Molecular and Cellular MR Imaging

Molecular and Cellular MR Imaging

Edited by

Michel M. J. Modo

Jeff W. M. Bulte

CRC Press
Taylor & Francis Group
Boca Raton London New York

CRC Press is an imprint of the
Taylor & Francis Group, an informa business

CRC Press
Taylor & Francis Group
6000 Broken Sound Parkway NW, Suite 300
Boca Raton, FL 33487-2742

Printed in the United States of America on acid-free paper
10 9 8 7 6 5 4 3 2 1

International Standard Book Number-10: 0-8493-7252-6 (Hardcover)
International Standard Book Number-13: 978-0-8493-7252-0 (Hardcover)

Library of Congress Cataloging-in-Publication Data

Molecular and cellular MR imaging / edited by Michel M.J.J. Modo and Jeff W.M. Bulte.
p. ; cm.
Includes bibliographical references and index.
ISBN-13: 978-0-8493-7252-0 (hardcover : alk. paper)
ISBN-10: 0-8493-7252-6 (hardcover : alk. paper)
1. Magnetic resonance imaging. 2. Molecular biology. 3. Cytology. I. Modo, Michel M. J. J. II. Bulte, Jeff W. M.
[DNLM: 1. Magnetic Resonance Imaging--methods. 2. Cytological Techniques--methods. 3. Molecular Diagnostic Techniques--methods. WN 185 M718 2007]

RC78.7.N83M65 2007
616.07'548--dc22 2006035039

Visit the Taylor & Francis Web site at
http://www.taylorandfrancis.com

and the CRC Press Web site at
http://www.crcpress.com

Table of Contents

Chapter 1 What Is Molecular and Cellular Imaging? 1
Michel M.J. Modo and Jeff W.M. Bulte

PART I Contrast Agents for Molecular and Cellular Imaging

Chapter 2 Physicochemical Principles of MR Contrast Agents 13
Peter Caravan

Chapter 3 Paramagnetic Contrast Agents 37
Silvio Aime, Zsolt Baranyai, Eliana Gianolio, and Enzo Terreno

Chapter 4 Superparamagnetic Contrast Agents 59
Claire Corot, Marc Port, Irène Guilbert, Philippe Robert, Isabelle Raynal, Caroline Robic, Jean-Sebastien Raynaud, Philippe Prigent, Anne Dencausse, and Jean-Marc Idée

Chapter 5 Physical Mechanism and Applications of CEST Contrast Agents 85
Michael T. McMahon, Jinyuan Zhou, Assaf A. Gilad, Jeff W.M. Bulte, and Peter C.M. van Zijl

Chapter 6 PARACEST Contrast Agents 101
A. Dean Sherry and Mark Woods

Chapter 7 Genetic Approaches for Modulating MRI Contrast 123
Eric T. Ahrens, William F. Goins, and Clinton S. Robison

PART II Molecular MR Imaging

Chapter 8 Molecular and Functional MR Imaging of Cancer 141
Michael A. Jacobs, Kristine Glunde, Barjor Gimi, Arvind P. Pathak, Ellen Ackerstaff, Dmitri Artemov, and Zaver M. Bhujwalla

Chapter 9 Molecular Imaging of Atherosclerosis with Magnetic Resonance 161
Karen C. Briley-Saebo, Willem J. Mulder, Fabien Hyafil, Venkatesh Mani, Vardan Amirbekian, Juan Gilberto S. Aguinaldo, James F. Rudd, Silvia H. Aguiar, and Zahi A. Fayad

Chapter 10 Molecular Imaging of Apoptosis 183
Mikko I. Kettunen and Kevin M. Brindle

Chapter 11 Molecular Imaging of Reporter Genes 199
Keren Ziv, Dorit Granot, Vicki Plaks, Batya Cohen, and Michal Neeman

Chapter 12 Pharmacological MRI as a Molecular Imaging Technique 219
Y. Iris Chen and Bruce G. Jenkins

PART III Cellular MR Imaging

Chapter 13 Cellular MR Imaging of the Liver Using Contrast Agents 247
N. Cem Balci and Sukru Mehmet Erturk

Chapter 14 Cellular Imaging of Macrophage Activity in Infection and Inflammation 259
Martin Rausch, Markus Rudin, Peter R. Allegrini, and Nicolau Beckmann

Chapter 15 Cellular Imaging of Macrophage Activity in Stroke 273
Ralph Weber and Mathias Hoehn

Chapter 16 Magnetic Resonance Imaging in Developmental Biology 283
Cyrus Papan, J. Michael Tyszka, and Russell E. Jacobs

Chapter 17 Methods for Labeling Nonphagocytic Cells with MR Contrast Agents 295
Joseph A. Frank, Stasia A. Anderson, and Ali S. Arbab

Chapter 18 Cellular Imaging of Cell Transplants 325
Michel M.J. Modo and Jeff W.M. Bulte

Chapter 19 Cellular and Molecular Imaging of the Diabetic Pancreas 343
Zdravka Medarova and Anna Moore

Chapter 20 Functional Cellular Imaging with Manganese 369
Vincent Van Meir and Annemie Van der Linden

PART IV Future Perspectives for Molecular and Cellular Imaging

Chapter 21 Translating Promising Experimental Approaches to Clinical Trials 395
Adrian D. Nunn

Chapter 22 An Outlook on Molecular and Cellular MR Imaging 405
Michel M.J. Modo and Jeff W.M. Bulte

Index 411

Preface

Molecular and cellular magnetic resonance (MR) imaging have recently emerged as novel technologies for the noninvasive assessment of biological processes in living organisms. The possibility to track the survival, migration, and differentiation of cells *in vivo*, as well as to be able to monitor particular gene or protein expression in living subjects, is not only becoming of great interest to scientists investigating fundamental aspects of health and disease, but is now also finding a translation into clinical settings.

The interdisciplinary nature of molecular and cellular MR imaging mandates various backgrounds in molecular and cell biology, chemistry, physics, image analysis, and drug discovery. In this book, a selected group of internationally recognized authors, each drawing on their specific expertise, highlight the diversity of skills necessary to further advance the field of molecular and cellular MR imaging. A constant dialog between these disciplines is vital to develop and translate promising approaches into reliable scientific applications and viable clinical diagnostic tools. This book provides a state-of-the-art overview of the various approaches to date that have been described to visualize cells and molecules by MR imaging and illustrates the application of these to interrogate specific biological processes in both animals and humans.

The Editors

Michel M.J. Modo, Ph.D., a Luxembourg native, is a Research Council of the United Kingdom (RCUK) fellow and Wolfson lecturer in stem cell imaging at the Centre for the Cellular Basis of Behaviour and the Medical Research Council (MRC) Centre for Neurodegeneration Research at the Institute of Psychiatry (IoP), King's College London. Dr. Modo graduated from Royal Holloway University of London with a degree in psychology and in 1995 spent 1 year as an undergraduate in the psychology department at McGill University in Montreal, Canada. In 2001, he earned his Ph.D. at the IoP and has since been interested in the application of molecular and cellular imaging to understand how stem cells promote functional recovery after brain damage.

Jeff W.M. Bulte, M.D., also a native from the Benelux (the Netherlands), is a professor of radiology in the Division of MR Research and is director of the cellular imaging section at the Institute for Cell Engineering, Johns Hopkins University School of Medicine. In 1991, Dr. Bulte graduated summa cum laude in medicine/immunoloy from the University of Groningen, and he spent 10 years in the Laboratory of Diagnostic Radiology Research at the National Institutes of Health before moving to Hopkins in 2001. His research specializes in molecular and cellular MR imaging.

Contributors

Ellen Ackerstaff
Department of Radiology and
Sidney Kimmel Comprehensive Cancer Center
Johns Hopkins University School of Medicine
Baltimore, Maryland

Silvia H. Aguiar
Department of Radiology
Mount Sinai School of Medicine
New York, New York

Juan Gilberto S. Aguinaldo
Department of Radiology
Mount Sinai School of Medicine
New York, New York

Eric T. Ahrens
Carnegie Mellon University
Pittsburgh, Pennsylvania

Silvio Aime
Department of Chemistry
IFM
University of Torino
Torino, Italy

Peter R. Allegrini
Novartis Institute for Biomedical Research
Novartis Pharma AG
Basel, Switzerland

Vardan Amirbekian
Department of Radiology
Mount Sinai School of Medicine
New York, New York

Stasia A. Anderson
National Institutes of Health
Bethesda, Maryland

Ali S. Arbab
Henry Ford Health System
Detroit, Michigan

Dmitri Artemov
Department of Radiology and
Sidney Kimmel Comprehensive Cancer Center
Johns Hopkins University School of Medicine
Baltimore, Maryland

N. Cem Balci
Department of Radiology
Saint Louis University
Saint Louis, Missouri

Zsolt Baranyai
Department of Chemistry
IFM and Molecular Imaging Center
University of Torino
Torino, Italy

Nicolau Beckmann
Novartis Institutes for BioMedical Research
Basel, Switzerland

Zaver M. Bhujwalla
Department of Radiology and
Sidney Kimmel Comprehensive Cancer Center
John Hopkins University School of Medicine
Baltimore, Maryland

Karen C. Briley-Saebo
Department of Radiology
Mount Sinai School of Medicine
New York, New York

Kevin M. Brindle
Department of Biochemistry
University of Cambridge
Cambridge, England

Jeff W.M. Bulte
Institute for Cell Engineering
Johns Hopkins University School of Medicine
Baltimore, Maryland

Peter Caravan
Epix Pharmaceuticals
Cambridge, Massachusetts

Y. Iris Chen
Athinoula A. Martinos Center
Massachusetts General Hospital
Charlestown, Massachusetts

Batya Cohen
Department of Biological Regulation
Weizmann Institute of Science
Rehovot, Israel

Claire Corot
Guerbet Research
Roissy, France

Anne Dencausse
Guerbet Research
Roissy, France

Sukru Mehmet Erturk
Department of Radiology
Brigham and Women's Hospital
Harvard Medical School
Boston, Massachusetts

Zahi A. Fayad
Mount Sinai School of Medicine
New York, New York

Joseph A. Frank
Laboratory of Diagnostic Radiology
Clinical Center
Bethesda, Maryland

Eliana Gianolio
Department of Chemistry
IFM and Molecular Imaging Center
University of Torino
Torino, Italy

Assaf A. Gilad
School of Medicine
Institute of Cell Engineering
Johns Hopkins University School of Medicine
Baltimore, Maryland

Barjor Gimi
Department of Radiology
University of Texas
Southwestern Medical Center
Dallas, Texas

Kristine Glunde
Department of Radiology and
Sidney Kimmel Comprehensive Cancer Center
Johns Hopkins University School of Medicine
Baltimore, Maryland

William F. Goins
Department of Molecular Genetics
School of Medicine
University of Pittsburgh
Pittsburgh, Pennsylvania

Dorit Granot
Department of Biological Regulation
Weizmann Institute of Science
Rehovot, Israel

Irène Guilbert
Guerbet Research
Roissy, France

Mathias Hoehn
Max-Planck-Institute for Neurological Research
Cologne, Germany

Fabien Hyafil
Department of Radiology
Mount Sinai School of Medicine
New York, New York

Jean-Marc Idée
Guerbet Research
Roissy, France

Michael A. Jacobs
Department of Radiology and
Sidney Kimmel Comprehensive Cancer Center
Johns Hopkins University School of Medicine
Baltimore, Maryland

Russell E. Jacobs
Beckman Institute
California Institute of Technology
Pasadena, California

Bruce G. Jenkins
Athinoula A. Martinos Center
Massachusetts General Hospital
Charlestown, Massachusetts

Mikko I. Kettunen
Department of Biochemistry
University of Cambridge
Cambridge, England

Venkatesh Mani
Department of Radiology
Mount Sinai School of Medicine
New York, New York

Michael T. McMahon
Kennedy Krieger Institute
Baltimore, Maryland

Zdravka Medarova
Harvard Medical School
Massachusetts General Hospital
Charlestown, Massachusetts

Michel M.J. Modo
Centre for the Cellular Basis of Behavior
Institute of Psychiatry
King's College London
London, England

Anna Moore
MGM Martinos Center for Biomedical Imaging
Massachusetts General Hospital
Charlestown, Massachusetts

Willem J. Mulder
Biomedical NMR, Department of Biomedical Engineering
Eindhoven University of Technology
Eindhoven, Netherlands

Michal Neeman
Department of Biological Regulation
Weizmann Institute of Science
Rehovot, Israel

Adrian D. Nunn
Bracco Research USA, Ltd.
Princeton, New Jersey

Cyrus Papan
Institute of Bioengineering and Nanotechnology
Thenanos, Singapore

Arvind P. Pathak
Department of Radiology and Sidney Kimmel Comprehensive Cancer Center
Johns Hopkins University School of Medicine
Baltimore, Maryland

Vicki Plaks
Department of Biological Regulation
Weizmann Institute of Science
Rehovot, Israel

Marc Port
Guerbet Research
Roissy, France

Philippe Prigent
Guerbet Research
Roissy, France

Martin Rausch
Novartis Institutes for Biomedical Research, Analytical and Imaging Science
Basel, Switzerland

Isabelle Raynal
Guerbet Research
Roissy, France

Jean-Sebastien Raynaud
Guerbet Research
Roissy, France

Philippe Robert
Guerbet Research
Roissy, France

Caroline Robic
Guerbet Research
Roissy, France

Clinton S. Robison
Department of Biological Sciences
Carnegie Mellon University
Pittsburgh, Pennsylvania

James F. Rudd
Department of Radiology
Mount Sinai School of Medicine
New York, New York

Markus Rudin
Institute for Biomedical Engineering
ETH and University of Zurich
Zurich, Switzerland

A. Dean Sherry
Department of Chemistry
University of Texas at Dallas
UT-Southwestern Medical Center
Advanced Imaging Research Center
Dallas, Texas

Enzo Terreno
Department of Chemistry
Molecular Imaging Center
University of Torino
Torino, Italy

J. Michael Tyszka
California Institute of Technology
Pasadena, California

Annemie Van der Linden
Bio-Imaging Lab
University of Antwerp
Antwerp, Belgium

Vincent Van Meir
Bio-Imaging Lab
University of Antwerp
Antwerp, Belgium

Peter C.M. van Zijl
F.M. Kirby Center for
Functional Brain Imaging
Kennedy Krieger Institute
Baltimore, Maryland

Ralph Weber
Max-Planck Institute for Neurological Research
and
University Clinic of Essen
Cologne, Germany

Mark Woods
Macrocyclics, Inc.
Dallas, Texas

Jinyuan Zhou
School of Medicine
Institute for Cell Engineering
Johns Hopkins University
Baltimore, Maryland

Keren Ziv
Department of Biological Regulation
Weizmann Institute of Science
Rehovot, Israel

1 What Is Molecular and Cellular Imaging?

Michel M.J. Modo and Jeff W.M. Bulte

CONTENTS

1.1 Introduction 1
1.2 What Is Molecular and Cellular Imaging? 2
1.3 Is There a Need for Molecular and Cellular Imaging in Biomedical Research? 2
1.4 How Does MR Imaging Compare with Other Imaging Modalities? 3
1.5 The Need for MR Contrast Agents 5
References 7

1.1 INTRODUCTION

The development of life is a proficiently orchestrated process of a myriad of distinctive molecules. From DNA to cells and from cells to organs, it is molecules that are the building blocks of all life.[1] Being able to understand how molecules and cells develop into animals also reveals how aberrant molecular or cellular processes contribute to the degeneration of physiological systems. The most powerful medical interventions are therefore deemed to intervene at the earliest stage when a molecular aberration can be detected. It is thought that this "molecular medicine" will not only treat symptoms of disease, but also lead to the prevention of symptoms and stop disease before it can harm the patient.[2,3]

Until recently, the study of these molecules and cells was mainly confined to invasive and irreversible histological and molecular biological techniques. Histological studies utilize a panoply of antibodies that detect highly specific molecules and allow, for instance, the differentiation between a variety of cellular phenotypes. Molecular biological techniques, such as polymerase chain reaction (PCR), can even describe the constituent parts of biological molecules and define "molecular fingerprints" of disease.[4] However, the disadvantage of both techniques is that they cannot easily be used on an intact living specimen. Histological techniques suffer from light scattering, and therefore have a very poor tissue penetration, whereas molecular biological techniques typically require a disintegration of tissue.

Invasive biopsies are needed to pursue these techniques. However, biopsies can cause tissue damage. Consequently, in many circumstances, biopsies are ethically unacceptable as they can result in iatrogenic complications. For instance, molecular changes in the hippocampus that could indicate early pathogenic events in Alzheimer's disease or epilepsy would require deep brain penetration that would injure surrounding tissue. Additionally, the accuracy of biopsies also depends on adequate sampling. If the disease is very localized in a large organ, such as the liver, the biopsy might sample a part of the organ that is not affected by the disease and lead to a false negative. Even if this approach is justified based on peripheral biomarkers, such as increases of a particular protein in the blood, tissue retrieval could only be considered once due to the inflicted damage to the organ. For that reason, biopsies have considerable limitations to present a potential early diagnosis of a disease. It is hence impossible to use this technique to monitor an organ for the

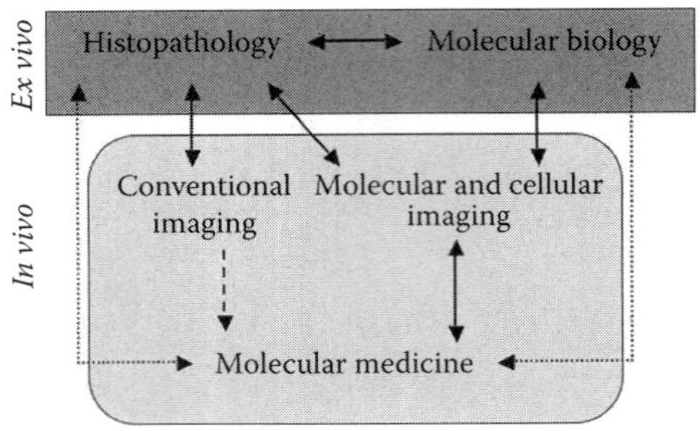

FIGURE 1.1 Both histopathology and molecular biology are predominantly *ex vivo* assessment techniques. In contrast, conventional and molecular and cellular imaging are *in vivo* analytical techniques that can be used to bridge the gap between the need for specific molecular information and its use in patient management in molecular medicine. However, molecular and cellular imaging is very dependent on both histopathology and molecular biology to identify imaging targets. Molecular biology is also dependent on histopathology to identify regions for further molecular analysis, whereas histopathology is dependent on molecular biology to devise probes that allow the localization of particular molecules in tissue sections. The interdependence of different approaches illustrates that particular techniques should not be used in a vacuum, as information derived from another analytical technique will contribute to a faster development. To realize the potential of molecular imaging, these different techniques should complement each other to ensure a rapid progression of technological innovation.

potential emergence of aberrant molecules. These techniques are therefore very limited to study biological or pathological processes in living organisms.

1.2 WHAT IS MOLECULAR AND CELLULAR IMAGING?

The application of histology and molecular biology in humans is very restricted. Neither can provide a satisfactory noninvasive deep tissue visualization of molecules or cells in living organisms. The development of molecular and cellular imaging aims to bridge this gap (Figure 1.1) and provide methods that allow the detection of molecules and their interaction in living organisms over time. To be able to visualize the presence and evolution of molecules or cells noninvasively in an intact animal will form an essential assessment if we are to unravel how life develops and pathology emerges. It is therefore possible to define molecular and cellular imaging as follows:

Molecular imaging — The visualization of specific molecules in an intact animal.
Cellular imaging — The visualization of specific cells in an intact animal.

The visualization of molecules or cells in intact animals distinguishes molecular and cellular imaging from histology and molecular biology that typically require a disintegration of the organisms. The specificity to particular molecules or cells differentiates molecular and cellular imaging from more conventional imaging techniques, such as anatomical and functional magnetic resonance (MR) imaging, which mainly describe gross morphology and organ blood flow. It is thus the target and not necessarily the technique that differentiates molecular or cellular imaging from conventional imaging.

1.3 IS THERE A NEED FOR MOLECULAR AND CELLULAR IMAGING IN BIOMEDICAL RESEARCH?

Although molecular biology might provide the targets for molecular interventions, it is a very poor diagnostic tool, as it lacks application in a living organism and cannot provide deep tissue information that would allow a spatial localization of an emerging pathogenesis. Over the past few years, it has been highlighted that there is a pressing need for more specific imaging techniques to enhance the *in vivo* study of molecules that are crucial for the development of molecular medicine.[5–7] Because the aim of molecular medicine is to treat molecular aberrations as early as possible, it will be important to develop molecular imaging as a diagnostic tool to allow clinicians to visualize aberrant disease molecules at an early stage in the living subject[8] (Figure 1.1). The realization of molecular medicine is therefore dependent on the development of molecular imaging as a reliable diagnostic tool to provide the bridge between molecular biology and molecular medicine.[5,9]

Apart from its potential application in clinical medicine, molecular and cellular imaging in experimental settings will provide an integrative technology to study biological processes and their relevance to behavior *in vivo*. The integration of behavior, histology, pharmacology, and molecular biology through imaging of these various elements will be essential for a more holistic development of novel approaches. Developments of molecular and cellular imaging will be central to the further development of pharmaceutical and biotechnological research as targets become more and more specific.[6,10] Imaging technology permits researchers to monitor a whole animal to determine effects on multiple organ systems *in vivo*.[11–13] Fine-tuning of pharmacological agents can go beyond current pharmacogenetic matching,[14] by accounting for specific gene expression profiles in particular brain regions. It is foreseeable that this could, for instance, allow the development of pharmacological agents that will target distinct regional neurochemical imbalances in psychiatric disease.

Developments in molecular and cellular imaging will allow the investigation of the elemental constituents of organs, and hence introduce a means to interrogate everything from gene expression to functional circuitries. The ability to link behavior to functional connectivity in the brain and tease out the molecular and anatomical changes underlying the changes through molecular imaging[15] will truly provide a powerful integration of different organizational levels from gene expression to its effect on behavior. As disease-related symptoms are but a modification of normal behavior, it is potentially possible to move beyond symptom-based diagnosis and focus on the specific underlying molecular changes. Although many medical disciplines already base their diagnosis on molecular pathology, this is mainly the case for easily accessible organs, such as the skin, and currently cannot be used for diseases pertaining to the brain or heart. It is therefore likely that medical disciplines concerned with internal organs will have the most to gain from molecular and cellular imaging, and at the same time are likely to see more change in clinical practice than existing approaches.[16] Molecular medicine will not only lead to earlier diagnosis and treatment, but also might redraw the definition of what we consider a disease. However, these predictions and promises are largely dependent on the advances and limits of technological developments in imaging.[9,17–20]

1.4 HOW DOES MR IMAGING COMPARE WITH OTHER IMAGING MODALITIES?

Many different techniques have the capability to visualize molecules *in vivo*.[19] Apart from their physical basis (i.e., the detection of resonant magnetic frequencies, radionuclides, emitted light, etc.), these techniques differ in many other aspects, from their invasiveness to their cost-effectiveness.[6] The choice of the technique will not only be dependent on the molecules or cells one wishes to detect in living subjects, but also be influenced by considering if this approach is to be translated into a clinical setting or if the subject is meant to undergo many repeated assessments. Repeated assessment of invasive procedures complicates clinical translation,[18] but in certain cases some invasiveness might be considered an acceptable risk if it outweighs the information that can be gained in order to help the patient.

Of all the existing molecular imaging techniques, positron emission tomography (PET) and MR imaging are the most advanced and available in both experimental and clinical environments. In contrast to light-dependent techniques, such as bioluminescent imaging, both also have excellent

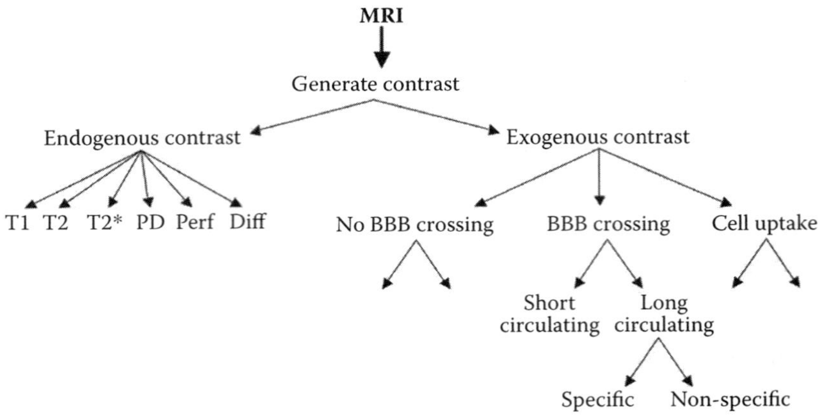

FIGURE 1.2 The basic principle behind MR imaging is to generate contrast between different tissues. Depending on the distribution and movement of water molecules (i.e., hydrogen atoms or protons), different endogenous contrast methods can, for instance, distinguish white from grey matter. However, in some cases exogenously administered contrast can help to highlight particular aspects. For example, the rather large MR contrast agents generally do not cross the intact blood–brain barrier (BBB). Leakage of contrast agents into the brain can therefore be used to assess damage to the BBB. Engineering of contrast agents with peptides that cause an active transport across the blood–brain barrier (e.g., putrescine) can be used to visualize targets that normally cannot be accessed by these agents. Contrast agents can either be cleared rapidly by the body's reticuloendothelial system (RES) or escape detection and generate MR contrast for prolonged episodes. Contrast particles can also either bind to specific targets, such as molecules, or be fairly unspecific and, for instance, merely be used as blood pool agents. Further engineering of MR contrast media will result in ever more sophisticated particles that can cross the intact BBB and be specifically taken up in one particular type of cell.

tissue penetration and, in principle, can visualize a whole subject. PET is a very powerful molecular imaging technique,[21] as the radioligands used to detect particular molecules are minute and easily cross the intact blood–brain barrier. These radioisotopes can easily be attached to other compounds that target particular molecules.[22–25] PET is therefore currently unbeatable in its ability to visualize specific molecules in the living brain.[26,27] However, the dependence on radioisotopes to produce an image in PET is a drawback, as it limits the number of times a single subject can be exposed to this activity.

In contrast, MR imaging mainly relies on detecting the nuclear magnetic resonance (NMR) signal of hydrogen (^1H) atoms after the application of a radiofrequency pulse. This noninvasiveness of MR imaging and the lack of radioactivity make it adept for serial studies of the same subject, even with short intervals between imaging sessions. The versatility of MR imaging is also interesting, as there are other MR techniques that complement molecular and cellular MR imaging. Functional/pharmacological MR imaging (f/phMRI), MR spectroscopy (MRS), and interventional MRI (iMRI) are but a few methods that can be achieved with the same hardware that is complementary to molecular/cellular MR imaging. The resolution achieved with MR imaging is largely dependent on the field strength of the magnet as it affects the signal-to-noise ratio (SNR), which determines how well a scan can contrast between different types of tissue (Figure 1.2). The tissue contrast, however, is a function of the distribution and chemical microenvironment of hydrogen atoms (see Chapter 2 for a basic overview of the physics and chemistry of MR imaging). The most commonly used field strengths for clinical scanners are 1.5 and 3.0 tesla (T) (64 and 128 MHz, respectively), whereas experimental studies using animals typically use field strengths at and above 4.7 T (170 MHz), as the target volume is smaller. Depending on the strength of the magnet and the sequences used to scan a subject, different aspects, such as grey or white matter in the brain, can be highlighted.[28]

The high spatial resolution (>10 times higher than PET), excellent tissue contrast, noninvasiveness for serial studies, and versatility make MR imaging a very attractive tool for molecular and cellular imaging that sets it apart from other techniques. Nevertheless, apart from large molecular complexes (e.g., N-acetyl-asparte, choline) that can be detected by MR spectroscopy,[29] MR imaging is not specific to particular molecules or cells. Similar to PET, to achieve specific detection of molecules or cells, MR imaging needs to increase its specificity and sensitivity by means of exogenous tracers or contrast agents.

1.5 THE NEED FOR MR CONTRAST AGENTS

Increasing sensitivity and specificity proves to be the challenge for molecular and cellular MR imaging. The detection of molecular or cellular events needs to exhibit *specificity* for the particular biological event. Specificity is therefore mainly reflected in the high fidelity and reliability of discriminating a particular molecule or cell from noise and other molecules. To achieve this high specificity, an antibody system targeting particular antigen, for instance, will selectively bind to the molecule/cell of interest.[30,31] By combining this antibody with a magnetic contrast agent, it will be possible to provide a selective detection of the molecule or cell of interest with MR imaging. However, other systems exist to specialize MR contrast agents to provide high specificity (Chapters 3 to 7). The properties of the MR contrast agent will determine its binding characteristics to the molecule of interest, its tissue penetration and circulation, and potentially its cellular uptake[32] (Figure 1.2). Modifying MR contrast agents into multifunctional entities (e.g., crossing the blood–brain barrier and selectively detecting amyloid plaques[33]) improves their attractiveness to molecular and cellular imaging.

Based on the functionalization of MR contrast agents, significant advances have been achieved in both cellular and molecular MR imaging (Figure 1.3). Notably, MR contrast agents can be shuttled into different types of cells *in vitro* or *in vivo* to track these by MR imaging[34] (Chapter 18), or they can be engineered to attach to particular molecules on tissues, such as blood vessels.[35] Increasing sophistication in the generation of these particles leads to ever more refined methods to detect specific molecules. For instance, even targeted MR contrast agents will produce a signal change if they are not bound to the molecule of interest. By engineering contrast agents to only produce a signal change when bound to a molecule of interest,[36,37] it is possible to scan the subject sooner, as there is no need to wait for unbound contrast agent to be washed out. These so-called smart MR contrast agents are but the start of contrast agents that change their properties depending on the environment[38] (Chapter 7). Environment sensing agents can be used for MR measurements as diverse as pH,[39] temperature,[40] or molecular interactions.[41] In the context of genetic studies, environmentally influenced agents can be biologically regulated to reflect an upregulation of a gene. Notably, the gene responsible for ferritin transport into cells can be used as an MR reporter[42] by increasing the intracellular iron load that can be detected by T_2*-weighted sequences (Chapters 7 and 11). As the molecular targets for visualization get sparser and more sophisticated, a greater emphasis needs to be placed on being able to detect fewer molecules in a smaller space.

The hardware systems used for visualization of a specific biological event hence need to achieve sufficient *sensitivity* to detect the effect of these MR "reporters." Although increasing the field strength of the MR scanner can improve some of these detection issues, the different physical characteristics of metal particles can facilitate the detection of sparse elements even within a relatively low magnetic field. Of the most commonly used MR contrast agents (gadolinium, ferric iron, and manganese), only iron particles are ferrimagnetic and produce a blooming effect that involves an area substantially greater than its localization.[43–45] Even small quantities of iron oxide particles can therefore be detected on MR scans. In some cases, even a single particle or cell can be detected.[46–48] In other cases, however, iron oxide particles might not be desirable, as the molecule might be too widely distributed throughout the organ and the use of ferumoxides might create a large signal void that no longer allows the localization of these events. Contrast agents based on

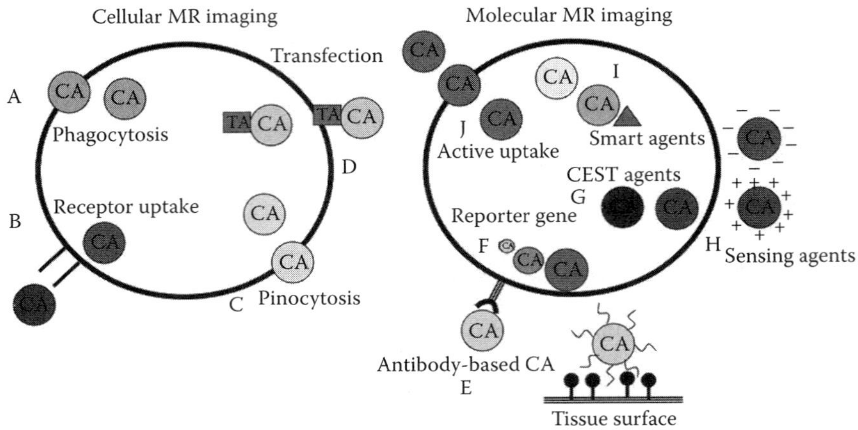

FIGURE 1.3 (please see color insert following page 210) Schematic overview of different approaches used to detect cells or molecules by MR imaging. For cellular imaging, contrast agent can be taken up by cells through phagocytosis (A), receptors (B), or pinocytosis (C). However, generally cells do not readily take up large compounds, but coating of particles with transfection agent can significantly improve uptake into cells (D). Specific molecules can be detected by MR imaging by conjugating an antibody with the contrast particle (E). Although antibodies generally recognize antigens, such as proteins, which are downstream products of a gene, it is also possible to directly study gene expression by genetically engineered cells to express ferritin that leads to a cellular increase in iron that can be detected by MR imaging (F). This reporter gene can then be linked to any gene of interest to study its expression *in vitro* or *in vivo*. Changes in the cell or tissue environment can also be detected by means of contrast agents. Chemical exchange saturation transfer (CEST) agents can be selectively activated by very narrow radiofrequency bands (G), whereas sensing agents change their relaxivity depending on the environmental conditions they encounter (H). However, sensing agents typically have a residual relaxivity, whereas smart agents will only induce a change in MR relaxivity when they bind to a particular molecule (I). Some contrast agents, such as manganese, can also be used to study, for instance, Ca^{2+} uptake into cells, as it normally enters through the Ca^{2+} channel when cells are activated (J).

gadolinium or manganese, producing hyperintensities in T_1-weighted images, might serve as alternatives in these cases. Contrast agents producing hyperintense signals are generally preferable to compounds causing hypointensities, as a positive signal is easier to interpret than the loss of a signal. Nevertheless, T_1 agents also often induce hypointensities on T_2-weighted scans, and therefore might not entirely circumvent this issue. Ideally, contrast agents use alternative atoms, such as fluorine,[49,50] to allow the detection of the molecular target or cell in a scan that does not affect the anatomical hydrogen-based MR image. At present, however, in most experiments iron oxide particles are the preferred agents, as they provide sufficient relaxivity to reliably detect even minute concentrations of contrast agent. Engineering of contrast agents based on their physicochemical properties therefore greatly influences *in vivo* MR detection. Meticulous considerations to these characteristics will ensure significant improvements in the application of molecular imaging agents.[45]

Although contrast agent design and engineering are developing rapidly and progressing the frontiers of molecular imaging, one of the greatest challenges over the coming years will be to ensure that these exciting advances find their translation into clinical applications. Not only will preclinical studies need to determine the feasibility and reliability of these novel contrast agents, but also prior to implementation in human subjects the safety of the newly engineered agents will need to be evaluated[32] (Chapter 21 deals with the clinical implementation of molecular and cellular MR agents). MR contrast agents so far have an excellent record of safety;[51] care must be taken to ensure that the procedures for translation of molecular and cellular imaging follow similar stringent tests and analyses that vouch for the agents' biocompatibility. The use of contrast agents to visualize intracellular targets or the use of contrast agents for cellular MR imaging especially needs to be thoroughly assessed, as they will remain localized to the compartments for considerably longer time frames than agents that are used for current conventional MR imaging.[34] The further engineering and addition of particular molecules to MR contrast agent demand also further refinement to biocompatibility testing. For instance, immunogenecity of peptides targeting particular molecules might arise as an issue. To date, only a few examples of dextran allergies prevent the use of contrast agents in humans.[52–54] However, as more specific elements enter contrast agent development, more of these compounds might elicit an immune response. Often these particular effects cannot be thoroughly assessed in preclinical experiments. In the future, the extensive testing for biocompatibility in human subjects might become a more complex procedure for the implementation of novel compounds prior to the testing of their specificity for particular molecules. Although many of these issues to date are mainly theoretical considerations, as clinical translations progress, the methods will find further refinement and improve the use of these agents.

There is no question that considerable progress has been achieved in MR contrast agent design. The improved specificity of these compounds and their more targeted application increase the potential of MR imaging as an analytical platform. The possibility to develop MR imaging methods in preclinical models and easily translate this approach to clinical application is central to this rapid progress. The versatility of MR imaging, its complementarities, and integration with other techniques, such as PET and single photon emission computer tomography (SPECT),[55,56] further enhance the attractiveness of MR imaging as a core integrative technique for cellular and molecular imaging. The development of multimodal agents further promises to expand the utility and versatility of MR imaging.[57]

Molecular and cellular MR imaging are interdisciplinary fields of study that require highly specialized expertise in contrast agent chemistry, MR physics, image analysis, and biological disciplines. It is this successful collaboration between different specialties that will progress MR imaging into a molecular and cellular imaging tool central to diagnostic analyses required for molecular medicine to flourish. This book aims to provide an integrated overview of these emerging fields.

REFERENCES

1. Gawad, C., Towards molecular medicine: a case for a biological periodic table, *Am. J. Pharmacogenomics*, 5, 207–211, 2005.
2. Dietel, M. and Sers, C., Personalized medicine and development of targeted therapies: the upcoming challenge for diagnostic molecular pathology. A review, *Virchows Arch.*, 448(6), 744–755, 2006.
3. Steel, M., Molecular medicine: promises, promises? *J. R. Soc. Med.*, 98, 197–199, 2005.
4. Bailey, W.J. and Ulrich, R., Molecular profiling approaches for identifying novel biomarkers, *Expert Opin. Drug Saf.*, 3, 137–151, 2004.
5. Heckl, S., Pipkorn, R., Nagele, T., Vogel, U., Kuker, W., and Voight, K., Molecular imaging: bridging the gap between neuroradiology and neurohistology, *Histol. Histopathol.*, 19, 651–668, 2004.
6. Rudin, M. and Weissleder, R., Molecular imaging in drug discovery and development, *Nat. Rev. Drug Discov.*, 2, 123–131, 2003.
7. Weissleder, R., Molecular imaging: exploring the next frontier, *Radiology*, 212, 609–614, 1999.
8. Calvo, B.F. and Semelka, R.C., Beyond anatomy: MR imaging as a molecular diagnostic tool, *Surg. Oncol. Clin. N. Am.*, 8, 171–183, 1999.
9. Schwaiger, M. and Weber, W., Molecular imaging: dream or reality? *Ernst Schering Res. Found. Workshop*, 48, 1–18, 2004.
10. Herschman, H.R., Molecular imaging: looking at problems, seeing solutions, *Science*, 302, 605–608, 2003.
11. Piwnica-Worms, D., Schuster, D.P., and Garbow, J.R., Molecular imaging of host-pathogen interactions in intact small animals, *Cell. Microbiol.*, 6, 319–331, 2004.
12. Gheysens, O. and Gambhir, S.S., Studying molecular and cellular processes in the intact organism, *Prog. Drug Res.*, 62, 117–150, 2005.

13. Massoud, T.F. and Gambhir, S.S., Molecular imaging in living subjects: seeing fundamental biological processes in a new light, *Genes Dev.*, 17, 545–580, 2003.
14. Staddon, S., Arranz, M.J., Mancama, D., Mata, I., and Kerwin, R.W., Clinical applications of pharmacogenetics in psychiatry, *Psychopharmacology (Berl.)*, 162, 18–23, 2002.
15. Jasanoff, A., Functional MRI using molecular imaging agents, *Trends Neurosci.*, 28, 120–126, 2005.
16. Ryan, J.M., Loy, R., and Tariot, P.N., Impact of molecular medicine on neuropsychiatry: the clinician's perspective, *Curr. Psychiatry Rep.*, 3, 355–360, 2001.
17. Jager, P.L., de Korte, M.A., Lub-de Hooge, M.N., van Waarde, A., Koopmans, K.P., Perik, P.J., and de Vries, E.G., Molecular imaging: what can be used today, *Cancer Imaging*, 5, S27–S32, 2005.
18. Pomper, M.G., Translational molecular imaging for cancer, *Cancer Imaging*, 5, S16–S26, 2005.
19. Levin, C.S., Primer on molecular imaging technology, *Eur. J. Nucl. Med. Mol. Imaging*, 32 (Suppl. 2), S325–S345, 2005.
20. Frost, J.J., Molecular imaging of the brain: a historical perspective, *Neuroimaging Clin. N. Am.*, 13, 653–658, 2003.
21. Hoh, C.K., Schiepers, C., Seltzer, M.A., Gambhir, S.S., Silverman, D.H., Czernin, J., Maddahi, J., and Phelps, M.E., PET in oncology: will it replace the other modalities? *Semin. Nucl. Med.*, 27, 94–106, 1997.
22. Conti, P.S., Introduction to imaging brain tumor metabolism with positron emission tomography (PET), *Cancer Invest.*, 13, 244–259, 1995.
23. Gibson, R.E., Burns, H.D., Hamill, T.G., Eng, W.S., Francis, B.E., and Ryan, C., Non-invasive radiotracer imaging as a tool for drug development, *Curr. Pharm. Design*, 6, 973–989, 2000.
24. Halldin, C., Gulyas, B., Langer, O., and Farde, L., Brain radioligands: state of the art and new trends, *Q. J. Nucl. Med.*, 45, 139–152, 2001.
25. Kegeles, L.S. and Mann, J.J., *In vivo* imaging of neurotransmitter systems using radiolabeled receptor ligands, *Neuropsychopharmacology*, 17, 293–307, 1997.
26. Phelps, M.E., PET: the merging of biology and imaging into molecular imaging, *J. Nucl. Med.*, 41, 661–681, 2000.
27. Czernin, J. and Phelps, M.E., Positron emission tomography scanning: current and future applications, *Annu. Rev. Med.*, 53, 89–112, 2002.
28. Sasaki, M., Inoue, T., Tohyama, K., Oikawa, H., Ehara, S., and Ogawa, A., High-field MRI of the central nervous system: current approaches to clinical and microscopic imaging, *Magn. Reson. Med. Sci.*, 2, 133–139, 2003.
29. Kwock, L., Localized MR spectroscopy: basic principles, *Neuroimaging Clin. N. Am.*, 8, 713–731, 1998.
30. Guccione, S., Li, K.C., and Bednarski, M.D., Molecular imaging and therapy directed at the neovasculature in pathologies. How imaging can be incorporated into vascular-targeted delivery systems to generate active therapeutic agents, *IEEE Eng. Med. Biol. Mag.*, 23, 50–56, 2004.
31. Artemov, D., Molecular magnetic resonance imaging with targeted contrast agents, *J. Cell. Biochem.*, 90, 518–524, 2003.
32. Lorusso, V., Pascolo, L., Fernetti, C., Anelli, P.L., Uggeri, F., and Tiribelli, C., Magnetic resonance contrast agents: from the bench to the patient, *Curr. Pharm. Design*, 11, 4079–4098, 2005.
33. Podulso, J.F., Wengenack, T.M., Curran, G.L., Wisniewski, T., Sigurdsson, E.M., Macura, S.I., Borowski, B.J., and Jack, C.R., Molecular targeting of Alzheimer's amyloid plaques for contrast-enhanced magnetic resonance imaging, *Neurobiol. Dis.*, 11, 315–329, 2002.
34. Modo, M., Hoehn, M., and Bulte, J.W., Cellular MR imaging, *Mol. Imaging*, 4, 143–164, 2005.
35. Delikatny, E.J. and Poptani, H., MR techniques for *in vivo* molecular and cellular imaging, *Radiol. Clin. N. Am.*, 43, 205–220, 2005.
36. Louie, A.Y., Huber, M.M., Ahrens, E.T., Rothbacher, U., Moats, R., Jacobs, R.E., Fraser, S.E., and Meade, T.J., *In vivo* visualization of gene expression using magnetic resonance imaging, *Nat. Biotechnol.*, 18, 321–325, 2000.
37. Li, W.H., Parigi, G., Fragai, M., Luchinat, C., and Meade, T.J., Mechanistic studies of a calcium-dependent MRI contrast agent, *Inorg. Chem.*, 41, 4018–4024, 2002.
38. Lowe, M.P., Activated MR contrast agents, *Curr. Pharm. Biotechnol.*, 5, 519–528, 2004.
39. Aime, S., Delli Castelli, D., and Terreno, E., Novel pH-reporter MRI contrast agents, *Angew. Chem. Int. Ed. Engl.*, 41, 4334–4336, 2002.

40. Aime, S., Botta, M., Fasano, M., Terreno, E., Kinchesh, P., Calabi, L., and Paleari, L., A new ytterbium chelate as contrast agent in chemical shift imaging and temperature sensitive probe for MR spectroscopy, *Magn. Reson. Med.*, 35, 648–651, 1996.
41. Perez, J.M., Josephson, L., O'Loughlin, T., Hogemann, D., and Weissleder, R., Magnetic relaxation switches capable of sensing molecular interactions, *Nat. Biotechnol.*, 20, 816–820, 2002.
42. Cohen, B., Dafni, H., Meir, G., and Neeman, M., Ferritin as novel MR-reporter for molecular imaging of gene expression, *Proc. Int. Soc. Magn. Reson. Med.*, 11, 1707, 2004.
43. Bonnemain, B., Superparamagnetic agents in magnetic resonance imaging: physicochemical characteristics and clinical applications. A review, *J. Drug Target*, 6, 167–174, 1998.
44. Bjornerud, A. and Johansson, L., The utility of superparamagnetic contrast agents in MRI: theoretical consideration and applications in the cardiovascular system, *NMR Biomed.*, 17, 465–477, 2004.
45. Reichert, D.E., Lewis, J.S., and Anderson, C.J., Metal complexes as diagnostic tools, *Coordination Chem. Rev.*, 184, 3–66, 1999.
46. Shapiro, E.M., Sharer, K., Skrtic, S., and Koretsky, A.P., *In vivo* detection of single cells by MRI, *Magn. Reson. Med.*, 55, 242–249, 2006.
47. Heyn, C., Ronald, J.A., Mackenzie, L.T., MacDonald, I.C., Chambers, A.F., Rutt, B.K., and Foster, P.J., *In vivo* magnetic resonance imaging of single cells in mouse brain with optical validation, *Magn. Reson. Med.*, 55, 23–29, 2006.
48. Shapiro, E.M., Skrtic, S., Sharer, K., Hill, J.M., Dunbar, C.E., and Koretsky, A.P., MRI detection of single particles for cellular imaging, *Proc. Natl. Acad. Sci. U.S.A.*, 101, 10901–10906, 2004.
49. Ahrens, E.T., Flores, R., Xu, H., and Morel, P.A., *In vivo* imaging platform for tracking immunotherapeutic cells, *Nat. Biotechnol.*, 23, 983–987, 2005.
50. Schwarz, R., Schuurmans, M., Seelig, J., and Kunnecke, B., 19F-MRI of perfluorononane as a novel contrast modality for gastrointestinal imaging, *Magn. Reson. Med.*, 41, 80–86, 1999.
51. Runge, V.M., Safety of approved MR contrast media for intravenous injection, *J. Magn. Reson. Imaging*, 12, 205–213, 2000.
52. Li, A., Wong, C.S., Wong, M.K., Lee, C.M., and Au Yeung, M.C., Acute adverse reactions to magnetic resonance contrast media: gadolinium chelates, *Br. J. Radiol.*, 79, 368–371, 2006.
53. Beaudouin, E., Kanny, G., Blanloeil, Y., Guilloux, L., Renaudin, J.M., and Moneret-Vautrin, D.A., Anaphylactic shock induced by gadoterate meglumine (DOTAREM), *Allerg. Immunol. (Paris)*, 35, 382–385, 2003.
54. Chu, W.C., Lam, W.W., and Metreweli, C., Incidence of adverse events after I.V. injection of MR contrast agents in a Chinese population. A comparison between gadopentetate and gadodiamide, *Acta Radiol.*, 41, 662–666, 2000.
55. Marsden, P.K., Strul, D., Keevil, S.F., Williams, S.C., and Cash, D., Simultaneous PET and NMR, *Br. J. Radiol.*, 75, S53–S59, 2002.
56. Jacobs, R.E. and Cherry, S.R., Complementary emerging techniques: high-resolution PET and MRI, *Curr. Opin. Neurobiol.*, 11, 621–629, 2001.
57. Roberts, T.P., Chuang, N., and Roberts, H.C., Neuroimaging: do we really need new contrast agents for MRI? *Eur. J. Radiol.*, 34, 166–178, 2000.

Part I

Contrast Agents for Molecular and Cellular Imaging

2 Physicochemical Principles of MR Contrast Agents

Peter Caravan

CONTENTS

2.1 Introduction 13
2.2 A Few Basic Principles of NMR 14
2.3 T_1, T_2, and T_2* Contrast Agents and Relaxivity 15
2.4 Chemistry of T_1 Agents 16
2.5 Paramagnetic Enhanced Nuclear Relaxation 17
2.6 T2 Agents 23
2.7 Imaging Physics: T_1 and T_2 Agents 25
2.8 PARACEST Agents 28
2.9 Direct Observation of Nonhydrogen Nuclei 31
2.10 Common Features and Concerns 32
2.11 Conclusion 32
References 33

2.1 INTRODUCTION

The magnetic resonance (MR) image in clinical and biological systems is typically an image of the hydrogen atoms in water and fat. Water hydrogen is chosen because it is very abundant; tissue is about 90 *M* (molar) in water hydrogen concentration. The ^{1}H isotope (the proton) is almost 100% naturally abundant and is the second most sensitive nucleus, behind tritium, ^{3}H, for nuclear magnetic resonance (NMR) detection. There are many sources of contrast in an MR image. The simplest is proton density, where tissue containing more water will give a greater signal. Tissue contrast can also be achieved by weighting the imaging sequence to display differences in proton relaxation rates ($1/T_1$ and $1/T_2$); exploiting differences in chemical shift or water diffusion; or the effect of flowing blood; or using magnetization transfer techniques. By utilizing one or more of these techniques, high-resolution images can be obtained providing excellent anatomical content, delineation of diseased tissue, and often valuable physiological information.

Sometimes additional contrast is required and exogenous materials are given that can alter the MR signal. These materials are called contrast agents. Contrast agents can act by changing the relaxation rates of neighboring water molecules and giving positive or negative contrast on a T_1- or T_2-weighted imaging sequence, respectively. A different class of contrast agent relies on magnetization transfer to provide negative contrast. Magnetization transfer and T_1- and T_2-weighted agents alter some property of water in a catalytic way, but it is still the water that is imaged. Other contrast agents use alternative nuclei such as fluorine or hyperpolarized nuclei such as carbon, helium, or xenon, and these nuclei are imaged directly.

There is an active research effort to extend MR imaging (MRI) beyond anatomy and physiology to the cellular and molecular level. Contrast agents are used to provide this information. The aim of this chapter is to provide an overview of the different contrast mechanisms and the relevant chemistry and biophysics for each class of contrast agent. Issues common to all contrast agents, such as formulation, speciation, stability, targeting, and excretion, are also discussed. Each specific class of contrast agent and its application to molecular and cellular imaging are described in greater detail in subsequent chapters in this book. This chapter assumes the reader has a basic knowledge of MRI and its terminology. Textbooks on the basic principles of MRI should be consulted for more detail than is given here.[1,2]

2.2 A FEW BASIC PRINCIPLES OF NMR

Atomic nuclei have magnetic moments that are proportional to their nuclear spin, I. The hydrogen atom has a spin of $^1/_2$. When an external magnetic field is applied, the nuclear moments align themselves with only certain allowed orientations; for $I = {}^1/_2$, there are only two possible orientations, denoted by the magnetic quantum number m_I, which has values of $+^1/_2$ or $-^1/_2$. In the case of hydrogen, the spins align either with or against the applied magnetic field. The spins can transition between these two states if the appropriate resonant energy (ΔE) is applied:

$$\Delta E = h\nu = \frac{\gamma h B_0}{2\pi} \tag{2.1}$$

Here γ is the magnetogyric ratio, a property specific to the nucleus in question, ν is the applied frequency (sometimes called Larmor frequency), B_0 is the external applied field, and h is Planck's constant. The frequency required will depend directly on the applied field and the magnetogyric ratio. If there is a difference in the population of spins between the $+^1/_2$ and $-^1/_2$ states, there will be a net absorption of energy when frequency ν is applied. The ratio of the population ($N_{-1/2}/N_{+1/2}$) between these two states is given by the Boltzmann equation:

$$\begin{aligned} \frac{N_{-1/2}}{N_{+1/2}} &= \exp(-h\nu / kT) \\ &= 1 - h\nu / kT; \qquad \text{since } h\nu << kT \end{aligned} \tag{2.2}$$

Since NMR deals with frequencies in the megahertz range, the excess population of spins is only about 1 in 100,000. This is the fundamental reason for the low sensitivity of NMR — only 0.001% of the hydrogen is detected.

Consideration of Equations 2.1 and 2.2 suggests that to increase sensitivity, one should work with a nucleus of high γ and at high applied fields, since this results in the greatest frequency required. Frequency is directly proportional to sensitivity. This is why hydrogen is often used (second highest γ of all nuclei) and why there is a push to higher-field MR imagers. Water is typically imaged because it is the most concentrated of all hydrogen-containing molecules. Fluorine also has a high γ, and highly concentrated perfluoro compounds have been imaged directly. Sensitivity could also be increased if the ratio in Equation 2.2 could be made much smaller. For certain nuclei (^{13}C, ^{129}Xe, ^{3}He), it is possible to polarize the material at low temperature, where kT is small, and then warm the material to room temperature and yet maintain this hyperpolarization.

When nuclei are placed in a magnetic field, it takes a certain time for them to align with or against the field. The time constant for this rate of alignment is called T_1. When radiofrequency (rf) is applied, spins absorb energy and undergo transitions between the $+^1/_2$ and $-^1/_2$ states. After the radiofrequency pulse is switched off, the spins emit energy and return to their initial equilibrium state. It is this emission of energy that is detected in the imaging experiment. The rate at which the nuclei return to their initial state is termed relaxation: for the component of magnetization parallel to the external field, the time constant is T_1, the longitudinal (also called spin-lattice) relaxation time; for the component of magnetization perpendicular to the external field, the time constant is T_2, the transverse (also called spin-spin) relaxation time. Transverse relaxation occurs because of local magnetic field inhomogeneities that are caused by (1) microscopic effects caused by magnetic interactions between neighboring molecules (chemistry) and (2) macroscopic effects related to the spatial variation of the external field (physics), e.g., through differences in magnetic susceptibility between air and liquid. The aggregate effect is termed T_2^*, while relaxation just due to molecular effects is termed T_2. Although typically it is only water that is detected, water in different tissues has different relaxation times. By making the image acquisition sensitive to differences in T_1, T_2, and T_2^*, contrast can be generated.

2.3 T_1, T_2, AND T_2^* CONTRAST AGENTS AND RELAXIVITY

In a T_1-weighted image, the repetition time (TR) is set short relative to T_1. Under these conditions, water hydrogens with long T_1 are not given enough time to relax (emit energy) before the next pulse of radiofrequency energy, and so the signal detected from these hydrogens is low. If T_1 is short, then relaxation is fast and most of the signal can be detected. Short T_1 results in positive image contrast.

In a T_2-weighted image, the echo repetition time (TE) is long relative to T_2. Here, fast transverse relaxation leads to signal loss. In a T_2-weighted image, tissue with long T_2 gives positive contrast, while regions with short T_2 will appear dark. Similarly, in T_2^*-weighted images (typically gradient echo images), tissue with short T_2^* will appear dark.

All contrast agents shorten T_1, T_2, and T_2^*. However, it is useful to classify MRI contrast agents into two broad groups based on whether the substance increases the transverse relaxation rate ($1/T_2$) by roughly the same amount that it increases the longitudinal relaxation rate ($1/T_1$) or whether $1/T_2$ is altered to a much greater extent. The first category is referred to as T_1 agents because, on a percentage basis, these agents alter $1/T_1$ of tissue more than $1/T_2$, owing to the fast endogenous transverse relaxation in tissue. With most pulse sequences, this dominant T_1 lowering effect gives rise to increases in signal intensity; these are positive contrast agents. The T_2 agents largely increase the $1/T_2$ of tissue selectively and cause a reduction in signal intensity; these are negative contrast agents. Paramagnetic gadolinium- and manganese-based contrast agents are examples of T_1 agents, while ferromagnetic and superparamagnetic iron oxide particles are examples of T_2 agents.

There are many mechanisms by which contrast agents shorten T_1 and T_2, but in many cases the effect of these mechanisms can be reduced to a single constant, called relaxivity. The simple way to quantify this effect is to consider the rate of relaxation, $1/T_1$ (sometimes denoted R_1). For most cases in medical imaging, the contrast agent increases the relaxation rate proportional to the amount of contrast agent:

$$\frac{1}{T_1} = \frac{1}{T_{1o}} + r_1 \,[\text{CA}] \tag{2.3}$$

where T_1 is the observed T_1 with contrast agent in the tissue, T_{1o} is the T_1 prior to addition of the contrast agent, [CA] is the concentration of contrast agent, and r_1 is the longitudinal relaxivity, often just relaxivity. The conventional units for r_1 are $mM^{-1}sec^{-1}$ (per millimolar per second, sometimes written as $l{\cdot}mol^{-1}sec^{-1}$). Thus, the slope of $1/T_1$ as a function of contrast agent concentration reveals the relaxivity. Transverse, or T_2, relaxivity is defined in an analogous way:

$$\frac{1}{T_2} = \frac{1}{T_{2o}} + r_2 \,[\text{CA}] \tag{2.4}$$

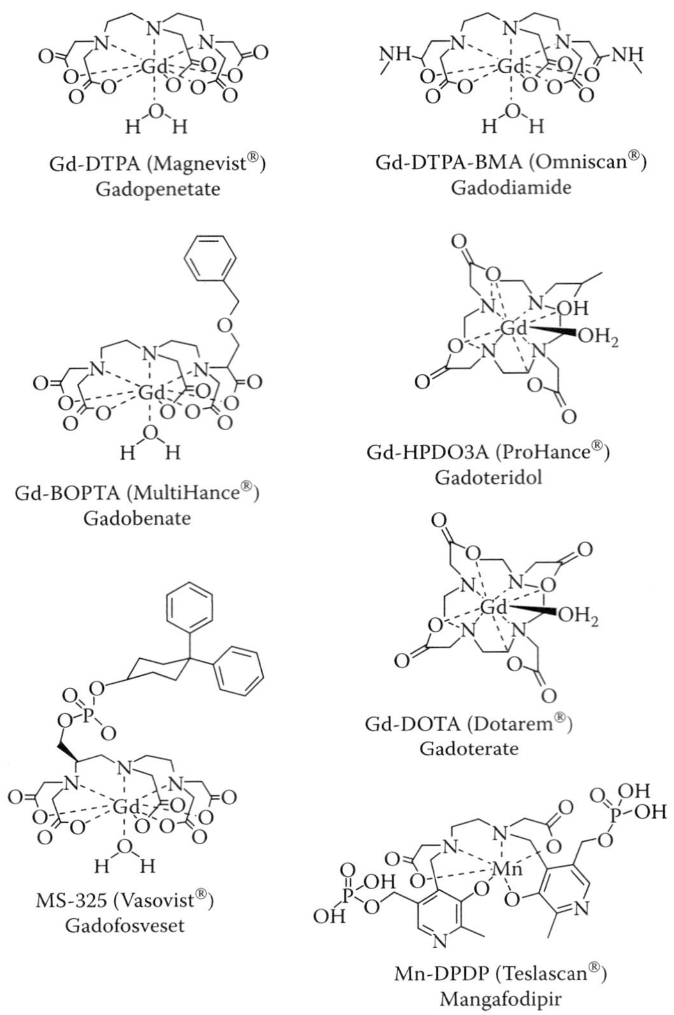

FIGURE 2.1 Some approved (U.S. or Europe) MRI contrast agents with trade name, generic name, and chemical abbreviation.

Relaxivity is a useful parameter that allows an *in vitro* ranking of various contrast agents. Increased relaxivity typically equates to greater contrast at an equivalent dose, or equivalent contrast at a lower dose. However, *in vivo*, signal change is more complex than the simple linear relationship implied by Equations 2.3 and 2.4. There are both physical and chemical reasons for this that will be described below. First, the chemistry of these contrast agents will be discussed.

2.4 CHEMISTRY OF T_1 AGENTS

MRI contrast agents must be biocompatible pharmaceuticals in addition to nuclear relaxation probes. Because of the relatively high doses used, they should have good water solubility. They should be nontoxic at the dose required to give the required imaging effect; $1/T_1$ changes as small as 10 to 20% can be detected by MRI.

T_1 agents are typically gadolinium(III) complexes, manganese(II) complexes, or, in several animal studies, just the Mn^{2+} cation[3,4] (given as $MnCl_2$; see also Chapter 20 for another Mn^{2+} application). Figure 2.1 shows some contrast agents approved for use in the U.S. and Europe.

Chemically, the gadolinium (Gd) compounds exhibit similar features: an eight-coordinate ligand binding to Gd and a single water molecule coordinated to Gd. The multidentate ligand is required for safety.[5] The ligand encapsulates the gadolinium, resulting in a high thermodynamic stability and kinetic inertness with respect to metal loss. This enables the contrast agent to be excreted intact — an important property since these contrast agents tend to be much less toxic than their individual components. For example, the DTPA ligand and gadolinium chloride both have an LD_{50} of 0.5 mmol/kg in rats (LD_{50} = dose that causes death in 50% of the animals), while the Gd-DTPA complex has nearly a factor of 20 higher safety margin, with an LD_{50} of 8 mmol/kg for the Gd-DTPA complex.[6]

Metal complex stability can be assessed by determining the metal-ligand stability constant (also called the formation constant).[7] Stability constants are typically very high for Gd complexes used as contrast agents,[5] $K > 10^{17}\ M^{-1}$. Stability constant determination requires knowledge of the protonation constants, pKa values, of the ligand. A less rigorous but still practical approach is to determine the relative stability compared to that of a known agent. For example, is the complex more or less stable than DTPA or EDTA?

Kinetics is also important. How fast will the complex release the gadolinium? Metal ion release is catalyzed by acid, competing metal ions such as zinc, or other coordinating anions like phosphate. Dissociation rates can be measured absolutely using, for instance, radiochemical[8] or optical methods[9] for detection. Alternately, some relative rate can be measured and compared to a benchmark compound like Gd-DTPA or Gd-DOTA. Vander Elst and colleagues have monitored the change in P-31 relaxation rate as a function of time when a Gd complex is subjected to a cocktail of phosphate groups,[10,11] and ranked the relative inertness of various contrast agents. As a general rule, macrocyclic ligands like DOTA tend to give more kinetically inert complexes than acyclic ligands like DTPA.

To illustrate the importance of both thermodynamic stability and kinetic inertness, consider Gd-DTPA, Gd-DTPA-BMA, and Gd-EDTA. The stability constants[5] for Gd-EDTA and Gd-DTPA-BMA are similar ($K = 10^{17}$) and much lower than for Gd-DTPA ($K = 10^{22.5}$). However, Gd-EDTA has a greater than 10-fold lower LD_{50} than the other two compounds.[12] Biodistribution studies[13] showed that there was more than 25 times as much Gd deposited in the femur (indicative of Gd loss from the complex) of a rat 14 days after Gd-EDTA was injected than when Gd-DTPA-BMA was administered. What may rationalize these findings is that the rate of transmetallation with other metal ions for Gd-DTPA-BMA is about the same as that for Gd-DTPA.[9] Gd-EDTA, on the other hand, is much more labile.[13] This indicates that complexes of lower stability can be used *in vivo* provided that they are kinetically inert.

There is a great deal of literature on the kinetics and thermodynamics of metal complex stability for gadolinium complexes. An excellent review is given by Brücher and Sherry[14] in a book on the chemistry of MR contrast agents. Lanthanide coordination chemistry tends to be very similar, so observations about stability and inertness for other lanthanides, like samarium-153, lutetium-177, and yttrium-90, used in nuclear medicine will also apply to gadolinium.

2.5 PARAMAGNETIC ENHANCED NUCLEAR RELAXATION

The paramagnetic ion and coordinated water molecule are essential to providing contrast. The gadolinium(III) ion has a high magnetic moment and a relatively slow electronic relaxation rate, properties that make it an excellent relaxer of water protons. The proximity of the coordinated water molecule leads to efficient relaxation. The coordinated water molecule is in rapid chemical exchange (10^6 exchanges per second) with solvating water molecules.[15] This rapid exchange leads to a catalytic effect whereby the Gd complex effectively shortens the relaxation times of the bulk solution.

The approved manganese agent, Mn-DPDP (Teslascan, Mangafodipir) does not have a site for coordinating water.[16] However, this compound dissociates *in vivo* and the manganese is taken up

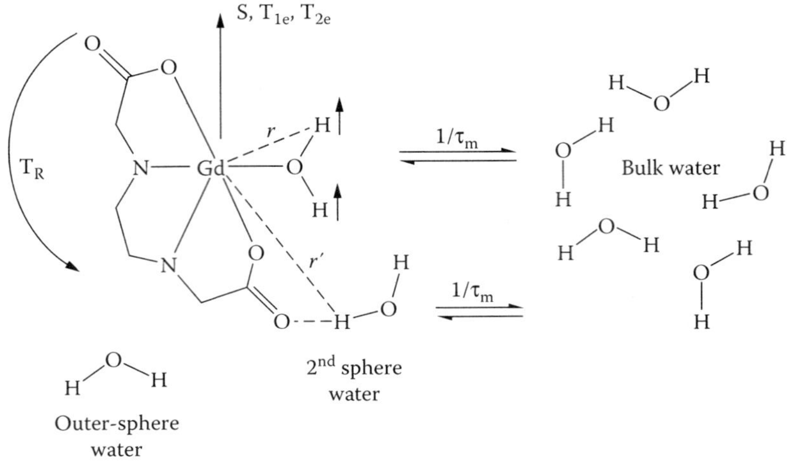

FIGURE 2.2 Molecular parameters that influence inner- and second-sphere relaxivity.

by hepatocytes. In the liver and gall bladder, the Mn(II) ion is bound to macromolecules, resulting in increased relaxivity.[17]

There are two pathways by which the paramagnetic complex enhances water relaxation. There is an inner-sphere effect whereby the metal-bound water is relaxed efficiently and this water undergoes rapid chemical exchange with other solvent water. This relaxation and fast water exchange catalytically enhances the relaxation rate of the bulk water. In addition, water not contained in the first coordination sphere can also be relaxed by the ion. This water is classified into two groups: second-sphere water, which denotes water molecules that directly hydrate the complex and have a lifetime in the second sphere longer than the time constant for water diffusion, and outer-sphere water, which is not associated with the complex but is diffusing nearby. Second- and outer-sphere water for this discussion are termed outer sphere. The different classes of water and the molecular parameters that determine relaxivity are shown in Figure 2.2. Relaxivity can be separated into inner- and outer-sphere relaxivity, r_1^{IS} and r_1^{OS}, respectively, and this is useful for understanding the biophysics behind the relaxation enhancement.

$$r_1 = r_1^{IS} + r_1^{OS} = \frac{q/[H_2O]}{T_{1m} + \tau_m} + r_1^{OS} \tag{2.5}$$

Here q is the number of water molecules in the inner sphere, T_{1m} is the relaxation time of these inner-sphere water protons, and τ_m is the lifetime of these waters in the inner sphere (the reciprocal of τ_m is the water exchange rate, $k_{ex} = 1/\tau_m$). Relaxivity depends directly on how many water molecules are coordinated, and inversely on the relaxation time of the bound water and how long it is bound. The relaxation rate at typical imaging fields for water protons bound to Gd or Mn is given by Equation 2.6:

$$\frac{1}{T_{1m}} = \frac{2}{15}\frac{\gamma_H^2 g_e^2 \mu_B^2 S(S+1)}{r_{M-H}^6}\left[\frac{3\tau_c}{1+\omega_H^2\tau_c^2}\right] \tag{2.6}$$

$$\frac{1}{\tau_c} = \frac{1}{\tau_R} + \frac{1}{T_{1e}} + \frac{1}{\tau_m} \tag{2.7}$$

Paramagnetic relaxation ($1/T_{1m}$) occurs via a dipolar mechanism. Relaxation depends on the spin quantum number (S), some fundamental constants (magnetogyric ratio, Bohr magneton, electronic g factor, $g_e = 2$ for Gd(III) and Mn(II)), the metal-to-hydrogen (M-H) distance, $r_{M\text{–}H}$, the proton Larmor frequency ω_H (in rad/sec), and a correlation time τ_c. The product $S(S + 1)$ is proportional to the magnetic moment. All other factors being equal, the higher the magnetic moment, the more efficient the relaxation. This is why Gd^{3+} ($S = 7/2$) is preferred to an ion such as copper (Cu^{2+}, $S = 1/2$). The dipolar effect depends on the distance between the ion and the hydrogen nucleus, r_{MH}, to the inverse sixth power. The inner-sphere water is critical; it has the shortest metal-to-hydrogen distance of water hydrating the metal complex. Mn^{2+} has a lower spin number ($S = 5/2$) than Gd^{3+}, but Mn^{2+} is a small ion and has a shorter Mn-H distance. Curiously, the $S(S + 1)/r^6$ term is approximately equal for Mn^{2+} and Gd^{3+}.

Fluctuating magnetic dipoles can induce spin transitions and cause spin relaxation. A correlation time is a time constant for characterizing these fluctuations; its reciprocal, $1/\tau_c$, is the average rate constant for these fluctuating dipoles. The closer this rate is to the Larmor frequency, the more efficient the relaxation. There are several processes that lead to fluctuating magnetic dipoles. Electronic relaxation ($1/T_{1e}$) at the Gd(III) ion creates a fluctuating field. Rotational diffusion ($1/\tau_R$) of the complex creates a fluctuating field. Water exchange ($1/\tau_m$) in and out of the coordination sphere creates a fluctuating field for the hydrogen nucleus. It is the fastest rate (shortest time constant) that determines the extent of relaxation (Equation 2.7). For most Gd(III) and Mn(II) complexes at imaging field strengths, it is rotational diffusion that is the dominant correlation time. The terms in square brackets are sometimes referred to as dispersive because once $\omega^2\tau_c^2 > 1$, the relaxation rate becomes slower and disperses with increasing frequency (field).

Mn(II) and Gd(III) are chosen as relaxation agents because electronic relaxation ($1/T_{1e}$) is slow, the magnetic moment is large, and water exchange is typically fast. To illustrate the effect of rotational diffusion and the field dependence on relaxivity, consider two similar compounds shown in Figure 2.1: Gd-DTPA and MS-325. MS-325 has the same Gd-binding ligand but also has a lipophilic group that enables it to bind to serum albumin. Small molecules like Gd-DTPA tumble very fast, in the gigahertz range (1 GHz = 1000 MHz), but the Larmor frequency for protons at imaging fields is much slower. For example, at 1.5 tesla, the Larmor frequency is about 65 MHz, so relaxation is not as efficient as it could be. Larger molecules like proteins tumble much more slowly. When contrast agents are made to tumble more slowly, relaxivity is increased. Lauffer[18] pointed out that if small contrast agents could be made to bind noncovalently to protein targets, then their relaxivity would be increased upon binding because the contrast agent would take on the rotational characteristics of the protein. This was termed receptor-induced magnetization enhancement (RIME). MS-325 is an example of a contrast agent designed to exploit the RIME effect.[19] In the absence of albumin, the relaxivity of MS-325 is about 50% greater than Gd-DTPA because its larger size results in a slower tumbling rate, but in the presence of albumin, the relaxivity is about 600% greater than Gd-DTPA at 1.5 tesla. This is illustrated in Figure 2.3, where the magnetic field dependence on relaxivity is plotted for MS-325 (circles) and Gd-DTPA (squares) in either serum albumin solution (filled symbols) or buffered saline (open symbols).

Figure 2.3 also shows that the relaxivity is rather field independent for Gd-DTPA and MS-325 in buffer, but the relaxivity of MS-325 bound to protein first increases and then decreases with field. At high fields, the inequality $\omega_H^2\tau_c^2 > 1$ is reached and T_{1m} will become longer (and relaxivity lower) with increasing field. The relaxivity first increases because the correlation time can also change with field. At low fields, electronic relaxation is very fast and the correlation time, τ_c, is approximately T_{1e}. The electronic relaxation rate for Gd(III) and Mn(II) decreases with the square of the magnetic field, so as field is increased, τ_c is getting longer and relaxivity increases. At some point, the rate of rotational diffusion is the fastest process and τ_c becomes τ_R. The field at which T_{1e} no longer dominates the correlation time will depend on the complex and the rotational correlation time. However, it appears safe to say that at 1.5 tesla and above, rotational motion is the correlation time that defines relaxivity.

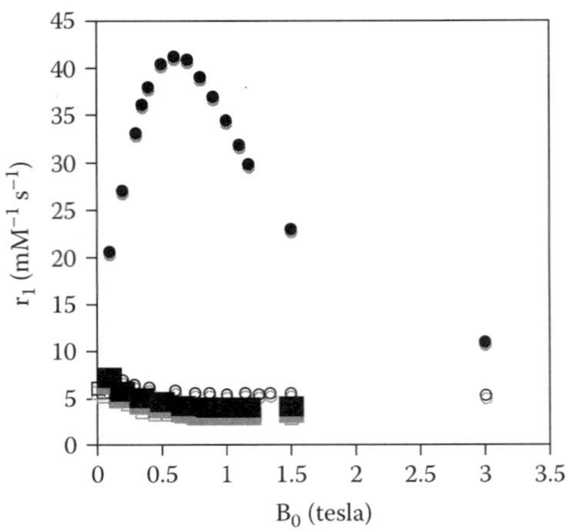

FIGURE 2.3 Magnetic field dependence on relaxivity for MS-325 (circles) and Gd-DTPA (squares) in either serum albumin solution (filled symbols) or buffered saline (open symbols) at 37°C.

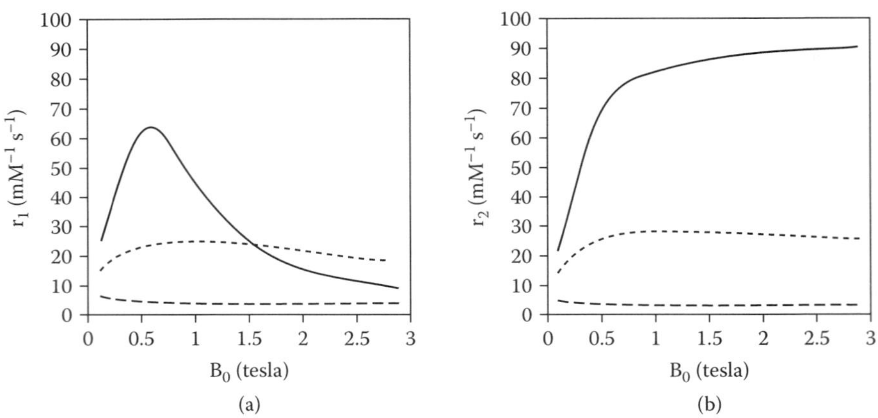

FIGURE 2.4 Effect of rotational correlation time on longitudinal (r_1) (a) and transverse (r_2) (b) relaxivities as a function of field strength. Long correlation time (τ_R = 10 nsec, —) typical of albumin binding gives high r_1 that decreases with increasing field and high r_2; intermediate correlation time (τ_R = 1 nsec, ---) shows relaxivity maximum for r_1 pushed out to higher field; short correlation time (τ_R = 0.1 nsec, ---) typical of ECF agents shows low, roughly field independent r_1, r_2. Simulations with other parameters typical of Gd-based agents.[22]

Transverse relaxivity can also be factored into inner- and outer-sphere terms. For Gd(III) and Mn(II), T_{2m} is predominantly governed by a dipolar mechanism (at very high fields, another mechanism called Curie spin relaxation[20,21] also shortens T_2). T_{2m} is given by Equation 2.8, which is very similar to that of Equation 8.6 for longitudinal relaxation, except that there is also a nondispersive term inside the square brackets.

$$\frac{1}{T_{2m}} = \frac{1}{15}\frac{\gamma_H^2 g_e^2 \mu_B^2 S(S+1)}{r_{M-H}^6}\left[4\tau_c + \frac{3\tau_c}{1+\omega_H^2\tau_c^2}\right] \tag{2.8}$$

Figure 2.4 further illustrates the effect of correlation time and field strength on relaxivity. In Figure 2.4, r_1 and r_2 are simulated over a range of fields encountered in MRI for correlation times of 0.1 nsec (typical of extracellular fluid, ECF, agents), 1 nsec (intermediate motion), and 10 nsec

TABLE 2.1
Relaxivities[68] ($mM^{-1}sec^{-1}$) of Selected Contrast Media (0.25 m*M*) in Plasma at 0.47, 1.5, and 3 Tesla at 37°

Compound		0.47 Tesla		1.5 Tesla		3 Tesla	
Chemical/Code	Commercial	r_1	r_2	r_1	r_2	r_1	r_2
Gd-DTPA	Magnevist	3.8	4.1	4.1	4.6	3.7	5.2
Gd-DTPA-BMA	Omniscan	4.4	4.6	4.3	5.2	4.0	5.6
Gd-HPDO3A	Prohance	4.8	6.1	4.1	5.0	3.7	5.7
Gd-DOTA	Dotarem	4.3	5.5	3.6	4.3	3.5	4.9
Gd-EOB-DTPA	Primovist	8.7	13	6.9	8.7	6.2	11.0
Gd-BOPTA	MultiHance	9.2	12.9	6.3	8.7	5.2	11.0
MS-325[a]	Vasovist	47.2	57.6	27.7	72.6	9.9	73
Gadomer		19	23	16.0	19.0	13.0	25
AMI-25	Endorem Feridex	NA	NA	4.5	33	2.7	45
SHU-555A	Resovist	15	101	7.4	95	3.3	160
SHU-555C	Supravist	22.3	99	10.7	38	5.6	95

Note: NA = not available.
[a] Data from Eldredge et al.[23]

(typical of albumin-bound agents). Figure 2.4 shows that the benefits of very slow rotation are seen at lower field strengths. Note also that r_1 does not go to zero because there is an outer-sphere component to relaxivity[22] and the correlation times that govern outer-sphere relaxivity are quite short. r_2 is always greater than r_1, and for very slow tumbling systems, the r_2/r_1 ratio becomes large at high fields. Electronic and nuclear relaxation are described in greater detail in various reviews and books.[5,20]

Figure 2.4 suggests that slow tumbling T_1 agents become less effective at high fields, but one must also recall that relaxation times for tissue are longer at high fields and that signal-to-noise ratio (SNR) increases with increased field. These factors and the choice of sequence mean that a contrast agent with a lower relaxivity at 3 T than 1.5 T may still provide greater contrast at 3 T than at 1.5 T. Table 2.1 lists relaxivities for some widely studied gadolinium complexes in plasma at 0.5, 1.5, and 3 tesla. The ECF agents show little field dependence, while the slow tumbling compounds Gadomer and MS-325 (albumin bound in plasma) and the iron oxide particles show strong field dependence.

The MS-325 example clearly shows the importance of chemical speciation on observed relaxation rates. Obviously, if the fraction of the contrast agent bound to the macromolecule is lower, then the observed relaxivity will also be lower. Contrast agents such as Gd-BOPTA (MultiHance, gadobenate) are only about 10% bound to plasma proteins and have relaxivities intermediate between albumin-bound MS-325 and ECF agents (Table 2.1). For compounds with reversible protein binding, the relaxivity observed will no longer be independent of concentration,[23] because changes in concentration will shift the equilibrium between free and bound. As a result, the fraction of MS-325 bound to albumin immediately after a bolus injection will be lower than that after the compound has distributed because the high concentration present in the bolus will saturate the albumin.[24]

Other chemical effects can also alter relaxivity *in vivo*. For instance, increasing q promises to increase relaxivity, and this is generally true in pure water.[25,26] However, increasing the number of waters bound also opens up a cleft around the metal ion that can allow other ligands to bind. Endogenous citrate, phosphate, and bicarbonate have high affinity for gadolinium.[27] When these ligands bind, they displace the bound water molecules and actually decrease relaxivity. This effect is typically only seen in $q \geq 2$ complexes, presumably because when there is only one water bound, there is not enough space near the metal ion to accommodate the larger bicarbonate or phosphate ion.

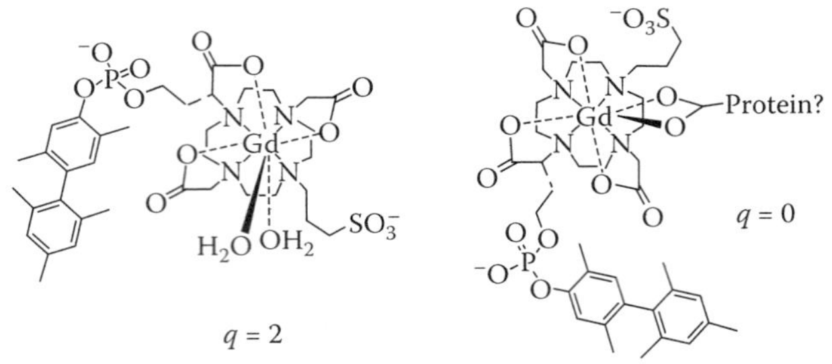

(a) PBS, pH 7.4, 37°C. $r_1 = 9.1\ mM^{-1}\ s^{-1}$ at 20 MHz

(b) PBS, pH 7.4, 37°C, 0.66 mM HSA, 0.1 mM Gd. Compound > 99% bound to HSA, $r_1 = 12.3\ mM^{-1}\ s^{-1}$ at 20 MHz

FIGURE 2.5 Gadolinium complex where $q = 2$ in buffered solution but $q = 0$ when bound to a protein. Hydration state suggested by relaxivity and confirmed by ^{1}H ENDOR.

Protein binding can also have an effect on q. For instance, Zech et al.[25] showed that the albumin binding $q = 2$ derivative shown in Figure 2.5 had high relaxivity in phosphate-buffered saline (PBS), but the expected relaxivity boost was missing in the presence of human serum albumin (HSA), even though the complex had high affinity to albumin. Electron-nuclear double resonance (ENDOR) spectroscopy showed that the two water molecules were displaced when the complex was bound to serum albumin, accounting for the lower than expected relaxivity. Presumably a protein side chain (Asp, Glu?) with high local concentration coordinated the Gd ion and displaced the waters. When europium is used as a surrogate for gadolinium, fluorescence lifetime measurements can also reveal hydration number changes.[27,28]

Chemical exchange of water in and out of the first coordination sphere is another parameter that can have a significant effect on relaxivity. The water residency time, τ_m, appears in the denominator of Equation 2.5 as $(T_{1m} + \tau_m)$. For fast tumbling molecules like Gd-DTPA, relaxation is less efficient because of rotational motion and $T_{1m} > \tau_m$. As a result, the first generation of clinical contrast agents all have similar relaxivities because their size is about the same, meaning T_{1m} is similar. However, the compounds in Figure 2.1 all have different water exchange rates. When T_{1m} is reduced, τ_m can become important and may limit relaxivity. For instance, when DTPA is converted into the bis(amide) DTPA-BMA, the water exchange rate at gadolinium is reduced by a factor of 10. For Gd-DTPA and Gd-DTPA-BMA, fast rotation means that T_{1m} is on the order of 10 μsec at 1.5 tesla. If T_{1m} is reduced by slowing down tumbling, say by protein binding, then T_{1m} is reduced to 0.7 μsec. In this scenario, the slow water exchange at the Gd-DTPA-BMA chelate would significantly limit its relaxivity.

Interaction with a protein target can also affect water exchange. Eldredge et al.[23] found that the relaxivity of MS-325 was different when bound to serum albumins of different species. For instance, relaxivity was almost twice as high when MS-325 was bound to human serum albumin than when bound to rabbit serum albumin. Based on variable temperature and variable field relaxivity measurements, they postulated that water exchange at MS-325 was slower when MS-325 was bound to rabbit serum albumin than when MS-325 was bound to human serum albumin.

The molecular factors that serve to increase or decrease r_1 will affect r_2 in the same way. However, T_2-weighted agents are typically used because of their susceptibility (T_2*) effect, rather than the pure T_2 shortening, as will be discussed below. It should be clear that relaxivity is not a constant, but depends strongly on environment. It is important to measure r_1 and r_2 under the conditions where the contrast agent will be used (magnetic field, physiological temperature, tissue) in order to better understand its effect *in vivo*.

When manganese is given as the simple salt $MnCl_2$, its relaxivity in water will be quite different than in the *in vivo* situation. Manganese is known to accumulate in the liver, where it has a high

TABLE 2.2
Relaxivities ($mM^{-1}sec^{-1}$) of Mn^{2+} in the Presence of Various Proteins[29] and Chelators[30] (Figure 2.6) at 0.47 Tesla and 25°C

Chelator Type	Protein	r_1	Hydration Number, q
None	None	9.0	6
None	Pyruvate kinase	275	?
None	Concanavalin A	96	?
None	Carboxypeptidase	43	?
EDTA derivative	None	6.4	1
DTPA derivative	None	3.5	0
EDTA derivative	Human serum albumin	55.9	1
DTPA derivative	Human serum albumin	4.9	0

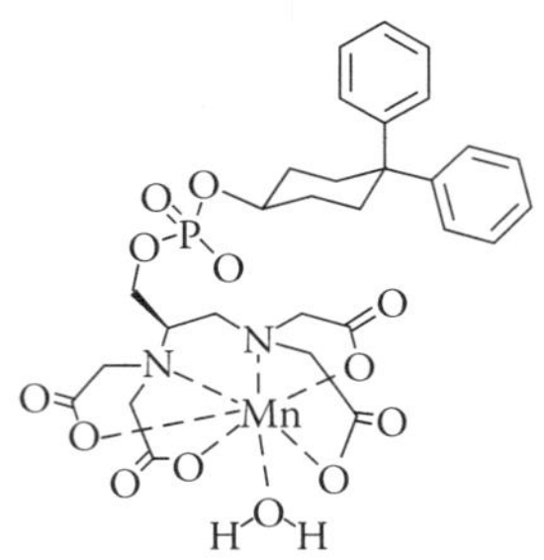

Mn-EDTA derivative
$q = 1$

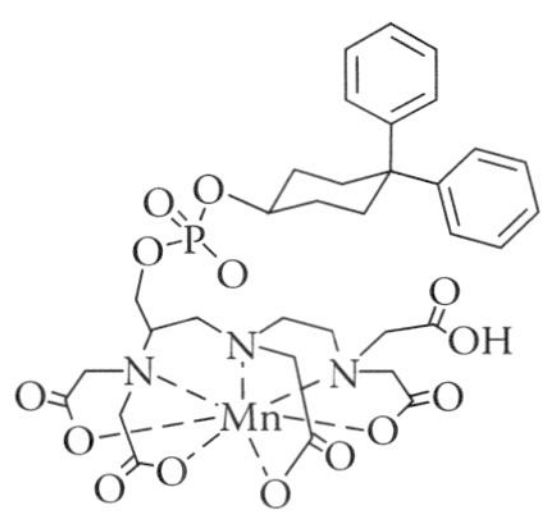

Mn-DTPA derivative
$q = 0$

FIGURE 2.6 Serum albumin-binding derivatives of manganese(II). The EDTA derivative (left) has one inner-sphere water while the DTPA derivative has no site available for direct water coordination, $q = 0$

relaxivity due to binding to macromolecules. The mechanism of action of Mn-DPDP is dissociation of the complex releasing free Mn^{2+} into the hepatocytes, resulting in manganese bound to macromolecules and increased relaxivity.[17] In his excellent review on contrast agents,[29] Lauffer listed relaxivities of Mn(II) bound to several proteins, and relaxivity can vary over two orders of magnitude. The speciation of manganese is critical to understanding its contrast-enhancing behavior. Table 2.2 illustrates this point. The relaxivity of the aqua ion is increased in the presence of proteins, but the choice of protein is also critical. Troughton et al.[30] recently described Mn(II) complexes of EDTA or DTPA ligands that were derivatized with the same albumin-binding group used with MS-325 (Figure 2.6). The Mn-EDTA derivative has one inner-sphere water, while the Mn-DTPA derivative does not have a water bound. As expected, protein binding has a large impact on the Mn-EDTA derivative but not on the $q = 0$ Mn-DTPA compound; these relaxivities are also listed in Table 2.2.

2.6 T_2 AGENTS

Paramagnetism generally involves the magnetism of small isolated ions that only behave as local magnets in the presence of an external magnetic field. For paramagnetic materials that contain multiple ions, the total magnetic susceptibility is directly proportional to the number of ions in the material. Therefore, the molar magnetic susceptibility (magnetic susceptibility divided by the number of ions) is constant. There are other materials that exhibit ferromagnetism and superparamagnetism. For certain materials such as ferrite (iron oxide) the individual spins of each iron cooperatively, via quantum mechanical interactions, build up to give the crystal a very large total spin, resulting in a very large molecular magnetic susceptibility that is a function of the number of spins. Such a material is called ferromagnetic, and its magnetism persists outside the external magnetic field. A weaker form of this is superparamagnetism: small particles of iron oxide with aligned spins in a magnetic field. Since the particles are small (submicron), the magnetic susceptibility effect is smaller than for large crystals of ferrites. Superparamagnets are no longer magnetic outside of the external field. These iron oxide particles represent an important class of contrast agents.[31]

The iron oxide particles consist of a core of one or more magnetic crystals of Fe_3O_4 embedded in a coating. Because these are materials, there is a distribution of sizes. Ultrasmall particles of iron oxide (USPIOs) have a single crystal core and a submicron diameter (e.g., ferumoxtran (Sinerem, Combidex, or AMI-227) has a crystal diameter of 4.3 to 4.9 nm and a global particle diameter of ca. 50 nm).[32] Small particles of iron oxide (SPIOs) have cores containing more than one crystal of Fe_3O_4 and are larger than USPIOs but still submicron (e.g., ferumoxide (Endorem, Feridex, or AMI-25) has a crystal diameter of 4.3 to 4.8 nm and a global particle diameter of ca. 200 nm).[32] USPIOs and SPIOs are small enough to form a stable suspension and can be administered intravenously. The size differences result in differences in pharmacokinetic behavior, which will be described below. There are also large particles that are used for oral applications (e.g., Abdoscan, 50-nm crystals making up a 3-μm particle).[33] USPIOs are also referred to as microcrystalline iron oxide nanoparticles (MIONs).

There are no inner-sphere water molecules in iron particles, and relaxation of water arises from the water molecules diffusing near the particle. However, the mechanism of outer-sphere relaxation is different than described above. One feature is that the crystals have a net magnetization, and as the external field is increased, this magnetization is increased (this is true as well for Gd, but the effect is much smaller). The modulation of this net magnetization can cause proton relaxation (so-called Curie spin relaxation). The theories describing the field dependence of iron oxide relaxivity have been reported.[33] Solvent relaxation induced by iron oxide systems is complex. Bulte et al.[34,35] used a combination of variable field T_2 measurements (T_2 relaxometry), electron paramagnetic resonance (EPR), and magnetization measurements to fully characterize one such USPIO, MION-46L. To fully explain their experimental findings, they proposed the existence of three different magnetic phases for this USPIO: a superparamagnetic core, an antiferromagnetic ferritin-like phase of incompletely converted iron oxyhydroxide, and a paramagnetic surface effect of ferric ions.

There are some generalities about relaxivity in these particles. For the USPIOs, longitudinal relaxivity (r_1) can be quite high and these can function as effective T_1 agents. The r_2/r_1 ratio for USPIOs is significantly larger than for gadolinium complexes, and r_2/r_1 increases with increasing magnetic field. When there is aggregation of crystals, which is the case in SPIOs, longitudinal relaxivity tends to decrease (r_1 drops) and transverse relaxivity increases (r_2 increases). Thus, for both the particles themselves and aggregates of particles the ratio of r_2/r_1 typically increases as the size of the particles or aggregates increases, though the T_2 relaxivity as a function of particle size can be quite complicated. See, for example, Weisskoff et al.[36] and Hardy and Henkelman.[37] The effect of aggregation of crystals is that the aggregate itself can be considered a large magnetized sphere whose magnetic moment increases with increasing field strength. This gives rise to susceptibility effects and the SPIOs can act as T_2* relaxation agents. This has important consequences when considering the effects of contrast agent compartmentalization on imaging (see below).

There is also a speciation effect on relaxivity for iron particles. When these nanoparticles cluster together, transverse relaxivity, r_2, increases. This phenomenon has been exploited by Perez et al., who have created sensors based on assembling and disassembling these nanoparticle clusters.[38,39]

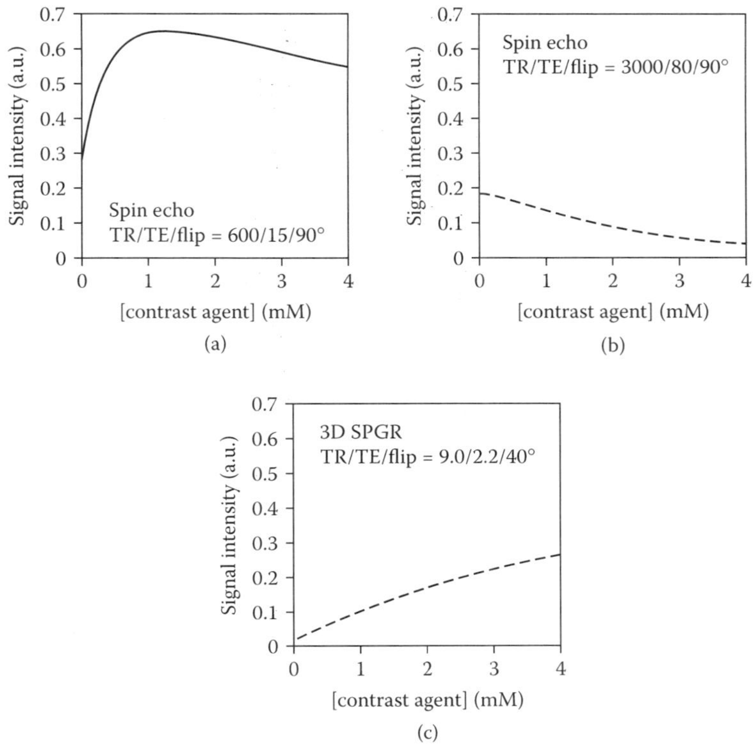

FIGURE 2.7 Effect of contrast agent on image intensity (baseline T_1, T_2 typical of muscle) on T_1- and T_2-weighted scans. (a) T_1-weighted spin echo (TR = 600 msec) shows linear increase of signal only for contrast agent concentration less than 0.5 m*M*. (b) T_2-weighted spin echo images (TR = 3000 msec) shows only T_2 signal loss effects due to contrast agent with no T_1 enhancement because of long TR. (c) Typical short-TR fast spoiled gradient echo sequence. The very short TE and short TR give monotonically increased image intensity across the entire range of contrast agent concentrations typically found in clinical scans.

2.7 IMAGING PHYSICS: T_1 AND T_2 AGENTS

Contrast agent behavior *in vivo* is quite complex. Even in the simple case of a single compartment with pure linear relaxation, the effect of the contrast agent on the MR image is generally nonlinear. In traditional spin echo sequences nonlinearity can be a result of T_1 saturation or T_2 signal loss. Once the contrast agent shortens $T_1 < TR/2$, increasing the contrast agent concentration will have little effect on increasing the available longitudinal magnetization because the tissue will have nearly fully recovered the magnetization before the next rf pulse. Because contrast agents affect both T_1 and T_2 relaxation, at high enough concentration the contrast agent will reduce T_2 to the order of TE, and will then decrease MR image intensity. These effects are illustrated in Figure 2.7, where signal intensity is plotted vs. contrast agent concentration for T_1- and T_2-weighted spin echo sequences. Figure 2.7 was generated assuming a contrast agent relaxivity of 4 $mM^{-1}sec^{-1}$, typical of most commercial ECF gadolinium agents, and tissue relaxation times typical of muscle (T_1 = 1200 msec, T_2 = 50 msec). For the T_1-weighted sequence (TR/TE = 600/15) (Figure 2.7a), signal intensity begins to level out at a contrast agent concentration between 0.5 and 1.0 m*M*; this is the range where the T_1 has dropped to around 300 msec, or TR/2. At concentrations above 1 m*M*, the T_1 effect is saturated, and the only *imaging* effect of the contrast agent is to make T_2 shorter and cause signal loss, even on this T_1-weighted sequence. Signal is lost because even a T_1-weighted sequence has a finite TE, and T_2 effects enter when T_2 is short enough.

The signal intensity plateau on the T_2-weighted scan (TR/TE = 3000/80) (Figure 2.7b) occurs at much lower contrast agent concentration. Because TR is so long, the only real effect of the contrast agent is to reduce (rather than increase) signal intensity on this T_2-weighted scan. T_2 agents create negative contrast exactly by providing enhanced T_2 relaxation, and thus darker images on T_2-weighted scans.

Increasing the relaxivity (r_1 or r_2) will have the effect of pushing the simulated curves in Figure 2.7a to the left, that is, peak signal and subsequent signal loss will occur at lower contrast agent concentrations. A more linear response of signal to contrast agent can be achieved with a fast three-dimensional spoiled gradient echo (3D SPGR) sequence. This is illustrated in Figure 2.7c, where signal intensity is plotted vs. contrast agent concentration using the same tissue relaxation times and relaxivities as in Figure 2.7a and b for a typical fast 3D SPGR sequence, TR/TE/flip = 9.0/2.2/40°. The short TR and very short TE ensure that signal intensity increases across the entire concentration range. At high concentration the effect of the contrast agent is becoming nonlinear, but the signal is still increasing with increasing contrast agent concentration.

However, in tissue, relaxation itself is usually nonlinear. The extent to which a metal complex influences tissue relaxation rates depends on three factors:

1. The chemical environment encountered by the metal complex. Binding of the agent to macromolecules can cause significant relaxivity enhancement. Similarly, clustering of nanoparticles can strongly alter r_2. This was discussed in detail above for both T_1 and T_2 agents.
2. Compartmentalization of the metal complex in tissue. Generally, tissue water is compartmentalized into intravascular, interstitial (fluid space between cells and capillaries), and intracellular space constituting roughly 5, 15, and 80% of total water, respectively. Cellular organelles further subdivide the intracellular component. If water exchange between any of these compartments is slow relative to the relaxation rate in the compartment with the longest T_1, multiexponential relaxation may result. This can decrease the effective tissue relaxivity of an agent because not all of the tissue water is encountering the paramagnetic center.
3. The magnetic susceptibility of the contrast agent. The contrast agent causes a microscopic field inhomogeneity on a biological scale of 10 to 1000 nm rather than on the chemical scale of 0.1 to 1 nm. This results in a reduction in apparent T_2.

Chemical speciation was discussed above. For molecular and cellular imaging it is useful to consider physical compartmentalization and magnetic susceptibility in more detail. Physical compartmentalization makes it more difficult to predict tissue relaxivity. With the exception of opsonization of iron oxide particles,[40] the liver-specific agents,[41] and specific cell-labeling preparations (see, e.g., Chapter 18), most contrast agents are designed to stay out of cells. Often the contrast agent will be localized to extracellular spaces. As a result, the simple relaxivity equations do not necessarily hold. For a Gd-based ECF agent in a test tube, it takes about 3 μsec for water to diffuse between Gd molecules;[42] in the time of a typical imaging TR, a given water molecule may interact with thousands or millions of Gd molecules, and all water molecules will interact with approximately the same number of Gd ions. However, if that same ECF agent is compartmentalized solely within the microvasculature, it takes between 2 and 20 sec for most of the water in the tissue (85% of it is extravascular) to physically diffuse into the microvasculature; most of the water in the tissue does not have the opportunity to be relaxed by the Gd within the TR of an imaging acquisition, resulting in a lower signal enhancement than that predicted by Equation 2.3 and assuming a uniform distribution of contrast agent throughout the tissue.

To deal with compartmentalization, the concept of water exchange and exchange time, τ, between compartments is often used.[43,44] The water exchange rate and the size of the compartments will determine the effect of the contrast agent on MRI signal. To illustrate this phenomenon, the two limiting cases of exchange will be described. For more detail, the reader should consult reviews on this topic.[42,44] In one extreme, water moves so fast between the biological compartments that the net effect is as if the contrast agent were uniformly spread throughout the whole tissue. This regime, called fast exchange, occurs whenever the exchange rate, $1/\tau$, between the compartments is much faster than the difference in relaxation rates between the compartments.[45] This occurs in blood where the red cell has a short water exchange time, on the order of 5 to 10 msec.[46] The intact red cell prevents most MR contrast agents from entering the cell, but as long as the plasma T_1 is longer than 20 msec, the two compartments of the blood (plasma + red cells) are in fast exchange and blood behaves for MR purposes as if the contrast agent were spread uniformly through the blood. In this case, the effective relaxation rate will be the weighted average of the relaxation in the two compartments. That is, if for compartment i (where $i = a, b$) the volume fraction is f_i, the initial T_1 is T_{1i}, and the concentration of agent is C_i (which could be zero), the whole tissue together will behave like

$$\frac{1}{T_1} = f_a\left(\frac{1}{T_{1a}} + r_{1a}C_a\right) + f_b\left(\frac{1}{T_{1b}} + r_{1b}C_b\right) \tag{2.9}$$

In slow exchange, the water exchange rate is much slower than the difference in relaxation rates between the compartments. In this case, a single relaxation time, and thus a single relaxivity, is meaningless, because the two microscopic compartments will relax with their own relaxation times. Very few biological compartments show true slow exchange, except at extremely high concentrations of contrast agent. The intermediate case, when exchange is neither slow nor fast (intermediate exchange), occurs very commonly. With intermediate exchange, relaxation behavior appears biexponential, although both the apparent compartment size and the effective T_1 of the two compartments will vary from their true biological size and T_1. It is possible to model the signal intensity behavior as a function of contrast agent concentration to estimate water exchange times *in vivo*. Although characterizing human tissue as having only one or two compartments is an oversimplification, these types of models have proved useful for explaining the effects of biological water mobility on contrast-enhanced scans.[47]

Biological compartmentalization also results in susceptibility contrast. The contrast agent causes microscopic field inhomogeneities sometimes called mesoscopic inhomogeneities.[48] Water diffusion causes the protons to dephase from one another due to the different magnetic fields that they experience during their random walks. Even in the absence of water diffusion, the field inhomogeneity causes intravoxel dephasing, and thus signal loss on gradient echo images due to the different microscopic magnetic fields within the voxel. The strength of the perturbing magnetic field is directly proportional to contrast agent concentration and its molar magnetic susceptibility (χ). The actual magnitude and even direction of the magnetic field shifts depend strongly on the size and shape of the biological compartment in which the contrast agent resides;[48] the size of the susceptibility contrast effect depends on how the water diffuses through the tissue.

The susceptibility T_2 effect is not limited by compartmentalization. For example, first-pass brain perfusion imaging[49] relies on the susceptibility effect of currently approved extracellular Gd-based agents. The blood volume in the brain is very small (4% in gray matter, 2% in white matter), and slow water exchange between the extravascular and intravascular spaces in the brain limits the size of signal changes due to any T_1-based contrast agent at acceptable doses. The susceptibility-based T_2 and T_2* effects can be much larger — as much as a 50% signal drop in normal gray matter at the same dose — due to the "action at a distance" effect possible with the outer-sphere effect. Thus, in cases of slow exchange and when only small compartments are available for the contrast agent, susceptibility contrast may be the medically relevant contrast mechanism of choice. Iron oxide particles with their much higher magnetic susceptibility are more potent susceptibility agents.

Physical compartmentalization and magnetic susceptibility influence how relaxation manifests itself in labeled cells. For T_1 agents, compartmentalization plays an important role. Terreno et al.[50]

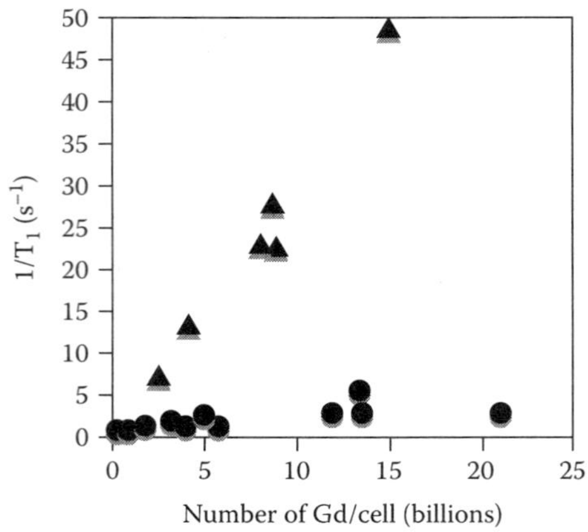

FIGURE 2.8 Effect of compartmentalization on observed T_1. Gd-HPDO3A introduced into cells via electroporation is distributed throughout the cytosol, resulting in efficient relaxation of intracellular water (triangles). Gd-HPDO3A introduced via pinocytosis sees the Gd localized in vesicles. Slow water exchange in and out of the vesicle limits the T_1 lowering effect of the contrast agent (circles).

recently showed that when Gd-HPDO3A was incorporated into rat hepatocarcinoma cells, the method of internalization influenced the observed water relaxation rates. If the gadolinium complex is internalized by pinocytosis, the complex is trapped inside intracellular vesicles. On the other hand, if electroporation is used, then the gadolinium complex is distributed throughout the cytoplasm. When the compound was trapped in vesicles, the observed relaxation rate increased with increasing Gd/cell and then reached a plateau (Figure 2.8 circles). This behavior is typical of intermediate to slow water exchange as the contrast agent concentration increases. Increased gadolinium concentration no longer has any effect on T_1 since water exchange through the vesicle is too slow to affect the other intracellular water. On the other hand, when the compound was distributed throughout the cytoplasm, the relaxation rate increased more quickly (greater slope, Figure 2.8 triangles) with increasing Gd/cell and did not reach a plateau, since the gadolinium was relaxing most of the intracellular water.

When cells are labeled with iron particles, the distribution of the iron particles does not affect the contrast because this is a through-space susceptibility-based relaxation mechanism.

2.8 PARACEST AGENTS

A more recent approach to providing contrast is a magnetization transfer technique termed chemical exchange saturation transfer (CEST) (Chapters 5 and 6). The magnetization transfer (MT) effect that is used clinically exploits a pool of hidden water in some tissues. Water protons associated with macromolecules (e.g., hydrogen bonded to protein and membrane surfaces) have restricted mobility and, because of this, have short T_2. This short T_2 results in a very broad line width of several kilohertz. Mobile water, which makes up most of the tissue, has a relatively long T_2 and a narrow line width. This is illustrated in Figure 2.9a, where magnetization is plotted as a function of frequency. If an rf pulse is applied at a frequency significantly different from the liquid water resonance (e.g., >1 kHz), then the hidden water can become saturated. This hidden water exchanges magnetization with the mobile water via chemical exchange and dipolar coupling. As a result, part of this saturation is transferred to the mobile water peak. This loss of magnetization results in signal loss, as shown in Figure 2.9b. The MT effect can provide contrast since different tissues exhibit MT effects of different magnitudes.

Similarly, magnetization (or saturation) transfer can also be used with contrast agents that have exchangeable hydrogens. Exchangeable hydrogens are typically N-H or O-H hydrogens on molecules

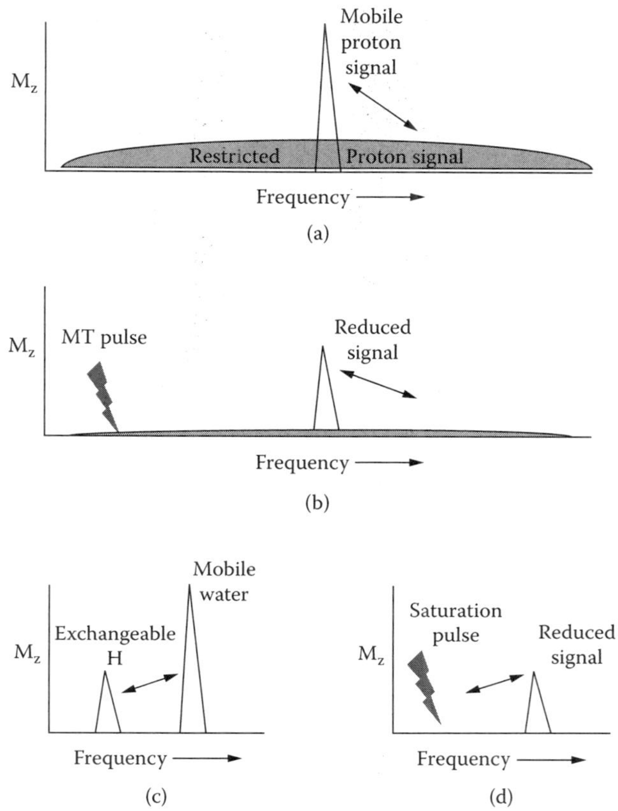

FIGURE 2.9 Magnetization transfer and the CEST effect. (a) Mobile water protons have a relatively long T_2 and resonate over a narrow range of frequencies. There is also a pool of protons with restricted mobility that have a short T_2 and resonate over a wide frequency range. (b) Application of an off-resonance MT pulse saturates the restricted pool, and some of this saturation is transferred to the mobile pool because of chemical exchange and dipolar coupling, resulting in a reduction in magnetization of the mobile pool — this is the MT effect. (c) NMR spectrum of an exchangeable hydrogen with a long T_2 and mobile water. (d) When a saturation pulse is applied at the frequency of the exchangeable hydrogen, this resonance is saturated; chemical exchange results in loss of magnetization of the mobile water — this is the CEST effect.

or exchangeable water molecules from metal complexes. The exchangeable hydrogen has a relatively long T_2 and a narrow line width, but because it is chemically different than water, it resonates at a different frequency (Figure 2.9c). If an rf pulse is applied at the frequency of the exchangeable hydrogen, this resonance becomes saturated (Figure 2.9d). If the hydrogen is undergoing chemical exchange with water, then some of this saturation is transferred to the water, generating negative contrast (Figure 2.9d). This is the CEST effect.[51] CEST differs from the general MT effect because it is only observed when the saturating rf pulse is at the frequency of the exchangeable water. If a different frequency is used, no effect is seen. In the MT effect, the hidden water resonance is so broad that it is excited over a broad range of frequencies. The CEST effect is attractive because it offers the possibility of a contrast agent that can only be observed if the correct pulse sequence is used, in other words a contrast agent that can be turned on and off.

It is intuitive that the rate of chemical exchange should be as fast as possible to maximize the CEST effect. However, the exchangeable hydrogen must resonate at a frequency different from that of water. In order for this to occur, the rate of exchange must be in the slow-exchange regime and the inequality $\Delta\omega\tau > 1$ must be met. Here $\Delta\omega$ is the chemical shift difference in frequency units between the exchangeable hydrogen (labeled ω_A in Figure 2.10) and the pure water resonance, ω_B.

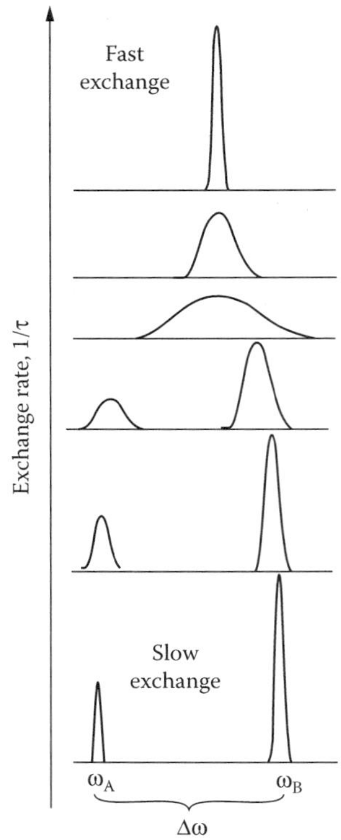

FIGURE 2.10 Effect of chemical exchange rate on the NMR spectrum of an exchangeable hydrogen with chemical shift ω_A and water (ω_B). In slow exchange, $\omega\tau > 1$, two peaks are observed. As $1/\tau$ increases, the peaks coalesce, and at fast exchange, $\omega\tau << 1$, there is a single peak resonating at a frequency that is the population-weighted average between ω_A and ω_B.

τ is the average time that the exchangeable hydrogen spends on its molecule; $1/\tau$ is the rate of hydrogen exchange. If this slow-exchange condition is not met, then only one resonance is observed in the spectrum at a frequency somewhere between the resonance of the water and the exchangeable hydrogen; the exact frequency depends on the relative population of the two groups of hydrogens. The effect of chemical exchange on the resonances is shown in Figure 2.10.

To increase the CEST effect, the number of exchangeable hydrogens can be increased. In this manner, polymeric agents with repeatable units of exchangeable hydrogens have been used.[52,53] The CEST efficiency can be improved by moving to faster exchanging systems that still meet the $\Delta\omega\tau > 1$ requirement. $\Delta\omega$ can be increased by going to higher fields since chemical shift in frequency units (hertz or radians/second) is directly proportional to applied field. Another way to increase $\Delta\omega$ is to incorporate a chemical shift reagent into the molecule.[54] Other paramagnetic lanthanides, but not gadolinium, are capable of inducing large chemical shifts. Several studies have been reported on europium, ytterbium, and other lanthanide complexes of DOTA tetraam (DOTAM) derivatives[54,55] (Figure 2.11) that contain two sources of exchangeable hydrogen — the coordinated water and the amide N-H. With the DOTAM ligand, water exchange is so slow that the gadolinium analog is not useful as a T_1 agent. However, the slow water exchange does mean that the $\Delta\omega\tau > 1$ requirement is met.

Water or proton exchange is obviously critical for this class of contrast agents. NMR is an ideal method for measuring this type of exchange. Depending on the rate of exchange, different techniques are used. Line shape analysis of the exchangeable proton can be done if the system is in slow exchange to intermediate exchange. Saturation transfer itself can be used to determine slow exchange rates. When line broadening is not apparent, two-dimensional exchange correlation

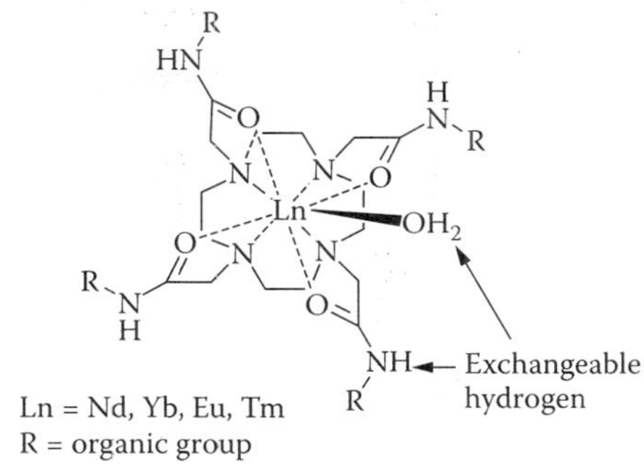

FIGURE 2.11 DOTA tetraamide (DOTAM) derivatives used as PARACEST agents have two sources of exchangeable hydrogen: the N-H amide hydrogen and the hydrogens on the exchangeable water molecule.

spectroscopy (EXSY) is useful. In the case of fast exchange with paramagnetic systems, it is often possible to measure the relaxation rates of the unbound water to determine the exchange rate. Because there is so much water in excess of the paramagnetic ion, the paramagnetic effect on relaxation is diluted. By extrapolating the observed rate to that of a single metal-bound water, estimation of very fast relaxation rates can be made. Varying the temperature or magnetic field is an excellent means of moving the system from slow to fast exchange and enables good estimation of this important parameter. Bertini and Luchinat give an excellent review of relaxation in the presence of chemical exchange.[56]

2.9 DIRECT OBSERVATION OF NONHYDROGEN NUCLEI

The lack of sensitivity in MRI stems from the very small degree of polarization among the nuclear spins. In a magnetic field there is a net magnetization, but this is small; about 0.0006% of the spins are polarized. A technique called spin exchange using a high-powered laser (also called optical pumping) can increase the polarization by four to five orders of magnitude (hyperpolarization).[57] Isotopes possessing long T_1 values can be hyperpolarized and used as contrast agents. The long T_1 is necessary to maintain the contrast medium in the hyperpolarized state long enough to image before the spins relax back to the equilibrium value.

Gases often have long T_1 values, and isotopes of the noble gases helium (^{3}He) and xenon (^{129}Xe) have been used for imaging. Recently, contrast agents with hyperpolarized carbon-13 were reported. Svensson and coworkers[58] described a ^{13}C-enriched water-soluble compound, (bis-1,1-(hydroxymethyl)-1-^{13}C-cyclopropane-d_8), that had long relaxation times (*in vitro*: T_1 = 82 sec, T_2 = 18 sec; *in vivo*: T_1 = 38 sec, T_2 = 1.3 sec). This could be formulated at a ^{13}C concentration of 200 m*M* and hyperpolarized to 15%. The authors used this material for a contrast-enhanced magnetic resonance angiography (CE-MRA) in rats. A major benefit of hyperpolarized contrast media is the excellent sensitivity and no background (high SNR). Challenges include the distribution and availability of the hyperpolarization equipment and imaging hardware compatibility for imaging nonhydrogen nuclei (not available on all clinical scanners).

The fluorine-19 isotope is 100% naturally abundant, and ^{19}F possesses a high magnetogyric ratio, giving ^{19}F a sensitivity that is 83% that of ^{1}H. For other spin = 1/2 nuclei, such as ^{13}C or ^{31}P, the sensitivity is only 0.02 or 6.6% that of the proton. Fluorine imaging has no background since there is very little fluorine in the body. There has been renewed interest in using fluorine as a probe for MRI. Ahrens et al.[59] used perfluorocarbon agents to load dendritic cells and then used ^{19}F MRI to track the cells *in vivo*. The Washington University group have a perfluorocarbon emulsion-based particle as a platform for targeted T_1 agents. Recently this group has demonstrated direct ^{19}F imaging of the particle.[60] Fluorine-containing gases have also been proposed as lung imaging agents.[61,62]

2.10 COMMON FEATURES AND CONCERNS

The various types of contrast agents described in this chapter try to make the most out of a signal-starved technique. To overcome this, high doses are often required, e.g., up to 300 μmol/kg or 200 mg/kg Gd-DTPA for angiography studies. The contrast agents should obviously be soluble at the high concentrations required to administer high doses in reasonable volumes. As these are diagnostic agents, the compounds should be as biologically inert as possible and should be thoroughly eliminated from the body in a reasonable period of time. The metal-based contrast agents should be inert with respect to metal ion loss. The compounds must be nontoxic at the doses used and ultimately must pass a series of acute and subacute dose toxicity studies if the compound is to advance into human studies. In order to understand the effect observed, it is important to know the speciation (composition) of the contrast agent in the tissue of interest.

Particles and very large macromolecular contrast agents have variable plasma lifetimes and are eventually cleared through the reticuloendothelial system (RES) and often end up concentrated in the liver.[31] The approved iron particle formulations are not excreted; rather, the iron is eventually resorbed into the body's iron pool.[63] For nonendogenous ions like gadolinium, it is critical that the contrast agent be completely eliminated from the body.

The pharmacokinetics of the agent should also be considered. The compound needs to accumulate sufficiently at the target (or have its relaxivity enhanced sufficiently) and clear from the blood enough to show an observable effect. Rapid uptake is desirable, as it allows one to easily visualize the image pre- and postadministration of the contrast agent. Localization and clearance can take minutes to days.[64,65] Macromolecular and particle-based contrast agents are restricted by size to intravascular targets. These compounds can also passively target the RES and report on functional lymph nodes and macrophage. The pharmacokinetics of these large agents tends to be slow. For instance, the use of iron particles for lymph node imaging requires waiting a day for sufficient nodal uptake.[64]

Small molecules distribute much more rapidly and have access to the extravascular space. Small molecules are also capable of being cleared by the renal and hepatobiliary systems. However, small molecules may not necessarily deliver sufficient contrast-enhancing payload to detect sparse targets.[66,67] The hyperpolarized compounds obviously require applications where the hyperpolarized agent rapidly occupies the region of interest because these compounds have an imaging half-life limited by their inherent relaxation times.

In developing a new agent or application, the choice of the contrast-enhancing group should be dictated by the problem addressed. Ideally, the compound should localize quickly, give bright positive contrast, and then be completely eliminated. T_1 agents give bright positive contrast, but they can be limited by compartmentalization (water exchange) effects. Small-molecule T_1 agents can have good pharmacokinetics but are limited in their sensitivity. Susceptibility-based contrast agents like iron particles show good sensitivity but often take a long time to localize; they provide negative contrast, which may be difficult to detect and can be confused with imaging artifacts, especially when trying to detect sparse targets. A practical issue to consider is how to link the targeting group to the contrast-enhancing group and still maintain the necessary affinity to the target. This may be more feasible with one contrast agent technology than with another. For animal studies, the bar is set lower with respect to elimination and toxicology, but these features must be addressed if the goal is translational research to humans.

2.11 CONCLUSION

There are many ways to create contrast in an MR image. Exogenous contrast agents can function in many ways: altering T_1, T_2, providing a source for magnetization transfer, or providing a new nucleus to detect. The chemistry and physics of these compounds are somewhat complex because of issues like field dependence and water exchange. However, the complex behavior offers many levers to adjust to create improved contrast agents for molecular and cellular imaging. The chapters that follow give specifics to do just that.

REFERENCES

1. M.A. Brown and R.C. Semelka, *MRI: Basic Principles and Applications*, John Wiley & Sons, New York, 1999.
2. J.P. Mugler, Basic principles, in *Clinical Magnetic Resonance Imaging*, 3rd ed., R.R. Edelman, J.R. Hesselink, M.B. Zlatkin, and J.V. Crues, Eds., Saunders, Philadelphia, 2005, pp. 357–375.
3. I. Aoki, S. Naruse, and C. Tanaka, Manganese-enhanced magnetic resonance imaging (MEMRI) of brain activity and applications to early detection of brain ischemia, *NMR Biomed.*, 17, 569–580, 2004.
4. A.C. Silva, J.H. Lee, I. Aoki, and A.P. Koretsky, Manganese-enhanced magnetic resonance imaging (MEMRI): methodological and practical considerations, *NMR Biomed.*, 17, 532–543, 2004.
5. P. Caravan, J.J. Ellison, T.J. McMurry, and R.B. Lauffer, Gadolinium(III) chelates as MRI contrast agents: structure, dynamics, and applications, *Chem. Rev.*, 99, 2293–2352, 1999.
6. H. Gries, Extracellular MRI contrast agents based on gadolinium, *Top. Curr. Chem.*, 221, 1–24, 2002.
7. A.E. Martell and R.J. Motekaitis, *Determination and Use of Stability Constants*, 2nd ed., John Wiley & Sons, New York, 1992.
8. P. Caravan, C. Comuzzi, W. Crooks, T.J. McMurry, G.R. Choppin, and S.R. Woulfe, Thermodynamic stability and kinetic inertness of MS-325, a new blood pool agent for magnetic resonance imaging, *Inorg. Chem.* 40, 2170–2176, 2001.
9. L. Sarka, L. Burai, R. Kiraly, L. Zekany, and E. Brücher, Studies on the kinetic stabilities of the Gd^{3+} complexes formed with the N-mono(methylamide), N'-mono(methylamide) and N,N"-bis(methylamide) derivatives of diethylenetriamine-N,N,N',N",N"-pentaacetic acid, *J. Inorg. Biochem.*, 91, 320–326, 2002.
10. L. Vander Elst, F. Maton, S. Laurent, F. Seghi, F. Chapelle, and R.N. Muller, A multinuclear MR study of Gd-EOB-DTPA: comprehensive preclinical characterization of an organ specific MRI contrast agent, *Magn. Reson. Med.*, 38, 604–614, 1997.
11. L. Vander Elst, Y. Van Haverbeke, J.-F. Goudemant, and R.N. Muller, Stability assessment of gadolinium complexes by P-31 and H-1 relaxometry, *Magn. Reson. Med.*, 31, 437–444, 1994.
12. W.P. Cacheris, S.C. Quay, and S.M. Rocklage, The relationship between thermodynamics and the toxicity of gadolinium complexes, *Magn. Reson. Imaging*, 8, 467–481, 1990.
13. P. Wedeking, K. Kumar, and M.F. Tweedle, Dissociation of gadolinium chelates in mice: relationship to chemical characteristics. *Magn. Reson. Imaging*, 10, 641–648, 1992.
14. E. Brücher and A.D. Sherry, Stability and toxicity of contrast agents, in *Chemistry of Contrast Agents in Medical Magnetic Resonance Imaging*, E. Toth and A.E. Merbach, Eds., Wiley, New York, 2001, pp. 45–119.
15. D.H. Powell, O.M. Ni Dhubhghaill, D. Pubanz, L. Helm, Y.S. Lebedev, W. Schlaepfer, and A.E. Merbach, High-pressure NMR kinetics. Part 74. Structural and dynamic parameters obtained from ^{17}O NMR, EPR, and NMRD studies of monomeric and dimeric Gd^{3+} complexes of interest in magnetic resonance imaging: an integrated and theoretically self-consistent approach, *J. Am. Chem. Soc.*, 118, 9333–9346, 1996.
16. S.M. Rocklage, W.P. Cacheris, S.C. Quay, F.E. Hahn, and K.N. Raymond, Manganese(II) N,N′-dipyridoxylethylenediamine-n,n'-diacetate 5,5'-bis(phosphate). Synthesis and characterization of a paramagnetic chelate for magnetic resonance imaging enhancement, *Inorg. Chem.*, 28, 477–485, 1989.
17. B. Gallez, G. Bacic, and H.M. Swartz, Evidence for the dissociation of the hepatobiliary MRI contrast agent Mn-DPDP, *Magn. Reson. Med.*, 35, 14–19, 1996.
18. R.B. Lauffer, Targeted relaxation enhancement agents for MRI, *Magn. Reson. Med.*, 22, 339, 1991.
19. R.B. Lauffer, D.J. Parmelee, S.U. Dunham, H.S. Ouellet, R.P. Dolan, S. Witte, T.J. McMurry, and R.C. Walovitch, MS-325: albumin-targeted contrast agent for MR angiography, *Radiology*, 207, 529–538, 1998.
20. I. Bertini and C. Luchinat, Relaxation, *Coord. Chem. Rev.*, 150, 77–110, 1996.
21. M. Gueron, Nuclear relaxation in macromolecules by paramagnetic ions: a novel mechanism, *J. Magn. Reson.*, 19, 58–66, 1975.

22. P. Caravan, N.J. Cloutier, M.T. Greenfield, S.A. McDermid, S.U. Dunham, J.W.M. Bulte, J.C. Amedio, Jr., R.J. Looby, R.M. Supkowski, W.D. Horrocks, Jr., T.J. McMurry, and R.B. Lauffer, The interaction of MS-325 with human serum albumin and its effect on proton relaxation rates, *J. Am. Chem. Soc.*, 124, 3152–3162, 2002.
23. H.B. Eldredge, M. Spiller, J.M. Chasse, M.T. Greenfield, and P. Caravan, Species dependence on plasma protein binding and relaxivity of the gadolinium-based MRI contrast agent MS-325, *Invest. Radiol.*, 41, 229–243, 2006.
24. M. Port, C. Corot, X. Violas, P. Robert, I. Raynal, and G. Gagneur, How to compare the efficiency of albumin-bound and nonalbumin-bound contrast agents *in vivo*: the concept of dynamic relaxivity, *Invest. Radiol.*, 40, 565–573, 2005.
25. S.G. Zech, W.C. Sun, V. Jacques, P. Caravan, A.V. Astashkin, and A.M. Raitsimring, Probing the water coordination of protein-targeted MRI contrast agents by pulsed ENDOR spectroscopy, *ChemPhysChem*, 6, 2570–2577, 2005.
26. X. Zhang, C.A. Chang, H.G. Brittain, J.M. Garrison, J. Telser, and M.F. Tweedle, Ph dependence of relaxivities and hydration numbers of gadolinium(III) complexes of macrocyclic amino carboxylates, *Inorg. Chem.*, 31, 5597–5600, 1992.
27. R.M. Supkowski and W.D. Horrocks, Jr., Displacement of inner-sphere water molecules from Eu^{3+} analogues of Gd^{3+} MRI contrast agents by carbonate and phosphate anions: dissociation constants from luminescence data in the rapid-exchange limit, *Inorg. Chem.*, 38, 5616–5619, 1999.
28. W.D. Horrocks, Jr. and D.R. Sudnick, Lanthanide ion probes of structure in biology. Laser-induced luminescence decay constants provide a direct measure of the number of metal-coordinated water molecules, *J. Am. Chem. Soc.*, 101, 334–340, 1979.
29. R.B. Lauffer, Paramagnetic metal complexes as water proton relaxation agents for NMR imaging: theory and design, *Chem. Rev.*, 87, 901–927, 1987.
30. J.S. Troughton, M.T. Greenfield, J.M. Greenwood, S. Dumas, A.J. Wiethoff, J. Wang, M. Spiller, T.J. McMurry, and P. Caravan, Synthesis and evaluation of a high relaxivity manganese(II)-based MRI contrast agent, *Inorg. Chem.*, 43, 6313–6323, 2004.
31. A. Bjornerud and L. Johansson, The utility of superparamagnetic contrast agents in MRI: theoretical consideration and applications in the cardiovascular system, *NMR Biomed.*, 17, 465–477, 2004.
32. C.W. Jung and P. Jacobs, Physical and chemical properties of superparamagnetic iron oxide MR contrast agents: ferumoxides, ferumoxtran, ferumoxsil, *Magn. Reson. Imaging*, 13, 661–674, 1995.
33. R.N. Muller, A. Roch, J.-M. Colet, A. Ouakssim, and P. Gillis, Particulate magnetic contrast agents, in *Chemistry of Contrast Agents in Medical Magnetic Resonance Imaging*, E. Toth and A.E. Merbach, Eds., Wiley, New York, 2001, pp. 417–435.
34. J.W. Bulte, R.A. Brooks, B.M. Moskowitz, L.H. Bryant, Jr., and J.A. Frank, Relaxometry and magnetometry of the MR contrast agent MION-46L, *Magn. Reson. Med.*, 42, 379–384, 1999.
35. J.W. Bulte, J. Vymazal, R.A. Brooks, C. Pierpaoli, and J.A. Frank, Frequency dependence of MR relaxation times. II. Iron oxides, *J. Magn. Reson. Imaging*, 3, 641–648, 1993.
36. R.M. Weisskoff, C.S. Zuo, J.L. Boxerman, and B.R. Rosen, Microscopic susceptibility variation and transverse relaxation: theory and experiment, *Magn. Reson. Med.*, 31, 601–610, 1994.
37. P. Hardy and R.M. Henkelman, On the transverse relaxation rate enhancement induced by diffusion of spins through inhomogeneous fields, *Magn. Reson. Med.*, 17, 348–356, 1991.
38. J.M. Perez, L. Josephson, T. O'Loughlin, D. Hogemann, and R. Weissleder, Magnetic relaxation switches capable of sensing molecular interactions, *Nat. Biotechnol.*, 20, 816–820, 2002.
39. J.M. Perez, T. O'Loughin, F.J. Simeone, R. Weissleder, and L. Josephson, DNA-based magnetic nanoparticle assembly acts as a magnetic relaxation nanoswitch allowing screening of DNA-cleaving agents, *J. Am. Chem. Soc.*, 124, 2856–2857, 2002.
40. A. Moore, R. Weissleder, and A. Bogdanov, Jr., Uptake of dextran-coated monocrystalline iron oxides in tumor cells and macrophages, *J. Magn. Reson. Imaging*, 7, 1140–1145, 1997.
41. O. Clement, N. Siauve, M. Lewin, E. de Kerviler, C.A. Cuenod, and G. Frija, Contrast agents in magnetic resonance imaging of the liver: present and future, *Biomed. Pharmacother.*, 52, 51–58, 1998.
42. K.M. Donahue, R.M. Weisskoff, and D. Burstein, Water diffusion and exchange as they influence contrast enhancement, *J. Magn. Reson. Imaging*, 7, 102–110, 1997.
43. C.F. Hazlewood, D.C. Chang, B.L. Nichols, and D.E. Woessner, Nuclear magnetic resonance transverse relaxation times of water protons in skeletal muscle, *Biophys. J.*, 14, 583–606, 1974.

44. X. Li, W.D. Rooney, and C.S. Springer, Jr., A unified magnetic resonance imaging pharmacokinetic theory: intravascular and extracellular contrast reagents, *Magn. Reson. Med.*, 54, 1351–1359, 2005.
45. A.C. McLaughlin and J.S. Leigh, Relaxation times in systems with chemical exchange, *J. Magn. Reson.*, 9, 296–304, 1973.
46. G.A. Wright, B.S. Hu, and A. Macovski, Estimating oxygen saturation of blood *in vivo* with MR imaging at 1.5 T, *J. Magn. Reson. Imaging*, 1, 275–283, 1991.
47. K.M. Donahue, R.M. Weisskoff, D.A. Chesler, K.K. Kwong, A.A. Bogdanov, Jr., J.B. Mandeville, and B.R. Rosen, Improving MR quantification of regional blood volume with intravascular T1 contrast agents: accuracy, precision, and water exchange, *Magn. Reson. Med.*, 36, 858–867, 1996.
48. A.L. Sukstanskii and D.A. Yablonskiy, Gaussian approximation in the theory of MR signal formation in the presence of structure-specific magnetic field inhomogeneities. Effects of impermeable susceptibility inclusions, *J. Magn. Reson.*, 167, 56–67, 2004.
49. B.R. Rosen, J.W. Belliveau, J.M. Vevea, and T.J. Brady, Perfusion imaging with NMR contrast agents, *Magn. Reson. Med.*, 14, 249–265, 1990.
50. E. Terreno, S. Geninatti Crich, S. Belfiore, L. Biancone, C. Cabella, G. Esposito, A.D. Manazza, and S. Aime, Effect of the intracellular localization of a Gd-based imaging probe on the relaxation enhancement of water protons, *Magn. Reson. Med.*, 55, 491–497, 2006.
51. K.M. Ward, A.H. Aletras, and R.S. Balaban, A new class of contrast agents for MRI based on proton chemical exchange dependent saturation transfer (CEST), *J. Magn. Reson.*, 143, 79–87, 2000.
52. K. Snoussi, J.W. Bulte, M. Gueron, and P.C. van Zijl, Sensitive CEST agents based on nucleic acid imino proton exchange: detection of poly(rU) and of a dendrimer-poly(rU) model for nucleic acid delivery and pharmacology, *Magn. Reson. Med.*, 49, 998–1005, 2003.
53. N. Goffeney, J.W. Bulte, J. Duyn, L.H. Bryant, Jr., and P.C. van Zijl, Sensitive NMR detection of cationic-polymer-based gene delivery systems using saturation transfer via proton exchange, *J. Am. Chem. Soc.*, 123, 8628–8629, 2001.
54. S. Zhang, M. Merritt, D.E. Woessner, R.E. Lenkinski, and A.D. Sherry, PARACEST agents: modulating MRI contrast via water proton exchange, *Acc. Chem. Res.*, 36, 783–790, 2003.
55. E. Terreno, D.D. Castelli, G. Cravotto, L. Milone, and S. Aime, Ln(III)-DOTAMGly complexes: a versatile series to assess the determinants of the efficacy of paramagnetic chemical exchange saturation transfer agents for magnetic resonance imaging applications, *Invest. Radiol.*, 39, 235–243, 2004.
56. I. Bertini and C. Luchinat, Hyperfine shift and relaxation in the presence of chemical exchange, *Coord. Chem. Rev.*, 150, 111–130, 1996.
57. H.E. Moller, X.J. Chen, B. Saam, K.D. Hagspiel, G.A. Johnson, T.A. Altes, E.E. de Lange, and H.U. Kauczor, MRI of the lungs using hyperpolarized noble gases, *Magn. Reson. Med.*, 47, 1029–1051, 2002.
58. J. Svensson, S. Mansson, E. Johansson, J.S. Petersson, and L.E. Olsson, Hyperpolarized ^{13}C MR angiography using trueFISP, *Magn. Reson. Med.*, 50, 256–262, 2003.
59. E.T. Ahrens, R. Flores, H. Xu, and P.A. Morel, *In vivo* imaging platform for tracking immunotherapeutic cells, *Nat. Biotechnol.*, 23, 983–987, 2005.
60. S.D. Caruthers, A.M. Neubauer, F.D. Hockett, R. Lamerichs, P.M. Winter, M.J. Scott, P.J. Gaffney, S.A. Wickline, and G.M. Lanza, *In vitro* demonstration using ^{19}F magnetic resonance to augment molecular imaging with paramagnetic perfluorocarbon nanoparticles at 1.5 tesla, *Invest. Radiol.*, 41, 305–312, 2006.
61. R.E. Jacob, Y.V. Chang, C.K. Choong, A. Bierhals, D. Zheng Hu, J. Zheng, D.A. Yablonskiy, J.C. Woods, D.S. Gierada, and M.S. Conradi, ^{19}F MR imaging of ventilation and diffusion in excised lungs, *Magn. Reson. Med.*, 54, 577–585, 2005.
62. J.M. Perez-Sanchez, R. Perez de Alejo, I. Rodriguez, M. Cortijo, G. Peces-Barba, and J. Ruiz-Cabello, *In vivo* diffusion weighted ^{19}F MRI using SF_6, *Magn. Reson. Med.*, 54, 460–463, 2005.
63. P. Reimer and T. Balzer, Ferucarbotran (Resovist): a new clinically approved res-specific contrast agent for contrast-enhanced MRI of the liver: properties, clinical development, and applications, *Eur. Radiol.*, 13, 1266–1276, 2003.
64. M.G. Harisinghani, W.T. Dixon, M.A. Saksena, E. Brachtel, D.J. Blezek, P.J. Dhawale, M. Torabi, and P.F. Hahn, MR lymphangiography: imaging strategies to optimize the imaging of lymph nodes with ferumoxtran-10, *Radiographics*, 24, 867–878, 2004.
65. J. Barkhausen, W. Ebert, C. Heyer, J.F. Debatin, and H.J. Weinmann, Detection of atherosclerotic plaque with gadofluorine-enhanced magnetic resonance imaging, *Circulation*, 108, 605–609, 2003.

66. M.F. Tweedle, P. Wedeking, J. Telser, C.H. Sotak, C.A. Chang, K. Kumar, X. Wan, and S.M. Eaton, Dependence of MR signal intensity on Gd tissue concentration over a broad dose range, *Magn. Reson. Med.*, 22, 191–194, 1991.
67. P. Wedeking, C.H. Sotak, J. Telser, K. Kumar, C.A. Chang, and M.F. Tweedle, Quantitative dependence of MR signal intensity on tissue concentration of Gd(HP-DO3A) in the nephrectomized rat, *Magn. Reson. Imaging*, 10, 97–108, 1992.
68. M. Rohrer, H. Bauer, J. Mintorovitch, M. Requardt, and H.J. Weinmann, Comparison of magnetic properties of MRI contrast media solutions at different magnetic field strengths, *Invest. Radiol.*, 40, 715–724, 2005.

3 Paramagnetic Contrast Agents

Silvio Aime, Zsolt Baranyai, Eliana Gianolio, and Enzo Terreno

CONTENTS

3.1 Introduction 37
3.2 Routes to Improve the Relaxivity of Gd-Chelates 39
3.2.1 Gd Complexes Containing Two Coordinated Water Molecules (q = 2) 40
3.2.2 Control of the Molecular Tumbling 40
3.2.3 Improved Contribution from Water Molecules in the Second Coordination Sphere 41
3.2.4 Gd-Loaded Carbon Nanostructures 42
3.3 Macromolecular Paramagnetic Systems 42
3.3.1 Supramolecular Adducts for the Accumulation of Gd Chelates at the Targeting Site 42
3.3.2 Targeting Human Serum Albumin (HSA) 43
3.3.3 Covalent Macromolecular Paramagnetic Systems 46
3.4 Responsive Agents 46
3.4.1 pH Sensitive 46
3.4.2 Agents Sensitive to Redox Potential 48
3.4.3 Enzyme Responsive 49
3.4.4 Concentration-Independent Responsive Probes 51
3.5 Effects on the Relaxivity When Going from Aqueous Solutions to Cellular Systems 52
3.6 Conclusions 54
References 54

3.1 INTRODUCTION

As it was early established that the main determinants of the contrast in a magnetic resonance (MR) image are the proton relaxation times T_1 and T_2, the search for MR imaging (MRI) contrast agents was oriented toward paramagnetic metal complexes because unpaired electrons display remarkable ability to reduce T_1 and T_2 of their solutions.[1,2] On this basis, the metals of choice have been identified among those ones having the higher number of unpaired electrons, namely, Mn(II) and Fe(III) (five unpaired electrons) among the transition metals and Gd(III) (seven unpaired electrons) among the lanthanides. Besides the number of unpaired electrons, an important parameter is represented by electronic relaxation time (τ_S), which has to be as long as possible.[3] The distribution of five and seven electrons in five 3d or seven 4f orbitals, respectively, provides a good basis for the attainment of long τ_S values. As their dominant effect is on T_1, paramagnetic MRI agents are often classified as positive agents because their effect causes hyperintensity in the region where they distribute. In fact, the shortening of T_1 allows the accumulation of a much higher number of scans per time unit without causing saturation of the nuclear magnetization. Conversely, iron oxide particles, whose dominant effect is on T_2, are usually classified as negative agents. Currently, several

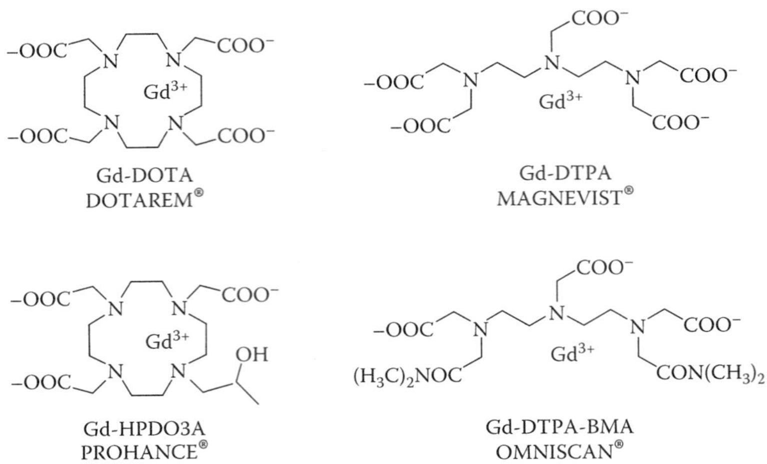

FIGURE 3.1 Structures of some Gd(III)-based MRI contrast agents currently used in clinical practice.

Gd-based agents and one Mn-based agent are available for clinical applications. Their use has led to remarkable improvements in medical diagnosis in terms of higher specificity, better tissue characterization, reduction of image artifacts, and functional information. Besides acting as catalyst for the relaxation of water protons, an MRI contrast agent has to possess several additional properties to guarantee the safety issues required for *in vivo* applications at the administered doses, namely, high thermodynamic (and possibly kinetic) stability, good solubility, and low osmolality.[4,5]

The first contrast agent (CA) introduced in clinical practice was Gd-DTPA (Magnevist®), in which the octa-coordinating DTPA ligand (diethylen triamine penta acetic acid; Figure 3.1) wraps around the Gd(III) ion. This bonding scheme yields a high stability ($\text{LogK}_{\text{Gd-L}} = 22$) and leaves one coordination site on the Gd(III) ion available for one water molecule ($q = 1$ complex). Several other CA based on Gd chelates with polyamminocarboxylate ligands with diagnostic indications analogue to those of Gd-DTPA have been made available for clinical applications (Figure 3.1).[6]

All display an extracellular vascular and extravascular distribution and are sometimes indicated as T_1-general agents. One of their main clinical indications deals with the assessment of the loss of the blood–brain barrier in the presence of tumor lesions or other diseases affecting the central nervous system (CNS). They are also used to investigate the occurrence of abnormalities in the vascular permeability. In this case, one makes use of the dynamic properties of the CA in the sense that the region of interest is imaged repeatedly within the first few minutes of the administration of the contrast agent and a graph reporting the changes of the contrast enhancement as a function of time is acquired. Neo-formed vessels display much higher permeability than normal capillaries, thus affecting the signal intensity of the interested regions accordingly. Currently, about one third of the MRI scans recorded in clinical settings take advantage of the use of paramagnetic contrast agents.

The relaxivity is the property by which a paramagnetic chelate is evaluated *in vitro* in order to be considered a potential contrast agent for MRI applications.[7] It represents the relaxation enhancement of water protons in the presence of the paramagnetic complex at 1 m*M* concentration. As detailed in Chapter 1, the relaxivity may vary with the field strength, and its value is determined by the sum of three contributions:[8]

$$r_1 = r_1^{is} + r_1^{2nd} + r_1^{os} \tag{3.1}$$

where r_1^{is} is the contribution arising from the exchange of water molecules in the inner coordination sphere of the paramagnetic ion and r_1^{2nd} and r_1^{os} represent the contributions arising from water

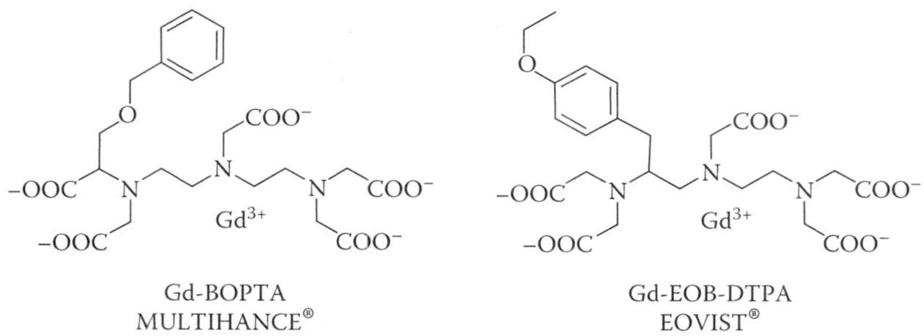

FIGURE 3.2 Structures of two hepatotropic Gd(III)-based MRI contrast agents currently used in clinical practice.

molecules in the second coordination sphere and from those ones diffusing in the proximity of the paramagnetic chelate. The inner sphere term scales up with the number of coordinated water molecules (q), and when not quenched by a long τ_M, it is basically determined by the value of the correlation time τ_C:

$$r_1^{is} = \frac{q[C]}{55.56(T_{1M} + \tau_M)} \tag{3.2}$$

$$(T_{1M})^{-1} = Kf(\tau_C) \tag{3.3}$$

$$(\tau_C)^{-1} = (\tau_R)^{-1} + (\tau_M)^{-1} + (\tau_S)^{-1} \tag{3.4}$$

where τ_R corresponds to the reorientational time of the magnetic vector joining the proton of the coordinated water and the paramagnetic ion, τ_M is the exchange lifetime of the coordinated water, and τ_S is the electronic relaxation time.

The relaxivity of the Gd chelates reported in Figure 3.1, at the imaging fields (i.e., >0.5 tesla), are very similar (ca. 4.7 ± 0.1 mM^{-1}sec^{-1} at 0.47 tesla and 298 K) because their value is determined by τ_R, which in turn, for these isotropically tumbling systems, is related to their (similar) molecular size.

By introducing a substituted aromatic moiety on the surface of a Gd-DTPA, two new contrast agents, Gd-BOPTA (Multihance®)[9] and Gd-EOB-DTPA (Eovist®),[10] were generated (Figure 3.2).

These chemical changes have profound effects on the *in vivo* relaxivity and biodistribution of these agents with respect to the parent one. First, the hydrophobic substituent promotes a weak reversible binding to serum albumin, thus yielding an increase of τ_R, which results in doubling the relaxivity shown by these agents in water. Then they are recognized by hepatocytes (more dramatically for Gd-EOB-DTPA), and this yields the activation of a liver excretion pathway in addition to the kidney one (which is the only mechanism used by the agents reported in Figure 3.1).

3.2 ROUTES TO IMPROVE THE RELAXIVITY OF Gd-CHELATES

As MRI is a relatively insensitive modality, the doses of required contrast agents are rather high; for instance, the currently used contrast agents have to be administered at doses of 0.1 mmol/kg, i.e., several grams per patient. To reduce the doses and, moreover, to pursue targeting protocols, it is necessary to have Gd complexes characterized by high relaxivities. The theory of paramagnetic relaxation foresees the possibility to markedly increase the relaxivity of Gd-based systems with respect to the currently available ones.[3,7] Therefore, many efforts have been made (and are currently

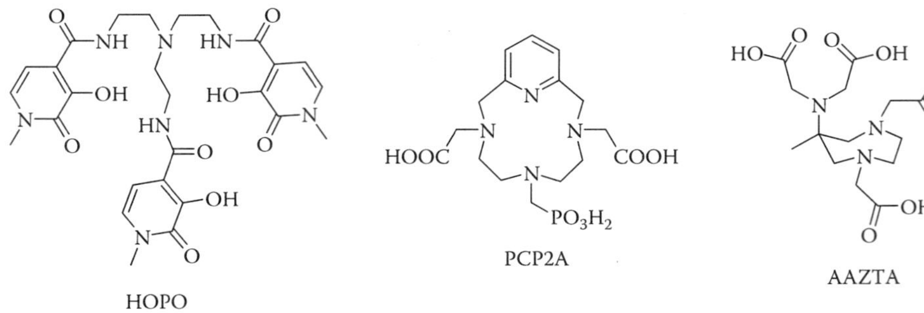

FIGURE 3.3 Structures of three heptadentate ligands that form $q = 2$ complexes with Gd(III).

made) to design new systems whose structural, electronic, and dynamic properties correspond at best to the characteristics the theory anticipates for high relaxivity agents.[11] Attention has been mainly focused on the following lines of activities.

3.2.1 Gd Complexes Containing Two Coordinated Water Molecules (q = 2)

According to Equation 3.2, doubling the value of q yields a doubling of r_1^{is}, but the increase of q may be at the expense of a decrease of the overall stability of the metal complex. In fact, on passing from the octa-coordinating ligands of Figure 3.1 and Figure 3.2 that form $q = 1$ complexes to heptadentate ligands that yield $q = 2$ systems, one has to pay much attention to the maintenance of a sufficiently high thermodynamic stability and low tendency to transmetalation with endogeneous metal ions such as Cu^{2+} and Zn^{2+}. The maintenance of a high stability is a requisite of overwhelming importance in molecular imaging applications, as it has been shown[12] that the cell membranes may act as sponges toward Gd^{3+} ions, thus shifting the dissociation equilibrium with a consequent large release of the toxic metal ions from the metal complexes. Moreover, the two water molecules may be replaced by bidentate anions present in solution, such as phosphate and carbonate and carboxylate moieties (e.g., aspartate and glutamate) present on the surface of proteins, thus leading to a decrease of the relaxivity. For these reasons, Gd-DO3A and its derivatives, although they show $q = 2$ in aqueous solutions, did not prove to be suitable for the purpose of the attainment of high relaxivities.[13] $q = 2$ systems that appear particularly interesting are those whose ligands are reported in Figure 3.3.

Gd-PCP2A consists of a ligand based on a pyridine-containing macrocycle bearing two acetic arms and one methylene-phosphonic arm. The presence of the phosphonic coordinating moiety appears to further improve the stability without reducing the number of the coordinated waters with respect to the parent Gd-PCTA.[14] Still, in the field of polyaminopolycarboxylate ligands a particularly valuable system is represented by Gd-AAZTA that shows a relaxivity of 7.1 $mM^{-1}sec^{-1}$ (at 20 MHz and 298 K) and good stability ($LogK_{Gd\text{-}L} = 20.1$), and it is inert toward interaction with bidentate anions present in solution as well as toward trasmetallation with Zn^{2+} and Cu^{2+}. AAZTA ligand is easily prepared and functionalized for targeting applications.[15] Gd-HOPO is the prototype of an interesting class developed by Hajela and co-workers in Berkeley. In this highly stable system, the Gd^{3+} ion is coordinated only by six oxygen donor atoms, and therefore, there is room for the additional coordination of two, or even three, water molecules. The peculiar coordination geometry of Gd-HOPO complexes does not allow the replacement of the two water molecules by endogenous ligands.[16]

3.2.2 Control of the Molecular Tumbling

As anticipated above, an important parameter in the determination of the relaxivity is represented by the reorientational time of the magnetic vector joining the paramagnetic center and the protons

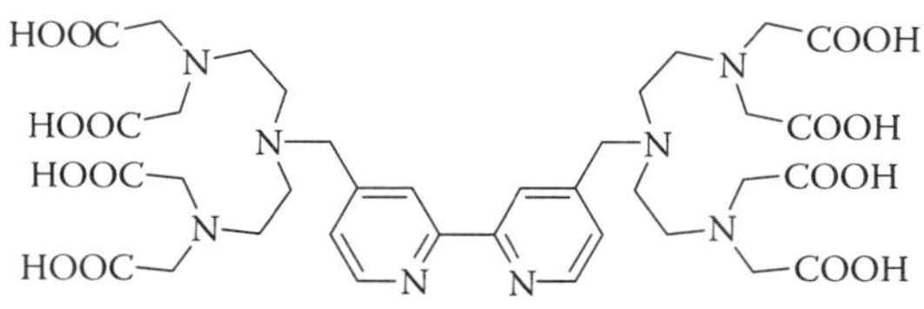

FIGURE 3.4 Structure of a heterotripodic ligand for binding Fe(II) and Gd(III) ions.

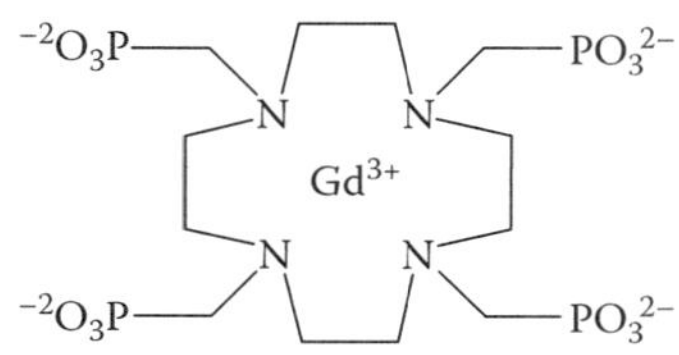

FIGURE 3.5 Structure of Gd-DOTP complex, a $q = 0$ complex, characterized by an extended second coordination hydration sphere.

of the coordinated water molecule. Although this approach has been widely exploited by designing covalent and noncovalent macromolecular systems, only recently have interesting achievements been obtained with small- and medium-sized agents. For instance, Gd-DOTA-like structures bearing bulk substituents on each of the four acetate arms display relaxivities that are significantly higher than those expected on the basis of values reported for systems characterized by similar molecular weight. This occurs because the coordinated water lies at the baricenter and along the principal axis of rotation of the complex.[17–19] Another interesting example deals with the formation of a tight assembly of up to six Gd chelates organized through the coordination around a central Fe(II) ion. The so-called metallo-star $[Fe\{Gd_2L(H_2O)_4\}_3]^{4-}$ formed with the ligand reported in Figure 3.4 shows a relaxivity value of 33.6 mM^{-1}sec^{-1} at 40 MHz and 298 K due also to the presence of two water molecules in the inner sphere of the Gd^{3+} ions.[20]

3.2.3 Improved Contribution from Water Molecules in the Second Coordination Sphere

The presence of coordinating groups like phosphonates, carboxylates, carboxo-amide, etc., introduces a number of polar atoms around the coordination cage that promote the organization of water molecules in the second coordination sphere.[21] These water molecules bind through hydrogen bonds to donor atoms of the first coordination sphere that bring them to a relatively short distance from the paramagnetic center. Borel et al.[22] carried out molecular docking (MD) calculations on Gd-DOTP (Figure 3.5) ($q = 0$) and showed the presence of 4.3 second-sphere water molecules with an average residence lifetime of 56 psec, which is one order of magnitude longer than the diffusion time of solvent water molecules. This second-sphere hydration shell is responsible for the higher relaxivity of the $q = 0$ Gd-DOTP complex (r_1 = 4.7 mM^{-1}sec^{-1}), which is the same as that shown by the $q = 1$ systems reported in Figure 3.1.

Another route to generate a large contribution from water molecules in the second coordination sphere has been pursued by forming supramolecular adducts between Gd complexes bearing hydrophobic substituents and β-cyclodextrin-containing systems. In such host–guest adducts, one may envisage the formation of chaltrates made of several water molecules whose mobility is slowed down by the constraints of the hydrophobic interaction.[23] Such a contribution is invariantly present in the noncovalent adducts between Gd chelates and proteins, and it is often responsible for a relevant part of the observed relaxation enhancement. In this context, an interesting example is provided by apoferritin loaded with Gd-HPDO3A.[24] The latter, commercial agent with $q = 1$ can

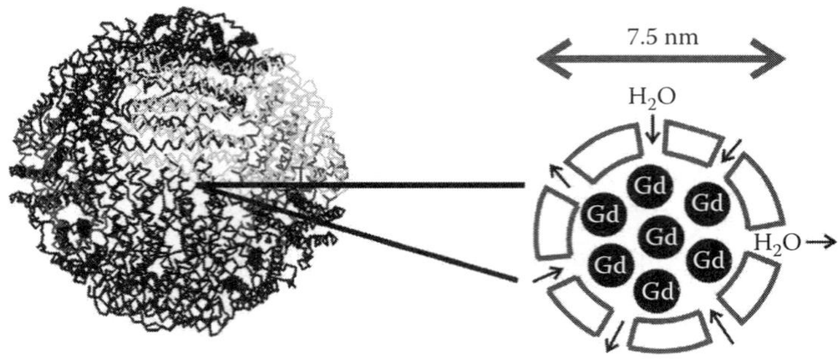

FIGURE 3.6 Schematic representation of Gd-loaded Apoferritin, a system showing outstandingly high relaxivity.

be entrapped into the protein cavity (up to 8 to 10 complex units) by a cleaving/reforming process determined by changing the pH of the solution from neutral to acidic and back to neutral. The relaxivity of the Gd-loaded apoferritin system is very high (ca. 80 $mM^{-1}sec^{-1}$/Gd), i.e., ca. 20 times higher than the relaxivity of Gd-HPDO3A in water. It has been shown that a large part of this relaxation enhancement arises from a sort of relaxation sink made from exchangeable protons and water molecules hydrogen bonded to the inner surface of the protein. This relaxation scheme is similar to the one proposed (*vide infra*) for describing the relaxation enhancement of supramolecular adducts on the outer surface of proteins, but here the effect is amplified by the spherical surface of the inner cavity (Figure 3.6).

3.2.4 Gd-Loaded Carbon Nanostructures

Interesting examples are represented by Gd-fullerenol (Gd@C60(OH)x),[25] for which a relaxivity of 38.5 $mM^{-1}sec^{-1}$ (40 MHz, 25°C) has been reported and accounted for in terms of an electronic transfer from Gd to the fullerene cage, followed by the dipolar coupling to a large number of water molecules hydrogen bonded to the external OH groups of the cage. More recently, an outstanding relaxivity (173 $mM^{-1}sec^{-1}$ at 20 MHz and 298 K) has been reported for Gd-loaded carbon nanotubes (20 to 80 nm in length and 1.4 nm in diameter).[26] It has been suggested that in this system Gd is organized under the form of small supramolecular clusters (8 to 10 atoms) that, through a sort of proximity effect, transfer their magnetic effect to the extended carbon surface of the nanotubes and, in turn, likely through hydroxyl or carboxylate functionalities, to the solvent water molecules. Although the rationale for the very high relaxivity shown by this system has to be further addressed, this result is highly promising in the search of innovative routes to attain relaxivities much higher than the ones currently available.

3.3 MACROMOLECULAR PARAMAGNETIC SYSTEMS

Under this category fall systems that tackle different issues, all of outstanding importance for the development of the use of paramagnetic contrast agents in MRI.

3.3.1 Supramolecular Adducts for the Accumulation of Gd Chelates at the Targeting Site

One of the hot spots deals with the setup of targeting strategies for molecular imaging protocols. The visualization of low-concentration epitopes requires the delivery at the targeting sites of a large number of Gd chelates to induce a detectable contrast in the MR image. We estimate that the MR visualization of a cell requires a number (N) of Gd chelates, given by $N = 10^9/r_1$. This yields $N = 10^7$ to 10^8 Gd/cell according to the relaxivity of the bound paramagnetic agent.[27]

Morawski and Caruthers developed nanosized particles, based on a microemulsion of fluorinated aliphatic chains, containing up to 10^5 Gd chelates.[28,29] The relaxivity per Gd is in the low-range values for immobilized systems (e.g., ca. 19 mM^{-1}sec^{-1}), but the relaxivity per particle is of the order of 1,800,000 mM^{-1}sec^{-1}. By endowing these Gd-containing particles with suitable targeting vectors (e.g., peptides), it has been possible to visualize the presence of several epitopes, like integrin, fibrin, etc.

Other Gd delivery systems include the Gd-chitosan nanoparticles (430 nm) containing 9.3% (w/w) of Gd(III),[30] the distearylamide gadopentetic acid microcapsules (106 to 149 μM) containing 5.13% (w/w) of Gd(III),[31] the gadolinium hexanedione (GdH) nanoparticles containing 2.5 mg/ml of GdH,[32] and the Gd-incorporated lipid nanoemulsions (100 nm) containing 3 mg/ml of Gd(III).[33]

Gd-loaded liposomes for MRI applications have also been extensively investigated, as they can encapsulate a large variety of both hydrophilic and hydrophobic contrast agents. Often the potential relaxation enhancement attainable by a given paramagnetic chelate entrapped in their inner cavities is partially quenched by the occurrence of a slow water exchange rate across the liposome membrane. On the other hand, the release of the Gd chelates from the delivery system is very limited, thus allowing the design of targeting experiments with such systems.[34] For the latter experiments, a prolonged circulation time in blood is a critical factor because it has been shown that there is a strong correlation between the residence time in the blood and its uptake by tumors in mice. The prolongation of the circulation time is often obtained by coating the liposomes with polyethyleneglycol (PEG), as the hydrophilic cover reduces the binding to plasma protein (opsonization) and enables the liposomes to escape recognition by liver and spleen.[35]

An interesting example based on the use of liposomes bearing Gd chelates bound to their membrane has been reported by Sipkins et al.[36] They succeeded in the visualization of $\alpha_v\beta_3$ receptors on tumor endothelium by a two-step procedure consisting of (1) targeting the epitopes of interest with a properly biotinylated antibody (Ab) and (2) recognition of the Ab by means of an avidin moiety bound on the surface of the paramagnetic liposomes. This approach provided enhanced and detailed detection of rabbit carcinoma through the imaging of the angiogenic vasculature.

Still in the context of delivering enough Gd(III) ions at the targeting site, Bhujwalla and co-workers were able to functionalize avidin with Gd-DTPA (ca. 12.5 units/protein) and used this system as a reporter of the recognition of HER-2/scan receptors previously targeted with suitably biotinylated monoclonal Ab (m-Ab).[37] The targeted epitope is a member of the epidermal growth factor family, and it is amplified in multiple cancer. As the expression level of the receptor was estimated at 7×10^5 receptors/cell, the visualization of a breast carcinoma in an experimental mouse model likely implies the binding of more than one avidin molecule per m-Ab.

3.3.2 Targeting Human Serum Albumin (HSA)

Targeting HSA represents another topic that has been widely investigated in the last decade, and it has provided a number of insights that can also be used to understand the binding of paramagnetic chelates to other proteins.[38]

Besides the attainment of high relaxivities, a high binding affinity to HSA enables Gd chelates to attain a long intravascular time, which is the property required for a good blood pool agent for MR angiography. Two systems, MS-325[39] and B22956 (Figure 3.7), have been developed with diagnostic indications of assessing peripheral[40] and coronary[41,42] (Figure 3.8) angiography, respectively.

Several other $q = 1$ systems have been investigated, yielding systems that show similar binding affinities and relaxation enhancements. Clearly further enhanced relaxivities are expected with $q = 2$ Gd chelates. We have recently investigated the binding of a lipophilic Gd-AAZTA complex (Figure 3.9) with fatted and defatted HSA. Gd-AAZTAC17 displays a better affinity for fatted in respect to defatted HSA (Ka = 7.1×10^4 M^{-1} and 2.4×10^4 M^{-1}, respectively), whereas the relaxivities of the supramolecular adducts go in the opposite direction (r_1^b = 63 and 84 mM^{-1}sec^{-1}). The relaxivity

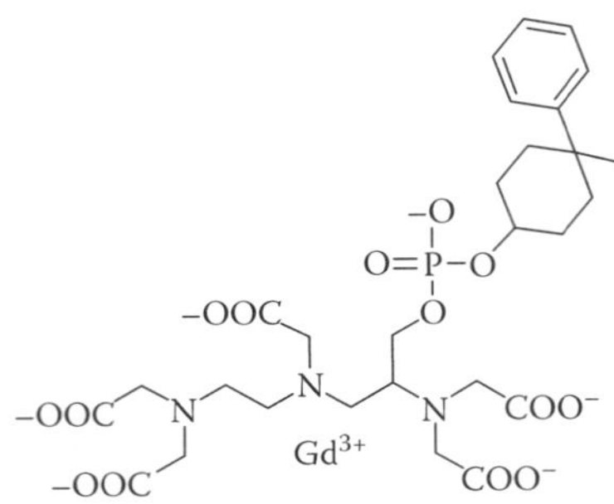

MS325

$K_A = 9.0 \times 10^4 \, M^{-1}$

$r_1^b = 37.2 \, mM^{-1} \, s^{-1}$

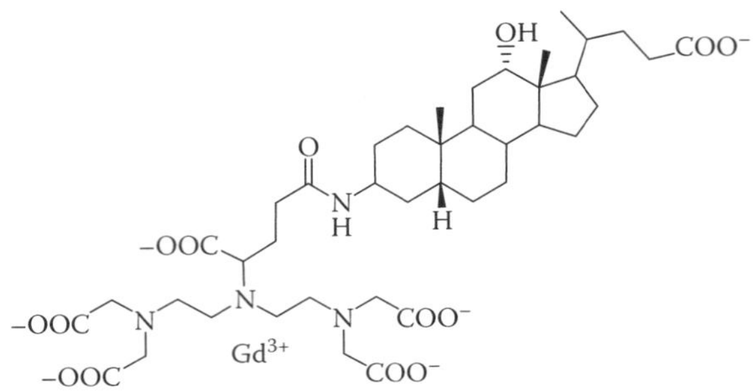

B22956

$K_A = 2.6 \times 10^4 \, M^{-1}$

$r_1^b = 27.6 \, mM^{-1} \, s^{-1}$

FIGURE 3.7 Structures of two Gd(III) complexes that show a high affinity toward HSA and are used in MR angiography investigations.

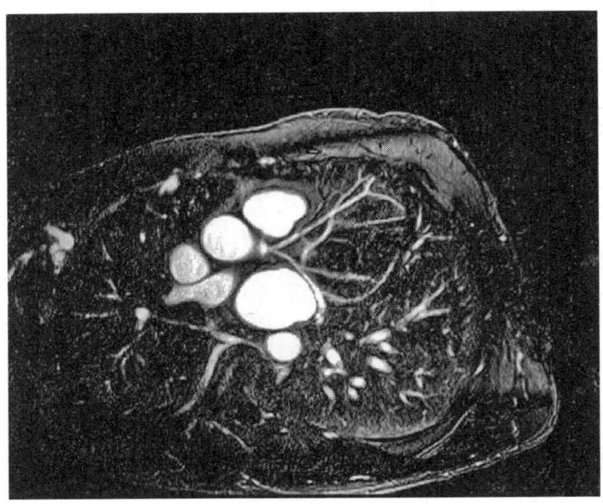

FIGURE 3.8 Inversion recovery, three-dimensional, gradient-recalled echo magnetic resonance coronary angiography (IR-3D-GRE-MRCA), performed after administration of B22956 at 1.5T.

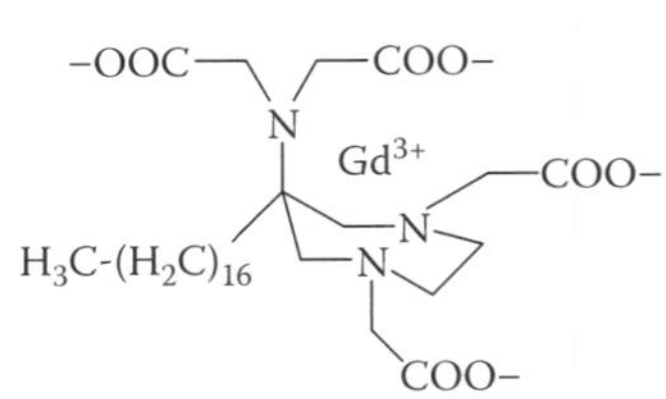

FIGURE 3.9 Structure of the lipophilic complex Gd-AAZTAC17.

shown by Gd-AAZTA/defatted HSA is by far the highest relaxivity until now reported for noncovalent adducts with slowly moving substrates.

Interestingly, the analysis of the nuclear magnetic relaxation dispersion (NMRD) profiles shows that the observed relaxivity receives large contributions from water molecules in the second coordination sphere. As elucidated by molecular docking calculations, the Gd complex enters more extensively in the hydrophobic pocket present in fatted HSA (therefore yielding a larger K_A and a

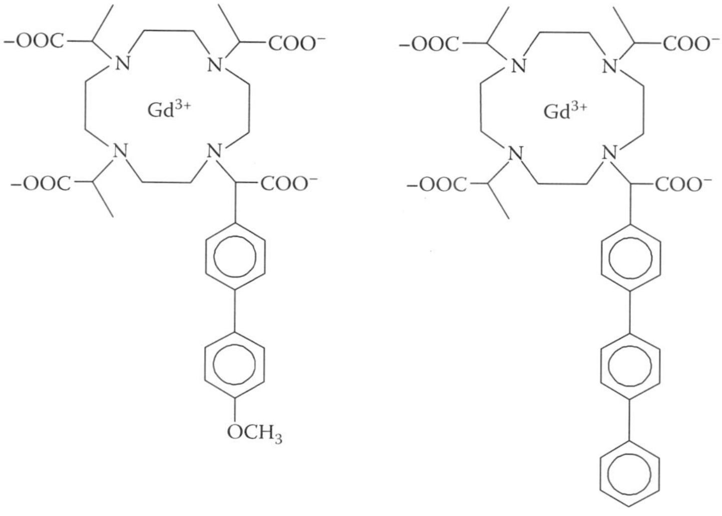

FIGURE 3.10 Two Gd-DOTMA derivatives synthesized for the binding interaction with HAS.

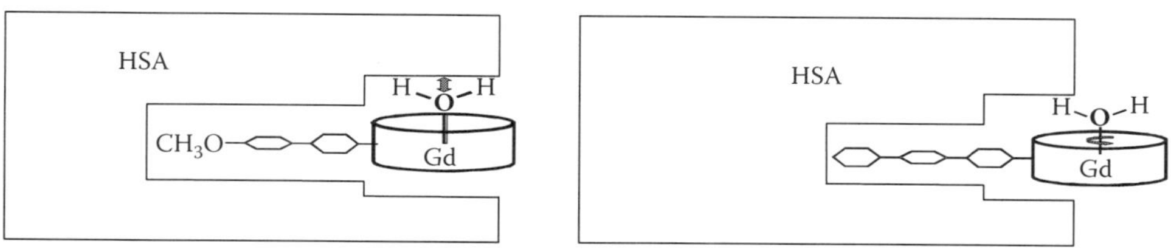

FIGURE 3.11 Complex A is deeply embedded in the hydrophobic cavity provided by HAS, whereas the longer substituent in complex B allows it to escape the interaction with donor atoms at the surface of the protein.

lower relaxivity, due to a limited contribution from the second coordination sphere) with respect to the corresponding adduct with defatted HSA.

A related study was recently undertaken on the interaction of two Gd-DOTMA derivatives (A and B in Figure 3.10) with HSA.[43] Both A and B complexes bind strongly to HAS, yielding K_a values of $2.7 \times 10^3\ M^{-1}$ and $9.5 \times 10^4\ M^{-1}$, respectively. r_1^b of complex A/HSA adduct, at 298 K and 20 MHz, is equal to 35 mM^{-1}sec^{-1}. On the basis of ^{17}O measurements we surmise that the expected relaxation enhancement has been quenched by a direct interference of the donor groups on the surface of the protein with the inner hydration sphere of the Gd(III) ion. We could not determine whether it results in a replacement of the inner sphere water or simply in a dramatic elongation of its exchange lifetime. However, this result clearly shows that the biphenyl moiety does not appear long enough to protrude the chelate moiety outside the interference of the residues on the surface of the protein in proximity of the binding site. It is worth noting that the observed relaxation enhancement has to be ascribed to second-sphere water molecules and mobile protons on the surface of the protein in the proximity of the paramagnetic center. Conversely, complex B contains a binding synthon made of three phenyl groups which resulted long enough to avoid such interference from the protein donor atoms (Figure 3.11): the observed r_1^b for complex B/HSA adduct is 43.5 mM^{-1}sec^{-1}. In spite of the recovered inner-sphere contribution, the limited increase of the observed relaxivity outlines the loss of the second-sphere term once the paramagnetic chelate is no longer embedded in the hydration layers of the protein.

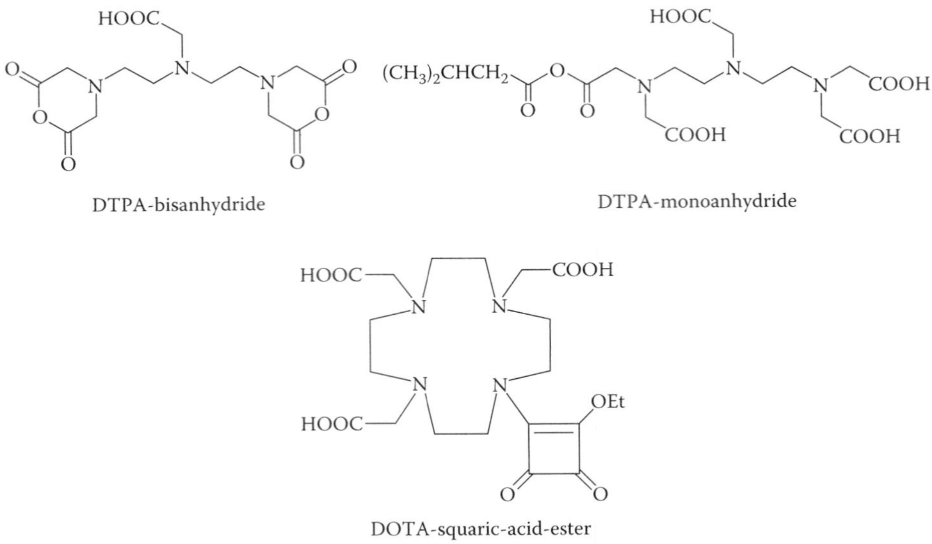

FIGURE 3.12 Examples of bifunctional ligands that easily conjugate with amino groups and are used to form macromolecular contrast agents.

3.3.3 Covalent Macromolecular Paramagnetic Systems

The most straightforward method to form covalent linkages deals with the reaction between amine groups on a macromolecular substrate and bifunctional ligands like those reported in Figure 3.12. Of course, a number of other coupling reactions with ligands bearing on their surface highly reactive functionalities such as isothyiocyanate, amino, carboxylate, sulfydril, etc., have also been reported. In this field, the synthesis[44] of [albumin-$(DTPA\text{-}Gd)_{30\text{–}34}$] was reported early and has been widely used to investigate vascular anatomy and physiology in tumor animal models.[45–47]

Besides albumin or other proteins, the macromolecular substrate may be a simple polyaminoacid like polylysine[48] or polyornithine[49] (whose polymerization degree can be controlled at will), dextran,[50] or polyethyleneglycol,[51] as well as a variety of copolymers.[52]

Another approach is based on the use of dendrimers, which are a class of highly branched synthetic spherical polymers (Figure 3.13) whose surface may be functionalized with Gd chelates. Kobayashi et al.[53] synthetized several dendrimer-based MRI contrast agents with different molecular weights ranging from 29 to 3850 KDa to change blood retention, tissue perfusion, and rate and pathway.

3.4 RESPONSIVE AGENTS

One of the major advantages of paramagnetic metal chelates with respect to iron oxide particles relies on the possibility of modulating their relaxivity as a function of specific parameter of the microenvironment in which the CA distributes. Therefore, to be a responsive agent, a paramagnetic metal complex has to be designed in such a way that at least one of the determinants of its relaxivity is significantly affected by the changes of the parameters of interest. Although essentially limited to *in vitro* investigation, several systems reporting changes in pH, temperature, redox potential, and enzymatic activity have already been reported.

3.4.1 pH Sensitive

Many pathological events are associated with alterations of pH homeostasis, including tumors and cardiovascular diseases.[54–57] The task of developing noninvasive methods for measuring pH *in vivo*

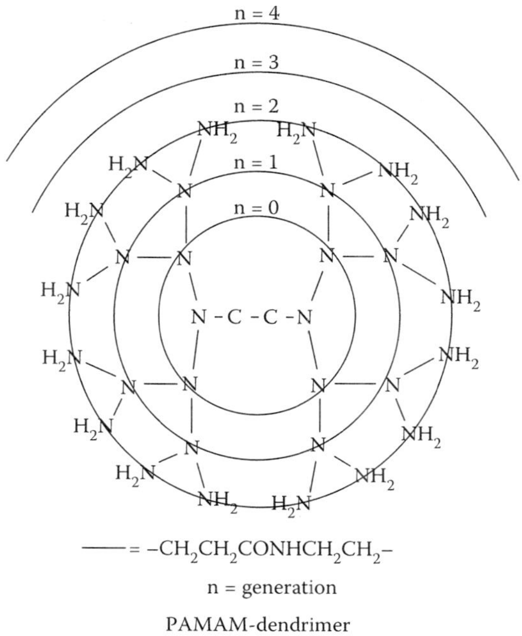

FIGURE 3.13 Scheme of polyamido amine (PAMAM) dendrimer used as case for the preparation of macromolecular contrast agents.

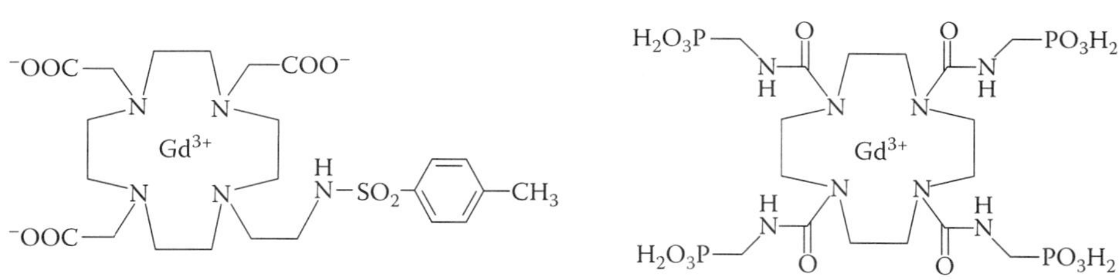

FIGURE 3.14 Schematic structure of two pH-responsive Gd chelates: complex A (left) is responsive via the change in the number of water molecules in the first hydration sphere, whereas complex B (right) is responsive via a change in the second hydration shell.

has therefore stimulated several studies aimed at designing probes sensitive to the hydrogenionic concentration.[58–61] The design of a Gd(III)-based complex whose relaxivity is pH dependent requires that at least one of the structural or dynamic parameters determining its relaxivity is made pH dependent.

For instance, it was found that the relaxivity of a series of macrocyclic Gd(III) complexes bearing β-arylsulfonamide groups[62] (Figure 3.14A) is markedly pH dependent on passing from about 8 $sec^{-1}mM^{-1}$ at pH < 4 to ca. 2.2 $sec^{-1}mM^{-1}$ at pH > 8. It has been demonstrated that the observed decrease (about fourfold) of r_1 is the result of a switch in the number of water molecules coordinated to the Gd(III) ion from 2 (at low pH values) to 0 (at basic pHs). This corresponds to a change in the coordination ability of the β-arylsulfonamide arm that binds the metal ion only when it is in the deprotonated form.

In some cases the pH dependence of the relaxivity is associated with changes in the structure of the second hydration shell. Zhang et al. reported on a case that deals with a macrocyclic tetramide derivative of DOTA (DOTA-4AmP; Figure 3.14B) displaying an unusual r_1 vs. pH dependence.[63] In fact, the relaxivity of this complex increases from pH 4 to pH 6, decreases up to pH 8.5, remains constant up to pH 10.5, and then increases again. The authors suggested that this behavior is related to the formation/disruption of the hydrogen bond network between the pendant phosphonate groups and the water bound to the Gd(III) ion. The deprotonation of phosphonate occurring at pH > 4 promotes the formation of the hydrogen bond network that slows down the exchange of the metal-bound water protons. On the contrary, the behavior observed at pH > 10.5 was accounted for in terms of a shortening of τ_M catalyzed by OH^- ions. Successively, it has been reported that this complex can be successfully used *in vivo* for mapping renal and systemic pH, on the assumption that its biodistribution is equal to that of the reference compound Gd-DOTP, a complex whose relaxivity is not affected by pH.[64]

Another useful route to pH-sensitive probes involves the inclusion of Gd(III) complexes into pH-sensitive liposomes.[65,66] The main component of these liposomes is represented by unsaturated phosphatidylethanolamine that is usually doped with negatively charged amphiphilic components in order to stabilize the lamellar phase necessary for the formation of the bilayer structure of liposomes. If the acidic group of the amphiphilic component is protonated, then the liposomes destabilize (transition phase) with the consequent increase of water exchange rate between the inner and outer compartments. On this basis, a relaxometric off–on pH switch can be designed by incorporating a Gd(III) complex into intact liposomes endowed with a very low water permeability. In this way, the relaxivity of the system will be very low (about 0.5 $sec^{-1}mM Gd^{-1}$ at 37°C and 10 MHz), but the protonation of the acidic components determines a significant relaxation enhancement (almost 4 $sec^{-1}mM Gd^{-1}$). This system has been successively tested in human blood where, unfortunately, it proved to be quite unstable owing to the presence of blood components (Na^+, Ca^{2+}, Mg^{2+}, and also HSA) that interfere with the phase transition of the liposomes.[67]

3.4.2 Agents Sensitive to Redox Potential

A diagnostic MRI probe sensitive to the *in vivo* redox potential would be very useful for detecting regions with a reduced oxygen partial pressure (P_{O2}), a typical symptom of several pathologies, including strokes.

Very few Gd(III) chelates sensitive to the tissue oxygenation have been so far reported. Our group has investigated the potential ability of Gd-DOTP to act as an allosteric effector of hemoglobin.[68] In fact, it has been observed that this chelate binds specifically to the T-form of the protein, which is characterized by a lower affinity toward oxygen. The interaction is driven by electrostatic forces and leads to a significant relaxivity enhancement (ca. fivefold) owing to the restricted molecular tumbling of the paramagnetic complex once it is bound to the protein. Although hemoglobin can be considered an excellent indirect target for detecting P_{O2}, the practical applicability of the method suffers from the inability of Gd-DOTP to enter red blood cells.

Another approach deals with the use of liposomes containing in their membrane an amphiphilic Gd(III) complex that has been designed to have a radical-sensitive disulfide bridge between the chelate and the lipid moiety[69] (Figure 3.15). The relaxivity (at 20 MHz and 25°C) of the liposomal paramagnetic agent is 13.6 $sec^{-1}mM^{-1}$, i.e., two-fold higher than that of the free Gd(III) complex (r_1 of 6.5 $sec^{-1}mM^{-1}$). Likely, the limited relaxivity enhancement of the bound form is due to the rotational flexibility of the complex on the surface of the liposome. This system has been tested *in vitro* by inducing the cleavage of the S–S bond with chemical (by dithiothreitol) or physical (by γ-rays) means. In both cases the relaxivity decreased from the value of the liposome-bound form to that of the free Gd(III) complex.

Though not based on the Gd(III) ion, it is worth recalling other paramagnetic probes acting on the relaxation rates of water protons, which have been proposed as responsive agents toward oxygen concentration. Basically, such systems rely on designing probes containing a metal that exists in two redox states endowed with quite different relaxation enhancements.

A first example is represented by the Mn(III)/Mn(II) redox switch. The complexes of Mn(II) and Mn(III) with the water-soluble tetraphenylsulfonate porphyrin (Figure 3.16) display significantly

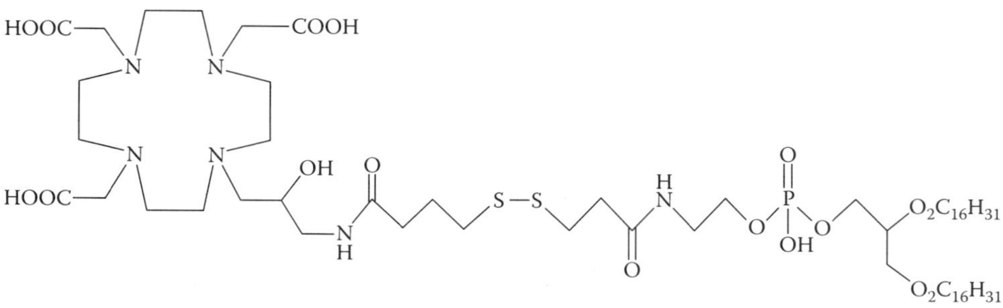

FIGURE 3.15 The Gd complex of this ligand is easily incorporated in the liposome membrane yielding a relaxivity typical of slowly tumbling systems. The cleavage of the S–S bond causes a decrease of relaxivity as a consequence of the loss of the characteristics of the supramolecular adduct.

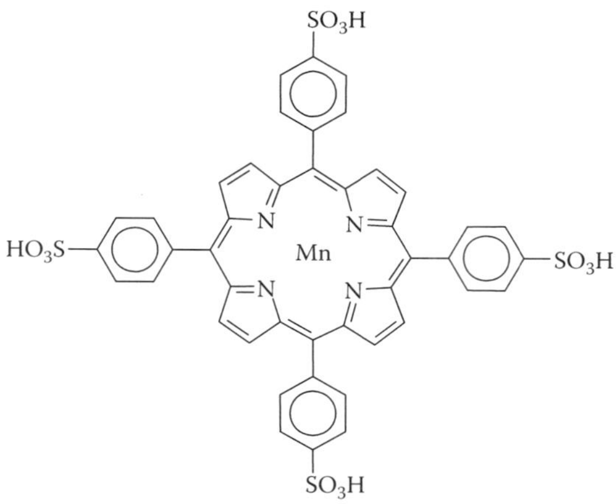

FIGURE 3.16 Mn can switch between oxidation states II and III, characterized by different relaxometric properties.

different r_1 values at low magnetic field strength (lower than 1 MHz), but very similar values at the fields used in clinical practice (>10 MHz).[70] However, the longer electronic relaxation rates of the Mn(II) complex make its relaxivity dependent on the rotational mobility of the chelate. In fact, upon interacting with a poly-β-cyclodextrin, a fourfold enhancement of the relaxivity of Mn(II)-TPPS at 20 MHz was detected, whereas little effect was observed for the Mn(III) complex. The ability of the Mn(II)/Mn(III) system to respond to changes in the partial pressure of oxygen has been demonstrated *in vitro*.

Another interesting redox switch deals with the Eu(III)/Eu(II) couple. Here, Eu(II) has the same electronic S_8 ground state of the Gd(III) ion, whereas Eu(III) ion is a poor relaxing probe owing to its very short electronic relaxation times. Unfortunately, Eu(II) is a very unstable cation. Therefore, efforts have been addressed to get more insight into the determinants of thermodynamic and kinetic stability of Eu(II) complexes.[71,72] Interestingly, it has been shown that the relaxivity of Eu(II)-DOTA is very similar to that of Gd(III)-DOTA, but the residence lifetime of the coordinated water in the former is noticeably shorter (τ_M of ca. 0.4 vs. 200 nsec at 25°C).[73]

3.4.3 Enzyme Responsive

One possible route to design enzyme-responsive agents is to synthesize paramagnetic complexes acting as substrate for a specific enzyme, whose binding to the active site of the protein can be signaled by the consequent variation in the relaxivity. Along this line of reasoning, Lauffer et al.

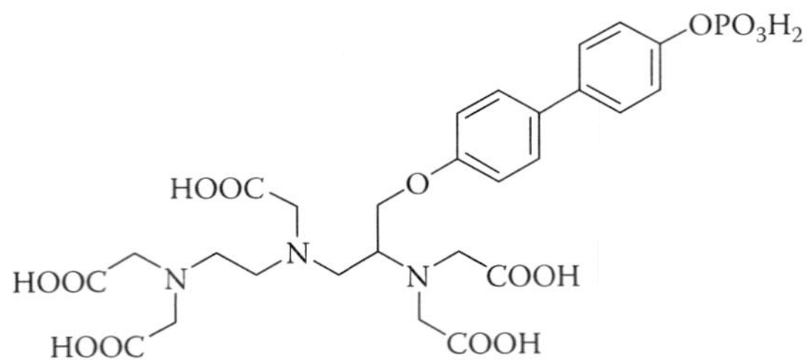

FIGURE 3.17 Enzyme-responsive agent: the action of serum alkaline phosphatase causes the release of phosphoric acid and a hydrophobic complex that displays high affinity to HSA.

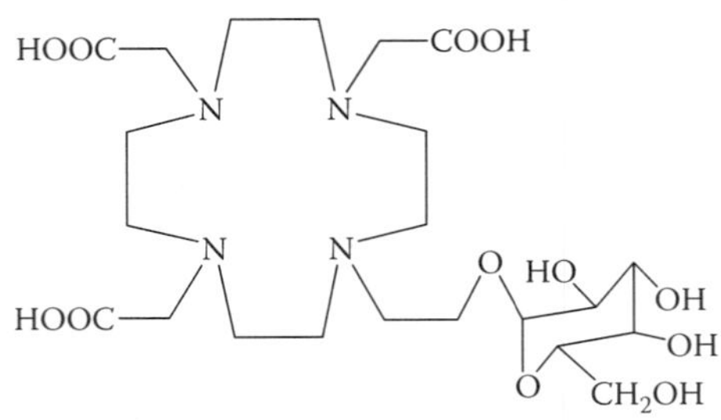

FIGURE 3.18 The action of β-galactosidase causes a change in the hydration state of the Gd^{3+} ion with a consequent increase of the observed relaxivity.

prepared a Gd(III) chelate containing a phosphoric ester (Figure 3.17) sensitive to the attack of the serum alkaline phosphatase.[74] The hydrolysis yields the exposure of a hydrophobic moiety well suitable to bind to HSA. Upon binding, there is an increase of the relaxivity as a consequence of the lengthening of the molecular reorientational time. This approach was used by the same research group for designing Gd(III) complexes sensitive to thrombin-activatable fibrinolysis inhibitor (TAFI), a carboxypeptidase B involved in clot degradation.[75]

Another example of enzyme-responsive agents has been provided by Moats et al. They synthesized a Gd-DOTA-like system containing a galactopyranosyl substituent capping the metal ion (Figure 3.18). Hence, the complex shows a low relaxivity (typical $q = 0$ complex). In the presence of β-galactosidase, an enzyme widely used as a marker of gene transfection, the sugar moiety is hydrolyzed and a water molecule can have free access to the paramagnetic center.[76] This complex has been successfully used for monitoring *in vivo* the gene expression during the development of embryos of *Xenopus laevis*.[77]

A procedure aimed at being of general applicability has been reported by Bodganov et al.[78] It relies on enzyme-mediated polymerization of paramagnetic substrates into oligomers of higher relaxivity. The paramagnetic substrate is represented by a Gd-DOTA-like complex bearing a cathecol functionality on its surface. In the presence of peroxydase, the monomers undergo rapid condensation into paramagnetic oligomers. To give support to the view of a wide applicability of this procedure in MRI, two interesting applications have been reported: (1) visualization of the occurrence of peroxydase activity at nanomolar concentration and (2) imaging of E-selectin on the surface of endothelial cells through a prior targeting with an anti-E-selectin peroxydase conjugate.

The development of Gd-based agents responsive to Factor XIII transglutaminase activity has been tackled independently by Weissleder's[79] and Neeman's[80] groups. Factor XIII is a key player in the final stages of blood coagulation, as it catalyzes the formation of a covalent bond between the glutamine residue in one fibrin chain and the ε-amino group of a lysine residue in a different chain. Thus, this transglutaminase reaction causes the formation of the fibrin network. A fibrinogen-specific peptide sequence, extracted from α2-antiplasmin (α2AP), the primary inhibitor of plasmin-mediated

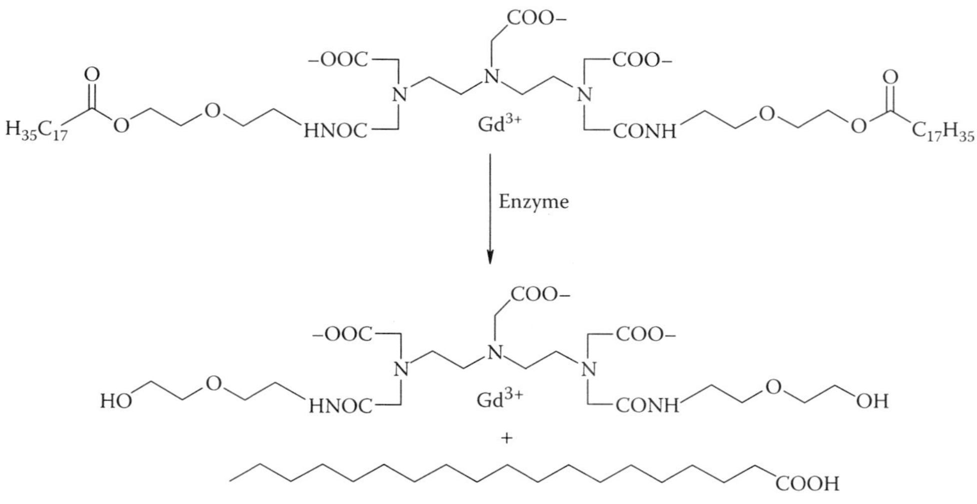

FIGURE 3.19 Particles made of insoluble Gd chelates. The hydrolysis of the ester bond by cellular enzymes causes the release of soluble Gd chelates with a consequent enhancement of cellular water protons.

fibrinolysis, was used. Thus, this peptide functionalized with Gd-DTPA cross-links with fibrin to an extent that depends on the local concentration of Factor XIII. Factor XIII activity is relevant to many pathologies, such as thrombotic disorder, coronary artery disease, myocardial infarction, and cerebrovascular diseases. Moreover, elevated activity of transglutaminase was shown at the boundaries of invading tumors, in association with angiogenesis.

Also, particles have been considered as enzymatic responsive agents. To this purpose, insoluble Gd(III) chelates have been synthesized by introducing, on the ligand surface, long aliphatic chains via an ester or a peptidic bond. The particles can be internalized into cells having phagocytic activity and then degraded by the action of the proper enzyme, which cleaves the bond between the Gd(III) chelate and the insolubilizing moiety (Figure 3.19). Thus, the increase of intracellular relaxivity becomes a function of the activity of the enzyme of choice. In principle, such an approach provides a representative example of responsive particles. Immediately after the internalization, the particles act as negative T_2 agents because they affect the bulk magnetic susceptibility. Then, their dissolution yields soluble Gd(III) chelates, which act as positive T_1 agents.[81]

3.4.4 Concentration-Independent Responsive Probes

In Section 3.4.3 it was shown how the relaxivity of a paramagnetic metal complex can be made responsive to a given parameter of its microenvironment. However, the *in vivo* application of responsive agents is still rather far off, mainly because in order to assign the observed relaxation rate to the actual value of the parameter of interest, one needs to know the local concentration of the contrast medium. So far, this problem has been tackled indirectly by using a reference compound whose relaxivity is not dependent on the parameter of interest.[64] From the measured proton relaxation enhancement induced by the reference compound one computes the local concentration in the region of interest and then uses it in the successive experiment with the responsive agent on the assumption that the two systems have the same biodistribution. Clearly it would be highly beneficial to rely on contrast agents whose responsiveness is independent of their actual concentration.

We have recently found[82] that this goal may be attained by applying a ratiometric approach that consists of measuring the ratio of the transverse and longitudinal paramagnetic contribution to the water proton relaxation rate promoted by the responsive agent. In fact, it has been demonstrated that at fields higher than 0.5 T and with responsive agents endowed with τ_R longer than 0.5 nsec,

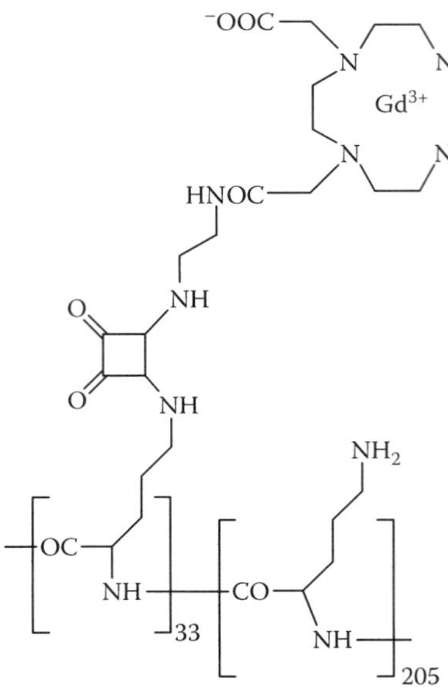

FIGURE 3.20 A macromolecular pH-responsive agent. The differential effect of pH and R_2 and R_1 has been exploited for the development of a concentration-independent procedure for the assessment of the solution pH.

the R_{2p}/R_{1p} ratio is independent of the concentration of the agent. In other words, one has to design a slowly moving system whose reorientational time is modulated by the pH of the solution but affecting to a different extent the longitudinal and transverse relaxation rates. To validate this approach, we investigated the pH responsiveness of a macromolecular adduct based on poly-L-ornithine, whose terminal amino groups have been partly functionalized with a macrocyclic Gd(III) complex by using a squarate moiety as a linker ((Gd-DOTAam)$_{33}$-Orn$_{205}$) (Figure 3.20). The pH responsiveness, in terms of R_{1p}, of an analogous paramagnetic macromolecule was already assessed[83] as a consequence of a change in molecular reorientational time caused by a reversible, pH-dependent transition between α-helical and random coil conformation.

The latter structure, caused by the electrostatic repulsion between the positively charged amino groups of the side chains, is stable at pH > 9, whereas at higher pH values the deprotonation of the amino groups promotes the formation of the α-helical conformation. This structural change has a different effect on the values of R_{2p} and R_{1p}, but their ratio is not affected by the concentration of the agent (Figure 3.21).

This procedure for making the pH responsiveness of a Gd-based probe independent of its concentration can be extended to the assessment of other parameters by the proper design of paramagnetic agents able to cause different effects on T_1 and T_2 of their solutions as a function of the parameter of interest.

3.5 EFFECTS ON THE RELAXIVITY WHEN GOING FROM AQUEOUS SOLUTIONS TO CELLULAR SYSTEMS

In this chapter it has been shown how the structure and dynamics of a paramagnetic metal chelate determine its ability to enhance water proton relaxation rates and that relaxivity is the key parameter in the evaluation of a paramagnetic chelate for MRI applications. In this section we focus on issues that may be relevant in determining the relaxivity of Gd chelates (and more generally of paramagnetic chelates) in *in vivo* applications.

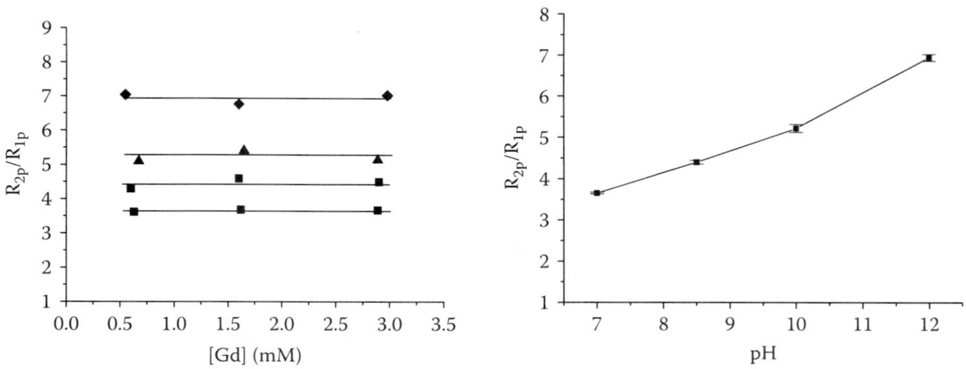

FIGURE 3.21 Left: Dependence of the relaxometric ratio on the concentration of Gd(III) for ((Gd-DOT-Aam)$_{33}$-Orn$_{205}$) at four pH values: pH 7 (squares), pH 8.5 (circles), pH 10 (triangles), and pH 12 (diamonds) (600 MHz, 25°C). Right: Corresponding pH dependence of the relaxometric ratio calculated from the data point reported on the left.

In general, for systems that distribute in vascular and extravascular space, it is possible to simulate the *in vivo* characteristics of biological fluids, and therefore to foresee the *in vivo* behavior of a given paramagnetic contrast agent. Also, for systems targeting epitopes on the cellular surface, one may translate the *in vitro* results to what is expected from *in vivo* experiments. More tricky is the situation encountered when the paramagnetic agent is internalized into cells. One may envisage two cases: (1) the contrast agent is entrapped into endosomes and (2) the contrast agent distributes in the cytoplasm. Case 1 is frequently met in the experiments of cellular labeling when the internalization occurs either by pinocytosis or by receptor-mediated endocytosis. It has been noted that when the paramagnetic metal complexes are confined in endosomes, the observed relaxation enhancement is quenched in the presence of relatively high amounts of internalized chelates. In Figure 3.22 such a saturation effect, measured for Gd-HPDO3A internalized in hepatoma (HTC) cells, leads to an observed proton relaxation rate of ca. 3 sec^{-1} that does not change when the number of chelates per cell increases from 1×10^{10} to 4.5×10^{10}. An example of the case 2 internalization route can be set up by use of the electroporation procedure. This technique causes the formation of transient hydrophilic pores on the cell membrane upon application of suitable electric pulses between two electrodes placed in the cell suspension. Upon internalization via the electroporation route, the contrast agent molecules enter the cytoplasm compartment, and no quenching effect on the observed relaxation rates has been observed, i.e., the relaxation rate increases upon increasing the amount of internalized Gd.

To account for the observed behavior, one may consider the cellular pellets (and a portion of tissue as well) at a multisite system where the water molecules are distributed in the extra- and intracellular (or cytosolic) compartments. Such compartments are separated by the cellular membrane, whose water permeability is crucial for determining the relaxometric behavior of the whole pellet. The behavior observed in Figure 3.22 may be explained in terms of a three-site water exchange model when the imaging probe is entrapped into endosomes (extracellular/cytoplasm/endosome compartments) and in terms of a two-site exchange model when the paramagnetic agent is only dispersed into the cytoplasm. On this basis, the quenching effect of the exchange on the relaxation rate of cytosolic water protons is the responsible factor for the saturation of the relaxation rate observed at high concentrations of the internalized probe for HTC pellets labeled by pinocytosis.[84] The obtained results mark the importance of the procedure used for labeling cells and demonstrate that the cytosol confinement of the probe yields higher relaxing efficiency, in turn allowing the MRI detection of a smaller number of cells with respect to the entrapment in endosomes.

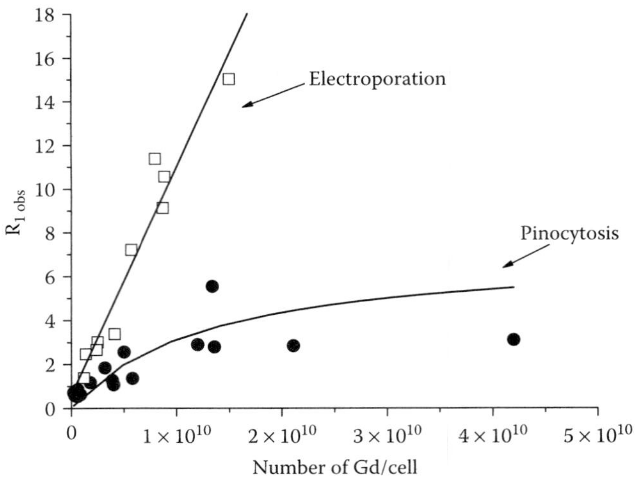

FIGURE 3.22 Effect of the intracellular localization of Gd-HPDO3A on the relaxation enhancement of water protons in HTC cellular pellets.

3.6 CONCLUSIONS

In the past two decades, Gd(III) chelates have been under intense scrutiny for their potential application as MRI contrast agents. This huge work has led to the synthesis of a number of hepta- and octa-coordinating ligands that efficiently wrap around the Gd^{3+} ion and yield chelates characterized by very high thermodynamic stabilities. An in-depth knowledge has been acquired about the relationships among structure, dynamics, and relaxometric properties of Gd(III) chelates. The control of the various determinants of the relaxivity has allowed the design of systems endowed with a relaxation enhancement capability much higher than that shown by the Gd base contrast agent currently used in clinical settings. Moreover, it has been shown that a proper design of the ligand characteristics may yield systems whose relaxivity acts as a reporter of a given physicochemical parameter of the microenvironment of the contrast agent. In this context, current research activities are focused on the possibility of developing responsive agents whose effect can be assessed independently from the knowledge of the absolute concentration.

Finally, the need for setting down MR molecular imaging protocols has prompted a great deal of work in the field of targeting agents. On one hand, this task has provided innovative work for the synthesis of highly sensitive Gd(III) chelates conjugated to suitable synthons that deliver the contrast agent to the targeting sites. On the other hand, in some instances, it has been found to be more convenient to design systems able to form highly stable supramolecular adducts as a route for the local amplification of the MRI response.

The search for systems endowed with a further enhanced sensitivity is still very active in many laboratories worldwide, and one may foresee that the achievements likely to come from this work will further strengthen the role of MRI in the armory of tools available for molecular imaging applications.

REFERENCES

1. Engelstadt, B.L., Wolf, G.L., in *Magnetic Resonance Imaging*, Stark, D.D., Bradley, W.G., Jr., Eds., V.C. Mosby Company, St. Louis, 1988, p. 161.
2. Rinck, P.A., *Magnetic Resonance in Medicine*, Blackwell Scientific Publications, Oxford, 1993.
3. Caravan, P. et al., Gadolinium(III) chelates as MRI contrast agents: structure, dynamics, and applications, *Chem. Rev.*, 99, 2293, 1999.

4. Tweedle, M.F., in *Lanthanide Probes in Life, Chemical and Earth Sciences: Theory and Practice*, Bünzli, J.-C.G., Choppin, G.R., Eds., Elsevier, Amsterdam, 1989, p. 127.
5. Brücher, E., Sherry, A.D., in *The Chemistry of Contrast Agents in Medical Magnetic Resonance Imaging*, Merbach, A.E., Tóth, E., Eds., Wiley, Chichester, U.K., 2001, p. 243.
6. Weinmann, H.J., Mühler, A., Radüchel, B., in *Biomedical Magnetic Resonance Imaging and Spectroscopy*, Young, I.R., Ed., John Wiley & Sons Ltd., Chichester, U.K., 2000, p. 705.
7. Aime, S. et al., Lanthanide(III) chelates for NMR biomedical applications, *Chem. Soc. Rev.*, 27, 19, 1998.
8. Banci, L., Bertini, I., Luchinat, C., *Nuclear and Electronic Relaxation*, VCH, Weinheim, 1991, p. 91.
9. Schmitt-Willich, H. et al., Synthesis and physicochemical characterization of a new gadolinium chelate: the liver-specific magnetic resonance imaging contrast agent Gd-EOB-DTPA, *Inorg. Chem.*, 38, 1134, 1999.
10. Uggeri, F. et al., Novel contrast agents for magnetic resonance imaging. Synthesis and characterization of the ligand BOPTA and its Ln(III)complexes (Ln = Gd, La, Lu). X-ray structure of disodium (TPS-9-145337286-C-S)-[4-carboxy-5,8,11-tris(carboxymethyl)-1-phenyl-2-oxa-5,8,11-triazatridecan-13-oato(5-)]gadolinate(2-) in a mixture with its enantiomer, *Inorg. Chem.*, 34, 633, 1995.
11. Aime, S. et al., Gd(III)-based contrast agents for MRI, *Adv. Inorg. Chem.*, 57, 173, 2005.
12. Cabella, C. et al., Cellular labeling with Gd(III)chelates: only high thermodynamic stabilities prevent the cells acting as "sponges" of Gd^{3+} ions, *Contrast Med. Mol. Imaging*, 1, 23, 2006.
13. Aime, S. et al., Ternary Gd(III)L-HSA adducts: evidence for the replacement of inner-sphere water molecules by coordinating groups of the protein. Implications for the design of contrast agents for MRI, *J. Biol. Inorg. Chem.*, 5, 488, 2000.
14. Aime, S. et al., [GdPCP2A(H2O)(2)](-): a paramagnetic contrast agent designed for improved applications in magnetic resonance imaging, *J. Med. Chem.*, 43, 4017, 2000.
15. Aime, S. et al., [Gd-AAZTA](-): a new structural entry for an improved generation of MRI contrast agents, *Inorg. Chem.*, 43, 7588, 2004.
16. Hajela, S. et al., A tris-hydroxymethyl-substituted derivative of Gd-TREN-Me-3,2-HOPO: an MRI relaxation agent with improved efficiency, *J. Am. Chem. Soc.*, 122, 11228, 2000.
17. Fulton, D.A. et al., Efficient relaxivity enhancement in dendritic gadolinium complexes: effective motional coupling in medium molecular weight conjugates, *Chem. Commun.*, 4, 474, 2005.
18. Vander Elst, L. et al., Physicochemical characterization of P760, a new macromolecular contrast agent with high relaxivity, *Eur. J. Inorg. Chem.*, 13, 2495, 2003.
19. Vander Elst, L. et al., *In vitro* relaxometric and luminescence characterization of P792 (Gadomelitol, Vistarem((R))), an efficient and rapid clearance blood pool MRI contrast agent, *Eur. J. Inorg. Chem.*, 6, 1142, 2005.
20. Livramento, J.B. et al., High relaxivity confined to a small molecular space: a metallostar-based, potential MRI contrast agent, *Angew. Chem. Int. Ed.*, 44, 1480, 2005.
21. Botta, M. et al., Second coordination sphere water molecules and relaxivity of gadolinium(III) complexes: implications for MRI contrast agents, *Eur. J. Inorg. Chem.*, 3, 399, 2000.
22. Borel, A. et al., Molecular dynamics simulations of MRI-relevant Gd-III chelates: direct access to outer-sphere relaxivity, *Chem. Eur. J.*, 7, 600, 2001.
23. Aime, S. et al., High-relaxivity contrast agents for magnetic resonance imaging based on multisite interactions between a beta-cyclodextrin oligomer and suitably functionalized Gd-III chelates, *Chem. Eur. J.*, 7, 5261, 2001.
24. Aime, S. et al., Compartmentalization of a gadolinium complex in the apoferritin cavity: a route to obtain high relaxivity contrast agents for magnetic resonance imaging, *Angew. Chem. Int. Ed.*, 41, 1017, 2002.
25. Toth, E. et al., Water-soluble gadofullerenes: toward high-relaxivity, pH-responsive MRI contrast agents, *J. Am. Chem. Soc.*, 127, 799, 2005.
26. Sitharaman, B. et al., Superparamagnetic gadonanotubes are high-performance MRI contrast agents, *Chem. Commun.*, 31, 3915, 2005.
27. Aime, S. et al., Insights into the use of paramagnetic Gd(III) complexes in MR-molecular imaging investigations, *J. Magn. Reson. Imaging*, 16, 394, 2002.
28. Morawski, A.M. et al., Targeted nanoparticles for quantitative imaging of sparse molecular epitopes with MRI, *Magn. Reson. Med.*, 51, 480, 2004.
29. Caruthers, S.D. et al., *In vitro* demonstration using F-19 magnetic resonance to augment molecular imaging with paramagnetic perfluorocarbon nanoparticles at 1.5 tesla, *Invest. Radiol.*, 41, 305, 2006.

30. Tokumitsu, H. et al., Chitosan-gadopentetic acid complex nanoparticles for gadolinium neutron-capture therapy of cancer: preparation by novel emulsion-droplet coalescence technique and characterization, *Pharm. Res.*, 16, 1830, 1999.
31. Jono, K. et al., Preparation of lecithin microcapsules by a dilution method using the Wurster process for intraarterial administration in gadolinium neutron capture therapy, *Chem. Pharm. Bull.*, 47, 54, 1999.
32. Oyewumi, M.O. et al., Comparison of cell uptake, biodistribution and tumor retention of folate-coated and PEG-coated gadolinium nanoparticles in tumor-bearing mice, *J. Control Release*, 95, 613, 2004.
33. Watanabe, T. et al., Tumor accumulation of gadolinium in lipid-nanoparticles intravenously injected for neutron-capture therapy of cancer, *Eur. J. Pharm. Biopharm.*, 54, 119, 2002.
34. Bertini, I. et al., Persistent contrast enhancement by sterically stabilized paramagnetic liposomes in murine melanoma, *Magn. Reson. Med.*, 52, 669, 2004.
35. Allen T.M. et al., Long-circulating (sterically stabilized) liposomes for targeted drug-delivery, *Trends Pharmacol. Sci.*, 15, 215, 1994.
36. Sipkins, D.A. et al., Detection of tumor angiogenesis *in vivo* by alpha(v)beta(3)-targeted magnetic resonance imaging, *Nat. Med.*, 4, 623, 1998.
37. Bhujwalla, Z.M. et al., Magnetic resonance molecular imaging of the HER-2/neu receptor, *Cancer Reson.*, 63, 2723, 2003.
38. Aime, S., Botta, M., Fasano, M., et al., in *The Chemistry of Contrast Agents in Medical Magnetic Resonance Imaging*, Merbach, A.E., Toth, E., Eds., John Wiley & Sons Ltd., Chichester, U.K., 2001, p. 193.
39. Caravan, P. et al., The interaction of MS-325 with human serum albumin and its effect on proton relaxation rates, *J. Am. Chem. Soc.*, 124, 3152, 2002.
40. Lahti, K.M. et al., Magnetic resonance angiography at 0.3 T using MS-325, *MAGMA*, 12, 88, 2001.
41. Zheng, J. et al., Single-session magnetic resonance coronary angiography and myocardial perfusion imaging using the new blood pool compound B-22956 (gadocoletic acid): initial experience in a porcine model of coronary artery disease, *Invest. Radiol.*, 40, 604, 2005.
42. De Haen, C. et al., Gadocoletic acid trisodium salt (B22956/1): a new blood pool magnetic resonance contrast agent with application in coronary angiography, *Invest. Radiol.*, 41, 279, 2006.
43. Aime, S. et al., New insights for pursuing high relaxivity MRI agents from modelling the binding interaction of Gd-III chelates to HSA, *ChemBioChem*, 6, 818, 2005.
44. Ogan, M.D. et al., Albumin labeled with GdDTPA. An intravascular contrast-enhancing agent for magentic resonance blood pool imaging: preparation and characterization, *Invest. Radiol.*, 22, 665, 1987.
45. Cohen, F.M. et al., Contrast-enhanced magnetic resonance imaging estimation of altered capillary permeability in experimental mammary carcinomas after X-irradiation, *Invest. Radiol.*, 29, 970, 1994.
46. Brash, R. et al., Assessing tumor angiogenesis using macromolecular MR imaging contrast media, *J. Magn. Reson. Imaging*, 7, 68, 1997.
47. Wikstrom, M.G. et al., Contrast-enhanced MRI of tumors. Comparison of Gd-DTPA and a macromolecular agent, *Invest. Radiol.*, 24, 609, 1989.
48. Schuhmann-Giampieri, G. et al., *In vivo* and *in vitro* evaluation of Gd-DTPA polylisine as a macromolecular contrast agent for magnetic resonance imaging, *Invest. Radiol.*, 26, 969, 1991.
49. Bremerich, J. et al., Slow clearance gadolinium-based and intravascular contrast media for three-dimensional MR angiography, *J. Magn. Reson. Imaging*, 13, 588, 2001.
50. Wang, S.C. et al., Evaluation of Gd-DTPA labeled dextran as an intravascular MR contrast agent: imaging characteristics in normal rat-tissues, *Radiology*, 175, 483, 1990.
51. Desser, T.S. et al., Dynamics of tumor imaging with Gd-DTPA polyethylene-glycol polymers: dependence on molecular-weight, *J. Magn. Reson. Imaging*, 4, 467, 1994.
52. Weissleder, R. et al., Size optimization of synthetic graft copolymers for *in vivo* angiogenesis imaging, *Bioconjug. Chem.*, 12, 213, 2001.
53. Kobayashi H. et al., Dendrimer-based molecular MRI contrast agents: characteristics and application, *Mol. Imaging*, 2, 1, 2003.
54. Gerweck, L.E. et al., Cellular pH gradient in tumor versus normal tissue: potential exploitation for the treatment of cancer, *Cancer Res.*, 56, 1194, 1996.
55. Tannock, I.F. et al., Heterogeneity of intracellular pH and of mechanisms that regulate intracellular pH in populations of cultured cells, *Cancer Res.*, 49, 4373, 1998.

56. Naghavi, M. et al., pH heterogeneity of human and rabbit atherosclerotic plaques: a new insight into detection of vulnerable plaque, *Atherosclerosis*, 164, 27, 2002.
57. Aime, S. et al., Paramagnetic complexes as novel NMR pH indicators, *Chem. Commun.*, 11, 1265, 1996.
58. Zhou, J.Y. et al., Using the amide proton signals of intracellular proteins and peptides to detect pH effects in MRI, *Nat. Med.*, 9, 1085, 2003.
59. Pilatus, U. et al., Real-time measurements of cellular oxygen consumption, pH, and energy metabolism using nuclear magnetic resonance spectroscopy, *Magn. Reson. Med.*, 45, 749, 2001.
60. Sun, Y. et al., Simultaneous measurements of temperature and pH *in vivo* using NMR in conjunction with $TmDOTP^{5-}$, *NMR Biomed.*, 13, 460, 2000.
61. Ward, K.M. et al., Determination of pH using water protons and chemical exchange dependent saturation transfer (CEST), *Magn. Reson. Med.*, 44, 799, 2000.
62. Lowe, M.P. et al., pH-dependent modulation of relaxivity and luminescence in macrocyclic gadolinium and europium complexes based on reversible intramolecular sulfonamide ligation, *J. Am. Chem. Soc.*, 123, 7601, 2001.
63. Zhang, S.R. et al., A novel pH-sensitive MRI contrast agent, *Angew. Chem. Int. Ed.*, 38, 3192, 1999.
64. Raghunand, N. et al., Renal and systemic pH imaging by contrast-enhanced MRI, *Magn. Reson. Med.*, 49, 249, 2003.
65. Lokling, K.E. et al., Tuning the MR properties of blood-stable pH-responsive paramagnetic liposomes, *Int. J. Pharm.*, 274, 75, 2004.
66. Lokling, K.E. et al., pH-sensitive paramagnetic liposomes for MRI: assessment of stability in blood, *Magn. Reson. Imaging*, 21, 531, 2003.
67. Lokling, K.E. et al., pH-sensitive paramagnetic liposomes as MRI contrast agents: *in vivo* feasibility studies, *Magn. Reson. Imaging*, 19, 731, 2001.
68. Aime, S. et al., Molecular recognition of r-states and t-states of human adult hemoglobin by a paramagnetic Gd(III) complex by means of the measurement of solvent water proton relaxation rates, *J. Am. Chem. Soc.*, 117, 9365, 1995.
69. Glogard, C. et al., Novel radical-responsive MRI contrast agent based on paramagnetic liposomes, *Magn. Reson. Chem.*, 41, 585, 2003.
70. Aime, S. et al., A p(O-2)-responsive MRI contrast agent based on the redox switch of manganese(II/III): porphyrin complexes, *Angew. Chem. Int. Ed.*, 39, 747, 2000.
71. Seibig, S. et al., Unexpected differences in the dynamics and in the nuclear and electronic relaxation properties of the isoelectronic [Eu-II(DTPA)(H2O)](3-) and [Gd-III(DTPA)(H2O)](2-) complexes (DTPA = diethylenetriamine-pentaacetate), *J. Am. Chem. Soc.*, 122, 5822, 2000.
72. Burai, L. et al., High-pressure NMR kinetics. Part 95. Solution and solid-state characterization of Eu-II chelates: a possible route towards redox responsive MRI contrast agents, *Chem. Eur. J.*, 6, 3761, 2000.
73. Burai, L. et al., Novel macrocyclic Eu-II complexes: fast water exchange related to an extreme M-O-water distance, *Chem. Eur. J.*, 9, 1394, 2003.
74. Lauffer, R.B., McMurry, T.J., Dunham, S.O., Scott, D.M., Parmelee, D.J., Dumas, S., PCT Int. Appl. WO9736619, 1997.
75. Nivorozhkin, A.L. et al., Enzyme-activated Gd^{3+} magnetic resonance imaging contrast agents with a prominent receptor-induced magnetization enhancement, *Angew. Chem. Int. Ed.*, 40, 2903, 2001.
76. Moats, R.A. et al., A "smart" magnetic resonance imaging agent that reports on specific enzymatic activity, *Angew. Chem. Int. Ed.*, 36, 726, 1997.
77. Louie, A.Y. et al., *In vivo* visualization of gene expression using magnetic resonance imaging, *Nat. Biotechnol.*, 18, 321, 2000.
78. Bodganov, A. et al., Oligomerization of paramagnetic substrates results in signal amplification and can be used for MR imaging of molecular targets, *Mol. Imaging*, 1, 16, 2002.
79. Tung, C.H. et al., Novel factor XIII. Probes for blood coagulation imaging, *ChemBioChem*, 4, 897, 2003.
80. Mazooz, G. et al., Development of magnetic resonance imaging contrast material for *in vivo* mapping of tissue transglutaminase activity, *Cancer Res.*, 65, 1369, 2005.
81. Aime, S. et al., Innovative magnetic resonance imaging diagnostic agents based on paramagnetic Gd(III) complexes, *Biopolymers*, 66, 419, 2002.

82. Aime, S. et al., *J. Am. Chem. Soc.*, 128, 11326–11327, 2006.
83. Aime, S. et al., A macromolecular Gd(III) complex as pH-responsive relaxometric probe for MRI applications, *Chem. Commun.*, 16, 1577, 1999.
84. Terreno, E. et al., Effect of the intracellular localization of Gd-based imaging probe on the relaxation enhancement of water protons, *Magn. Reson. Med.*, 55, 491, 2006.

4 Superparamagnetic Contrast Agents

Claire Corot, Marc Port, Irène Guilbert, Philippe Robert, Isabelle Raynal, Caroline Robic, Jean-Sebastien Raynaud, Philippe Prigent, Anne Dencausse, and Jean-Marc Idée

CONTENTS

4.1 Introduction 60
4.2 Physicochemical Aspects 60
4.2.1 Synthesis of Magnetic Nanoparticles 60
4.2.1.1 Synthesis of Iron Oxide Crystalline Structures 60
4.2.1.2 Surface Modification of Magnetic Nanoparticles 62
4.2.1.3 Surface Modification with Targeting Ligands 62
4.2.2 Magnetic Properties 63
4.2.2.1 Superparamagnetism 63
4.2.2.2 Relaxivity of Superparamagnetic Materials 64
4.2.3 Physicochemical Characterization 65
4.2.3.1 Size, Shape, and Distribution of the Iron Oxide Crystal 65
4.2.3.2 Magnetic and Relaxometric Properties 67
4.2.3.3 Hydrodynamic Particle Size 67
4.2.3.4 Charge 67
4.3 Biological Aspects 67
4.3.1 *In Vitro* Endocytosis 67
4.3.1.1 Mechanism of Internalization 69
4.3.1.2 Nanoparticle Uptake by Nonprofessional Phagocytic Cells 70
4.3.1.3 Technical Note 70
4.3.2 Pharmacokinetics, Biodistribution, and Metabolism 70
4.3.2.1 Pharmacokinetics and Biodistribution 70
4.3.2.2 Species Differences 72
4.3.2.3 Metabolism: Persistence of the Cell Label 72
4.3.3 Biocompatibility 74
4.4 Imaging Modalities 74
4.4.1 Positive or Negative Contrast? 74
4.4.2 Spatial Resolution 75
4.5 Clinical Applications 75
4.5.1 Liver Imaging 76
4.5.2 Metastatic Lymph Node Imaging 76
4.5.3 Macrophage Imaging 76
4.5.4 Blood Pool Agent Imaging 77
4.5.4.1 Tumor Imaging 77

4.5.4.2 MR Angiography (MRA) 77
4.5.4.3 Cerebral Blood Volume Imaging 77
4.6 Conclusion 77
References 78

4.1 INTRODUCTION

The introduction of superparamagnetic nanoparticle contrast agents in magnetic resonance imaging (MRI) has constituted a major improvement of the range of tools available to clinicians. Due to the efficacy of these agents, they are now proposed in the gastrointestinal (GI) tract and liver imaging.[1,2] The prospects for new applications in lymph node and functional imaging are very promising. The capacity of nanoparticles to target inflammatory lesions via macrophage labeling opens up major and very exciting prospects for the characterization of numerous inflammatory and degenerative diseases.[3] By the addition of targeted ligands, these agents are also potential disease-specific products.[4,5]

There are currently two distinct classes of superparamagnetic nanoparticles, depending on particle size: superparamagnetic iron oxide (SPIO) particles with a mean particle diameter of more than 50 nm and ultrasmall superparamagnetic iron oxide (USPIO) particles with a smaller diameter.[1,6]

Two compounds in the SPIO family are commercialized for intravenous use: ferumoxides (Endorem, Europe; Feridex, U.S. and Japan) and ferucarbotran (Resovist, Europe and Japan). In both cases, the clinical targets are liver tumors. These nanoparticles are medium sized (70 to 150 nm) and coated with dextran (ferumoxides) or carboxydextran (ferucarbotran).[7]

Several USPIO particles have been investigated in humans such as ferumoxtran-10 (dextran),[8,9] VSOP (very small iron oxide particle; citrate),[10] feruglose (pegylated starch),[11] and SHU555C (carboxydextran).[12]

Some USPIO nanoparticles have a similar composition to SPIO nanoparticles, but have a smaller total diameter. It is generally accepted that larger particles are more rapidly and preferentially taken up by macrophages, but the nature of the hydrophilic coating material is also very important.[13]

This chapter reviews the main physicochemical characteristics as well as the biological properties useful for the imaging applications of superparamagnetic contrast agents.

4.2 PHYSICOCHEMICAL ASPECTS

Iron oxide nanoparticles composed of maghemite and magnetite (Fe_2O_3, Fe_3O_4) stabilized by various coating agents are characterized by a large magnetic moment in the presence of a static external magnetic field, which makes them suitable as MR contrast agents. This large magnetic moment is caused by a crystal ordering (spinels) that induces a cooperativity between the individual paramagnetic ions constituting the crystal. These small superparamagnetic crystals are smaller than a magnetic domain (approximately 30 nm), and they consequently do not show any magnetic remanence (i.e., restoration of the induced magnetization to zero upon removal of the external magnetic field), unlike ferromagnetic materials. Several classes of iron oxide nanoparticles are investigated. Structure–activity relationship programs are based on optimization of blood clearance, biocompatibility, tissue accessibility, and cellular targeting.

4.2.1 Synthesis of Magnetic Nanoparticles

4.2.1.1 Synthesis of Iron Oxide Crystalline Structures

Numerous chemical methods can be used to synthesize magnetic nanoparticles for medical imaging applications. The first main chemical challenge probably consists of defining experimental conditions leading to a monodisperse population of magnetic grains of suitable size. The second critical point is to select a reproducible process that can be industrialized without any complex purification procedure such as ultracentrifugation,[14] size exclusion chromatography,[15] magnetic filtration,[16] or flow field gradient.[17]

Various processes have been adapted to produce magnetic nanoparticles from solution techniques or from the aerosol or vapor phase.

Solution Technique

The solution technique is probably the simplest and most efficient chemical pathway to obtain magnetic particles with appreciable control over size and shape. Iron oxides (either Fe_3O_4 or γFe_2O_3) are usually prepared by an aging stoichiometric mixture of ferrous and ferric salts in aqueous medium. The chemical reaction of Fe_3O_4 formation may be written as

$$Fe^{2+} + 2\ Fe^{3+} + 8\ OH^{-} \rightarrow Fe_3O_4 + 4\ H_2O \qquad (4.1)$$

According to the thermodynamics of this reaction, complete precipitation of Fe_3O_4 should be expected between pH 9 and 14 with a stoichiometric ratio of 2/1 (Fe^{3+}/Fe^{2+}) in a nonoxidizing oxygen environment.[18]

However, magnetite (Fe_3O_4) is very sensitive to oxidation and is transformed into maghemite (γFe_2O_3) in the presence of oxygen, according to the following:

$$Fe_3O_4 + 2\ H^{+} \rightarrow \gamma Fe_2O_3 + Fe^{2+} + H_2O \qquad (4.2)$$

Oxidation in air is not the only way to transform magnetite (Fe_3O_4) into maghemite (γFe_2O_3). Various electron or ion transfers, depending on the pH of the suspension, are involved. Under acidic and anaerobic conditions, surface Fe^{2+} ions are desorbed as hexa-aqua complexes in solution, whereas, under basic conditions, oxidation of magnetite involves oxidation-reduction of the surface of magnetite. Oxidation of ferrous ions is always correlated with migration of cations through the lattice framework, creating cationic vacancies in order to maintain the charge balance, explaining the structure of maghemite. Consequently, maghemite has a spinel structure inverse to that of magnetite (in maghemite, iron ions are distributed in the octahedral (Oh) and tetrahedral (Td) sites of the spinel structure (Formula 4.1)), but differs from magnetite by the presence of cationic vacancies within the octahedral site. The vacancies ordering scheme is closely related to the sample preparation method and results in symmetry lowering and possibly superstructures. The vacancies can be completely random or partially or totally ordered. It has been shown, essentially from combined infrared (IR) spectroscopy and x-ray diffraction, that vacancy ordering occurs only for particles exceeding 5 nm.[19]

Formula 4.1: Structure of Magnetite and Maghemite

$$Fe_3O_4\text{: } [Fe^{3+}]_{Td}\ [Fe^{3+}\ Fe^{2+}]Oh\ O_4$$

$$\gamma Fe_2O_3\text{: } 0.75\ [Fe^{3+}]_{Td}\ [Fe^{3+}{}_{5/3}\ V_{1/3}]Oh\ O_4$$

Although the classical coprecipitation method allows modification of the mean size of nanoparticles by adjusting pH, ionic strength, temperature, or the ratio of Fe(II) over Fe(III) (range from 4 to 15 nm), control of particle size distribution is limited with this synthetic method. On the other hand, uniform and quasi-homodisperse nanoparticles can be obtained by coprecipitation in reverse micelles.[20] However, these particles are only soluble in organic solvents.

A microemulsion synthesis providing water-soluble nanoparticles with precise control of size dispersity has recently been described using reactive metal salts, as magnetite nanoparticles with a diameter of 4 nm were prepared by controlled hydrolysis of an aqueous solution of ferrous and ferric chloride with ammonium hydroxide in microemulsion formed by sodium bis(2-ethylhexyl) sulfosuccinate (AOT) as the surfactant and heptane as the continuous phase.[21]

High-Temperature Decomposition of Organic Precursor

Nanoparticles with a high level of monodispersity and size control can be obtained by high-temperature decomposition of iron organic precursors such as $Fe(Cup)_3$, $Fe(CO)_5$, or $Fe(acac)_3$. However, this type of process must be improved to be suitable for industrial preparation, especially in terms of safety of the reactant and the high temperature required.

4.2.1.2 Surface Modification of Magnetic Nanoparticles

After synthesis of the magnetic nanoparticle, coating is required to prevent destabilization and agglomeration of the colloidal suspension and to make the nanoparticles soluble in aqueous or biological media. The coating may also provide a certain chemical functional to conjugate targeted ligands. A very high density of coating is often needed to effectively stabilize iron oxide nanoparticles, and polymeric or monomeric coatings have been used. Coating chemistry can be assisted by sonochemistry.[22]

Many polymeric coating materials have been used, such as dextran, carboxymethylated dextran, carboxydextran, starch, polyethyleneglycol (PEG), arabinogalactan, glycosaminoglycan, organic siloxane, and sulfonated styrene-divinylbenzene.[23]

Carboxylated polyamino amine (PAMAM) dendrimers (generation 4.5) have been used as a stabilizing iron oxide coating. These SPIO nanocomposites (magnetodendrimers) have been optimized for *in vivo* tracking of stem cells.[24] Oxidation of Fe(II) at slightly elevated pH and temperature resulted in the formation of highly water soluble magnetodendrimers with an overall hydrodynamic size of 20 to 30 nm.

Mornet et al.[25] developed an original synthetic route to obtain versatile ultrasmall superparamagnetic iron oxide (VUSPIO) in a multistep procedure consisting of colloidal maghemite synthesis, surface modification by silanation of the iron core with aminopropylsylane groups, and conjugation with partially oxidized dextran and subsequent reduction of the shiff base.

The process engineered by Massart[26] for rapid synthesis of homogeneous γFe_2O_3 nanoparticles allowed coating by a wide range of monomeric species, such as amino acids, α-hydroxyacids (citric, tartaric, gluconic), hydroxamate (arginine hydroxamate), or dimercaptosuccinic acid (DMSA).

VSOP C184 is under clinical investigation stabilized by a monomeric coating (citric acid). Process optimization of VSOP C184 has led to nanoparticles with an iron core diameter of 4 nm and a hydrodynamic diameter of 8.6 nm.[10] Two processes have been developed to coat the iron core: *in situ* coating and postsynthesis coating. For example, ferumoxtran-10 and ferumoxides are prepared by the Molday coprecipitation method with *in situ* coating by dextran. In contrast, nanoparticles derived from the Massart technology are obtained by a postsynthesis coating process. Another coating approach has been proposed consisting of encapsulating iron oxide nanoparticles in liposomes to obtain magnetoliposomes. De Cuyper and Joniau[27] and Bulte et al.[28] developed this process to synthesize magnetoliposomes containing one to six crystals per vesicle with a hydrodynamic diameter of 40 nm. These magnetoliposomes were pegylated to prolong their blood half-life.

4.2.1.3 Surface Modification with Targeting Ligands

For molecular imaging, biovectors able to recognize a biological target must be grafted onto the surface of iron nanoparticles. Many biovectors are used in molecular imaging, such as antibodies, proteins, peptides, peptidomimetics, and small targeting ligands. Various processes have been used to couple these biovectors onto nanoparticles.

The first process is based on an oxidative conjugation strategy, which produces aldehydes on a carbohydrate coating such as dextran. This oxidative process using periodate was employed to covalently couple human serum albumin (HSA) or transferrin onto a carboxydextran nanoparticle or to couple a monoclonal antibody or a wheat germ agglutinin lectin onto a dextran nanoparticle.[29–31]

However, in the case of the transferrin biovector, a substantial loss of the biological activity of the protein was observed.[32] To minimize this type of detrimental effect, a new versatile nonoxidative technology was developed allowing the introduction of various chemical linkers. First, a dextran nanoparticle was cross-linked by epichlorohydrin and ammonia. The resulting amine-terminated cross-linked iron oxide (CLIO) nanoparticle is a powerful platform to conjugated biovectors with a wide range of heterobifunctional linkers.[33] Weissleder's group used this technology to graft transferrin, annexin V, anti-VCAM (vascular cellular adhesion molecule) or anti-E-selectin antibodies, oligonucleotides, and twin arginine translocase (TAT) peptides to CLIO nanoparticles.[34–38] This technology has recently been used to develop a nanoparticle library that recognizes apoptotic cells comprising 146 nanoparticles decorated with different synthetic small ligands.[39,40] Although this CLIO technology has provided interesting targeted USPIO particles, which have been used to reach proof of principle in molecular imaging, industrialization of it raises a major problem, as the cross-linking agent, epichlorohydrin, is classified as a carcinogenic, mutagenic, and reprotoxic substance.

Oxidative processes have also been used by Sonrico et al. to couple a monoamine PEG-folic acid to a VUSPIO and by Zhang and Zhang to obtain pegylated USPIO coated by folic acid.[41,42]

The 2,3-DMSA technology, described previously, allowed coupling of biovectors via S–S bonds using SPDP (N-succimidyl 3-(2-pyridyl thio) propionate) as the heterobifunctional linker or C–S bonds using MBS (maleinidobenzoyl-N-hydroxysuccinimide ester) as the heterobifunctional linker. This technology has been used to couple antibodies, lectins, and annexin V to DMSA nanoparticles.[43,44]

4.2.2 Magnetic Properties

4.2.2.1 Superparamagnetism[45,46]

Bulk ferromagnetic materials are composed of fully magnetized domains of micron size. In the absence of an external magnetic field, a ferromagnetic material is not magnetized, as the magnetization of these ferromagnetic domains is oriented in several directions. Superparamagnetism occurs when the size of the crystals is smaller than ferromagnetic domains (approximately 30 nm) and, consequently, they do not show any magnetic remanence (i.e., restoration of the induced magnetization to zero upon removal of the external magnetic field), unlike ferromagnetic materials. Each crystal is then considered to be a fully magnetized single magnetic monodomain, and thus can be considered to be a monomagnet. This monomagnet is characterized by a supermagnetic spin larger than the sum of the individual paramagnetic spins due to a cooperativity effect. At the atomic level, this magnetic cooperativity is a direct consequence of the spinel structure of the crystal, which allows strong magnetic coupling and consequently perfect alignment of the individual magnetic spins.

The crystalline nature of superparamagnetic nanoparticles is anisotropic, as the crystal comprises an anisotropic field and, consequently, the magnetic moment of the nanomagnet tends to align along privileged axes called axes of easy magnetization. At each point in time, the magnetization is therefore not equal to zero, but is almost saturated even in the absence of an external magnetic field. However, the magnetic moment jumps from one easy direction to another, which cancels the time-averaged magnetization. This magnetic fluctuation is called the Néel relaxation process and is characterized by a correlation time τ_N. When a static external field Bo is applied, the magnetic moment is forced to align with Bo and the correlation time τ_N has an infinite value.

As a result, superparamagnetic crystals are characterized by a large magnetic moment in the presence of an external magnetic field Bo, but no remnant magnetic moment when the field is

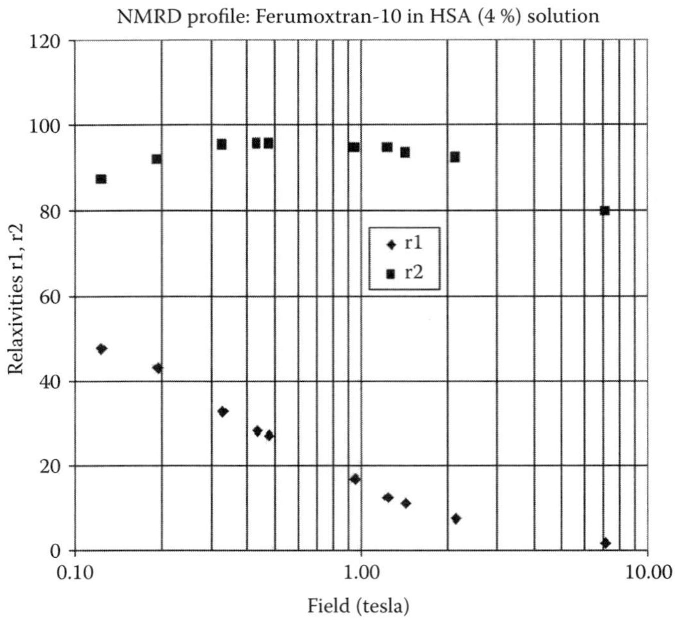

FIGURE 4.1 NMRD profile of ferumoxtran-10 in 4% human serum albumin (HSA).

zero, contrary to ferromagnetic substances, which have a remnant magnetic moment at zero field once magnetized.

These colloids of magnetic iron oxide are composed of crystals measuring 4 to 6 nm. When these crystals are placed in an external magnetic field, they align and create very high local magnetic field gradients inducing water proton spin dephasing and, as a result, reducing the T1 and T2 relaxation times of the surrounding water.

4.2.2.2 Relaxivity of Superparamagnetic Materials[45,46]

The efficiency by which a contrast agent can enhance the proton relaxation rate in a homogeneous medium is called relaxivity of the agent and is defined by

$$R_{1,2} = + r_{1,2}\, C \tag{4.3}$$

where $R_{1,2}$ (unit sec^{-1}) is the respective T_1 or T_2 proton relaxation rate in the presence of the contrast agent, $R^{\circ}_{1,2}$ is the relaxation rate in the absence of contrast agent, and C is the contrast agent concentration (unit m*M*). The constant of proportionality $r_{1,2}$ (unit $mM^{-1}sec^{-1}$) is called relaxivity and is a measure of how much the proton relaxation rate is increased per unit of concentration of contrast medium.

The dipolar interaction between surrounding water protons and the high magnetic moment (super spin) of superparamagnetic particles results in high longitudinal r_1 and transverse r_2 relaxivity.

The theory describing the magnetic interaction of superparamagnetic compounds with water protons has been described by different theoretical models derived from the classical outer-sphere paramagnetic relaxation model. In the classical outer-sphere theory, the dipolar interaction fluctuates due to the translational diffusion time of the water molecule and the Néel relaxation process. The classical outer-sphere theory has been modified to take into account the high Curie relaxation of superparamagnetic crystals. This theory can be used to interpret and fit the field dependence of the proton relaxation rate (nuclear magnetic relaxation dispersion (NMRD) profile; Figure 4.1) by using:

- A Freed (low field fitting, characteristic correlation times τ_D and τ_N) and an Ayant (high field fitting, characteristic correlation time τ_D) models of fitting weighted by magnetization of the sample (Langevin function). This basic theory has been improved by introducing the influence of the anisotropy energy of the superparamagnetic crystal, allowing the magnetization moment to proceed around the easy direction rather than being blocked along the axis of easy magnetization.
- A classical outer-sphere theory completed by a secular term (predominant term at high field, proportional to τ_D) to fit the NMRD r_2 data. The secular term explains the increase of r_2 relaxivity at high field according to the size of the superparamagnetic crystal.

However, in most situations, it is the significant capacity of superparamagnetic nanoparticles to reduce the so-called T_2* relaxivity that is used in MR imaging. This r_2* relaxivity is called susceptibility effect and describes an increase of T_2 relaxation rates due to a magnetization difference, ΔM, between different voxels in the MRI image. A large ΔM occurs as a result of the nonhomogeneous distribution of superparamagnetic particles, which gives rise to local field gradients that accelerate the loss of phase coherence of the spins contributing to the MR signal. This process is much more important for superparamagnetic particles than for paramagnetic species, as the induced magnetization of a superparamagnetic particle is high due to the high susceptibility of iron oxide. Although the dipolar coupling described by the outer-sphere theory is a close-range effect requiring an interaction between water protons and the magnetic center, the susceptibility effect affects protons much farther away from the magnetic centers. The magnitude of this susceptibility effect also depends on many factors, such as compartmentalization of the contrast agent, type of imaging sequence, and aggregation of the contrast agent.

It should be noted that any aggregation has an important impact on the T_1, T_2, or T_2* efficiency of a superparamagnetic particle. At the clinical field used in MRI (1 to 3 T), agglomeration tends to slightly decrease r_1, but markedly increases r_2 and particularly r_2* (Table 4.1).

4.2.3 Physicochemical Characterization

The synthesis of superparamagnetic nanoparticles is a complex process because of their colloidal nature. Consequently, a full set of analytical methods should be used to characterize the efficacy (in terms of magnetization and relaxivity), purity of nanoparticles, and reproducibility of the synthesis process. Moreover, because the size, geometry, heterogeneity, composition of the crystals, charge of the particles, and nature of the coating strongly influence the physicochemical and biological behavior of the particles, an accurate description of the physicochemical properties of these nanoparticles is crucial to fully understand the system and its efficiency, particularly for structure–activity relationship programs.

4.2.3.1 Size, Shape, and Distribution of the Iron Oxide Crystal

The size of the crystals varies from agent to agent, but also depends on the measurement technique.[6,52–54] The core size is generally between 4 and 10 nm (Table 4.2). The size distribution and the shape of the crystal can be appreciated by transmission electronic microscopy (TEM) and need to be measured on a statistically significant number of crystals. Moreover, the sample preparation can induce aggregation of the colloids, and the TEM measurements may consequently not reflect the crystal size and distribution in solution. The size of the crystals can also be measured by x-ray diffraction (XRD) by analyzing x-ray line broadening. Moreover, XRD can provide information about the crystal composition and structure. Composition of the crystal can also be assessed by Mössbauer spectroscopy to distinguish between magnetite and maghemite phases. Mössbauer spectroscopy also provides information about the magnitude of the Néel relaxation time, τ_N.

TABLE 4.1
Relaxivities r_1, r_2 in Water, at 37°C ($mM^{-1}sec^{-1}$)

Medium: Water	Hydrodynamic Diameter (nm)	0.47 T (20 MHz) r_1	0.47 T (20 MHz) r_2	1.42 T (60 MHz) r_1	1.42 T (60 MHz) r_2	3 T (128 MHz)[a] r_1	3 T (128 MHz)[a] r_2	7.1 T (300 MHz) r_1	7.1 T (300 MHz) r_2
Ferumoxides AMI25[b]	120–180	20.2	101	10.1	120	5	130	1.8	132
Ferucarbotran A SHU555A[b]	65	23.6	179	9.7	189	4.5	200	1.6	205
Ferumoxtran-10 AMI227[b]	15–30	23.3	65	9.9	65	5	66	1.4	71
Ferumoxytol C7228[b]	35	38	83	15	89	7.5	92	2	95
Ferucarbotran C SHU555C[47]	20	24	60	—	—	—	—	—	—
Feruglose NC100150[48,49]	15–20	21.8	35.3	—	—	—	—	—	—
VSOP C184[50]	8.6	18.7	30	14	33.4	8	34	3.5	34.2
CLIO-TAT peptide[51]	30	22.3	51.9	—	—	—	—	—	—
P904B[b]	25–30	34	86	14	87	7.5	89	2	91

Note: —, not available.

[a] Values extrapolated from the NMRD profile.
[b] Guerbet Research data.

TABLE 4.2
Physicochemical Parameters

	Size of Crystal Core (nm)	Magnetization	Coating	Hydrodynamic Diameter (nm)
Ferumoxides AMI25	5[55]	94 emu/g Fe (50 kg)[55]	Dextran T10 kDa	120–180[56]
Ferumoxtran-10 AMI227	6[55]	95 emu/g Fe (50 kg)[55]	Dextran T10 kDa, T1 kDa	15–30[56]
Ferumoxytol C7228	6.7[56]		Carboxymethyl dextran T10 kDa	35[56]
Ferucarbotran A SHU555A	4[57]		Carboxydextran T1.8 kDa	65[57]
Ferucarbotran C SHU555C	4[57]		Carboxydextran T1.8 kDa	20[58]
Feruglose NC100150	6[59]	86 emu/g Fe Msat[54]	Methylcellulose, PEG	15–20[59]
VSOP C184	4[10]		Citrate	8.6[10]
CLIO	8.7[60]		Dextran periodate	30[60]

Note: Msat = magnetization at saturation.

4.2.3.2 Magnetic and Relaxometric Properties

The magnetic properties of nanoparticles are classically studied by the magnetization behavior in response to the applied magnetic field (magnetometry), which confirms the superparamagnetic property and indicates magnetization at saturation (MS) (Table 4.2). Interpretation of magnetic behavior according to the Langevin function also indicates the magnetic diameter of the crystal. Analysis of nuclear magnetic relaxation dispersion (NMRD) curves is very informative, as fitting of NMRD curves provides information about relaxometric diameter, magnetization at saturation, the Néel relaxation time, τ_N, and anisotropy energy (Figure 4.1). The r_2/r_1 ratio is also an indicator of the size of the nanoparticles and possibly an agglomeration process.

4.2.3.3 Hydrodynamic Particle Size

The hydrodynamic size of the nanoparticles (i.e., the global size of the particle comprising one or several magnetic crystals surrounded by the coating molecules) is usually measured by photon correlation spectroscopy (PCS), which analyzes the quasi-elastic light diffusion of nanoparticles when illuminated by a monochromatic laser beam. The intensity of the diffused light is modulated by the Brownian motion of the particles in solution. Various mathematical analyses of PCS data can provide an estimation of the hydrodynamic size and polydispersity of the nanoparticles. However, the values obtained by PCS intimately depend on the mathematical models and weighting parameters used (unimodal or multimodal distribution, number, volume or intensity distribution). It is therefore very difficult to compare hydrodynamic size measurements described in the literature for different nanoparticles, and any relationship between size and biodistribution must be proposed with caution (Table 4.2).

In general, it is hazardous to compare nanoparticle sizes measured by different methods, such as iron oxide diameter obtained by TEM or XRD, relaxometric diameter (obtained by fitting of NMRD curves), magnetometric diameter (obtained by superconducting quantum interface device [SQUID]), or hydrodynamic diameter (obtained by PCS).

4.2.3.4 Charge

The charge on the surface of the nanoparticles is usually determined by the zeta potential measurement deduced from measurement of electrophoretic mobility. Unfortunately, the value of zeta potential of superparamagnetic nanoparticles is rarely described in the literature. Nevertheless, the effect of charge on the biodistribution of superparamagnetic nanoparticles is a key point. Positive and negative charges both have been found to decrease in blood half-life, but via different mechanisms of interaction,[61] and they may also influence the biocompatibility of these nanoparticles.

4.3 BIOLOGICAL ASPECTS

After intravenous injection, nanoparticles are preferentially taken up by professional phagocytic cells. The size and nature of the coating material influence cellular uptake as well as nanoparticle intracellular trafficking and metabolism. The impact of intracellular compartmentalization, modification of nanoparticle structure, and degradation on their relaxometric behavior and biocompatibility profiles must be considered (Figure 4.2).

4.3.1 *In Vitro* Endocytosis

Macrophage uptake of nanoparticles is dependent on dose, incubation time, and cell type. Moore et al. showed that, under the same incubation conditions (1 mg Fe/1 million cells, 1 h), iron dextran nanoparticle (microcrystalline iron oxide nanoparticle (MION)) uptake was 310 ng Fe/million cells in J-774 macrophage-like cells, whereas it was 970 ng Fe/million cells in murine peritoneal

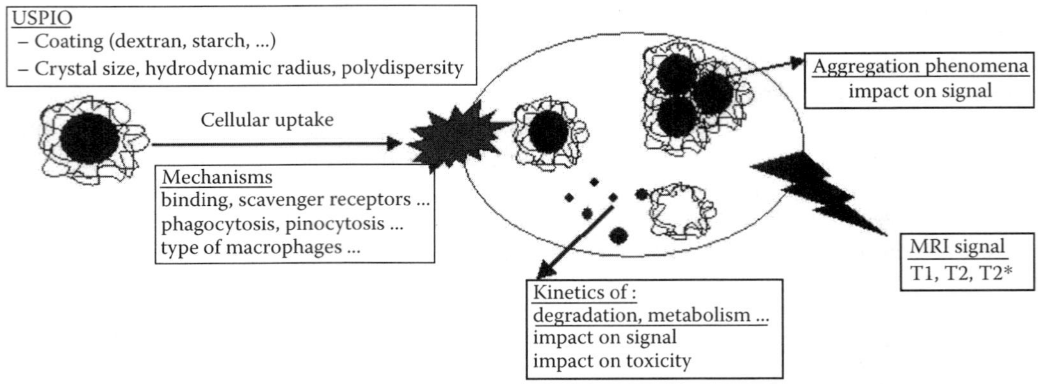

FIGURE 4.2 Cellular interactions and metabolic parameters relevant for MR efficacy of iron oxide nanoparticles.

TABLE 4.3
***In Vitro* Macrophage Uptake of USPIO and SPIO**

	Incubation Concentration (μg Fe/ml)	Incubation Time (h)	Human Promonocytic Cell Line THP-1 (pg Fe/cell)[a]	Adherent Human Monocytes (pg Fe/cell)[64]
Ferumoxides	100	4	5.0–7.0	9
AMI25	200	24		
Ferumoxtran	100	4	0.5–1.8	2
AMI227	200	24		
Ferumoxytol	200	24	1.0–1.5	
C7228				
Ferucarbotran A	100	4	8.0–12.0	28
SHU555A	200	24		
Ferucarbotran C	100	4		4
SHU555C				
P904	200	24	1.8–2.3	

[a] Guerbet Research data.

macrophages.[62] A dose-dependent relationship is observed with saturation of uptake at very high concentrations. Furthermore, the higher the uptake rate, the shorter the time required to achieve maximum uptake.[13,63,64] Quantitative comparison of uptake among various nanoparticles must be performed under the same experimental conditions (Table 4.3).

SPIO shows a higher uptake than USPIO when taking into account the same classes of compounds, i.e., dextran- (ferumoxides vs. ferumoxtran-10) or carboxydextran- (SHU555A vs. SHU555C) coated nanoparticles. As these agents have an identical composition, the larger particle size of SPIO must be considered to be responsible for the higher rate of macrophage uptake (Table 4.3).

Ionic dextran nanoparticles such as ferucarbotran (SHU555A and SHU555C) have a higher uptake than nonionic dextran nanoparticles (ferumoxides and ferumoxtran-10). A similar difference is observed for carboxymethyldextran USPIO ferumoxytol compared to nonionic dextran USPIO ferumoxtran-10. Metz et al. have shown that, following incubation of ferumoxides and ferucarbotran, at plasma concentrations calculated according to the human injected doses of 17 and 9 μg/ml, with adherent human monocytes, similar uptakes were observed for the two compounds (3.78 ± 0.74 and 4.47 ± 0.84 pg Fe/cell, respectively).[64]

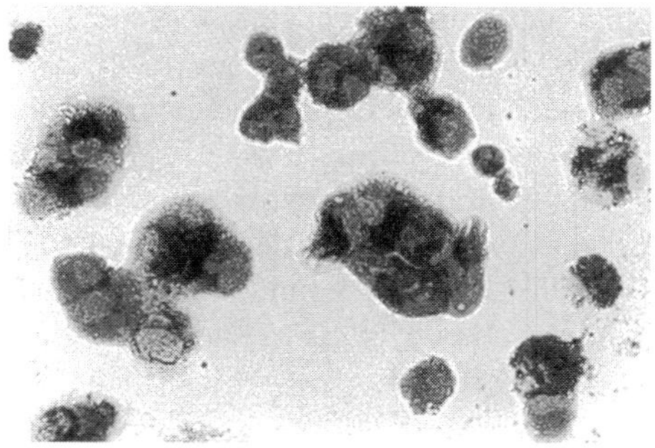

FIGURE 4.3 (please see color insert following page 210) Iron staining and haematoxilin Eosin Safran (HES) coloration of THP-1 cells (human monocytic cell line) 24 h after incubation of 100 μg Fe/ml of ferumoxides.

The influence of the nanoparticle anionic charge in the macrophage is also important for small organic coating materials. Uptake of the citrate-coated VSOP C125 nanoparticles was much greater than that of DDM 43/34/103 dextran-coated nanoparticles in a P388 macrophage cell line,[63] and similar findings have been reported for anionic magnetic nanoparticle (AMNP) nanoparticles (meso-2,3-dimercaptosuccinic acid) compared to ferumoxtran-10 in a RAW macrophage cell line.[65]

Improvement of hydrophilicity by the addition of hydroxyl groups (P904; Table 4.3) or PEG groups limits phagocyte uptake.[42]

4.3.1.1 Mechanism of Internalization

Internalization of foreign materials by cells has classically been divided into uptake of large materials (phagocytosis) and uptake of soluble materials or nanoparticles (pinocytosis). Pinocytosis can be further divided into clathrin-dependent fluid-phase endocytosis and macropinocytosis. Phagocytosis and receptor-mediated endocytosis share many characteristics, although the extent of overlap between these processes has not been fully described. Using uncoated polystyrene in the size range of 100 to 1000 nm, no apparent size-related discontinuity was observed between pinocytic and phagocytic uptake by rat peritoneal macrophages.[66] Since iron oxide nanoparticles have hydrodynamic diameters between 9 and 180 nm, pinocytosis is mainly involved (Figure 4.3).

Macrophages actively take up a wide range of negatively charged macromolecules. Under physiological conditions, scavenger receptors (SRs) clear anionic nanoparticles. Under pathological conditions, scavenger receptors mediate the recruitment, activation, and transformation of macrophages and other cells that may be related to the development of atherosclerosis and disorders caused by accumulation of denatured materials.

Polynucleotides such as poly-inosinic acid or polysaccharides such as fucoidan are ligands of scavenger receptors. Raynal et al. observed dose-related inhibition of ferumoxides uptake by poly-inosinic acid and fucoidan, which indicates that an SR-mediated endocytosis pathway is involved for this dextran-coated SPIO. By improving the sensitivity of quantification, the same type of inhibition was observed with ferumoxtran-10.[13]

In the presence of cytochalasin B, which blocks the cytoskeleton with subsequent inhibition of phagocytosis, uptake of citrate-coated nanoparticles by P388 macrophages was significantly reduced, whereas no change was observed in the presence of colchicine, which blocks intracellular structures and thereby suppresses nonspecific pinocytosis. Opposite results were observed for dextran-coated nanoparticles.[63] Moore et al. also showed that N-ethylmaleimide (NEM), an inhibitor of vesicular traffic, caused a dramatic decrease in uptake of dextran-coated USPIO (–72%) in mouse peritoneal macrophages, and localization of these nanoparticles within macropinosomes or phagosomes was demonstrated by fluorescent microscopy.[62] This compartmentalization was also observed in several electron microscopy studies, which have shown that iron oxide nanoparticles are predominantly found in large phagolysosomes.

Because scavenger receptors and complement receptors are prominent on macrophage membranes, it is also likely that opsonized nanoparticles are recognized by these receptors and are actively taken up by receptor-mediated endocytosis. Moore et al. showed that the uptake of opsonized nanoparticles by mouse peritoneal macrophages was dramatically increased compared to nonopsonized nanoparticles. In the presence of an excess of plasma proteins, uptake of opsonized nanoparticles was reduced, whereas excess of free dextran had no effect.[62]

Rogers and Basu showed that the uptake of ferumoxides by activated macrophages is regulated by endogenous cytokines and serum components as well as exogenous lovastatin. In cultured murine macrophage-like cells (J774A.1) pretreated with 1.0 μM lovastatin and incubated with ferumoxides, SPIO uptake was significantly reduced to 61% of control levels. In the same study, interferon-gamma (1000 U/ml) increased the SPIO uptake to 163% of control levels. Interleukin-4 (40 ng/ml) also increased uptake (178% of control). In cells incubated with SPIO in the absence of serum proteins, SPIO uptake decreased to 57% of control.[67] MRI signal changes after SPIO administration may therefore reflect macrophage phagocytic capacity as well as the presence of macrophages.

Altogether, these results indicate that several mechanisms could be involved in macrophage uptake, with intensity varying according to the nanoparticle characteristics and the level of cellular activation.

4.3.1.2 Nanoparticle Uptake by Nonprofessional Phagocytic Cells

It has also been observed that the uptake of iron oxide nanoparticles is not limited to professional phagocytes. USPIO and SPIO uptake has been described in many other cell types, such as endothelial cells and tumor cells. For example, the uptake of anionic AMNP nanoparticles by HeLa cells (tumor cell line) exceeds the uptake by RAW macrophages.[65] For dextran-coated nanoparticles, Moore et al. observed an uptake by 9L gliosarcoma, C6 glioma, or LX-1 small cell lung carcinoma, which were significantly lower than in macrophages.[62] Cell labeling with magnetic nanoparticles is an increasingly investigated method for *in vitro* cell separation and *in vivo* imaging.

4.3.1.3 Technical Note

Relaxation methods may be used to accurately quantify the concentration of iron oxide particles within cells. However, several criteria must be met to use relaxation methods to evaluate the behavior of iron oxide nanoparticles in cells: well-defined relaxivity in the cell, complete homogenization of cell samples, and no intracellular aggregation of iron oxide nanoparticles, which can modify the relaxometric properties of the magnetic nanoparticle. In cells, the effective relaxivity may differ from that measured in solution due to compartmentalization of nanoparticles. To eliminate the effects of compartmentalization on relaxivity, the tissue samples need to be completely homogenized.

4.3.2 Pharmacokinetics, Biodistribution, and Metabolism

4.3.2.1 Pharmacokinetics and Biodistribution

Nanoparticles are usually taken up by macrophages in the liver, spleen, and bone marrow or lymph nodes. The different uptake rates of various nanoparticles result in longer blood half-lives, avoiding phagocytosis in the liver and spleen, consequently allowing them to reach other targets.

It is generally accepted that larger particles are preferentially taken up more rapidly by macrophages, but the nature of the hydrophilic coating material is also of major importance. The blood half-lives of the various iron oxide nanoparticles administered in patients vary from 1 to 36 h (Table 4.4). An inverse relationship between the potency of human macrophage uptake *in vitro* and human blood half-life is in favor of the role of the liver and spleen for the blood clearance of these SPIO/USPIO particles, as these tissues have a high content of macrophages (Table 4.4). In the case of iron oxide nanoparticles with the same coating material, USPIO particles are less prone to liver uptake due to their smaller size (ferumoxtran-10 vs. ferumoxides and SHU555C vs. SHU555A).

TABLE 4.4
Human and Rat Blood Half-Lives and Macrophage Uptake of SPIO and USPIO

Human	Dose (μg Fe/kg)	Human Half-Life (h)	Rat Half-Life (h)	Macrophage Uptake
Ferumoxides AMI25	30	2[68]	0.1[74]	High
Ferumoxtran AMI227	45	24–36[69]	2.7[a]	Low
Ferumoxytol C7228	18–70	10–14[70]	1.1[a]	Low
Ferucarbotran A SHU555A	8–12	2.4–3.6[71]	0.7[75]	High
Ferucarbotran C SHU555C	40	6[72]	1[76]	Medium
Feruglose NC100150	36	6[73]	No data available	No data available
VSOP	15–75	0.6–1.3[50]	0.25–0.50[77]	High

[a] Guerbet Research data.

As for *in vitro* macrophage uptake, the blood half-life is lower for ionic dextran (carboxy and carboxymethyl) than for nonionic dextran.

Size is not the only factor involved, as very small USPIO (8.7 nm) composed of citrated-coated iron oxide nanoparticles (VSOP) present the shortest blood half-life in humans (approximately 1 h).

Blood half-life is dose dependent for all iron oxide nanoparticles. This is a well-known phenomenon that has already been demonstrated for various particle systems and is related to progressive saturation of macrophage uptake in the liver or other macrophage-rich organs, such as the spleen and bone marrow. At high doses, a progressive saturation of macrophage uptake in these organs leads to an increase of the free form in the plasma. However, the slight increase of half-life found in the range of clinical doses is not considered to have any relevant clinical impact in terms of the global pharmacokinetic profile of the compound.

Although accumulation of USPIO particles in macrophages has been clearly established (Table 4.3), their pathway of transport to tissue macrophages has not been fully elucidated. Several mechanisms have been proposed:[3,78,79]

1. USPIO are endocytosed by activated blood monocytes that migrate into pathological tissues. For example, the long blood half-life of ferumoxtran-10 allows sufficient time for blood monocytes to endocytose nanoparticles and for progressive migration of these cells.
2. Transcytosis of USPIO across the endothelium and migration of USPIO particles into the tissues, followed by progressive endocytosis of these USPIO particles by *in situ* macrophages.
3. Transport of USPIO particles into the pathological tissue, in some cases via the inflammatory neovasculature (vasa vasorum) irrigating the media and adventitia in atherosclerotic lesions.

Vascular permeability to ferumoxtran-10 was demonstrated in a cultured endothelial cell monolayer model. While endothelial cell clearance was observed, the increased permeability in the presence of bradykinin indicates that ferumoxtran-10 transport is similar to albumin endothelial transport via vesicular and junctional pathways.[80] Electron microscopy has demonstrated the presence of USPIO in endothelial cells in several animal models.

TABLE 4.5
Dose Species Differences in Pharmacokinetic Parameters of Ferumoxtran-10 and Comparison with Cardiac Output in the Same Range of Injected Dose

Ferumoxtran-10	Dose (μmol Fe/kg)	Blood Half-Life (min)	Lymph Node Uptake (× Endogenous Fe)[a]	Cardiac Output (l/min/kg)
Mouse	60	18	× 1.8	0.6
Rat	50	168	× 5.0	0.4
Rabbit	45	228	—	0.25
Monkey	30	342	—	0.16
Human	45	2160	—	0.09

[a] × Endogenous Fe, ratio of iron concentration measured postinjection of ferumoxtran-10 to the iron concentration before injection.

Source: Guerbet Research data.

All three pathways (endocytosis, transcytosis, and inflammatory neovessels) are assumed to be involved, with their respective contribution dependent on the patient's pathophysiological status.

4.3.2.2 Species Differences

The blood half-life of USPIO differs between humans and animal species. For example, the blood half-life in rats is three- to tenfold lower than in humans for various iron oxide nanoparticles, but the same ranking between products is observed (Table 4.4). At doses of 30 or 45 μmol Fe/kg, the blood-half life of ferumoxtran-10 is 2 to 3 h in rats and 24 to 36 h in humans. Comparison between mice and rats showed that the increase in blood half-life leads to an increase in lymph node uptake (Table 4.5). As in humans, increasing the dose in animals resulted in a longer blood half-life due to progressive saturation of uptake by the liver and spleen.[3,10] Since the access of USPIO particles to deep compartments is facilitated by prolonged blood residence time, animal imaging experiments are generally performed using high doses of USPIO particles (200 to 1000 μmol Fe/kg) compared to the human clinical dose of 45 μmol Fe/kg.

The blood half-life, which increases according to the size of the animal species, is also dependent on hemodynamic parameters such as cardiac output (Table 4.5). A high blood circulation rate enhanced turnover of circulating nanoparticles due to an increased probability of contact with macrophage-rich tissues. Nevertheless, this type of relationship must be interpreted cautiously, as other major factors must also be taken into account, such as the total number of accessible macrophages and their functional status. For example, the blood half-life of ferumoxtran-10 is very low in dogs and pigs. This is attributed to the large number of liver macrophages in dogs compared to other species, whereas in pigs, nanoparticles may be captured by pulmonary intravascular macrophages (PIMs).[81]

4.3.2.3 Metabolism: Persistence of the Cell Label

Sun et al. demonstrated a reduction of cellular iron oxide content, evaluated by histologic staining and iron content determination using inductively coupled plasma-mass spectrometry (ICP-MS) 4 days after reculture of labeled cells (human fibroblasts, immortalized rat progenitor cells, and HEP-G2-hepatoma cells) for the SPIO and USPIO SHU555A and SHU555C. This effect was attributed to cell division, but could also be due to cellular degradation of the compounds. R2 changes were observed more than 6 days after incubation.[82]

In vitro metabolism studies of NC100150 Injection showed that the nanoparticles are completely solubilized within 4 to 7 days of incubation at pH 4.5 (10 m*M* citrate). However, when the pH values were increased (5 to 5.5), the rate of solubilization was markedly decreased. The pH of lysosomes in the liver is 3.5 to 4, sufficiently low to solubilize iron oxide nanoparticles. In addition, the Kupffer cells and endothelial cells of the liver contain ferritin and transferrin that may facilitate the metabolism of nanoparticles.[83,84]

Dextran-coated iron oxide nanoparticles are biodegradable, and therefore do not have any long-term toxicity: intracellular dextranase cleaves the dextran moiety and iron oxide is solubilized into iron ions, which are progressively incorporated into the hemoglobin pool.[3] Van Beers et al.[85] showed that the distribution of ferumoxtran-10 in the liver is fairly complex, as it can be located in both the vascular and interstitial spaces as well as inside cells. The predominant site of uptake is Kupffer cells; negligible uptake is observed experimentally in hepatocytes only at very high dose levels. The highest concentrations per gram of tissue, much higher than those observed in the liver, are found in the spleen and lymph nodes. The distribution and elimination data obtained with ^{14}C-ferumoxtran-10 are investigated, in comparison with ^{59}Fe data, to understand the metabolism of nanoparticles. In view of the differences between the outcome of ^{59}Fe- and ^{14}C-linked radioactivities, the dextran coating appears to undergo progressive degradation after uptake by macrophages and is almost exclusively eliminated in the urine (89% in 56 days) due to the low molecular weight of the dextran used. The remaining dextran is excreted in the feces. The iron contained in ferumoxtran-10 is incorporated into the body's iron store and is progressively found in the red blood cells (hemoglobin). Like endogenous iron, it is eliminated very slowly, as only 16 to 21% of the iron injected is eliminated after 84 days in the feces (negligible urinary excretion $< 1\%$). The same behavior has been described for ferumoxides. The degradation of iron oxide has been described to occur in the lysosomes of macrophages. Similar rates of erythrocyte incorporation of iron have been reported for radiolabeled ferritin.[86]

The results of a rat biodistribution study showed that feruglose was taken up and distributed equally in liver endothelial cells and Kupffer cells following a single 5 mg Fe/kg bolus injection. Liver endothelial cells and Kupffer cells exhibited similar uptake patterns and metabolism of feruglose, whereas no uptake was observed in liver parenchymal cells, suggesting that these nanoparticles may be taken up by adsorptive or receptor-mediated endocytosis rather than fluid-phase endocytosis. However, light microscopy indicated an increased iron load within hepatocytes 24 h postinjection.[83,84]

The increased iron content found in hepatocytes 3 days postinjection supports the hypothesis that Kupffer cells and endothelial cells release iron metabolized from nanoparticles, presumably as ferritin and transferrin. Since hepatocytes contain ferritin receptors, it is likely that a fraction of the ferritin released from Kupffer cells and liver endothelial cells is immediately taken up by hepatocytes via receptor-mediated endocytosis.

Feruglose exhibits prolonged $1/T2^*$ enhancement in rat liver. The liver enhancement persisted at time points when the concentration of iron oxide particles present in the liver was below the limit of detection of the method. The prolonged $1/T2^*$ enhancement is likely a result of particle breakdown products and induction of ferritin and hemosiderin with increasing iron core/loading factors.[83,84]

Compartmentalization of ferritin/hemosiderin may cause a significant decrease in the MRI signal intensity of the liver. The combined results of this study imply that the prolonged imaging effect (in terms of signal loss) significantly exceeds the half-life of feruglose in liver.

Phase II clinical trials have revealed that, after a single bolus injection of feruglose, the liver remains hypointense on T1-weighted images for several months postinjection.[83,84]

Murine liver $R2^*$ and R2 were quantified longitudinally after administration of 2.5 mg Fe/kg of ferumoxides, ferumoxytol, or feruglose. All three contrast agents significantly increased liver $R2^*$ and R2 4 h after injection. After 10 days, $R2^*$ and R2 for both the ferumoxides and ferumoxytol groups had recovered to saline control levels, whereas the increase of $R2^*$ and R2 of the feruglose group was significantly sustained compared to the control at day 50. Histology revealed feruglose in both Kupffer cells and endothelial cells, while ferumoxides and ferumoxytol were only found in Kupffer cells.[87]

Biodistribution to endothelial cells is an important aspect, as these cells are considered to be less effective at metabolizing and degrading particulate iron than Kupffer cells. It is therefore not always possible to predict the pharmacokinetics, metabolic profile, or subcellular distribution for one type of iron oxide nanoparticles based on the results from similar iron oxide compounds due to the major variations in their coating composition.

4.3.3 Biocompatibility

USPIO particles have a good biocompatibility profile,[86,88] and their uptake by macrophages is not associated with cell activation. Interleukin-1 is not released during *in vitro* endocytosis of ferumoxtran-10 by macrophages.[13]

VSOPs, which are taken up by RAW macrophages *in vitro* via endocytosis, result in a significant increase in oxidative stress compared to control levels. One day after incubation, the decrease in oxidative stress, compared with control levels, indicated that the initial increase in oxidative stress was only transient. The incorporated particles do not appear to significantly affect cellular proliferation. The initial transient oxidative stress can be inhibited by desferrioxamine, an iron chelating agent, or by an antioxidant, and this effect may be related to intracellular iron released from VSOPs in the cells.[89]

The systemic safety of several iron oxide nanoparticles has been evaluated after injection in humans (Table 4.4), indicating that these products have a satisfactory safety profile according to standard toxicological and pharmacological tests.

4.4 IMAGING MODALITIES

The superparamagnetic properties of iron oxide particles induce strong magnetic field distortions around the particles. On T2*-weighted sequences, this effect generally results in marked, focal signal defects. Image contrast enhancement by iron oxide nanoparticles is highly dependent on the biodistribution, tissue sequestration, and resulting spatial distribution of magnetic cores. Clustering of the crystal is more important for the T2* relaxation effect than the homogeneous biodistribution. Large clusters of SPIO particles in the liver produce a more marked signal decrease on gradient echo (GRE) pulse sequences than finely clustered SPIO in the spleen and blood. Relaxation effects of clustered iron oxide nanoparticles *in vivo* may allow investigation of reticuloendothelial system (RES) cell function and may influence the choice of the best-suited MR sequences for the diagnosis of liver, spleen, and lymph node diseases.[2]

4.4.1 Positive or Negative Contrast?

The term *magnetic susceptibility imaging* has been used to describe MRI sequences that are sensitive to magnetic susceptibility, mainly gradient echo sequences. While it offers lower signal-to-noise ratio (SNR) and spatial resolution than spin echo (SE) sequence, T2*-weighted GRE sequences are more sensitive. Generally, long TR and TE are used (TR > 100 msec and TE > 10 to 20 msec) associated with a low flip angle (FA < 20°). Nevertheless, large variation in TR, TE, and FA are observed. A combination of multiple echos improves SNR with an increase in the bandwidth and a shortening of the acquisition time without alteration of the spatial resolution (multi echo data image combination [MEDIC] sequence).[90]

In addition to the T2 and T2* effects, USPIO particles also have inherent T1-shortening properties.[91,92] In view of the r_1 and r_2 relaxivity properties of USPIO, the T1-shortening effects could be expected to be intimately dependent on the local concentration of the particle: at low concentrations, a T1-weighted positive contrast can be observed, but at high concentrations, susceptibility effects are predominant, resulting in irreversible destruction of the MR signal around the particle. This effect also extends locally to the neighboring voxels, resulting in a higher detection sensitivity.[93]

Promising new sequences have recently been developed to provide positive contrast with USPIO. Several approaches have been investigated, such as:

- Positive contrast due to a basic T1-weighted acquisition: At low concentrations of USPIO, the T1 effect is prevalent (high sensitivity, but fast switch between T1 enhancement and T2 cancellation of the signal).[93,94]
- Positive contrast due to very short T2 relaxation: Sequences are based on ultrashort TE (UTE).[95,96] The difference between ultrashort TE (μsec) and medium TE (msec) acquisition leads to suppression of long T2 values (low sensitivity, but good quantification).
- Positive contrast due to a dephasing effect: A dephasing gradient on the slice axis is added to the spoil signal across the sample, except in regions where the local gradients surrounding the superparamagnetic marker present an appropriate amplitude and orientation to refocus the lost signal.[97,98]
- Positive contrast due to susceptibility effects: Saturation of on-resonance water leads to tissue signal saturation, except for the signal from off-resonance positive lobes of susceptibility (high contrast, but low spatial resolution).[99–101]

4.4.2 Spatial Resolution

Spatial resolution is a key parameter for the detection of low concentrations of contrast agent. This was clearly demonstrated by Dodd et al. in an *in vitro* study.[102] This experiment was performed at high magnetic field (7 tesla) to achieve a high spatial resolution while maintaining a satisfactory signal-to-noise ratio (SNR). In this study, detection of USPIO-labeled T-lymphocytes was performed with a T2*-weighted MR sequence at variable spatial resolutions. Measurements of signal intensity decrease showed that detection is a function of voxel size. Johansson et al. subsequently confirmed these results *in vitro* and *in vivo* with an arginine–glycine–aspartate (RGD)-labeled USPIO designed for thrombus visualization.[103] This study was performed at 1.5 T and demonstrated the critical importance of spatial resolution for good contrast detection.

To avoid a partial volume effect, spatial resolution must ideally fit the object size. To achieve this goal, several authors have performed MR microimaging (voxels smaller than $(100\ \mu m)^3$) and demonstrated the feasibility of single-cell visualization.[104,105] Imaging has been performed on *ex vivo* samples at conventional 1.5 T or very high field, resulting in a limit of detection of iron oxide particles less than 1 μ*M* on gradient echo imaging.

4.5 CLINICAL APPLICATIONS

SPIO particles have been introduced as contrast agents for magnetic resonance imaging because of their strong T2 and T2* relaxivities, which induce a marked decrease in signal intensity, i.e., negative enhancement of various target organs on T2- and T2*-weighted images. The relaxation effects of iron oxide nanoparticles are influenced by their local concentration as well as the applied field strength and the environment in which these agents interact with surrounding protons. Nanoparticle contrast agents not only are distributed in the intravascular extracellular space, but also undergo subsequent intracellular uptake by macrophages. This intracellular uptake of iron oxide has recently been recognized as an important characteristic of phagocyte-associated pathological processes.

In addition, these particles, especially USPIO, also have a high T1 relaxivity, which results in an additional increase in signal intensity (positive enhancement) on T1-weighted images. Applications of iron oxide contrast agents as blood pool agents have been investigated to overcome the limits of time window and spatial resolution imaging related to the use of contrast agents with a short intravascular half-life and the rapid extracellular distribution of extracellular gadolinium chelates.

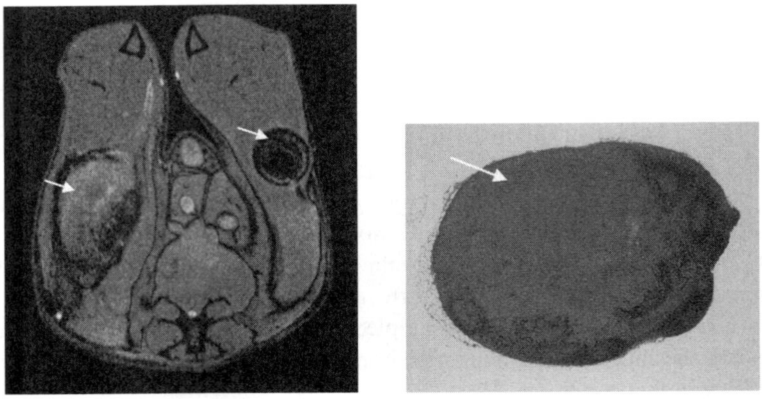

(a) (b)

FIGURE 4.4 (a) MatLyLu metastatic lymph node in rat, True Fisp sequence 24 h postinjection of 300 μmol Fe/kg of ferumoxtran-10. Left arrow: Metastatic lymph node, no uptake of ferumoxtran-10 in the major part of the node where macrophage is absent due to the high tumoral cellular content. The metastasis remains bright 24 h post ferumoxtran-10 injection. Right arrow: Normal lymph node. Note the dark signal due to the susceptibility of ferumoxtran-10 internalized into macrophages present in the normal lymph node. (b) Histology of a metastatic lymph node (arrow).

4.5.1 Liver Imaging

SPIO particles predominantly shorten T2 and T2* relaxation times of normal hepatic parenchyma, resulting in a decreased MR signal intensity of normal liver. Tumors lack a permanent decrease in signal intensity after administration of iron oxide, as they are largely devoid of macrophages. Hepatic tumors or metastases as small as 2 to 3 mm can be detected. SPIO-enhanced imaging allows both increased lesion conspicuity and increased lesion detection compared to nonenhanced imaging[106,107] (see Chapter 13).

4.5.2 Metastatic Lymph Node Imaging

Superparamagnetic agents that can be taken up by normal nodes after intravenous or subcutaneous injection would be very useful clinically. After intravenous administration, ferumoxtran-10 slowly extravasates from the vascular space into the interstitial space and is then transported to lymph nodes via lymphatics. Direct uptake by lymph nodes from blood vessels has also been demonstrated. Once inside lymph nodes, these nanoparticles bind to macrophages, producing a decrease in signal intensity on T2- and T2*-weighted images. If part of the lymph node or the entire node is infiltrated by tumor, a defect of ferumoxtran-10 uptake is observed due to the absence of macrophages, and these areas continue to retain their high signal intensity after administration of the contrast material (Figure 4.4). Harisinghani et al. reported a sensitivity of 100% with a specificity of 95.7% for characterization of lymph nodes in patients with prostate cancer.[9,108]

4.5.3 Macrophage Imaging

Magnetic resonance (MR) imaging applications are being developed for the detection of macrophage phagocytic activity. The concept of *in vivo* macrophage-specific MR imaging is based on the fact that USPIO particles are sufficiently small, after a certain vascular circulation period, to be selectively ingested by macrophages, which are located in the organs of the mononuclear phagocyte system (i.e., bone marrow, lymph nodes, liver, and spleen), or by migrating macrophages, which are mobilized by an inflammatory disease process in other parts of the body. The ability of MR imaging to depict *in vivo* phagocytosis of USPIO particles by macrophages has been demonstrated in animal models and in proof-of-concept clinical studies in atherosclerotic plaques, renal graft rejection, multiple sclerosis, stroke, brain tumors, soft-tissue infection, etc.[109–127] (see Chapters 14 and 15).

4.5.4 Blood Pool Agent Imaging

4.5.4.1 Tumor Imaging

The differentiation of benign and malignant lesions still constitutes a problem with standard MR imaging techniques. It has been shown in animal studies and a proof-of-concept clinical study that the kinetic enhancement profiles, based on a dynamic contrast-enhanced T1 sequence, of benign and malignant lesions were significantly different on USPIO-enhanced images. However, this technique presents a number of limitations: (1) the low tumor enhancement observed is partly due to the smaller distribution volume compared to small gadolinium chelates, which diffuse into the global tumor extracellular space, and (2) the long MR acquisition time of 60 min to detect a sufficient signal is difficult to tolerate, especially for older patients.[128]

4.5.4.2 MR Angiography (MRA)

Various USPIO particles have been evaluated clinically in different territories, such as coronary, pulmonary, or peripheral angiography. Although these studies confirm the feasibility of using USPIO in first-pass and equilibrium-phase MRA, some limitations were detected:

1. The SNR is higher for first-pass MRA than for equilibrium-phase MRA, as the contrast agent is less diluted in the blood and the reduction in T1 relaxation is greater.
2. Equilibrium-phase images acquired over many minutes show both arterial and venous contrast enhancement, which complicates interpretation of the vasculature and requires specific postprocessing image analysis to distinguish these two territories.
3. At high concentrations, the T2* effects predominate even on sequences with ultrashort echo times, which limits the range of dose.[70,71,129,130]

4.5.4.3 Cerebral Blood Volume Imaging

The use of an iron particle blood pool contrast agent for functional MRI (fMRI), proposed in 1996 by Berry,[131] improves the sensitivity of detection of changes of cerebral blood volume activation compared to the blood oxygen level dependent (BOLD) effect (the sensitivity is increased by a factor of 2 to 12).

This technique has found potential applications in neurosciences.[131–133] cerebral blood flow (CBF) and cerebral blood volume (CBV) measurements by USPIO bolus tracking and vessel size index for neovascular characterization have also been described.[134,135]

4.6 CONCLUSION

Superparamagnetic iron oxide nanoparticles have been used for several years as negative MRI contrast agents. They have high T_2 relaxivity and a high magnetic moment, which generates microscopic field inhomogeneities. Consequently, they produce a marked decrease in signal intensity of the organs in which they accumulate through macrophage uptake. In terms of the effect on signal, the use of the dose-dependent effect of nanoparticles on T1 relaxation time will undoubtedly be the subject of increasing research in the near future. The importance of the r_1/r_2 ratio as a function of the magnetic field is also emphasized in studies establishing the most appropriate sequences for the molecules under clinical development. Iron oxide nanoparticles can also behave as blood pool contrast agents. In terms of pharmacokinetics, research into preventing the uptake of iron oxide particles by the reticuloendothelial system by an appropriate selection of particle charge, size, and coating is of major interest. The chemical coating of these nanoparticles may also allow them to be linked to molecules capable of specifically targeting a specific area, such as an organ, a disease, or a particular biological system. Furthermore, *ex vivo* labeling using iron oxide nanoparticles of progenitor and stem cells that can be subsequently tracked *in vivo* with MRI is a major research subject.

REFERENCES

1. Bonnemain, B., Superparamagnetic agents in magnetic resonance imaging, physicochemical characteristics and clinical applications. A review, *J. Drug Target*, 6, 167, 1998.
2. Weissleder, R. and Papisov, M., Pharmaceutical iron oxides for MR imaging, *Rev. Magn. Reson. Med.*, 4, 1, 1992.
3. Corot, C. et al., Macrophage imaging in central nervous system and in carotid atherosclerotic plaque using ultrasmall superparamagnetic iron oxide in magnetic resonance imaging, *Invest. Radiol.*, 39, 619, 2004.
4. Weissleder, R., Bogdanov, A., and Papisov, M., Drug targeting in magnetic resonance imaging, *Magn. Reson. Q.*, 8, 55, 1992.
5. Bulte, J.W. and Kraitchman, D.L., Iron oxide MR contrast agents for molecular and cellular imaging, *NMR Biomed.*, 17, 484, 2004.
6. Benderbous, S. et al., Superparamagnetic agents, physicochemical characteristics and preclinical imaging evaluation, *Acad. Radiol.*, 3 (Suppl. 2), S292, 1996.
7. Reimer, P. and Tombach, B., Hepatic MRI with SPIO, detection and characterization of focal liver lesions, *Eur. Radiol.*, 8, 1198, 1998.
8. Clement, O. and Luciani, A., Imaging the lymphatic system, possibilities and clinical applications, *Eur. Radiol.*, 14, 1498, 2004.
9. Harisinghani, M.G. et al., Noninvasive detection of clinically occult lymph-node metastases in prostate cancer, *N. Engl. J. Med.*, 348, 2491, 2003.
10. Wagner, S. et al., Monomer-coated very small superparamagnetic iron oxide particles as contrast medium for magnetic resonance imaging, preclinical *in vivo* characterization, *Invest. Radiol.*, 37, 167, 2002.
11. Taylor, A.M. et al., Safety and preliminary findings with the intravascular contrast agent NC100150 injection for MR coronary angiography, *J. Magn. Reson. Imaging*, 9, 220, 1999.
12. Tombach, B. et al., First-pass and equilibrium MRA of the aortoiliac region with a superparamagnetic iron oxide blood pool MR contrast agent SH U 555 C, results of a human pilot study, *NMR Biomed.*, 17, 500, 2004.
13. Raynal, I. et al., Macrophage endocytosis of superparamagnetic iron oxide nanoparticles, mechanisms and comparison of ferumoxides and ferumoxtran-10, *Invest. Radiol.*, 39, 56, 2004.
14. Sjogren, C.E. et al., Crystal size and properties of superparamagnetic iron oxide SPIO particles, *Magn. Reson. Imaging*, 15, 55, 1997.
15. Nunes, A.C. and Yu, Z.C., Fractionation of a water-based ferrofluid, *J. Magnetism Magn. Mater.*, 65, 265, 1987.
16. Babes, L. et al., Synthesis of iron oxide nanoparticles used as MRI contrast agents, a parametric study, *J. Colloid Interface Sci.*, 2122, 474, 1999.
17. Thurm, S. and Odenbach, S., Magnetic separation of ferrofluids, *J. Magnetism Magn. Mater.*, 252, 247, 2002.
18. Jolivet, J.P., Chaneac, C., and Tronc, E., Iron oxide chemistry: from molecular clusters to extended solid networks, *Chem. Commun.*, 5, 481, 2004.
19. Morales, M.P. et al., Surface and internal spin canting in γ-Fe_2O_3 nanoparticles, *Chem. Mater.*, 11, 3058, 1999.
20. LaConte, L., Nitin, N., and Gang, B., Magnetic nanoparticules probes, *Nanotoday*, 32, 38, 2005.
21. Rivas, J. et al., Production and characterization of iron, boron amorphous particles, *J. Magnetism Magn. Mater.*, 122, 1, 1993.

22. Shafi, K.V. et al., Magnetic enhancement of γ-Fe2O3 nanoparticles by sonochemical coating, *Chem. Mater.*, 14, 1778, 2002.
23. Lawaczeck, R., Menzel, M., and Pietsch, H., Superparamagnetic iron oxide particles, contrast media for magnetic resonance imaging, *Appl. Organometal. Chem.*, 18, 506, 2004.
24. Bulte, J.W.M. et al., Magnetodendrimers allow endosomal magnetic labelling and *in vivo* tracking of stem cells, *Nat. Biotechnol.*, 19, 1141, 2001.
25. Mornet, S., Portier, J., and Duguet, E., A method for synthesis and functionalization of ultrasmall superparamagnetic covalent carriers based on maghemite and dextran, *J. Magnetism Magn. Mater.*, 293, 127, 2005.
26. Massart, R., Preparation of aqueous magnetic liquids in alkaline and acid media, *IEE Trans. Magn.*, 17, 1247, 1981 (abstract).
27. De Cuyper, M. and Joniau, M., Magnetoliposomes: formation and structural characterization, *Eur. Biophys. J.*, 15, 311, 1988.
28. Bulte, J.W.M. et al., Preparation, relaxometry, and biokinetics of PEGylated magnetoliposomes as MR contrast agent, *J. Magnetism Magn. Mater.*, 194, 204, 1999.
29. Weissleder, R. et al., Antimyosin-labeled monocrystalline iron oxide allows detection of myocardial infarct, MR antibody imaging, *Radiology*, 182, 381, 1992.
30. Remsen, L.G. et al., MR of carcinoma specific monoclonal antibody conjugated to monocrystalline iron oxide nanoparticules, the potential for noninvasive diagnosis, *Am. J. Neuroradiol.*, 17, 411, 1996.
31. Kresse, M. et al., Targeting of ultrasmall superparamagnetic iron oxide USPIO particles to tumor cells *in vivo* by using transferring receptor pathways, *MRM*, 40, 236, 1998.
32. Savellano, D.H. et al., The transferring receptor, a potential molecular imaging marker for human cancer, *Neoplasia*, 5, 495, 2003.
33. Wunderbaldinger, P., Josephson, L., and Weissleder, R., Crosslinked iron oxides CLIO, a new platform for the development of targeted MR contrast agents, *Acad. Radiol.*, 9 (Suppl. 2), S304, 2002.
34. Schellenberger, E.A. et al., Annexin V-CLIO, a nanoparticule for detecting apoptosis by MRI, *Mol. Imaging*, 1, 102, 2002.
35. Tsourkas, A. et al., *In vivo* imaging of activated endothelium using an anti-VCAM1 magnetooptical probe, *Bioconjugate Chem.*, 16, 576, 2005.
36. Kang, H.W., Jr. et al., Magnetic resonance imaging of inducible E-selectin expression in human endothelial cell culture, *Bioconjugate Chem.*, 13, 122, 2002.
37. Josephson, L. et al., High-efficiency intracellular magnetic labelling with novel superparamagnetic tat peptide conjugates, *Bioconjugate Chem.*, 10, 186, 1999.
38. Högemann, D. et al., Improvement of MRI probes to allow efficient detection of gene expression, *Bioconjugate Chem.*, 11, 941, 2000.
39. Weissleder, R. et al., Cell-specific targeting of nanoparticules by multivalent attachment of small molecules, *Nat. Biotechnol.*, 11, 1418, 2005.
40. Schellenberger, E.A., Surface-functionalized nanoparticle library yields probes for apoptotic cells, *ChemBioChem*, 5, 275, 2004.
41. Sonvico, F. et al., Folate-conjugated iron oxide nanoparticles for solid tumor targeting as potential specific magnetic hyperthermia mediators, synthesis, physicochemical characterization, and *in vitro* experiments, *Bioconjugate Chem.*, 16, 1181, 2005.
42. Zhang, Y. and Zhang, J., Surface modification of monodisperse magnetite nanoparticles for improved intracellular uptake to breast cancer cells, *J. Colloid Interface Sci.*, 2832, 352, 2005.
43. Fauconnier, N. et al., Thiolation of maghemite nanoparticles by dimercaptosuccinic acid, *J. Colloid Interface*, 194, 427, 1997.
44. Roger, J. et al., Some biomedical applications of ferrofluids, *Eur. Phys. J. AP*, 5, 321, 1999.
45. Roch, A., Muller, R.N., and Gillis, P., Theory of proton relaxation induced by superparamagnetic particles, *J. Chem. Phys.*, 110, 5403, 1999.
46. Muller, R.N. et al., Transverse relaxivity of particulate MRI contrast media: from theories to experiments, *Magn. Reson. Med.*, 222, 178, 1991.
47. Clarke, S.E. et al., Comparison of two blood pool contrast agents for 0.5-T MR angiography, experimental study in rabbits, *Radiology*, 214, 787, 2000.
48. Fan, X. et al., Differentiation of non-metastatic and metastatic rodent prostate tumors with high spectral and spatial resolution MRI, *Magn. Reson. Med.*, 45, 1046, 2001.

49. Kellar, K.E. et al., Important considerations in the design of iron oxide nanoparticles as contrast agents for T1-weighted MRI and MRA, *Acad. Radiol.*, 9 (Suppl. 1), S34, 2002.
50. Taupitz, M. et al., Phase I clinical evaluation of citrate-coated monocrystalline very small superparamagnetic iron oxide particles as a new contrast medium for magnetic resonance imaging, *Invest. Radiol.*, 39, 394, 2004.
51. Wunderbaldinger, P., Josephson, L., and Weissleder, R., Crosslinked iron oxides CLIO, a new platform for the development of targeted MR contrast agents, *Acad. Radiol.*, 9 (Suppl. 2), S304, 2002.
52. Jung, C.W. and Jacobs, P., Physical and chemical properties of superparamagnetic iron oxide MR contrast agents, ferumoxides, ferumoxtran, ferumoxsil, *Magn. Reson. Imaging*, 13, 661, 1995.
53. Cheng, F.Y. et al., Characterization of aqueous dispersions of Fe_3O_4 nanoparticles and their biomedical applications, *Biomaterials*, 26, 729, 2005.
54. Kellar, K.E. et al., 'NC100150,' a preparation of iron oxide nanoparticles ideal for positive-contrast MR angiography, *MAGMA*, 8, 207, 1999.
55. Jung, C., Surface properties of superparamagnetic iron oxide MR contrast agents: Ferumoxide, Ferumoxtran, Ferumoxsil, *Magn. Reson. Imaging*, 13, 675, 1995.
56. Idee, J.M. et al., The superparamagnetic iron oxides nanoparticles for magnetic resonance imaging applications, in *Nanomaterials for Cancer Therapy and Diagnosis*, Wiley-VCH, Verlag GmBH Weinheim, G., 2006, chap. 6.
57. Bremer, C. et al., RES specific imaging of the liver and spleen with iron oxide particles designed for blood pool MR-angiography, *J. Magn. Reson. Imaging*, 10, 461, 1999.
58. Knollmann, F.D. et al., Differences in predominant enhancement mechanisms of superparamagnetic iron oxide and ultrasmall superparamagnetic iron oxide for contrast-enhanced portal magnetic resonance angiographies, *Invest. Radiol.*, 339, 637, 1998.
59. Kellar, K.E. et al., NC100150 injection, a preparation of optimized iron oxide nanoparticules for positive-contrast MR angiography, *J. Magn. Reson. Imaging*, 11, 488, 2000.
60. Reynolds, F. et al., Method of determining nanoparticle core weight, *Anal. Chem.*, 773, 814, 2005.
61. Weissleder, R. et al., Long circulating iron oxides for MR imaging, *Adv. Drug Delivery Rev.*, 13, 321, 1995.
62. Moore, A., Weissleder, R., and Bogdanov, A., Uptake of dextran-coated monocrystalline iron oxides in tumor cells and macrophages, *J. Magn. Reson. Imaging*, 7, 1140, 1997.
63. Fleige, G. et al., *In vitro* characterization of two different ultrasmall iron oxide particles for magnetic resonance cell tracking, *Invest. Radiol.*, 37, 482, 2002.
64. Metz, S. et al., Capacity of human monocytes to phagocytose approved iron oxide MR contrast agents *in vitro*, *Eur. Radiol.*, 14, 1851, 2004.
65. Wilhelm, C. et al., Intracellular uptake of anionic superparamagnetic nanoparticles as a function of their surface coating, *Biomaterials*, 24, 1001, 2003.
66. Koval, M. et al., Size of IgG-opsonized particles determines macrophage response during internalization, *Exp. Cell Res.*, 242, 265, 1998.
67. Rogers, W.J. and Basu, P., Factors regulating macrophage endocytosis of nanoparticles, implications for targeted magnetic resonance plaque imaging, *Atherosclerosis*, 178, 67, 2005.
68. Clément, O. et al., Liver imaging with ferumoxides: Feridex, fundamentals, controversies and practical aspects, *Top. Magn. Reson. Imaging*, 9, 167, 1998.
69. McLachlan, S.J. et al., Phase I clinical evaluation of a new iron oxide MR contrast agent, *J. Magn. Reson. Imaging*, 43, 301, 1994.
70. Li, W. et al., First-pass contrast-enhanced magnetic resonance angiography in humans using ferumoxytol, a novel ultrasmall superparamagnetic iron oxide USPIO-based blood pool agent, *J. Magn. Reson. Imaging*, 21, 46, 2005.
71. Reimer, P. et al., SPIO-enhanced 2D-TOF MR angiography of the portal venous system, results of an intraindividual comparison, *J. Magn. Reson. Imaging*, 7, 945, 1997.
72. Simon, G.H. et al., Ultrasmall supraparamagnetic iron oxide-enhanced magnetic resonance imaging of antigen-induced arthritis, a comparative study between SHU 555C, ferumoxtran-10 and ferumoxytol, *Invest. Radiol.*, 41, 45, 2006.
73. Daldrup-Link, H.E. et al., Macromolecular contrast medium feruglose versus small molecular contrast medium gadopentetate enhanced magnetic resonance imaging, differentiation of benign and malignant breast lesions, *Acad. Radiol.*, 10, 1237, 2003.

74. Weissleder, R. et al., Ultrasmall superparamagnetic iron oxide, characterization of a new class of contrast agents for MR imaging, *Radiology*, 1752, 489, 1990.
75. Lawaczeck, R. et al. Magnetic iron oxide particles coated with carboxydextran for parenteral administration and liver contrasting: preclinical profile of SHU 555A, *Acta Radiol.*, 38, 584, 1997.
76. Turetschek, K. et al., MRI monitoring of tumor response following angiogenesis inhibition in an experimental human breast cancer model, *Eur. J. Nucl. Med. Mol. Imaging*, 30, 448, 2003.
77. Taupitz, M. et al., New generation of monomer-stabilized very small superparamagnetic iron oxide particles (VSOP) as contrast medium for MR angiography, preclinical results in rats and rabbits, *J. Magn. Reson. Imaging*, 126, 905, 2000.
78. Dousset, V. et al., *In vivo* macrophage activity imaging in the central nervous system detected by magnetic resonance, *Magn. Reson. Med.*, 41, 329, 1999.
79. Rausch, M. et al., MRI-based monitoring of inflammation and tissue damage in acute and chronic relapsing EAE, *Magn. Reson. Med.*, 50, 309, 2003.
80. Martin-Chouly, C.A. et al., *In vitro* evaluation of vascular permeability to contrast media using cultured endothelial cell monolayers, *Invest. Radiol.*, 34, 663, 1999.
81. Brain, J.D. et al., Pulmonary intravascular macrophages, their contribution to the mononuclear phagocyte system in 13 species, *Am. J. Physiol.*, 276, 146, 1999.
82. Sun, R. et al., Physical and biological characterization of superparamagnetic iron oxide, and ultrasmall superparamagnetic iron oxide-labeled cells, a comparison, *Invest. Radiol.*, 40, 504, 2005.
83. Briley-Saebo, K. et al., Hepatic cellular distribution and degradation of iron oxide nanoparticles following single intravenous injection in rats, implications for magnetic resonance imaging, *Cell Tissue Res.*, 316, 315, 2004.
84. Briley-Saebo, K. et al., Long-term imaging effects in rat liver after a single injection of an iron oxide nanoparticle based MR contrast agent, *J. Magn. Reson. Imaging*, 20, 622, 2004.
85. Van Beers, B.E. et al., Biodistribution of ultrasmall iron oxide particles in the rat liver, *J. Magn. Reson. Imaging*, 134, 594, 2001.
86. Bourrinet, P. et al., Preclinical safety and pharmacokinetic profile of ferumoxtran-10, an ultrasmall superparamagnetic iron oxide magnetic resonance contrast agent, *Invest. Radiol.*, 413, 313, 2006.
87. Kalber, T.L. et al., A longitudinal study of R2* and R2 magnetic resonance imaging relaxation rate measurements in murine liver after a single administration of 3 different iron oxide-based contrast agents, *Invest. Radiol.*, 40, 784, 2005.
88. Leenders, W., Ferumoxtran-10 advanced magnetics, *Idrugs*, 6, 987, 2003.
89. Stroh, A. et al., Iron oxide particles for molecular magnetic resonance imaging cause transient oxidative stress in rat macrophages, *Free Radic. Biol. Med.*, 36, 976, 2004.
90. Torabi, M., Aquino, S.L., and Harisinghani, M.G., Current concepts in lymph node imaging, *J. Nucl. Med.*, 45, 1509, 2004.
91. Chambon, C. et al., Superparamagnetic iron oxides as positive MR contrast agents, *in vitro* and *in vivo* evidence, *Magn. Reson. Imaging*, 11, 509, 1993.
92. Canet, E. et al., Superparamagnetic iron oxide particles and positive enhancement for myocardial perfusion studies assessed by subsecond T1-weighted MRI, *Magn. Reson. Imaging*, 11, 1139, 1993.
93. Daldrup-Link, H.E. et al., Targeting of hematopoietic progenitor cells with MR contrast agents, *Radiology*, 2283, 760, 2003.
94. Ruehm, S.G. et al., Magnetic resonance imaging of atherosclerotic plaque with ultrasmall superparamagnetic particles of iron oxide in hyperlipidemic rabbits, *Circulation*, 1033, 415, 2001.
95. Robson, M.D. et al., Magnetic resonance, an introduction to ultrashort TE UTE imaging, *J. Comput. Assist. Tomogr.*, 276, 825, 2003.
96. Crowe, L.A. et al., *Ex vivo* MR imaging of atherosclerotic rabbit aorta labelled with USPIO: enhancement of iron loaded regions in UTE imaging, in *ISMRM*, Miami, 2005, p. 115.
97. Seppenwoolde, J.H., Viergever, M.A., and Bakker, C.J., Passive tracking exploiting local signal conservation, the white marker phenomenon, *Magn. Reson. Med.*, 504, 784, 2003.
98. Coristine, A.J. et al., Positive contrast labelling of SPIO loaded cells in cell samples and spinal cord injury, in *ISMRM*, Hawaii, 2004, p. 163.
99. Cunningham, C.H. et al., Positive contrast magnetic resonance imaging of cells labeled with magnetic nanoparticles, *Magn. Reson. Med.*, 535, 999, 2005.

100. Pintaske, J., Martirosian, P., and Schick, F., Bright visualization of samples containing SPIO-labeled cells, improvements in MR sensitivity, in *ISMRM*, Miami, 2005, p. 405.
101. Stuber, M. et al., Shedding light on the dark spot with IRON, a method that generates positive contrast in the presence of superparamagnetic nanoparticles, *ISNRN*, Miami, 2005, p. 2608.
102. Dodd, S.J. et al., Detection of single mammalian cells by high-resolution magnetic resonance imaging, *Biophysics*, 761, 103, 1999.
103. Johansson, L.O. et al., A targeted contrast agent for magnetic resonance imaging of thrombus, implications of spatial resolution, *J. Magn. Reson. Imaging*, 134, 615, 2001.
104. Foster-Gareau, P. et al., Imaging single mammalian cells with a 1.5 T clinical MRI scanner, *Magn. Reson. Med.*, 495, 968, 2003.
105. Shapiro, E.M. et al., MRI detection of single particles for cellular imaging, *Proc. Natl. Acad. Sci. U.S.A.*, 10130, 10901, 2004.
106. Bellin, M.F. et al., Liver metastases, safety and efficacy of detection with superparamagnetic iron oxide in MR imaging, *Radiology*, 193, 657, 1994.
107. Semelka, R.C. and Helmberger, T.K., Contrast agents for MR imaging of the liver, *Radiology*, 218, 27, 2001.
108. Harisinghani, M.G. et al., MR lymphangiography, imaging strategies to optimize the imaging of lymph nodes with ferumoxtran-10, *Radiographics*, 24, 867, 2004.
109. Kooi, M.E. et al. Accumulation of ultrasmall superparamagnetic particles of iron oxide in human atherosclerotic plaques can be detected by *in vivo* magnetic resonance imaging, *Circulation*, 10719, 2453, 2003.
110. Schmitz, S. et al., Magnetic resonance imaging of atherosclerotic plaques using superparamagnetic iron oxide particles, *J. Magn. Reson. Imaging*, 14, 355, 2001.
111. Litovsky, S. et al., Superparamagnetic iron oxide-based method for quantifying recruitment of monocytes to mouse atherosclerotic lesions *in vivo*, enhancement by tissue necrosis factor-alpha, interleukin-1beta, and interferon-gamma, *Circulation*, 107, 1545, 2003.
112. Hyafil, F. et al., Ferumoxtran-10-enhanced MRI of the hypercholesterolemic rabbit aorta, relationship between signal loss and macrophage infiltration, *Arterioscler. Thromb. Vasc. Biol.*, 26, 176, 2006.
113. Trivedi, R. et al., Noninvasive imaging of carotid plaque inflammation, *Neurology*, 63, 187, 2004.
114. Rausch, M. et al., Dynamic patterns of USPIO enhancement can be observed in macrophages after ischemic brain damage, *Magn. Reson. Med.*, 465, 1018, 2001.
115. Varallyay, P. et al. Comparison of two superparamagnetic viral-sized iron oxide particles ferumoxides and ferumoxtran-10 with a gadolinium chelate in imaging intracranial tumors, *Am. J. Neuroradiol.*, 23, 510, 2002.
116. Enochs, S. et al., Improved delineation of human brain tumors on MR images using a long-circulating-superparamagnetic iron oxide agent, *J. Magn. Reson. Imaging*, 9, 228, 1999.
117. Dousset, V. et al., Comparison of ultrasmall particles of iron oxide USPIO enhanced T2-weighted, conventional T2-weighted, and gadolinium enhanced T1-weighted MR images in rats with experimental autoimmune encephalomyelitis, *Am. J. Neuroradiol.*, 202, 223, 1999.
118. Dousset, V. et al., Dose and scanning delay using USPIO for central nervous system macrophage imaging, *MAGMA*, 8, 185, 1999.
119. Xu, S. et al., Study of relapsing remitting experimental allergic encephalomyelitis SJL mouse model using MION-46L enhanced *in vivo* MRI, early histopathological correlation, *J. Neurosci. Res.*, 525, 549, 1998.
120. Hauger, O. et al., Nephrotoxic nephritis and obstructive nephropathy, evaluation with MR imaging enhanced with ultrasmall superparamagnetic iron oxide, preliminary findings in a rat model, *Radiology*, 2173, 819, 2000.
121. Saleh, A. et al., *In vivo* MRI of brain inflammation in human ischaemic stroke, *Brain*, 127, 1670, 2004.
122. Saleh, A. et al., Central nervous system inflammatory response after cerebral infarction as detected by magnetic resonance imaging, *NMR Biomed.*, 17, 163, 2004.
123. Le Duc, G. et al., Use of T2-weighted susceptibility contrast MRI for mapping the blood volume in the glioma-bearing rat brain, *Magn. Reson. Med.*, 42, 754, 1999.
124. Simon, G.H. et al., Ultrasmall superparamagnetic iron-oxide-enhanced MR imaging of normal bone marrow in rodents, original research, *Acad. Radiol.*, 12, 1190, 2005.

125. Daldrup-Link, H.E. et al., Iron oxide-enhanced MR imaging of bone marrow in patients with non-Hodgkin's lymphoma, differentiation between tumor infiltration and hypercellular bone marrow, *Eur. Radiol.*, 12, 1557, 2002.
126. Lutz, A.M. et al., Detection of synovial macrophages in an experimental rabbit model of antigen-induced arthritis, ultrasmall superparamagnetic iron oxide-enhanced MR imaging, *Radiology*, 233, 149, 2004.
127. Kaim, A.H. et al., MR imaging with ultrasmall superparamagnetic iron oxide particles in experimental soft-tissue infections in rats, *Radiology*, 2253, 808, 2002.
128. Daldrup-Link, H.E. et al., Quantification of breast tumor microvascular permeability with feruglose-enhanced MR imaging, initial phase II multicenter trial, *Radiology*, 229, 885, 2003.
129. Bjornerud, A., Johansson, L.O., and Ahlstrom, H.K., Pre-clinical results with Clariscan NC100150 injection, experience from different disease models, *MAGMA*, 12, 99, 2001.
130. Klein, C. et al., Improvement of image quality of non-invasive coronary artery imaging with magnetic resonance by the use of the intravascular contrast agent Clariscan NC100150 injection in patients with coronary artery disease, *J. Magn. Reson. Imaging*, 17, 656, 2003.
131. Berry, I., Benderbous, S. et al., Contribution of Sinerem used as blood-pool contrast agent, detection of cerebral blood volume changes during apnea in the rabbit, *Magn. Reson. Med.*, 363, 415, 1996.
132. Dubowitz, D.J. et al., Enhancing fMRI contrast in awake-behaving primates using intravascular magnetite dextran nanoparticles, *Neuroreport*, 1211, 2335, 2001.
133. Leite, F.P. et al., Repeated fMRI using iron oxide contrast agent in awake, behaving macaques at 3 tesla, *Neuroimage*, 62, 283, 2002.
134. Simonsen, C.Z. et al., CBF and CBV measurements by USPIO bolus tracking, reproducibility and comparison with Gd-based values, *J. Magn. Reson. Imaging*, 9, 342, 1999.
135. Troprès, I. et al., *In vivo* assessment of tumoral angiogenesis, *Magn. Reson. Med.*, 51, 533, 2004.

5 Physical Mechanism and Applications of CEST Contrast Agents

Michael T. McMahon, Jinyuan Zhou, Assaf A. Gilad, Jeff W.M. Bulte, and Peter C.M. van Zijl

CONTENTS

5.1 Introduction 85
5.2 Contrast Enhancement Mechanism 87
5.3 Chemical Exchange Rates for These Agents Can Be Measured Using QUEST and QUESP Experiments 89
5.4 pH Dependence of the Exchange Rate for Two Example CEST Agents, PLL and SPD-5 93
5.5 Changes in Endogenous Protein Content Can Be Used To Detect Rat Brain Tumors *In Vivo* via CEST Imaging 94
5.6 Conclusion 96
References 97

5.1 INTRODUCTION

The imaging of cells with magnetic resonance imaging (MRI) requires some magnetic labeling strategy that allows the discrimination between labeled cells and their surroundings. To date, there are two major classes of MR contrast agents, paramagnetic agents, producing large positive signal enhancement from decreasing T_1, and superparamagnetic iron oxide particles, which produce large negative T_2 contrast. Both of these have been described in depth in the preceding chapters. In addition, a third class of contrast agents has been developed recently, so-called chemical exchange saturation transfer (CEST) agents,[1–5] which can be switched on and off externally with an applied radio frequency (rf) field.

In this chapter, we will summarize the previous MR studies of chemical exchange and describe the basic mechanism for chemical exchange image contrast using the Bloch equations to relate the contrast to the concentration of the agent and its exchange and relaxation properties. In addition, we will describe some of the unique features of these MRI agents, which include the ability to label cells with different frequencies to distinguish between these cells.

There are many different types of chemical exchange that can influence MR spectra. Intramolecular exchange is one category and includes helix–coil transitions of nucleic acids, folding/unfolding of proteins, and other types of conformational equilibria for a molecule. A second category is intermolecular exchange, which includes the binding of small molecules to macromolecules or organometallic complexes, protonation/deprotonation equilibria, isotope exchange processes, and enzyme-catalyzed reactions. Chemical exchange effects on nuclear magnetic resonance (NMR) spectra have been observed since 1951,[6,7] and a quantitative description of these processes was presented in many papers[8–15] in the 1950s and 1960s. Initial attempts at deducing exchange time constants used the exchangeable resonance line shape directly,[9,11] although this approach was limited in studying slow-exchange constants because the line shape had to be characterized very accurately both with exchange and in the absence of this exchange. Many techniques are now available to study these exchange processes in great detail, especially with the advent of two-dimensional NMR spectroscopy.[16,17] These techniques have been reviewed elsewhere,[18–25] including as topics in many NMR textbooks,[26–30] and as such are not the subject of this chapter. One technique that is relevant was developed by Forsen and Hoffman,[10,31,32] which was later modified for FT NMR experiments.[33] Their experiment to measure the exchange rate was to apply a saturation pulse on resonance with one of the exchangeable peaks and measure the result of this pulse on the other sites.[10] This basic technique can be used to quantify the semisolid proton pool present in tissue (conventional magnetization transfer (MT))[34–36] and also obtain chemical exchange information on metabolites,[1–5] both shown by Balaban and coworkers.

In addition to these studies, many investigations have been carried out to characterize chemical exchange in both proteins and nucleic acids.[37–45] This characterization has been shown to be an important tool for studying protein folding.[39] In addition to two-dimensional or higher-dimensionality exchange spectroscopy (EXSY) methods[46–48] that could create resolution between the many exchangeable sites by the added dimensions, techniques were developed by Mori et al.[49,50] to study just the solvent-exchangeable protons (intermolecular exchange) in macromolecules with one-dimensional NMR experiments, which they termed water exchange (WEX) filters. After the exchange rates in the macromolecules were measured, the data could then be used to determine the folding/unfolding of a macromolecule. In one particularly thorough study, Bai et al.[38] produced an empirical formula relating a protein's amino acid sequence and its random coil (completely unfolded) chemical exchange rates. This was based on the measurement of dipeptide exchange rates and provided a sound fundamental basis for understanding chemical exchange in biological macromolecules as it relates to the secondary and tertiary structure of proteins.

These experiments inspired van Zijl and coworkers to look for biological macromolecules that could give very large CEST effects through protonation/deprotonation and that might be used as contrast agents. They were the first to show that some special proteins and nucleic acids that have multiple similar amide groups[51] or imino groups[52] could produce large CEST contrast (within the micromolar concentration range), opening the door for use of these macromolecules as exogenous contrast agents. In addition, as has been shown by Sherry, Aime, and coworkers, these exogenous agents can also be made up of small chelates of paramagnetic lanthanides,[53–55] which have water hopping on/off the complex instead of just protons, and which have been termed PARACEST agents, the subject of the following chapter. Both CEST and PARACEST agents provide the unique property to MRI probes that they can have multiple colors because the exchangeable peak can have different frequencies compared to water, which should provide a big advantage for imaging multiple cell populations over other agents, enabling the distinction of different cell lines. The concept of multiple-color PARACEST agents has been demonstrated by Aime and coworkers.[56] The large contrast seen *in vitro* may allow these contrast agents to also be detectable as cell tracking agents *in vivo*.

For MRI using CEST, we are interested in proton exchange processes that involve water (intermolecular exchange) because of the large abundance of water protons in biological tissue and the enhancement of solute molecule signal possible from rapid chemical exchange of their protons with water. Balaban and coworkers were the first to come up with the idea of using the saturation transfer experiment for imaging molecules with chemically exchangeable groups1 and then demonstrated that chemical exchange between protons on metabolites and water could be detected sensitively with MRI *ex vivo*[2] and *in vivo*[3] on endogenous metabolites such as urea or ammonia. In the so-called amide proton transfer (APT) imaging, Zhou et al.[57,58] demonstrated that this CEST effect could in fact be used to image the pH effect in rat brain during ischemia, as well as highlight regions with large endogenous protein content, such as is the case in tumors *in vivo*.

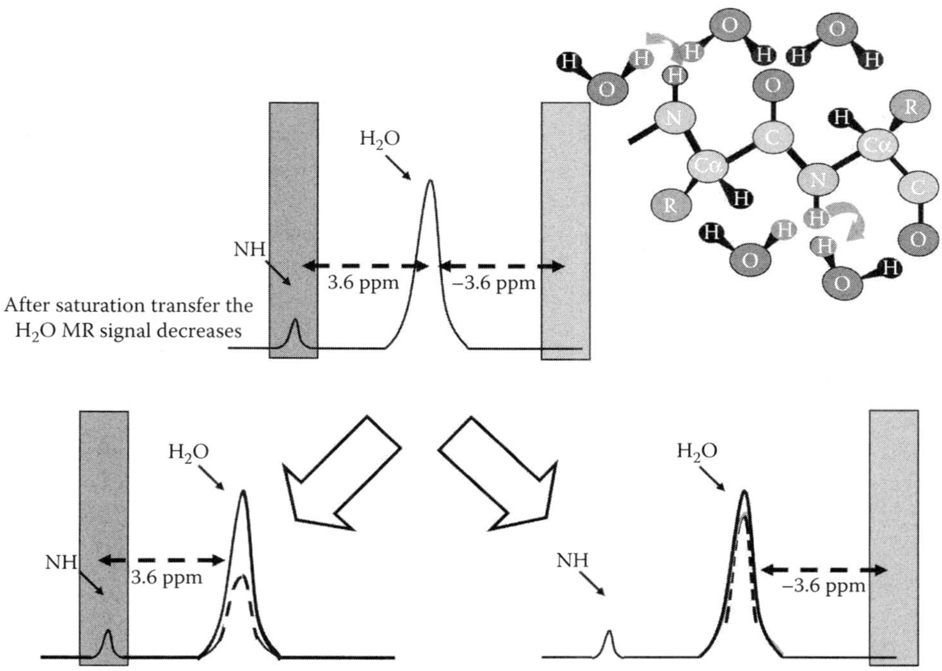

FIGURE 5.1 Contrast mechanism for CEST agents. Saturation pulses are applied at two positions, depicted by rectangles, with the initial and resulting signal intensity superpositioned. Proton exchange between amide protons and water is depicted by arrows. Because the exchange rate is fast, amide protons transfer to multiple water molecules, resulting in an amplification of signal loss, where each amide may transfer its magnetization to hundreds or thousands of water protons, depending on the exact exchange rate. Signal loss due to direct saturation of the waterline is shown as well, from $\Delta\omega = -3.6$ ppm.

5.2 CONTRAST ENHANCEMENT MECHANISM

Rapid exchange between CEST agent protons and water is necessary for these agents to be detectable at low concentrations. This exchange produces a proton/proton sensitivity enhancement factor allowing micromolar detection of polymers,[51,52] and potentially this enhancement could be used to image nanomolar concentrations of agents by using the appropriate polymers/imaging sequence. The mechanism and an example MRI sequence by which CEST agents produce contrast are displayed in Figure 5.1 and Figure 5.2 and are described as follows. An rf pulse is applied on the solute exchangeable peak, and provided this pulse saturates the peak sufficiently before the solute proton hops to water, chemical exchange of the proton leads to loss of water signal, allowing the exchangeable site to be detected indirectly. This signal loss is distinguished from the direct saturation of the waterline produced by this pulse and also from solid-like macromolecular saturation transfer[34–36] by applying a saturation pulse on the opposite side of the waterline and comparing the resulting images. For good CEST agents, the exchange should be rapid, producing a sensitivity enhancement because many water protons can be saturated from a single exchangeable site on the agent. In the case of one such agent, poly-L-lysine, the exchange rate is ~400 Hz at pH 7.3, allowing enhancements larger than 500,000 depending on size of the polymer.[51,59] Assuming a small pool of solute protons (s) and a large pool of water protons (w), it is helpful to define the asymmetric magnetization transfer ratio (MTR_{asym}) as the difference in water signal between two experiments, one with the rf field on resonance with the exchangeable peak $S_w^{+\Delta\omega_{sw}}$ and the other with the rf field at the same distance from water but on the opposite side ($S_w^{-\Delta\omega_{sw}}$) (Figure 5.1) normalized by $S_w^{-\Delta\omega_{sw}}$:[59]

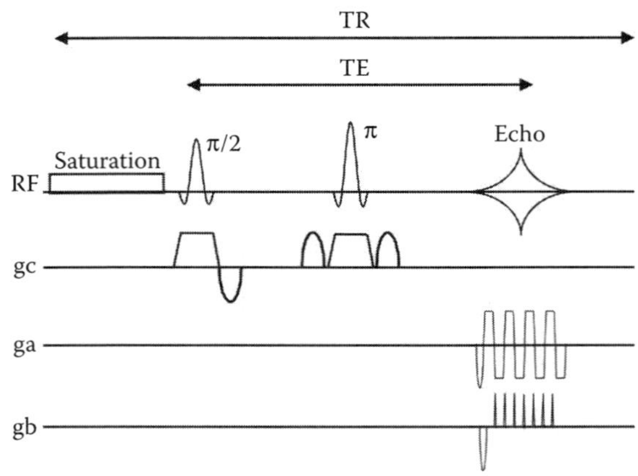

FIGURE 5.2 A saturation transfer imaging pulse sequence suitable for CEST imaging, using a spin-echo EPI (echo planar imaging) acquisition and a cw saturation pulse. (Reproduced from Zhou, J. et al., *Magn. Reson. Med.*, 50, 1120–1126, 2003. With permission.)

$$MTR_{asym} = \frac{S_w^{-\Delta\omega_{sw}} - S_w^{+\Delta\omega_{sw}}}{S_w^{-\Delta\omega_{sw}}} \approx \frac{S_{0w} - S_w^{+\Delta\omega_{sw}}}{S_{0w}} \tag{5.1}$$

where S_{0w} is the water signal without a saturation pulse. If the conventional MT effects were symmetric with respect to water, any contribution due to proton exchange would produce a positive MT difference. By obtaining images using a saturation pulse and calculating MTR_{asym}, CEST contrast can be obtained, which has been shown previously.[3,57,58,60,61]

The MR contrast produced by exchangeable peaks on an agent depends on a multitude of parameters, including concentration, number of exchangeable protons, proton exchange rate, T_1, T_2, saturation time, and saturation efficiency.[4,51–53,62–64] Of these, the chemical exchange rate is often the parameter of interest that reflects tissue pH and the molecular environment, such as salt or metal content. Chemical exchange processes can be described by modified Bloch equations[10,11,46,53,63,65–67] with exchange terms. While applying B_1 along the x axis, the Bloch equations for a two-pool proton exchange model are:[63]

$$\frac{dM_{xs}}{dt} = -\Delta\omega_s M_{ys} - R_{2s} M_{xs} - k_{sw} M_{xs} + k_{ws} M_{xw} \tag{5.2}$$

$$\frac{dM_{ys}}{dt} = \Delta\omega_s M_{xs} + \omega_1 M_{zs} - R_{2s} M_{ys} - k_{sw} M_{ys} + k_{ws} M_{yw} \tag{5.3}$$

$$\frac{dM_{zs}}{dt} = -\omega_1 M_{ys} - R_{1s}(M_{zs} - M_{0s}) - k_{sw} M_{zs} + k_{ws} M_{zw} \tag{5.4}$$

$$\frac{dM_{xw}}{dt} = -\Delta\omega_w M_{yw} - R_{2w} M_{xw} + k_{sw} M_{xs} - k_{ws} M_{xw} \tag{5.5}$$

$$\frac{dM_{yw}}{dt} = \Delta\omega_w M_{xw} + \omega_1 M_{zw} - R_{2w} M_{yw} + k_{sw} M_{ys} - k_{ws} M_{yw} \tag{5.6}$$

$$\frac{dM_{zw}}{dt} = -\omega_1 M_{yw} - R_{1w}(M_{zw} - M_{0w}) + k_{sw}M_{zs} - k_{ws}M_{zw} \tag{5.7}$$

in which $\omega_0 = \gamma B_0$ and $\omega_1 = \gamma B_1$; $\Delta\omega_s$ and $\Delta\omega_w$ are the chemical shift differences between the saturation pulse and the solute and water resonance frequencies, respectively; and M_0 is the equilibrium magnetization. Proton exchange between the two pools occurs with rates k_{sw} (solute water), k_{ws} (water solute), and $k_{sw}M_{os} = k_{ws}M_{ow}$ at equilibrium. In a recent paper,[63] a concise expression based on Equations 5.2 to 5.7 was derived for the signal intensities of the water protons (S_w) in the CEST experiment. This expression was for the proton transfer ratio (PTR), which in the limit of small direct saturation and solid saturation transfer is equal to MTR_{asym}, defined in Equation 5.1. The resulting fractional signal reduction is given by

$$PTR = \frac{S_{0w} - S_w(t_{sat}, \alpha)}{S_{0w}} = \frac{k_{sw} \cdot \alpha \cdot x_{CA}}{R_{1w} + k_{sw} \cdot x_{CA}} \left[1 - e^{-(R_{1w} + k_{sw} \cdot x_{CA})t_{sat}}\right] \tag{5.8}$$

in which k_{sw} is the chemical exchange rate, x_{CA} is the fractional concentration of exchangeable protons of the contrast agent, t_{sat} is the saturation time, α is the saturation efficiency, and the term $k_{sw}x_{CA}$ accounts for back exchange of saturated water protons to the solute, which will occur when the exchange rate or the concentration of exchangeable protons for the CEST agent is very high. The water signal intensity depends on the pulse power ($\omega_1 = \gamma B_1$) via[63]

$$\alpha = \frac{\omega_1^2}{\omega_1^2 + pq} \tag{5.9}$$

in which

$$p = R_{2s} + k_{sw} - k_{sw}^2 \cdot x_{CA} / (R_{2w} + k_{sw} \cdot x_{CA})$$

$$q = R_{1s} + k_{sw} - k_{sw}^2 \cdot x_{CA} / (R_{1w} + k_{sw} \cdot x_{CA})$$

These equations are valid for saturation transfer experiments under the assumptions of sufficiently separate resonances for these two types of protons (i.e., no direct saturation of the other resonance when saturating one). As can be seen, to a first approximation, PTR = MTR_{asym}. These expressions show that the contrast will increase as the exchange rate increases, although the condition PTR = MTR_{asym} breaks down as the exchange rate approaches the separation between the exchangeable peak and water. Because the frequency separation between groups will increase with magnetic field, this breakdown, along with the relaxation dependence, makes these agents more effective at higher fields.

5.3 CHEMICAL EXCHANGE RATES FOR THESE AGENTS CAN BE MEASURED USING QUEST AND QUESP EXPERIMENTS

It is possible to obtain quantitative exchange information on either endogenous CEST-producing macromolecules or exogenous CEST contrast agents similar to T_1 or T_2 maps using saturation transfer imaging. Two new methods to monitor local exchange rate based on CEST were introduced recently by McMahon et al.[59] These two MRI-compatible approaches are quantifying exchange using saturation time (QUEST) dependence and quantifying exchange using saturation power

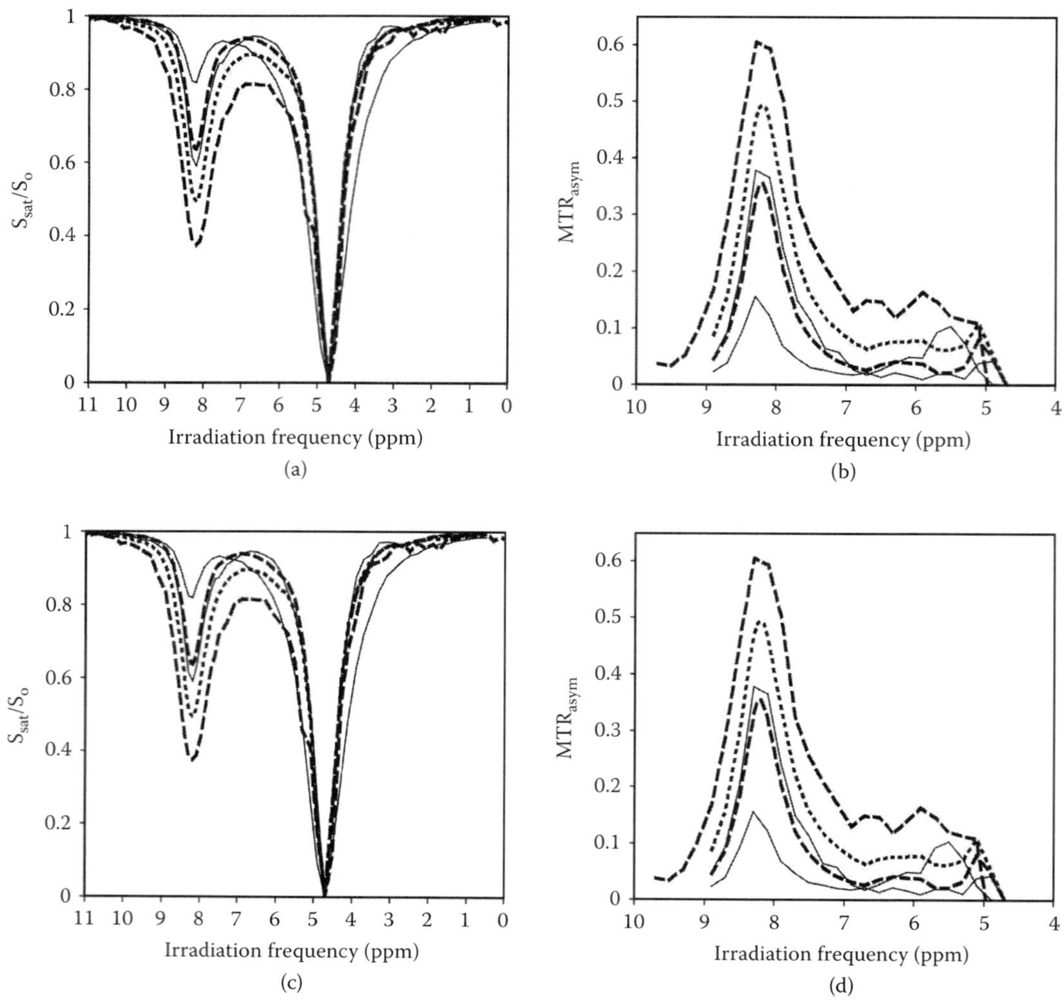

FIGURE 5.3 Saturation transfer as a function of saturation frequency (z-spectra) at 37°C and resulting MTR_{asym} spectra calculated using Equation 5.8. (a) Z-spectra of SPD-5 as a function of pH: 5.6 (—), 6.7 (– –), 7.0 (—), 7.6 (…), 8.1 (– –). (b) MTR_{asym} for SPD-5 as a function of pH, frequency. (c) Experimental z-spectra of PLL as a function of pH: 6.0 (—), 6.5 (…), 7.3 (—), 7.7 (…), 7.9 (– –). (d) Resulting MTR_{asym} for PLL as a function of frequency. (Reproduced from McMahon, M.T. et al., *Magn. Reson. Med.*, 55, 836–847, 2006. With permission.)

(QUESP) dependence. QUEST and QUESP experiments both use the same saturation transfer pulse sequence shown in Figure 5.2 with a continuous wave (cw) saturation pulse with variable offset. QUEST data are acquired by simply varying the saturation time and keeping the power constant, while QUESP data are collected by varying the saturation power while keeping the saturation time constant. The techniques were applied to poly-L-lysine (PLL) and a generation 5 polyamidoamine dendrimer (SPD-5) to measure the pH dependence of amide proton exchange rates in the physiological range *in vitro*. Figure 5.3 shows the frequency dependence of the saturation transfer effect on water (z-spectra; a and c) and the calculated MTR_{asym} (b and d) for these two contrast agents, displaying essentially one site that is exchangeable with water at the saturation fields used. For all of the QUEST data collected, MTR_{asym} reached a steady state by ~10 sec, which was therefore used for the saturation time for QUESP. One-dimensional NMR spectra were acquired.

Two saturation powers ($\omega_1/2\pi$) of 100 or 200 Hz were used in collecting the QUEST data. The MTR_{asym} dependence is shown as a function of saturation time at various pH values in Figure 5.4

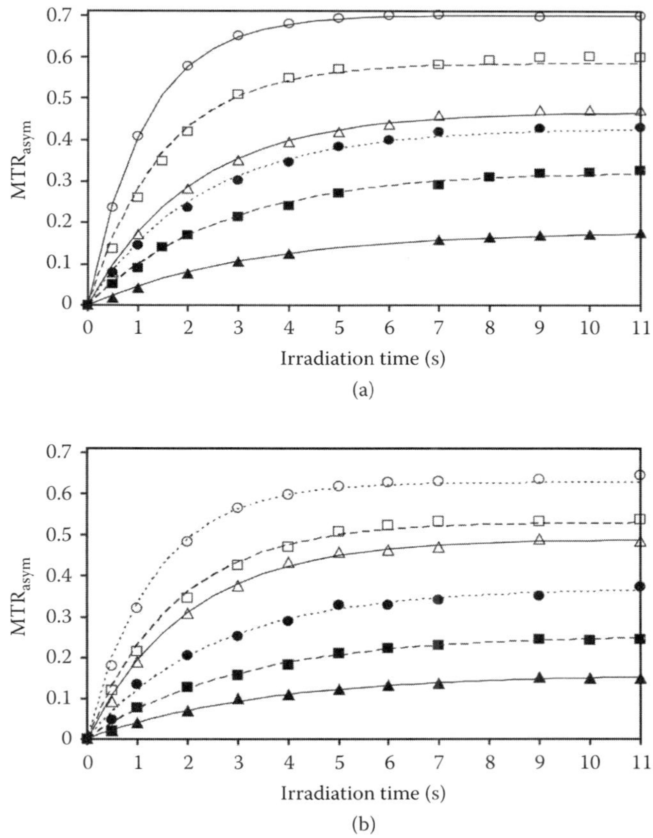

FIGURE 5.4 QUEST results. Plot of MTR_{asym} as a function of saturation time at various pH values. (a) SPD-5 data: ▲, pH 5.6; ■, pH 6.3; ●, pH 6.7; △, pH 7.0; □, pH 7.6; ○, pH 8.1; lines are best fits to Equation 5.4. (b) PLL data: ▲, pH 6.0; ■, pH 6.5; ●, pH 6.7; △, pH 7.3; □, pH 7.7; ○, pH 7.9; lines are best fits to Equation 5.4. (Reproduced from McMahon, M.T. et al., *Magn. Reson. Med.*, 55, 836–847, 2006. With permission.)

for SPD-5 (a) and PLL (b). Analytical fits to this data were obtained through Equation 5.8, and numerical fits were obtained using Equations 5.2 to 5.7. For all experimental data Equation 5.1 was used to calculate MTR_{asym} and relate this to PTR. More details on the numerical solutions and the nonlinear fitting parameters are provided by McMahon et al.[59] For exchange rates slower than 400 Hz, $\omega_1/2\pi = 100$ Hz, the best fits shown are the analytical solutions (Equation 5.8), which fit the data reliably in this range. This equation can be used successfully at this power level because the spillover is less than 1% and the weak rf saturation field approximation holds. For exchange rates faster than 400 Hz, the experimental data were collected with a 200-Hz saturation field. The analytical fits in Figure 5.4 are still excellent for these exchange rates, but the determined rates are incorrect, as judged from comparison with the numerical and line width results in Table 5.1. The analytical fits systematically underestimate the exchange rate by more than 10%. For the fastest exchange rate, PLL at pH 7.9, there is very little change in MTR_{asym} with exchange rates of 1200 Hz or higher. As such, the 95% confidence interval for this exchange rate determination is much broader than the lower exchange rates.

The QUESP data were collected with saturation field strengths of 50, 75, 100, 150, 200, and 250 Hz. Again, the data were fit to both Equation 5.8 analytically and Equations 5.2 to 5.7 numerically. In Figure 5.5, the analytical QUESP fits are shown for both SPD-5 (a) and PLL (b). Here also the exchange rates are accurate below 400 Hz, as judged from comparison with the line width and numerical measurements.

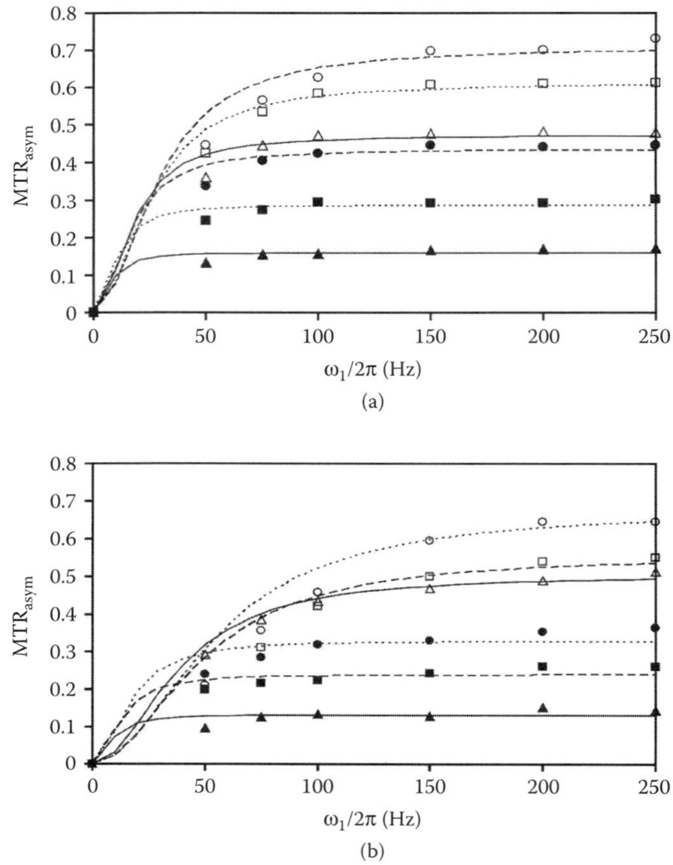

FIGURE 5.5 QUESP results. Plot of saturation power dependence of MTR_{asym} at various pH values. (a) SPD-5 data: ▲, pH 5.6; ■, pH 6.3; ●, pH 6.7; △, pH 7.0; □, pH 7.6; ○, pH 8.1; lines are best fits to Equation 5.4. (b) PLL data: ▲, pH 6.0; ■, pH 6.5; ●, pH 6.7; △, pH 7.3; □, pH 7.7; ○, pH 7.9; lines are best fits to Equation 5.4. (Reproduced from McMahon, M.T. et al., *Magn. Reson. Med.*, 55, 836–847, 2006. With permission.)

The experimental measurements in Table 5.1 and Figure 5.4 and Figure 5.5 show that the QUEST and QUESP MRI-based approaches provide exchange rates that are on the same order of magnitude as those determined by established spectroscopic methods. For the SPD-5 CEST agent in the pH range from 5.7 to 7.6, the agreement is well within experimental error with the more established line shape[8,68] and WEX filter[49,50,69] measurements. Even though the exchange rates produced by the analytical fits to the experimental data are systematically lower than those obtained numerically, the overall agreement between the exchange rates obtained by both numerical and analytical estimates is quite acceptable within this range. For the PLL CEST agent, the data obtained on the slow rates show good agreement using analytical fits to either QUEST or QUESP, while for the larger exchange rates, it is necessary to use numerical fits to these experiments to obtain accurate rates. MTR_{asym} depends on $k_{sw}x_{CA}$, and so it is difficult to decouple these effects without explicitly measuring x_{CA} separately, and this measurement was found to be the main source of error. For this reason, the sharpest resonances on the CEST agents were chosen to measure x_{CA} experimentally.

While the QUEST and QUESP MRI techniques are suitable for determining proton transfer rates of CEST agents, additional caution must be exercised with regard to the interpretation of these rates. This is due to the existence of competing processes that also saturate the water signal. For the *in vitro* examples described, direct water saturation is the only effect, but when working *in vivo*, conventional MT effects also contribute. The magnitude of direct water saturation depends on the

TABLE 5.1
Exchange Constants for the Various pH Values for PLL and the Dendrimer Obtained from Line Width, WEX, QUEST, and QUESP

pH G5 PAMAM	x_{CA} (×10³) Exp.	LW k_{sw} (Hz)	WEX k_{sw}[a] (Hz)	QUEST k_{sw}[a] (Hz) Numerical	QUEST k_{sw}[a] (Hz) Analytical	QUESP k_{sw}[a] (Hz) Numerical[c]	QUESP k_{sw}[a] (Hz) Analytical
8.1	1.368	580	—	521 ± 24	454 ± 13	550 + 200/–150	436 + 268/–131
7.6	1.307	382	305 + 98/–73	415 + 85/–104	316 + 54/–41	350 ± 50	300 + 73/–54
7.0	1.488	165	202 + 59/–48	161 ± 17	155 ± 17	175 ± 25	151 ± 26
6.7	1.456	115	128 + 37/–31	153 ± 15	133 ± 15	150 ± 25	132 ± 20
6.3	1.634	90	83 ± 13	78 ± 6	74 ± 6	75 ± 25	64 ± 11
5.6	1.346	27	39 ± 4	45 ± 3	41 ± 2	37 ± 12.5	37 ± 3
				PLL			
7.9	0.656	1211	—	1257 + 500/–400	782 ± 47	1250 ± 550	788 + 482/–378
7.7	0.579	639	—	660 + 143/–100	552 ± 17	650 ± 150	534 + 186/–124
7.3	0.674	398	404 + 153/–106	410 ± 70	379 ± 14	450 ± 150	379 + 98/–77
6.7[b]	1.044	133	122 ± 15	151 ± 24	140 ± 8	125 ± 50	121 ± 11
6.5	1.035	71	89 ± 10	81 ± 10	83 ± 7	75 ± 25	79 ± 6
6.0	0.907	39	48 ± 1	49.8 ± 4	52 ± 3	50 ± 12.5	45 ± 8

[a] Errors were obtained using the F statistic and the 95% confidence limits.
[b] PLL = 30 kD.
[c] Fit using Equations 5.2 to 5.7 to a grid of solutions every 25 Hz.

relationships among the frequency difference between the solute proton and water ($\Delta\omega_{sw}$), the exchange rate (k_{sw}), the saturation bandwidth, and the water line width (proportional to the transverse relaxation rate, R_{2w}). Analysis of the Bloch equations over a series of conditions highlighted three key points: (1) the analytical solution for MTR_{asym} is accurate with these relaxation measurements at 11.7 T for exchange rates lower than 550 Hz, (2) the reliability of the analytical solution degrades as R_{2w} increases above a threshold of 3.0 Hz, and (3) the effect that the exchange rate has on MTR_{asym} levels off as this approaches $\Delta\omega$.[59] In addition, the proper choice of ω_1 field is particularly important for this experiment. For exchange rates of 600 Hz or higher, 100-Hz saturation fields would not allow accurate measurement of the exchange rate. Therefore, $\omega_1/2\pi$ must be raised to measure these exchange rates. If a 200-Hz saturation field is used, measurement of rates up to 1400 Hz is feasible. The QUEST experiment is less desirable to carry out *in vivo* than QUESP due to its higher dependence on R_{2w}. There are also larger errors produced using QUESP when measuring exchange rates greater than 500 Hz due to the reduced change in MTR_{asym}.

5.4 PH DEPENDENCE OF THE EXCHANGE RATE FOR TWO EXAMPLE CEST AGENTS, PLL AND SPD-5

In addition to measuring exchange rates, it would also be interesting to determine the pH from CEST data, and the results above determined that both SPD-5 and PLL CEST contrast agents have a pronounced pH dependence over the physiological range of these rates. Because PLL has a single resonance and a faster exchange rate, it may be more practical than SPD-5 for *in vivo* applications. The pH dependence of the amide proton exchange rates for PLL and SPD-5 can be calibrated by fitting the function first mentioned in Gregory et al.:[70]

$$k_{sw} = k_0 + k_a \times 10^{-pH} + k_b \times 10^{pH-pK_w} \qquad (5.10)$$

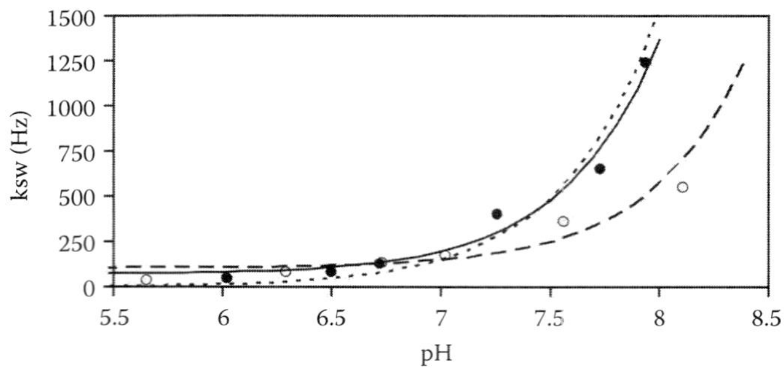

FIGURE 5.6 Dependence of the proton exchange constant on pH at T = 37°C for the two different contrast agents: PLL (●) and SPD-5 (○). Exchange rates are the average of those obtained by the four different measurements, and lines are the best fits using Equation 5.9. For PLL, the solid line represents our fit and the dashed line (…) represents the fit using constants found in Bai et al.[33] translated to 37°C, as described in this reference. (Reproduced from McMahon, M.T. et al., *Magn. Reson. Med.*, 55, 836–847, 2006. With permission.)

to the data in Table 5.1, in which k_0 (spontaneous exchange), k_a (acid-catalyzed exchange), and k_b (base-catalyzed exchange) were used as fitting constants. This expression is nice, because it can be translated to different temperatures, as mentioned in Bai et al.[38] The results of this fitting are given in Figure 5.6. For PLL, the best fit was $k_0 = 68.9$ Hz, $k_a = 1.21$ Hz, and $k_b = 1.92 \times 10^9$ Hz, and for SPD-5, the best fit was $k_0 = 106.4$ Hz, $k_a = 25.8$ Hz, and $k_b = 5.45 \times 10^8$ Hz. As is apparent, this fitting was dominated by the base-catalyzed term. It is interesting to compare these measurements to the empirical estimate provided in Bai et al.,[38] which used a set of peptides (including PLL) at low salt conditions. Their estimate is $k_0 = 0.004$ Hz, $k_a = 0.965$ Hz, and $k_b = 2.28 \times 10^9$ Hz, which is quite close, even though they used different salts/concentrations. If their calibration were to be used, the agreement with the new calibration only starts to really deviate from the data below pH 6.7. This could be due to the presence of phosphates in the phosphate-buffered saline (PBS) buffer, which also catalyze proton exchange.[71]

If either PLL or SPD-5 were to be used *in vivo*, its effect would compete with MT and other effects, such as APT effects of endogenous mobile proteins and peptides. For these endogenous compounds, which have not yet been identified, Zhou et al.,[57] using just the base-catalyzed term, found $k_b = 3.28 \times 10^8$ (adjusting k_b so that $pK_w = 14.17$, the value that was used by McMahon et al.[59] and Bai et al.[38]). Even neglecting the other two terms, for PLL the best fit was obtained with $k_b = 2.13 \times 10^9$, which is a factor of ~6.5 higher. Since CEST contrast goes in proportion to $e^{-(k_{sw} * x_{CA})t_{sat}}$, these agents should be visible *in vivo* as well. Using these expressions, the pH could be mapped for CEST agents using either the QUEST or QUESP experiments proposed.

Recently CEST imaging of endogenous proteins has been extended to the imaging of acute ischemic stroke,[57,60] taking advantage of the pH change that occurs in the tissue and produces a change in the proton exchange properties of proteins. pH provides an indicator of cellular metabolism, and therefore should allow CEST images to complement other MR images and provide additional information. QUEST or QUESP experiments might also prove useful for further quantification of these effects.

5.5 CHANGES IN ENDOGENOUS PROTEIN CONTENT CAN BE USED TO DETECT RAT BRAIN TUMORS *IN VIVO* VIA CEST IMAGING

Zhou and coworkers have demonstrated the possibility of producing MR image contrast that reflects endogenous cellular protein content and amide proton exchange properties in the 9L rat brain tumor model.[58] These experiments used standard MR imaging hardware, a horizontal-bore 4.7-T GE CSI

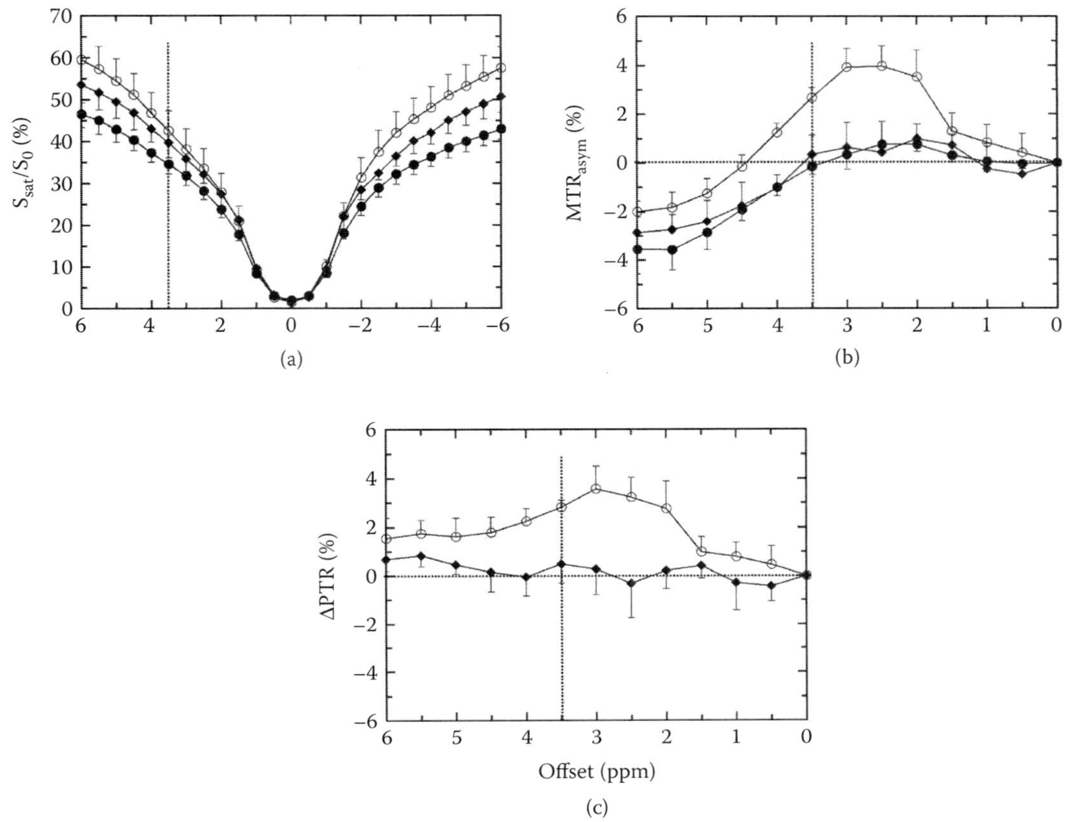

FIGURE 5.7 Plots of signal intensity as a function of saturation pulse offset: (a) Z-spectra, (b) MTR_{asym} spectra, (c) ΔPTR spectra. The data were acquired on a tumor-implanted rat brain (10 or 11 days; $N = 5$). Contralateral region data were represented as solid circles; peritumoral tissue data, diamonds; and tumor data, open circles. The z-spectra show the saturation of the water resonance as a function of rf irradiation frequency with respect to water. Signal attenuation is due mainly to direct water saturation close to the water frequency and the solid-like MT effect over the whole spectral range. Notice that the z-spectra are asymmetric, and when one compares the positive and negative offsets in the 2 to 3.5 ppm region, there are noticeable drops on the positive side. The MTR_{asym} and ΔPTR spectra show that the tumor changes are only visible in the 2 to 3.5 ppm offset range from water, corresponding to the exchangeable proton range in the spectra. (Reproduced from Zhou, J. et al., *Magn. Reson. Med.*, 50, 1120–1126, 2003. With permission.)

animal imager with a 3-cm surface coil for rf transmission and detection. The pulse sequence they utilized is shown schematically in Figure 5.2, with the exception that the cw saturation pulse was replaced by a train of 400 Gaussian pulses (length = 6.6 msec, flip angle = 180°, delay = 3.4 msec, total duration = 4 sec, average rf power ~ 50 Hz).

As was described in the theory section, the saturation pulses shown in Figure 5.2 can produce a signal reduction in water via a direct effect, by some type of saturation transfer to water from exchangeable peaks on solutes, or from solid-like macromolecules. In order to tease out the desired CEST contrast from the other two effects, images were acquired with the saturation pulses swept from +6 to –6 ppm from the waterline. This technique has been routinely applied in conventional MT experiments[34–36] with the plots of these rf saturation effects as a function of saturation frequency called z-spectra. In Figure 5.7 the z-spectra for tumor, peritumoral tissue, and contralateral normal tissue for the 9L brain tumor model in rats are plotted. As can be seen, the signal intensities for all the tissues are substantially reduced over this frequency range (max signal is 60%), which is

TABLE 5.2
Typical Shifts from the Water Peak (ppm) of Exchangeable Groups in Peptides and Proteins

Group	Shift (ppm)
His	5.3, 4.3
Amide	3.6
Asn	3.1
Arg	1.8
Thr	0.6

due mostly to the effects of direct saturation and also conventional MT. The CEST effect on the z-spectra is asymmetric and highly local, and distinguishable from the effects of solid tissue as seen in Figure 5.7b, by plotting MTR_{asym} for tumor tissue. In the 2 to 3.5 ppm region there is a large, sharp asymmetry, which can be assigned to exchangeable peaks on soluble proteins. This range of chemical shifts corresponds to the range expected for different exchangeable sites in proteins and peptides, as shown in Table 5.1, with amide protons being the most abundant. The data in Table 5.1 were taken from the BioMagResBank database.[72]

Based on these z-spectra, images can be generated to visualize the peritumoral tissue using saturation offsets of ±3.5 ppm, as shown in Figure 5.8. This figure also compares the APT-weighted images with several other common MRI types and with histology, in which the high signal-to-noise ratio (SNR) APT images were acquired using frequency-labeling offsets of ±3.5 ppm (TR = 10 sec, 16 scans). The increased MTR_{asym} or APTR in tumor means that there is increased protein/peptide content or increased intracellular pH with respect to normal tissue or increased water content or even differing relaxation rates. However, the increased protein/peptide content is the most likely explanation due to the fact that only a small pH increase (<0.1 unit) is often detected in tumor. This is also supported by the idea that tumors have an increased peptide/protein content compared to normal tissue.[73] Figure 5.8 shows water content and diffusion-weighted images that also support this notion; the tumor does not deviate sharply from other areas of the brain. The water content image was calculated based on T_{1w}, as described in Zhou et al.,[58] because it has been shown to be inversely proportional to the content. In addition, this is supported by multiple spectroscopic studies in which a large backbone amide peak was found resonating at 3.6 ppm from water.[69,72,74,75] It is important to note that the APT images have the potential capability to differentiate the peritumoral tissue from tumor and show a clear boundary of the tumor, which agrees well with histology.

5.6 CONCLUSION

The CEST contrast mechanism is a very promising new tool for studying biological phenomena. Endogenous small metabolites and larger biological macromolecules can be highlighted using the saturation transfer imaging sequence, and two new methods have been developed to measure their exchange rates. In addition, special macromolecules have been produced that can be used as contrast agents for CEST MR imaging. A systematic study of the pH dependence of the exchange rate for two of these new polycationic CEST contrast agents has been carried out using the QUEST and QUESP CEST imaging experiments. The contrast is highly dependent on pH, an effect that was also used to image ischemic strokes with endogenous proteins. Imaging of macromolecules with multiple exchangeable groups should be useful for many future molecular imaging studies.

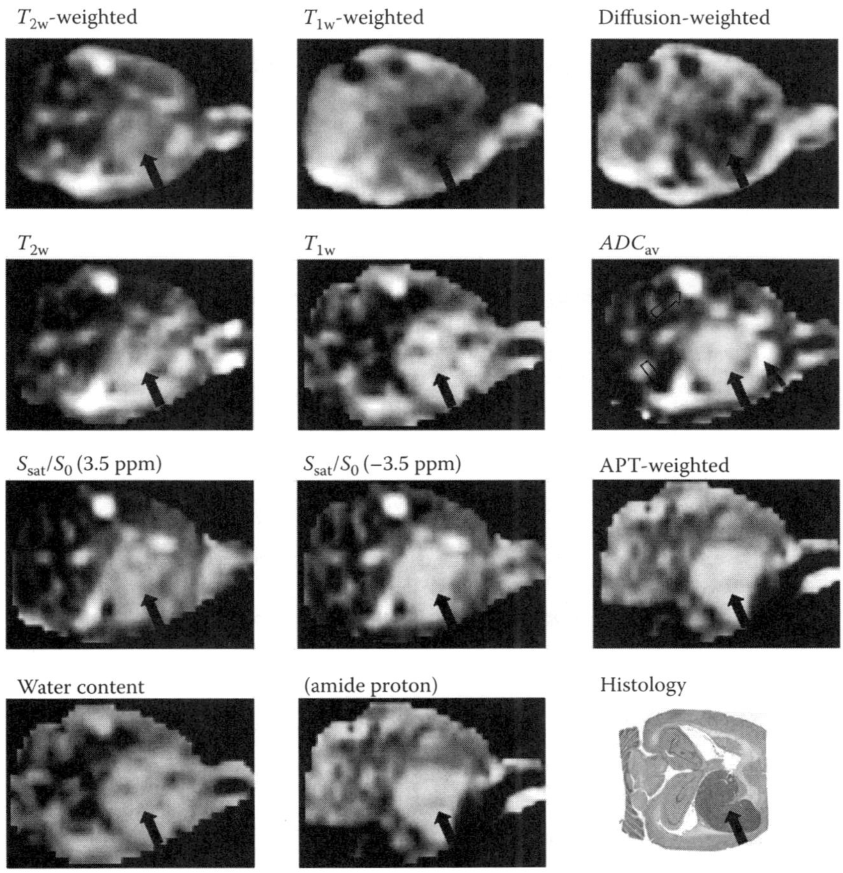

FIGURE 5.8 Comparison of APT images with several common types of MR images and histology for a rat brain tumor. The conventional MR images include a T_2-weighted image, T_{2w} map, T1-weighted image, T_{1w} map, isotropic diffusion-weighted image, and ADC_{av} map. The APT images include the MTR_{asym} (3.5 ppm) (i.e., APT-weighted) image and an amide proton content map. The APT-weighted image was obtained by subtracting the S_{sat}/S_0 images acquired at frequency labeling offsets of ±3.5 ppm (16 scans), which are also displayed. Notice that the hyperintensity peritumoral tissue (arrowhead) and cerebral spinal fluid (CSF) (open arrow) in the ADC_{av} map (also the S_{sat}/S_0 (±3.5 ppm) images) become normal in the APT-weighted image. The tumor is visible in all of the MR images (arrow), but its contour is much clearer in the APT images. (Reproduced from Zhou, J. et al., *Magn. Reson. Med.*, 50, 1120–1126, 2003. With permission.)

REFERENCES

1. Wolff, S.D. and Balaban, R.S., NMR imaging of labile proton-exchange, *J. Magn. Reson.*, 86, 164–169, 1990.
2. Guivel-Scharen, V., Sinnwell, T., Wolff, S.D., and Balaban, R.S., Detection of proton chemical exchange between metabolites and water in biological tissues, *J. Magn. Reson.*, 133, 36–45, 1998.
3. Dagher, A.P., Aletras, A., Choyke, P., and Balaban, R.S., Imaging of urea using chemical exchange-dependent saturation transfer at 1.5 T, *J. Magn. Reson. Imaging*, 12, 745–748, 2000.
4. Ward, K., Aletras, A., and Balaban, R., A new class of contrast agents for MRI based on proton chemical exchange dependent saturation transfer (CEST), *J. Magn. Reson.*, 143, 79–87, 2000.
5. Ward, K. and Balaban, R.S., Determination of pH using water protons and chemical exchange dependent saturation transfer (CEST), *Magn. Reson. Med.*, 44, 799–802, 2000.

6. Arnold, J.T. and Packard, M.E., Variations in absolute chemical shift of nuclear induction signals of hydroxyl groups of methyl and ethyl alcohol, *J. Chem. Phys.*, 19, 1608–1609, 1951.
7. Liddel, U. and Ramsey, N.F., Temperature dependent magnetic shielding in ethyl alcohol, *J. Chem. Phys.*, 19, 1608–1608, 1951.
8. Gutowsky, H.S. and Holm, C.H., Rate Processes and nuclear magnetic resonance spectra. 2. Hindered internal rotation of amides, *J. Chem. Phys.*, 25, 1228–1234, 1956.
9. Gutowsky, H.S. and Saika, A., Dissociation, chemical exchange, and the proton magnetic resonance in some aqueous electrolytes, *J. Chem. Phys.*, 21, 1688–1694, 1953.
10. Forsen, S. and Hoffman, R.A., Study of moderately rapid chemical exchange reactions by means of nuclear magnetic double resonance, *J. Chem. Phys.*, 39, 2892–2901, 1963.
11. McConnell, H.M., Reaction rates by nuclear magnetic resonance, *J. Chem. Phys.*, 28, 430–431, 1958.
12. Arnold, J.T., Magnetic resonances of protons in ethyl alcohol, *Phys. Rev.*, 102, 136–150, 1956.
13. Allerhand, A. and Gutowsky, H.S., Spin-echo NMR studies of chemical exchange. 1. Some general aspects, *J. Chem. Phys.*, 41, 2115–2126, 1964.
14. Gutowsky, H.S., Vold, R.L., and Wells, E.J., Theory of chemical exchange effects in magnetic resonance, *J. Chem. Phys.*, 43, 4107–, 1965.
15. Woessner, D.E., Nuclear transfer effects in nuclear magnetic resonance pulse experiments, *J. Chem. Phys.*, 35, 41–48, 1961.
16. Jeener, J., Ampere Summer School, Basko Polje, Yugoslavia, 1971.
17. Aue, W.P., Bartholdi, E., and Ernst, R.R., 2-Dimensional spectroscopy: application to nuclear magnetic-resonance, *J. Chem. Phys.*, 64, 2229–2246, 1976.
18. Johnson, C.S., Chemical rate processes and magnetic resonance, *Adv. Magn. Reson.*, 1, 33–102, 1965.
19. Willem, R., 2D NMR Applied to dynamic stereochemical problems, *Prog. NMR Spectr.*, 20, 1–94, 1987.
20. Desvaux, H. and Berthault, P., Study of dynamic processes in liquids using off-resonance rf irradiation, *Prog. NMR Spectrosc.*, 35, 295–340, 1999.
21. Dempsey, C.E., Hydrogen exchange in peptides and proteins using NMR spectroscopy, *Prog. NMR Spectrosc.*, 39, 135–170, 2001.
22. Bain, A.D., Chemical exchange in NMR, *Prog. NMR Spectrosc.*, 43, 63–103, 2003.
23. Alger, J.R. and Shulman, R.G., Nmr-studies of enzymatic rates *invitro* and *invivo* by magnetization transfer, *Q. Rev. Biophys.*, 17, 83–124, 1984.
24. Morris, P.G., *Annual Reports on NMR Spectroscopy*, Academic Press, London, 1988, pp. 1–60.
25. Leibfritz, D. and Dreher, W., Magnetization transfer MRS, *NMR Biomed.*, 14, 65–76, 2001.
26. Ernst, R.R., Bodenhausen, G., and Wokaun, A., *Principles of Nuclear Magnetic Resonance in One and Two Dimensions*, Clarendon Press, Oxford, 1990 (reprint with corrections).
27. Jackman, L.M., Cotton, F.A., and Adams, R.D., *Dynamic Nuclear Magnetic Resonance Spectroscopy*, Academic Press, New York, 1975.
28. Kaplan, J.I. and Fraenkel, G., *NMR of Chemically Exchanging Systems*, Academic Press, New York, 1980.
29. Sandström, J., *Dynamic NMR Spectroscopy*, Academic Press, London, 1982.
30. Cavanagh, J., Fairbrother, W.J., Palmer, A.G., III, and Skelton, N.J., *Protein NMR Spectroscopy: Principles and Practice*, Academic Press, San Diego, 1996.
31. Forsen, S. and Hoffman, R.A., A new method for study of moderately rapid chemical exchange rates employing nuclear magnetic double resonance, *Acta Chem. Scand.*, 17, 1787–1788, 1963.
32. Forsen, S. and Hoffman, R.A., High resolution nuclear magnetic double and multiple resonance, *Prog. NMR Spectrosc.*, 1, 15–204, 1966.
33. Mann, B.E., Application of Forsen-Hoffman method of measuring rates of exchange to C-13 NMR-spectrum of Cis-decalin, *J. Magn. Reson.*, 21, 17–23, 1976.
34. Wolff, S.D. and Balaban, R.S., Magnetization transfer contrast (MTC) and tissue water proton relaxation *in vivo*, *Magn. Reson. Med.*, 10, 135–144, 1989.
35. Henkelman, R.M., Stanisz, G.J., and Graham, S.J., Magnetization transfer in MRI: a review, *NMR Biomed.*, 14, 57–64, 2001.
36. Bryant, R.G., The dynamics of water-protein interactions, *Annu. Rev. Biophys. Biomol. Struct.*, 25, 29–53, 1996.
37. Englander, S.W. and Kallenbach, N.R., Hydrogen exchange and structural dynamics of proteins and nucleic acids, *Q. Rev. Biophys.*, 16, 521–655, 1984.

38. Bai, Y., Milne, J.S., Mayne, L., and Englander, S.W., Primary structure effects on peptide group hydrogen exchange, *Protein Struct. Funct. Genet.*, 17, 75–86, 1993.
39. Krishna, M.M.G., Hoang, L., Lin, Y., and Englander, S.W., Hydrogen exchange methods to study protein folding, *Methods*, 34, 51–64, 2004.
40. Molday, R.S., Englander, S.W., and Kallen, R.G., Primary structure effects on peptide group hydrogen exchange, *Biochemistry*, 11, 150–158, 1972.
41. Englander, S.W. and Poulsen, A., Hydrogen-tritium exchange of the random chain polypeptide, *Biopolymers*, 7, 379–393, 1969.
42. Maity, H., Lim, W.K., Rumbley, J.N., and Englander, S.W., Protein hydrogen exchange mechanism: local fluctuations, *Protein Sci.*, 12, 153–160, 2002.
43. Connolly, G.P., Bai, Y., Jeng, M.-F., and Englander, S.W., Isotope effects in peptide group hydrogen exchange, *Protein Struct. Funct. Genet.*, 17, 87–92, 1993.
44. Roder, H., Wagner, G., and Wuthrich, K., Individual amide proton exchange rates in thermally unfolded basic pancreatic trypsin inhibitor, *Biochemistry*, 24, 7407–7411, 1985.
45. Kim, P.S. and Baldwin, R.L., Influence of charge on the rate of amide proton exchange, *Biochemistry*, 21, 1–5, 1982.
46. Jeener, J., Meier, B.H., Bachmann, P., and Ernst, R.R., Investigation of exchange processes by 2-dimensional NMR-spectroscopy, *J. Chem. Phys.*, 71, 4546–4553, 1979.
47. Perrin, C.L. and Dwyer, T.J., Application of 2-dimensional NMR to kinetics of chemical-exchange, *Chem. Rev.*, 90, 935–967, 1990.
48. Macura, S., Westler, W.M., and Markley, J.L., 2-Dimensional exchange spectroscopy of proteins, *Nucl. Magn. Reson. C*, 239, 106–144, 1994.
49. Mori, S., Abeygunawardana, C., vanZijl, P.C.M., and Berg, J.M., Water exchange filter with improved sensitivity (WEX II) to study solvent-exchangeable protons. Application to the consensus zinc finger peptide CP-I, *J. Magn. Reson. Ser. B*, 110, 96–101, 1996.
50. Mori, S., Johnson, M.O., Berg, J.M., and Vanzijl, P.C.M., Water exchange filter (Wex filter) for nuclear-magnetic-resonance studies of macromolecules, *J. Am. Chem. Soc.*, 116, 11982–11984, 1994.
51. Goffeney, N., Bulte, J.W., Duyn, J., Bryant, L.H., Jr., and van Zijl, P.C., Sensitive NMR detection of cationic-polymer-based gene delivery systems using saturation transfer via proton exchange, *J. Am. Chem. Soc.*, 123, 8628–8629, 2001.
52. Snoussi, K., Bulte, J.W., Gueron, M., and van Zijl, P.C., Sensitive CEST agents based on nucleic acid imino proton exchange: detection of poly(rU) and of a dendrimer-poly(rU) model for nucleic acid delivery and pharmacology, *Magn. Reson. Med.*, 49, 998–1005, 2003.
53. Zhang, S., Merritt, M., Woessner, D.E., Lenkinski, R.E., and Sherry, A.D., PARACEST agents: modulating MRI contrast via water proton exchange, *Acc. Chem. Res.*, 36, 783–790, 2003.
54. Aime, S., Barge, A., Delli Castelli, D., Fedeli, F., Mortillaro, A., Nielsen, F.U., and Terreno, E., Paramagnetic lanthanide(III) complexes as pH-sensitive chemical exchange saturation transfer (CEST) contrast agents for MRI applications, *Magn. Reson. Med.*, 47, 639–648, 2002.
55. Zhang, S., Winter, P., Wu, K., and Sherry, A.D., A novel europium(III)-based MRI contrast agent, *J. Am. Chem. Soc.*, 123, 1517–1518, 2001.
56. Aime, S., Carrera, C., Castelli, D.D., Crich, S.G., and Terreno, E., Tunable imaging of cells labeled with MRI-PARACEST agents, *Angew. Chem. Int. Ed.*, 44, 1813–1815, 2005.
57. Zhou, J., Payen, J.-F., Wilson, D.A., Traystman, R.J., and van Zijl, P.C., Using the amid proton signals of intracellular proteins and peptides to detect pH effects in MRI, *Nat. Med.*, 9, 1085–1090, 2003.
58. Zhou, J., Lal, B., Wilson, D.A., Laterra, J., and van Zijl, P.C., Amide proton transfer (APT) contrast for imaging of brain tumors, *Magn. Reson. Med.*, 50, 1120–1126, 2003.
59. McMahon, M.T., Gilad, A.A., Zhou, J., Sun, P.Z., Bulte, J.W.M., and vanZijl, P.C.M., Quantifying exchange rates in CEST agents using the saturation time and saturation power dependencies of the magnetization transfer effect on the MRI signal (QUEST and QUESP): pH calibration for poly-L-lysine and a starburst dendrimer, *Magn. Reson. Med.*, 55, 836–847, 2006.
60. Sun, P.Z., Zhou, J., Sun, W., Huang, J., and van Zijl, P.C., Suppression of lipid artifacts in amide proton transfer imaging, *Magn. Reson. Med.*, 54, 222–225, 2005.
61. Gilad, A.A., McMahon, M.T., Winnard, P.T.J., Raman, V., Bulte, J.W.M., and van Zijl, P.C.M., Developing a New Class of CEST Reporter Genes, paper presented at the Fourth Annual Meeting of the Society for Molecular Imaging Conference, Cologne, Germany, September 7–10, 2005.

62. Aime, S., Delli Castelli, D., and Terreno, E., Novel pH-reporter MRI contrast agents, *Angew. Chem. Int. Ed.*, 41, 4334–4336, 2002.
63. Zhou, J., Wilson, D.A., Sun, P.Z., Klaus, J.A., and van Zijl, P.C.M., Quantitative description of proton exchange processes between water and endogenous and exogenous agents for WEX, CEST, and APT experiments, *Magn. Reson. Med.*, 51, 945–952, 2004.
64. Terreno, E., Castelli, D.D., Cravotto, G., Milone, L., and Aime, S., Ln(III)-DOTAMGIY complexes: a versatile series to assess the determinants of the efficacy of paramagnetic chemical exchange saturation transfer agents for magnetic resonance imaging applications, *Invest. Radiol.*, 39, 235–243, 2004.
65. McConnell, B.M. and von Hippell, P.H., Hydrogen exchange as a probe of the dynamic structure of DNA. I. General acid-base catalysis, *J. Mol. Biol.*, 50, 297–316, 1970.
66. Kingsley, P.B. and Monahan, W.G., Effects of off-resonance irradiation, cross-relaxation, and chemical exchange on steady-state magnetization and effective spin-lattice relaxation times, *J. Magn. Reson.*, 143, 360–375, 2000.
67. Woessner, D.E., Zhang, S., Merritt, M.E., and Sherry, A.D., Numerical solution of the Bloch equations provides insights into the optimum design of PARACEST agents for MRI, *Magn. Reson. Med.*, 53, 790–799, 2005.
68. Gutowsky, H.S., McCall, D.W., and Slichter, C.P., Nuclear magnetic resonance multiplets in liquids, *J. Chem. Phys.*, 21, 279–292, 1953.
69. van Zijl, P.C., Zhou, J., Mori, N., Payen, J.F., Wilson, D., and Mori, S., Mechanism of magnetization transfer during on-resonance water saturation. A new approach to detect mobile proteins, peptides, and lipids, *Magn. Reson. Med.*, 49, 440–449, 2003.
70. Gregory, R.B., Crabo, L., Percy, A.J., and Rosenburg, A., Water catalysis of peptide hydrogen isotope exchange, *Biochemistry*, 22, 910–917, 1983.
71. Liepinsh, E. and Otting, G., Proton exchange rates from amino acid side chains: implications for image contrast, *Magn. Reson. Med.*, 35, 30–42, 1996.
72. Seavey, B.R., Farr, E.A., Westler, W.M., and Markley, J.L., A relational database for sequence-specific protein NMR data, *J. Biomol. NMR*, 1, 217–236, 1991.
73. Howe, F.A., Barton, S.J., Cudlip, S.A., Stubbs, M., Saunders, D.E., Murphy, M., Wilkins, P., Opstad, K.S., Doyle, V.L., McLean, M.A., Bell, B.A., and Griffiths, J.R., Metabolic profiles of human brain tumors using quantitative *in vivo* 1H magnetic resonance spectroscopy, *Magn. Reson. Med.*, 49, 223–232, 2003.
74. Cavanagh, J., Fairbrother, W.J., Palmer, A.G., III, and Skelton, N.J., *Protein NMR Spectroscopy: Principles and Practice*, Academic Press, San Diego, 1996.
75. Wüthrich, K., *NMR of Proteins and Nucleic Acids*, Wiley, New York, 1986.

6 PARACEST Contrast Agents

A. Dean Sherry and Mark Woods

CONTENTS

6.1 Paramagnetic CEST (PARACEST) Agents ... 101
6.2 Water Exchange in Lanthanide Complexes ... 101
6.3 Lanthanide Complexes as PARACEST Agents ... 104
6.4 CEST Spectra: Fine-Tuning the Choice of Proton Antenna and Lanthanide Ion ... 106
6.5 PARACEST Agents as Sensors ... 110
6.6 Other Responsive PARACEST Agents ... 112
6.6.1 Temperature ... 112
6.6.2 Metabolites ... 113
6.7 A General Model for the Design of Responsive PARACEST Agents ... 114
6.8 Amplifying CEST with Supramolecular PARACEST Agents ... 117
6.9 Summary ... 119
Acknowledgments ... 120
References ... 121

6.1 PARAMAGNETIC CEST (PARACEST) AGENTS

The previous chapter introduced the concept of using chemical exchange saturation transfer (CEST) methods to produce contrast in magnetic resonance (MR) images. Since MR image intensity is largely determined by the water content of tissues and the relaxation characteristics of that water, any molecule that is in chemical exchange with water can potentially be used to transfer saturated spins into the pool of water protons, and hence used to initiate CEST contrast. In biology, this includes any molecule, large or small, that contains an exchangeable proton such as –NH, –OH, –SH, $-PO_4H$, or, in unusual circumstances, $-CO_2H$. The protons in such groups typically have nuclear magnetic resonance (NMR) chemical shifts that differ from tissue water by <5 ppm. This makes it difficult, if not impossible, to saturate one of these exchangeable spins without partial saturation of at least some of the bulk water spins as well. More will be presented on this later. One could expand the choices of exchangeable spins from simple proton systems (–NH, –OH, etc.) to a water molecule (H_2O) if the chemical shift degeneracy between water molecules in different chemical environments or compartments could be lifted. As we shall see, this is the role fulfilled by paramagnetic metal ions and, in particular, the trivalent lanthanide (III) ions.

6.2 WATER EXCHANGE IN LANTHANIDE COMPLEXES

Paramagnetic lanthanide ions first made an impact in NMR during the early 1970s with the discovery of lanthanide shift reagents.[1] The first shift reagents were organic-soluble, Lewis acid type complexes that were used to increase the chemical shifts of the proton resonances of organic molecules, essentially turning low-field, non-first-order NMR spectra into the equivalent of high-field, first-order spectra.[2] Later, it was found that the hyperfine shifts induced by the lanthanide aqua ions were also useful in refining the solution structures of molecules such as nucleic acids, peptides, proteins, and other biological molecules.[3] The magnitude of the shifts induced by lanthanide ions in the proton resonances of a molecule can be quite large, even several hundred ppm, and this is especially true of protons that are part of the lanthanide complex itself. In these cases, lanthanide-induced hyperfine shifts are not reduced by weak binding interactions and subsequent equilibrium averaging.

In the early 1980s lanthanide ions had a substantially larger impact on NMR with the introduction of gadolinium complexes as contrast agents for MR imaging (MRI).[4] The gadolinium(III) ion is unique in the lanthanide series of ions in that it has seven unpaired electrons equally distributed into the 7f orbitals. Since these electrons are distributed isotropically, they cannot induce hyperfine NMR shifts in nearby protons as observed with other lanthanide ions. Gadolinium does, however, have a profound effect on the relaxation rates of any nucleus in close proximity to the ion, and this is the foundation for the use of gadolinium(III) as a contrast agent for medical MRI.[5] The factors that govern the effectiveness of these gadolinium-based agents are discussed in detail in Chapter 3, but one important factor is the rate of water exchange in such complexes. For gadolinium-based agents it has long been understood that the rate of this water exchange must be rapid; however, when these complexes were first introduced into clinical medicine, there was no way to measure this exchange rate. By drawing parallels with transition metal chemistry, it was assumed that the rate of water exchange in complexes such as $GdDPTA^{2-}$ and $GdDOTA^{-}$, the first MRI contrast agents, was rapid like that of the Gd^{3+} aqua ion. It was not until a decade after the introduction of $GdDPTA^{2-}$ into clinical medicine that the pioneering work of Merbach allowed the first measurement of these exchange rates in 1993.[6] Surprisingly, the rates obtained for these complexes were much slower than had previously been assumed and three orders of magnitude slower than that of the Gd^{3+} aqua ion. This discovery sparked an upsurge into research directed at controlling the water exchange rate in lanthanide complexes.

The variable-temperature ^{17}O NMR work of Merbach's group showed that the water residence lifetime (τ_M) of $GdDPTA^{2-}(H_2O)$ is 303 nsec and for $GdDOTA^{-}(H_2O)$, 244 nsec.[6] At the same time, it was discovered that replacing two of the anionic carboxylate ligating groups in DTPA with neutral ligands such as amides caused a significant drop in the water exchange rate such that the Gd^{3+} complex of bis-methylamide-DTPA, GdDTPA-BMA, was found to have a ~7-fold longer τ_M value (τ_M = 2.2 μsec) than in $GdDTPA^{2-}$ itself.[6] This observation is comparatively easy to understand if the mechanism of water exchange in these monohydrated complexes is considered. Water exchange occurs through a dissociative process; the coordinated water molecule leaves the complex to form an eight-coordinate intermediate that rapidly reacts with a new water molecule from bulk solvent.[7] Clearly, the rate-limiting step in this mechanism is the dissociation of the coordinated water molecule. Empirically, the more electron density that is placed on the metal ion, the smaller the demand for electron density from the coordinated water molecule and the weaker the interaction between the two. Thus, a ligand that contains several good electron donors such as carboxylates will render the metal ion relatively electron rich, and so water exchange will be rapid. Substitution of a carboxylate side-chain ligand by a ligand donor like an amide that donates less electron density to the lanthanide leads to a stronger metal–water interaction, and consequently slower water exchange.[8] The potential of this approach for controlling water exchange rates is nicely demonstrated by the series of complexes shown in Figure 6.1. The acetate arms of $GdDOTA^{-}$ are gradually replaced by acetamide pendant arms having a methyl phosphonate substituent on the amide. With the removal of each acetate arm, the observed water exchange rate increases such that by the time all the acetates have been replaced by amides, the exchange rate is two orders of magnitude slower than that of GdDOTA.

So swapping acetate pendant arms for amides slows the rate of water exchange, but it has also been found that the nature of the amide substituent has a profound effect on the rate of water exchange.[8] In a study that examined the water exchange rate of several lanthanide tetra-amide–DOTA complexes as a function of amide substituent, it was found that water exchange is

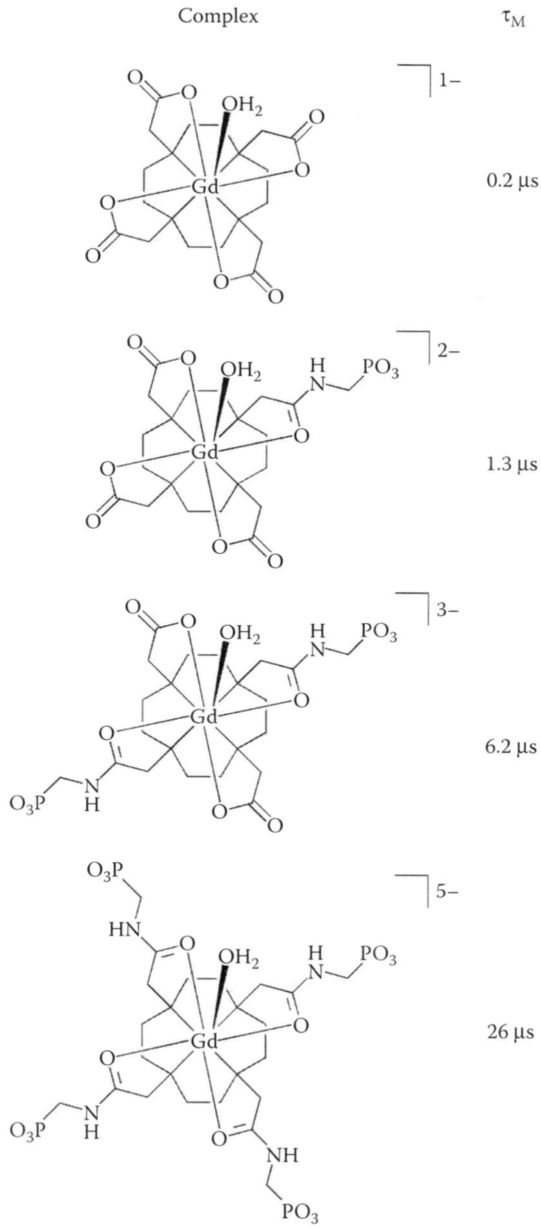

FIGURE 6.1 Water exchange rates for gadolinium(III) complexes of DOTA and three DOTA-amide ligands varies dramatically with an increase in the number of amide pendant arms.

slower in Eu^{3+} complexes that have large hydrophobic amide substituents that hinder access of an approaching water molecule (Figure 6.2).[9] One can easily extend τ_M from 100 μsec to over 1 msec by altering the amide substituent. This means that a resonance from coordinated water molecule in EuDOTAM can only be observed in acetonitrile solution because the rate of water exchange (τ_M = 120 μsec, for the square antiprismatic coordination geometry) in aqueous solution is too fast to observe by NMR.[10] However, at the other end of the spectrum, the rate of water exchange in EuDOTA-4AmCE is so slow that a resonance from the coordinated water molecule can be observed at +50 ppm in the proton NMR spectrum, even in pure water at room temperature.[11]

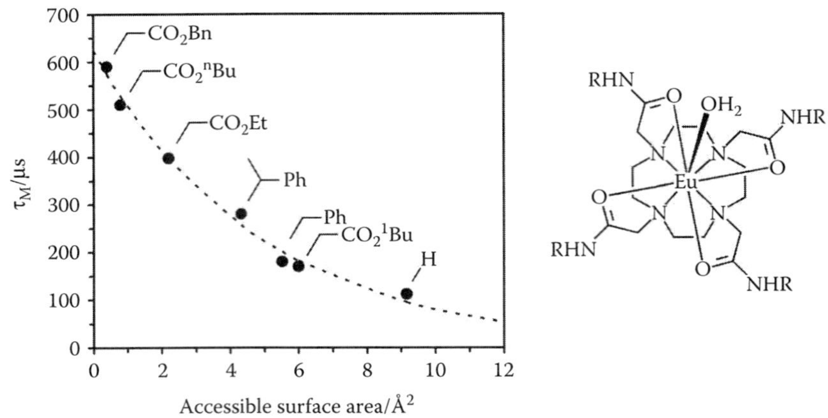

FIGURE 6.2 The effect on the water exchange rate of changing the amide substituent in LnDOTA–tetra-amide complexes. (Data from Aime, S. et al., *Chem. Commun.*, 10, 1120–1121, 2002.)

6.3 LANTHANIDE COMPLEXES AS PARACEST AGENTS

The observation of a resonance arising from the coordinated water molecule in the NMR spectrum of these complexes has opened an entirely new avenue for generating image contrast. The technique of chemical exchange saturation transfer (CEST), also sometimes referred to as magnetization transfer (MT), is not a new concept in NMR spectroscopy. In 1963, Forsen and Hoffman[12] introduced the idea of saturating a single resonance in one molecule that is in chemical exchange with a second molecule as a method to measure kinetic rate constants. This basic technique has since been widely applied in chemistry and biology to determine unidirectional rate constants of simple reactions. However, it was not until 2000 that Ward first proposed using CEST as a mechanism for generating image contrast using exogenous agents.[13] Their first example of exogenous CEST agents employed a class of diamagnetic molecules that contain exchangeable –NH or –OH protons. MR contrast was then initiated by switching on radio frequency (RF) irradiation at the NMR frequency of an exchanging proton. Chemical exchange of these –NH or –OH protons with those of the solvent water transfers the chemical saturation to the bulk water pool. This has the effect of reducing the total solvent water signal collected during MR image acquisition, hence darkening an image. More detailed descriptions of the mechanism of CEST contrast are discussed in Chapter 5 and elsewhere.[14]

One limitation for use of diamagnetic molecules as exogenous CEST agents is that the chemical shifts of those exchangeable proton resonances typically lie very close to that of bulk solvent water, typically <5 ppm. This leads to two problems: (1) a substantial amount of off-resonance direct saturation of the solvent water will occur during the presaturation phase of the experiment, and (2) the presaturation frequencies applied to these diamagnetic agents will also activate CEST arising from endogenous species. This is discussed in Chapter 5. It would be preferable then to use exogenous CEST agents that have as large of a chemical shift difference between the chemically exchanging species and bulk water ($\Delta\omega$) as possible. Paramagnetic species can induce much larger chemical shifts, and we have already seen that the resonance of the coordinated water molecule in europium(III) tetra-amide complexes is shifted to around 50 ppm, about 10 times farther away from the solvent water as the exchangeable protons of diamagnetic compounds. Since the rate of water exchange in these lanthanide DOTA–tetra-amide complexes is moderately slow and CEST requires intermediate-to-slow water exchange, these complexes afford an important opportunity to improve exogenous CEST agents.

The factors that govern the CEST effect are well described by a simple chemical exchange Bloch model,[15] and so it should be possible to calculate the parameters required to maximize CEST efficiency. To illustrate this technique, let us consider an idealized system in which some of the protons on a PARACEST agent (Pool B) exchange with those of the solvent water (Pool A). If the spins of Pool B are selectively saturated by a frequency-selective RF pulse and undergo chemical exchange with those in Pool A during saturation, then the observed intensity of the signal of Pool A will decrease. At steady state the signal intensity of Pool A will be given by

$$\frac{M_{A\infty}}{M_{A0}} = \frac{1}{\left(1 + k_2 T_{1A}\right)} \tag{6.1}$$

where $M_{A\infty}$ is the intensity of Pool A after prolonged presaturation of Pool B spins, M_{A0} is the initial intensity of Pool A, k_2 is the unidirectional rate constant for the spins moving from Pool A to Pool B, and T_{1A} is the spin-lattice relaxation time of Pool A spins. This indicates that a CEST effect will be observed if the T_1 of the solvent water (Pool A in this model) is long compared to the rate, k_2.

$$\Delta\omega = k_{ex} \tag{6.2}$$

CEST is also subject to the limitations of the slow exchange boundary condition (Equation 6.2), so larger chemical shift differences render faster exchange rates permissible. In practical terms, this means that faster proton exchanging sites can be used for CEST contrast whenever $\Delta\omega$ is large, an inherent feature of PARACEST. In practice, however, other parameters, such as relaxation rates of spins and the power and duration of presaturation pulse, also affect CEST. To help understand how these parameters will affect CEST, let us return to a prototype experiment in which $\Delta\omega$ is sufficiently large that irradiation of the protons of Pool B does not result in any direct saturation of the protons in Pool A. In tissue, the concentration of water protons (Pool A) is on the order of ~80 *M*, whereas the concentration of the PARACEST agent must be no more than millimolar, preferably lower. However, an NMR experiment detects nuclear magnetization, not the nuclei themselves, but since the magnetization of each pool at equilibrium is given by the Boltzmann distribution (Equation 6.1), the magnetization, *M*, of each pool will be proportional to their concentrations, $[H]_a = M_0^a$ and $[H]_b = M_0^b$. This allows one to relate the exchange of nuclear magnetization between the two pools to the actual exchange of protons. When exchange occurs, protons move from Pool A to Pool B at a rate k_a ($k_a = 1/\tau_a$), where τ_a is the residence lifetime of bulk water protons (Pool A). As long as the exchangeable protons on the CEST agent are completely saturated, and some early publications make this assumption, then the protons that leave Pool A having a *Z*-magnetization, M_Z^a, will be replaced by protons from Pool B with no net magnetization. This will cause the bulk magnetization of Pool A to decrease at the rate $k_a M_Z^a$. Of course, spin-lattice relaxation will simultaneously work to return the *Z*-magnetization of Pool A to its equilibrium value at a rate $[M_0^a - M_Z^a]/T_{1a}$, where T_{1a} is the longitudinal relaxation time of bulk water protons. Once the system has reached steady state, these rates will be equal, such that:

$$k_a M_Z^a = \frac{\left[M_0^a - M_Z^a\right]}{T_{1a}} \tag{6.3}$$

This can be rearranged to give the normalized steady-state value of M_Z^a, often referred to as the *Z*-value. Note that smaller *Z*-values indicate larger CEST effects.

$$Z = \frac{M_Z^a}{M_0^a} = \frac{\tau_a}{\left(T_{1a} + \tau_a\right)} \tag{6.4}$$

From Equation 6.4, one sees that the observed CEST effect will be larger when the residence lifetime of a proton in Pool A, τ_a, is short and the longitudinal relaxation time of this pool, T_{1a}, is long. T_{1a} is an intrinsic property of bulk water and is typically longer than 1 sec unless a relaxation agent is present. It should be noted that the T_{1a} of bulk water tends to be longer at higher fields, so this implies, all else being equal, that CEST will be greater at higher magnetic fields. τ_a, on the other hand, is a feature of interactions between the solvent water and the PARACEST agent, and this can be manipulated by chemical design so as to position it in an optimal range. Mass balance dictates that $k_b M_0{}^b = k_a M_0{}^a$ and $\tau_a = \tau_b\ (M_0{}^b/M_0{}^a)$, so Equation 6.4 can be written in terms of agent concentration and lanthanide-bound water lifetime (τ_b).

$$Z = \frac{M_Z^a}{M_0^a} = \frac{\tau_b}{\left(T_{1a}(M_0^a / M_0^b) + \tau_b\right)} \qquad (6.5)$$

Equation 6.5 shows that a maximum CEST effect is obtained when the residence lifetime of a proton on a PARACEST agent, τ_b, is as short as possible and the concentration of the agent ($M_0{}^b$) is as high as possible. In addition, the CEST also depends upon the extent of saturation of Pool B; less than complete saturation will result in lower CEST contrast. The extent of saturation of Pool B is determined by the power of the applied presaturation pulse, B_1 ($\omega_1 = 2\pi B_1$), as well as other physical characteristics of the nuclear spin system, including relaxation times, chemical shifts, and exchange rates.[8] So, the dichotomy in this experiment is that CEST will nearly always increase with larger applied B_1, yet this (power) is the very parameter that we need to minimize for *in vivo* work. To do this, there are essentially three parameters that can be modified: the residence lifetime of the protons on the agent, τ_b, the chemical shift of the exchangeable protons, $\Delta\omega$, and the relaxation properties of the agent. Clearly, $\Delta\omega$ should be as large as possible so that one does not indirectly saturate bulk water protons while saturating the exchanging Pool B protons. Equation 6.5 shows that shorter values of τ_b result in larger CEST if the protons of Pool B are satisfactorily saturated. However, if τ_b is too short, then the protons of Pool B will be poorly saturated, and this ultimately limits the observed CEST. Since these two effects run counter to one another, an optimal value of τ_b must be calculated. A numerical analysis that involves solution of the Bloch equations formulated to include exchange between two pools of protons shows that this optimal value is dependent upon the saturation power, B_1.[8]

$$k_b = \omega_1 = 2\pi B_1 \quad \text{or} \quad \tau_b = \frac{1}{\omega_1} = \frac{1}{2\pi B_1} \qquad (6.6)$$

This fundamental concept is graphically illustrated in Figure 6.3, which shows that for a relatively small B_1 of 50 Hz, an optimal CEST effect will be observed for an agent with a bound water lifetime, τ_b, of 3 msec, much longer than reported for any PARACEST agent to date. For B_1 values of 100, 200, or 500 Hz, the optimal bound water lifetimes decrease from 1.5 msec, 735 μsec, and 296 μsec, respectively. Thus, higher presaturation power allows one to choose an agent undergoing faster water exchange, a useful guiding principle for the design of PARACEST agents.

6.4 CEST SPECTRA: FINE-TUNING THE CHOICE OF PROTON ANTENNA AND LANTHANIDE ION

One can most easily identify all protons (or water molecules) that exchange with bulk water at a rate that yields CEST by recording what is commonly called a *Z*-spectrum or a CEST spectrum. This involves applying a frequency-selective presaturation pulse (typically of 1 to 2 sec duration), followed immediately by a non-frequency-selective observe pulse (typically a hard 90° pulse) to

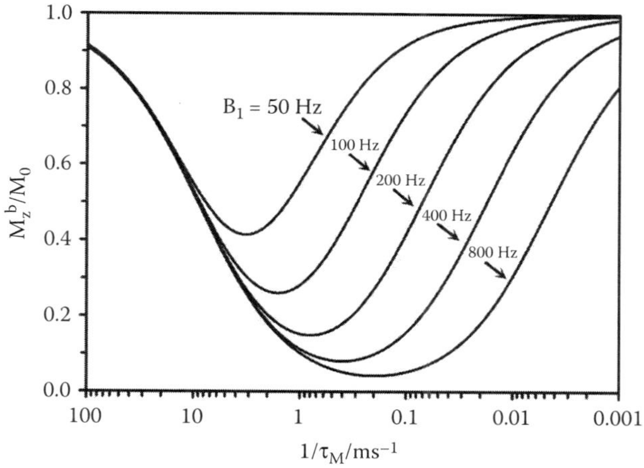

FIGURE 6.3 The magnitude of the CEST effect calculated for typical parameters for a 20 m*M* solution of a PARACEST agent (T_{1bulk} = 2.5 sec, T_{1bound} = 1.0 sec, $\Delta\omega$ = 50 ppm at 200 MHz) as a function of the water exchange rate ($1/\tau_M$) for different presaturation powers (B_1). The advantage of selecting an appropriate water exchange rate is evident.

sample the steady-state bulk water signal intensity. If one repeats this experiment many times over a range of presaturation frequencies that cover the entire chemical shift range, one might expect to locate the exchanging proton species; a plot of residual water intensity vs. frequency of the presaturation pulse yields a CEST spectrum. Three examples are illustrated in Figure 6.4. The CEST spectrum of EuDOTA-4AmCE^{3+} (Figure 6.4a) shows that a substantial reduction in the solvent water intensity is achieved when a presaturation pulse is applied at 50 ppm,[16] which corresponds to the chemical shift of the exchanging Eu^{3+}-bound water molecule in this complex. CEST spectra such as these have now been recorded for many structurally related DOTA–tetra-amide complexes,[8] and in general, the chemical shift of the Eu^{3+}-coordinated water molecule does not differ much from this 50 ppm example from one complex to another. It should be pointed out, however, that the protons of the Eu^{3+}-bound water molecule are not the only exchangeable protons in this complex. The amide protons are not shifted as strongly as those of the coordinated water molecule, and although it is difficult to see in the spectrum of the Eu^{3+} complexes, they are easily detected in the spectra of other lanthanide complexes, such as those formed with ytterbium or dysprosium.[17] The CEST spectrum of DyDOTAM^{3+}, for example (Figure 6.4b), shows a strong CEST peak from amide proton exchange near 80 ppm that seems considerably larger than the CEST peak from the Dy^{3+}-coordinated water molecule near –720 ppm. There are three reasons why the amide protons give rise to larger CEST effect in this complex. First, the bound water molecule is situated relatively near to the highly paramagnetic Dy^{3+} ion, and hence those protons relax rather quickly (short T_1). The amide protons, on the other hand, are situated farther away from the paramagnetic center, and therefore experience less relaxation by the Dy^{3+} ion. We have already seen how relaxation of the exchanging protons is detrimental to CEST. Second, there are eight exchanging amide protons in this complex, compared to only two water protons, so this would translate into doubling of CEST for the amide protons relative to water if all other factors are identical. Finally, the rate of proton exchange between the amides of this complex and bulk water is considerably slower than the rate of water exchange between the Dy^{3+}-bound water molecule and bulk water, and in this example, amide proton exchange is more advantageous for CEST. Thus, all three factors must be considered when comparing CEST peaks in any quantitative way.

Another strategy for increasing the number of exchangeable protons available for CEST is to replace the amide side chains in cyclen-based complexes with hydroxyethyl side chains. Hydroxy-ethyl groups are known to retain their protons upon coordination to a lanthanide ion, and this then

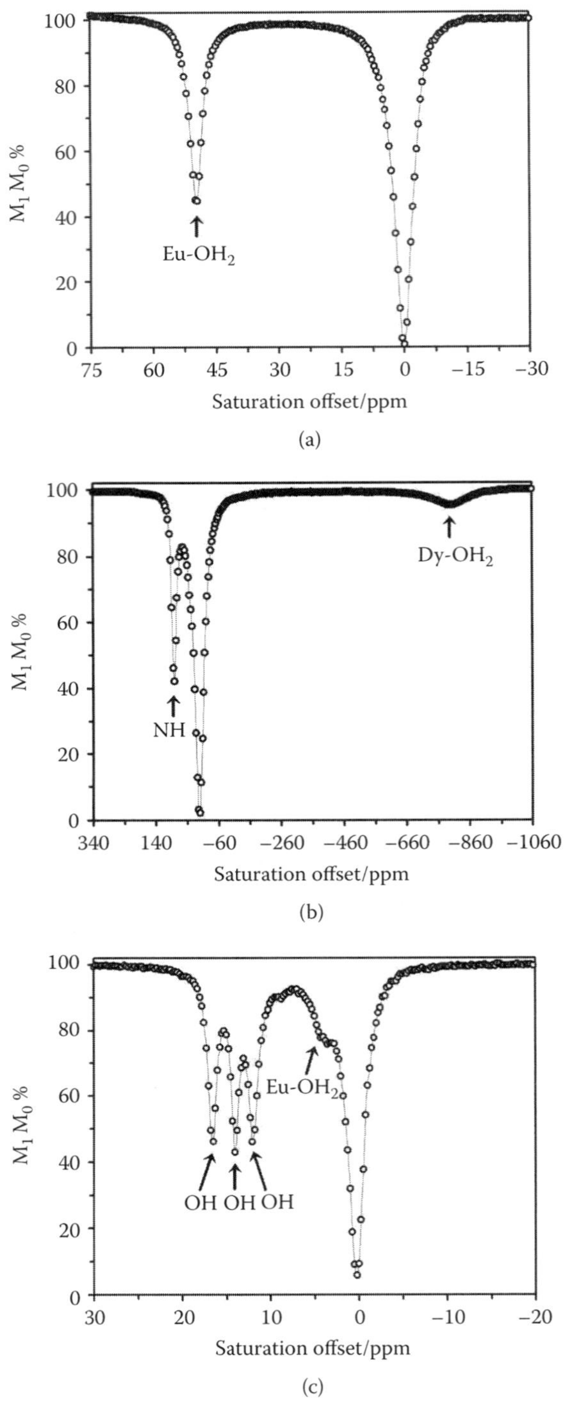

FIGURE 6.4 CEST spectra. (a) A 30 m*M* solution of EuDOTA-4AmCE^{3+} recorded at 270 MHz and 25°C; the peak at +50 ppm reflects exchange with the coordinated water molecule. (b) DyDOTAM^{3+} recorded at 400 MHz and 25°C; the peak at +80 ppm reflects exchange with the amide protons while the peak at −720 ppm reflects the coordinated water molecule. (c) A 35 m*M* solution of EuCNPHC^{3+} recorded at 270 MHz and 25°C; the three peaks at 16, 14, and 12 ppm reflect exchange with each of the hydroxyl protons.

TABLE 6.1
C_J Values as Reported by Bleaney et al.[32] for Each Lanthanide Ion and the Observed Chemical Shifts of the Ln^{3+}-Bound Water Protons in Complexes of DOTA-4AmCE[33]

Ln^{3+}	Ce	Pr	Nd	Pm	Sm	Eu	Gd	Tb	Dy	Ho	Er	Tm	Yb
C_J	–6.3	–11.0	–4.2	2.0	–0.7	4.0	0.0	–86	–100	–39	33	53	22
Ln^{3+}-bound H_2O (ppm)	—	–60	–32	—	–4	50	—	–600	–720	–360	200	500	200

Note: The magnitude of C_J predicts the magnitude of the pseudocontact shift contribution by the various lanthanide ions.

provides another type of chemical exchange site for CEST activation. Interestingly, the rate of proton exchange in Eu-*S*-THP^{3+} has been found to be too fast (relative to $\Delta\omega$) to detect a CEST effect.[14] However, when one hydroxyethyl group was replaced by an amide pendant arm, as in EuCNPHC^{3+}, then not only was CEST from the pendant hydroxyethyl groups observed (with the complex dissolved in wet acetonitrile), but each of the three magnetically nonequivalent exchange sites also transferred saturated spins to bulk water (Figure 6.4c).[14] This suggests that replacing one ligand side chain from a hydroxyethyl group to an amide results in slower exchange for the remaining three hydroxyethyl protons. Although interesting from a structural point of view, the utility of such hydroxyethyl-based systems appears to be limited by the small magnitude of $\Delta\omega$ in these systems. As weak donors, the hydroxyl groups do not impart a strong ligand field on the Eu^{3+} ion, and since ligand field is one of the major factors in governing the magnitude of the hyperfine shift in such systems, a weak ligand field gives rise to small hyperfine shifts (small $\Delta\omega$), and hence proton exchange must be much slower to observe a CEST effect in such systems. Interestingly, the rate of proton exchange between the coordinated hydroxyethyl groups and bulk water increases dramatically as water is added to this system, and eventually exceeds that permitted by the CEST requirements; consequently, CEST is not detected from any of the exchangeable sites with the complex dissolved in pure water. Thus, it appears from the limited number of hydroxyethyl-based ligand systems examined so far that these are not viable platforms for PARACEST.

These systems do highlight one further advantage of using lanthanide-based agents for CEST. Because the coordination chemistry of the lanthanides is virtually identical along the series, isostructural complexes can be prepared with any lanthanide ion and any given ligand system. This means that the hyperfine shift characteristics and relaxation properties of a PARACEST agent can be tuned by the choice of lanthanide ion. As can be seen from Table 6.1, both the magnitude and sign of the chemical shift of protons in lanthanide complexes can be altered considerably by lanthanide ion substitution. This means that more than one PARACEST agent could be administered simultaneously, and each could be activated separately by an appropriate choice of CEST frequency. This concept was nicely demonstrated both in phantoms and in living cells by Aime and coworkers.[18] A schematic of this experiment is illustrated in Figure 6.5 with four samples: one containing EuDOTA-4AmC$^-$ ($\Delta\omega$ = 50 ppm downfield), one containing TbDOTA-4AmC$^-$ ($\Delta\omega$ = 600 ppm upfield), one containing a mixture of both complexes, and a control containing only water. When a presaturation pulse is applied at +50 ppm prior to image acquisition, a change in total water signal intensity is observed only in those samples containing the europium complex. No change in water intensity is observed in samples that do not contain the europium complex. Conversely, a presaturation pulse applied at –600 ppm activates only the terbium complex, and a decrease in water intensity is observed only in those samples containing the terbium complex. As illustrated for the sample containing both complexes, the presence of a second complex does not affect the performance of the first complex as long as the saturation frequencies are well separated from one another. There are many imaging applications where the use of multiple PARACEST agents could be envisioned; for example, cell-tracking experiments might allow more than one

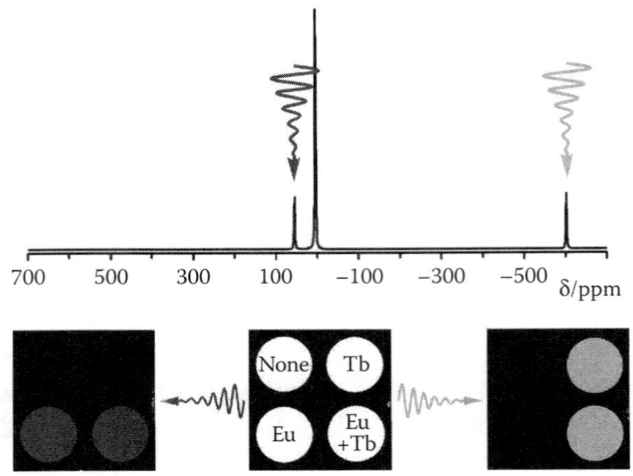

FIGURE 6.5 (please see color insert following page 210) A simulated ^{1}H NMR spectrum showing the coordinated water proton resonances of EuDOTA-4AmC$^-$ (+50 ppm) and TbDOTA-4AmC$^-$ (–600 ppm) in water (resonances arising from the ligand were omitted for clarity). Below this is a schematic representation of the result of selectively irradiating these resonances on samples containing EuDOTA-4AmC$^-$, TbDOTA-4AmC$^-$, and a mixture of the two. It is seen that each complex may be activated selectively regardless of the presence or absence of the other complex.[18]

type of cell to be tracked simultaneously, or two types of PARACEST agents could be given at equal concentrations, one for reporting a physiological measure such as pH and another simply acting as a concentration marker.

6.5 PARACEST AGENTS AS SENSORS

Two emerging areas of MRI contrast agent development are responsive agents and targeted agents. Contrast agents that respond to changes in the concentration of biologically important species, sometimes known as smart agents, are of particular interest since MRI is already so widely used for anatomical imaging. Many Gd^{3+}-based agents that exhibit changes in relaxivity with changes in pH, metal ion concentration, endogenous anions, and enzyme activity have been reported.[14] One of the main difficulties in using such agents as a platform for responsive imaging contrast agents is that the T_1 effects of Gd^{3+} are never completely silent due to a large outer-sphere relaxation effect. Thus, for quantitative measures of pH, metal ion concentration, or enzyme activity, for example, one must account for all forms of Gd^{3+} in solution, not an intractable problem but certainly a difficult one. Consider the example of a pH-responsive agent that exhibits a doubling of its relaxivity in response to a change in pH from 7.4 to 6. If the pH is 7.4 in one tissue volume element and 6 in a second, nearby volume element, and the concentration of the agent is the same in both elements, then the voxel at pH 6 will appear brighter than that at pH 7.4. However, if the pH is 7.4 in both voxels, but the concentration of the agent is twice as high in one voxel vs. the other, then similar image contrast would be observed in both voxels. So, to generate an accurate pH map using a Gd^{3+}-based T_1 agent, it is necessary to account for the local concentration of the agent in each image voxel. Although this has been accomplished in pH mapping studies,[19,20] the experimental protocol is more cumbersome than one would desire for translation of the technique to clinical medicine.

We have seen that CEST originates from chemical exchange of protons or water molecules and that the magnitude of the CEST effect is related to the rate of this exchange. Since proton exchange in these systems is normally a pH-dependent process, so is the magnitude of CEST. This feature, recognized from the very beginning of the development of exogenous CEST agents,[13] allows a direct measure of pH if the concentration of the CEST agent is known, as is the case for traditional

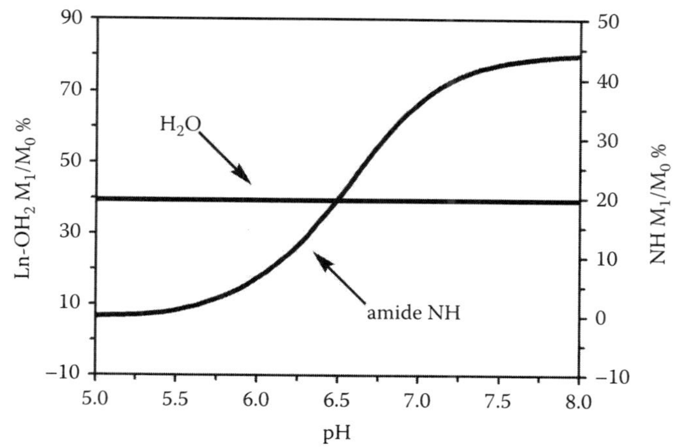

FIGURE 6.6 The ratio between the CEST effects arising from the coordinated water molecule (straight line) and the amide protons (curved line) of PrDOTA-4AmC⁻ may be used to determine pH without knowing the concentration of the complex in solution.[33]

Gd^{3+}-based T_1 contrast agents. But unlike Gd^{3+}-based agents, it is possible to administer and independently visualize two CEST agents in the same experiment (Figure 6.5), and this feature could be used to develop a concentration-independent method of determining pH. For example, the exchange of the amide protons of YbDOTA-4AmC⁻, shifted 20 ppm upfield, responds differently to changes in pH than does the exchange of protons of the coordinated water molecule of EuDOTA-4AmC⁻, shifted 50 ppm downfield. By selectively saturating first one pool of protons followed by the other, it is possible to acquire two different CEST pH profiles for a solution containing both complexes.[21] The ratio of the two CEST profiles is therefore concentration independent and can be used to determine pH directly without knowing the analytical concentration of either agent.

Of course it would be preferable to administer a single compound rather than a cocktail of compounds, so single-agent ratiometric approaches have also been explored.[22] It can be seen from the CEST spectrum of $DyDOTAM^{3+}$ (Figure 6.4) that if the conditions are optimal, CEST can be observed from both the coordinated water molecule and amide protons of a single complex. If the CEST pH response arising from the amide protons differs from that of the coordinated water molecule, then a ratiometric approach could also be used after injection of a single complex. Although CEST arising from the coordinated water molecule of EuDOTA-4AmC⁻ is reasonably high, CEST from the amide protons in this same complex is relatively small, because, as noted above, $\Delta\omega$ is relatively small for the –NH protons in this complex.[22] Conversely, the –NH protons of YbDOTA-4AmC⁻ display a much larger $\Delta\omega$ due to differences in the hyperfine shifting properties of Yb^{3+} vs. Eu^{3+}, and hence CEST from the –NH protons is quite favorable. Unfortunately, CEST from the exchanging water molecule in this complex is not detected because water exchange is too fast in Yb^{3+} complexes.[21,22] This illustrates the importance of understanding the relationships between $\Delta\omega$ (as determined by lanthanide-induced hyperfine shifts and molecular geometry), proton exchange rates, and CEST efficiency.

Complexes of DOTA-4AmC formed with lanthanide ions from the early part of the series, and in particular Pr^{3+}, show reasonable CEST from both the –NH protons and the exchange water molecule, an advantage for designing a pH sensor that can act as its own internal concentration marker.[22,23] CEST arising from the coordinated water molecule of PrDOTA-4AmC⁻ is independent of pH between 5 and 8, while CEST arising from the amide protons exhibits a marked pH dependence (Figure 6.6).[22,23] Thus, absolute pH can be determined with this single complex by using the CEST effect from the coordinated water molecule protons as a concentration marker and using the ratio of the two CEST effects as a direct readout of pH.[24]

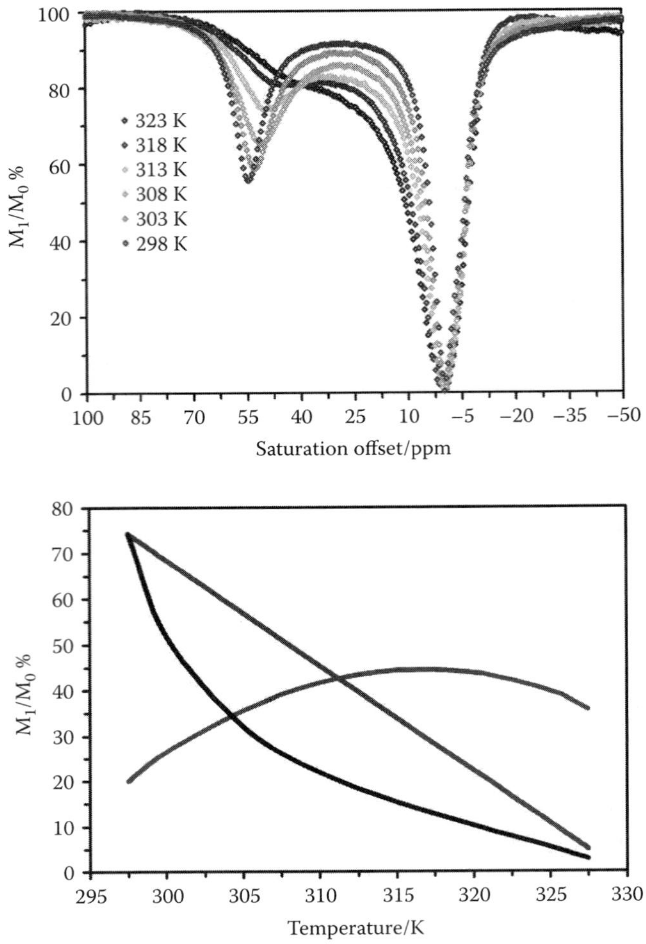

FIGURE 6.7 (please see color insert) The CEST spectra of 10 m*M* solution of EuDOTA-4AmCE^{3+} recorded at 400 MHz, pH 7.3, $B_1 = 714$ Hz and different temperatures (top)[25] and the ratiometric determination of temperature using PrDOTA-4AmC$^-$ comparing the water (blue) and amide (red) protons; the ratio ST(H_2O)/ST(NH) is shown in black (bottom).[24]

6.6 OTHER RESPONSIVE PARACEST AGENTS

6.6.1 Temperature

In addition to changes in pH, the rate of proton exchange is also affected by temperature. The rate of exchange accelerates with increasing temperature according to the Arrhenius equation, and this has a profound effect upon the magnitude and shape of the peaks in the CEST spectrum (Figure 6.7). A peak in the CEST spectrum arising from a slowly exchanging Eu^{3+}-bound water molecule is relatively sharp (see peak labeled 298°C), but as the exchange rate increases, the peak broadens, shifts toward the bulk water peak, and decreases in apparent intensity. Furthermore, the hyperfine shift induced by the paramagnetic Eu^{3+} is also highly temperature dependent and decreases rather sharply with increasing temperature. Thus, the combined effect of smaller hyperfine shift *plus* a shift of the bound water peak toward the bulk water peak with increasing temperatures makes this an extremely sensitive method to measure absolute temperature. Zhang et al. used the same PrDOTA-4AmC$^-$ complex described above and a ratiometric approach to measure temperature.[22]

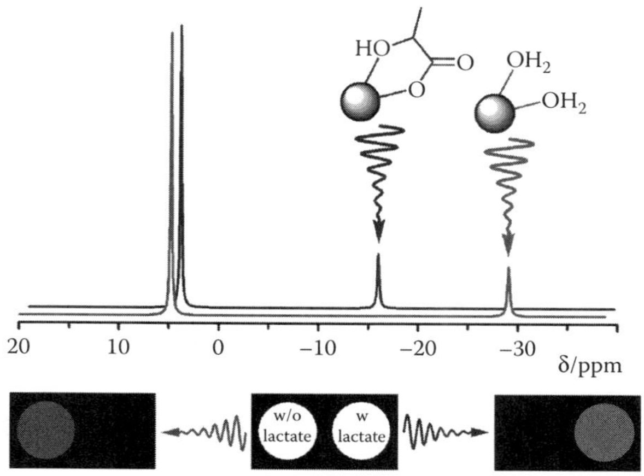

FIGURE 6.8 (please see color insert) A schematic illustration of the NMR spectra of YbMB-DO3AM^{3+} (a q = 2 complex) showing only the –NH protons of the complex in the presence and absence of lactate. Upon formation of a ternary complex between YbMB-DO3AM^{3+} and lactate (lactate displaces both inner-sphere water molecules), the hyperfine shifted –NH protons appear at a different chemical shift near –16 ppm. Thus, two CEST experiments, one with preirradiation of a sample containing both YbMB-DO3AM^{3+} and lactate at –16 ppm, followed by another with preirradiation at –29 ppm, provide a direct readout of the amount of lactate-bound vs. free complex in solution.[24] (Illustration derived from method described in Aime et al.[26])

Acceleration of the coordinated water exchange rate with increasing temperature resulted in the linear decrease in CEST, while CEST arising from the amide protons exhibits a more complex temperature behavior due to the interplay of changing $\Delta\omega$, k_{ex}, and T_1. Initially, CEST from the amide protons increases with temperature until a maximum is reached, and then falls at higher temperatures. Although CEST from both pools exhibits a temperature response, so neither pool can be viewed as a simple concentration marker, the CEST ratio of the two pools still affords a concentration-independent measure of temperature (Figure 6.7).

Zhang et al. took a different approach to measure temperature with PARACEST agents.[25] Rather than using the magnitude of the CEST effect, they monitored the frequency of the CEST peak with changing temperature. In an imaging context, this means acquiring a collection of images in which the frequency of the presaturation pulse has been varied over the range of interest. In phantom systems using EuDOTA-4AmC^{-}, it was possible to determine the temperature of the sample using this method with sufficient accuracy that even slight temperature gradients in the sample could be observed. Since it is the frequency of the presaturation pulse that is monitored in this method, this technique is also independent of the concentration of the CEST agent.

6.6.2 Metabolites

Changes in $\Delta\omega$ have also been utilized as a method for detecting the presence of lactate by CEST imaging.[26] Many endogenous anions, including lactate, are known to displace the two inner-sphere water molecules in heptadentate DO3A-triamide lanthanide complexes such as YbMB-DO3AM^{3+}.[26,27] Displacement of water by lactate induces a substantial change in the chemical shift of the amide proton resonances in the Yb^{3+} complex, from –29 to –16 ppm (Figure 6.8). CEST activation of the free complex (in the absence of lactate) by using a presaturation pulse at –29 ppm resulted in a 60% reduction in the bulk water signal intensity. Addition of lactate to the sample gradually reversed this CEST until the complex was completely saturated with lactate. Conversely, CEST by presaturation at –16 ppm was somewhat different because even in the absence of lactate there is still a reduction in the solvent water signal after presaturation. Nonetheless, the magnitude of the CEST effect observed after presaturation at –16 ppm did increase with increasing lactate concentrations.

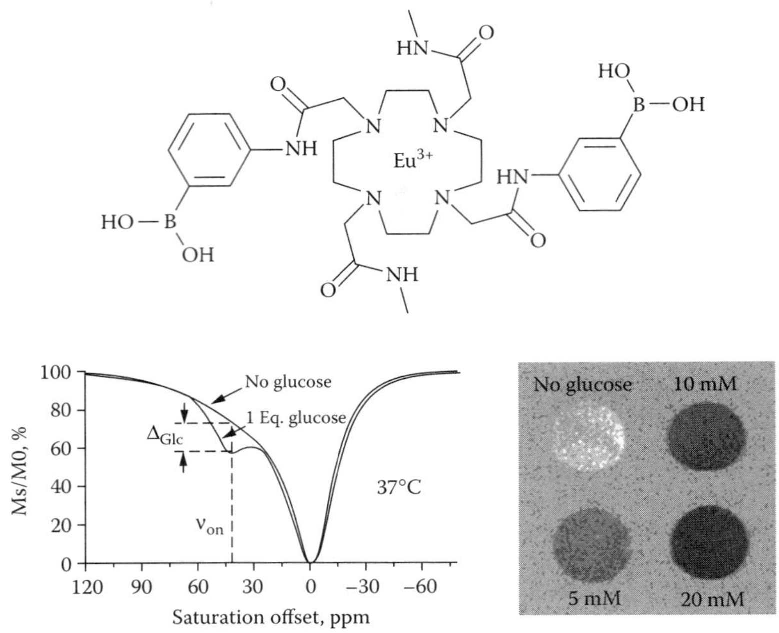

FIGURE 6.9 Chemical structure of EuDTMA-2PB and *Z*-spectra of the complex in the absence and presence of one equivalent of glucose. The *Z*-spectra were collected on samples containing 10 m*M* EuDTMA-2PB, pH 7.4, 37°C. The images show four tubes containing 10 m*M* EuDTMA-2PB with the indicated concentrations of glucose.[25]

Although not proposed by the authors at the time, the ratio of CEST by activation at –29 vs. –16 ppm could be used to give a concentration-independent ratiometric method of determining the concentration of lactate in solution. It is not clear, however, to what extent other endogenous anions would interfere with the operation of this agent *in vivo*.

6.7 A GENERAL MODEL FOR THE DESIGN OF RESPONSIVE PARACEST AGENTS

There are, of course, other aspects of biology other than pH, temperature, or metabolite levels that one might like to monitor in tissue, so it is of interest to have a general platform for responsive PARACEST agents. One general approach is to use known molecular recognition systems that either obstruct water exchange in a PARACEST complex or catalytically enhance proton exchange from a slow-exchange Ln^{3+}-bound water molecule. Examples of each are illustrated below. The first example is given by the Eu^{3+} complex of the tetra-amide ligand shown in Figure 6.9. The tetra-amide side arms of this cyclen-based macrocycle almost ensure that water exchange will be in an appropriate regime for the Eu^{3+} complex to act as a PARACEST system. However, the actual water exchange rate is still difficult to predict without experimental verification. The two phenylboronate groups appended to the *trans*-amides of this ligand were included knowing that phenylboronates such as these condense with *cis*-hydroxyl groups of sugars, so this design was based upon the expectation that the two phenylborates would condense with two different *cis*-hydroxyl groups of glucose and thereby position the sugar directly over the Eu^{3+}-bound water molecule. It was unknown whether this type of molecular recognition would result in a slowing of water exchange by limiting solvent access to the Ln^{3+} coordination site or catalyze water exchange by closely positioned –OH groups of the bound sugar. The complex, EuDTMA-2PB, was synthesized and characterized by usual chemical methods, and sugar binding studies revealed that this agent had, rather surprisingly,

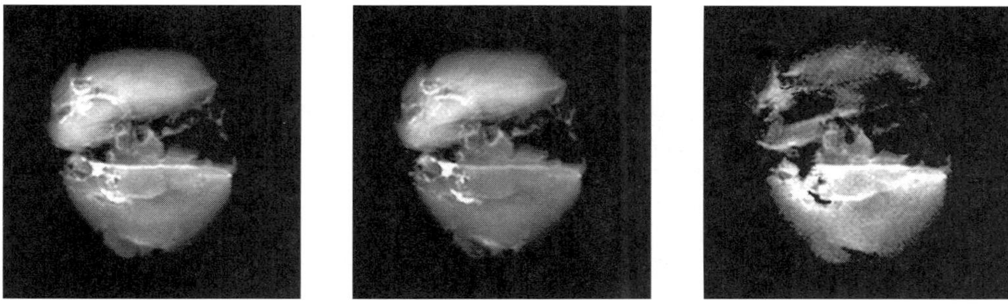

2 s presaturation pulse applied at −42 ppm | 2 s presaturation pulse applied at +42 ppm | Difference CEST image

FIGURE 6.10 *Ex vivo* MR images of mouse livers (matrix = 256 × 256, field of view (FOV) = 40 × 40 mm) after a bench perfusion with 10 m*M* EuDTMA-2PB at 37°C. The top liver in each panel was perfused without glucose while the liver in the bottom of each panel was perfused with 10 m*M* glucose. The images were collected using a 4.7-T Varian Inova horizontal-bore MR system with livers positioned over a 2.5-cm surface coil so that both livers could be imaged simultaneously. A spin-echo sequence (TR/TE = 4 sec/12 msec) with a Gaussian-shaped presaturation pulse (B_1 peak power = 500 Hz, duration = 2 sec) was used for imaging. (Jimin Ren, unpublished data.)

a rather significant binding preference for glucose over fructose.[28] As other organic-based bis-phenylboronate structures of this type typically display the highest affinity for fructose, this surprising result indicates that phenylboronate rings of EuDTMA-2PB must be oriented in such a way that condensation with the *cis*-hydroxy groups of glucose is more favorable than condensation with the *cis*-hydroxy groups of fructose. Although crystals of this adduct have not been isolated to confirm this hypothesis, a variety of binding studies using ultraviolet absorption and circular dichroism CD spectroscopy verified that a 1:1 EuDTMA-2PB:glucose is the predominant structure in solution, and glucose is likely complexed via the 1,2-*cis*-hydroxy groups to one phenylboronate and the 4,6-*cis*-hydroxy groups to another phenylboronate. Molecular modeling indicates that the resulting EuDTMA-2PB:glucose adduct would position the glucose molecule approximately over the Eu^{3+}-OH_2, as originally hypothesized.

Z-spectra of EuDTMA-2PB in the absence and presence of one equivalent of glucose are shown in Figure 6.9. The *Z*-spectrum in the absence of glucose at 37°C shows only a broad featureless water peak centered at 0 ppm, representing bulk water, while the spectrum collected in the presence of glucose shows, in addition, a well-defined CEST peak near 42 ppm, corresponding to exchange from a Eu^{3+}-bound water molecule. These features are consistent with slowing of water exchange by the glucose cap in the EuDTMA-2PB:glucose adduct, as we had anticipated. A fitting of CEST spectra such as these collected at different temperatures indicates that water exchanges slows ~2-fold upon binding of glucose to EuDTMA-2PB, yet this relatively modest alteration of water exchange is easily detected by CEST imaging.

Further binding studies of this PARACEST sensor with various sugars provided more interesting insights. First, CD binding studies verified that EuDTMA-2PB forms a 1:1 adduct with glucose where both ends of the glucose molecule are complexed by phenylboronates.[28] Both UV and CD titrations indicated an unusual affinity order, D-glucose > D-fructose > D-galactose, whereas the *cis*-isomer of EuDTMA-2PB, EuDTMA-1PB (the monophenylboronic acid derivative), and even the free ligand, DTMA-2PB, all formed complexes with sugars in the stability order D-fructose > D-galactose > D-glucose. This indicates that the spatial orientation of the *trans*-phenylboronate rings in EuDTMA-2PB determines the sugar binding specificity in this case. This unusual binding selectivity for glucose, at least compared to other phenylboronate-based sugar sensors, suggested that this sensor might be selective enough to allow imaging of glucose in tissues. Some early results using isolated, perfused mouse livers are illustrated in Figure 6.10. Since EuDTMA-2PB is weakly

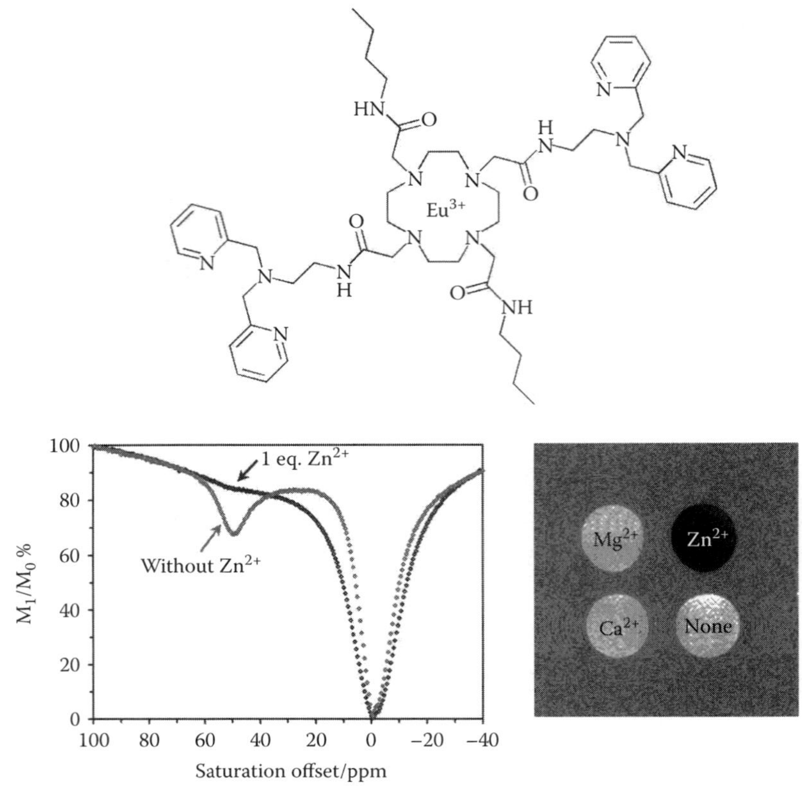

FIGURE 6.11 Chemical structure of a Zn^{2+}-sensitive PARACEST agent (top). CEST spectra of 20 mm of EuDOTAMpy in the absence and presence of 20 mm of Zn^{2+} at pH 8, 25°C, using a 2-sec frequency-selective soft pulse (B_1 = 1000 Hz) was followed by a non-frequency-selective 60° readout hard pulse for each data point (bottom, left).[28] The CEST images compare the effects of adding 1 equivalent of Mg^{2+} versus Zn^{2+} (bottom, right). Please note that the black/white image intensities of each sample are inverted for clarity in this figure.

paramagnetic and has little influence on tissue relaxation rates, the images shown in the first panel are essentially identical to those collected in the absence of the agent (not shown). Furthermore, both livers have similar intensities throughout the tissue, indicating that the presence or absence of glucose has no effect on the images. However, the images collected following a 2-sec presaturation pulse on the EuDTMA-2PB-bound water peak near 42 ppm were quite different (middle panel). Now, the liver with glucose was substantially darker than the liver without glucose. A subtraction of these two images yielded the CEST image shown in the right panel. Here, one finds essentially complete cancellation of the image intensities in the liver without glucose, while the image of the liver with glucose is bright throughout. Given the assumption that EuDTMA-2PB remains extracellular and is uniformly distributed in liver, the uniformity of the CEST image intensities of the glucose perfused liver suggests that glucose is also equally distributed throughout. Interestingly, the vascular areas appear brighter in this image, consistent with higher concentrations of both EuDTMA-2PB and glucose in these volume elements.

A second example of where molecular recognition was used in the design of responsive PARACEST agents is given by the Eu^{3+} complex of the ligand shown in Figure 6.11. Here, the four pyridine ligands appended to a tetra-amide macrocycle are a known recognition system for Zn^{2+}. Again, molecular modeling indicated that Zn^{2+} might form a tetrahedral complex with the four pyridine ligands directly above the Eu^{3+}-bound water molecule, and perhaps result in slowing of its exchange with bulk solvent. Surprisingly, quite the opposite was observed (Figure 6.11).[28]

EuDOTAMpy alone shows a CEST peak near 50 ppm, typical of an exchanging Eu^{3+}-bound water molecule in slow to intermediate exchange with bulk water. Note that this CEST spectrum was collected at 25°C, not 37°C, as reported for the glucose sensor in Figure 6.9. Thus, one would expect water exchange to be somewhat slower at 25°C than at 37°C, and the Eu^{3+}-bound water exchanging CEST peak is centered at a more highly shifted position. In this case, addition of Zn^{2+} to EuDOTAMpy resulted in disappearance of the exchanging CEST peak near 50 ppm, indicating that water exchange had either slowed drastically ($k_{ex} << \Delta\omega$, too slow for CEST) or increased substantially ($k_{ex} >> \Delta\omega$, too fast for CEST). These could not be differentiated without additional NMR and potentiometric titrations, which revealed that the Zn^{2+} adduct had an additional pK_a near 7, compared to EuDOTAMpy alone, apparently due to a Zn^{2+}-OH species.[28] This led to a working hypothesis that this nearby Zn^{2+}-OH species catalyzes proton exchange between the Eu^{3+}-bound water molecule and bulk solvent such that exchange becomes too fast for CEST. Such prototropic exchange has been shown to enhance water relaxivity in the Gd^{3+}-based, pH-sensitive contrast agent, GdDOTA-4AmP^{5-}, but was unreported for PARACEST complexes until this example was discovered. CEST images of phantoms (Figure 6.11) and potentiometric pH data both showed that EuDOTAMpy is quite selective for Zn^{2+} over other typical biological divalent ions.

6.8 AMPLIFYING CEST WITH SUPRAMOLECULAR PARACEST AGENTS

The examples described above illustrate the versatility of PARACEST as a platform for biologically responsive imaging agents. Since the MRI signal generated by a PARACEST agent is based upon chemical exchange, a fundamental parameter in biological systems, it is relatively easy to imagine that responsive agents could be designed to detect a wide variety of metabolic events in tissue. An important question that remains, however, is the sensitivity limits of PARACEST as imaging agents for MRI. Although it is difficult to compare the sensitivity of PARACEST and typical T_1 agents, since their mechanisms for producing contrast in an image are so different, one can estimate on the basis of exchange theory that PARACEST agents such as those described above have about the same contrast sensitivity as typical low molecular weight, extracellular, nonspecific, Gd^{3+}-based T_1 agents. For imaging bulk metabolic indices such as tissue pH or distribution of glucose, this concentration requirement may be acceptable for eventual clinical use. The glucose sensor described in Figure 6.9, for example, easily detects glucose by CEST imaging using ~1 to 2 mM agent, similar to a typical clinical dose of Gd^{3+}. However, for molecular imaging applications involving detection of cellular apoptosis or cancer markers, one clearly would like to amplify the CEST signal. One logical way to achieve this is to increase the number of exchangeable protons on a PARACEST agent. For example, supramolecular assemblies could be envisioned to multiply the number of exchangeable protons compared to that possible in simple, low molecular weight, paramagnetic complexes such as those described above. Goffeney and coworkers have demonstrated that CEST can be amplified by as much as 5.8×10^5 by using polyamides or dendrimers that have a large number of chemically equivalent, exchangeable NH groups.[29] The problem with diamagnetic polymers again is that the antenna frequency for CEST activation differs little from that of the solvent water. One could potentially overcome this problem by adding a suitable lanthanide shift reagent to increase the $\Delta\omega$ of the exchanging protons in the polymer. Aime and coworkers used TmDOTP^{5-} as a hyperfine shift reagent to shift the exchangeable guanidine protons in polyarginine well away from the bulk water signal.[30] Here, ionic interactions between TmDOTP^{5-} and the positively charged polymer, polyarginine, resulted in a downfield shift of ~25 ppm in the exchangeable guanidino protons, and this provided a measurable CEST effect (defined as a 5% reduction in bulk water signal intensity), with only 30 μM TmDOTP^{5-} and 1.7 μM polyarginine, a considerable improvement over 1 to 2 mM.

Even greater sensitivity can be achieved by encapsulating a paramagnetic shift reagent in the inner core of a liposome.[30] Since water exchange across a lipid bilayer is relatively slow, the presence

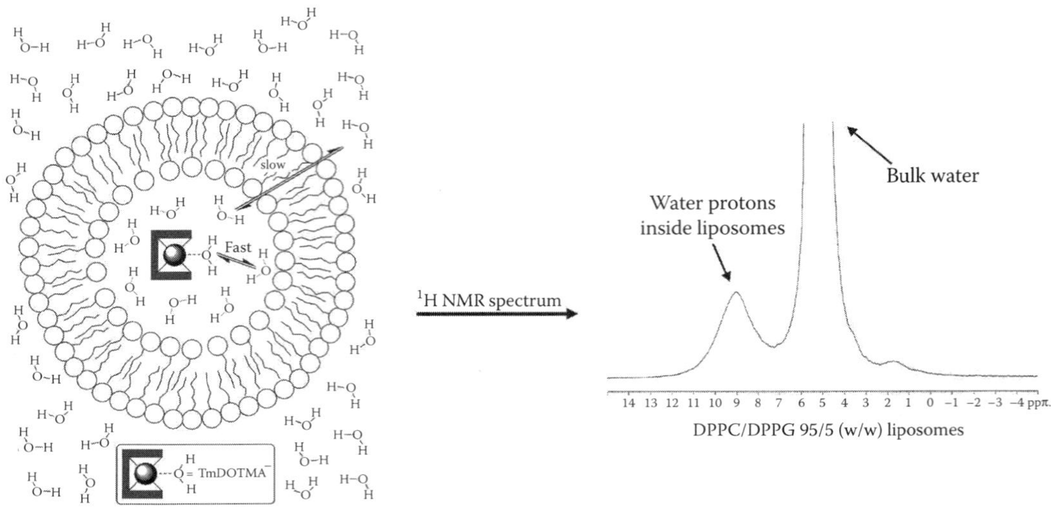

FIGURE 6.12 A schematic of a LIPOCEST particle. $TmDOTMA^-$ has one inner-sphere water molecule in rapid exchange with bulk water, and hence is not a PARACEST agent on its own. However, encapsulation of this complex within the inner core of a liposome containing a fixed amount of trapped water molecules results in a shifted water resonance that can be used for CEST activation. Given an adequate liposomal water exchange rate, this can amplify CEST in proportion to the amount of trapped water inside the liposome. (^1H NMR spectrum compliments of Silvio Aime and Enzo Terreno, University of Torino.)

of a shift reagent entrapped within the liposome gives rise to two signals in the NMR spectrum — one from bulk water and another from water entrapped within the liposome (Figure 6.12). Application of a frequency-selective presaturation pulse at the frequency of the shifted water peak results in CEST to bulk water just as before, but in this case the water exchange rate is determined by the characteristics of the liposomal membrane, not the lanthanide complex, hence the name LIPOCEST.[31] Liposomes made up from phospholipids and cholesterol with ~0.12 *M* $TmDOTMA^-$ entrapped within their inner core show a separate water resonance near 3.1 ppm, which, upon CEST activation, leads to a 5% reduction in bulk water signal intensity with only 90 p*M* LIPOCEST agent (B_1 = 12 μT). Although this equates to 42 μ*M* in $TmDOTMA^-$, it remains a remarkably low sensitivity level for CEST contrast. Such LIPOCEST systems will likely prove sensitive enough to allow molecular imaging of apoptosis or signatures of cancer, clearly an important advance for MRI.

PARACEST type complexes have also been added to the surface of nanoparticles for molecular targeting and imaging purposes. An example of such a particle is illustrated in Figure 6.13. Here, a Eu^{3+} complex of a DOTA-tetra-amide derivative with a phospholipid side chain on one ligand arm was incorporated onto the surface of a perfluorocarbon-filled nanoparticle along with antifibrin antibodies to direct the nanoparticle to fibrin. Each nanoparticle had ~130,000 Eu^{3+} chelates, and each chelate had one slowly exchanging water molecule. A CEST spectrum of the resulting nanoparticle shows a broad but symmetrical peak near 52 ppm, characteristic of the exchanging Eu^{3+}-bound molecules. This result suggests that all surface PARACEST agents are in a similar chemical environment on the surface of the nanoparticle, and each is equally exposed to bulk solvent. *In vitro* plasma clots were treated with targeted nanoparticles formulated with or without the PARACEST chelate, and CEST images were generated by subtraction of images collected with RF presaturation at ±52 ppm. The resulting CEST images clearly show CEST contrast at the clot surface in the presence of PARACEST particles but none in the presence of nanoparticles lacking the PARACEST chelate. This illustrates that PARACEST sensitivity can be amplified considerably, making this technology available to a wide variety of molecular imaging applications.

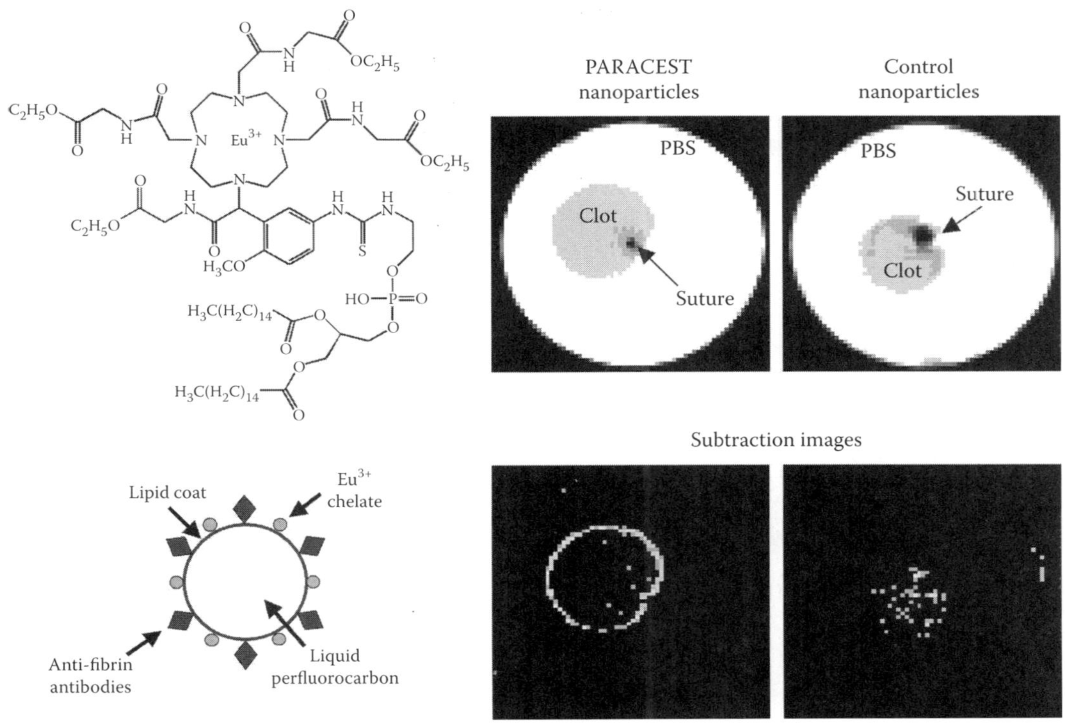

FIGURE 6.13 Molecular imaging of fibrin with antibody-targeted PARACEST nanoparticles. The CEST images were collected at 4.7 T. (Unpublished data compliments of Patrick Winter and Greg Lanza, Washington University, St. Louis.)

6.9 SUMMARY

Paramagnetic CEST agents were discovered only recently, yet there has been great progress in moving these interesting systems toward viable clinical molecular imaging tools. One could argue that progress toward their development has been rapid because so much has been learned about the factors that control water exchange in Gd^{3+}-based systems over the past decade. Most of this prior knowledge has been directly translatable to PARACEST, even though optimization of Gd^{3+}-based agents requires faster water exchange while PARACEST agents require slower-exchanging species. Examples of responsive Gd^{3+} and PARACEST agents both exist; Gd^{3+} systems normally rely on either a change in q (the number of inner-sphere water exchange sites) or molecular rotation, while PARACEST designs so far have been based on changes in water or proton exchange. One could argue that the PARACEST platform is more versatile because it is based on a fundamental property of biological systems — chemical exchange. PARACEST also has the advantage in that virtually any of the lanthanide ions can be used, so that one can optimize any particular ligand system by choosing a paramagnetic lanthanide ion that induces a large vs. small hyperfine shift or exchanges a single inner-sphere water molecule at some particular rate. One could envision using a mixture of PARACEST agents to report multiple biological variables such as pH, tissue redox, or enzyme activity in a single imaging experiment. It is difficult to envision how this could be done with T_1 or T_2 relaxation agents. One final feature of PARACEST that was not emphasized in the above discussion but may prove extremely important is their ability to be switched on and off by the MRI sequence. This may allow pretargeting of silent antibodies or nanoparticles, for example, during a normal clinical MR imaging procedure, followed by a subsequent scan with CEST contrast to identify those molecular markers or physiological responses of interest.

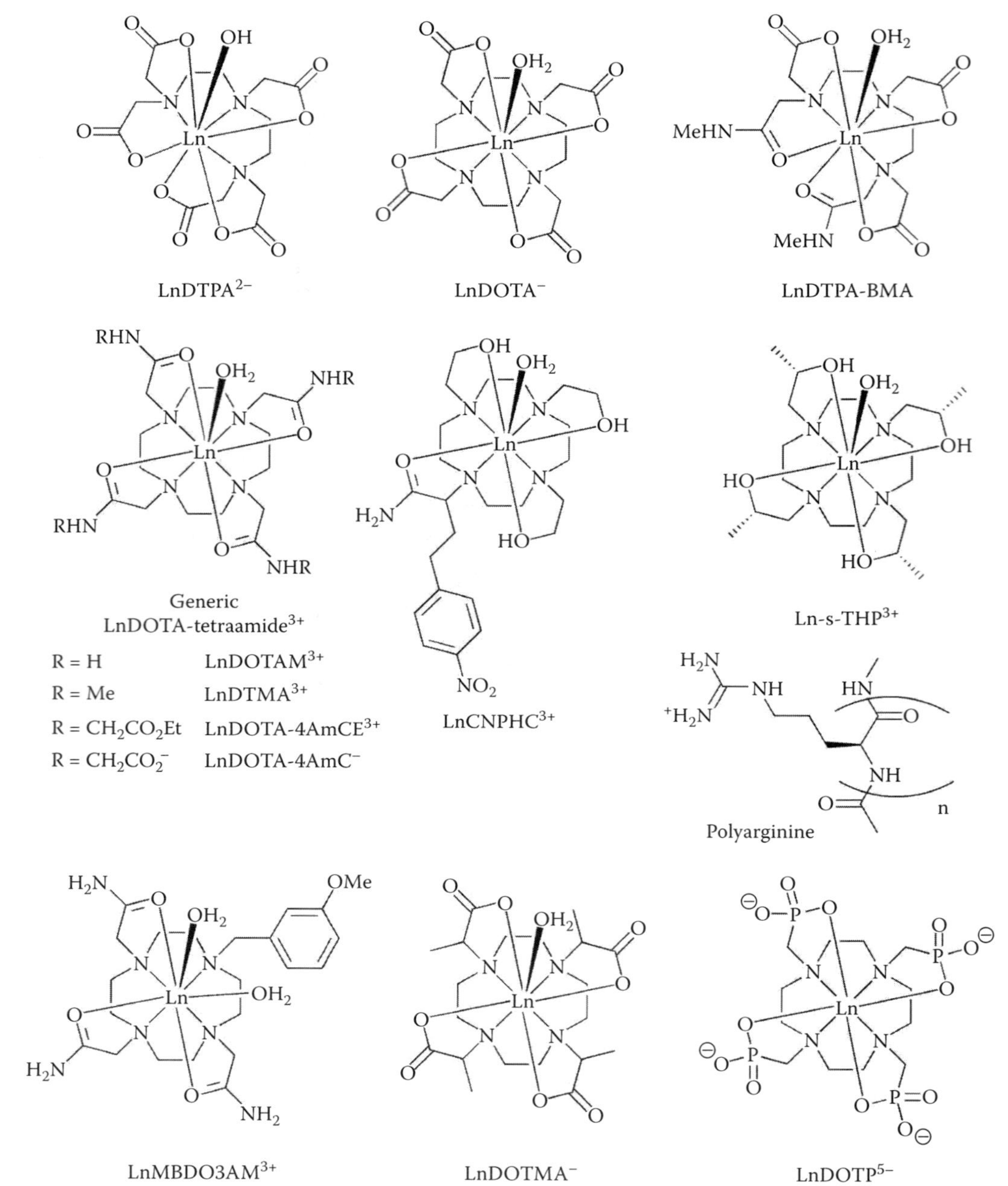

CHART 6.1 The structures of various lanthanide complexes discussed in this chapter.

ACKNOWLEDGMENTS

The authors thank Drs. Jimin Ren and Donald Woessner of UT Southwestern Medical Center for the liver CEST spectra shown in Figure 6.10 and for helping with CEST theory, respectively. The authors also thank the National Institutes of Health (EB-04285, M.W.; CA-115531 and RR-02584, A.D.S.) and the Robert A. Welch Foundation (AT-584, A.D.S.) for financial assistance.

REFERENCES

1. Hinckley, C.C., Lanthanide shift reagents, *Proc. Rare Earth Res. Conf.*, 1, 362–371, 1973.
2. Hinckley, C.C., Applications of lanthanide shift reagents, *Mod. Methods Steroid Anal.*, 1, 265–279, 1973.
3. Peters, J.A., Huskens, J., and Raber, D.J., Lanthanide induced shifts and relaxation rate enhancements, *Prog. Nucl. Magn. Reson. Spectrosc.*, 28, 283–350, 1996.
4. Lauffer, R.B., Paramagnetic metal-complexes as water proton relaxation agents for NMR imaging: theory and design, *Chem. Rev.*, 87, 901–927, 1987.
5. Caravan, P., Ellison, J.J., McMurry, T.J., and Lauffer, R.B., Gadolinium(III) chelates as MRI contrast agents: structure, dynamics, and applications, *Chem. Rev.*, 99, 2293–2352, 1999.
6. Micskei, K., Helm, L., Brucher, E., and Merbach, A.E., Oxygen-17 NMR study of water exchange on gadolinium polyaminopolyacetates [Gd(DTPA)(H2O)]2- and [Gd(DOTA)(H2O)]- related to NMR imaging, *Inorg. Chem.*, 32, 3844–3850, 1993.
7. Merbach, A.E. and Toth, E., *The Chemistry of Contrast Agents in Medical Magnetic Resonance Imaging*, Wiley, New York, 2001.
8. Zhang, S., Merritt, M., Woessner, D.E., Lenkinski, R.E., and Sherry, A.D., PARACEST agents: modulating MRI contrast via water proton exchange, *Acc. Chem. Res.*, 36, 783–790, 2003.
9. Aime, S., Barge, A., Batsanov, A.S., Botta, M., Castelli, D.D., Fedeli, F., Mortillaro, A., Parker, D., and Puschmann, H., Controlling the variation of axial water exchange rates in macrocyclic lanthanide(III) complexes, *Chem. Commun.*, 10, 1120–1121, 2002.
10. Dunand, F.A., Aime, S., and Merbach, A.E., First 17O NMR observation of coordinated water on both isomers of $[Eu(DOTAM)(H_2O)]^{3+}$: a direct access to water exchange and its role in the isomerization, *J. Am. Chem. Soc.*, 122, 1506–1512, 2000.
11. Zhang, S., Wu, K., Biewer, M.C., and Sherry, A.D., 1H and 17O NMR detection of a lanthanide-bound water molecule at ambient temperatures in pure water as solvent, *Inorg. Chem.*, 40, 4284–4290, 2001.
12. Forsen, S. and Hoffman, A., Study of moderately rapid chemical exchange reactions by means of nuclear magnetic resonance double resonance, *J. Chem. Phys.*, 39, 2892–2901, 1963.
13. Ward, K.M., Aletras, A.H., and Balaban, R.S., A new class of contrast agents for MRI based on proton chemical exchange dependent saturation transfer (CEST), *J. Magn. Reson.*, 143, 79–87, 2000.
14. Woods, M., Kovacs, Z., Zhang, S., and Sherry, A.D., Towards the rational design of magnetic resonance imaging contrast agents: isolation of the two coordination isomers of lanthanide DOTA-type complexes, *Angew. Chem. Int. Ed.*, 42, 5889–5892, 2003.
15. Bloch, F., Nuclear induction, *Phys. Rev.*, 70, 460–474, 1946.
16. Zhang, S., Winter, P., Wu, K., and Sherry, A.D., A novel europium(III)-based MRI contrast agent, *J. Am. Chem. Soc.*, 123, 1517–1518, 2001.
17. Zhang, S., Michaudet, L., Burgess, S., and Sherry, A.D., The amide protons of an ytterbium(III) dota tetraamide complex act as efficient antennae for transfer of magnetization to bulk water, *Angew. Chem. Int. Ed.*, 41, 1919–1921, 2002.
18. Aime, S., Carrera, C., Delli Castelli, D., Geninatti Crich, S., and Terreno, E., Tunable imaging of cells labeled with MRI-PARACEST agents, *Angew. Chem. Int. Ed.*, 44, 1813–1815, 2005.
19. Raghunand, N., Howison, C., Sherry, A.D., Zhang, S., and Gillies, R.J., Renal and systemic pH imaging by contrast-enhanced MRI, *Magn. Reson. Med.*, 49, 249–257, 2003.
20. Garcia-Martin, M.L., Martinez, G.V., Raghunand, N., Sherry, A.D., Zhang, S., and Gillies, R.J., High resolution pH_e imaging of rat glioma using pH-dependent relaxivity, *Magn. Reson. Med.*, 55, 309–315, 2006.
21. Aime, S., Barge, A., Castelli, D.D., Fedeli, F., Mortillaro, A., Nielsen, F.U., and Terreno, E., Paramagnetic lanthanide(III) complexes as pH-sensitive chemical exchange saturation transfer (CEST) contrast agents for MRI applications, *Magn. Reson. Med.*, 47, 639–648, 2002.
22. Zhang, S., Kovacs, Z., Burgess, S., Aime, S., Terreno, E., and Sherry, A.D., {DOTA-bis(amide)}lanthanide complexes: NMR evidence for differences in water-molecule exchange rates for coordination isomers, *Chem. Eur. J.*, 7, 288–296, 2001.

23. Dickins, R.S., Aime, S., Batsanov, A.S., Beeby, A., Botta, M., Bruce, J.I., Howard, J.A.K., Love, C.S., Parker, D., Peacock, R.D., and Puschmann, H., Structural, luminescence, and NMR studies of the reversible binding of acetate, lactate, citrate, and selected amino acids to chiral diaqua ytterbium, gadolinium, and europium complexes, *J. Am. Chem. Soc.*, 124, 12697–12705, 2002.
24. Terreno, E., Castelli Daniela, D., Cravotto, G., Milone, L., and Aime, S., Ln(III)-DOTAMGly complexes: a versatile series to assess the determinants of the efficacy of paramagnetic chemical exchange saturation transfer agents for magnetic resonance imaging applications, *Invest. Radiol.*, 39, 235–243, 2004.
25. Zhang, S., Malloy, C., and Sherry, A.D., MRI thermometry based on PARACEST agents, *J. Am. Chem. Soc.*, 127, 17572–17573, 2005.
26. Aime, S., Delli Castelli, D., Fedeli, F., and Terreno, E., A paramagnetic MRI-CEST agent responsive to lactate concentration, *J. Am. Chem. Soc.*, 124, 9364–9365, 2002.
27. Dickins, R.S., Gunnlaugsson, T., Parker, D., and Peacock, R.D., Reversible anion binding in aqueous solution at a cationic heptacoordinate lanthanide center: selective bicarbonate sensing by time-delayed luminescence, *Chem. Commun.*, 16, 1643–1644, 1998.
28. Zhang, S., Trokowski, R., and Sherry, A.D., A paramagnetic CEST agent for imaging glucose by MRI, *J. Am. Chem. Soc.*, 125, 15288–15289, 2003.
28. Goffeney, N., Bulte, J.W.M., Duyn, J., Bryant, L.H., and van Zijl, P.C.M., Sensitive NMR detection of cationic-polymer-based gene delivery systems using saturation transfer *via* proton exchange, *J. Am. Chem. Soc.*, 123, 8628–8629, 2001.
30. Aime, S., Delli Castelli, D., and Terreno, E., Supramolecular adducts between poly-L-arginine and [TmIIIdotp]: a route to sensitivity-enhanced magnetic resonance imaging-chemical exchange saturation transfer agents, *Angew. Chem. Int. Ed.*, 42, 4527–4529, 2003.
31. Aime, S., Castelli, D.D., and Terreno, E., Highly sensitive MRI chemical exchange saturation transfer agents using liposomes, *Angew. Chem. Int. Ed.*, 44, 5513–5515, 2005.
32. Bleaney, B., Dobson, C.M., Levine, B.A., Martin, R.B., Williams, R.J.P., and Xavier, A.V., Origin of lanthanide nuclear magnetic resonance shifts and their uses, *Chem. Commun.*, 791–793, 1972.
33. Zhang, S. and Sherry, A.D., Physical characteristics of lanthanide complexes that act as magnetization transfer (MT) contrast agents, *J. Solid State Chem.*, 171, 38–43, 2003.

7 Genetic Approaches for Modulating MRI Contrast

Eric T. Ahrens, William F. Goins, and Clinton S. Robison

CONTENTS

7.1 Introduction 123
7.2 Current Approaches 124
7.3 MRI Reporters Based on Metalloproteins 126
7.4 Design of Metalloprotein-Based Reporter cDNAs 127
7.5 Gene Therapy and Viral Vectors as Delivery Vehicles 129
7.6 Applications to Cellular Therapeutics 130
7.7 Conclusions 131
Acknowledgments 132
References 132

7.1 INTRODUCTION

Significant strides have been made in understanding the subcellular mechanisms of human disease, ushering in the age of molecular medicine. Molecular medicine encompasses diagnostics and treatments that are focused specifically on the proteomic and genetic levels. Although this field is still in its infancy, scientists have become increasingly adept at integrating clinical observations with the molecular basis of disease and treatment. Much of the remarkable progress in developing detailed insights into the molecular aspects of disease can be attributed to the rapid growth in biotechnology. The design and application of new research tools and methods have helped to bridge the gap between medical practice and basic science. Recombinant DNA methods, reverse genetics, gene amplification, and fluorescence imaging are examples of *in vitro* approaches to understand the human organism. These technologies have made possible the discovery of defective genes and their phenotypes, have improved methods for characterizing tissue pathology at a molecular level, and have aided in the design of transgenic animal models of human disease. The development of molecular imaging tools for use in the intact organism will surely play a vital role in our goal of understanding the genetic and biochemical basis of disease in its native environment.

An understanding of the molecular basis of disease will also open up new opportunities for treatment. For example, recombinant proteins have been produced in bioreactors for the treatment of metabolic disease, immune deficiencies, anemia, and cancer. Moreover, new technologies have been developed for treating genetic defects or modifying the function of tissues in order to achieve a therapeutic outcome. The introduction of engineered genes into somatic cells provides a way to alter the course of pathologic processes. This strategy has been generally referred to as gene therapy. Here again, *in vivo* molecular imaging can play a vital role by indicating where and when therapeutic gene delivery has occurred and how long expression of gene products persists.

The central nervous system (CNS) represents a particularly important target for gene therapy. Because of the blood–brain barrier (BBB) and the fact that neurons are postmitotic, more standard biological therapies such as enzyme administration are often ineffective in the CNS. Moreover, the BBB limits the effectiveness of treating malignancies in the CNS using cellular therapeutics (e.g., immunotherapy). Thus, it is apparent that successful gene delivery to the brain will require vectors that can transduce postmitotic cells.

The development of noninvasive *in vivo* imaging methods that can track vector biodistribution and gene expression is urgently needed in this emerging field. Magnetic resonance imaging (MRI) works superbly in the brain; it yields vivid, high-resolution images, and there is intrinsic contrast among various CNS tissues. Although MRI is the imaging modality of choice for the CNS, it still has limited utility for molecular imaging due to the lack of suitable reporters.

This chapter describes recent developments in molecular imaging using MRI to detect gene activity. We emphasize *in vivo* approaches utilizing MRI reporters that do not require exogenous metal-complexed agents. Instead, we highlight emerging concepts in the development of probeless MRI reporter genes. We review modern recombinant viral vectors used in recent gene therapy and imaging studies, particularly in the CNS. Finally, we describe the potential use of MRI reporters in the emerging field of cellular therapeutics.

7.2 CURRENT APPROACHES

Gene expression can be monitored by incorporating a marker gene that is expressed along with one or more genes of interest. These may be combined as either separate or multicistronic messages or as fusion proteins. Visualization of the marker gene products is most commonly achieved by histological preparations (e.g., by using a β-galactosidase or β-Gal assay), immune detection, or fluorescence microscopy (e.g., using green fluorescent protein (GFP)). These methods often have limited use for noninvasive imaging of tissues, as they typically involve histological preparations, requiring the sacrifice and processing of the subject material. Fluorescent markers are routinely imaged in living cells, but generally it is not possible to image at tissue depths exceeding approximately 500 μm at common excitation and emission wavelengths.[1] This limitation rules out fluorescence imaging for detecting gene activity in the large animal disease models that are frequently required by the Food and Drug Administration (FDA) for preclinical studies supporting Phase I trials. Other methods, such as positron emission tomography (PET)[2,3] and bioluminescent imaging (BLI) reporters (i.e., luciferases),[4–6] are used to detect gene expression *in vivo* and are often sufficient for a variety of molecular imaging applications. However, bioluminescence has tissue depth limitations, particularly in the CNS, where light must pass through the skull. Longitudinal studies using PET require redosing of the subject with ionizing radiation. Both PET and BLI have limited spatial resolution, often on the order of cubic millimeters or larger.

MRI offers advantages over other modalities in its ability to provide three-dimensional images at reasonably high spatial resolution as well as longitudinal information. Several pioneering papers have used metal-complexed MRI agents as probes to detect gene activity *in vivo*;[7,8] these approaches have been reviewed extensively.[9] In one of these approaches, the relaxivity of a novel Gd complex, called EgadMe, is increased in response to local enzymatic activity expressed from a β-Gal reporter gene.[7,10] The mechanism for this increase stems from a steric blocking of water access to the inner coordination sphere of the Gd^{3+} using a galactopyranose unit attached to the tetraazamacrocycle chelate. When the agent is exposed to the enzyme β-Gal, the galactopyranose ring is cleaved from the macrocycle, making an inner coordination site of the Gd^{3+} available for water exchange; this availability effectively turns on the agent. This idea was tested in living frog embryos,[7] and Figure 7.1 shows representative results.

EgadME has generated much interest and provides a glimpse of the future possibilities of molecular imaging using probe-based MRI. However, key challenges remain. A large molecular weight (i.e., EgadMe) complex must be delivered intracellularly. In Figure 7.1, the agent was injected directly into the cells. The delivery of metal complexes to the cells of interest is challenging since they generally have poor penetrance into tissues and organs. Alternative approaches have been

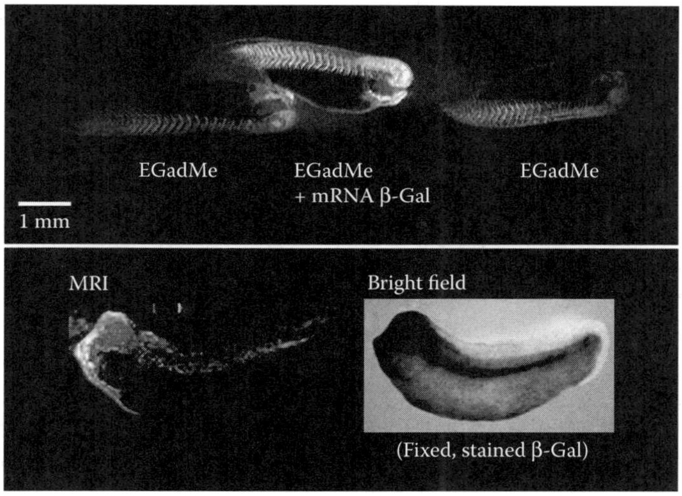

FIGURE 7.1 (please see color insert following page 210) MRI of living *Xenopus laevis* embryos showing β-Gal expression. Top panel: MRI of three embryos microinjected with EgadMe at the two-cell stage. The embryo in the center was also injected with β-Gal mRNA, resulting in higher-intensity regions. Bottom panel: Embryos were prepared in a similar fashion as above, except that plasmids encoding for lacZ were microinjected into one side of the embryo only. Hyperintense regions, which can be interpreted as regions of enzyme cleavage of EgadMe, were found on one side of the embryo only. Using whole-mount cytochemistry (right panel), the enzyme expression pattern following X-Gal staining correlates with the MRI results. These figures are adaptations of the MRI data sets presented in Louie et al.[7]

proposed to detect β-Gal activity. For example, Yu and Mason[11] have recently investigated novel fluorinated galactopyranoside-based substrates *in vitro* for detecting β-Gal enzyme activity by using ^{19}F nuclear magnetic resonance (NMR).

Historically, methods to detect specific endogenous gene expression by MRI employ metal-tagged ligands as probes to target specific cell surface receptors. The use of antibody-conjugated contrast agents has been contemplated for some time,[12,13] although *in vivo* results often lack robust contrast. More recently, the receptor-targeting approach has been revisited using super-loaded Gd-based agents having significantly improved relaxivity.[14] High-relaxivity Gd-chelate-perfluorocarbon nanoparticles have been investigated for imaging cells with sparse surface receptor protein expression.[15,16]

Another receptor-based approach described by Weissleder et al.[8] utilized an engineered glioma cell line to overexpress the transferrin receptor (TfR-1). These cells were grown as tumors in mice. The tumors selectively bound an MRI-potent, microcrystalline iron oxide nanoparticle (MION)-transferrin conjugated probe after intravenous administration. The binding occurred preferentially to cells expressing high levels of the TfR-1 transgene, thus imparting exogenous contrast. Overall, the receptor–ligand approach may have broad applicability for transgene detection if an exogenous or unique cell surface marker can serve as a receptor and the contrast agent probe is able to bind to this receptor with adequate specificity. As with EgadME, delivery of the probes to certain tissues *in vivo*, such as the CNS, may present significant challenges.

Another interesting intracellular approach employs a ternary probe complex comprised of a Gd chelate, a peptide nucleic acid (PNA) sequence, and a transmembrane carrier peptide.[17] This complex was used to highlight tumor cells upregulating c-myc RNA.[17] Once internalized in the cell via the peptide, the PNA annealed with c-myc transcripts, resulting in selective accumulation and hyperintensity in cancer cells having a c-myc positive phenotype.

In vivo ^{31}P magnetic resonance spectroscopy (MRS) has also been used extensively in certain tissue types (e.g., liver and muscle) to detect transgene expression using creatine kinase[18] and arginine kinase[19] as marker genes. Recent *in vivo* studies have employed creatine kinase as a marker gene for quantifying adenovirus-mediated gene transfer into hepatocytes.[20] ^{31}P MRS has been used to detect and quantify inorganic polyphosphate (poly-P) accumulated in yeast cells designed to overexpress the vacuolar genes encoding for the vacuolar transporter chaperone (Vtc) complex and the vacuolar H+-ATPase (V-ATPase).[21] How these yeast-derived transgenes could be applied to *in vivo* imaging of mammalian gene expression remains to be investigated.

Overall, this is a rapidly emerging field, and there has been numerous creative solutions proposed for reporters and probes for molecular MRI of gene expression. With the exception of the MRS approaches, these methodologies have all relied on inorganic metal–ion complexes to detect the gene expression products.

7.3 MRI REPORTERS BASED ON METALLOPROTEINS

Molecular medicine would greatly benefit from an MRI reporter that could monitor *in vivo* gene expression without relying on the delivery of exogenous contrast agents to cells. Ideally, such a reporter would be genetically encoded to produce detectable image contrast and would cause no cytotoxicity or immune response. Existing delivery solutions could be employed (i.e., vectors) that are suitable for tagging cells. We note that early investigations into detecting gene expression using MRI in the absence of exogenous metal–ion probes contemplated the use of tyrosinase as a reporter gene and melanin complexes created from its expression.[22,23]

Many organisms produce intracellular, biomineralized, superparamagnetic nanocrystals incorporating iron. The magnetosome structure of the magnetotactic bacteria is a dramatic example.[24] Assembling magnetosomes within the bacteria creates a magnetic dipole moment in the cell and a biomagnetic compass. However, the precise genetic control of magnetosome formation is complex and involves many loci.[25,26] There are many examples of paramagnetic metalloproteins, particularly ferritins, found in nature. Ferritins are part of a superfamily of iron storage proteins that are found in virtually all animals, plants, fungi, and bacteria.

Recently, several groups have reported the use of metalloproteins from the ferritin (FT) family as MRI reporter molecules.[27–29] The idea is to induce the expression and assembly of the primary FT MRI reporter in cells using molecular-genetic methods. Gene expression can then be followed as the FT protein is made superparamagnetic by sequestering iron that is naturally present. Thus, the reporter protein is synthesized as an intracellular MRI contrast agent, without the requirements of exogenous metal agents. The temporal formation of the contrast agent is governed via the molecular-genetic design of the system. The delivery of the gene reporter can be achieved using existing recombinant viral vectors or transgenic animal technologies, thereby integrating current vehicles with MRI detection. By combining the MRI reporter with other recombinant materials of interest (e.g., a therapeutic gene, vector, or cell), a large number of applications for this technology may be possible.

Ferritin is a natural prototype for developing MRI reporter proteins. It is ubiquitous and structurally conserved throughout the kingdoms.[30] Ferritin is primarily responsible for the storage of intracellular iron in a nontoxic form and also for the physiological homeostasis of iron metabolism. The quaternary structure of eukaryotic ferritins and certain bacterial ferritins consists of polymers assembled from 24 subunits to form a protein shell having an outer diameter of 12 nm and an internal diameter of 9 nm. Within this hollow sphere is where the iron oxyhydroxide core is formed. The two predominant mammalian isoforms of monomeric ferritin, the heavy chain (H) subunit and light chain (L) subunit, preferentially combine to form heteropolymeric shells[31] called apoferritin (i.e., nascent ferritin polymer lacking iron). Previous reports suggest that most mammalian cells preferentially form H-L heteropolymers rather than homopolymers.[31] H and L chains act synergistically, where the L subunits enhance stability of the ferrihydrite cores and the H subunits provide the catalytic center for the ferroxidase activity.[32] Additionally, H-L heteropolymers acquire iron more readily than homopolymers.[33]

Because the Fe^{3+} stored in FT does not participate in the Fenton reaction, it is not expected to be toxic, as demonstrated by many *in vitro* studies.[34,35] In fact, several reports indicate that FT overexpression leads to an increased resistance to chemically induced oxidative stress.[34,36] Interestingly, many bacteria detoxify metals encountered in their environment using ferritin and ferritin-like molecules.[37]

Ferritin is a natural source of intrinsic MRI contrast in various tissues and organs. Due to its crystalline ferrihydrite core, FT exhibits an anomalously high superparamagnetism[38] and a marked effect on solvent NMR relaxation rates.[38–41] For example, in anatomical brain scans ferritin is a major contributor to the T_1- and T_2-weighted contrast variations seen in gray matter.[40] T_2-weighted brain scans of primates reveal large regions of hypointensity as a consequence of the normal aging process, for example, in the *substantia nigra*,[42,43] that have been attributed to the accumulation of holoferritin. Iron physiology of the CNS and its concentration heterogeneity are active areas of research.[44–46]

Ferritin is also a natural by-product of the intracellular degradation of many iron-based MRI contrast agents that are currently used for cellular imaging. The use of superparamagnetic iron oxide (SPIO) or MION agents is of great interest for a wide range of *in vivo* cellular MRI tracking studies.[47–55] When used intracellularly, these agents can degrade within low pH vesicles, resulting in labile iron release. Consequently, cells compensate by upregulating ferritin to detoxify and store this excess Fe.[56]

7.4 DESIGN OF METALLOPROTEIN-BASED REPORTER cDNAS

Ideally, effective MRI gene reporter systems should be highly sensitive and have a rapid response time. The reporter gene "cassette" should be kept to a minimal nucleotide length for incorporation into vectors with insert size constraints. Furthermore, the gene products should be relatively nontoxic to cells even when expressed at high levels. From within the FT superfamily there are an immense number of different ferritin molecules that could be investigated for their efficacy as MRI gene reporters. These possibilities include FTs from various animals, plants, and bacteria. The nucleotide length (nt) of ferritin subunits is modest (~540 nt) compared to many common gene reporters (GFP, ~750 nt; LacZ, ~1020 nt).

Our initial studies involved the coexpression of both the H and L subunits of FT using two separate vectors.[27] To evaluate the efficacy of these subunits as reporters for *in vivo* MRI, we visualized their expression in the mouse brain. We injected adenovirus (AdV) encoding the FT subunits stereotactically into the striatum and then imaged the mice at 5, 11, and 39 days postinoculation.[27] After 5 days, transduced cells displayed robust contrast in T_2-weighted images (left arrows, Figure 7.2a), and the contrast could be seen out to 39 days.[27] No significant contrast could be detected on the contralateral side injected with the AdV-lacZ control vector (right arrows, Figure 7.2a). At 5 days posttransduction, we performed histology in selected brains. The X-Gal staining pattern of the AdV-lacZ inoculation mimicked the AdV-FT-induced MRI contrast (Figure 7.2b). In the same brains used for MRI, immunohistochemistry (IHC) was performed to detect human FT expression in the striatum (Figure 7.2c). The spatial pattern of recombinant FT expression was consistent with the MRI (Figure 7.2a).

We envision that next-generation reporter cDNAs will be engineered to be more versatile and provide optimal contrast within specific cell types. This design process is intimately related to cellular Fe physiology, which is still not fully elucidated and an active area of research. Cellular mechanisms may limit the rate of iron storage into the FT reporter, and contrast could possibly be optimized by including accessory molecules, or co-reporters, that accelerate the amount of iron transported into cells and loaded into FT. For example, the transferrin receptor (TfR-1 or CD71)[57] can potentially be used as a co-reporter, as described by Deans et al.[29] In many cell types, TfR-1 plays an important role in the cellular uptake of iron from the circulation. At the cell surface, TfR-1

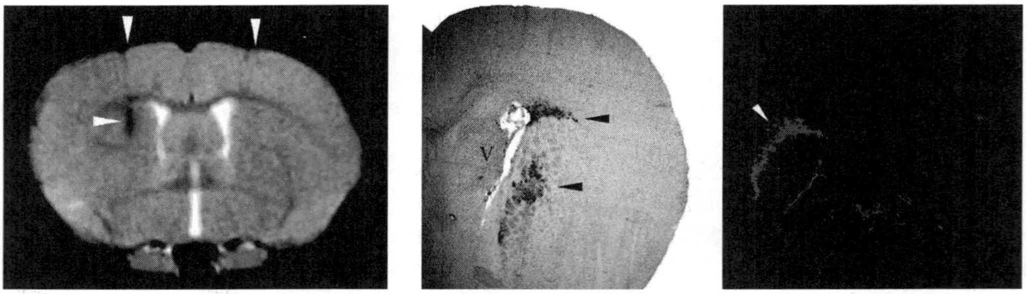

(a) (b) (c)

FIGURE 7.2 (please see color insert) *In vivo* MRI of ferritin-based reporter expression in the mouse brain. AdV carrying transgenes encoding the ferritin L and H subunits was inoculated into the striatum. (a) Injection sites in a T_2-weighted image 5 days postinoculation (arrows; left, MRI reporters; right, AdV-lacZ control). The MRI reporter-transduced cells appear hypointense, and the cells with LacZ show no contrast. (b) X-Gal stained AdV-lacZ transduced pattern at 5 days posttransduction. The staining pattern, mostly in white matter (top arrow) and striatum (bottom arrow), is comparable to the MRI pattern; "V" denotes ventricle. (c) Immunohistochemistry (IHC) results of ferritin transgene expression in mouse brain sections 5 days posttransduction. AdV with MRI reporters were inoculated into the left striatum and show immunoreactivity; AdV-lacZ was injected in the contralateral side and shows negative immunoreactivity over background. *In vivo* images were obtained in anesthetized mice at 11.7 T, at a resolution of 98 μm in-plane with 0.75-mm-thick slices. For additional details, see Genove et al.[27]

molecules capture iron–transferrin complexes from the extracellular milieu, followed by iron uptake into cells via endocytosis.[58] Multicistronic reporter cassettes could include the FT reporter, together with co-reporters, which serves to widen the hypothetical bottleneck in cellular and physiological iron acquisition into tissues. At the same time, because the MRI reporter cassettes will ultimately become incorporated into vector or cellular genomes that may carry therapeutic genes, the design of a practical reporter system will also focus on reducing the amount of genetic material and the number of regulated cistrons necessary for optimal reporter response.

In designing metalloprotein reporters, steps should be taken to minimize toxicity of the reporter transgenes. Overexpression of FT stores Fe in the 3+ state, which is nontoxic. However, the potential toxicity of the FT reporter system may be coupled to how different cell types regulate iron. For example, in the CNS it is known that increased cellular Fe accompanies several neurological disorders, such as Parkinson's and Alzheimer's, but it also occurs during the course of normal aging.[59] It is interesting to note that neuropathologic conditions are also accompanied by a defective FT regulation and an increase in the reactive Fe^{2+} species. Evidence points toward a scenario where Fe is toxic if it is not oxidized and stored in FT. If the FT signaling pathway is in some way defective in the cells, resulting in an insufficient supply of FT, or perhaps aberrant (L or H) subunit stoichiometries, toxicity can follow.[59] A detailed discussion of this interesting area is beyond the scope of this chapter. However, MRI gene reporter studies may be able to contribute to the basic understanding of cellular Fe physiology in the CNS.

Ultimately, reporter toxicity will be related to the ability to regulate levels of transgene products. Sufficient levels are needed to produce meaningful image contrast, but excessive levels are undesirable. Furthermore, one only needs reporter expression for a short time window before the actual MRI scan. Achieving appropriate levels of transgene expression at the appropriate times via molecular biological or pharmacological manipulations is not unique to imaging technologies. The challenge of regulating gene expression is a common theme to the field of gene therapy. Methods to manipulate transgene expression levels are an active area of research,[60] and focus on these techniques will only intensify as the field of molecular medicine further develops.

7.5 GENE THERAPY AND VIRAL VECTORS AS DELIVERY VEHICLES

Incorporating MRI reporters into viral vectors could potentially have a major impact in the field of gene therapy, particularly in the preclinical phase. Non-MRI imaging reporters are commonly incorporated into viral gene therapy studies. Over 500 human clinical trials have been performed to date using a variety of delivery vehicles. Viral vectors represent the ideal gene delivery platform due to their natural ability to infect eukaryotic cells. The permissivity of a tissue or cell type not only determines the mechanism of vector delivery, but also dictates a suitable vector for a particular application. Numerous viral vector systems have been employed in gene therapy clinical trials.[61,62] Retroviruses and AdV account for the majority (35 and 28%, respectively) of the approved trials, while adeno-associated virus (AAV) and herpes simplex virus (HSV) (3 and 1%, respectively) have only recently been used.[63] Every vector system possesses its own pros and cons with regard to (1) vector genome stability, (2) virus-associated toxicity, (3) virus/transgene immunogenicity, (4) vector payload capacity, (5) ease and scalability of production, (6) safety in humans, and (7) transduction efficiency. Although many viral vector systems have been employed for gene transfer/therapy studies, only those vectors that have been extensively used to express imaging reporter genes are described in this chapter.

The earliest gene transfer studies were performed with retroviral vectors. Although these vectors achieve stable long-term transgene expression, the vector payload is limited to one gene or potentially two smaller genes. These vectors have proven to be nonimmunogenic and display high efficacy; however, their utility has been limited by the fact that they are unable to transduce nondividing cells[64] (e.g., neurons). *Ex vivo* approaches where cells are first transduced *in vitro* and then reintroduced back into the host represent an active niche for retroviruses. For example, retroviral vectors expressing fluorescent proteins such as eGFP have been used to transduce stem cell populations prior to transplantation for imaging follow-up.[65,66] Other studies have employed vectors expressing an eGFP-HSV thymidine kinase (tk) fusion protein for micro-PET imaging of vector-transduced tumor cells[67,68] or stem cells.[69] Vectors expressing luciferase (luc) have also been used to image retroviral transduced tumor cells *in vivo*.[67] Retroviral vector-mediated expression of the sodium iodide symporter (NIS) enabled imaging of transduced tumor cells in mouse glioma tumor models.[70,71] Finally, retroviral vectors were employed to express the norepinephrine transporter as a reporter gene.[72]

Lentiviruses (LVs) possess all of the features of standard retroviruses except they can transduce both dividing and nondividing cells.[73,74] However, they still possess the ability to integrate into the host cell genome,[75,76] suggesting that safety issues may limit their utility in clinical trials. LV vectors expressing eGFP, RFP, or luc imaging reporters have been employed in stem cell gene marking studies[77] and in *in vivo* transduction of the eye.[78,79] LV vectors expressing an HSVtk-eGFP-luc chimera were employed in micro-PET and BLI to monitor vector-transduced lymphocyte migration.[80]

AAV is a small DNA virus that is not associated with any disease in humans. It integrates into the host chromosome in a site-specific manner as part of its natural infectious cycle, suggesting that AAV will be amenable to long-term therapeutic applications. AAV vectors are relatively nontoxic and thus have been proven efficacious in a variety of preclinical applications.[81,82] However, the small genome size has limited the vector payload to single genes of 4 kb in size. Low titers, the difficulty of production, and the inability to readminister the vector due to the host immune response[83,84] complicate the use of this vector system. *In vivo* fluorescence imaging of eGFP-expressing AAV vectors showed expression for many weeks.[85,86] Expression of a luc reporter using an AAV vector resulted in signals within the target tissues, but was confounded by high-background signal in peritoneum and liver.[87,88] PET or SPECT imaging of AAV vectors expressing either HSVtk[89] or the dopamine transporter (DAT)[90] yielded images of higher resolution, yet it was difficult to identify the regions of vector transduction.

HSV has numerous features that are attractive for gene delivery to the nervous system. The virus possesses an extremely broad host range[91,92] and is capable of efficiently infecting most cell types, tissues, and animals. HSV does not integrate into the host genome, yet persists as a circular nonintegrated episome[93] within neurons. Although early replication-defective HSV vectors displayed residual toxicity,[94,95] the combined elimination of multiple genes has reduced the cytotoxicity[96,97] and enabled persistence in other cell types besides neurons. Replication-defective HSV vectors in which 40 kb of viral sequences has been removed allow for the incorporation of multiple[98] or very large[99] therapeutic genes. HSV vectors can readily be prepared to high titer and purity.[100,101] HSV possesses a unique natural promoter system that displays long-term transgene expression in sensory neurons of the PNS[102–104] and striatal neurons of the CNS.[105] HSV vectors expressing either eGFP or RFP reporters have been employed in tumors[106,107] and normal rodent brain[108] to evaluate vector spread. Vectors expressing luc have been used for live imaging studies of virus distribution in animals,[109–111] and in most instances, very low doses of vector (10^3 to 10^5 particles) were required to produce a distinct signal above background. Since HSV encodes the tk gene product, it already possesses a native imaging reporter that has been employed in micro-PET studies.[112–114] Finally, HSV vectors expressing TfR-1 have been employed in preliminary MRI studies.[115]

AdV possesses many attractive features for short-term applications. First-generation AdV vectors deleted for the E1 region were among the earliest DNA viruses developed and implemented for gene therapeutic applications, along with retroviruses. These vectors have been shown to persist as episomal molecules and mediate high-level therapeutic gene expression in a variety of dividing and nondividing cell types.[116,117] However, the level of transgene expression from the vector falls at least tenfold over the first several weeks following inoculation, suggesting that these vectors are best suited to short-term therapeutic applications. AdV vectors can accept moderate-sized transgene inserts (~5 to 10 kb) and thus have a greater payload capacity than AAV, retrovirus, and LV vectors. AdV can be constructed with relative ease using commercially available kits that can yield batches of high purity and titer. A major disadvantage of the first-generation AdV vectors is that they continue to express viral gene products in addition to the transgene, which can result in a significant immune response against the vector;[118–120] this may account in part for the rapid fall-off in transgene expression.[121] Later-generation "gutless" AdV vectors display a reduced inflammatory reaction and increased transgene capacity;[122] however, they are difficult to grow and purify to high enough titers for *in vivo* studies.

AdV expressing eGFP or RFP has been employed in imaging studies of vector injection into tumor xenografts;[123,124] however, since signal detection is limited by tissue depth, this technique is only applicable to flank tumor xenograft models. Although BLI of luc reporter gene expression shows better penetrance than eGFP,[125–128] nonspecific signal in tissues such as liver and poor resolution have complicated the use of this imaging system. HSVtk has been used as an imaging reporter in the background of AdV vectors in micro-PET.[129–132] There has also been extensive use of the NIS reporter system in AdV vectors for imaging.[133–135] Relatively high vector doses of 10^9 to 10^{12} were required to obtain a detectable signal above background. Finally, PET imaging with AdV expressing the dopamine D2 receptor,[136,137] SSTR2,[138,139] and NET[140] has shown promise, yet some issues concerning resolution and high background in tissues in which these receptors are naturally expressed need to be resolved.

7.6 APPLICATIONS TO CELLULAR THERAPEUTICS

Therapeutic cells have enormous potential for treating disease because of their ability to carry out complex functions and their responsiveness to surrounding tissues. In the simplest mode, cells can be isolated, cultured *ex vivo* to achieve therapeutic quantities. Cells are then implanted to produce soluble factors or perform complex tasks such as reconstitution of tissues, organs, or immune responses.

Cellular therapeutics are currently applied to various diseases, including cancer, cardiovascular, neurological, hematological, and immunological disorders.[141–147] A wide variety of cell types are being examined for future therapeutic uses, including blood cells, myoblasts, bone marrow cells, cardiomyocytes, chondrocytes, dendritic cells, fibroblasts, hepatocytes, pancreatic islets, keratinocytes, and stem cells.

Visualizing therapeutic cells noninvasively can be difficult, and approaches are needed that hasten testing of these treatments. In the future, imaging will be closely aligned with cellular therapeutics. Imaging will be used to help calibrate dosage and delivery efficacy and the biodistribution of products when introduced into the patient. The success of cellular therapeutics, such as immunotherapy and stem cell transfers, can be highly variable among patients, and imaging can provide a means for real-time monitoring of therapeutic progression.

Labeling cells for MRI can be an additional *ex vivo* treatment to the cells prior to implantation into the patient. *Ex vivo* labeling with SPIO nanoparticles has been shown to be effective in visualizing immune cells and other cell types. However, there are challenges with using SPIO particles as intracellular agents. For example, the mean intracellular agent concentration is diluted by cell mitosis, tending to reduce the cell's contrast over time. Also, the SPIO particles can degrade over time inside cells. Furthermore, when SPIO-labeled cells die, the SPIO nanoparticles may be taken up by resident phagocytic cells, resulting in nonspecific macrophage labeling. Alternatives to SPIO labeling agents, such as fluorocarbon-based nanoparticles,[148] do not undergo cell degradation; however, dilution due to cell division and nonspecific labeling of resident phagocytes may occur.

There are potential advantages for using nucleic acid-based reporters for *ex vivo* labeling of therapeutic cells. The challenges associated with delivering a high molecular weight complex intracellularly are alleviated, as only a transgene needs to be delivered to the cells *ex vivo* via efficient vectors, as described above. The labeled cell and progeny can be designed to carry the reporter transgenes for extended periods, and thus the reporter can be perpetual. Loss of cell contrast due to mitosis or lysosomal degradation is mitigated. Furthermore, if the labeled cell dies, the reporter metalloprotein is degraded rapidly by proteases, and the contrast agent will not be transferred to nonspecific phagocytes.

However, there are also potential pitfalls associated with metalloprotein-based MRI reporters. In addition to the concerns mentioned above, such as potential toxicity, it is important that the cell labeling scheme does not cause changes in phenotype that would significantly alter its therapeutic value. For example, many immunotherapeutic strategies rely on specific immune cell subsets (e.g., mature vs. immature); thus, it is essential that the label does not significantly alter the cell's phenotype. Similarly, stem cell therapies that rely on the cell's pluripotent or multipotent potential should not have this ability compromised by the reporter.

7.7 CONCLUSIONS

MRI offers advantages over other modalities in its ability to provide longitudinal three-dimensional images at high resolution. Gene-sensing MRI using nucleic acid-based reporters is still in a nascent state; however, the need for a versatile platform is overwhelmingly clear. Due to the inherent lack of sensitivity of MRI, a practical gene-sensing approach may require new generations of engineered reporters providing rapid and robust contrast. Metalloproteins, particularly within the family of ferritins, offer intriguing avenues for development. Fundamentally, the choice of imaging technique used is largely dependent upon the currently available technology, the experimental design, and the primary questions the researchers ask. Multimodality gene expression reporter modules will be a driving factor in the development and validation of new generations of molecular and cellular therapeutics.

ACKNOWLEDGMENTS

We acknowledge support from the National Institutes of Health grants R01-EB005740, P01-HD047675, P41-EB001977, and P50-ES012359.

REFERENCES

1. Lichtman, J.W. and Fraser, S.E., The neuronal naturalist: watching neurons in their native habitat, *Nat. Neurosci.*, 4, 1215, 2001.
2. Gambhir, S.S. et al., Imaging adenoviral-directed reporter gene expression in living animals with positron emission tomography, *Proc. Natl. Acad. Sci. U.S.A.*, 96, 2333, 1999.
3. Min, J.J. and Gambhir, S.S., Gene therapy progress and prospects: noninvasive imaging of gene therapy in living subjects, *Gene Ther.*, 11, 115, 2004.
4. Contag, C.H. and Bachmann, M.H., Advances in *in vivo* bioluminescence imaging of gene expression, *Annu. Rev. Biomed. Eng.*, 4, 235, 2002.
5. Costa, G.L. et al., Adoptive immunotherapy of experimental autoimmune encephalomyelitis via T cell delivery of the IL-12 p40 subunit, *J. Immunol.*, 167, 2379, 2001.
6. Hardy, J. et al., Extracellular replication of Listeria *monocytogenes* in the murine gall bladder, *Science*, 303, 851, 2004.
7. Louie, A.Y. et al., *In vivo* visualization of gene expression using magnetic resonance imaging, *Nat. Biotechnol.*, 18, 321, 2000.
8. Weissleder, R. et al., *In vivo* magnetic resonance imaging of transgene expression, *Nat. Med.*, 6, 351, 2000.
9. Basilion, J.P., Yeon, S., and Botnar, R., Magnetic resonance imaging: utility as a molecular imaging modality, in *In Vivo Cellular and Molecular Imaging*, Ahrens, E.T., Ed., Elsevier, San Diego, 2005, chap. 1.
10. Moats, R.A., Fraser, S.E., and Meade, T.J., A "smart" magnetic resonance imaging agent that reports on specific enzymatic activity, *Angew. Chem. Int. Ed.*, 36, 726, 1997.
11. Yu, J.X. and Mason, R.P., Synthesis and characterization of novel lacZ gene reporter molecules: detection of beta-galactosidase activity by F-19 nuclear magnetic resonance of polyglycosylated fluorinated vitamin B-6, *J. Med. Chem.*, 49, 1991, 2006.
12. Shreve, P. and Aisen, A.M., Monoclonal antibodies labeled with polymeric paramagnetic ion chelates, *Magn. Reson. Med.*, 3, 336, 1986.
13. Unger, E.C. et al., Magnetic resonance imaging using gadolinium labeled monoclonal antibody, *Invest. Radiol.*, 20, 693, 1985.
14. Wickline, S.A. and Lanza, G.M., Molecular imaging, targeted therapeutics, and nanoscience, *J. Cell. Biochem. Suppl.*, 39, 90, 2002.
15. Morawski, A.M. et al., Targeted nanoparticles for quantitative imaging of sparse molecular epitopes with MRI, *Magn. Reson. Med.*, 51, 480, 2004.
16. Morawski, A.M. et al., Quantitative "magnetic resonance immunohistochemistry" with ligand-targeted F-19 nanoparticles, *Magn. Reson. Med.*, 52, 1255, 2004.
17. Heckl, S. et al., Intracellular visualization of prostate cancer using magnetic resonance imaging, *Cancer Res.*, 63, 4766, 2003.
18. Koretsky, A.P. et al., NMR detection of creatine-kinase expressed in liver of transgenic mice: determination of free ADP levels, *Proc. Natl. Acad. Sci. U.S.A.*, 87, 3112, 1990.
19. Walter, G., Barton, E.R., and Sweeney, H.L., Noninvasive measurement of gene expression in skeletal muscle, *Proc. Natl. Acad. Sci. U.S.A.*, 97, 5151, 2000.
20. Li, Z. et al., Creatine kinase, a magnetic resonance-detectable marker gene for quantification of liver-directed gene transfer, *Hum. Gene Ther.*, 16, 1429, 2005.
21. Ki, S. et al., A novel magnetic resonance-based method to measure gene expression in living cells, *Nucl. Acids Res.*, 34, e51, 2006.
22. Enochs, W.S. et al., Paramagnetic metal scavenging by melanin: MR imaging, *Radiology*, 204, 417, 1997.

23. Weissleder, R. et al., MR imaging and scintigraphy of gene expression through melanin induction, *Radiology*, 204, 425, 1997.
24. Stolz, J.F., Chang, S.B.R., and Kirschvink, J.L., Magnetotactic bacteria and single-domain magnetite in hemipelagic sediments, *Nature*, 321, 849, 1986.
25. Grunberg, K. et al., A large gene cluster encoding several magnetosome proteins is conserved in different species of magnetotactic bacteria, *Appl. Environ. Microbiol.*, 67, 4573, 2001.
26. Matsunaga, T. et al., Cloning and characterization of a gene, mpsA, encoding a protein associated with intracellular magnetic particles from *Magnetospirillum* sp. strain AMB-1, *Biochem. Biophys. Res. Commun.*, 268, 932, 2000.
27. Genove, G. et al., A new transgene reporter for *in vivo* magnetic resonance imaging, *Nat. Med.*, 11, 450, 2005.
28. Cohen, B. et al., Ferritin as an endogenous MRI reporter for noninvasive imaging of gene expression in C6 glioma tumors, *Neoplasia*, 7, 109, 2005.
29. Deans, A.E. et al., Cellular MRI contrast via coexpression of transferrin receptor and ferritin, *Magn. Reson. Med.*, 56, 51, 2006.
30. Theil, E.C., Ferritin: structure, gene regulation, and cellular function in animals, plants, and microorganisms, *Annu. Rev. Biochem.*, 56, 289, 1987.
31. Santambrogio, P. et al., Production and characterization of recombinant heteropolymers of human ferritin H-chain and L-chain, *J. Biol. Chem.*, 268, 12744, 1993.
32. Levi, S. et al., Evidence that H-chains and L-chains have cooperative roles in the iron-uptake mechanism of human ferritin, *Biochem. J.*, 288, 591, 1992.
33. Kim, H.J. et al., Expression of heteropolymeric ferritin improves iron storage in *Saccharomyces cerevisiae*, *Appl. Environ. Microbiol.*, 69, 1999, 2003.
34. Epsztejn, S. et al., H-ferritin subunit overexpression in erythroid cells reduces the oxidative stress response and induces multidrug resistance properties, *Blood*, 94, 3593, 1999.
35. Corsi, B. et al., Transient overexpression of human H- and L-ferritin chains in COS cells, *Biochem. J.*, 330, 315, 1998.
36. Cozzi, A. et al., Overexpression of wild type and mutated human ferritin H-chain in HeLa cells: *in vivo* role of ferritin ferroxidase activity, *J. Biol. Chem.*, 275, 25122, 2000.
37. Andrews, S.C., Robinson, A.K., and Rodriguez-Quinones, F., Bacterial iron homeostasis, *FEMS Microbiol. Rev.*, 27, 215, 2003.
38. Bulte, J.W. et al., Magnetoferritin: characterization of a novel superparamagnetic MR contrast agent, *J. Magn. Reson. Imaging*, 4, 497, 1994.
39. Gottesfeld, Z. and Neeman, M., Ferritin effect on the transverse relaxation of water: NMR microscopy at 9.4 T, *Magn. Reson. Med.*, 35, 514, 1996.
40. Vymazal, J. et al., The relation between brain iron and NMR relaxation times: an *in vitro* study, *Magn. Reson. Med.*, 35, 56, 1996.
41. Vymazal, J. et al., Iron uptake by ferritin: NMR relaxometry studies at low iron loads, *J. Inorg. Biochem.*, 71, 153, 1998.
42. Dhenain, M. et al., T2-weighted MRI studies of mouse lemurs: a primate model of brain aging, *Neurobiol. Aging*, 18, 517, 1997.
43. Dhenain, M. et al., Age dependence of the T2-weighted MRI signal in brain structures of a prosimian primate (*Microcebus murinus*), *Neurosci. Lett.*, 237, 85, 1997.
44. Connor, J.R. et al., Iron and iron management proteins in neurobiology, *Pediatr. Neurol.*, 25, 118, 2001.
45. Rouault, T.A., Systemic iron metabolism: a review and implications for brain iron metabolism, *Pediatr. Neurol.*, 25, 130, 2001.
46. Connor, J.R., What we have learned about the role of iron in neurobiology from neurological diseases, *J. Neurochem.*, 81, 80, 2002.
47. Yeh, T.C. et al., *In-vivo* dynamic MRI tracking of rat T-cells labeled with superparamagnetic iron-oxide particles, *Magn. Reson. Med.*, 33, 200, 1995.
48. Schulze, E. et al., Cellular uptake and trafficking of a prototypical magnetic iron oxide label *in vitro*, *Invest. Radiol.*, 30, 604, 1995.
49. Moore, A., Weissleder, R., and Bogdanov, A., Uptake of dextran-coated monocrystalline iron oxides in tumor cells and macrophages, *J. Magn. Reson. Imaging*, 7, 1140, 1997.

50. Weissleder, R. et al., Magnetically labeled cells can be detected by MR imaging, *J. Magn. Reson. Imaging*, 7, 258, 1997.
51. Schoepf, U. et al., Intracellular magnetic labeling of lymphocytes for *in vivo* trafficking studies, *Biotechniques*, 24, 642, 1998.
52. Josephson, L. et al., High-efficiency intracellular magnetic labeling with novel superparamagnetic-tat peptide conjugates, *Bioconj. Chem.*, 10, 186, 1999.
53. Ahrens, E.T. et al., Receptor-mediated endocytosis of iron-oxide particles provides efficient labeling of dendritic cells for *in vivo* MR imaging, *Magn. Reson. Med.*, 49, 1006, 2003.
54. Kircher, M.F. et al., *In vivo* high resolution three-dimensional imaging of antigen-specific cytotoxic T-lymphocyte trafficking to tumors, *Cancer Res.*, 63, 6838, 2003.
55. Bulte, J.W.M. et al., Preparation of magnetically labeled cells for cell tracking by magnetic resonance imaging, *Methods Enzymol.*, 386, 275, 2004.
56. Okon, E. et al., Biodegradation of magnetite dextran nanoparticles in the rat: a histologic and biophysical study, *Lab. Invest.*, 71, 895, 1994.
57. Welch, S., *Transferrin: The Iron Carrier*, CRC Press, Boca Raton, FL, 1992.
58. Klausner, R.D. et al., Receptor-mediated endocytosis of transferrin in K562 cells, *J. Biol. Chem.*, 258, 4715, 1983.
59. Moos, T. and Morgan, E.H., The metabolism of neuronal iron and its pathogenic role in neurological disease, in *Redox-Active Metals in Neurological Disorders*, Levine, S.M., Conor, J.R., and Schipper, H.M., Eds., New York Academy of Sciences, New York, 2004, chap. 1.
60. Fussenegger, M., The impact of mammalian gene regulation concepts on functional genomic research, metabolic engineering, and advanced gene therapies, *Biotechnol. Prog.*, 17, 1, 2001.
61. Davidson, B.L. and Breakefield, X.O., Viral vectors for gene delivery to the nervous system, *Nat. Rev. Neurosci.*, 4, 353, 2003.
62. Kay, M.A., Glorioso, J.C., and Naldini, L., Viral vectors for gene therapy: the art of turning infectious agents into vehicles of therapeutics, *Nat. Med.*, 7, 33, 2001.
63. Lotze, M.T. and Kost, T.A., Viruses as gene delivery vectors: application to gene function, target validation, and assay development, *Cancer Gene Ther.*, 9, 692, 2002.
64. Miller, D.G., Adam, M.A., and Miller, A.D., Gene transfer by retrovirus vectors occurs only in cells that are actively replicating at the time of infection, *Mol. Cell. Biol.*, 10, 4239, 1990.
65. Hirschmann, F. et al., Vital marking of articular chondrocytes by retroviral infection using green fluorescence protein, *Osteoarthr. Cartilage*, 10, 109, 2002.
66. Niyibizi, C. et al., The fate of mesenchymal stem cells transplanted into immunocompetent neonatal mice: implications for skeletal gene therapy via stem cells, *Mol. Ther.*, 9, 955, 2004.
67. Ponomarev, V. et al., A novel triple-modality reporter gene for whole-body fluorescent, bioluminescent, and nuclear noninvasive imaging, *Eur. J. Nucl. Med. Mol. Imaging*, 31, 740, 2004.
68. Serganova, I. et al., Molecular imaging of temporal dynamics and spatial heterogeneity of hypoxia-inducible factor-1 signal transduction activity in tumors in living mice, *Cancer Res.*, 64, 6101, 2004.
69. Hung, S.C. et al., Mesenchymal stem cell targeting of microscopic tumors and tumor stroma development monitored by noninvasive *in vivo* positron emission tomography imaging, *Clin. Cancer Res.*, 11, 7749, 2005.
70. Cho, J.Y., A transporter gene (sodium iodide symporter) for dual purposes in gene therapy: imaging and therapy, *Curr. Gene Ther.*, 2, 393, 2002.
71. Cho, J.Y. et al., *In vivo* imaging and radioiodine therapy following sodium iodide symporter gene transfer in animal model of intracerebral gliomas, *Gene Ther.*, 9, 1139, 2002.
72. Anton, M. et al., Use of the norepinephrine transporter as a reporter gene for non-invasive imaging of genetically modified cells, *J. Gene Med.*, 6, 119, 2004.
73. Blomer, U. et al., Applications of gene therapy to the CNS, *Hum. Mol. Genet.*, 5, 1397, 1996.
74. Naldini, L. et al., Efficient transfer, integration, and sustained long-term expression of the transgene in adult rat brains injected with a lentiviral vector, *Proc. Natl. Acad. Sci. U.S.A.*, 93, 11382, 1996.
75. Mitchell, R.S. et al., Retroviral DNA integration: ASLV, HIV, and MLV show distinct target site preferences, *PLoS Biol.*, 2, E234, 2004.
76. Schroder, A.R. et al., HIV-1 integration in the human genome favors active genes and local hotspots, *Cell*, 110, 521, 2002.

77. Okada, S. et al., *In vivo* imaging of engrafted neural stem cells: its application in evaluating the optimal timing of transplantation for spinal cord injury, *FASEB J.*, 19, 1839, 2005.
78. Balaggan, K.S. et al., Stable and efficient intraocular gene transfer using pseudotyped EIAV lentiviral vectors, *J. Gene Med.*, 8, 275, 2006.
79. Loewen, N. et al., Long-term, targeted genetic modification of the aqueous humor outflow tract coupled with noninvasive imaging of gene expression *in vivo*, *Invest. Ophthalmol. Vis. Sci.*, 45, 3091, 2004.
80. Kim, Y.J. et al., Multimodality imaging of lymphocytic migration using lentiviral-based transduction of a tri-fusion reporter gene, *Mol. Imaging Biol.*, 6, 331, 2004.
81. Grieger, J.C. and Samulski, R.J., Adeno-associated virus as a gene therapy vector: vector development, production and clinical applications, *Adv. Biochem. Eng. Biotechnol.*, 99, 119, 2005.
82. Lu, Y., Recombinant adeno-associated virus as delivery vector for gene therapy: a review, *Stem Cells Dev.*, 13, 133, 2004.
83. Peden, C.S. et al., Circulating anti-wild-type adeno-associated virus type 2 (AAV2) antibodies inhibit recombinant AAV2 (rAAV2)-mediated, but not rAAV5-mediated, gene transfer in the brain, *J. Virol.*, 78, 6344, 2004.
84. Zaiss, A.K. and Muruve, D.A., Immune responses to adeno-associated virus vectors, *Curr. Gene Ther.*, 5, 323, 2005.
85. Bennett, J. et al., Real-time, noninvasive *in vivo* assessment of adeno-associated virus-mediated retinal transduction, *Invest. Ophthalmol. Vis. Sci.*, 38, 2857, 1997.
86. Enger, P.O. et al., Adeno-associated viral vectors penetrate human solid tumor tissue *in vivo* more effectively than adenoviral vectors, *Hum. Gene Ther.*, 13, 1115, 2002.
87. Lipshutz, G.S. et al., *In utero* delivery of adeno-associated viral vectors: intraperitoneal gene transfer produces long-term expression, *Mol. Ther.*, 3, 284, 2001.
88. Ogura, T. et al., Utility of intraperitoneal administration as a route of AAV serotype 5 vector-mediated neonatal gene transfer, *J. Gene Med.*, 8, 990, 2006.
89. Hajitou, A. et al., A hybrid vector for ligand-directed tumor targeting and molecular imaging, *Cell*, 125, 385, 2006.
90. Auricchio, A. et al., *In vivo* quantitative noninvasive imaging of gene transfer by single-photon emission computerized tomography, *Hum. Gene Ther.*, 14, 255, 2003.
91. Glorioso, J.C. et al., Gene transfer to brain using herpes simplex virus vectors, *Ann. Neurol.*, 35 (Suppl.), S28, 1994.
92. Wolfe, D. et al., Engineering herpes simplex virus vectors for CNS applications, *Exp. Neurol.*, 159, 34, 1999.
93. Mellerick, D.M. and Fraser, N., Physical state of the latent herpes simplex virus genome in a mouse model system: evidence suggesting an episomal state, *Virology*, 158, 265, 1987.
94. Samaniego, L., Webb, A., and DeLuca, N., Functional interaction between herpes simplex virus immediate-early proteins during infection: gene expression as a consequence of ICP27 and different domains of ICP4, *J. Virol.*, 69, 5705, 1995.
95. Wu, N. et al., Prolonged gene expression and cell survival after infection by a herpes simplex virus mutant defective in the immediate-early genes encoding ICP4, ICP27, and ICP22, *J. Virol.*, 70, 6358, 1996.
96. Krisky, D.M. et al., Deletion of multiple immediate-early genes from herpes simplex virus reduces cytotoxicity and permits long-term gene expression in neurons, *Gene Ther.*, 5, 1593, 1998.
97. Samaniego, L.A., Neiderhiser, L., and DeLuca, N.A., Persistence and expression of the herpes simplex virus genome in the absence of immediate-early proteins, *J. Virol.*, 72, 3307, 1998.
98. Krisky, D.M. et al., Development of herpes simplex virus replication-defective multigene vectors for combination gene therapy applications, *Gene Ther.*, 5, 1517, 1998.
99. Akkaraju, G.R. et al., Herpes simplex virus vector-mediated dystrophin gene transfer and expression in MDX mouse skeletal muscle, *J. Gene Med.*, 1, 280, 1999.
100. Jiang, C. et al., Immobilized cobalt affinity chromatography provides a novel, efficient method for herpes simplex virus type 1 gene vector purification, *J. Virol.*, 78, 8994, 2004.
101. Ozuer, A. et al., Evaluation of infection parameters in the production of replication-defective HSV-1 viral vectors, *Biotechnol. Prog.*, 18, 476, 2002.
102. Goins, W.F. et al., Herpes simplex virus type 1 vector-mediated expression of nerve growth factor protects dorsal root ganglion neurons from peroxide toxicity, *J. Virol.*, 73, 519, 1999.

103. Goins, W.F. et al., A novel latency-active promoter is contained within the herpes simplex virus type 1 UL flanking repeats, *J. Virol.*, 68, 2239, 1994.
104. Goins, W.F. et al., Herpes simplex virus mediated nerve growth factor expression in bladder and afferent neurons: potential treatment for diabetic bladder dysfunction, *J. Urol.*, 165, 1748, 2001.
105. Puskovic, V. et al., Prolonged biologically active transgene expression driven by HSV LAP2 in brain *in vivo*, *Mol. Ther.*, 10, 67, 2004.
106. Adusumilli, P.S. et al., Imaging and therapy of malignant pleural mesothelioma using replication-competent herpes simplex viruses, *J. Genet. Med.*, 8, 603, 2006.
107. Rueger, M.A. et al., Variability in infectivity of primary cell cultures of human brain tumors with HSV-1 amplicon vectors, *Gene Ther.*, 12, 588, 2005.
108. Wang, S., Petravicz, J., and Breakefield, X.O., Single HSV-amplicon vector mediates drug-induced gene expression via dimerizer system, *Mol. Ther.*, 7, 790, 2003.
109. Luker, G.D. et al., Noninvasive bioluminescence imaging of herpes simplex virus type 1 infection and therapy in living mice, *J. Virol.*, 76, 12149, 2002.
110. Luker, G.D. et al., Bioluminescence imaging reveals systemic dissemination of herpes simplex virus type 1 in the absence of interferon receptors, *J. Virol.*, 77, 11082, 2003.
111. Luker, K.E. et al., Transgenic reporter mouse for bioluminescence imaging of herpes simplex virus 1 infection in living mice, *Virology*, 347, 286, 2006.
112. Jacobs, A. et al., Functional coexpression of HSV-1 thymidine kinase and green fluorescent protein: implications for noninvasive imaging of transgene expression, *Neoplasia*, 1, 154, 1999.
113. Jacobs, A.H. et al., Improved herpes simplex virus type 1 amplicon vectors for proportional coexpression of positron emission tomography marker and therapeutic genes, *Hum. Gene Ther.*, 14, 277, 2003.
114. Mullerad, M. et al., Use of positron emission tomography to target prostate cancer gene therapy by oncolytic herpes simplex virus, *Mol. Imaging Biol.*, 8, 30, 2006.
115. Ichikawa, T. et al., MRI of transgene expression: correlation to therapeutic gene expression, *Neoplasia*, 4, 523, 2002.
116. McConnell, M.J. and Imperiale, M.J., Biology of adenovirus and its use as a vector for gene therapy, *Hum. Gene Ther.*, 15, 1022, 2004.
117. Rein, D.T., Breidenbach, M., and Curiel, D.T., Current developments in adenovirus-based cancer gene therapy, *Future Oncol.*, 2, 137, 2006.
118. Hackett, N.R. et al., Antivector and antitransgene host responses in gene therapy, *Curr. Opin. Mol. Ther.*, 2, 376, 2000.
119. Yang, Y. et al., Cellular and humoral immune responses to viral antigens create barriers to lung-directed gene therapy with recombinant adenoviruses, *J. Virol.*, 69, 2004, 1995.
120. Yang, Y. et al., Cellular immunity to viral antigens limits E1-deleted adenoviruses for gene therapy, *Proc. Natl. Acad. Sci. U.S.A.*, 91, 4407, 1994.
121. Thomas, C.E. et al., Preexisting antiadenoviral immunity is not a barrier to efficient and stable transduction of the brain, mediated by novel high-capacity adenovirus vectors, *Hum. Gene Ther.*, 12, 839, 2001.
122. Palmer, D.J. and Ng, P., Helper-dependent adenoviral vectors for gene therapy, *Hum. Gene Ther.*, 16, 1, 2005.
123. Le, L.P. et al., Dynamic monitoring of oncolytic adenovirus *in vivo* by genetic capsid labeling, *J. Natl. Cancer Inst.*, 98, 203, 2006.
124. Ono, H.A. et al., Noninvasive visualization of adenovirus replication with a fluorescent reporter in the E3 region, *Cancer Res.*, 65, 10154, 2005.
125. Ilagan, R. et al., Imaging androgen receptor function during flutamide treatment in the LAPC9 xenograft model, *Mol. Cancer Ther.*, 4, 1662, 2005.
126. Lee, C.T. et al., *In vivo* imaging of adenovirus transduction and enhanced therapeutic efficacy of combination therapy with conditionally replicating adenovirus and adenovirus-p27, *Cancer Res.*, 66, 372, 2006.
127. Rehemtulla, A. et al., Molecular imaging of gene expression and efficacy following adenoviral-mediated brain tumor gene therapy, *Mol. Imaging*, 1, 43, 2002.
128. Sato, M. et al., Optimization of adenoviral vectors to direct highly amplified prostate-specific expression for imaging and gene therapy, *Mol. Ther.*, 8, 726, 2003.

129. Anton, M. et al., Coexpression of herpesviral thymidine kinase reporter gene and VEGF gene for noninvasive monitoring of therapeutic gene transfer: an *in vitro* evaluation, *J. Nucl. Med.*, 45, 1743, 2004.
130. Bengel, F.M. et al., Noninvasive imaging of transgene expression by use of positron emission tomography in a pig model of myocardial gene transfer, *Circulation*, 108, 2127, 2003.
131. Green, L.A. et al., A tracer kinetic model for 18F-FHBG for quantitating herpes simplex virus type 1 thymidine kinase reporter gene expression in living animals using PET, *J. Nucl. Med.*, 45, 1560, 2004.
132. Richard, J.C. et al., Repetitive imaging of reporter gene expression in the lung, *Mol. Imaging*, 2, 342, 2003.
133. Barton, K.N. et al., GENIS: gene expression of sodium iodide symporter for noninvasive imaging of gene therapy vectors and quantification of gene expression *in vivo*, *Mol. Ther.*, 8, 508, 2003.
134. Dwyer, R.M. et al., Sodium iodide symporter-mediated radioiodide imaging and therapy of ovarian tumor xenografts in mice, *Gene Ther.*, 13, 60, 2006.
135. Lee, K.H. et al., Accuracy of myocardial sodium/iodide symporter gene expression imaging with radioiodide: evaluation with a dual-gene adenovirus vector, *J. Nucl. Med.*, 46, 652, 2005.
136. Chen, I.Y. et al., Micro-positron emission tomography imaging of cardiac gene expression in rats using bicistronic adenoviral vector-mediated gene delivery, *Circulation*, 109, 1415, 2004.
137. Umegaki, H. et al., Longitudinal follow-up study of adenoviral vector-mediated gene transfer of dopamine D2 receptors in the striatum in young, middle-aged, and aged rats: a positron emission tomography study, *Neuroscience*, 121, 479, 2003.
138. Rogers, B.E. et al., MicroPET imaging of gene transfer with a somatostatin receptor-based reporter gene and (94m)Tc-Demotate 1, *J. Nucl. Med.*, 46, 1889, 2005.
139. Verwijnen, S.M. et al., Molecular imaging and treatment of malignant gliomas following adenoviral transfer of the herpes simplex virus-thymidine kinase gene and the somatostatin receptor subtype 2 gene, *Cancer Biother. Radiopharm.*, 19, 111, 2004.
140. Buursma, A.R. et al., The human norepinephrine transporter in combination with 11C-m-hydroxyephedrine as a reporter gene/reporter probe for PET of gene therapy, *J. Nucl. Med.*, 46, 2068, 2005.
141. Reya, T. et al., Stem cells, cancer, and cancer stem cells, *Nature*, 414, 105, 2001.
142. Gage, F.H., Mammalian neural stem cells, *Science*, 287, 1433, 2000.
143. Weissman, I.L., Stem cells: units of development, units of regeneration, and units in evolution, *Cell*, 100, 157, 2000.
144. McKay, R., Stem cells in the central nervous system, *Science*, 276, 66, 1997.
145. Emerson, S.G., *Ex vivo* expansion of hematopoietic precursors, progenitors, and stem cells: the next generation of cellular therapeutics, *Blood*, 87, 3082, 1996.
146. Armitage, J.O., Bone-marrow transplantation, *N. Engl. J. Med.*, 330, 827, 1994.
147. Banchereau, J. et al., Immunobiology of dendritic cells, *Annu. Rev. Immunol.*, 18, 767, 2000.
148. Ahrens, E.T. et al., *In vivo* imaging platform for tracking immunotherapeutic cells, *Nat. Biotechnol.*, 23, 983, 2005.

Part II

Molecular MR Imaging

8 Molecular and Functional MR Imaging of Cancer

Michael A. Jacobs, Kristine Glunde, Barjor Gimi, Arvind P. Pathak, Ellen Ackerstaff, Dmitri Artemov, and Zaver M. Bhujwalla

CONTENTS

8.1 Introduction 141
8.2 Molecular MR Imaging of Cellular Targets in Cancer 142
8.3 Characterizing the Extracellular Matrix, the Physiological Environment, and Their Role in the Malignant Phenotype 146
8.4 Clinical Molecular and Functional MR Imaging 148
8.5 Multimodality Molecular Imaging 153
Acknowledgments 154
References 154

8.1 INTRODUCTION

Noninvasive imaging has become a powerful tool in the investigation of various disease processes, including cancer. Advances in magnetic resonance (MR), positron emission tomography (PET), single positron emission computed tomography (SPECT), ultrasound, and optical imaging techniques, as well as the development of novel imaging agents, have revolutionized our understanding of diseases. As our understanding of cancer advances, so does our recognition of the complexities of this disease, with new discoveries serving to emphasize these complexities. While a decade ago the focus in cancer research was primarily on genetic alterations, it is now apparent that the tumor's physiological microenvironment, the interaction between host cells, including stem cells and cancer cells, the extracellular matrix (ECM), and a multitude of secreted factors and cytokines influence progression, aggressiveness, and response of the disease to treatment. Therefore, identifying central targets that act across these levels of multiplicity is essential for the successful treatment of this disease. Multiparametric molecular and functional MR methods have several key roles to play in the treatment of cancer — they can be used to (1) reveal key targets for therapy, (2) visualize delivery of the therapy, and (3) assess the outcome of treatment. Altered choline metabolism is one example of the identification of a common feature of cancer revealed by MR spectroscopy, which can be exploited for diagnosis and potentially for treatment. Similarly, the use of vascular imaging to detect the effects of antiangiogenic and antivascular drugs is another example of the applications of MR imaging (MRI) in cancer treatment. Image-guided incorporation of nano- and microdevices into tumors for slow release of therapeutic agents and gene delivery is another area of promise in cancer treatment.

One of the most exciting developments in the application of MR methods in cancer has arisen from the development of smart contrast agents to detect receptor expression and specific molecular

TABLE 8.1
Some of the Currently Available Applications of MR Methods in Cancer

MR Method	Applications
Vascular imaging and angiogenesis	Vascular volume
	Permeability surface area product
	Extraction fraction
	Vessel size
	Blood oxygenation level dependent (BOLD) contrast imaging
	ECM remodeling
Metabolic imaging	Total choline
	Lactate
	Lipid
	NAA
	pH
	Energy (ATP, PCr, Pi)
	Drug pharmacokinetics
	Labeled substrate utilization
Targeted contrast imaging	Receptor expression on epithelial and endothelial cells
Diffusion imaging	Cell death
	Edema
	Fiber mapping
Chemical exchange saturation transfer (CEST), amide proton transfer (APT)	pH
	Receptor expression
	Gene expression

targets. Mechanisms for generating contrast include the use of paramagnetic and superparamagnetic agents, chemical exchange saturation transfer (CEST) and paramagnetic CEST (PARACEST), and spectroscopic imaging of specific agents or compounds. Targeted contrast can be generated for cell surface receptors using antibodies, or by substrate accumulation/activation of the probe, or through endogenous expression of MR-detectable reporters.

Clinically, MR methods have already made a significant impact in cancer diagnosis as well as in detecting tumor recurrence. While many important advances have been made in generating molecular, physiological, and metabolic contrast using MR methods, several challenges remain. The development of novel strategies to detect and image specific pathways and targets continues to be at the forefront of current challenges in MR contrast. The low concentration of receptors and molecular targets and the inherent insensitivity of MR detection have led to amplification strategies to improve detection limits, and a major challenge for the future continues to be increasing the sensitivity of detection. Several critical pathways and molecules require intracellular access of reporter molecules, and the internalization of MR reporters is another challenge for MR contrast. Finally, as the acquisition of multiparametric and multimodality images becomes routinely available, an integrated multimodal approach presents unique computing challenges as well as exciting opportunities.

In this chapter, we have reviewed MR applications in molecular and functional imaging of cancer. Some of these applications are summarized in Table 8.1. We have used examples of work performed in our program to demonstrate some of the capabilities of MR methods in cancer. Since several chapters in this volume cover specific aspects of MR applications in cancer, those aspects are not discussed here.

8.2 MOLECULAR MR IMAGING OF CELLULAR TARGETS IN CANCER

MR molecular imaging of cellular targets remains challenging because of the inherently low sensitivity of MRI and spectroscopy and the low concentrations of target molecules in living cells.

Reliable MR detection of these targets requires efficient methods of contrast enhancement. Current applications of MR in molecular imaging can be broadly categorized as (1) MR imaging of cells preloaded with an MR contrast agent such as superparamagnetic iron oxide (SPIO) nanoparticles or expressing an endogenous reporter gene under the control of a specific promoter, and (2) MR imaging of cell surface targets such as receptors using exogenous contrast agents like paramagnetic and superparamagnetic probes or MR spectroscopic imaging detection of an enzyme-coupled reaction with an appropriate substrate.

A detailed review of the first group of MRI applications, especially for imaging stem cells, can be found in Chapter 3. Briefly, the ability to load cells with a high concentration of the imaging probes, such as SPIO nanoparticles or fluorinated tracers, *ex vivo* has been shown to generate sufficient MR contrast for detection *in vivo* in preclinical systems and also in pilot clinical studies.[1] The expression of a significant number of reporter molecules such as CEST reporters (discussed separately in Chapter 1)[2] or ferritin protein[3] in the target cell is another approach to cell imaging. The large number of protein molecules expressed under the control of an efficient promoter provides label amplification. Contrast generated by CEST affects the large signal of bulk water, while overexpression of ferritin results in increased intracellular iron content by the target cell with a corresponding significant decrease of T_2 and T_2* relaxation times. These approaches require the use of engineered cancer cells and can provide important information in preclinical models of cancer. They can also be used to monitor the efficiency of gene expression or the modulation of gene expression following gene therapy. Contrast agents that become MR detectable only upon enzymatic activation through reporter gene expression, such as β-galactosidase, can also be used to detect the expression of these reporter genes.[4]

The second group of MR applications for molecular imaging, which comprises exogenous contrast agents to detect cell surface targets, requires (1) a highly efficient and specific targeting mechanism that recognizes cell surface markers and (2) efficient effector molecules or MR probes that can generate strong MR contrast from a small amount of probe targeted to the cell. Specific targeting can be achieved by monoclonal antibodies (mAbs), mAb fragments, or specially designed minibodies, diabodies, or synthetic targeting peptides.[5,6] An alternative approach is to use proteins that have a high binding affinity to the cell marker of choice, such as the transferrin/ferritin receptor[7] or annexin V (or synaptotagmin I)/phosphatidylserine residues in apoptotic cells.[8] While complete antibodies provide the highest binding affinity to their antigen in most cases, their large molecular size may restrict efficient delivery to the interstitium of solid tumors. On the other hand, these antibodies can provide excellent targeting properties for accessible receptors, such as those expressed in the lumen of blood vessels.[9,10]

Various mechanisms of contrast generation have been explored for MR imaging of cell surface receptors. Paramagnetic T_1 contrast agents such as gadolinium are effective in reducing the T_1 relaxation time of multiple water molecules in the immediate milieu of the gadolinium (Gd) ion. At high magnetic fields, the detection threshold of water in the presence of Gd is in the range of 10 μM Gd. To image cells expressing a million receptors on the plasma membrane, each receptor should be labeled with at least several dozen gadolinium ions to provide sufficient sensitivity for MR detection. Attempts to directly label mAb with Gd–chelate groups have not been successful in solid tumors, partly due to the limited number of groups that can be conjugated to the mAb without diminishing its binding affinity and partly due to limited extravasation and diffusion of the large molecular complexes in the tumor microenvironment. On the other hand, $\alpha_v\beta_3$ integrins expressed in the angiogenic endothelium of tumors have been imaged successfully. The accessibility of these integrins permits the use of large blood pool nanocomplexes, such as paramagnetic-polymerized liposomes[10] or Gd-perfluorocarbon nanoparticles,[11] that can contain several thousand Gd ions per particle.

Macromolecular imaging platforms such as protein or dendrimer Gd conjugates or liposomes that can carry a large number of derivatized GdDOTA or GdDTPA substrates can also be used to image cell surface receptors. These imaging agents can be administered independently from the targeting antibodies following a multistep labeling approach,[12] with better delivery compared to a single large molecular weight targeted MR imaging agent. This approach was demonstrated *in vivo* for imaging of HER-2/neu receptors in a preclinical model of breast cancer using biotinylated primary anti-HER-2/neu mAb and avidin(GdDTPA) conjugate.[13] The combination of high receptor expression levels, relatively small molecular sizes of the components, and amplification arising from multiple Gd ions attached to a single avidin molecule in conjunction with multiple avidin-binding biotins per mAb resulted in detectable positive T_1 MR contrast in HER-2/neu-positive tumors.[13] Interestingly, our recent observations also suggest that cross-linking of multiple biotinylated mAb with avidin results in rapid internalization of HER-2/neu receptors together with the attached probe. This provides a novel strategy for internalization of the contrast agent as well as for delivery of therapeutic cargo. It is also possible that the higher sensitivity of this approach resulted from efficient loading of cells with the internalized contrast agent.

Several studies have also used targeted superparamagnetic nanoparticles for MR imaging of tumor cell surface receptors *in vivo*. Nanoparticles typically consist of a single polycrystalline iron oxide magnetic core coated with a polymer for biocompatibility. The large magnetic moment of the superparamagnetic core results in a high T_2/T_2^* relaxivity that allows imaging of a single cell labeled with iron oxide nanoparticles (micron size).[14,15] The high contrast generated by this approach makes nanoparticles an ideal contrast agent for prelabeling cells *ex vivo* for cell tracking MR applications.[16] We have used this approach to track endothelial cell motility, invasive potential, and network response in the presence and absence of cancer cells.[17] Human vascular endothelial cells (HUVECs) labeled with the superparamagnetic iron oxide T_2 contrast agent Feridex® were cocultured with MDA-MB-231 breast cancer cells in the presence of an extracellular matrix (ECM) gel. HUVECs responded to paracrine factors secreted by the cancer cells by invading and migrating through the ECM gel toward the cancer cells (Figure 8.1). This assay can be used to observe HUVEC motility and network formation in response to proangiogenic or antiangiogenic stimuli, and we are currently using it to understand the response of endothelial cells to physiological environments, such as hypoxia, typically found in tumors.

Transferrin conjugated with monocrystalline iron oxide nanoparticles (MIONs) has also been used to image cancer cells overexpressing engineered transferrin receptor (ETR) that rapidly internalizes MIONs, producing strong negative T_2^* contrast in the ETR-transfected tumors.[7] The use of mAb conjugated to iron oxide nanoparticles was also reported for imaging tumor cell surface markers such as underglycosylated mucin-1 antigen[18] or phosphatidylserine residues in apoptotic cancer cells.[8]

Spectroscopic imaging can also be used to generate receptor- or gene expression-specific contrast. One such example is the enzymatic conversion of the prodrug 5-fluorocytosine (5-FC) to 5-fluorouracil (5-FU) by the enzyme cytosine deaminase (CD). Stegman et al.[19] transfected cells with bacterial or yeast CD to convert 5-FC to 5-FU and detected the conversion *in vivo*. In a study by Aboagye et al.,[20] CD was covalently bound to a mAb specific to the L6 antigen expressed in tumors derived from human lung adenocarcinoma H2981 cells. An efficient amplification of the probe and ^{19}F chemical shifts of both 5-FU and 5-FC were detected in cultured cells and in animal tumor models.[20] Yu and Mason have developed a ^{19}F-containing substrate for the standard reporter enzyme LacZ, which was used to detect the cleavage of this construct by LacZ-expressing MCF-7 cells.[21] A potential problem of MR spectroscopy, however, is its limited sensitivity, and a threshold concentration of the product in the range of ~1 m*M* is required for detection.

It is very likely that future translational applications in molecular imaging will depend heavily upon interventional radiology for image-guided delivery of viral vectors, prodrug therapy, and smart contrast agents using novel imaging platforms. MRI of targeted liposomes may ultimately be used to deliver image-guided molecular-based therapy to cancer patients.[22,23] The use of microdevices containing slow-release polymers that can be implanted within tumors under image guidance is another exciting area of development. Recently, Gimi et al.[24] created microcontainers coated with a very thin gold layer, which can be detected with MRI (see Figure 8.2). As shown

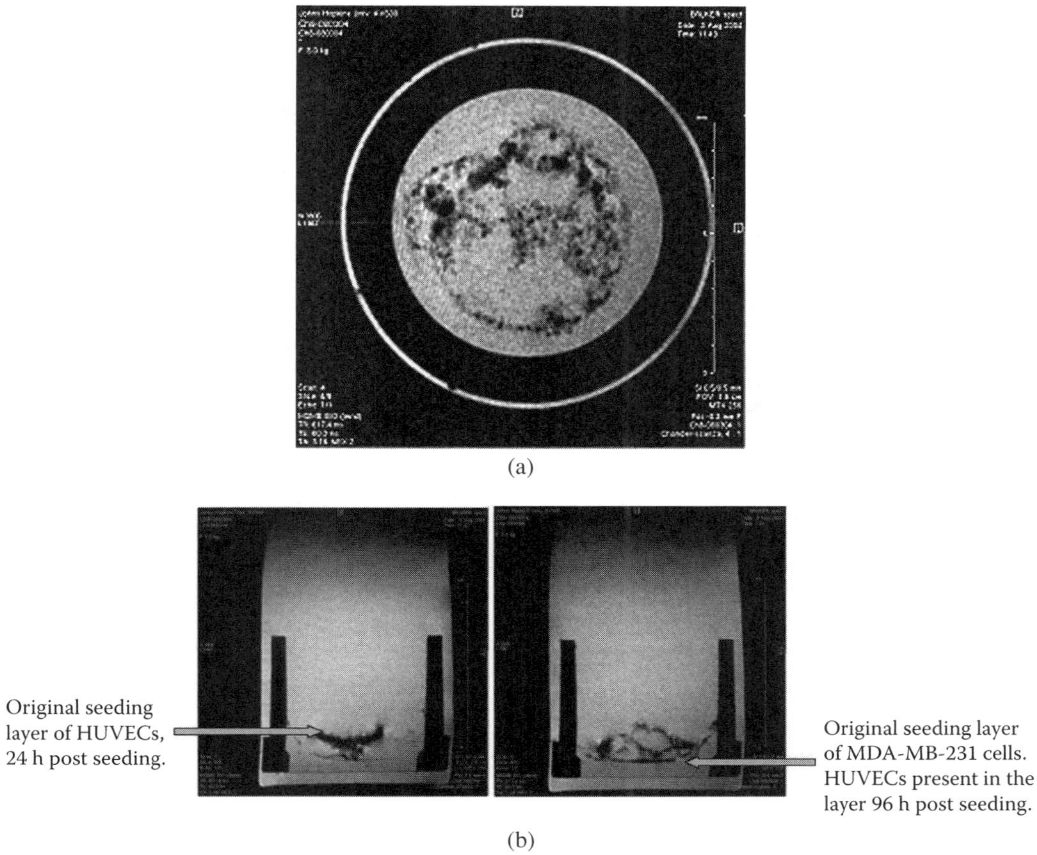

FIGURE 8.1 (a) A representative axial image of a chamber containing Feridex-labeled HUVECs, showing a detailed network-like structure 24 h postseeding. The T_2-weighted spin echo MR image was obtained with a single acquisition over a 0.5-mm slice, 1.6-cm field of view, 256 × 256 acquisition matrix, TR = 857 msec, and TE = 90 msec. (b) Coronal images show HUVECs in the upper seed layer 24 h postseeding, whereas an increased presence of HUVECs is observed in the MDA-MB-231 cancer cell layer 96 h postseeding, along with ECM gel remodeling. No invasion was observed in control chambers (not shown). (Adapted from Gimi, B. et al., *Neoplasia*, 8, 207–213, 2006. With permission.)

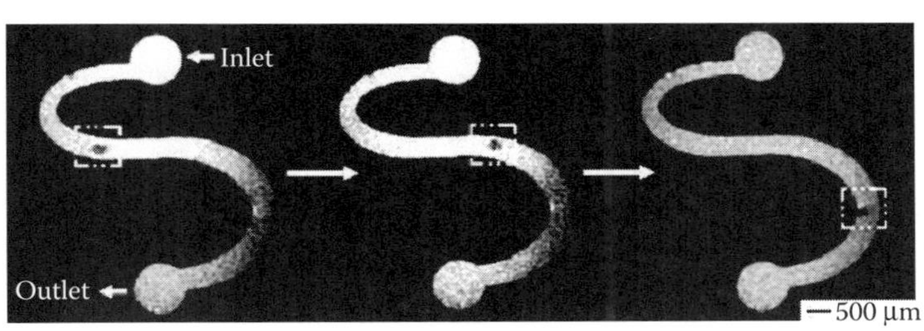

FIGURE 8.2 MR tracking of a biocontainer in a fluidic channel. MR images of the container at different time points taken under pressure-driven flow of the fluid. (Adapted from Gimi, B. et al., *Biomed. Microdevices*, 7, 341–345, 2005. With permission.)

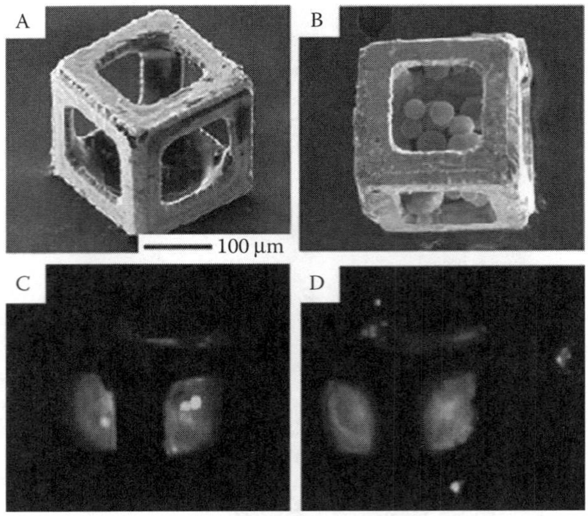

FIGURE 8.3 Scanning electron microscope (SEM) image of (A) a hollow, open-surface biocontainer and (B) a device loaded with glass microbeads. (C) Fluorescence microscopy images of a biocontainer loaded with cell-ECM-agarose with the cell viability stain Calcein-AM. (D) Release of viable cells from the biocontainer. (Adapted from Gimi, B. et al., *Biomed. Microdevices*, 7, 341–345, 2005. With permission.)

in Figure 8.3, these microcontainers can be used to encapsulate cells or drug-releasing polymers in cancer treatment.

8.3 CHARACTERIZING THE EXTRACELLULAR MATRIX, THE PHYSIOLOGICAL ENVIRONMENT, AND THEIR ROLE IN THE MALIGNANT PHENOTYPE

Traditionally, perfusion imaging with MRI has played an important role in understanding tumor vasculature in preclinical models as well as in diagnosis in the clinic. There are several excellent reviews on vascular imaging with MRI, and therefore here we have focused on the cancer cell–ECM continuum and the application of MR techniques in understanding the interaction between cancer cells, ECM, and endothelial cells. MR methods can be translated from bench to bedside, and therefore can be used to study isolated cells, human tumor xenografts, and clinical tumors.

Tumors display abnormal physiological environments that primarily arise from the aberrant and chaotic vasculature that develops through neoangiogenesis. Since the discovery of hypoxia-inducible factor and the identification of hypoxia response elements as transcriptional controls in multiple genes,[25] it has become increasingly evident that the physiological environment, and hypoxia in particular, plays an important role in the cancer phenotype. The ECM presents one of the earliest lines of defense against the invading and disseminating cancer cell. Cancer cells invade through the ECM and into surrounding tissues by secreting proteolytic enzymes such as serine proteases and matrix-degrading metalloproteinases, or by inducing tumor stromal cells to secrete these enzymes.[26–28] Figure 8.4 shows an example of how MR methods can be adapted to study the ability of cancer cells to degrade ECM, an important step in the metastatic cascade. This MR-compatible invasion assay was designed to understand if physiological environments typically found in tumors promoted or prevented invasion. The advantages of this system are that it provides careful control of physiological conditions, as well as quantitation of ECM degradation over time and metabolic information of the cells.[29] Perfluorotripropylamine (FTPA)-doped alginate beads in the MR tube and FTPA embedded into the ECM layer are used to measure oxygen tensions from the T_1 relaxation time of FTPA. In this system, we can regulate oxygen tensions to less than 1.5%, which is necessary to evaluate the impact of the hypoxic tumor environment on cancer cell invasion.

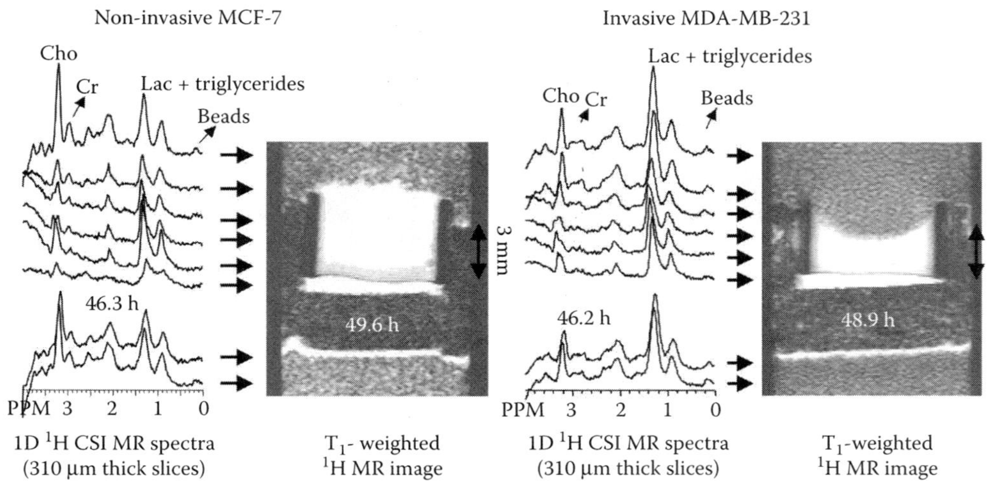

FIGURE 8.4 Expanded ^{1}H MR image of sample showing the ECM gel layer for MCF-7 and MDA-MB-231 cells at comparable time points. Images were obtained with TR = 1 sec, TE = 30 msec, FOV = 40 mm, slice thickness = 2 mm, and in-plane resolution = 78 μm. Corresponding ^{1}H MR spectra are from 310-μm-thick localized slices from within the sample. Cho denotes the signal from total choline-containing compounds (Cho + PC + GPC); Cr denotes the signal from total creatine-containing compounds; Lac+Triglycerides denotes the signal from lactate and triglyceride peak. With permission.

Typical invasion and metabolic data obtained with this assay for a noninvasive (MCF-7) and an invasive (MDA-MB-231) breast cancer cell line are shown in Figure 8.4. A significant degradation of the ECM layer by MDA-MB-231 cells, but not MCF-7 cells, is observed as early as 24 h. Such a system can be used to detect the effects of anti-invasive treatments and is especially useful for testing the efficacy of treatments based on metabolic interventions, such as the downregulation of choline kinase, discussed later.

It is also possible that proteolytic enzymes reduce the integrity of the tumor stroma and promote tumoral lymphatic development, as well as the movement of macromolecules through the ECM.[27] Understanding the contribution of factors that influence delivery, movement, and clearance of macromolecules through the ECM and supporting stroma of solid tumors is important for delineating critical mechanisms in cancer invasion and metastasis. We therefore developed a noninvasive method to quantify the extravascular transport of macromolecules through the ECM of solid tumors *in vivo* using MRI.[30] We have used this method to characterize differences in vascularization (vascular volume, permeability surface area product) and lymphatic-convective transport (macromolecular fluid transport rates/volumes, fraction of draining/pooling voxels) *in vivo* in two human breast cancer xenografts (MCF-7 and MDA-MB-231) preselected for their differences in invasiveness (as shown in Figure 8.4). The more invasive tumor model, MDA-MB-231, exhibited a significantly higher number of draining voxels, as well as larger extravascular fluid clearance, than MCF-7 tumors, which may reflect the integrity of its ECM (Figure 8.5). A higher incidence of lymph node metastases was observed in MDA-MB-231 tumors than in MCF-7 tumors, which was most likely a consequence of their differential invasiveness, altered ECM integrity resulting in increased drainage areas, and increased lymphatic vessel area.[31]

Human tumor xenograft models provide an important preclinical bridge to translational studies. The combination of molecular imaging strategies and multimodality imaging has provided novel opportunities to further characterize the tumor microenvironment. A critical response to hypoxia is the induction of the hypoxia-inducible factor (HIF-1). HIF-1 acts as a transcriptional activator for, among others, vascular endothelial growth factor (VEGF), inducible nitric oxide synthase, heme oxygenase 1, glucose transporter 1 (GLUT1), and the glycolytic enzymes: aldolase A, enolase 1,

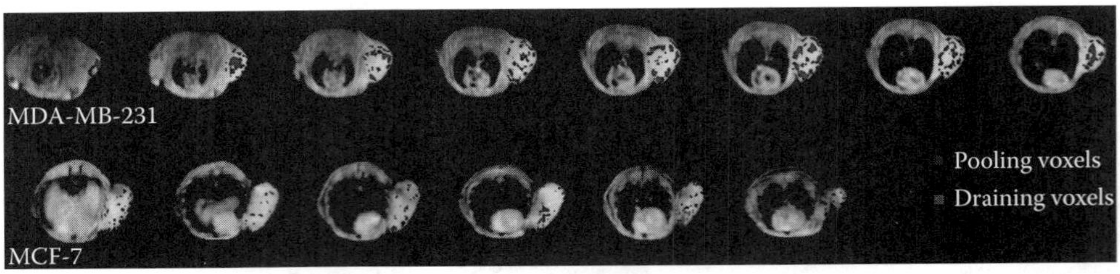

FIGURE 8.5 (please see color insert following page 210) Representative multislice functional MRI maps of pooling and draining voxels for an MDA-MB-231-tumor-bearing animal (top) and an MCF-7-tumor-bearing animal (bottom). (Adapted from Pathak, A.P. et al., *Cancer Res.*, 66, 5151–5158, 2006. With permission.)

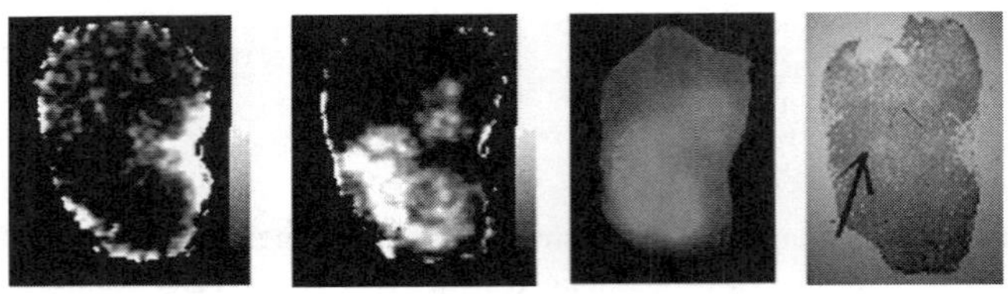

FIGURE 8.6 (please see color insert) Map of (a) vascular volume and (b) permeability surface area product (PSP) obtained from an HRE-GFP PC-3 tumor (180 mm^3). Vascular volume ranged from 0 to 344 μl/g and PSP from 0 to 24 μl/g-min. (c) Fluorescent microscopy of a tissue section obtained from the imaged slice, using a Nikon TS100-F microscope (1× objective) with a wavelength of 512 nm. (d) H&E-stained 5-μm-thick section from this slice. The region containing the GFP expression consisted of viable cells. The less dense staining in the upper part of the section is due to uneven sectioning. The only area of dying cells was in a small focus shown by the arrow. (Adapted from Raman, V. et al., Combined and Co-registered MRI, MRS, and Optical Imaging Characterization of Tumor Vascularization, Metabolism and Hypoxia, paper presented at International Society for Magnetic Resonance in Medicine, Honolulu, 2002. With permission.)

lactate dehydrogenase A, and phosphoglycerate kinase 1.[25] These target genes contain hypoxia response elements (HREs), which include one or more HIF-1 binding sites. We have exploited the HRE to study the relationship between hypoxia and vascularization using *in vivo* MRI and optical imaging of a human prostate cancer xenograft model derived from cells stably expressing green fluorescent protein (GFP) under the regulation of an HRE promoter.[32] MRI of the macromolecular contrast agent, albumin-GdDTPA, was performed to obtain colocalized maps of vascular volume and vascular permeability,[32] after which the GFP distribution was determined in freshly excised tissue slices corresponding to the slices imaged with MR. Distinct distributions of fluorescence and vascular parameters were detected in the tumors (Figure 8.6). Regions of low vascular volume measured by MRI were associated with regions exhibiting high fluorescence and vice versa. Regions of high permeability were coincident with regions of low vascular volume and high fluorescence (Figure 8.6). These observations are consistent with the possibility that regions of low vascular volume are hypoxic, and with hypoxia acting as a transcriptional activator of VEGF, a potent vascular permeability factor. We are currently investigating the role of hypoxia in ECM integrity using the HRE-GFP system in MDA-MB-231 tumors. Such studies can also be extended to understanding the role of extracellular pH (pH_e) in ECM integrity, since it is also possible to measure pH_e with proton spectroscopic imaging.[33,34]

8.4 CLINICAL MOLECULAR AND FUNCTIONAL MR IMAGING

While MR molecular imaging of cellular targets is still at the stage of preclinical applications, several MR methods are playing an increasingly important role in cancer diagnosis and treatment

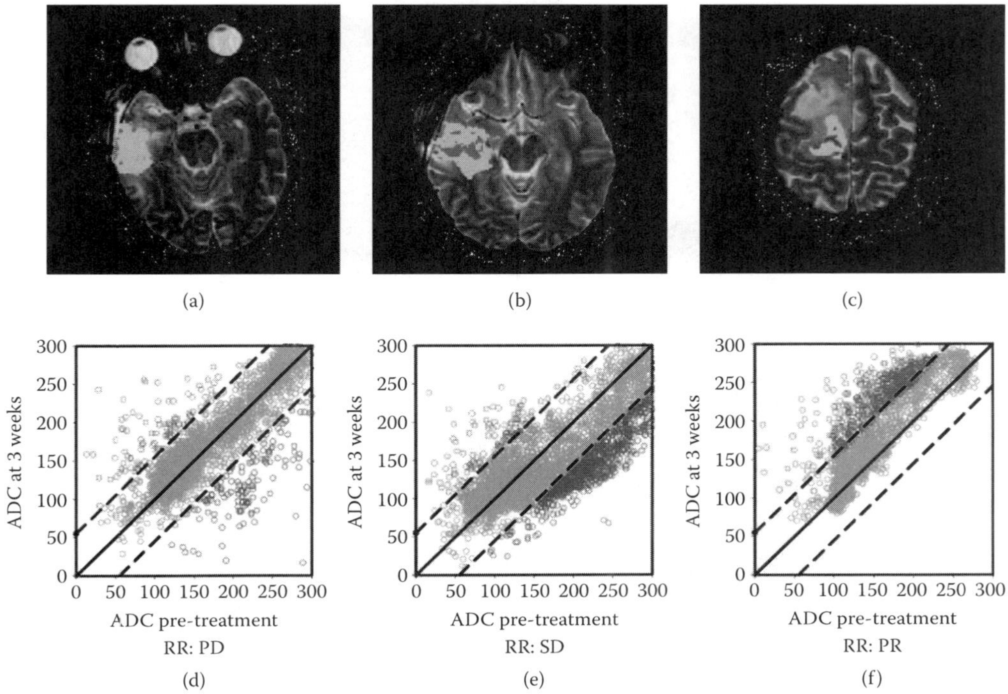

FIGURE 8.7 (please see color insert) Representative MR images of three patients with glioblastoma multiforme. MR images shown are at 3 weeks into a 7-week fractionated radiation regimen. Regions of interests were drawn for each tumor image by using anatomical images. (a–c) The regional spatial distributions of ADC changes in a single slice from each tumor are shown as color overlays for progressive disease (a), stable disease (b), and partial response (c) patients. The red pixels indicate areas of increased diffusion, whereas the blue and green pixels indicate regions of decreased and unchanged ADC, respectively. (d–f) The scatter plots quantitatively show the distribution of ADC (1×10^5 mm^2/sec) changes for the entire three-dimensional tumor volume, whereas radiographic response (RR) labels are given at the bottom, indicating each patient's WHO-based classification at week 10. (Adapted from Hamstra, D.A. et al., *Proc. Natl. Acad. Sci. U.S.A.*, 102, 16759–16764, 2005. With permission.)

in the clinic. Methods that combine different parameters into a single diagnostic index are being developed[35,36] to characterize the complexity of cancerous tissue. Techniques such as T_1, T_2, and contrast-enhanced MR imaging, proton magnetic resonance spectroscopic imaging (MRSI), and sodium MR imaging, when used in combination, can provide a comprehensive data set with potentially more power to diagnose cancer than any single measure alone.[36] In addition, the use of novel contrast and specific therapeutic agents has led to the ability to interrogate the molecular environment of the tumor using MRI. Several studies have shown the utility of using diffusion-weighted imaging (DWI) for characterizing brain tumors (Figure 8.7).[37,38] DWI is sensitive to subvoxel changes in the diffusion of water within the intra- and intercellular environments.[39] These changes in water diffusion result in changes in the MR signal intensity in DWI.[37,39,40] Although the precise mechanisms for these signal intensity changes are still being investigated, there is evidence that factors such as the movement of water from the extracellular space to the intracellular space and vice versa, increased tortuosity of the diffusion pathways, restriction of the cellular membrane permeability, cellular density, and disruption of cellular membrane depolarization contribute to these signals.[41–43] DWI can be used to derive a quantitative biophysical parameter, called the apparent diffusion coefficient (ADC) of water. The ADC provides an average value of the distance a water molecule has moved within a biological tissue and is related to the state of tissue, for example, during the evolution of cerebral ischemia[43] and tumor progression.[44]

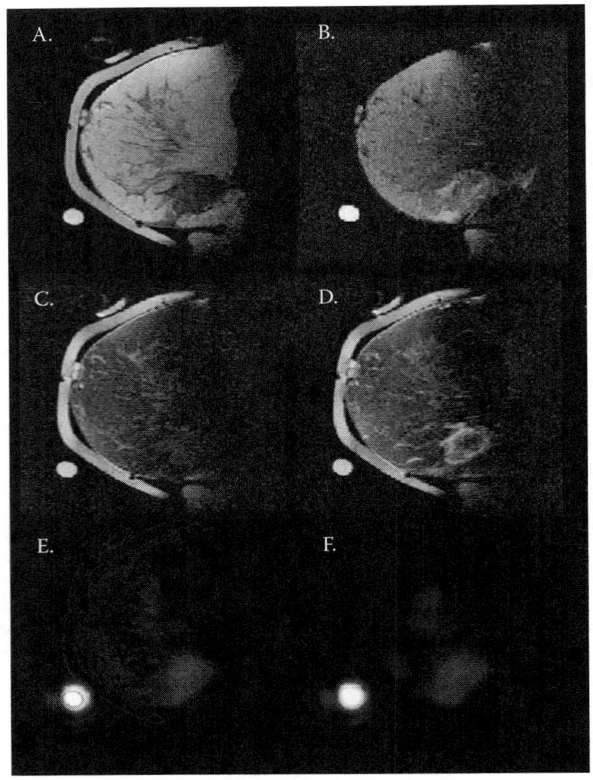

FIGURE 8.8 (please see color insert) (A) Sagittal T_1-weighted MR image of a 63-year-old woman with invasive carcinoma in the lower inner breast. (B) Fat-suppressed fast spin echo T_2-weighted MR image. Fat-suppressed T_1-weighted pre- (C) and post- (D) gadolinium contrast-enhanced MR image. (E–F) Sodium MR images demonstrating increased sodium signal within the breast lesion. A blue contour plot generated on (D) is shown superimposed on the sodium MR image in (E). (Adapted from Jacobs, M.A. et al., *Technol. Cancer Res. Treat.*, 3, 543–550, 2004. With permission.)

While imaging techniques such as contrast-enhanced MRI provide high sensitivity for the detection of cancerous lesions in brain, breast, liver, and prostate,[45–48] there is a need to increase specificity and provide metabolic information. An imaging modality gaining importance in the clinical setting is sodium MR imaging. Sodium is abundant in most tissues and has important biological implications. Changes in sodium concentration can occur through disruption of the sodium-potassium pump within cells or through metabolic changes that alter sodium exchange across the cell membrane. Although intracellular sodium (ICS) concentration cannot be imaged separately from extracellular sodium (ECS) concentration without shift reagents or sophisticated multiquantum MR methods,[49] the total sodium concentration (TSC) can be determined with MR.[50] Since normal tissue TSC is heavily dependent on the ratio of intracellular to extracellular volumes,[50,51] increased interstitial space through a change in cellular organization or increased vascular volume will lead to an increase in TSC. The altered Na^+/H^+ exchange kinetics and the acidic extracellular microenvironment of the tumor cells both cause an increase in intracellular Na^+ concentration. The use of sodium MR in breast cancer is shown in Figure 8.8. The increased TSC within the lesion is clearly visible and provides additional information about the molecular environment.

Currently, DWI, perfusion imaging, and MRS play an important role in cancer detection and treatment. Proton MRS has, in fact, emerged as a valuable technique for evaluating metabolite levels in different tissues in different organs.

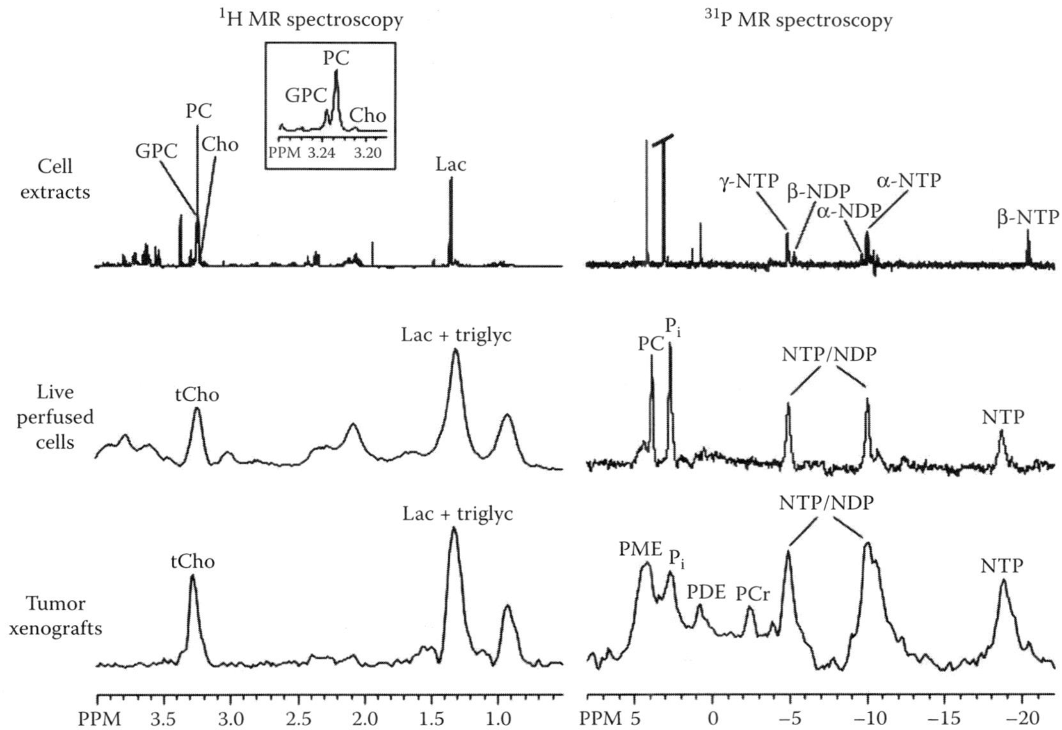

FIGURE 8.9 Proton (left) and phosphorus (right) MR spectra of MDA-MB-231 breast cancer xenografts (bottom), live perfused MDA-MB-231 breast cancer cells (middle), and MDA-MB-231 breast cancer cell extracts (top). Cho, free choline; GPC, glycerophosphocholine; Lac, lactate; NDP, nucleoside diphosphate; NTP, nucleoside triphosphate; PC, phosphocholine; PCr, phosphocreatine; PDE, phosphodiester; P_i, inorganic phosphate; PME, phosphomonoester; tCho, total choline-containing metabolites; Triglyc, triacylglycerides. (Adapted from Ackerstaff, E. et al., *J. Cell. Biochem.*, 90, 525–533, 2003. With permission.)

A high total choline-containing compound (tCho) signal appears to be a spectroscopic hallmark, not only of breast cancer,[52–54] prostate cancer,[55,56] and head and neck cancers,[57] but also in human brain tumors, where a correlation between tumor grade and choline concentration has been reported.[58,59] In the brain, the signal from tCho is compared to the signal from N-acetylaspartate (NAA),[59] whereas in the prostate, it is compared to normal tissue biomarkers such as citrate and creatine to detect malignancy.[55,56,60] Proton MRSI can therefore be used to detect the extent of invasion of brain,[61] prostate,[62] and breast[63] tumors, among others, and to differentiate between recurrence and necrosis following treatment[62,64] in cancer patients. Proton spectra of small volumes of about 0.2 to 2 cm^3 can be detected within a clinically reasonable time frame of about 15 min. Multivoxel MRSI techniques that provide two- or three-dimensional spatial localization by means of slice selection and phase encoding are also being developed for clinical studies.[65] The combined total choline signal at 3.2 to 3.3 ppm detected *in vivo* consists of the resonances from free choline (Cho) at 3.21 ppm, phosphocholine (PC) at 3.23 ppm, and glycerophosphocholine (GPC) at 3.24 ppm. Phosphorus MRS can detect ^{31}P-containing choline compounds such as PC at 3.4 ppm and GPC at 0.5 ppm. Figure 8.9 demonstrates spectral resolutions that can be achieved by applying ^{1}H and ^{31}P MRS to solid MDA-MB-231 breast tumor xenografts *in vivo*, to live perfused MDA-MB-231 breast cancer cells, or to cell extracts from the same cell line. While ^{31}P MRS has been applied in many preclinical studies, its limited sensitivity only provides coarse spectral and spatial resolution in the clinical setting. Phosphorus MRS can be useful *ex vivo*, as in human brain tumor

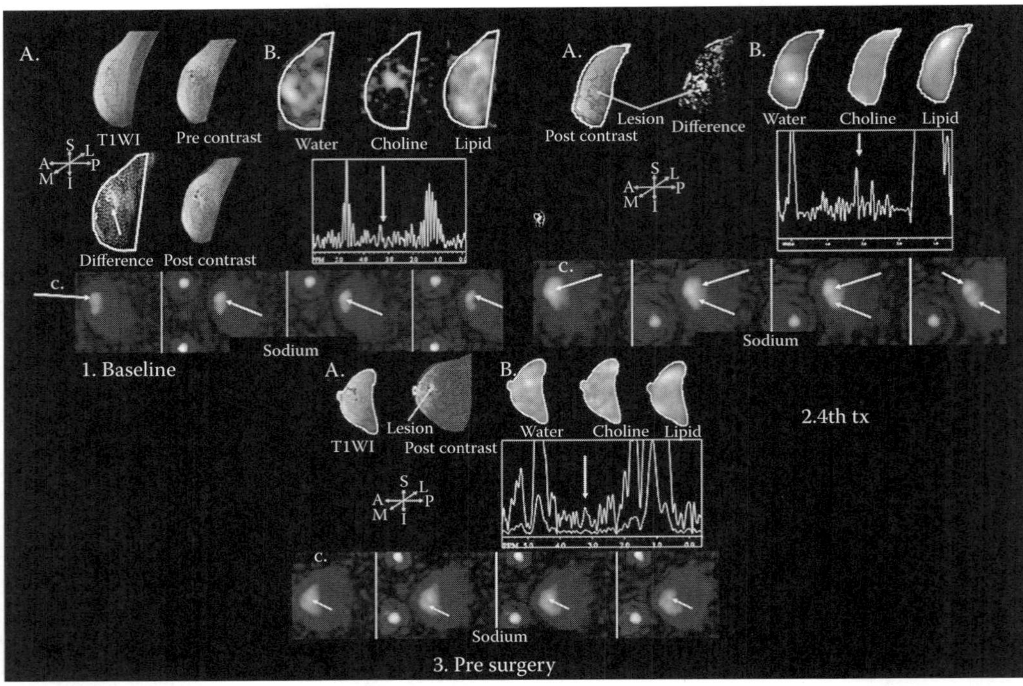

FIGURE 8.10 (please see color insert) Forty-nine-year-old female with infiltrating ductal carcinoma receiving preoperative chemotherapy. (1) Baseline MR volume was 2.9 cc with a visible tCho signal at 3.2 ppm and a sodium concentration of 58.7 m*M*/kg. (2) After the fourth treatment, the MR volume was 0.90 cc and the tCho signal was still visible; however, her sodium concentration had decreased to 48.9 m*M*/kg. (3) Before surgery, the MRI volume was 0.53 cc. The sodium concentration within the lesion remained constant after the fourth treatment.

extract samples.[66] Magic angle-spinning (MAS) ^{1}H MRS is currently emerging as a technology for examining intact biological tissue *ex vivo*, which can then undergo further histologic, biochemical, and genetic analyses.[67]

The ability to use molecular imaging for the detection of an early therapeutic response is under active research, especially with the development of molecular-based targeting agents that may control but not reduce tumor volume. Traditionally, volumetric changes of the tumor have been used (e.g., RECIST (Response Evaluation Criteria in Solid Tumors) or World Health Organization (WHO) criteria)[68,69] to gauge response to treatment. In a recent study, a decrease in the tCho signal in proton spectra of breast cancer was shown to be effective in predicting a response.[70,71] The use of water and fat ratios also detected response with good success.[71,72] Using proton MRI, MRSI, and sodium MRI, we have also demonstrated similar findings in the breast.[36] The use of combined proton, MRS, and sodium MRI is novel in the clinical neoadjuvant setting of breast cancer (Figure 8.10) and can be extended to other organ systems. Recent work has shown that sodium MRI and DWI in experimental brain tumors can predict response.[73] Combining these methods provides a more thorough evaluation of the molecular environment of the tumor.[36,73,74] This is demonstrated in Figure 8.10, where changes of the TSC are observed and can be correlated to the clinical exam or biomarkers, such as estrogen or Ki-67 (a marker of proliferation) in breast cancer.

As mentioned in the introduction, MR molecular and functional imaging can be used to identify key targets, visualize delivery of therapy, and assess treatment outcome. The aberrant choline phospholipid metabolism of cancer cells is one example of the identification of a target for treatment (for review, see Ackerstaff et al.[75]). Human breast and prostate cancer cells in culture typically exhibit elevated PC levels.[76,77] A step-wise increase of PC and tCho was observed in immortalized, oncogene-transformed, or tumor-derived human mammary epithelial cells (HMECs), used as a model of breast cancer progression.[76] Treatments and alterations in molecular pathways that increase or decrease cancer aggressiveness resulted in consistent changes in PC and tCho.[78–81]

Choline phospholipid metabolism comprises a complex network of biosynthetic and breakdown pathways, with one or more enzymes acting per pathway. Choline kinase generates intracellular PC in the cytosine diphosphate choline (CDP choline) pathway, the major biosynthetic pathway for *de novo* phosphatidylcholine (PtdCho) synthesis in mammalian cells. PtdCho-specific phospholipase C produces PC from membrane PtdCho by a breakdown pathway. Cytosine triphosphate (CTP):phosphocholine cytidylyltransferase utilizes PC, and therefore its activity correlates inversely with intracellular PC levels. PC itself has been reported to be mitogenic and may be a second messenger or mediator for the mitogenic activity of several growth factors.[82,83] Recent studies have started exploring the possibility of altering the expression or activity of enzymes involved in choline phospholipid metabolism as novel therapeutic targets for cancer treatment. Since phosphatidylcholine-specific phospholipase D (PC-PLD) was shown to be involved in many aspects of cell proliferation and oncogenic signaling, it could prove valuable as a target for therapeutic intervention in cancers, particularly breast cancers.[84–87] Overexpression of choline kinase has been reported in several human tumor-derived cell lines of multiple origins, as well as biopsies of lung, colon, and prostate carcinomas, which were compared to matching normal tissue from the same patient.[88] The activity of choline kinase was shown to be higher in malignant tissue.[89–91] Figure 8.11a shows an example of this increase in choline kinase expression levels in malignant breast cancer cell lines compared to a nonmalignant human mammary epithelial cell line. *Ras* oncogene transformation has also been linked to choline kinase stimulation in cancers leading to elevated PC levels.[79,92–94] Ras GTPases are among the most important oncogenes in human carcinogenesis, and roughly 30% of all human tumors contain *ras* mutations.[95] Choline kinase inhibition[96–98] and downregulation by means of small interfering RNA (siRNA)[99] are currently being investigated as potential novel antitumor therapies. Downregulation of choline kinase in human breast cancer cells using siRNA specific to choline kinase was found to efficiently decrease cellular ^{1}H MR-detectable PC levels, while human mammary epithelial cells remained unaffected by this treatment,[99] as demonstrated in Figure 8.11. The emerging knowledge of the genetic and molecular regulation of choline kinase, as well as other enzymes in the choline phospholipid metabolic pathway, will have a significant impact on improving potential choline kinase-targeted cancer therapies and on identifying other potential targets in the choline cycle. Noninvasive MRS and MRSI of choline phospholipid metabolites will prove valuable in future studies to evaluate these novel therapies.

8.5 MULTIMODALITY MOLECULAR IMAGING

The use of multimodality imaging methods is in the early stages of becoming the *de facto* standard. Combined PET/CT imaging is increasingly utilized in colorectal,[100] breast,[101] and pancreatic cancer.[102] Recent advances in technology have spurred the development of combined PET/MR scanners.[103] These advances will provide an excellent means of combining the target-specific contrast agents currently available clinically for PET studies with the anatomical and functional capabilities of MR methods. We are currently investigating the use of proton MRI, MRS, and PET/CT in the evaluation of breast tumors before and after preoperative chemotherapy. This represents a significant advance for colocalization of imaging with pathological and functional information.[104] Figure 8.12 demonstrates the power of this approach; the fusion of the PET/CT images provides some anatomical detail, but by using MRI, vascularity as well as the TSC of the tumor can be determined. Such a multimodality approach will enable the clinician to better understand how the patient is responding to treatment (Figure 8.12). The scientific potential of these methods is only beginning to be tapped, and further studies are needed to characterize the combinations of modalities that will be most useful for specific treatments and organ sites.

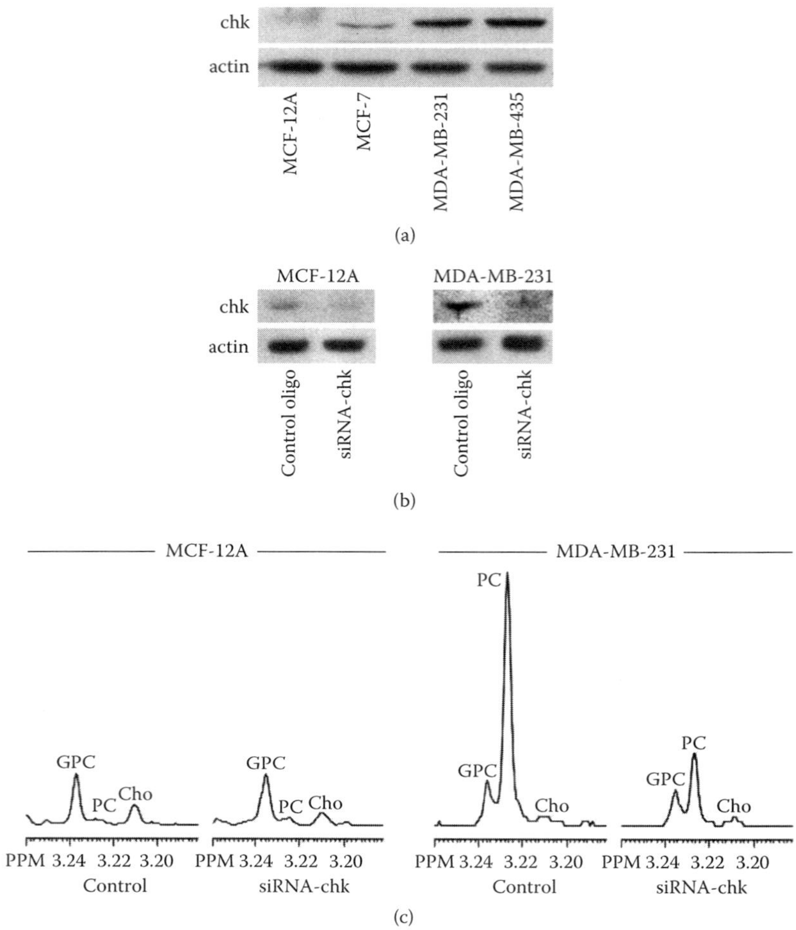

FIGURE 8.11 Western blots using choline kinase (chk) antibody (top) and the corresponding actin loading controls (bottom) of (a) a panel of HMECs — nonmalignant HMECs MCF-12A, weakly metastatic MCF-7 breast cancer cells, and highly metastatic MDA-MB-231 and MDA-MB-435 breast cancer cells — and of (b) MCF-12A and MDA-MB-231 cells treated with siRNA specific to choline kinase (siRNA-chk) compared to the respective oligofectamine controls (control oligo). (c) Corresponding ^{1}H MR spectra of siRNA-chk-treated MCF-12A and MDA-MB-231 cells compared to their respective controls. Cho, free choline; GPC, glycerophosphocholine; PC, phosphocholine. (Adapted from Glunde, K. et al., *Cancer Res.*, 65, 11034–11043, 2005. With permission.)

ACKNOWLEDGMENTS

Work from the authors' program was supported through funding by NIH grants R01 CA73850, R01 CA082337, P50 CA103175, R21 CA112216, R01 CA100184, R01 CA97310, and 3P30 CA006973-43S1.

REFERENCES

1. de Vries, I.J., Lesterhuis, W.J., Barentsz, J.O., Verdijk, P., van Krieken, J.H., Boerman, O.C., Oyen, W.J., Bonenkamp, J.J., Boezeman, J.B., Adema, G.J., Bulte, J.W., Scheenen, T.W., Punt, C.J., Heerschap, A., and Figdor, C.G., Magnetic resonance tracking of dendritic cells in melanoma patients for monitoring of cellular therapy, *Nat Biotechnol*, 23, 1407–1413, 2005.

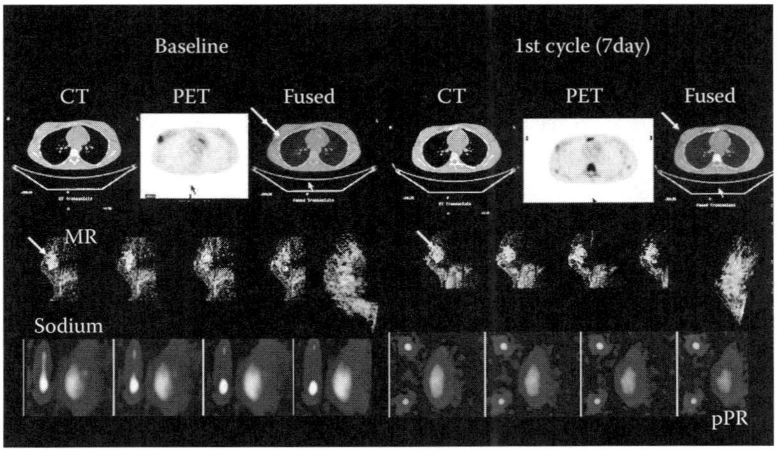

FIGURE 8.12 (please see color insert) Combined multimodality images on a 35-year-old female with invasive carcinoma. MR volume was 32 cc at baseline, 28 cc after cycle 1, and 12 cc before surgery. The sodium concentration within the lesion before treatment was 42.8 m*M*, whereas after cycle 1 it was 41 m*M*. The standardized uptake value (SUV) changed significantly from 6.7 to 3.15 after cycle 1.

2. McMahon, M.T., Gilad, A.A., Zhou, J., Sun, P.Z., Bulte, J.W., and van Zijl, P.C., Quantifying exchange rates in chemical exchange saturation transfer agents using the saturation time and saturation power dependencies of the magnetization transfer effect on the magnetic resonance imaging signal (QUEST and QUESP): Ph calibration for poly-L-lysine and a starburst dendrimer, *Magn Reson Med*, 55, 836–847, 2006.
3. Cohen, B., Dafni, H., Meir, G., Harmelin, A., and Neeman, M., Ferritin as an endogenous MRI reporter for noninvasive imaging of gene expression in C6 glioma tumors, *Neoplasia*, 7, 109–117, 2005.
4. Louie, A.Y., Huber, M.M., Ahrens, E.T., Rothbacher, U., Moats, R., Jacobs, R.E., Fraser, S.E., and Meade, T.J., *In vivo* visualization of gene expression using magnetic resonance imaging, *Nat Biotechnol*, 18, 321–325, 2000.
5. Li, L., Yazaki, P.J., Anderson, A.L., Crow, D., Colcher, D., Wu, A.M., Williams, L.E., Wong, J.Y., Raubitschek, A., and Shively, J.E., Improved biodistribution and radioimmunoimaging with poly(ethylene glycol)-DOTA-conjugated anti-CEA diabody, *Bioconjug Chem*, 17, 68–76, 2006.
6. Olafsen, T., Tan, G.J., Cheung, C.W., Yazaki, P.J., Park, J.M., Shively, J.E., Williams, L.E., Raubitschek, A.A., Press, M.F., and Wu, A.M., Characterization of engineered anti-p185HER-2 (scFv-CH3)2 antibody fragments (minibodies) for tumor targeting, *Protein Eng Design Selection*, 17, 315–323, 2004.
7. Weissleder, R., Moore, A., Mahmood, U., Bhorade, R., Benveniste, H., Chiocca, E.A., and Basilion, J.P., *In vivo* magnetic resonance imaging of transgene expression, *Nat Med*, 6, 351–355, 2000.
8. Zhao, M., Beauregard, D.A., Loizou, L., Davletov, B., and Brindle, K.M., Non-invasive detection of apoptosis using magnetic resonance imaging and a targeted contrast agent, *Nat Med*, 7, 1241–1244, 2001.
9. Winter, P.M., Caruthers, S.D., Kassner, A., Harris, T.D., Chinen, L.K., Allen, J.S., Lacy, E.K., Zhang, H., Robertson, J.D., Wickline, S.A., and Lanza, G.M., Molecular imaging of angiogenesis in nascent Vx-2 rabbit tumors using a novel alpha(nu)beta3-targeted nanoparticle and 1.5 tesla magnetic resonance imaging, *Cancer Res*, 63, 5838–5843, 2003.
10. Sipkins, D.A., Cheresh, D.A., Kazemi, M.R., Nevin, L.M., Bednarski, M.D., and Li, K.C., Detection of tumor angiogenesis *in vivo* by alphaVbeta3-targeted magnetic resonance imaging, *Nat Med*, 4, 623–626, 1998.
11. Anderson, S.A., Rader, R.K., Westlin, W.F., Null, C., Jackson, D., Lanza, G.M., Wickline, S.A., and Kotyk, J.J., Magnetic resonance contrast enhancement of neovasculature with alpha(v)beta(3)-targeted nanoparticles, *Magn Reson Med*, 44, 433–439, 2000.
12. Paganelli, G., Grana, C., Chinol, M., Cremonesi, M., De Cicco, C., De Braud, F., Robertson, C., Zurrida, S., Casadio, C., Zoboli, S., Siccardi, A.G., and Veronesi, U., Antibody-guided three-step therapy for high grade glioma with yttrium-90 biotin, *Eur J Nucl Med*, 26, 348–357, 1999.

13. Artemov, D., Mori, N., Ravi, R., and Bhujwalla, Z.M., Magnetic resonance molecular imaging of the HER-2/neu receptor, *Cancer Res*, 63, 2723–2727, 2003.
14. Dodd, S.J., Williams, M., Suhan, J.P., Williams, D.S., Koretsky, A.P., and Ho, C., Detection of single mammalian cells by high-resolution magnetic resonance imaging, *Biophys J*, 76 (Pt 1), 103–109, 1999.
15. Heyn, C., Ronald, J.A., Mackenzie, L.T., MacDonald, I.C., Chambers, A.F., Rutt, B.K., and Foster, P.J., *In vivo* magnetic resonance imaging of single cells in mouse brain with optical validation, *Magn Reson Med*, 55, 23–29, 2006.
16. Bulte, J.W., Arbab, A.S., Douglas, T., and Frank, J.A., Preparation of magnetically labeled cells for cell tracking by magnetic resonance imaging, *Methods Enzymol*, 386, 275–299, 2004.
17. Gimi, B., Mori, N., Ackerstaff, E., Frost, E.E., Bulte, J.W., and Bhujwalla, Z.M., Noninvasive MRI of endothelial cell response to human breast cancer cells, *Neoplasia*, 8, 207–213, 2006.
18. Moore, A., Medarova, Z., Potthast, A., and Dai, G., *In vivo* targeting of underglycosylated MUC-1 tumor antigen using a multimodal imaging probe, *Cancer Res*, 64, 1821–1827, 2004.
19. Stegman, L.D., Rehemtulla, A., Beattie, B., Kievit, E., Lawrence, T.S., Blasberg, R.G., Tjuvajev, J.G., and Ross, B.D., Noninvasive quantitation of cytosine deaminase transgene expression in human tumor xenografts with *in vivo* magnetic resonance spectroscopy, *Proc Natl Acad Sci USA*, 96, 9821–9826, 1999.
20. Aboagye, E.O., Artemov, D., Senter, P.D., and Bhujwalla, Z.M., Intratumoral conversion of 5-fluorocytosine to 5-fluorouracil by monoclonal antibody-cytosine deaminase conjugates: noninvasive detection of prodrug activation by magnetic resonance spectroscopy and spectroscopic imaging, *Cancer Res*, 58, 4075–4078, 1998.
21. Yu, J. and Mason, R.P., Synthesis and characterization of novel lacZ gene reporter molecules: detection of beta-galactosidase activity by ^{19}F nuclear magnetic resonance of polyglycosylated fluorinated vitamin B6, *J Med Chem*, 49, 1991–1999, 2006.
22. Mulder, W.J., Strijkers, G.J., Habets, J.W., Bleeker, E.J., van der Schaft, D.W., Storm, G., Koning, G.A., Griffioen, A.W., and Nicolay, K., MR molecular imaging and fluorescence microscopy for identification of activated tumor endothelium using a bimodal lipidic nanoparticle, *FASEB J*, 19, 2008–2010, 2005.
23. Viglianti, B.L., Abraham, S.A., Michelich, C.R., Yarmolenko, P.S., MacFall, J.R., Bally, M.B., and Dewhirst, M.W., *In vivo* monitoring of tissue pharmacokinetics of liposome/drug using MRI: illustration of targeted delivery, *Magn Reson Med*, 51, 1153–1162, 2004.
24. Gimi, B., Leong, T., Gu, Z., Yang, M., Artemov, D., Bhujwalla, Z.M., and Gracias, D.H., Self-assembled three dimensional radio frequency (RF) shielded containers for cell encapsulation, *Biomed Microdevices*, 7, 341–345, 2005.
25. Semenza, G.L., Expression of hypoxia-inducible factor 1: mechanisms and consequences, *Biochem Pharmacol*, 59, 47–53, 2000.
26. Liotta, L.A. and Kohn, E.C., The microenvironment of the tumour-host interface, *Nature*, 411, 375–379, 2001.
27. Egeblad, M. and Werb, Z., New functions for the matrix metalloproteinases in cancer progression, *Nat Rev Cancer*, 2, 161–174, 2002.
28. Bissell, M.J. and Radisky, D., Putting tumours in context, *Nat Rev Cancer*, 1, 46–54, 2001.
29. Pilatus, U., Ackerstaff, E., Artemov, D., Mori, N., Gillies, R.J., and Bhujwalla, Z.M., Imaging prostate cancer invasion with multi-nuclear magnetic resonance methods: the metabolic Boyden chamber, *Neoplasia*, 2, 273–279, 2000.
30. Pathak, A.P., Artemov, D., Ward, B.D., Jackson, D.G., Neeman, M., and Bhujwalla, Z.M., Characterizing extravascular fluid transport of macromolecules in the tumor interstitium by magnetic resonance imaging, *Cancer Res*, 65, 1425–1432, 2005.
31. Pathak, A.P., Artemov, D., Neeman, M., and Bhujwalla, Z.M., Lymph node metastasis in breast cancer xenografts is associated with increased regions of extravascular drain, lymphatic vessel area, and invasive phenotype, *Cancer Res*, 66, 5151–5158, 2006.
32. Raman, V., Artemov, D., Mironchik, E., and Bhujwalla, Z.M., Combined and Co-registered MRI, MRS, and Optical Imaging Characterization of Tumor Vascularization, Metabolism and Hypoxia, paper presented at International Society for Magnetic Resonance in Medicine, Honolulu, 2002.
33. van Sluis, R., Bhujwalla, Z.M., Raghunand, N., Ballesteros, P., Alvarez, J., Cerdan, S., Galons, J.P., and Gillies, R.J., *In vivo* imaging of extracellular pH using ^{1}H MRSI, *Magn Reson Med*, 41, 743–750, 1999.

34. Bhujwalla, Z.M., Artemov, D., Ballesteros, P., Cerdan, S., Gillies, R.J., and Solaiyappan, M., Combined vascular and extracellular pH imaging of solid tumors, *NMR Biomed*, 15, 114–119, 2002.
35. Jacobs, M.A., Ouwerkerk, R., Bottomley, P.A., Barker, P.B., Davidson, N.E., Wolff, A.C., Bhujwalla, Z.M., and Bluemke, D.B., Multiparameter Proton, Sodium, and Spectroscopic Imaging of Human Breast Cancer, paper presented at Proceedings of the International Society of Magnetic Resonance 11th Scientific Meeting, 20th Annual Meeting, Toronto, Canada, 2003.
36. Jacobs, M.A., Ouwerkerk, R., Wolff, A.C., Stearns, V., Bottomley, P.A., Barker, P.B., Argani, P., Khouri, N., Davidson, N.E., Bhujwalla, Z.M., and Bluemke, D.A., Multiparametric and multinuclear magnetic resonance imaging of human breast cancer: current applications, *Technol Cancer Res Treat*, 3, 543–550, 2004.
37. Chenevert, T.L., Stegman, L.D., Taylor, J.M., Robertson, P.L., Greenberg, H.S., Rehemtulla, A., and Ross, B.D., Diffusion magnetic resonance imaging: an early surrogate marker of therapeutic efficacy in brain tumors, *J Natl Cancer Inst*, 92, 2029–2036, 2000.
38. Hamstra, D.A., Chenevert, T.L., Moffat, B.A., Johnson, T.D., Meyer, C.R., Mukherji, S.K., Quint, D.J., Gebarski, S.S., Fan, X., Tsien, C.I., Lawrence, T.S., Junck, L., Rehemtulla, A., and Ross, B.D., Evaluation of the functional diffusion map as an early biomarker of time-to-progression and overall survival in high-grade glioma, *Proc Natl Acad Sci USA*, 102, 16759–16764, 2005.
39. Le Bihan, D., Breton, E., Lallemand, D., Grenier, P., Cabanis, E., and Laval Jeantet, M., MR imaging of intravoxel incoherent motions: application to diffusion and perfusion in neurologic disorders, *Radiology*, 161, 401–407, 1986.
40. Moseley, M.E., Kucharczyk, J., Mintorovitch, J., Cohen, Y., Kurhanewicz, J., Derugin, N., Asgari, H., and Norman, D., Diffusion-weighted MR imaging of acute stroke: correlation with T_2-weighted and magnetic susceptibility-enhanced MR imaging in cats, *Am J Neuroradiol*, 11, 423–429, 1990.
41. Mintorovitch, J., Moseley, M., Chileuitt, L., Shimizu, H., Cohen, Y., and Weinstein, P., Comparison of diffusion and T2 weighted MRI for the early detection of cerebral ischemia and reperfusion in rats, *Magn Reson Med*, 18, 39–50, 1991.
42. Minematsu, K., Li, L., Sotak, C., Davis, M., and Fisher, M., Reversible focal ischemic injury demonstrated by diffusion-weighted magnetic resonance imaging in rat brain, *Stroke*, 23, 1304–1311, 1992.
43. Knight, R.A., Ordidge, R., Helpern, J., Chopp, M., Rodolosi, L., and Peck, D., Temporal evolution of ischemic damage in rat brain measured by proton nuclear magnetic resonance imaging, *Stroke*, 22, 802–808, 1991.
44. Brunberg, J.A., Chenevert, T.L., McKeever, P.E., Ross, D.A., Junck, L.R., Muraszko, K.M., Dauser, R., Pipe, J.G., and Betley, A.T., *In vivo* MR determination of water diffusion coefficients and diffusion anisotropy: correlation with structural alteration in gliomas of the cerebral hemispheres, *Am J Neuroradiol*, 16, 361–371, 1995.
45. Carr, D.H., Brown, J., Bydder, G.M., Weinmann, H.J., Speck, U., Thomas, D.J., and Young, I.R., Intravenous chelated gadolinium as a contrast agent in NMR imaging of cerebral tumours, *Lancet*, 1, 484–486, 1984.
46. Orel, S.G., Differentiating benign from malignant enhancing lesions identified at MR imaging of the breast: are time-signal intensity curves an accurate predictor? *Radiology*, 211, 5–7, 1999.
47. Kamel, I.R. and Bluemke, D.A., MR imaging of liver tumors, *Radiol Clin North Am*, 41, 51–65, 2003.
48. Kim, J.K., Hong, S.S., Choi, Y.J., Park, S.H., Ahn, H., Kim, C.S., and Cho, K.S., Wash-in rate on the basis of dynamic contrast-enhanced MRI: usefulness for prostate cancer detection and localization, *J Magn Reson Imaging*, 22, 639–646, 2005.
49. Tanase, C. and Boada, F.E., Triple-quantum-filtered imaging of sodium in presence of B(0) inhomogeneities, *J Magn Reson*, 174, 270–278, 2005.
50. Ouwerkerk, R., Bleich, K.B., Gillen, J.S., Pomper, M.G., and Bottomley, P.A., Tissue sodium concentration in human brain tumors as measured with ^{23}Na MR imaging, *Radiology*, 227, 529–537, 2003.
51. Thulborn, K.R., Gindin, T.S., Davis, D., and Erb, P., Comprehensive MR imaging protocol for stroke management: tissue sodium concentration as a measure of tissue viability in nonhuman primate studies and in clinical studies, *Radiology*, 213, 156–166, 1999.
52. Gribbestad, I.S., Singstad, T.E., Nilsen, G., Fjosne, H.E., Engan, T., Haugen, O.A., and Rinck, P.A., *In vivo* ^{1}H MRS of normal breast and breast tumors using a dedicated double breast coil, *J Magn Reson Imaging*, 8, 1191–1197, 1998.

53. Roebuck, J.R., Cecil, K.M., Schnall, M.D., and Lenkinski, R.E., Human breast lesions: characterization with proton MR spectroscopy, *Radiology*, 209, 269–275, 1998.
54. Kvistad, K.A., Bakken, I.J., Gribbestad, I.S., Ehrnholm, B., Lundgren, S., Fjosne, H.E., and Haraldseth, O., Characterization of neoplastic and normal human breast tissues with *in vivo* ^{1}H MR spectroscopy, *J Magn Reson Imaging*, 10, 159–164, 1999.
55. Zakian, K.L., Sircar, K., Hricak, H., Chen, H.N., Shukla-Dave, A., Eberhardt, S., Muruganandham, M., Ebora, L., Kattan, M.W., Reuter, V.E., Scardino, P.T., and Koutcher, J.A., Correlation of proton MR spectroscopic imaging with Gleason score based on step-section pathologic analysis after radical prostatectomy, *Radiology*, 234, 804–814, 2005.
56. Kurhanewicz, J., Vigneron, D.B., Nelson, S.J., Hricak, H., MacDonald, J.M., Konety, B., and Narayan, P., Citrate as an *in vivo* marker to discriminate prostate cancer from benign prostatic hyperplasia and normal prostate peripheral zone: detection via localized proton spectroscopy, *Urology*, 45, 459–466, 1995.
57. Mukherji, S.K., Schiro, S., Castillo, M., Kwock, L., Muller, K.E., and Blackstock, W., Proton MR spectroscopy of squamous cell carcinoma of the extracranial head and neck: *in vitro* and *in vivo* studies, *Am J Neuroradiol*, 18, 1057–1072, 1997.
58. Jenkinson, M.D., Smith, T.S., Joyce, K., Fildes, D., du Plessis, D.G., Warnke, P.C., and Walker, C., MRS of oligodendroglial tumors: correlation with histopathology and genetic subtypes, *Neurology*, 64, 2085–2089, 2005.
59. Gill, S.S. et al., Proton MR spectroscopy of intracranial tumours: *in vivo* and *in vitro* studies, *J Comput Assist Tomogr*, 14, 497–504, 1990.
60. Yacoe, M.E., Sommer, G., and Peehl, D., *In vitro* proton spectroscopy of normal and abnormal prostate, *Magn Reson Med*, 19, 429–438, 1991.
61. Li, X., Lu, Y., Pirzkall, A., McKnight, T., and Nelson, S.J., Analysis of the spatial characteristics of metabolic abnormalities in newly diagnosed glioma patients, *J Magn Reson Imaging*, 16, 229–237, 2002.
62. Kurhanewicz, J., Vigneron, D.B., and Nelson, S.J., Three-dimensional magnetic resonance spectroscopic imaging of brain and prostate cancer, *Neoplasia*, 2, 166–189, 2000.
63. Katz-Brull, R., Lavin, P.T., and Lenkinski, R.E., Clinical utility of proton magnetic resonance spectroscopy in characterizing breast lesions, *J Natl Cancer Inst*, 94, 1197–1203, 2002.
64. Leach, M.O., Verrill, M., Glaholm, J., Smith, T.A., Collins, D.J., Payne, G.S., Sharp, J.C., Ronen, S.M., McCready, V.R., Powles, T.J., and Smith, I.E., Measurements of human breast cancer using magnetic resonance spectroscopy: a review of clinical measurements and a report of localized ^{31}P measurements of response to treatment, *NMR Biomed*, 11, 314–340, 1998.
65. Kurhanewicz, J., Swanson, M.G., Nelson, S.J., and Vigneron, D.B., Combined magnetic resonance imaging and spectroscopic imaging approach to molecular imaging of prostate cancer, *J Magn Reson Imaging*, 16, 451–463, 2002.
66. Loening, N.M., Chamberlin, A.M., Zepeda, A.G., Gonzalez, R.G., and Cheng, L.L., Quantification of phosphocholine and glycerophosphocholine with ^{31}P edited ^{1}H NMR spectroscopy, *NMR Biomed*, 18, 413–420, 2005.
67. Cheng, L.L., Chang, I.W., Louis, D.N., and Gonzalez, R.G., Correlation of high-resolution magic angle spinning proton magnetic resonance spectroscopy with histopathology of intact human brain tumor specimens, *Cancer Res*, 58, 1825–1832, 1998.
68. James, K., Eisenhauer, E., Christian, M., Terenziani, M., Vena, D., Muldal, A., and Therasse, P., Measuring response in solid tumors: unidimensional versus bidimensional measurement, *J Natl Cancer Inst*, 91, 523–528, 1999.
69. Therasse, P., Arbuck, S.G., Eisenhauer, E.A., Wanders, J., Kaplan, R.S., Rubinstein, L., Verweij, J., Van Glabbeke, M., van Oosterom, A.T., Christian, M.C., and Gwyther, S.G., New guidelines to evaluate the response to treatment in solid tumors, *J Natl Cancer Inst*, 92, 205–216, 2000.
70. Meisamy, S., Bolan, P.J., Baker, E.H., Bliss, R.L., Gulbahce, E., Everson, L.I., Nelson, M.T., Emory, T.H., Tuttle, T.M., Yee, D., and Garwood, M., Neoadjuvant chemotherapy of locally advanced breast cancer: predicting response with *in vivo* ^{1}H MR spectroscopy — a pilot study at 4 T, *Radiology*, 233, 424–431, 2004.
71. Manton, D.J., Chaturvedi, A., Hubbard, A., Lind, M.J., Lowry, M., Maraveyas, A., Pickles, M.D., Tozer, D.J., and Turnbull, L.W., Neoadjuvant chemotherapy in breast cancer: early response prediction with quantitative MR imaging and spectroscopy, *Br J Cancer*, 94, 427–435, 2006.

72. Jagannathan, N.R., Kumar, M., Raghunathan, P., Coshic, O., Julka, P.K., and Rath, G.K., Assessment of the therapeutic response of human breast carcinoma using *in vivo* volume localized proton magnetic resonance spectroscopy, *Curr Sci*, 76, 777–782, 1999.
73. Schepkin, V.D., Ross, B.D., Chenevert, T.L., Rehemtulla, A., Sharma, S., Kumar, M., and Stojanovska, J., Sodium magnetic resonance imaging of chemotherapeutic response in a rat glioma, *Magn Reson Med*, 53, 85–92, 2005.
74. Jacobs, M.A., Barker, P.B., Argani, P., Ouwerkerk, R., Bhujwalla, Z.M., and Bluemke, D.A., Combined dynamic contrast enhanced breast MR and proton spectroscopic imaging: a feasibility study, *J Magn Reson Imaging*, 21, 23–28, 2005.
75. Ackerstaff, E., Glunde, K., and Bhujwalla, Z.M., Choline phospholipid metabolism: a target in cancer cells? *J Cell Biochem*, 90, 525–533, 2003.
76. Aboagye, E.O. and Bhujwalla, Z.M., Malignant transformation alters membrane choline phospholipid metabolism of human mammary epithelial cells, *Cancer Res*, 59, 80–84, 1999.
77. Ackerstaff, E., Pflug, B.R., Nelson, J.B., and Bhujwalla, Z.M., Detection of increased choline compounds with proton nuclear magnetic resonance spectroscopy subsequent to malignant transformation of human prostatic epithelial cells, *Cancer Res*, 61, 3599–3603, 2001.
78. Bhujwalla, Z.M., Aboagye, E.O., Gillies, R.J., Chacko, V.P., Mendola, C.E., and Backer, J.M., Nm23-transfected MDA-MB-435 human breast carcinoma cells form tumors with altered phospholipid metabolism and pH: a 31P nuclear magnetic resonance study *in vivo* and *in vitro*, *Magn Reson Med*, 41, 897–903, 1999.
79. Ronen, S.M., Jackson, L.E., Beloueche, M., and Leach, M.O., Magnetic resonance detects changes in phosphocholine associated with Ras activation and inhibition in NIH 3T3 cells, *Br J Cancer*, 84, 691–696, 2001.
80. Sterin, M., Cohen, J.S., Mardor, Y., Berman, E., and Ringel, I., Levels of phospholipid metabolites in breast cancer cells treated with antimitotic drugs: a ^{31}P-magnetic resonance spectroscopy study, *Cancer Res*, 61, 7536–7543, 2001.
81. Natarajan, K., Mori, N., Artemov, D., and Bhujwalla, Z.M., Exposure of human breast cancer cells to the anti-inflammatory agent indomethacin alters choline phospholipid metabolites and Nm23 expression, *Neoplasia*, 4, 409–416, 2002.
82. Cuadrado, A., Carnero, A., Dolfi, F., Jimenez, B., and Lacal, J.C., Phosphorylcholine: a novel second messenger essential for mitogenic activity of growth factors, *Oncogene*, 8, 2959–2968, 1993.
83. Chung, T., Huang, J.S., Mukherjee, J.J., Crilly, K.S., and Kiss, Z., Expression of human choline kinase in NIH 3T3 fibroblasts increases the mitogenic potential of insulin and insulin-like growth factor I, *Cell Signal*, 12, 279–288, 2000.
84. Foster, D.A. and Xu, L., Phospholipase D in cell proliferation and cancer, *Mol Cancer Res*, 1, 789–800, 2003.
85. Rodriguez-Gonzalez, A., Ramirez de Molina, A., Benitez-Rajal, J., and Lacal, J.C., Phospholipase D and choline kinase: their role in cancer development and their potential as drug targets, *Prog Cell Cycle Res*, 5, 191–201, 2003.
86. Steed, P.M. and Chow, A.H., Intracellular signaling by phospholipase D as a therapeutic target, *Curr Pharm Biotechnol*, 2, 241–256, 2001.
87. Pai, J.K., Frank, E.A., Blood, C., and Chu, M., Novel ketoepoxides block phospholipase D activation and tumor cell invasion, *Anticancer Drug Design*, 9, 363–372, 1994.
88. Ramirez de Molina, A., Rodriguez-Gonzalez, A., Gutierrez, R., Martinez-Pineiro, L., Sanchez, J., Bonilla, F., Rosell, R., and Lacal, J., Overexpression of choline kinase is a frequent feature in human tumor-derived cell lines and in lung, prostate, and colorectal human cancers, *Biochem Biophys Res Commun*, 296, 580–583, 2002.
89. Ramirez de Molina, A., Gutierrez, R., Ramos, M.A., Silva, J.M., Silva, J., Bonilla, F., Sanchez, J.J., and Lacal, J.C., Increased choline kinase activity in human breast carcinomas: clinical evidence for a potential novel antitumor strategy, *Oncogene*, 21, 4317–4322, 2002.
90. Nakagami, K., Uchida, T., Ohwada, S., Koibuchi, Y., and Morishita, Y., Increased choline kinase activity in 1,2-dimethylhydrazine-induced rat colon cancer, *Jpn J Cancer Res*, 90, 1212–1217, 1999.
91. Nakagami, K., Uchida, T., Ohwada, S., Koibuchi, Y., Suda, Y., Sekine, T., and Morishita, Y., Increased choline kinase activity and elevated phosphocholine levels in human colon cancer, *Jpn J Cancer Res*, 90, 419–424, 1999.

92. Ratnam, S. and Kent, C., Early increase in choline kinase activity upon induction of the H-*ras* oncogene in mouse fibroblast cell lines, *Arch Biochem Biophys*, 323, 313–322, 1995.
93. Teegarden, D., Taparowsky, E.J., and Kent, C., Altered phosphatidylcholine metabolism in C3H10T1/2 cells transfected with the Harvey-*ras* oncogene, *J Biol Chem*, 265, 6042–6047, 1990.
94. Ramirez de Molina, A., Penalva, V., Lucas, L., and Lacal, J.C., Regulation of choline kinase activity by Ras proteins involves Ral-GDS and PI3K, *Oncogene*, 21, 937–946, 2002.
95. Adjei, A.A., Blocking oncogenic Ras signaling for cancer therapy, *J Natl Cancer Inst*, 93, 1062–1074, 2001.
96. Hernandez-Alcoceba, R., Saniger, L., Campos, J., Nunez, M.C., Khaless, F., Gallo, M.A., Espinosa, A., and Lacal, J.C., Choline kinase inhibitors as a novel approach for antiproliferative drug design, *Oncogene*, 15, 2289–2301, 1997.
97. Rodriguez-Gonzalez, A., de Molina, A.R., Fernandez, F., Ramos, M.A., Nunez Mdel, C., Campos, J., and Lacal, J.C., Inhibition of choline kinase as a specific cytotoxic strategy in oncogene-transformed cells, *Oncogene*, 22, 8803–8812, 2003.
98. Hernandez-Alcoceba, R., Fernandez, F., and Lacal, J.C., *In vivo* antitumor activity of choline kinase inhibitors: a novel target for anticancer drug discovery, *Cancer Res*, 59, 3112–3118, 1999.
99. Glunde, K., Raman, V., Mori, N., and Bhujwalla, Z.M., RNA interference-mediated choline kinase suppression in breast cancer cells induces differentiation and reduces proliferation, *Cancer Res*, 65, 11034–11043, 2005.
100. Saif, M.W., Management of colorectal cancer in pregnancy: a multimodality approach, *Clin Colorectal Cancer*, 5, 247–256, 2005.
101. Smith, J.A. and Andreopoulou, E., An overview of the status of imaging screening technology for breast cancer, *Ann Oncol*, 15 (Suppl 1), I18–I26, 2004.
102. Yang, G.Y., Wagner, T.D., Fuss, M., and Thomas, C.R., Jr., Multimodality approaches for pancreatic cancer, *CA Cancer J Clin*, 55, 352–367, 2005.
103. Seemann, M.D., Whole-body PET/MRI: the future in oncological imaging, *Technol Cancer Res Treat*, 4, 577–582, 2005.
104. Jacobs, M.A., Stearns, V., Wolff, A., Ouwerkerk, R., Bluemke, D.A., and Wahl, R.L., Multimodality (MR/PET/CT) monitoring preoperative systemic therapy in operable breast cancer, presented at International Society for Magnetic Resonance in Medicine, 14, 2006.

9 Molecular Imaging of Atherosclerosis with Magnetic Resonance

Karen C. Briley-Saebo, Willem J. Mulder, Fabien Hyafil, Venkatesh Mani, Vardan Amirbekian, Juan Gilberto S. Aguinaldo, James F. Rudd, Silvia H. Aguiar, and Zahi A. Fayad

CONTENTS

9.1 Introduction 161
9.1.1 Atherosclerosis 161
9.1.2 Multicontrast MRI 162
9.1.3 Contrast-Enhanced MRI 164
9.2 Lipid-Based Nanoparticles 164
9.2.1 Lipid Properties 164
9.2.2 Targeting Strategies 165
9.2.3 Liposomes 166
9.2.4 Micelles 167
9.2.5 Quantum Dots 169
9.2.6 Microemulsions 169
9.2.7 Lipoproteins 169
9.3 Iron Oxide Particles 170
9.3.1 Properties 170
9.3.2 Detection of Macrophages with USPIOs 172
9.3.3 Alternative Iron Oxide Particles 172
9.3.4 Positive Contrast Imaging Techniques for the Detection of USPIO 174
9.3.5 Perspectives 175
Acknowledgment 176
References 176

9.1 INTRODUCTION

9.1.1 ATHEROSCLEROSIS

Despite important clinical advances in the prevention and treatment of atherosclerosis during the past 20 years, atherosclerotic disease remains the first cause of mortality in industrialized countries.[1,2] Atherosclerosis is a complex disease where cholesterol deposition, inflammation, and plaque formation play a major role. Although the causes of atherosclerosis are not completely understood, most atherosclerotic lesions are characterized by a thickening of the arterial intima and are typically composed of a lipid core and an overlying fibrous cap.[3–6] X-ray angiography remains the gold standard for diagnosis and quantification of atherosclerotic plaques. X-ray angiography is an invasive technique that is used to visualize flow-limiting arterial stenoses, allowing for an indirect measure of atherosclerotic burden. Positive remodeling of the arterial wall — a process in which the vessel dilates to limit the narrowing of the lumen in the presence of atherosclerotic plaques — leads to an underestimation of the true extension of atherosclerosis disease with x-ray angiography.[7]

High-resolution magnetic resonance imaging (MRI) allows a submillimeter resolution of the arterial wall and has emerged as one of the most promising techniques for the direct evaluation of the burden of atherosclerosis.[8,9] Imaging atherosclerosis with MRI has shown a much wider distribution of atherosclerosis than predicted with angiography. Additionally, studies with MRI have indicated the need to discriminate between vulnerable or high-risk plaques, which typically cause myocardial infarction or strokes, and more stable plaques that have lipid cores surrounded by a thick fibrous cap.[10,11]

Myocardial infarction or strokes are usually caused by the sudden endothelial disruption of an atherosclerotic plaque by superficial intimal erosion or fibrous cap rupture. Plaque rupture, with subsequent exposure of the thrombogenic extracellular matrix to circulating blood, may trigger the formation of an intraluminal thrombus. In some cases, the thrombus may become obstructive and cut the oxygen supply to the downstream organs, thereby leading to acute ischemia and irreversible necrosis of the heart (myocardial infarction) or the brain (stroke).[10] It has become clear that the cellular and extracellular composition of atherosclerotic plaque is the primary determinant of plaque stability.[12] Lesions with a large lipid core, thin fibrous cap, preponderance of inflammatory cells (macrophages), and few vascular smooth muscle cells (VSMCs) are at the highest risk of rupture.[13] Inflammatory cells, particularly macrophages, produce enzymes such as metalloproteinases that break down the matrix proteins in the fibrous cap. In addition, macrophages secrete inflammatory cytokines, in particular interferon (IFN-γ), which inhibit VSMC proliferation and collagen synthesis. Furthermore, VSMCs in the fibrous cap have a reduced proliferative capacity and a propensity to apoptosis.[14] Consequently, inflammation within the plaques tends toward destruction of the fibrous cap and subsequent thrombosis. There appears to be a dynamic balance within the plaque between macrophages, which promote erosion and rupture of the fibrous cap, and VSMCs, which nourish and repair it. Since these processes are independent of plaque size, small asymptomatic plaques that are typically not seen using conventional x-ray angiography can rupture. On the other hand, some large plaques that obstruct flow and produce symptoms such as angina may be stable and not life threatening. As a result, there is an urgent need to discriminate stable plaques from potentially unstable lesions in clinical practice.

9.1.2 Multicontrast MRI

By taking advantage of intrinsic differences in the relaxation properties and proton densities between atherosclerotic plaque components, MRI can provide high-resolution imaging of plaques without injecting any contrast agent (CA).[8,11,15] The noncontrast agent approach has been referred to as multicontrast MRI. The small size of the vessels and adjacent lumen requires the acquisition of high spatial and contrast resolutions for atherosclerosis imaging. Atherosclerotic plaques are usually imaged using high-resolution black-blood fast spin echo MR sequences. Black-blood sequences improve the contrast between atherosclerotic plaques and the lumen and offer a better delineation of the contours of the plaque. These sequences null the signal of the flowing blood by using preparatory pulses (double inversion recovery or parallel saturation bands).[16] New black-blood techniques have recently been introduced for the simultaneous acquisition of multiple slices and allow the analysis of a full-length arterial segment with a reduced total examination time.[17,18] Using these techniques, clinical studies demonstrated that MRI provides excellent quantitative capabilities for the measurement of total plaque volume with an error in vessel wall area measurement as low as 2.6% for the aorta[19] and 3.5% in the carotids.[20,21]

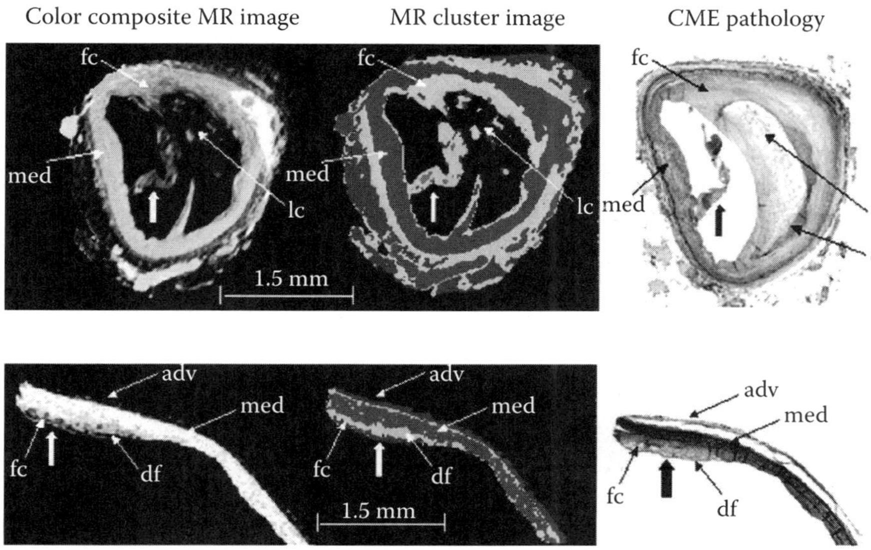

FIGURE 9.1 (please see color insert following page 210) Cluster analysis of human atherosclerotic plaque based upon the relative change in signal enhancement after application of T_1, T_2, and PDW spin echo sequences. Adv = adventia; nc = necrotic core; df = dense fibrous; fc = fibrous cap; med = media. (From Ref 15. With permission.)

TABLE 9.1
Representative Effect of Sequence Parameters on the Enhancement of Various Plaque Components

	Sequence Weighting		
Tissue	**T^1 Weighted**	**T^2 Weighted**	**PDW**
Media	–	–	±
Fibrocellular	+	+	+
Lipid	±	+	+
Necrotic core	±	+	+
Thrombus	–	–	–
Dense fibrous	±	–	±

Note: Signal intensities are expressed relative to adjacent muscle. + = hyperintense; ± = isointense; – = hypointense; PDW = protein density weighted.

Signal intensities detected with MRI are influenced by the relaxation times of protons (T_1 and T_2) and the proton density present in the different components of atherosclerotic plaques. The timing of the excitation pulses of an MR sequence will determine the weight of T_1 and T_2 relaxation times or proton density in the image contrast. Multicontrast MRI is based on successive T_1, T_2, and proton density weighted sequences. Analysis of signal intensities detected on each of these sequences allows differentiation of each of the atherosclerotic plaque components (lipid core, fibrous tissue, hemorrhage, calcification) by their different relaxation properties on MRI.[22] Development of dedicated software, which analyzes the signal intensities of multicontrast MRI on a pixel-by-pixel basis, has further improved the identification of atherosclerotic plaques components.[15] An example of multicontrast MRI showing different plaque components and automatic segmentation of these plaques using a k-means cluster algorithm is shown in Table 9.1.

Multicontrast MRI studies have primarily focused on carotid atherosclerotic plaques. The superficial location of carotid arteries and their relative absence of motion represent less of a technical challenge for imaging than the aorta or the coronary arteries. In this location, multicontrast MRI could also be compared to corresponding histology of atherosclerotic plaques from endarterectomy specimens. For example, studies have demonstrated that *in vivo* multicontrast MR of human carotid arteries had a sensitivity of 85% and specificity of 92% for the identification of a lipid core and acute intraplaque hemorrhage.[23] Recent histopathological studies suggest that intraplaque hemorrhage may play a role in plaque rupture and also represent a potent atherogenic stimulus. Preliminary studies have shown that[24,25] multicontrast MRI can also accurately detect intraplaque hemorrhages in carotid atherosclerotic plaques using T_2*-weighted sequences. Interestingly, a recent study[25] found that the detection of these hemorrhages in carotid atherosclerotic plaques with MRI was associated with an accelerated increase of plaque volume in the next 18 months.

In summary, multicontrast MRI allows for the detection of the different components of atherosclerotic plaques with high accuracy and is particularly promising for the study of carotid atherosclerotic plaques. However, due to low signal-to-noise ratios (SNRs) and partial voluming effects, application of multicontrast MRI is limited to large arteries such as the carotids or aorta. Contrast agents targeted to specific molecules present in atherosclerotic lesions are needed to further improve the detection and characterization of vulnerable plaques using MRI.

9.1.3 Contrast-Enhanced MRI

Contrast agents are used to enhance the relaxation properties of tissue and regions that contain the material. Contrast agents may be either paramagnetic (containing lanthanide metals) or superparamagnetic (containing particulate iron oxide particles). Generally, paramagnetic materials enhance the longitudinal magnetization of water protons, thereby inducing MR signal gain (positive signal). On the other hand, superparamagnetic materials enhance the transverse magnetization and induce MR signal loss. Currently, contrast agents are designed specifically for the detection and characterization of atherosclerotic plaque. These agents are based on either lipid- or iron oxide-based platforms. Most lipid-based materials contain paramagnetic gadolinium and generate signal gain upon uptake or association with the plaque. The advantage of lipid-based materials is related to the ease of associating antibodies or peptides with the surface of micelles and liposomes. Therefore, it is possible to target plaque components or molecules associated with atherosclerotic plaque vulnerability. Iron oxide particles that induce signal loss are typically passively targeted to functionalized macrophages within the lesion. Due to the low concentration of contrast agent that may be targeted to a specific receptor or cell, targeting of very specific molecules or activities within the plaque requires very efficient contrast agents in order to be detected by MRI. Therefore, this chapter will focus primarily on lipid-based contrast agents that have high longitudinal relaxivities and on iron oxide materials with highly effective transverse relaxivities.

9.2 LIPID-BASED NANOPARTICLES

9.2.1 Lipid Properties

Lipids are naturally occurring amphiphilic molecules that contain a hydrophilic head group and a hydrophobic tail, as shown in Figure 9.2a. Because of the dual character of the lipids and the energetically unfavorable contact between the lipid tails and water, amphiphiles self-associate into aggregates of different sizes and geometries, as shown in Figure 9.2b.[26,27] The major forces that direct the self-assembly of amphiphilic molecules into well-defined structures in water derive from the hydrophobic associative interactions of the lipid tails and the repulsive interactions between the hydrophilic head groups. A wide variety of lipid aggregate geometries occur, of which micelles, liposomes, and microemulsions are among the most relevant for MR-based contrast agents. The

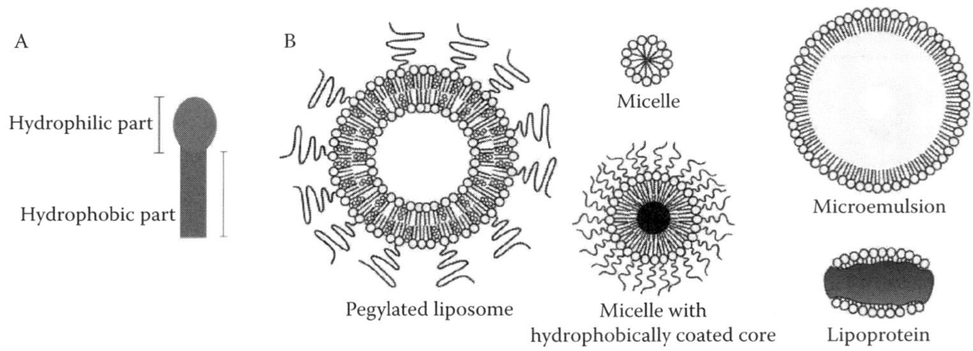

FIGURE 9.2 (a) Depiction of a naturally occurring amphiphilic molecule containing a hydrophilic head group and a hydrophobic tail. (b) Possible lipid aggregates for *in vivo* use. Whereas pegylated liposomes consist of a bilayer of lipids, micelles are made up of a monolayer. The vesicle of the liposome may be used to transport cytotoxic material or other drugs for targeted drug delivery. Micelles may contain hydrophobic solid nanoparticles in their corona. Microemulsions consist of a surfactant (amphiphile) monolayer covering an oil, fluid, or gas.

geometry of the aggregates that are formed in aqueous solution depends not only upon the structure of the amphiphilic molecules, but also on parameters such as pH, temperature, and concentration.[28] The length of the hydrophobic chains and the size of the head group (in relation to the chain) determine the curvature of the aggregate and whether a micelle-like structure or a bilayer (liposome) structure will be formed.[29]

Lipids may also be used as a shell to cover a solid core. In this way, iron oxide nanoparticles or nanocrystallic colloids (quantum dots),[30–32] as well as glass, silica, or mica, can be coated with biocompatible molecules. Another class of lipid-based nanoparticles is the lipoproteins. Lipoproteins are naturally occurring assemblies that contain both proteins (e.g., apolipoprotein AI) and lipids. Important nanosized lipoproteins are high-density lipoprotein (HDL) and low-density lipoprotein (LDL). Both lipoproteins have an important role in the transports of cholesterol and triglycerides. Since lipoproteins are endogenous, they can be used as natural nanoparticulate carriers for contrast-generating material.[33,34]

The ability of lipid-based materials to promote MR signal enhancement is related to the structure and composition of the material. The longitudinal relaxivity, r_1, associated with liposomes is largely dependent upon the location of the gadolinium lipid within the structure. If the gadolinium is contained within the vesicle or within the lipid bilayers, then the r_1 values may be reduced, relative to Gd-DTPA ($r_1 = 4\ \text{sec}^{-1}\text{m}M^{-1}$ at 20 MHz).[35] In addition, the relaxation mechanisms of these types of materials are greatly modulated by the rate of water exchange through the lipid bilayers.[36–39] For gadolinium mixed micelles and liposomes where the gadolinium is attached to the polar head of the lipid and thus directed outward toward the solvent, the classical inner-sphere/outer-sphere theory can be applied to describe the longitudinal relaxation.[40–43] As long as water molecules can freely exchange with the gadolinium ion (and are not shielded by other components associated with the particle), the r_1 values associated with mixed micelles and this class of liposome are much larger than that (>3 times) of Gd-DTPA.[42–47] The increase in the r_1 value associated with these materials is related to the decrease in the overall rate of molecular tumbling (large particles rotate slower than small gadolinium chelates).[40,48,49]

9.2.2 Targeting Strategies

Target specificity can be introduced to lipid-based nanoparticles via several routes. The endogenous lipoproteins are inherently target specific, but may be rerouted to other targets by enriching the

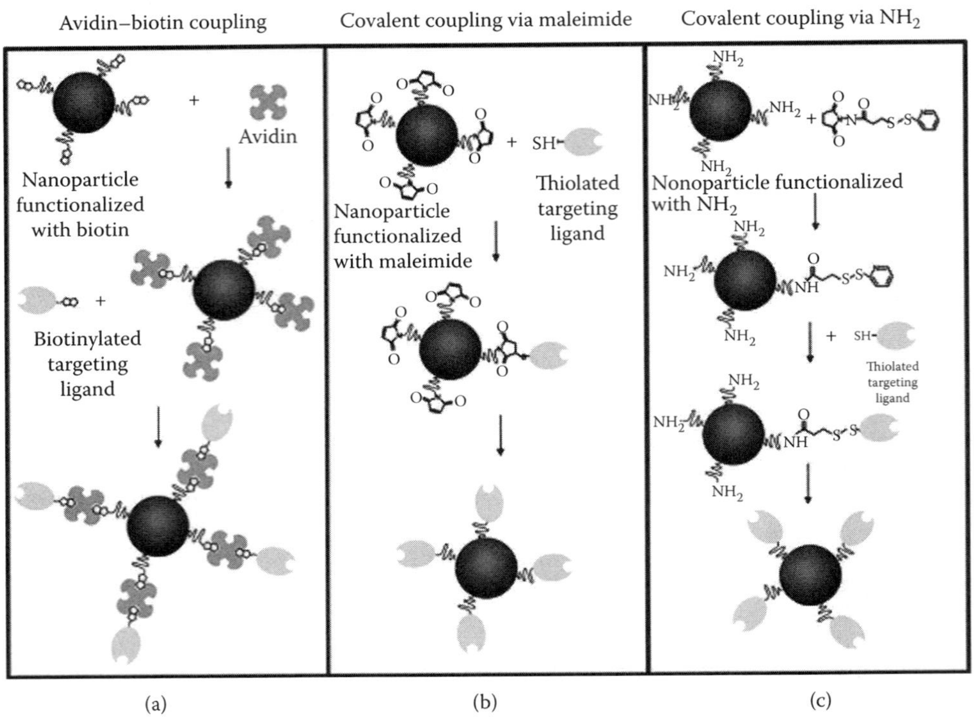

FIGURE 9.3 (a) Example of targeting nanoparticles via biotin–avidin interactions. (b) Example of a nanoparticle targeting strategy using a maleimide–thiol linkage. (c) Example of targeting nanoparticles using covalent coupling via NH_2 after activation with N-succinimidyl 3-[2-pyridyldithio]propionate.

protein with peptidic motives. Target specificity for the synthetic lipid-based nanoparticles can be introduced by conjugating targeting ligands (antibodies and peptides). Several chemical conjugation strategies have been described to link a variety of targeting ligands to MRI contrast agents.[28,44,50] These targeting ligands may include monoclonal antibodies (mAbs), antibody fragments (Fabs), (recombinant) proteins, peptides, peptidomimetics, sugars, and small molecules. Roughly, the ligand–nanoparticle coupling can be divided into noncovalent linkage and covalent linkage.

Noncovalent coupling is usually done with an avidin–biotin linkage. Avidin is a tetrameric protein with a molecular weight of 68 kDa and is capable of strongly binding four biotins ($KA \approx 1.7 \times 1015\ M^{-1}$). The biotin–avidin interaction has been exploited for conjugating different types of nanoparticulate MRI contrast agents with biotinylated proteins or peptides (Figure 9.3a, biotin–avidin linkage). Although this method is simple and effective, the introduction of avidin in the conjugate has certain drawbacks, one of which is the size of the conjugate, but more importantly, the immunogenic properties and the fast clearance of avidin by the liver need to be considered.[51] Covalently linking the ligand to the contrast agent directly leads to smaller conjugates, which may have more favorable pharmacokinetic properties. Although several methods have been described, covalent strategies can be divided into the formation of (1) an amide bond between activated carboxyl groups and amino groups, (2) a disulfide bond, and (3) a thioether bond between maleimide and thiol, as shown in Figure 9.3b. Furthermore, amines can be activated for conjugation to thiol-(SH) exposing ligands with N-succinimidyl 3-[2-pyridyldithio]propionate, as shown in Figure 9.3c.

9.2.3 Liposomes

Liposomes can be defined as spherical, self-closed structures formed by one or several concentric lipid bilayers with an aqueous phase inside and between the lipid bilayers.[44] They can vary in size

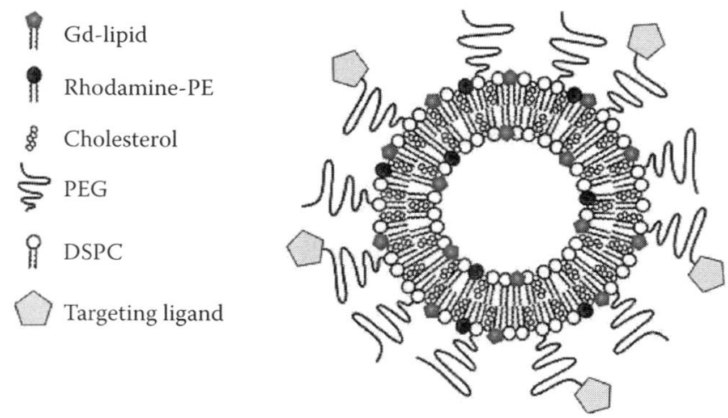

FIGURE 9.4 Schematic representation of a bimodal liposome.

and lamellarity and are therefore subdivided into multilamellar vesicles (MLVs; consisting of several concentric bilayers), large unilamellar vesicles (LUVs; in a size range of 200 to 800 nm), and small unilamellar vesicles (SUVs; in a size range of 50 to 150 nm). Paramagnetic properties are introduced by incorporating paramagnetic agents in the aqueous lumen or paramagnetic amphiphiles in the liposomal bilayer.

A polymerized liposomal MRI contrast agent has been developed[52] that is composed of a Gd-DTPA-based lipid, an amphiphilic carrier lipid, and a biotinylated amphiphile. The biotinylated lipid was used to introduce target specificity via an avidin–biotin linkage procedure. Liposomes coupled to v3-specific LM609 antibodies have been used to detect angiogenesis in tumor-bearing rabbits and to detect the expression of leukocyte adhesion molecules (ICAM-1) with MRI in the brain of a mouse model of multiple sclerosis.[53,54] Both the v3-integrin and ICAM-1 are important cell surface receptors associated with atherosclerosis. Many atherosclerotic lesions have neovascularization[55] of the vasa vasorum network, which is accompanied with an increased expression of v3 on the immature endothelium of these neovessels. ICAM-1, also expressed at the endothelium, plays an important role in the early atherosclerotic inflammatory process.

Recently, a bimodal targeted liposomal contrast agent was introduced (Figure 9.4, liposomal CAs), composed of a Gd-DTPA lipid, a fluorescent lipid, carrier lipids distearoyl phosphatidyl-choline [DSPC] and cholesterol), and a PEG lipid for improved stability and pharmacokinetics.[28,56] The PEG lipids were used to covalently link multiple antibodies specific for E-selectin; this contrast agent was made specifically for inflammatory processes present in the early phases of atherosclerosis. Angiogenesis, the formation of new blood vessels, is an important feature of atherosclerotic plaques.[13] Specificity for angiogenic blood vessels was realized by linking several v3-specific arginine-glycine-aspartic acid (RGD) peptides to this liposomal system.[57] The effectiveness for the detection of regions rich in neovessels was demonstrated in a tumor model. Furthermore, the bimodal character of this contrast agent allowed the detection of the contrast agent on histologic sections of the excised tumor tissue with fluorescence microscopy. Fluorescent microscopy confirmed the specific and exclusive association of the contrast agent with activated tumor vessels. This study demonstrated the interest of validating the MRI findings with a complementary technique like fluorescence microscopy.

9.2.4 Micelles

In contrast to liposomes, micelles are composed of aggregated amphiphiles, are single lipid layers, and do not enclose an aqueous lumen.[42,43] Therefore, the stability of the aggregate formed is different from bilayered vesicles and has to be considered when applied *in vivo*. An important feature of micelles is the critical micelle concentration (CMC). The CMC reflects at what concentration monomers of the amphiphile assemble into micellular aggregates. The lower the CMC, the more stable the micelles formed.

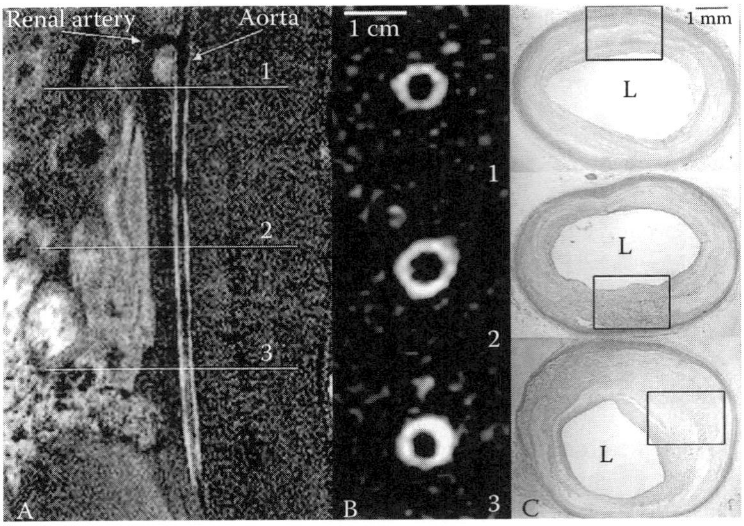

FIGURE 9.5 (please see color insert) (A) *In vivo* MR images of the aortic vessel wall upon injection of gadofluorine. (B) At three positions a tranverse MRI slice of the aorta in (A) is depicted. (C) The corresponding histopathological sections, stained with hematoxylin and eosin (H&E), show a good correlation with the MR images of the sections in (B). (D) Magnified areas are indicated by squares in (C). At level 1, enhancement is heterogeneous because of accumulation of lipids within the fibrous area. At level 2, the large lipid core corresponds to the highest enhancement within the plaque. Plaque at level 3 is mainly composed of lipids with perfect matching of the highest plaque enhancement. Ad, adventitia; F, loose fibrous; FC, fibrous cap; L, lumen; LC, lipid core; M, macrophages. (From Ref 59. With permission.)

Gadofluorine M (a preclinical agent developed by Schering, Germany) is a lipophilic micellular contrast agent that has been shown to improve the detection and characterization of atherosclerotic plaques.[58,59] Atherosclerotic plaques in the aortic arch of hyperlipidemic rabbits were studied with MRI after the injection of Gadofluorine M. Enhancement occurred in the aortic wall of hyperlipidemic rabbits as a function of time postinjection, as shown in Figure 9.5.[59] The MRI findings revealed a ring-shaped contrast enhancement of the aortic wall, which was in agreement with Sudan red stainings that matched *ex vivo* MR scans of the same specimen. This study also demonstrated that early and advanced lesions could be discriminated by Gadofluorine M-enhanced imaging.[59] The enhancement of advanced lesions was higher than in early lesions after the injection of Gadofluorine M and correlated with the density of the vasa vasorum, suggesting that the intensity of plaque enhancement with Gadofluorine M may be dependent on neovessel density.

Bimodal micelles may be composed of phospholipids, a surfactant, a fluorescent lipid, and an amphiphilic gadolinium contrast. Significant vessel wall enhancement of atherosclerotic mice (ApoE–/–) has been observed for untargeted mixed micelles.[60] The uptake was much higher in atherosclerotic plaque than in normal vessel wall of wild-type mice. Additionally, it has been shown that the blood clearance and biodistribution of mixed micelles are not influenced by micelle size (26 nm formed using dipalmitoyl-phosphatidyl-choline [DPPC] and 106 nm formed using palmitoyl-2-oleoyl-sn-glycerol-3-phosphocholine [POPC]), indicating that the efficacy and biodistribution of micelles are regulated by factors other than the hydrated particle size. The liver retention associated with the mixed micelles in mice is relatively low (<8% of the injected dose in mice), indicating that mixed micelles may prove an interesting platform for molecular imaging.[60] Additionally, targeted mixed micelles were recently developed for active targeting of functional macrophages in atherosclerotic plaque. Specific micelles were labeled with biotinylated antibodies specific for the macrophage scavenger receptor, via an avidin–biotin linkage. An improved uptake of the antibody-conjugated micelles, compared to untargeted micelles, was observed on MRI and fluorescence microscopy.[61]

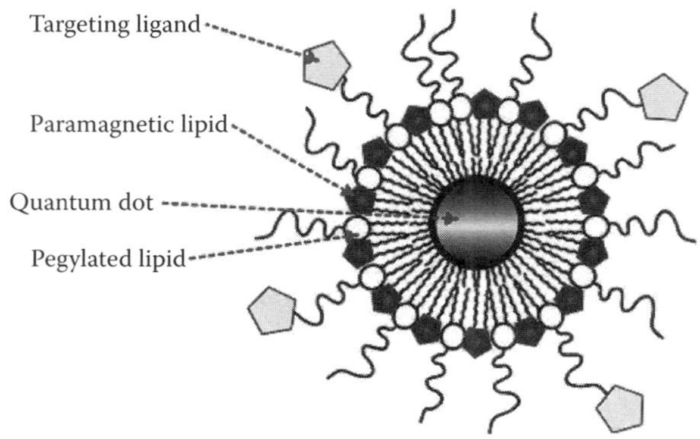

FIGURE 9.6 Schematic representation of a paramagnetic micelle quantum dot.

9.2.5 Quantum Dots

Quantum dots (QDs) are colloidal nanocrystals that have gained much interest for biological imaging purposes the past few years, because of their bright fluorescence, excellent photostability, high molar extinction coefficient, and narrow and tunable emission spectrum.[62] QDs capped in phospholipid micelles were among the first to be applied for *in vivo* imaging.[31]

Recently, QDs capped in paramagnetic micelles (pQDs) for parallel detection with optical methods and MRI have been developed.[32] The lipid coating of these QDs was comprised of a pegylated lipid and a paramagnetic Gd-DTPA-based lipid, as shown in Figure 9.6 (schematic representation of the nanoparticle). The ionic relaxivity r_1 was 12 mM^{-1}sec^{-1}, which is threefold higher than that of Gd-DTPA. Since the pQDs contain approximately 300 lipids, half of which is Gd-DTPA-BSA, the relaxivity per mM pQD is estimated to be ca. 2000 mM^{-1}sec^{-1}. This high relaxivity makes the pQD contrast agent an attractive candidate for molecular MRI purposes. For example, the nanoparticle may be functionalized by covalently linking RGD peptides and annexin A5 to allow the detection of cells expressing avb3 and apoptotic cells, respectively.

9.2.6 Microemulsions

Microemulsions are mixtures of water, oil, and an amphiphile that result in an optically isotropic and thermodynamically stable solution. Several studies have shown the utility of nanoparticles based on microemulsions composed of a perfluorocarbon core covered with a monolayer of (paramagnetic) lipids for MRI applications.[46,63–65] For molecular imaging of atherosclerosis, these nanoparticles were functionalized with antibodies specific for fibrin, allowing for enhancement of thrombus on *in vivo* MRI.[65,66] Perfluorocarbon nanoparticles targeted to the v3-integrin have been used for detecting angiogenesis in a rabbit model of atherosclerosis.[67] These nanoparticles may also be loaded with drugs, and the perfluorocarbon core of the nanoparticles enables particle detection with 19F nuclear magnetic resonance (NMR) spectroscopy and imaging, which makes quantification and hot spot imaging possible.[64]

9.2.7 Lipoproteins

Low-density lipoprotein (LDL) and high-density lipoprotein (HDL) are endogenous lipid nanoparticles that play an important role in the transport of cholesterol. These nanoparticles consist of a lipid core of cholesterol esters and triglycerides covered by a phospholipid monolayer that contains a large apolipoprotein.[68] The apolipoprotein B-100 (ApoB-100, 550 kDa) associated with LDL contains one or more clusters of cationic amino acids that naturally target LDL receptors (LDLRs) on cells. Uptake of LDL is via endocytosis, with subsequent uptake into cellular lysosomes that degrade the material. All LDL receptors are recycled so that many LDL particles may be accumulated

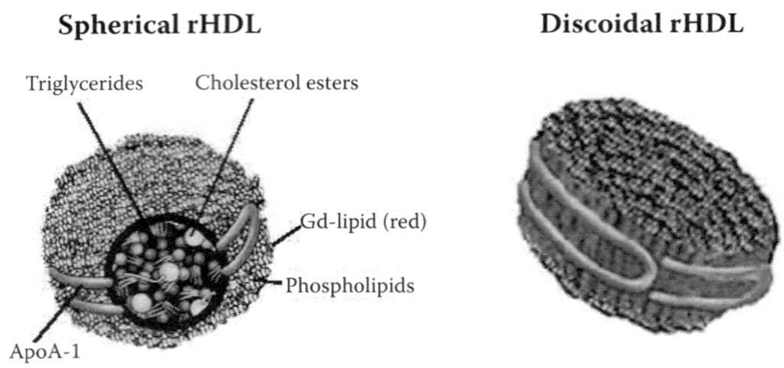

FIGURE 9.7 (please see color insert) Schematic representation of a gadolinium-labeled high-density lipoprotein (HDL) particle. (From Ref 33. With permission.)

within a given cell. LDL receptors are found on several normal tissue types (liver, adrenal glands, ovaries, etc.); however, macrophages in atherosclerotic plaques and some tumoral cells overexpress LDLR.[69,70] As a result, lanthanide-labeled LDL may be used as natural targets for atherosclerotic plaques and other cells that overexpress LDLR.[34,71–78] Specifically, recent studies have shown that manganese-mesoporphyrin-(MnMeso) labeled LDL may allow both the detection of atherosclerotic plaque and the assessment of lipoprotein kinetics in the vessel wall.[78]

HDLs may provide protection against cardiovascular events by decreasing cholesterol accumulation in atherosclerotic plaque.[79,80] Additionally, HDL also may reduce atherosclerosis by decreasing LDL oxidation, a critical step in foam cell formation.[79,80] The apolipoprotein (ApoA-I), associated with HDL, allows for receptor-mediated uptake of the material within the lesion. Once inside the plaque, the HDL sequesters the free cholesterol via ATP-binding cassette transporter 1 (ABACA1) and lecithin-cholesterol acyltransferase. Once HDL sequesters the cholesterol, the shape of the particle changes from a discoid form (β-form) to spherical α-HDL.[79,80] Due to the therapeutic advantages associated with HDL, contrast agents based upon this platform (as opposed to LDL) may be desirable, as shown in Figure 9.7. Studies using gadolinium-labeled α-LDL have shown that these particles penetrate the plaque, allowing for detection by MRI.[33] Additionally, these studies indicate that the labeled HDL was primarily localized within the intima layer, with some accumulation in macrophages, as shown in Figure 9.8.

9.3 IRON OXIDE PARTICLES

9.3.1 Properties

Unlike lipid-based nanoparticles that are flexible macromolecular structures, superparamagnetic iron oxide particles (SPIOs) are rigid crystalline structures. SPIOs are normally composed of an iron oxide core (usually magnetite $Fe_2^{3+}O_3Fe^{2+}O$) and a coating material that stabilizes the core and prevents aggregation or sedimentation.[81,82] In order for the iron oxide core to exhibit superparamagnetic properties, the size of the core (or crystal) must be smaller than a Weiss domain but large enough (>3 nm) to allow for spin coupling or interaction between the spins associated with Fe^{3+} and Fe^{2+}.[81] As a result, most iron oxide particles used as contrast agents for MRI have core sizes ranging from 4 to 5 nm. The coating material associated with the iron oxide particle generally reflects the matrix in which the iron oxides were originally formed. Currently, the only clinically approved iron oxide particles (Feridex and Combidex) for injection are prepared using dextran-based coating materials.[83,84] The iron oxide cores are formed within a dextran matrix that is broken down to form particles ranging from 20 to 100 nm in size.[83,84] Studies have shown that the pharmacokinetics and biodistribution of SPIOs are modulated by both the total hydrated size of

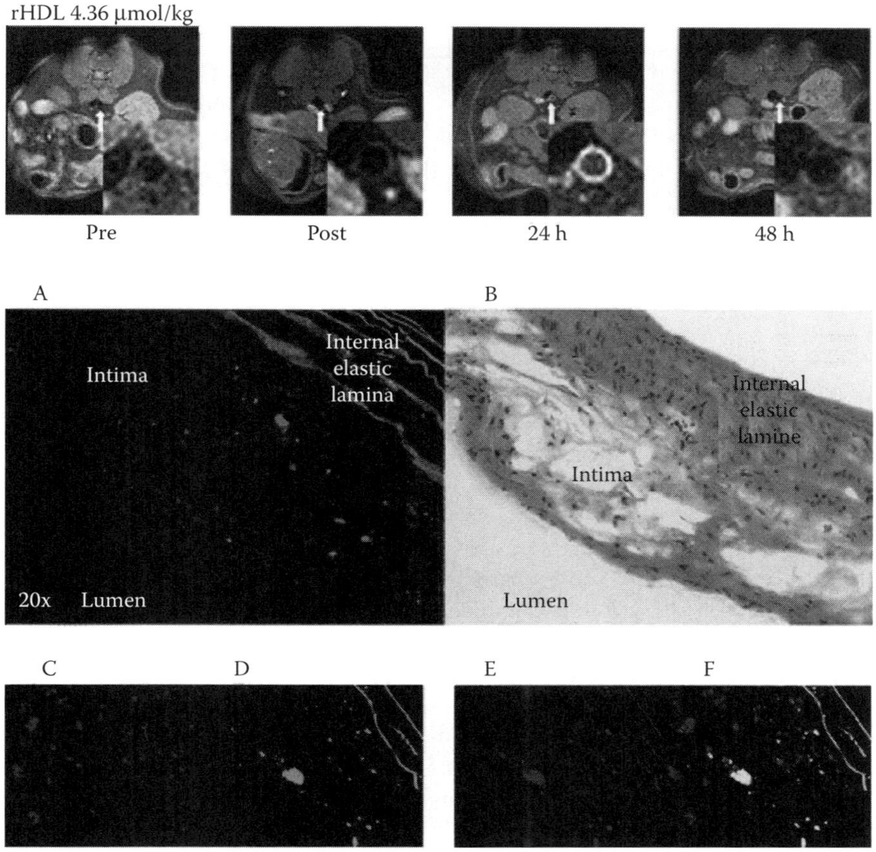

FIGURE 9.8 (please see color insert) Top: *In vivo* MRI at different time points (pre- and postinjection of contrast agent at 24, 48, 72, and 96 h) and dosages of Gd as determined by ICP-MS measurements. White arrows point to the abdominal aorta; the insets denote a magnification of the aorta region. Middle and bottom: (A) Confocal fluorescence microscopy of an atherosclerotic plaque. Blue denotes nuclei (DAPI staining), and green denotes rHDL-NBD labeled. (B) Histopathological section stained with hematoxylin and eosin (H&E). (C) DAPI staining of an atherosclerotic plaque. (D) rHDL-NBD staining. (E) Antibody selective for macrophages, CD68:RPE. (F) Merged images (yellow indicates colocalization). (From Reference 33. With permission.)

the particle and the coating material present.[85–90] For dextran-based SPIOs, the uptake of the particles into inflammatory cells associated with the mononuclear phagocytic system is controlled solely by the size of the particle.[89] Due to the large hydrated particle size of Feridex (known as Endorem, ferumoxide, and AMI-25). this material is primarily cleared by cells of the reticular endothelial system (RES) with significant liver uptake (>80% of the injected dose).[91,92] As a result, Feridex is not suitable for the imaging of atherosclerotic plaque since only a limited amount of the material is taken up by functionalized macrophages associated with the lesion.[93] On the other hand, Combidex (known as Sinerem, ferumoxtran-10, and AMI-227, available in Europe) has a mean hydrated particle diameter of only 20 nm.[83] Since Combidex is composed of single crystals coated with dextran, this material has been classified as an ultrasmall iron oxide (USPIO) particle.[81,83] Combidex has a neutral dextran T-10 coating associated with the surface of the iron oxide core and exhibits relatively long blood half-lives (>200 min) in rats and humans.[84,90,94] Due to the long circulation times, the particle is primarily taken up by bone marrow, lymphatic tissue, and other cells associated with the mononuclear phagocytic system.[89,94–97] Cells of the RES take approximately 6% of the injected dose. As a result, high dosages are required to obtain sufficient uptake of USPIO particles by macrophages in atherosclerotic plaques.

9.3.2 Detection of Macrophages with USPIOs

The compartmentalization of USPIO particles within tissue, such as a plaque macrophage, produces large susceptibility differences between the tissue containing the contrast agent and the surrounding tissue. The susceptibility differences create local magnetic field inhomogeneities (or gradients) that promote the dephasing of water protons. The dephasing of the water protons enhances the *effective* transverse magnetization (reduces T_2^*), causing signal loss. T_2^* relaxation (also termed susceptibility-induced relaxation or susceptibility effects) can be described based on models showing the dependence of T_2^* on the magnetic moment of iron oxide particles in solution.[98–101] Assuming a Gaussian field distribution, the free induction decay (FID = $1/T_2^* = R_2^*$) is proportional to the concentration of USPIO in the tissue (C) and the magnetization (M_l) generated by the particle ($1/T_2^* \approx CM_l$). For USPIO particles, the magnetization (defined by the Langevin equation) is normally reported as the saturation magnetization (M_s) that defines the maximum magnetization obtained at high field strengths.[101] The net magnetization created by the tissue containing the contrast agent is a sum of all the individual magnetic moments of each particle in the tissue. Due to the signal loss generated by T_2^* effects, signal loss reflects USPIO uptake in the macrophage-rich regions of atherosclerotic plaque. By increasing the concentration of USPIO in the macrophages (i.e., high dose of Combidex) or increasing M_s (by optimizing the iron core or increasing imaging field strengths), the efficacy of USPIO particles for plaque imaging can be increased.

Preliminary studies showed that Combidex-laden macrophages could be detected by MRI at sites of inflammatory cellular activity within the central nervous system.[102–104] Following this observation, studies were performed to evaluate the ability of Combidex to detect inflammation (or functional macrophages) in atherosclerotic plaque.[105–111] The preliminary results clearly indicate that Combidex is phagocytosed by functional macrophages in the atherosclerotic plaque of hyperlipidemic rabbits.[105,106,112] The uptake of Combidex into the macrophages resulted in significant signal loss that was used to identify inflammation or active plaque formation,[112] as shown in Figure 9.9 and Figure 9.10. Additionally, studies have revealed that Combidex is not taken up by all plaque macrophages (or homogeneously distributed) within the lesion. Uptake into atherosclerotic plaque is primarily restricted to small non-foam cell macrophages with colocalization in both the subendothelial layer and deep intima, as shown in Figure 9.9.[105,106,108,112] Although the majority of Combidex may be taken up via passive diffusion through the nonnormal endothelium, studies involving cytokine stimulation suggest that uptake is facilitated by both monocyte recruitment and activation of macrophage phagocytosis.[113]

Clinical studies in patients have confirmed the preclinical results. Preliminary studies, involving 11 symptomatic patients scheduled for carotid endarterectomy with USPIO-enhanced MRI, found a 24% decrease in signal intensity on corresponding T_2^*-weighted sequences.[114] Additionally, 75% of ruptured or rupture-prone lesions showed USPIO uptake in macrophages (as shown by histology).[114] Addtional studies expanded on this work and demonstrated that the optimum time for imaging symptomatic carotid plaque was between 24 and 36 h after injection of USPIO.[115,116] As demonstrated in the preclinical studies, uptake of Combidex within the human plaque macrophages was heterogeneous with uptake observed in active macrophages close to neovessels.

9.3.3 Alternative Iron Oxide Particles

Although most studies have been performed using Combidex, a few preliminary studies have been performed using alternative iron oxide formulations. Feromoxytrol (Advanced Magnetics) is a USPIO particle (30 nm) with a carbohydrate (nondextran) coating.[89] Due to differences in the coating material, the circulation time of Feromoxytrol is significantly less than that of Combidex. Studies in balloon-injured rabbits indicate that uptake into atherosclerotic plaque was substantially less for Feromoxytrol (no significant signal loss) relative to Combidex under equivalent experimental conditions.[89] The results of the study strongly suggest that plasma clearance of USPIO is more

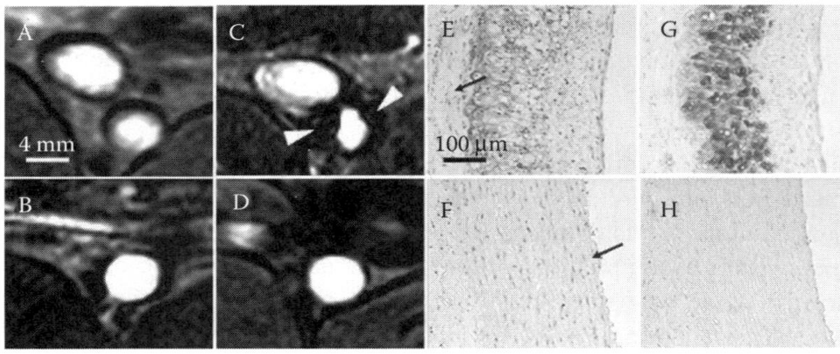

FIGURE 9.9 (please see color insert) Detection of macrophages in the aorta of a rabbit with ferumoxtran-10-enhanced MRI. MRIs were performed immediately (A and B) and 5 days (C and D) after the injection of ferumoxtran-10 (56 mg Fe/kg) in a model of focal inflammation induced by balloon injury in the aorta of hypercholesterolemic rabbits. The figure shows axial views of the injured (A, C) and noninjured (B, D) aorta of the same rabbit. No magnetic susceptibility artifact (MSA) is detected in the walls of injured (A) and noninjured (B) aortas immediately after ferumoxtran-10 injection. Five days later, strong MSAs (white arrowheads) are observed as signal voids only in the injured aorta (C). Cross-sections of the injured (E, G) and noninjured (F, H) aorta of this rabbit were obtained 5 days after the injection of ferumoxtran-10 and stained with Perl's reagent to detect iron (E, F) and monoclonal antibodies against RAM-11, a marker of rabbit macrophage (G, H). In the injured aorta, ferumoxtran-10 (blue stain) accumulated in the deeper layers of a thick neointima (E) and localized essentially in macrophage-rich areas (G), whereas no ferumoxtran-10 was found in the noninjured aorta (F). Black arrow: internal elastic lamina. Magnification: ×20. (Adapted from Hyafil, F. et al., *Arterioscler. Thromb. Vasc. Biol.*, 26, 176–181, 2006.)

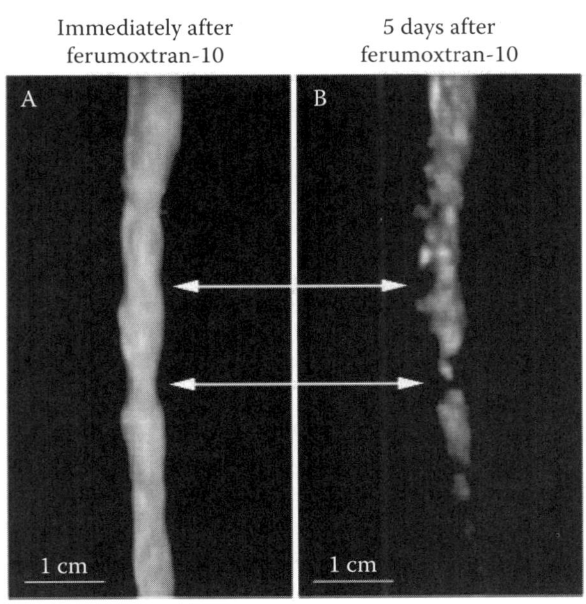

FIGURE 9.10 Ferumoxtran-10 (Combidex)-enhanced MRI in balloon-injured rabbit model. The figure shows longitudinal views of the aorta immediately after injection of a low USPIO dose (A, 2.8 mg Fe/kg) and 5 days after administration of a high USPIO dose (B, 56 mg Fe/kg). The small arrowhead indicates the location of the left renal artery. The large arrows present 5 days postinjection indicate luminal irregularities that were formed due to the signal loss generated by USPIO uptake in activated plaque macrophages. (Adapted from Hyafil, F et al., *Arterioscler Thromb Vasc Biol*, 26, 176–181, 2006.)

important for plaque labeling than phagocytosis. These results are not surprising, since the signal loss generated by T_2* effects is dependent upon the concentration of the material found within the plaque. The longer a material circulates (prior to clearance), the greater the potential uptake into active plaque macrophages.

Methods have also been described that allow for coating iron oxide particles with phospholipids or incorporation of iron into the lumen of liposomes.[117] The most promising lipid approach, with respect to plaque imaging, is related to the formation of iron oxide micelles that are composed of lipids or polymers. The micelle platform allows for the easy preparation of micelles that contain one or multiple iron oxide nanoparticles. For example, annexin V functionalized micellular iron oxide particles have been developed for the detection of apoptotic cells.[118] In cell cultures, the annexin V conjugated particles showed high affinity for apoptotic cells with subsequent T_2* effects and signal loss. Control cells showed almost no T_2 decrease, with limited reduction in signal intensity post annexin V treatment. The results of this study indicate that this material might be very useful for detecting apoptotic cells in atherosclerotic plaques.

9.3.4 Positive Contrast Imaging Techniques for the Detection of USPIO

Issues associated with partial volume effects and other artifacts have limited the potential use of T_2* techniques to detect USPIO atherosclerotic plaque. As shown in Figure 9.9, significant signal loss (due to perivascular artifacts) was observed at water–fat interfaces around the vessel wall. Since the negative contrast obtained using conventional T_2/T_2*-weighted sequences can be confused with signal loss caused by other sources, the sensitivity for detection of inflammation in plaque by USPIO is dramatically reduced.[112] Recently, several techniques have been developed that allow for the generation of positive signal in the presence of iron oxide particles, as summarized in Table 9.2. By utilizing such techniques, it may be possible to improve the differentiation between signal loss generated by USPIO and signal loss due to partial voluming effects or other artifacts.

The current positive contrast techniques can be summarized as follows:

1. Off-resonance methods that generate positive signal by shifting the excitation frequency to match the frequency shift caused by the presence of magnetic entities[119]
2. Spectral selective radio frequency (RF) pulse methods that selectively excite regions containing magnetic entities resulting in positive contrast[120,121]
3. Dephased MRI techniques that produce positive contrast by using the shift in k-space caused by magnetic entities,[122] including white marker imaging techniques that generate positive signal by completely rephasing only areas where magnetic entities are present[123,124]
4. Ultrashort TE imaging that evaluates the change in signal intensity obtained at various echo times (normally two TEs are used)

This method is a form of T2* mapping and assumes the susceptibility effects generated by magnetic entities are related to the change in signal obtained at very short echo times vs. the signal intensity obtained at long echo times. In the gradient echo acquisition for superparamagnetic particles susceptibility (GRASP) technique, conventional T_2*-weighted gradient echo (GRE) sequences are modified so that the rephasing z-gradient is reduced (typically 25 to 35% of normal rephasing is used). In tissues that do not contain dipolar fields generated by magnetic entities, no signal is observed (tissue signal is similar to that of noise) since the gradient moments are not rephased. However, in regions where local magnetic dipoles are present (regions containing iron oxide particles, ferritin/hemosiderin, etc.), the signal from regions containing dipoles is rephased, thereby generating a positive signal. By comparing the images obtained using conventional T_2*-weighted GRE sequences and GRASP, it is possible to identify regions of the plaque containing USPIO (fractionated Combidex), as shown in Figure 9.11.

TABLE 9.2
Comparison of Positive Contrast Techniques to Selectively Visualize Magnetic Entities such as Iron Laden Cells

Method	Off-Resonance Imaging	Spectral Selective RF Pulse Methods	Ultrashort TE Imaging	Dephased MRI/White Marker/GRASP
Principle	Signal from on-resonant water protons is suppressed, but off-resonance signal generated by magnetic entities is maintained, resulting in positive contrast	Uses spectrally selective RF saturation pulse to selectively excite only spins affected by magnetic entities	Subtracts the signal obtained using long echo times from the signal obtained using ultrashort echo times to generate positive contrast	Gradient rephasing is modified so that only regions where magnetic entities are present are completely rephased and produce positive signal
Sequence typically used	Spin echo	Spin echo	Gradient echo	Gradient echo
Contrast generated	Good contrast	Limited optimum contrast with gradual transition	Contrast generated by image subtraction	Robust contrast with sudden transition
Speed	Fast	Slow	Fast	Fast
Effect of B_0 inhomogeneity on positive contrast	Limited	Extreme (requires excellent shimming)	Limited	Limited (shimming preferred)
Ease of implementation	Complex	Difficult	Easy, but image subtraction can be problematic if motion is involved	Easy
Selectivity	Can be changed by using different bandwidth for prepulse	Controlled by bandwidth	Can only be controlled by changing TE values	Can be controlled by varying dephasing gradients

9.3.5 Perspectives

Molecular imaging of atherosclerotic plaques is very challenging because of the small size of the arterial wall, signal of the adjacent lumen, and artifacts due to breathing and systolic expansion of the arterial wall. Thanks to the absence of ionizing radiation and the high spatial resolution, MRI represents one of the most promising technologies for molecular imaging of atherosclerosis. For a long time molecular MRI was hampered by the need of high concentrations of contrast agents to obtain an effective signal. However, new contrast agents with high relaxivities or payloads of contrast moieties have changed the game and now offer to effectively detect changes in signal on MRI with much lower concentrations of contrast agent. As described in this chapter, these new platforms can be easily coupled with specific ligands and target receptors in atherosclerotic plaques. Furthermore, fluorescent probes have been combined with the MR contrast agent, allowing direct detection of the targets in atherosclerotic plaques using fluorescent lights. Hence, these new tools offer expanding combinations of contrast molecules and targets, which should clearly enhance the development of molecular MRI for atherosclerosis disease.

However, the key pathological steps that lead from a stable atherosclerotic plaque to an acute ischemic event remain poorly understood. The development of molecular MRI of atherosclerosis will clearly help to detect and follow the changes in atherosclerotic plaque composition *in vivo* and, hopefully, identify the most relevant markers of plaque instability. Moreover, recent clinical

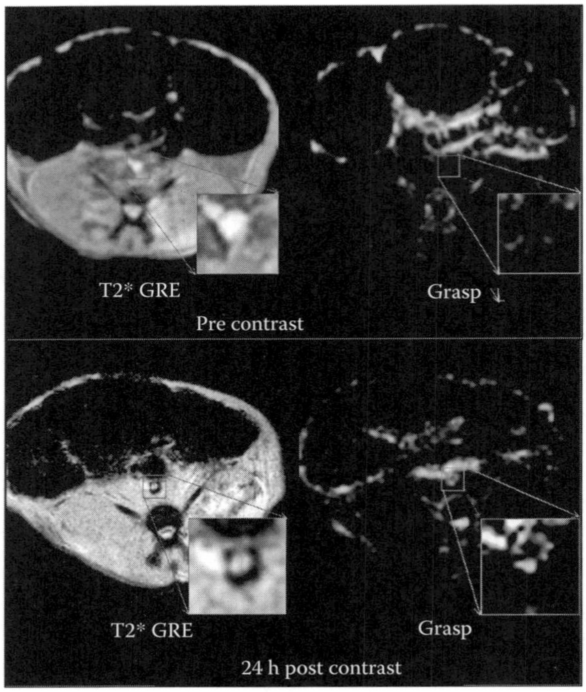

FIGURE 9.11 Positive contrast obtained using the white marker technique GRASP. Images obtained both precontrast and 24 h postinjection of a 4.7 mg Fe/kg dose in rabbits. The USPIO particle was made from fractionated Combidex (12 nm). At 24 h postinjection, signal loss was observed using traditional GRE sequences. The presence of iron was confirmed by the generation of a positive signal using GRASP.

studies have underscored the multiple locations of vulnerable and ruptured atherosclerotic plaques and the diffuse inflammation of the arterial tree in patients with acute ischemic events compared to stable patients.[125] Therefore, the concept of detecting infrequent vulnerable atherosclerotic plaques with imaging and treating them individually could move to a more global identification of vulnerable patients at high risk of acute clinical events, irrespective of the arterial location. Interestingly, extensive research has already been done with lipid aggregates as pharmaceutical carriers. This should make the shift from an MRI contrast agent to a therapeutic agent relatively easy and may allow not only detection but also targeting of specific treatments into these multiple vulnerable atherosclerotic plaques.

ACKNOWLEDGMENT

This work was partially supported by the Fédération Française de Cardiologie, Paris, France (F.H.), and NIH/NHLBI HL071021 and HL078667 (Z.A.F.).

REFERENCES

1. Naghavi M, Libby P, Falk E, et al. From vulnerable plaque to vulnerable patient: a call for new definitions and risk assessment strategies: part I. *Circulation* 2003;108:1664–1672.
2. Naghavi M, Libby P, Falk E, et al. From vulnerable plaque to vulnerable patient: a call for new definitions and risk assessment strategies: part II. *Circulation* 2003;108:1772–1778.
3. Fuster V, Moreno PR, Fayad ZA, et al. Atherothrombosis and high-risk plaque. I. Evolving concepts. *J Am Coll Cardiol* 2005;46:937–954.

4. Fuster V, Fayad ZA, Moreno PR, et al. Atherothrombosis and high-risk plaque. II. Approaches by noninvasive computed tomographic/magnetic resonance imaging. *J Am Coll Cardiol* 2005; 46:1209–1218.
5. Libby P. Current concepts of the pathogenesis of the acute coronary syndromes. *Circulation* 2001;104:365–372.
6. Libby P. Molecular bases of the acute coronary syndromes. *Circulation* 1995;91:2844–2850.
7. Glagov S, Bassiouny HS, Giddens DP, et al. Pathobiology of plaque modeling and complication. *Surg Clin North Am* 1995;75:545–556.
8. Yuan C, Kerwin WS. MRI of atherosclerosis. *J Magn Reson Imaging* 2004;19:710–719.
9. Fayad ZA, Fuster V, Fallon JT, et al. Noninvasive *in vivo* human coronary artery lumen and wall imaging using black-blood magnetic resonance imaging. *Circulation* 2000;102:506–510.
10. Faxon DP, Fuster V, Libby P, et al. Atherosclerotic vascular disease conference: writing group III: pathophysiology. *Circulation* 2004;109:2617–2625.
11. Choudhury RP, Fuster V, Fayad ZA. Molecular, cellular and functional imaging of atherothrombosis. *Nat Rev Drug Discov* 2004;3:913–925.
12. Davies MJ. Acute coronary thrombosis: the role of plaque disruption and its initiation and prevention. *Eur Heart J* 1995;16(Suppl L):3–7.
13. Virmani R, Kolodgie FD, Burke AP, et al. Atherosclerotic plaque progression and vulnerability to rupture: angiogenesis as a source of intraplaque hemorrhage. *Arterioscler Thromb Vasc Biol* 2005;25:2054–2061.
14. Bennett MR, Evan GI, Schwartz SM. Apoptosis of human vascular smooth muscle cells derived from normal vessels and coronary atherosclerotic plaques. *J Clin Invest* 1995;95:2266–2274.
15. Itskovich VV, Samber DD, Mani V, et al. Quantification of human atherosclerotic plaques using spatially enhanced cluster analysis of multicontrast-weighted magnetic resonance images. *Magn Reson Med* 2004;52:515–523.
16. Mani V, Itskovich VV, Aguiar SH, et al. Comparison of gated and non-gated fast multislice black-blood carotid imaging using rapid extended coverage and inflow/outflow saturation techniques. *J Magn Reson Imaging* 2005;22:628–633.
17. Itskovich VV, Mani V, Mizsei G, et al. Parallel and nonparallel simultaneous multislice black-blood double inversion recovery techniques for vessel wall imaging. *J Magn Reson Imaging* 2004;19:459–467.
18. Mani V, Itskovich VV, Szimtenings M, et al. Rapid extended coverage simultaneous multisection black-blood vessel wall MR imaging. *Radiology* 2004;232:281–288.
19. Summers RM, Andrasko-Bourgeois J, Feuerstein IM, et al. Evaluation of the aortic root by MRI: insights from patients with homozygous familial hypercholesterolemia. *Circulation* 1998;98:509–518.
20. Corti R, Fuster V, Fayad ZA, et al. Lipid lowering by simvastatin induces regression of human atherosclerotic lesions: two years' follow-up by high-resolution noninvasive magnetic resonance imaging. *Circulation* 2002;106:2884–2887.
21. Corti R, Fayad ZA, Fuster V, et al. Effects of lipid-lowering by simvastatin on human atherosclerotic lesions: a longitudinal study by high-resolution, noninvasive magnetic resonance imaging. *Circulation* 2001;104:249–252.
22. Toussaint JF, Gouya H, Glurton D, et al. A T2 classification for the discrimination of atheromatous primary and restenotic coronary plaques at 1.5T by high-resolution MRI. *Proc Int Soc Magn Reson Med* 1999;1:84.
23. Yuan C, Mitsumori LM, Beach KW, et al. Carotid atherosclerotic plaque: noninvasive MR characterization and identification of vulnerable lesions. *Radiology* 2001;221:285–299.
24. Chu B, Kampschulte A, Ferguson MS, et al. Hemorrhage in the atherosclerotic carotid plaque: a high-resolution MRI study. *Stroke* 2004;5:1079–1084.
25. Takaya N, Yuan C, Chu B, et al. Presence of intraplaque hemorrhage stimulates progression of carotid atherosclerotic plaques: a high-resolution magnetic resonance imaging study. *Circulation* 2005;111a:2768–2775.
26. Degiorgio V, Corti M. *Physics of Amphiphiles: Micelles, Vesicles and Microemulsions. Proceedings of the International School of Physics, Enricol Fermi, Course Xc*. Elsevier Science Ltd., The Netherlands, 1985.
27. Cevc G. *Phospholipids Handbook*. Marcel Dekker, New York, 1993.

28. Mulder WJ, Strijkers GJ, van Tilborg GA, et al. Lipid-based nanoparticles for contrast-enhanced MRI and molecular imaging. *NMR Biomed* 2006;19:142–164.
29. Belsito S, Bartucci R, Sportelli L. Lipid chain length effect on the phase behaviour of PCs/PEG:2000-PEs mixtures. A spin label electron spin resonance and spectrophotometric study. *Biophys Chem* 2001;93:11–22.
30. Nitin N, LaConte LE, Zurkiya O, et al. Functionalization and peptide-based delivery of magnetic nanoparticles as an intracellular MRI contrast agent. *J Biol Inorg Chem* 2004;9:706–712.
31. Dubertret B, Skourides P, Norris DJ, et al. *In vivo* imaging of quantum dots encapsulated in phospholipid micelles. *Science* 2002;298:1759–1762.
32. Mulder WJ, Koole R, Brandwijk RJ, et al. Quantum dots with a paramagnetic coating as a bimodal molecular imaging probe. *Nano Lett* 2006;6:1–6.
33. Frias JC, Williams KJ, Fisher EA, et al. Recombinant HDL-like nanoparticles: a specific contrast agent for MRI of atherosclerotic plaques. *J Am Chem Soc* 2004;126:16316–16317.
34. Zheng G, Li H, Zhang M, et al. Low-density lipoprotein reconstituted by pyropheophorbide cholesterol oleate as target-specific photosensitizer. *Bioconjug Chem* 2002;13:392–396.
35. Unger E, Shen DK, Wu GL, et al. Liposomes as MR contrast agents: pros and cons. *Magn Reson Med* 1991;22:304–308; discussion 313.
36. Fossheim SL, Colet JM, Mansson S, et al. Paramagnetic liposomes as magnetic resonance imaging contrast agents. Assessment of contrast efficacy in various liver models. *Invest Radiol* 1998;33:810–821.
37. Fossheim SL, Fahlvik AK, Klaveness J, et al. Paramagnetic liposomes as MRI contrast agents: influence of liposomal physicochemical properties on the *in vitro* relaxivity. *Magn Reson Imaging* 1999;17:83–89.
38. Barsky D, Putz B, Schulten K. Theory of heterogeneous relaxation in compartmentalized tissues. *Magn Reson Med* 1997;37:666–675.
39. Barsky D, Putz B, Schulten K, et al. Theory of paramagnetic contrast agents in liposome systems. *Magn Reson Med* 1992;24:1–13.
40. Koenig SH, Brown RD, III. Relaxation of solvent protons by paramagnetic ions and its dependence on magnetic field and chemical environment: implications for NMR imaging. *Magn Reson Med* 1984;1:478–495.
41. Kellar KE, Henrichs PM, Spiller M, et al. Relaxation of solvent protons by solute Gd3+-chelates revisited. *Magn Reson Med* 1997;37:730–735.
42. Kimpe K, Parac-Vogt TN, Laurent S, et al. Potential MRI contrast agents based on micellar incorporation of amphiphilic bis(alkylamide) derivatives of [(Gd-DTPA)(H2O)](2-). *Eur J Inorg Chem* 2003;16:3021–3027.
43. Parac-Vogt TN, Kimpe K, Laurent S, et al. Gadolinium DTPA-monoamide complexes incorporated into mixed micelles as possible MRI contrast agents. *Eur J Inorg Chem* 2004;17:3538–3543.
44. Torchilin VP. Recent advances with liposomes as pharmaceutical carriers. *Nat Rev Drug Discov* 2005;4:145–160.
45. Lanza GM, Wallace KD, Scott MJ, et al. A novel site-targeted ultrasonic contrast agent with broad biomedical application. *Circulation* 1996;94:3334–3340.
46. Lanza GM, Winter P, Caruthers S, et al. Novel paramagnetic contrast agents for molecular imaging and targeted drug delivery. *Curr Pharm Biotechnol* 2004;5:495–507.
47. Lanza GM, Abendschein DR, Yu X, et al. Molecular imaging and targeted drug delivery with a novel, ligand-directed paramagnetic nanoparticle technology. *Acad Radiol* 2002;9(Suppl 2):S330–331.
48. Caravan P, Ellison JJ, McMurry TJ, et al. Gadolinium(III) chelates as MRI contrast agents: structure, dynamics, and applications. *Chem Rev* 1999;99:2293–2352.
49. Aime S, Botta M, Fasano M, et al. Lanthanide(III) chelates for NMR biomedical applications. *Chem Soc Rev* 1998;27:19–29.
50. Torchilin VP, Lukyanov AN, Gao Z, et al. Immunomicelles: targeted pharmaceutical carriers for poorly soluble drugs. *Proc Natl Acad Sci USA* 2003;100:6039–6044.
51. Dafni H, Gilead A, Nevo N, et al. Modulation of the pharmacokinetics of macromolecular contrast material by avidin chase: MRI, optical, and inductively coupled plasma mass spectrometry tracking of triply labeled albumin. *Magn Reson Med* 2003;50:904–914.
52. Storrs RW, Tropper FD, Li HY, et al. Paramagnetic polymerized liposomes as new recirculating MR contrast agents. *J Magn Reson Imaging* 1995;5:719–724.

53. Sipkins DA, Gijbels K, Tropper FD, et al. ICAM-1 expression in autoimmune encephalitis visualized using magnetic resonance imaging. *J Neuroimmunol* 2000;104:1–9.
54. Sipkins DA, Cheresh DA, Kazemi MR, et al. Detection of tumor angiogenesis *in vivo* by alphaVbeta3-targeted magnetic resonance imaging. *Nat Med* 1998;4:623–626.
55. Kolodgie FD, Gold HK, Burke AP, et al. Intraplaque hemorrhage and progression of coronary atheroma. *N Engl J Med* 2003;349:2316–2325.
56. Mulder WJ, Strijkers GJ, Griffioen AW, et al. A liposomal system for contrast-enhanced magnetic resonance imaging of molecular targets. *Bioconjug Chem* 2004;15:799–806.
57. Mulder WJ, Strijkers GJ, Habets JW, et al. MR molecular imaging and fluorescence microscopy for identification of activated tumor endothelium using a bimodal lipidic nanoparticle. *FASEB J* 2005;19:2008–2010.
58. Barkhausen J, Ebert W, Heyer C, et al. Detection of atherosclerotic plaque with gadofluorine-enhanced magnetic resonance imaging. *Circulation* 2003;108:605–609.
59. Sirol M, Itskovich VV, Mani V, et al. Lipid-rich atherosclerotic plaques detected by gadofluorine-enhanced *in vivo* magnetic resonance imaging. *Circulation* 2004;109:2890–2896.
60. Briley-Saebo KCMV, Amirbekian V, Frias JC, Fayad ZA. Gadolinium Mixed Micelles: Effect of Size on *In Vitro* and *In Vivo* Efficacy. Paper presented at Society for Cardiovascular Magnetic Resonance, Miami, 2006.
61. Amirbekian V, Amirbekian S, Lipinski MJ, et al. Magnetic resonance imaging (*in-vivo*) of apolipoprotein-E knockout mice to detect atherosclerosis with gadolinium-containing micelles and immunomicelles molecularly targeted to macrophages via the macrophage scavenger receptor. *Circulation* 2005;112:U286–U286.
62. Michalet X, Pinaud FF, Bentolila LA, et al. Quantum dots for live cells, *in vivo* imaging, and diagnostics. *Science* 2005;307:538–544.
63. Lanza GM, Winter PM, Caruthers SD, et al. Magnetic resonance molecular imaging with nanoparticles. *J Nucl Cardiol* 2004;11:733–743.
64. Lanza GM, Winter PM, Neubauer AM, et al. (1)H/(19)F magnetic resonance molecular imaging with perfluorocarbon nanoparticles. *Curr Top Dev Biol* 2005;70:57–76.
65. Cyrus T, Winter PM, Caruthers SD, et al. Magnetic resonance nanoparticles for cardiovascular molecular imaging and therapy. *Expert Rev Cardiovasc Ther* 2005;3:705–715.
66. Flacke S, Fischer S, Scott MJ, et al. Novel MRI contrast agent for molecular imaging of fibrin: implications for detecting vulnerable plaques. *Circulation* 2001;104:1280–1285.
67. Winter PM, Morawski AM, Caruthers SD, et al. Molecular imaging of angiogenesis in early-stage atherosclerosis with alpha(v)beta3-integrin-targeted nanoparticles. *Circulation* 2003;108:2270–2274.
68. Chapman MJ. The potential role of HDL- and LDL-cholesterol modulation in atheromatous plaque development. *Curr Med Res Opin* 2005;21(Suppl 6):S17–S22.
69. Chun PW, Espinosa AJ, Lee CW, et al. Low density lipoprotein receptor regulation. Kinetic models. *Biophys Chem* 1985;21:185–209.
70. Wehr H. The 1985 Nobel Prize for achievements in the fields of physiology or medicine. *Postepy Biochem* 1986;32:395–399.
71. Zheng G, Li H, Yang K, et al. Tricarbocyanine cholesterol laurates labeled LDL: new near infrared fluorescent probes (NIRFs) for monitoring tumors and gene therapy of familial hypercholesterolemia. *Bioorg Med Chem Lett* 2002;12:1485–1488.
72. Li H, Zhang Z, Blessington D, et al. Carbocyanine labeled LDL for optical imaging of tumors. *Acad Radiol* 2004;11:669–677.
73. Li H, Gray BD, Corbin I, et al. MR and fluorescent imaging of low-density lipoprotein receptors. *Acad Radiol* 2004;11:1251–1259.
74. Wu SP, Lee I, Ghoroghchian PP, et al. Near-infrared optical imaging of B16 melanoma cells via low-density lipoprotein-mediated uptake and delivery of high emission dipole strength tris [(porphinato)zinc(II)] fluorophores. *Bioconjug Chem* 2005;16:542–550.
75. Zheng G, Chen J, Li H, et al. Rerouting lipoprotein nanoparticles to selected alternate receptors for the targeted delivery of cancer diagnostic and therapeutic agents. *Proc Natl Acad Sci USA* 2005;102:17757–17762.
76. Ponty E, Favre G, Benaniba R, et al. Biodistribution study of 99mTc-labeled LDL in B16-melanoma-bearing mice. Visualization of a preferential uptake by the tumor. *Int J Cancer* 1993;54:411–417.

77. Sauer I, Dunay IR, Weisgraber K, et al. An apolipoprotein E-derived peptide mediates uptake of sterically stabilized liposomes into brain capillary endothelial cells. *Biochemistry* 2005;44:2021–2029.
78. Mitsumori LM, Ricks JL, Rosenfeld ME, et al. Development of a lipoprotein based molecular imaging MR contrast agent for the noninvasive detection of early atherosclerotic disease. *Int J Cardiovasc Imaging* 2004;20:561–567.
79. Brewer HB, Jr. Increasing HDL cholesterol levels. *N Engl J Med* 2004;350:1491–1494.
80. Brewer HB, Jr. Focus on high-density lipoproteins in reducing cardiovascular risk. *Am Heart J* 2004;148(Suppl):S14–S18.
81. Wang YXJ, Hussain SM, Krestin GP. Superparamagnetic iron oxide contrast agents: physicochemical characteristics and applications in MR imaging. *Eur Radiol* 2001;11:2319–2331.
82. Weissleder R, Papisov M. Pharmaceutical iron oxides for MR imaging. *Rev Magn Reson Med* 1992;4:1–20.
83. Jung CW, Jacobs P. Physical and chemical properties of superparamagnetic iron oxide MR contrast agents: ferumoxides, ferumoxtran, ferumoxsil. *Magn Reson Imaging* 1995;13:661–674.
84. Jung CW. Surface properties of superparamagnetic iron oxide MR contrast agents: ferumoxides, ferumoxtran, ferumoxsil. *Magn Reson Imaging* 1995;13:675–691.
85. Pouliquen D, Perdrisot R, Ermias A, et al. Superparamagnetic iron oxide nanoparticles as a liver MRI contrast agent: contribution of microencapsulation to improved biodistribution. *Magn Reson Imaging* 1989;7:619–627.
86. Pouliquen D, Le Jeune JJ, Perdrisot R, et al. Iron oxide nanoparticles for use as an MRI contrast agent: pharmacokinetics and metabolism. *Magn Reson Imaging* 1991;9:275–283.
87. Chouly C, Pouliquen D, Lucet I, et al. Development of superparamagnetic nanoparticles for MRI: effect of particle size, charge and surface nature on biodistribution. *J Microencapsul* 1996;13:245–255.
88. Pouliquen D, Perroud H, Calza F, et al. Investigation of the magnetic properties of iron oxide nanoparticles used as contrast agent for MRI. *Magn Reson Med* 1992;24:75–84.
89. Yancy AD, Olzinski AR, Hu TC, et al. Differential uptake of ferumoxtran-10 and ferumoxytol, ultrasmall superparamagnetic iron oxide contrast agents in rabbit: critical determinants of atherosclerotic plaque labeling. *J Magn Reson Imaging* 2005;21:432–442.
90. Bonnemain B. Pharmacokinetic and hemodynamic safety of two superparamagnetic agents, Endorem and Sinerem, in cirrhotic rats. *Acad Radiol* 1998;5(Suppl 1):S151–S153; discussion S156.
91. Majumdar S, Zoghbi SS, Gore JC. Pharmacokinetics of superparamagnetic iron-oxide MR contrast agents in the rat. *Invest Radiol* 1990;25:771–777.
92. Hundt W, Petsch R, Helmberger T, et al. Signal changes in liver and spleen after Endorem administration in patients with and without liver cirrhosis. *Eur Radiol* 2000;10:409–416.
93. Clement O, Siauve N, Cuenod CA, et al. Liver imaging with ferumoxides (Feridex): fundamentals, controversies, and practical aspects. *Top Magn Reson Imaging* 1998;9:167–182.
94. Daldrup HE, Link TM, Blasius S, et al. Monitoring radiation-induced changes in bone marrow histopathology with ultra-small superparamagnetic iron oxide (USPIO)-enhanced MRI. *J Magn Reson Imaging* 1999;9:643–652.
95. Weissleder R, Elizondo G, Wittenberg J, et al. Ultrasmall superparamagnetic iron oxide: characterization of a new class of contrast agents for MR imaging. *Radiology* 1990;175:489–493.
96. Dupas B, Berreur M, Rohanizadeh R, et al. Electron microscopy study of intrahepatic ultrasmall superparamagnetic iron oxide kinetics in the rat. Relation with magnetic resonance imaging. *Biol Cell* 1999;91:195–208.
97. Seneterre E, Weissleder R, Jaramillo D, et al. Bone marrow: ultrasmall superparamagnetic iron oxide for MR imaging. *Radiology* 1991;179:529–533.
98. Hardy P, Henkelman RM. On the transverse relaxation rate enhancement induced by diffusion of spins through inhomogeneous fields. *Magn Reson Med* 1991;17:348–356.
99. Majumdar S, Gore JC. Studies of diffusion in random-fields produced by variations in susceptibility. *J Magn Reson* 1988;78:41–55.
100. Majumdar S, Zoghbi SS, Gore JC. The influence of pulse sequence on the relaxation effects of superparamagnetic iron-oxide contrast agents. *Magn Reson Med* 1989;10:289–301.
101. Yablonskiy DA, Haacke EM. Theory of NMR signal behavior in magnetically inhomogeneous tissues: the static dephasing regime. *Magn Reson Med* 1994;32:749–763.

102. Dousset V, Delalande C, Ballarino L, et al. *In vivo* macrophage activity imaging in the central nervous system detected by magnetic resonance. *Magn Reson Med* 1999;41:329–333.
103. Dousset V, Ballarino L, Delalande C, et al. Comparison of ultrasmall particles of iron oxide (USPIO)-enhanced T2-weighted, conventional T2-weighted, and gadolinium-enhanced T1-weighted MR images in rats with experimental autoimmune encephalomyelitis. *Am J Neuroradiol* 1999;20:223–227.
104. Dousset V, Gomez C, Petry KG, et al. Dose and scanning delay using USPIO for central nervous system macrophage imaging. *MAGMA* 1999;8:185–189.
105. Ruehm SG, Corot C, Vogt P, et al. Magnetic resonance imaging of atherosclerotic plaque with ultrasmall superparamagnetic particles of iron oxide in hyperlipidemic rabbits. *Circulation* 2001;103:415–422.
106. Ruehm SG, Corot C, Vogt P, et al. Ultrasmall superparamagnetic iron oxide-enhanced MR imaging of atherosclerotic plaque in hyperlipidemic rabbits. *Acad Radiol* 2002;9(Suppl 1):S143–S144.
107. Schmitz SA, Hansel J, Wagner S, et al. SPIO-enhanced MR angiography for the detection of venous thrombi in an animal model. *Rofo* 1999;170:316–321.
108. Schmitz SA, Coupland SE, Gust R, et al. Superparamagnetic iron oxide-enhanced MRI of atherosclerotic plaques in Watanabe hereditable hyperlipidemic rabbits. *Invest Radiol* 2000;35:460–471.
109. Schmitz SA, Winterhalter S, Schiffler S, et al. USPIO-enhanced direct MR imaging of thrombus: preclinical evaluation in rabbits. *Radiology* 2001;221:237–243.
110. Schmitz SA, Taupitz M, Wagner S, et al. Magnetic resonance imaging of atherosclerotic plaques using superparamagnetic iron oxide particles. *J Magn Reson Imaging* 2001;14:355–361.
111. Schmitz SA, Taupitz M, Wagner S, et al. Iron-oxide-enhanced magnetic resonance imaging of atherosclerotic plaques: postmortem analysis of accuracy, inter-observer agreement, and pitfalls. *Invest Radiol* 2002;37:405–411.
112. Hyafil F, Laissy JP, Mazighi M, et al. Ferumoxtran-10-enhanced MRI of the hypercholesterolemic rabbit aorta: relationship between signal loss and macrophage infiltration. *Arterioscler Thromb Vasc Biol* 2006;26:176–181.
113. Litovsky S, Madjid M, Zarrabi A, et al. Superparamagnetic iron oxide-based method for quantifying recruitment of monocytes to mouse atherosclerotic lesions *in vivo*: enhancement by tissue necrosis factor-alpha, interleukin-1beta, and interferon-gamma. *Circulation* 2003;107:1545–1549.
114. Kooi ME, Cappendijk VC, Cleutjens KB, et al. Accumulation of ultrasmall superparamagnetic particles of iron oxide in human atherosclerotic plaques can be detected by *in vivo* magnetic resonance imaging. *Circulation* 2003;107:2453–2458.
115. Trivedi RA, JM UK-I, Graves MJ, et al. *In vivo* detection of macrophages in human carotid atheroma: temporal dependence of ultrasmall superparamagnetic particles of iron oxide-enhanced MRI. *Stroke* 2004;35:1631–1635.
116. Trivedi RA, JM UK-I, Graves MJ, et al. Noninvasive imaging of carotid plaque inflammation. *Neurology* 2004;63:187–188.
117. de Cuyper M, Joniau M. Potentialities of magnetoliposomes in studying symmetric and asymmetric phospholipid transfer processes. *Biochim Biophys Acta* 1990;1027:172–178.
118. Van Tilborg GA, Mulder WJ, Strijkers G, et al. Imaging of Apoptosis Using Targeted Lipid-Based Bimodal Contrast Agents. Paper presented at SMI 2005, Cologne, Germany, 2005.
119. Stuber M, Gilson WD, Schaer M, et al. Shedding Light on the Dark Spot with IRON: A Method That Generates Positive Contrast in the Presence of Superparamagnetic Nanoparticles. Paper presented at Proceedings of the International Society of Magnetic Resonance Medicine, Miami Beach, FL, 2005.
120. Cunningham CH, Vigneron DB, Chen AP, et al. Design of symmetric-sweep spectral-spatial RF pulses for spectral editing. *Magn Reson Med* 2004;52:147–153.
121. Cunningham CH, Arai T, Yang PC, et al. Positive contrast magnetic resonance imaging of cells labeled with magnetic nanoparticles. *Magn Reson Med* 2005;53:999–1005.
122. Bakker CJ, Seppenwoolde JH, Vincken KL. Dephased MRI. *Magn Reson Med* 2005.
123. Mani V, Briley-Saebo KC, Itskovich VV, et al. Gradient echo acquisition for superparamagnetic particles with positive contrast (GRASP): sequence characterization in membrane and glass superparamagnetic iron oxide phantoms at 1.5T and 3T. *Magn Reson Med* 2006;55:126–135.
124. Seppenwoolde JH, Viergever MA, Bakker CJ. Passive tracking exploiting local signal conservation: the white marker phenomenon. *Magn Reson Med* 2003;50:784–790.
125. Rioufol G, Finet G, Ginon I, et al. Multiple atherosclerotic plaque rupture in acute coronary syndrome: a three-vessel intravascular ultrasound study. *Circulation* 2002;106:804–808.

10 Molecular Imaging of Apoptosis

Mikko I. Kettunen and Kevin M. Brindle

CONTENTS

10.1 Apoptosis 183
10.2 Molecular Imaging of Apoptosis 184
10.3 Imaging Changes in Cell and Tissue Architecture 186
10.4 Detecting Cell Death through Changes in Metabolism 187
10.4.1 Lipid Metabolism 187
10.4.2 Energy Metabolism 188
10.5 Imaging of Phosphatidylserine Exposure 189
10.6 Caspase Activity 191
10.7 Altered pH 192
10.8 Concluding Remarks 192
Acknowledgments 193
References 193

10.1 APOPTOSIS

Apoptosis, or programmed cell death, plays an important role in the control of development and in the maintenance of tissue homeostasis in multicellular organisms. Apoptosis progresses through a series of energy-requiring and tightly regulated steps that conclude with the engulfment of dying cells by neighboring phagocytic cells, in a process that avoids the inflammatory reaction caused by cellular necrosis.[1,2] Dysregulation of apoptosis is associated with a number of pathological states. Cancer, for example, is believed to result from inadequate apoptosis as well as unrestrained cell division,[3] whereas increased rates of apoptosis occur in pathological conditions, such as cardiac infarction and Alzheimer's disease (see, for example, references 4 to 8 and references cited therein). Apoptosis has been associated particularly with those regions that show less damage in the initial stages of disease (the penumbra), and subsequently in the chronic stages of disease, where the amount of apoptosis may vary depending on disease progression. In oncology, the early onset of apoptosis post-treatment is a good prognostic indicator for outcome and could potentially be used to monitor the success of therapy long before other signs of cell death are present, such as tumor shrinkage.[9–12] A noninvasive imaging technique for detecting apoptosis would be invaluable, therefore, in diagnosing disease and detecting responses to treatment.

Apoptosis is mediated by signaling pathways, in which the major players are a family of cysteine proteases called caspases, which cleave their substrate proteins at specific aspartic acid residues. These proteases are synthesized as inactive precursors, called procaspases, which are usually activated through cleavage by other caspases, resulting in an amplifying proteolytic cascade. There are two major apoptotic pathways in mammalian cells.[2] The death receptor pathway is triggered by binding of ligands, such as tumor necrosis factor (TNF), to their respective receptors, which results in activation of procaspase-8. Caspase-8 then goes on to cleave other apoptotic substrate proteins. The mitochondrial pathway also responds to extracellular signals and internal insults, such as DNA damage. Both pathways converge at the surface of the mitochondria, where pro- and antiapoptotic members of the Bcl-2 family of proteins compete to regulate the release of the electron carrier protein, cytochrome c. Cytochrome c, when released into the cytosol, associates with Apaf-1 and procaspase-9, and possibly other proteins, to form the apoptosome, which cleaves caspase-3. Caspase-3 is also activated by the caspase-8 activated in the death receptor pathway. Caspase-3 then mediates the downstream cleavage of other proteins in the cell, which causes the morphological changes that are characteristic of the process. The amplifying effect of the caspase proteolytic cascade, as well as the presence of positive feedback loops in the apoptotic program, ensures that once the process has started, the cell becomes committed to death. However, there are buffers or dampeners in the program that are thought to prevent inadvertent or accidental activation. The antiapoptotic Bcl-2 family members, for example, are thought to act as buffers that minimize accidental release of mitochondrial contents. At a cellular level, apoptosis is manifested as alterations in the cytoskeleton, disassembly of the nuclear envelope, and DNA fragmentation. Finally, the cell breaks up to form apoptotic bodies, which are then engulfed or phagocytosed by neighboring cells or macrophages.[13] Recognition of apoptotic cells by neighboring cells is mediated by phosphatidylserine (PS), which is transported from the inner leaflet of the plasma membrane to the cell surface in the early stages of the process.[14–16]

Apoptosis, however, is not the only form of cell death. The other classical form of cell death, necrosis, occurs following more acute damage, when cells cannot undergo apoptosis due, for example, to loss of adenosine triphosphate (ATP). It is now becoming clear that classical cell necrosis and caspase-mediated apoptosis are extremes of a continuum of related and interrelated processes.[2,17,18] Consequently, these different forms of cell death cannot easily be distinguished, as dying cells can display a wide range of characteristics, some of which are predominantly necrotic while others are symptomatic of apoptosis. This makes imaging of apoptosis, and its separation from other forms of cell death, difficult *in vitro* and nearly impossible *in vivo*. However, this is not necessarily a problem since it may be sufficient simply to diagnose cell death in it all its forms; for example, in oncology, a successful drug treatment is defined ultimately by the death of tumor tissue regardless of the mode of cell death.

10.2 MOLECULAR IMAGING OF APOPTOSIS

Recent studies have demonstrated the feasibility of imaging cell death or apoptosis *in vivo*. Radionuclide and optical methods for detecting apoptosis, through binding of an appropriately labeled annexin V molecule to the phosphatidylserine (PS) exposed on the surface of apoptotic cells (see below), have shown considerable promise and, in some cases, have already gone into clinical trials.[5] An indirect method for detecting cell death, positron emission tomography (PET) detection of [^{18}F] fluorodeoxyglucose (FDG) uptake in tumors, is already widely used clinically as a way of assessing tumor responses to therapy. FDG is a glucose analog that is taken up by cells and phosphorylated by hexokinase to produce FDG 6-phosphate. This cannot be metabolized further and is trapped in the cell by the phosphorylation. Hence, [^{18}F] FDG uptake is a surrogate marker of glycolytic activity, which can be reduced in those tumors that show a positive response to treatment.[19–21]

Magnetic resonance imaging (MRI) and localized magnetic resonance spectroscopy (MRS) could also be valuable tools for detecting apoptosis or cell death in the clinic. MR has relatively good spatial and temporal resolution, compared to PET, and is already in widespread clinical use. Endogenous tissue contrasts, which may change in apoptotic or necrotic tissue, can be exploited in ^{1}H MRI detection of cell death. These include contrasts that are based on the diffusion and relaxation properties of water protons as they move and interact with macromolecules and other cellular structures. The applicability of ^{1}H MRI can be further expanded through the introduction

TABLE 10.1
Potential Targets for Molecular Imaging of Apoptosis Using Magnetic Resonance

Target	Method of MR Detection	Marker Appearance	Principal Problem	References
Activation of caspases	Cleavable contrast agent for MRI, cleavable substrate for MRS	Early	Target access	Not yet been achieved
Externalization of PS	Contrast agent for MRI	Early	Access of probe to and washout from tissue	102, 107, 109
Altered pH	$^{31}P/^{1}H$ MRS, contrast agent	Early	Background	56, 127, 129
Altered lipid metabolism	^{1}H MRS	Early	Nonspecific	42, 69, 71
Altered energy metabolism	$^{31}P/^{13}C$ MRS, fMRI	Late	Background, sensitivity	56–58, 95
Cellular shrinkage, loss of cellularity	Diffusion MRI, relaxation MRI	Late	Nonspecific	32, 33, 46, 120

of exogenous contrast agents. MRS measurements of tissue metabolism also allow detection of cell death, since MR-observable metabolism is frequently perturbed during cell death. In addition to protons, several other nuclei, such as phosphorus (^{31}P) and carbon (^{13}C), can also be studied, which widens the range of metabolites that can be detected. This spectroscopic approach can be further expanded to give spatial information through spectroscopic imaging, although with low resolution, compared to ^{1}H MRI of tissue water protons. The markers of cell death that could be detected using noninvasive MRI or MRS methods *in vivo* are summarized in Table 10.1. These markers range from changes in plasma membrane phospholipid composition to gross morphological changes in tissue architecture resulting from apoptosis-specific changes, such as cell shrinkage. However, to serve as useful imaging targets for MRI, several requirements must be fulfilled. The insensitivity of the technique, compared to imaging modalities such as PET, means that the imaging target must be present in relatively high concentration. In the case of exogenous contrast agents, the biodistribution of the contrast agent can significantly affect its utility for detecting apoptosis. There must be unrestricted access of the agent to the target and sufficient washout of unbound material to minimize nonspecific background signal, and thus to generate tissue contrast. The progression of the apoptotic process also sets additional challenges for imaging the process. The levels of apoptosis are frequently relatively low *in vivo*, typically only a few percent, and this situation does not change much in pathological situations, where it may increase to a few tens of percent.[22] For example, significant levels of apoptosis have been observed in clinical trials of drugs designed to promote apoptosis. Lymphoma shows high levels of apoptosis following therapy (e.g., 40%[23]), and up to 15% apoptosis has been observed in treated breast tumors (3- to 6-fold increase above untreated)[24] and 19% apoptosis in treated melanoma (up from <1% in untreated).[25] This means that the imaging target has to be specific to apoptotic or dying cells and to produce a relatively large signal in order to be able to distinguish dying cells, which are in a minority, from the majority of cells, which remain viable. The low-steady-state levels of apoptosis are partially explained by the fact that apoptotic cells are rapidly removed by surrounding cells, which means that the imaging target is only temporarily available before the surrounding tissue removes it. This is not necessarily a problem since, in the case of a targeted contrast agent, this may amplify the imaging signal by accumulating the agent in the tissue. Finally, there is the issue of distinguishing between different forms of cell death, although given the interrelationship between necrosis and apoptosis, and the fact that just detecting cell death may be sufficient, this need not be a problem. In the following, we summarize the current state of MR-based methods for detecting tissue apoptosis and cell death in general. The use of MR to detect therapy response and apoptosis has been the subject of several recent reviews (for example, see references 26 to 30).

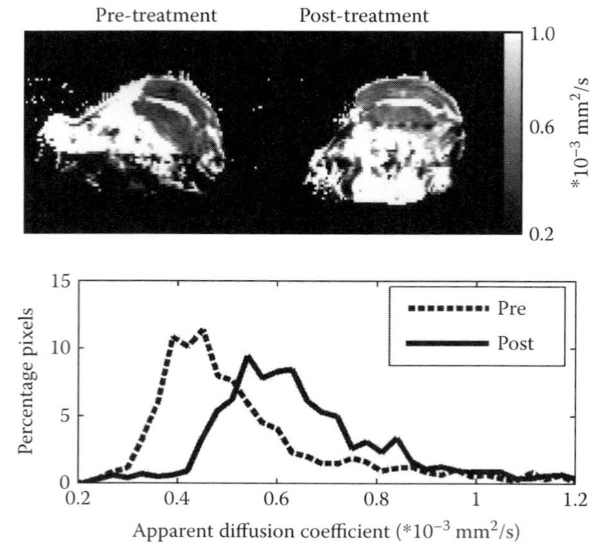

FIGURE 10.1 Water apparent diffusion coefficient (ADC) maps in EL-4 mouse lymphomas *in vivo*. An increase in ADC is observed following etoposide/cyclophosphamide chemotherapy.

10.3 IMAGING CHANGES IN CELL AND TISSUE ARCHITECTURE

Detection of cell death through changes in the MR properties of tissue water could, in principle, allow imaging of cell death at relatively high spatial resolution. This is likely to be important in detecting small regions of tissue undergoing apoptosis. Apoptotic alterations in cellular architecture include changes in the size of cellular organelles, rearrangement of the cytoskeleton, and increases in the permeability of membranes to water, all of which can potentially influence the diffusion and MR relaxation properties of tissue water. The advantage of exploiting an endogenous marker of apoptosis, such as the MR properties of tissue water, is that the method is completely noninvasive. The disadvantage, in this case, is that the gross cellular alterations that affect the MR properties of tissue water occur relatively late in the apoptotic process, and also their specificity for the process may be limited.

Diffusion-weighted MRI offers a powerful tool to study the movement of water in tissue. For example, in cerebral ischemia, diffusion-weighted MRI can be used to detect the lesion within the first hours of ischemia, when more conventional MR contrasts still appear relatively unchanged.[31] Altered water diffusion also appears to be a relatively early marker for successful therapy in tumors[32–35] (see Figure 10.1). A comparison of these systems, however, highlights some of the problems associated with the use of endogenous contrast. Due to technical limitations, most of the experiments have been performed using relatively low b-values (0 to 1000 sec/mm^2; the b-value defines the amount of applied diffusion weighting), causing the measured apparent diffusion coefficient to be a weighted average of water mobility in several different compartments, mainly the intra- and extracellular compartments. In cerebral ischemia, a marked reduction in water diffusion due to loss of energy metabolism is thought to reflect changes in the distribution of water between the intra- and extracellular spaces, primarily an influx of water into cells, as well as changes in the movement of water within these pools.[36] In the clinic, a lowered diffusion rate is usually associated with permanently damaged tissue, although this may not be the case in experimental settings. In tumors, on the other hand, successful treatment leads to increased tissue water diffusion. The major mechanism responsible for these early changes in water diffusion in tumors is thought to be cell loss and an increase in the extracellular space, reflecting a general increase in cell death.[37–40] Therefore, cell death can be detected through an increase or a decrease in the diffusion coefficient of water, depending on the tissue studied. Furthermore, measurements of the diffusion coefficient cannot currently give detailed information on the type of cell death. It is possible that the use of higher b-values, combined with a more detailed data analysis, could make these diffusion experiments more sensitive to intracellular changes, and therefore potentially more specific for the tissue changes that take place during apoptosis.[39,41] Alternatively, measurements on intracellular metabolite diffusion might be used to detect the cellular changes that accompany apoptosis.[42] Whether the sensitivity of these diffusion methods can be improved to allow detection of the low levels of apoptosis that are frequently found in the clinic remains to be seen. In the meantime, however, diffusion measurements appear to be a promising noninvasive tool for following therapy response in oncology and in neurological diseases.[33]

MR contrast based on the relaxation properties of tissue water appears to be less useful in the detection of apoptosis, primarily because the responses are delayed and usually relatively small.[33] Interestingly, recent studies have suggested that T_1 in the rotating frame ($T_{1\rho}$) may be a more sensitive marker for early tissue alterations in a variety of pathological states, when compared to measurements of other tissue water relaxation times.[43–45] In tumors, regions with early increases in $T_{1\rho}$ coincided with regions where a high apoptotic index was detected histologically.[45,46] The use of relaxation changes, however, for detecting tissue death is challenging since they are affected by numerous other physiological parameters, such as blood flow and oxygenation, which could make correlation with cell death ambiguous.

There has been some interest recently in using ^{23}Na MRI to monitor cell death, as increased sodium signals have been observed following cerebral ischemia,[47–50] tumor therapy,[51–53] and liver failure.[54] The increased signals may result from alterations in sodium distribution between the intra- and extracellular spaces, due to loss of functional plasma membrane ion channels or destruction of cell membranes. Sodium imaging could therefore allow direct assessment of ongoing cell death. However, the biggest limitation of sodium MRI, and also other nonproton imaging approaches, is its relatively low sensitivity, which makes imaging difficult.

To summarize, contrast based on changes in cell architecture, particularly measurements of water diffusion, shows some promise in detecting cell death. The biggest limitation of these methods currently is a lack of specificity. A range of different processes can affect these parameters to a varying extent, and therefore the observed changes may be difficult to interpret. The question remains whether these techniques could be made sensitive and specific enough for the clinical detection of small regions of tissue undergoing apoptosis.

10.4 DETECTING CELL DEATH THROUGH CHANGES IN METABOLISM

Several ^{31}P MRS-detectable metabolic markers of apoptosis have been identified in tumor cells *in vitro*, including the accumulation of fructose 1,6-biphosphate (FBP) and cytosine diphosphocholine (CDP choline),[55,56] as well as variable changes in the concentrations of phospholipid metabolites and NAD(H).[26,56–58] A general limitation of MRS of nonproton nuclei, however, is its relatively low sensitivity, which limits both spatial and temporal resolution. A further challenge is the limited spectral resolution available *in vivo*, which may prevent detection of the metabolites of interest.[26] While proton MRS also allows the detection of cell death-dependent changes in the concentrations of metabolites involved in phospholipid metabolism, namely, choline-containing compounds (choline, glycerophosphocholine, phosphocholine, as well as taurine and myo-inositol) at 3.2 ppm,[56,59–62] the most promising spectroscopic approach for detecting apoptosis has focused on the changes in signals from neutral lipids.

10.4.1 Lipid Metabolism

An association between NMR signals from mobile lipids and cell death, including apoptosis, was reported more than a decade ago.[63,64] Initially, the methylene ($–CH_2–$) signal at 1.3 ppm and the methyl ($-CH_3$) signal at 0.9 ppm were identified as potential markers for apoptosis.[65,66] Subsequent studies have also shown changes in the levels of polyunsaturated fatty acids (PUFAs; at 2.8 and 5.4 ppm).[67,68] These changes in lipid signals have been reported in a range of cell lines and following a variety of different apoptosis-inducing treatments *in vitro*, suggesting that altered lipid metabolism is a relatively common consequence of cell death.[69] Importantly, similar observations have also been made *in vivo*, mostly in gliomas[62,67,70] but also in a lymphoma tumor model.[71] It should be stressed, however, that increased lipid signals are not necessarily a specific marker of apoptosis since altered lipid levels are also associated with necrosis and the formation of scar tissue *in vivo*[42,72] and with cell stress *in vitro*.[73,74]

Despite the apparent success of apoptosis detection using ^{1}H MRS measurements of lipid accumulation, there are several issues that need to be resolved. Although there is good evidence that the signals arise from an accumulation of triacylglycerols and cholesterol esters,[69] the mechanisms responsible for this lipid accumulation are not yet fully understood. The accumulation of triacylglycerols could result from several pathways. Increased catabolism of membrane phospholipids, resulting from phospholipase activation, produces free fatty acids that can be converted into triacylglycerols.[60,73,75] Phospholipase activation, especially of phospholipase A_2, has been linked to increased lipid signals,[67] and the breakdown of mitochondrial membranes, in association with increased catabolic activity, has been suggested as a potential source of the mobile lipids.[76] The accumulation of diacylglycerols and triacylglycerols could also result from inhibition of phospholipid biosynthesis. Inhibition of phosphatidylcholine biosynthesis, which resulted in the accumulation of CDP choline, has been reported in cells undergoing apoptosis.[55,56] Triacylglycerol accumulation could also result from altered lipid uptake since cells readily take up lipids from the surrounding medium.[63,77] Lipid accumulation, however, does not appear to be an integral part of the cell death program since lipid accumulation in apoptotic HuT 78 cells was prevented using an acyl-CoA synthetase inhibitor, with only minor effects on the progress of the apoptotic program.[78]

Two major mobile lipid domains have been suggested to be responsible for the NMR-visible lipids: lipid bodies with a diameter of 0.5 to 10 μm in the cytosol or extracellular space[63,72] and smaller microdomains or rafts located in the plasma membrane.[79–82] The current belief is that the majority of the signal is produced by lipids located in the lipid bodies.[62,63,67,72,83,84] This is supported by the close correlation observed between the accumulation of lipid droplets and the increase in the MR signals,[63,72,74] and by the observation that the restricted diffusion of the lipids, measured by MR, is consistent with the size of the lipid droplets.[85,86]

The ^{1}H MRS(I) detection of these lipids can also be technically challenging. Fourier-based two-dimensional MRS imaging is inherently more time consuming than MRI due to the need for phase encoding in both directions. This, combined with the relatively low concentration of the lipids, significantly limits the achievable spatial resolution and makes detection of relatively low levels of cell death difficult, as the majority of cells inside a voxel may not be undergoing apoptosis. Another disadvantage of low resolution is the vulnerability to contamination from lipid signals from surrounding tissue. Finally, the inherent variation in cellular lipid content may limit the application of lipid spectroscopy to situations where either follow-up studies are possible, such as in monitoring response to therapy, or there are no significant amounts of endogenous lipid, such as in cerebral ischemia.[87]

10.4.2 Energy Metabolism

A casualty of the apoptotic program is mitochondrial function. Therefore, it seems reasonable to expect that apoptosis might be detected from ^{31}P MRS changes in cellular energy status, such as an increase in Pi concentration, a loss of ATP, or a change in the ADP/ATP ratio, which can be estimated from the phosphocreatine/ATP ratio in those tissues containing the enzyme creatine kinase. Changes in cellular energy status have been detected by ^{31}P MRS in cells undergoing apoptosis.[55,58,88] While ^{31}P MRS has shown great promise *in vitro*, there are important limitations

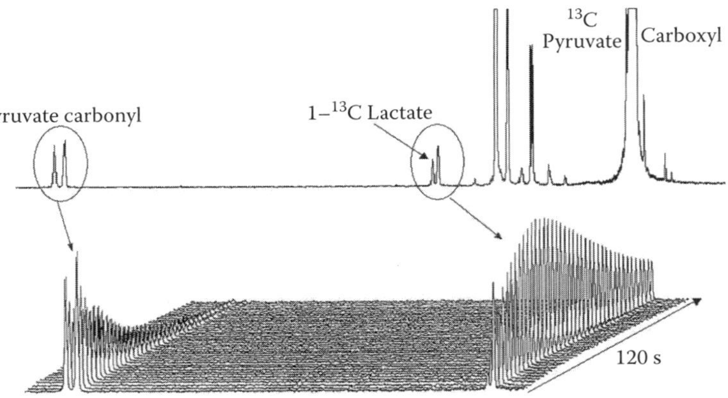

FIGURE 10.2 *In vitro* ^{13}C MR spectroscopy of hyperpolarized [1-^{13}C]-pyruvate in EL-4 cells. Series of low-flip angle single-scan spectra were collected every second over a period of 2 min. Data show the loss of hyperpolarization as evidenced by the disappearance of the natural abundance pyruvate carbonyl signal. The signal amplification obtained through hyperpolarization allows detection of the metabolic transfer of the ^{13}C label from pyruvate to lactate. (Data courtesy of Mr. Sam Day, University of Cambridge.)

for its use *in vivo*. Foremost among these limitations is the evidence that cellular energy status is maintained until relatively late during the process. Furthermore, increased leakiness and membrane damage, especially in necrotic cells, may lead to washout of metabolites.[89] These effects would make it difficult to detect small areas of apoptosis against a background from normal healthy tissue, rendering the method useful only in those situations where there is widespread cell death.

^{13}C spectroscopy has long been used to study glucose metabolism *in vitro* and *in vivo*,[88,90] and it could, in principle, be used to detect cell death in much the same way as [^{18}F] FDG has been used with PET. However, the biggest limitation of ^{13}C-based approaches has been sensitivity. The recent introduction of hyperpolarized ^{13}C-labeled substrates may overcome this problem to some extent[91] and offers the exciting possibility of being able to truly image tissue metabolism (see Figure 10.2).

Finally, alterations in cellular energy status could be estimated from tissue oxygenation by using direct methods, such as fluorine-based oxygen-sensitive probes,[92,93] or indirectly via functional MRI (fMRI). The latter method is based on blood deoxyhemoglobin concentration and has shown promise to detect alterations in energy metabolism in both normal and pathological states.[94,95] While offering good spatial resolution, the sensitivity of the methodology remains relatively low. Furthermore, it is not specific for cell death, but rather detects alterations in blood flow, blood volume, and tissue oxygenation, making the interpretation of results difficult.

10.5 IMAGING OF PHOSPHATIDYLSERINE EXPOSURE

Exposure of phosphatidylserine (PS) on the cell surface occurs relatively early in the apoptotic process.[96] PS is normally resident predominantly on the inner leaflet of the plasma membrane bilayer, and this asymmetry is maintained by an ATP-dependent translocase that transports aminophospholipids from the outer leaflet to the inner leaflet. During apoptosis, PS may be flipped to the outer leaflet, due to inhibition of this translocase and by activation of a Ca^{2+}-dependent scramblase, which transports lipids bidirectionally.[97] As a result, the number of surface PS molecules can increase 100- to 1000-fold.[98,99] For example, in Jurkat cells, the induction of apoptosis results in the exposure of ~240 pmol of PS/10^6 cells.[100] At the estimated cell densities found in tumors of ~10^9/ml,[101] this corresponds to a local PS concentration of ~200 μM. However, PS exposure can also occur during necrosis due to the leakiness of the plasma membrane under these conditions, and therefore is not specific to apoptosis per se.[18]

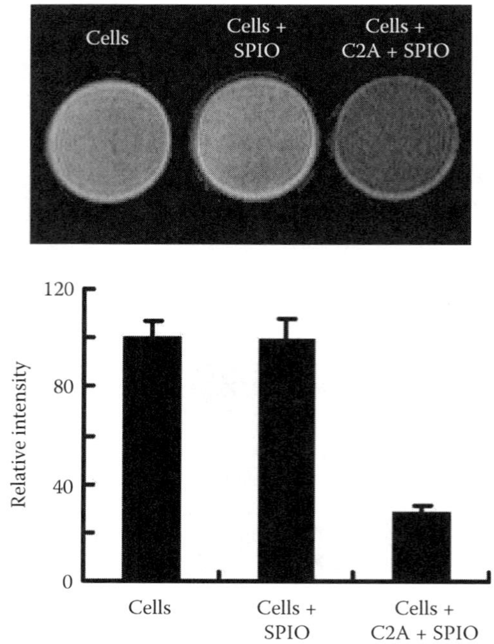

FIGURE 10.3 Imaging of phosphatidylserine exposure in apoptotic EL-4 cells. Apoptotic cells were incubated in the presence of streptavidin-SPIO (30 ng Fe/ml). T_2-weighted MRI showed a significant decrease in signal intensity only when the cells had also been incubated with biotin-C2A-GST (1.5 μ*M*) prior to addition of the streptavidin-SPIO. For further details, see Jung et al.[106] and Zhao et al.[107]

PS exposure can be targeted with the protein annexin V, which in the presence of Ca^{2+} binds to PS with nanomolar affinity. This feature has been exploited for apoptosis detection *in vitro* and *in vivo* using a range of different imaging modalities, including fluorescence, radionuclide, and MRI, and a range of different animal models, from graft rejection to tumor therapy (see, for example, references 5, 96, 102–104 and references cited therein). Recently, other molecules that also bind to PS have been introduced,[105] including the C2A domain of synaptotagmin I, which like annexin V binds to PS in a Ca^{2+}-dependent manner and with nanomolar affinity.[106] The protein has been conjugated to superparamagnetic iron oxide (SPIO) nanoparticles (Figure 10.3) and to Gd^{3+} chelates and used in the MRI detection of apoptosis *in vitro*[106,107] and *in vivo*.[107]

Imaging of PS exposure, using radiolabeled annexin V, has already undergone clinical trials.[5] Therefore, any new MRI-based method for detecting apoptosis via PS exposure needs to be evaluated in comparison with these methods. While MRI has much better image resolution than radionuclide imaging methods such as PET and single photon emission computed tomograpy (SPECT), it has much poorer sensitivity. Typical MR contrast agents are based on either SPIO nanoparticles or gadolinium (Gd) chelates. Currently, MR typically requires contrast agent concentrations in the range of 10^{-3} to 10^{-5} *M*, as compared to 10^{-10} to 10^{-11} *M* for SPECT and 10^{-11} to 10^{-12} *M* for PET.[108] This lack of sensitivity may limit the detection of relatively low levels of apoptosis by MR, which are nevertheless detectable using radionuclide or optical imaging techniques.

All *in vivo* MRI studies of apoptosis in heart and tumors using PS-targeted agents have so far been based on SPIO.[102,107,109] This is not surprising given their higher T_2 relaxivity, especially at the high magnetic fields used for preclinical research. Their increased sensitivity to detection is compromised to some extent by the fact that the contrast obtained is negative — the targeted regions become hypointense in the MR image. MRI images, particularly of tumors, are inherently heterogeneous and modest signal decreases may be difficult to separate from other processes such as hemorrhage, which are also likely to occur in damaged tissue. The biggest limitation, however, of these SPIO-based contrast agents is their large size (usually tens of nanometers in diameter), which limits their extravasation from the vascular space into the tissue and may also influence the specificity of the targeted agent. As important as access of the agent to the tissue is the washout of unbound material, which is responsible for the generation of specific contrast.[110] It should be noted, however, that pathological tissue may have increased permeability, which alleviates these problems to some extent.[111]

An alternative approach, using Gd^{3+}-based chelates, has so far been less successful for targeted imaging due to the much lower T_1 relaxitivies of these contrast agents. The minimum tissue concentration of Gd^{3+} required for MR detection has been estimated to be of the order of 10 to 100 μM,[112] which is high compared to other imaging modalities, where the limits are usually in the nanomolar range. This limits the applicability of Gd^{3+} chelates that would otherwise offer several advantages over SPIO-based agents. Most importantly, the contrast produced is positive in T_1-weighted images, which makes them easier to detect in a heterogeneous image background. The molecules are also usually smaller in size, so they permeate blood vessels more efficiently. The sensitivity limitations of Gd^{3+} chelates can be reduced, to some extent, by using targeted agents that carry a larger Gd^{3+} load per molecule. An added advantage of increasing the size of the Gd^{3+} carrier is that larger molecules have longer correlation times, which further improves their relaxivity. The increased relaxitivity comes at the price of increased molecular size, which may limit vascular permeability.

A number of different approaches have been suggested for generating high-relaxivity Gd^{3+}-based contrast agents. For example, avidin, which binds with very high affinity to biotin (Kd ~ 10^{-15} M), can be conjugated to multiple Gd^{3+} chelates and then bound to a biotin-labeled targeting agent, an approach that has been shown to work *in vivo*.[101] While this approach is relatively generic and allows the use of smaller molecules, which can be combined if necessary into a larger complex *in situ*, it nevertheless has some limitations. Avidin is immunogenic, which may limit its use in the clinic, and also shows nonspecific binding. These problems can be partially resolved by coating the molecule with polyethylene glycol groups.[113] An alternative to using avidin as the carrier molecule is to use synthetic carrier molecules, such as polyamidoamine (PAMAM) dendrimers, or nanoparticles loaded with Gd^{3+}-DTPA.[112,114] The full potential of MR-based PS imaging remains to be determined, particularly in relation to the already established radionuclide imaging methods.

10.6 CASPASE ACTIVITY

Possibly the best target for imaging apoptosis would be caspase activity, since increased caspase activity is among the earliest and most specific markers of apoptosis, and the basal activities of these enzymes are low. An enzyme-activatable contrast agent, in which there is an enzyme-mediated change in relaxivity or chemical shift, has the advantages, in principle, of signal amplification, specificity, and low nonspecific background signal.[115] The usefulness of this approach for molecular imaging has been demonstrated previously with detection of caspase-3 activity using bioluminescence.[115,116]

Enzyme-activated probes have been developed for MRI, although these have yet to achieve large signal increases. For example, Louie et al. described a Gd^{3+} chelate for which the relaxivity increased in the presence of β-galactosidase activity, an enzyme commonly used as a gene reporter in biological systems.[117] MRI agents have also been developed where there are target-dependent changes in relaxivity of the agent due to changes in agent size.[118] To image caspase activity, this would require that any agent is taken up by the cell. This might be achieved by the use of membrane-penetrating peptides, such as TAT,[119] provided that the agent was subsequently released into the cytosol rather than confined to an endocytic vesicle or some other cell compartment. It would also be an advantage if the nonreacted contrast agent could be cleared from the cell, thus minimizing any nonspecific background signal. An appealing alternative to relaxation-based reagents would be to exploit chemical shift. For example, it might be possible to design a probe molecule with an enzyme-cleavable structure, in which cleavage results in a change in the chemical shift of a probe resonance. In this case, the background signal would be minimal. Fluorine spectroscopy, for example, has been used successfully to follow enzyme activation of a pro-drug to generate an effective chemotherapeutic agent,[120] and one can envision similar approaches to image the activity of other enzymes. In fact, recent work has shown that β-galactosidase activity can be detected using fluorine MRS.[121] The question remains, however, whether it would be possible to load enough probe molecules into cells *in vivo* to produce a strong enough signal to allow some form of imaging.

10.7 ALTERED PH

Alterations in cellular pH are often associated with ongoing cell death and can be followed using MR. For example, cellular acidification of CHO cells induced to undergo programmed cell death in a hollow-fiber bioreactor was detected using ^{31}P MRS measurements of the chemical shift of the resonance from intracellular Pi.[56] Alternatively, one could try to measure extracellular pH in the expectation that this might also decrease, using exogenous pH probes such as the ^{1}H MRS-detectable molecule 2-imidazol-1-yl-3-ethoxycarbonyl-propionate (IEPA), in which the pH is estimated from the chemical shifts of the probe's resonances.[122] These approaches suffer the same limitations as ^{1}H MRS detection of mobile lipid, and it would be difficult to detect relatively low levels of tissue apoptosis by this method.

An alternative and more sensitive approach for detecting changes in extracellular pH would be to use lanthanide-based contrast agents that modify water ^{1}H signal intensity in the MR image in a pH-dependent manner.[123] This provides much higher spatial and temporal resolution than the spectroscopic methods. Two different approaches have been tried, one based on chemical exchange between the free water signal and the water signal shifted by binding of water molecules to the lanthanide, and the other based on pH-dependent relaxation of the water resonance by the lanthanide. The first approach relies on so-called chemical exchange saturation transfer (CEST), where saturation of the resonance from the bound water molecule alters the intensity of the signal from free water in an exchange process that depends on pH.[124,125] Recently, a similar approach was taken using an endogenous contrast agent to measure pH in cerebral ischemia. In this case, pH was estimated from measurements of proton exchange between free water and endogenous protein amide protons.[126] An alternative is to use a contrast agent that shows pH-dependent changes in relaxivity.[127] For both methods, the concentration of the contrast agent also needs to be known to calculate pH. This problem can be solved, for example, by using a second, non-pH-sensitive contrast agent, with similar biodistribution as the pH-sensitive contrast agent.[123,127,128]

10.8 CONCLUDING REMARKS

The ideal apoptosis detection method needs to be sensitive, in order to allow detection of relatively small regions of tissue cell death, and specific, so as to minimize background signal from surrounding healthy tissue. At the present time, none of the MR-based methods fulfill both criteria. ^{1}H MRS detection of cellular lipid accumulation and ^{1}H MRI measurements of water diffusion and relaxation have the virtue that they exploit endogenous contrast, and thus can be used in the clinic now. However, they lack both sensitivity and specificity, which is particularly true for the ^{1}H MRS measurements of lipid accumulation. Targeted contrast media, in principle, could be made that are both more sensitive and specific. However, their translation into the clinic requires considerable further development in preclinical animal models, where they will need to be carefully evaluated in comparison to existing, predominantly radionuclide based imaging methods for detecting tissue cell death.

ACKNOWLEDGMENTS

The work in KMB's laboratory is funded by grants from Cancer Research U.K. (CUK grant C197/A3514). We are members of the European Molecular Imaging Laboratory (EMIL), a "network of excellence" funded by the European Union.

REFERENCES

1. Alberts, B. et al., *Molecular Biology of the Cell*, Garland Science, New York, 2002, p. 1010.
2. Hengartner, M.O., The biochemistry of apoptosis, *Nature*, 407, 770, 2000.
3. Evan, G. and Littlewood, T., A matter of life and cell death, *Science*, 281, 1317, 1998.
4. Mattson, M.P., Culmsee, C., and Yu, Z.F., Apoptotic and antiapoptotic mechanisms in stroke, *Cell Tissue Res.*, 301, 173, 2000.
5. Blankenberg, F., Mari, C., and Strauss, H.W., Imaging cell death *in vivo*, *Q. J. Nucl. Med.*, 47, 337, 2003.
6. Garg, S., Narula, J., and Chandrashekhar, Y., Apoptosis and heart failure: clinical relevance and therapeutic target, *J. Mol. Cell. Cardiol.*, 38, 73, 2005.
7. Ghavami, S. et al., Apoptosis in liver diseases: detection and therapeutic applications, *Med. Sci. Monit.*, 11, RA337, 2005.
8. Lossi, L. et al., Apoptosis in the mammalian CNS: lessons from animal models, *Vet. J.*, 170, 52, 2005.
9. Chang, J. et al., Apoptosis and proliferation as predictors of chemotherapy response in patients with breast carcinoma, *Cancer*, 89, 2145, 2000.
10. Ellis, P.A. et al., Preoperative chemotherapy induces apoptosis in early breast cancer, *Lancet*, 349, 849, 1997.
11. Meyn, R.E. et al., Heterogeneity in the development of apoptosis in irradiated murine tumours of different histologies, *Int. J. Radiat. Biol.*, 64, 583, 1993.
12. Niemeyer, C.M. et al., Low-dose versus high-dose methotrexate during remission induction in childhood acute lymphoblastic leukemia (Protocol 81-01 update), *Blood*, 78, 2514, 1991.
13. Savill, J. and Fadok, V., Corpse clearance defines the meaning of cell death, *Nature*, 407, 784, 2000.
14. Fadok, V.A. et al., A receptor for phosphatidylserine-specific clearance of apoptotic cells, *Nature*, 405, 85, 2000.
15. Martin, S.J. et al., Early redistribution of plasma membrane phosphatidylserine is a general feature of apoptosis regardless of the initiating stimulus: inhibition by overexpression of Bcl-2 and Abl, *J. Exp. Med.*, 182, 1545, 1995.
16. Naito, M. et al., Phosphatidylserine externalization is a downstream event of interleukin-1 beta-converting enzyme family protease activation during apoptosis, *Blood*, 89, 2060, 1997.
17. Leist, M. and Jaattela, M., Four deaths and a funeral: from caspases to alternative mechanisms, *Nat. Rev. Mol. Cell Biol.*, 2, 589, 2001.
18. Corsten, M.F. et al., Counting heads in the war against cancer: defining the role of annexin A5 imaging in cancer treatment and surveillance, *Cancer Res.*, 66, 1255, 2006.
19. Romer, W. et al., Positron emission tomography in non-Hodgkin's lymphoma: assessment of chemotherapy with fluorodeoxyglucose, *Blood*, 91, 4464, 1998.
20. von Schulthess, G.K., Steinert, H.C., and Hany, T.F., Integrated PET/CT: current applications and future directions, *Radiology*, 238, 405, 2006.
21. Juweid, M.E. and Cheson, B.D., Positron-emission tomography and assessment of cancer therapy, *N. Engl. J. Med.*, 354, 496, 2006.
22. Meyn, R.E. et al., Apoptosis in murine tumors treated with chemotherapy agents, *Anticancer Drugs*, 6, 443, 1995.
23. Maciorowski, Z. et al., Comparative analysis of apoptosis measured by Hoechst and flow cytometry in non-Hodgkin's lymphomas, *Cytometry*, 32, 44, 1998.
24. Symmans, W.F. et al., Paclitaxel-induced apoptosis and mitotic arrest assessed by serial fine-needle aspiration: implications for early prediction of breast cancer response to neoadjuvant treatment, *Clin. Cancer Res.*, 6, 4610, 2000.

25. Jansen, B. et al., Chemosensitisation of malignant melanoma by BCL2 antisense therapy, *Lancet*, 356, 1728, 2000.
26. Brindle, K.M., Detection of apoptosis in tumors using magnetic resonance imaging and spectroscopy, *Adv. Enzyme Regul.*, 42, 101, 2002.
27. Brauer, M., *In vivo* monitoring of apoptosis, *Prog. Neuropsychopharmacol. Biol. Psychiatry*, 27, 323, 2003.
28. Hakumäki, J.M. and Liimatainen, T., Molecular imaging of apoptosis in cancer, *Eur. J. Radiol.*, 56, 143, 2005.
29. Kettunen, M.I. and Brindle, K.M., Apoptosis detection using magnetic resonance imaging and spectroscopy, *Prog. Nucl. Mag. Res. Sp.*, 47, 175, 2005.
30. Kettunen, M.I. and Gröhn, O.H., Tumour gene therapy monitoring using magnetic resonance imaging and spectroscopy, *Curr. Gene Ther.*, 5, 685, 2005.
31. Moseley, M.E. et al., Early detection of regional cerebral ischemia in cats: comparison of diffusion- and T2-weighted MRI and spectroscopy, *Magn. Reson. Med.*, 14, 330, 1990.
32. Zhao, M. et al., Early detection of treatment response by diffusion-weighted 1H-NMR spectroscopy in a murine tumour *in vivo*, *Br. J. Cancer*, 73, 61, 1996.
33. Kauppinen, R.A., Monitoring cytotoxic tumour treatment response by diffusion magnetic resonance imaging and proton spectroscopy, *NMR Biomed.*, 15, 6, 2002.
34. Ross, B.D. et al., Evaluation of cancer therapy using diffusion magnetic resonance imaging, *Mol. Cancer Ther.*, 2, 581, 2003.
35. Moffat, B.A. et al., Diffusion imaging for evaluation of tumor therapies in preclinical animal models, *MAGMA*, 17, 249, 2004.
36. Sotak, C.H., Nuclear magnetic resonance (NMR) measurement of the apparent diffusion coefficient (ADC) of tissue water and its relationship to cell volume changes in pathological states, *Neurochem. Int.*, 45, 569, 2004.
37. Sugahara, T. et al., Usefulness of diffusion-weighted MRI with echo-planar technique in the evaluation of cellularity in gliomas, *J. Magn. Reson. Imaging*, 9, 53, 1999.
38. Gupta, R.K. et al., Inverse correlation between choline magnetic resonance spectroscopy signal intensity and the apparent diffusion coefficient in human glioma, *Magn. Reson. Med.*, 41, 2, 1999.
39. Valonen, P.K. et al., Water diffusion in a rat glioma during ganciclovir-thymidine kinase gene therapy-induced programmed cell death *in vivo*: correlation with cell density, *J. Magn. Reson. Imaging*, 19, 389, 2004.
40. Schepkin, V.D. et al., Sodium magnetic resonance imaging of chemotherapeutic response in a rat glioma, *Magn. Reson. Med.*, 53, 85, 2005.
41. Minard, K.R. et al., Simultaneous 1H PFG-NMR and confocal microscopy of monolayer cell cultures: effects of apoptosis and necrosis on water diffusion and compartmentalization, *Magn. Reson. Med.*, 52, 495, 2004.
42. Hakumäki, J.M. et al., Quantitative 1H nuclear magnetic resonance diffusion spectroscopy of BT4C rat glioma during thymidine kinase-mediated gene therapy *in vivo*: identification of apoptotic response, *Cancer Res.*, 58, 3791, 1998.
43. Gröhn, O.H. et al., Quantitative magnetic resonance imaging assessment of cerebral ischemia in rat using on-resonance T(1) in the rotating frame, *Magn. Reson. Med.*, 42, 268, 1999.
44. Duvvuri, U. et al., Quantitative T1rho magnetic resonance imaging of RIF-1 tumors *in vivo*: detection of early response to cyclophosphamide therapy, *Cancer Res.*, 61, 7747, 2001.
45. Hakumäki, J.M. et al., Early gene therapy-induced apoptotic response in BT4C gliomas by magnetic resonance relaxation contrast T1 in the rotating frame, *Cancer Gene Ther.*, 9, 338, 2002.
46. Gröhn, O.H. et al., Novel magnetic resonance imaging contrasts for monitoring response to gene therapy in rat glioma, *Cancer Res.*, 63, 7571, 2003.
47. Thulborn, K.R. et al., Comprehensive MR imaging protocol for stroke management: tissue sodium concentration as a measure of tissue viability in nonhuman primate studies and in clinical studies, *Radiology*, 213, 156, 1999.
48. Lin, S.P. et al., Direct, longitudinal comparison of (1)H and (23)Na MRI after transient focal cerebral ischemia, *Stroke*, 32, 925, 2001.
49. Bartha, R. et al., Sodium T2*-weighted MR imaging of acute focal cerebral ischemia in rabbits, *Magn. Reson. Imaging*, 22, 983, 2004.

50. Jones, S.C. et al., Stroke onset time using sodium MRI in rat focal cerebral ischemia, *Stroke*, 37, 883, 2006.
51. Kline, R.P. et al., Rapid *in vivo* monitoring of chemotherapeutic response using weighted sodium magnetic resonance imaging, *Clin. Cancer Res.*, 6, 2146, 2000.
52. Babsky, A.M. et al., Application of 23Na MRI to monitor chemotherapeutic response in RIF-1 tumors, *Neoplasia*, 7, 658, 2005.
53. Schepkin, V.D. et al., Sodium and proton diffusion MRI as biomarkers for early therapeutic response in subcutaneous tumors, *Magn. Reson. Imaging*, 24, 273, 2006.
54. Maril, N. et al., Detection of evolving acute tubular necrosis with renal 23Na MRI: studies in rats, *Kidney Int.*, 69, 765, 2006.
55. Anthony, M.L., Zhao, M., and Brindle, K.M., Inhibition of phosphatidylcholine biosynthesis following induction of apoptosis in HL-60 cells, *J. Biol. Chem.*, 274, 19686, 1999.
56. Williams, S.N., Anthony, M.L., and Brindle, K.M., Induction of apoptosis in two mammalian cell lines results in increased levels of fructose-1,6-bisphosphate and CDP-choline as determined by 31P MRS, *Magn. Reson. Med.*, 40, 411, 1998.
57. Adebodun, F. and Post, J.F., 31P NMR characterization of cellular metabolism during dexamethasone induced apoptosis in human leukemic cell lines, *J. Cell Physiol.*, 158, 180, 1994.
58. Muruganandham, M. et al., Metabolic signatures associated with a NAD synthesis inhibitor-induced tumor apoptosis identified by 1H-decoupled-31P magnetic resonance spectroscopy, *Clin. Cancer Res.*, 11, 3503, 2005.
59. Gupta, R.K. et al., Relationships between choline magnetic resonance spectroscopy, apparent diffusion coefficient and quantitative histopathology in human glioma, *J. Neurooncol.*, 50, 215, 2000.
60. Milkevitch, M. et al., Increases in NMR-visible lipid and glycerophosphocholine during phenylbutyrate-induced apoptosis in human prostate cancer cells, *Biochim. Biophys. Acta*, 1734, 1, 2005.
61. Ronen, S.M. et al., Magnetic resonance detects changes in phosphocholine associated with Ras activation and inhibition in NIH 3T3 cells, *Br. J. Cancer*, 84, 691, 2001.
62. Lehtimäki, K.K. et al., Metabolite changes in BT4C rat gliomas undergoing ganciclovir-thymidine kinase gene therapy-induced programmed cell death as studied by 1H NMR spectroscopy *in vivo*, *ex vivo*, and *in vitro*, *J. Biol. Chem.*, 278, 45915, 2003.
63. Callies, R. et al., The appearance of neutral lipid signals in the 1H NMR spectra of a myeloma cell line correlates with the induced formation of cytoplasmic lipid droplets, *Magn. Reson. Med.*, 29, 546, 1993.
64. Blankenberg, F.G. et al., Detection of apoptotic cell death by proton nuclear magnetic resonance spectroscopy, *Blood*, 87, 1951, 1996.
65. Blankenberg, F.G. et al., Quantitative analysis of apoptotic cell death using proton nuclear magnetic resonance spectroscopy, *Blood*, 89, 3778, 1997.
66. Al-Saffar, N.M. et al., Apoptosis is associated with triacylglycerol accumulation in Jurkat T-cells, *Br. J. Cancer*, 86, 963, 2002.
67. Hakumäki, J.M. et al., 1H MRS detects polyunsaturated fatty acid accumulation during gene therapy of glioma: implications for the *in vivo* detection of apoptosis, *Nat. Med.*, 5, 1323, 1999.
68. Griffin, J.L. et al., Assignment of 1H nuclear magnetic resonance visible polyunsaturated fatty acids in BT4C gliomas undergoing ganciclovir-thymidine kinase gene therapy-induced programmed cell death, *Cancer Res.*, 63, 3195, 2003.
69. Hakumäki, J.M. and Kauppinen, R.A., 1H NMR visible lipids in the life and death of cells, *Trends Biochem. Sci.*, 25, 357, 2000.
70. Ross, B.D., Kim, B., and Davidson, B.L., Assessment of ganciclovir toxicity to experimental intracranial gliomas following recombinant adenoviral-mediated transfer of the herpes simplex virus thymidine kinase gene by magnetic resonance imaging and proton magnetic resonance spectroscopy, *Clin. Cancer Res.*, 1, 651, 1995.
71. Schmitz, J.E. et al., 1H MRS-visible lipids accumulate during apoptosis of lymphoma cells *in vitro* and *in vivo*, *Magn. Reson. Med.*, 54, 43, 2005.
72. Zoula, S. et al., Correlation between the occurrence of 1H-MRS lipid signal, necrosis and lipid droplets during C6 rat glioma development, *NMR Biomed.*, 16, 199, 2003.
73. Delikatny, E.J. et al., Modulation of MR-visible mobile lipid levels by cell culture conditions and correlations with chemotactic response, *Int. J. Cancer*, 65, 238, 1996.

74. Barba, I. et al., Mobile lipid production after confluence and pH stress in perfused C6 cells, *NMR Biomed.*, 14, 33, 2001.
75. Veale, M.F. et al., The generation of 1H-NMR-detectable mobile lipid in stimulated lymphocytes: relationship to cellular activation, the cell cycle, and phosphatidylcholine-specific phospholipase C, *Biochem. Biophys. Res. Commun.*, 239, 868, 1997.
76. Delikatny, E.J. et al., Nuclear magnetic resonance-visible lipids induced by cationic lipophilic chemotherapeutic agents are accompanied by increased lipid droplet formation and damaged mitochondria, *Cancer Res.*, 62, 1394, 2002.
77. Wright, L.C. et al., The origin of 1H NMR-visible triacylglycerol in human neutrophils. High fatty acid environments result in preferential sequestration of palmitic acid into plasma membrane triacylglycerol, *Eur. J. Biochem.*, 267, 68, 2000.
78. Iorio, E. et al., Triacsin C inhibits the formation of 1H NMR-visible mobile lipids and lipid bodies in HuT 78 apoptotic cells, *Biochim. Biophys. Acta*, 1634, 1, 2003.
79. Mountford, C.E. and Wright, L.C., Organization of lipids in the plasma membranes of malignant and stimulated cells: a new model, *Trends Biochem. Sci.*, 13, 172, 1988.
80. Le Moyec, L. et al., Lipid signals detected by NMR proton spectroscopy of whole cells are not correlated to lipid droplets evidenced by the Nile red staining, *Cell. Mol. Biol.*, 43, 703, 1997.
81. Ferretti, A. et al., Biophysical and structural characterization of 1H-NMR-detectable mobile lipid domains in NIH-3T3 fibroblasts, *Biochim. Biophys. Acta*, 1438, 329, 1999.
82. Wright, L.C. et al., Detergent-resistant membrane fractions contribute to the total 1H NMR-visible lipid signal in cells, *Eur. J. Biochem.*, 270, 2091, 2003.
83. Barba, I., Cabanas, M.E., and Arus, C., The relationship between nuclear magnetic resonance-visible lipids, lipid droplets, and cell proliferation in cultured C6 cells, *Cancer Res.*, 59, 1861, 1999.
84. Di Vito, M. et al., 1H NMR-visible mobile lipid domains correlate with cytoplasmic lipid bodies in apoptotic T-lymphoblastoid cells, *Biochim. Biophys. Acta*, 1530, 47, 2001.
85. Remy, C. et al., Evidence that mobile lipids detected in rat brain glioma by 1H nuclear magnetic resonance correspond to lipid droplets, *Cancer Res.*, 57, 407, 1997.
86. Perez, Y. et al., Measurement by nuclear magnetic resonance diffusion of the dimensions of the mobile lipid compartment in C6 cells, *Cancer Res.*, 62, 5672, 2002.
87. Gasparovic, C. et al., Magnetic resonance lipid signals in rat brain after experimental stroke correlate with neutral lipid accumulation, *Neurosci. Lett.*, 301, 87, 2001.
88. Seymour, A.M., Imaging cardiac metabolism in heart failure: the potential of NMR spectroscopy in the era of metabolism revisited, *Heart Lung Circ.*, 12, 25, 2003.
89. Tozer, G.M. and Griffiths, J.R., The contribution made by cell death and oxygenation to 31P MRS observations of tumour energy metabolism, *NMR Biomed.*, 5, 279, 1992.
90. Gruetter, R., *In vivo* 13C NMR studies of compartmentalized cerebral carbohydrate metabolism, *Neurochem. Int.*, 41, 143, 2002.
91. Ardenkjaer-Larsen, J.H. et al., Increase in signal-to-noise ratio of >10,000 times in liquid-state NMR, *Proc. Natl. Acad. Sci. U.S.A.*, 100, 10158, 2003.
92. Gillies, R.J. et al., MRI of the tumor microenvironment, *J. Magn. Reson. Imaging*, 16, 430, 2002.
93. Zhao, D., Jiang, L., and Mason, R.P., Measuring changes in tumor oxygenation, *Methods Enzymol.*, 386, 378, 2004.
94. Gröhn, O.H. and Kauppinen, R.A., Assessment of brain tissue viability in acute ischemic stroke by BOLD MRI, *NMR Biomed.*, 14, 432, 2001.
95. Rodrigues, L.M. et al., Tumor R2* is a prognostic indicator of acute radiotherapeutic response in rodent tumors, *J. Magn. Reson. Imaging*, 19, 482, 2004.
96. Lahorte, C.M. et al., Apoptosis-detecting radioligands: current state of the art and future perspectives, *Eur. J. Nucl. Med. Mol. Imaging*, 31, 887, 2004.
97. Zwaal, R.F. and Schroit, A.J., Pathophysiologic implications of membrane phospholipid asymmetry in blood cells, *Blood*, 89, 1121, 1997.
98. Bennett, M.R. et al., Binding and phagocytosis of apoptotic vascular smooth muscle cells is mediated in part by exposure of phosphatidylserine, *Circ. Res.*, 77, 1136, 1995.
99. Tait, J.F., Smith, C., and Wood, B.L., Measurement of phosphatidylserine exposure in leukocytes and platelets by whole-blood flow cytometry with annexin V, *Blood Cells Mol. Dis.*, 25, 271, 1999.

100. Borisenko, G.G. et al., Macrophage recognition of externalized phosphatidylserine and phagocytosis of apoptotic Jurkat cells: existence of a threshold, *Arch. Biochem. Biophys.*, 413, 41, 2003.
101. Artemov, D. et al., Magnetic resonance molecular imaging of the HER-2/neu receptor, *Cancer Res.*, 63, 2723, 2003.
102. Schellenberger, E.A. et al., Annexin V-CLIO: a nanoparticle for detecting apoptosis by MRI, *Mol. Imaging*, 1, 102, 2002.
103. Green, A.M. and Steinmetz, N.D., Monitoring apoptosis in real time, *Cancer J.*, 8, 82, 2002.
104. Kietselaer, B.L. et al., The role of labeled annexin A5 in imaging of programmed cell death. From animal to clinical imaging, *Q. J. Nucl. Med.*, 47, 349, 2003.
105. Hanshaw, R.G. and Smith, B.D., New reagents for phosphatidylserine recognition and detection of apoptosis, *Bioorg. Med. Chem.*, 13, 5035, 2005.
106. Jung, H.I. et al., Detection of apoptosis using the C2A domain of synaptotagmin I, *Bioconjug. Chem.*, 15, 983, 2004.
107. Zhao, M. et al., Non-invasive detection of apoptosis using magnetic resonance imaging and a targeted contrast agent, *Nat. Med.*, 7, 1241, 2001.
108. Massoud, T.F. and Gambhir, S.S., Molecular imaging in living subjects: seeing fundamental biological processes in a new light, *Genes Dev.*, 17, 545, 2003.
109. Sosnovik, D.E. et al., Magnetic resonance imaging of cardiomyocyte apoptosis with a novel magneto-optical nanoparticle, *Magn. Reson. Med.*, 54, 718, 2005.
110. Dreher, M.R. et al., Tumor vascular permeability, accumulation, and penetration of macromolecular drug carriers, *J. Natl. Cancer Inst.*, 98, 335, 2006.
111. McDonald, D.M. and Choyke, P.L., Imaging of angiogenesis: from microscope to clinic, *Nat. Med.*, 9, 713, 2003.
112. Aime, S. et al., Insights into the use of paramagnetic Gd(III) complexes in MR-molecular imaging investigations, *J. Magn. Reson. Imaging*, 16, 394, 2002.
113. Chinol, M. et al., Biochemical modifications of avidin improve pharmacokinetics and biodistribution, and reduce immunogenicity, *Br. J. Cancer*, 78, 189, 1998.
114. Esfand, R. and Tomalia, D.A., Poly(amidoamine) (PAMAM) dendrimers: from biomimicry to drug delivery and biomedical applications, *Drug Discov. Today*, 6, 427, 2001.
115. Jaffer, F.A. and Weissleder, R., Seeing within: molecular imaging of the cardiovascular system, *Circ. Res.*, 94, 433, 2004.
116. Laxman, B. et al., Noninvasive real-time imaging of apoptosis, *Proc. Natl. Acad. Sci. U.S.A.*, 99, 16551, 2002.
117. Louie, A.Y. et al., *In vivo* visualization of gene expression using magnetic resonance imaging, *Nat. Biotechnol.*, 18, 321, 2000.
118. Perez, J.M. et al., Magnetic relaxation switches capable of sensing molecular interactions, *Nat. Biotechnol.*, 20, 816, 2002.
119. Josephson, L. et al., High-efficiency intracellular magnetic labeling with novel superparamagnetic-Tat peptide conjugates, *Bioconjug. Chem.*, 10, 186, 1999.
120. Hamstra, D.A. et al., The use of 19F spectroscopy and diffusion-weighted MRI to evaluate differences in gene-dependent enzyme prodrug therapies, *Mol. Ther.*, 10, 916, 2004.
121. Yu, J. and Mason, R.P., Synthesis and characterization of novel lacZ gene reporter molecules: detection of beta-galactosidase activity by 19F nuclear magnetic resonance of polyglycosylated fluorinated vitamin B6, *J. Med. Chem.*, 49, 1991, 2006.
122. Gil, S. et al., Imidazol-1-ylalkanoic acids as extrinsic 1H NMR probes for the determination of intracellular pH, extracellular pH and cell volume, *Bioorg. Med. Chem.*, 2, 305, 1994.
123. Gillies, R.J. et al., pH imaging. A review of pH measurement methods and applications in cancers, *IEEE Eng. Med. Biol. Mag.*, 23, 57, 2004.
124. Ward, K.M., Aletras, A.H., and Balaban, R.S., A new class of contrast agents for MRI based on proton chemical exchange dependent saturation transfer (CEST), *J. Magn. Reson.*, 143, 79, 2000.
125. Aime, S. et al., Paramagnetic lanthanide(III) complexes as pH-sensitive chemical exchange saturation transfer (CEST) contrast agents for MRI applications, *Magn. Reson. Med.*, 47, 639, 2002.
126. Zhou, J. et al., Using the amide proton signals of intracellular proteins and peptides to detect pH effects in MRI, *Nat. Med.*, 9, 1085, 2003.

127. Garcia-Martin, M.L. et al., High resolution pH(e) imaging of rat glioma using pH-dependent relaxivity, *Magn. Reson. Med.*, 55, 309, 2006.
128. Aime, S., Delli Castelli, D., and Terreno, E., Novel pH-reporter MRI contrast agents, *Angew. Chem. Int. Ed. Engl.*, 41, 4334, 2002.
129. Gottlieb, R.A. et al., Apoptosis induced in Jurkat cells by several agents is preceded by intracellular acidification, *Proc. Natl. Acad. Sci. U.S.A.*, 93, 654, 1996.

11 Molecular Imaging of Reporter Genes

Keren Ziv, Dorit Granot, Vicki Plaks, Batya Cohen, and Michal Neeman

CONTENTS

11.1 Reporter Gene Applications in Biomedical Research 199
11.2 Methods of Gene Transfer for Molecular Imaging 200
11.3 Spatial- and Temporal-Specific Promoters 200
11.4 Genetically Modified Mouse Models 201
11.5 Fluorescent Proteins as Reporter Genes 202
11.6 Luciferases as Reporter Genes 202
11.7 β-Galactosidase as a Reporter Gene 203
11.8 The Suicide Genes Thymidine Kinase and Cytosine Deaminase as Reporters of Gene Expression 204
11.9 Creatine and Arginine Kinases as Reporter Genes 205
11.10 Application of Biotin Ligase for Constructing Reporter Genes 206
11.11 CEST Agents as Reporter Genes 207
11.12 Tyrosinase as a Reporter Gene 207
11.13 Iron Uptake and Storage Proteins as Reporter Genes 208
11.14 Transferrin Receptor as a Reporter Gene for MRI 208
11.15 Ferritin as a Reporter Gene 209
11.16 Summary and Future Perspective 211
Acknowledgments 211
References 211

11.1 REPORTER GENE APPLICATIONS IN BIOMEDICAL RESEARCH

Reporter genes have become an indispensable tool for contemporary biomedical research. Briefly, the transcriptional activation of gene expression is rendered visible for detection through construction of a transgene in which activation of the promoter of the gene of interest leads to the generation of an easily detectable signal. Widely used reporter genes, including fluorescent proteins such as the green fluorescent protein (GFP), enzymes catalyzing formation of detectable products such as β-galactosidase, and bioluminescence generating enzymes such as the firefly luciferase can now be detected in tissue culture as well as in intact animals. Dynamic spatial mapping of gene expression in whole animals is one of the central challenges of molecular imaging in general and magnetic resonance imaging (MRI) in particular. This chapter will review the exciting progress in design and application of novel candidate reporter genes that may allow detection of the transcriptional regulation of gene expression by MRI. Candidate reporter genes include enzymes, receptors, transporters, and peptides whose expression can be mapped by MRI either through endogenous contrast mechanisms or through their interaction with a specific reporter probe. Future challenges remain in evaluating the ability to exploit these reporters for actual monitoring of spatial and temporal changes in gene expression in transgenic animals, and in targeted cellular and molecular therapy.

Availability of the sequence for the human and mouse genomes provided the foundation for the yet greater challenge of deciphering the contribution of each gene to development and disease. The level at which it will be possible to explore gene function is dependent, in part, upon the ability to monitor and image gene expression *in vivo* in real time.

One way to monitor gene expression is to utilize reporter genes, which upon delivery to cells or animal models assist in following gene localization and transcriptional regulation. Transfection of reporter genes into eukaryotic cells is used widely to study *cis*-regulatory sequences or trans-acting factors that modulate the transcriptional activity of certain promoters. The introduction of reporter genes into animal genomes provides opportunities to investigate the activity of a given promoter in the context of a living organism. The most commonly used reporter genes are LUC (luciferase), which oxidizes luciferin; CAT (chloramphenicol acetyltransferase), which transfers acetyl groups to chloramphenicol; GAL (β-galactosidase) and GUS (β-glucuronidase), known to hydrolyze galactosides and glucuronides, respectively, to yield colored products; GFP; and the newly discovered RFP (red fluorescent protein), which gives fluorescence emission after excitation.

Reporter genes can be expressed either constitutively or inducibly, depending on the attached promoter. Attachment of promoters such as thymidine kinase (Tk), long terminal repeat (LTR), Rous sarcoma virus (RSV), cytomegalovirus (CMV), phosphoglycerate kinase (PGK), and elongation factor-1 (EF1), which induce constitutive expression, is usually used to monitor cell trafficking. Alternatively, coupling a reporter gene to an inducible promoter or enhancer (e.g., hypoxia response element (HRE), interferon stimulatory response element (ISRE)) will provide the possibility to control the induction of reporter expression. Inducible reporter genes can also be constructed to be sensitive to specific endogenous molecular processes, including the regulation of endogenous gene expression, the activity of specific signal transduction pathways, specific protein–protein interactions, and post-transcriptional regulation of protein expression.

Molecular imaging methods provide powerful means for studying gene function through the use of reporter genes, namely, nucleic acid sequences encoding easily detectable proteins, which upon delivery to cells or animal models assist in following the spatial and temporal transcriptional regulation of gene expression.[1] Multimodality reporter constructs can be generated as fusion genes containing cDNA from two or three different reporter genes, establishing a single fusion protein that retains functionality of each of the composite gene products and allowing cross-validation of results using various imaging methods.[2]

11.2 METHODS OF GENE TRANSFER FOR MOLECULAR IMAGING

Analysis of gene expression both *in vitro* and *in vivo* using reporter genes requires efficient delivery of genetic material into the cell or tissue of interest, in such a way that the construct will be expressed at the appropriate level, precise time, and for a sufficient duration. The limited efficiency of classical gene transfer methods, such as transfection and microinjection, can be overcome by viral vector gene delivery systems, which are widely used today for both *ex vivo* and *in vivo* gene transfer (Table 11.1).[3–7] Once a successful gene delivery is obtained, molecular imaging aims to monitor the expression level as well as to obtain spatial and temporal information on the regulation of transgene expression.

11.3 SPATIAL- AND TEMPORAL-SPECIFIC PROMOTERS

Spatial control depends on the availability of specific transcription regulatory elements or promoters that will limit the expression of transgene to the desired cell or tissue. A range of cellular promoters

TABLE 11.1
Viral Vectors for Delivery of Transgene Reporters

Virus	Genome	Transgene Capacity	Dividing Cells	Nondividing Cells	Duration of Expression	References
Adenoviruses	ds DNA	30 kb	+	+	Transient	3, 4
Adeno-associated viruses	ss DNA	4 kb	+	+	Stable	5
Herpes simplex virus 1	ds DNA	50 kb	+	+	Transient	6
Retro/lenti viruses	ss RNA	8 kb	+	+	Stable	7

have been identified for specific tissues, including the liver (albumin[8] and liver activator protein (LAP)[9]), muscle (myosin light chain 1[10]), endothelial cell tie-2,[11–13] VE-cadherin,[14] and skin.[15] Tissue-specific promoters allow a second level of control over transgene expression, in addition to that of selective transduction by the vector, and can be used to follow selective differentiation of progenitor stem cells.

Temporal control has also been pursued vigorously in the past years, and a variety of different regulatory systems have been developed. The inducing agents include heavy metal ions, heat shock, isopropyl β-D-thiogalactoside (β-gal), antibiotics, and steroid hormones.[16] Some of the inducers are toxic to mammalian cells, can influence the expression of endogenous genes, and in the absence of the inducer show high background expression of their target genes.

A popular approach for inducible gene expression, which was applied for the study of inducible expression of two MR reporter genes, tyrosinase[17] and ferritin,[18] includes the use of one of the tetracycline (tet)-inducible systems, tTA (Tet-Off) or rtTA (Tet-On). The tetracycline-controlled transactivator (tTA) was generated by fusing the DNA-binding domain of tetracycline resistance operon (TetR) from *Escherichia coli* with the transcription activation domain of virion protein 16 (VP16) of herpes simplex virus (HSV).[19] A second construct contains the target gene under the control of a minimal promoter sequence of the human cytomegalovirus promoter IE (P) combined with tet operator sequences of *E. coli* (*tetO*). In the absence of tetracycline, tTA binds to *tetO* and activates P, which in turn initiates transcription of the downstream target gene. In the presence of tetracycline, tTA dissociates from *tetO*, terminating transcription. The reverse tetracycline-controlled transcriptional activator system (rtTA), or Tet-On system, includes a mutant Tet repressor fused to VP16 to form rtTA and the responsive element *tetO* sequences linked to a P-driven target gene. In the absence of tetracycline, the target gene is not transcribed. However, in the presence of Tet, rtTA binds to *tetO* and P, which in turn activates transcription. Alternative inducible gene expression systems include the ecdysone receptor[20] and the estrogen receptor[21] systems.

11.4 GENETICALLY MODIFIED MOUSE MODELS

Generation of genetically modified mice carrying reporter genes facilitates the study of mammalian gene function. Technological advances expanded the choice of gene manipulation from straightforward gene inactivation or overexpression to selective modification of gene expression pattern, structure, and function in desired cell types, at specific times. Combining conventional/conditional, knockout/knock-in, inducible, and even reversible gene manipulation strategies provides the freedom to design an optimal model to study the function of a gene in a specific organ system during development or in postnatal and adult life.

One of the simplest ways to study gene function in a mouse is exogenous overexpression of a protein in some or all tissues. To generate a standard transgenic mouse, a vector containing the transgene and any desired reporters is injected into a fertilized mouse egg. The DNA usually integrates into one or more loci during the first few cell divisions of preimplantation development.

The number of copies of the transgenic fragment can vary from one to several hundred. A knock-in mouse is generated by targeted insertion of the transgene at a selected locus. The transgene is flanked by DNA from a noncritical locus, and homologous recombination allows the transgene to be targeted to that specific integration site. Site-specific knock-in results in a more consistent level of transgene expression from generation to generation. Elimination of a gene or the deletion of a functional domain of the protein can be achieved through specific gene targeting to generate a knockout mouse. Homologous recombination removes one or more exons from a gene, which results in the production of a mutated/truncated protein or, more often, in the absence of the protein (null mouse). The phenotypes of knockout mice can be very complex because all tissues of the mouse are affected. A way to avoid this is to ablate specific genes in selective tissues at certain stages of development using recombinases (e.g., Cre and Flp).

Site-specific recombinase (SSR) systems (Cre-loxp and Flp-FRP) led to major breakthroughs in advancing the knowledge of gene function. Cre (causes recombination of the bacteriophage P1 genome) and Flp (named for its ability to invert, or "flip," a DNA segment in *Saccharomyces cerevisiae*) are able to recombine specific sequences of DNA with high fidelity,[22,23] allowing the generation of gene deletions, insertions, inversions, and exchanges in exogenous systems.[24–26] Cre and Flp recombine DNA at defined target sites, termed loxP (locus of crossover (x) in P1)[27] and FRT (Flp recombinase recognition target),[28] respectively, in both actively dividing and postmitotic cells, as well as in most tissue types. These two systems have been combined with various inducible systems (e.g., Tet and ER) to enable the induction of genetic changes late in embryogenesis or adult tissues, with reporter genes, which are generated upon recombination.

11.5 FLUORESCENT PROTEINS AS REPORTER GENES

In vivo fluorescence imaging ranges in spatial resolution and penetration depth from high-resolution cellular imaging of superficial regions (intravital microscopy) to low-resolution whole-body imaging.[29] Applications include monitoring tumor growth and metastasis[30–33] as well as transcriptional regulation of gene expression.[34,35] Endogenously fluorescent proteins provide a powerful tool for tracing bacteria, viruses, and mammalian cells in animals.[36–40]

Aequorea GFP, one of the most widely used fluorescent proteins, has many useful characteristics.[41] It is heat stable and active as a fusion protein partner. GFP, like other fluorescent proteins, does not require cofactors or chemical staining for *in vivo* imaging. *In vitro*, GFP-labeled cells are easily distinguishable from nonfluorescent tissues. A major drawback of GFP for *in vivo* imaging is its short absorption and emission wavelength, which overlaps with the endogenous absorption and autofluorescence of many tissues. One of the key strategies for imaging deeper tissues has been to use compounds with high quantum yields that emit far-red (600 to 700 nm) and near-infrared (NIR) (700 to 900 nm) light where tissue absorption and autofluorescence are minimal.[29] A wide range of red-shifted fluorescent proteins was developed that are more favorable for *in vivo* imaging.[42] Recently, photoactivable fluorescent proteins were reported, which display little initial fluorescence under excitation at the imaging wavelength but increase their fluorescence after activation by irradiation at a different wavelength.[43] These molecular switches enable the storage of information on a molecular level, and their use as reporter genes allows protein tracking in living cells,[44,45] as photobleaching techniques do not allow direct visualization of protein movement routes within a living cell. Reversible photoswitching was recently reported for Dronpa, a mutant of a GFP-like fluorescent protein cloned from the coral Pectiniidae.[46]

11.6 LUCIFERASES AS REPORTER GENES

Bioluminescence provides another approach for whole-animal imaging of reporter genes.[47] Because of its simplicity and ease of generating cells expressing the luciferase gene (*luc*), its primary uses have been for tracking tumor cells,[48,49] stem cells, immune cells,[50] and bacteria,[51] as well as for imaging gene expression.

Firefly *Photinus pyralis* luciferase and its substrate D-luciferin (a benzothiazole) are the most commonly used enzyme–substrate pair for *in vivo* imaging[48,52–54] because of the long wavelength (562 nm) and high quantum yield. D-luciferin easily penetrates various organs and cell types upon systemic injection to catalyze reactions in live animals and is well tolerated even at high doses. More recently, imaging of bioluminescence in mice expressing *Renilla reniformis* after administration of coelenterazine has also been reported.[55]

Luciferase has been applied in experimental gene transfer studies,[56–58] as a screening tool for rapid identification of transgenic founder mice,[59] or to visualize activation of specific pathways in cancer formation.[60] Most recently, engineered luciferases have been used to image specific cellular processes such as image protein–protein interactions[61] and NF-κB degradation,[62] and a chimeric IκB alpha-firefly luciferase was applied as a reporter of IKK activation.[63]

11.7 β-GALACTOSIDASE AS A REPORTER GENE

One of the most abundant reporter genes in use today is the LacZ gene encoding for the *E. coli* β-galactosidase (β-gal) enzyme that catalyzes hydrolysis of β-D-galactosides. β-gal possesses broad substrate specificity; thus, a variety of compounds that bring about colored or fluorescent products are appropriate candidates for cleavage. This feature and its absence from eukaryotic cells, as well as its relative inertness, contribute to its wide use.

Real-time *in vivo* detection of β-gal activity was reported for molecular imaging of gene expression by nuclear magnetic resonance (NMR) spectroscopy and MRI, as well as by far-red fluorescence imaging. Fluorescence imaging of β-gal was made possible using a fluorescent substrate analog, 9H-(1,3-dichloro-9,9-dimethylacridin- 2-one-7-yl) β-D-galactopyranoside (DDAOG; excitation/emission at 465/608 nm), which is hydrolyzed to form DDAO, with red-shifted excitation and emission peaks at 646 and 659 nm.[64] Specifically, β-gal-expressing 9L gliomas were shown to be readily detectable by red fluorescence imaging in comparison with the native 9L gliomas. Furthermore, herpes simplex virus amplicon-mediated LacZ gene transfer into tumors could be serially visualized over time.

The early generation of MRI contrast agents that functioned as substrates for β-gal (Egad[65]) yielded a 20% increase in relaxivity after cleavage. This study established the ability of the enzyme to recognize the MRI substrate, hydrolyze it, and generate contrast, but the change in relaxivity was not sufficient for *in vivo* imaging.[66,67]

The next-generation contrast agent, EgadMe, consisted of a paramagnetic Gd^{3+} ion chelated by $DOTA^{2-}$ that interacts with eight of the nine coordination sites of the metal ion. A galactopyranose ring occupied the ninth coordination site, and thus blocked any access of water protons to the chelated Gd^{3+}. Conjugation of α-methyl group to an ethylenic carbon on the sugar linkage arm of the chelate increased the rigidity and lowered the rotation of galactopyranose. In this inactive conformation, the contrast material did not alter relaxation time of the water. However, once galactopyranose was cleaved from the macrocycle by β-gal, a free coordination site of Gd^{3+} was exposed, thereby activating the contrast agent.[68]

In vivo MRI studies were limited to microinjection of EgadMe. Fragments of mRNA encoding for β-gal were injected into a single cell of two-cell-stage *Xenopus laevis* embryos, along with nuclear-localized GFP (nGFP), to enable independent fluorescent detection of the engineered progeny, whereas EgadMe was microinjected into both cells. Consequently, while in cells expressing the β-gal enzyme EgadMe was activated, yielding an increase in the signal intensity (shortening T_1 of the water), it remained inactive in nonexpressing cells. Moreover, signal in the embryo head was depicted by MRI (and was verified by light microscopy of X-gal staining on the fixed embryo), while fluorescence of nGFP was attenuated, demonstrating the superiority of MRI in detection of signal from deeper tissues. Injection of DNA encoding the β-gal into blastomers preinjected with

EgadMe revealed enhanced MR signal intensity that correlated with the expression patterns obtained by staining with X-gal after fixation. Yet, there are a few drawbacks for this method. Among them is the mosaic expression of the injected DNA in *X. laevis* embryos,[69] or unbalanced distribution of the mRNA to descendants, due to rapid cell division at the blastomere stage, raising a subclone of the originally injected cell.[70] In addition, the EgadMe contrast agent not only is a low-affinity substrate of the enzyme (its cleavage by β-gal is a few orders of magnitude slower than the colorimetric biochemical indicator ONPG), but also does not penetrate cell membranes, thus requiring injection into cells.

Alternative substrate analogs were developed for detection of β-gal activity by NMR spectroscopy (MRS). 4-Fluoro-2-nitrophenyl-β-D-galactopyranoside (PFONPG) was synthesized by the addition of a fluorine atom to a biochemical indicator of β-gal activity.[71] As opposed to EgadMe, PFONPG is easily taken up by cells and can be detected as a single, narrow ^{19}F NMR resonance. The chemical shift of the compound is pH sensitive. Upon cleavage by β-gal, the signal of PFONPG decreased, while the signal due to the cleavage product (PFONP), with a chemical shift greater than 3.6 ppm, increased. Since the product is also pH sensitive, its chemical shift may extend over 9 ppm, but the chemical shifts of the substrate and product did not overlap.[72] The kinetic curves demonstrated rapid enzymatic activity, emphasizing the minor disturbance of the fluorine addition to the original substrate.

Cleavage of PFONPG was studied on two prostate cancer cell lines (PC-3 and LNCAP C4-2), expressing β-gal under the constitutive CMV promoter. In both cases the genetically engineered enzyme successfully hydrolyzed the contrast agent, yielding appropriate chemical shift in the ^{19}F NMR spectrum. To assess the qualities and strength of the selected reporter probe, β-gal was also cloned under the specific regulation of the bone sialo protein (BSP) promoter. A much lower signal (5%) was generated upon cleavage of PFONPG.

Six additional ^{19}F-conjugated β-galactosides showed a single, narrow line width in ^{19}F NMR that varied little in chemical shift and cleavage products with a minimal chemical shift of 3.56 ppm and a maximal chemical shift of 9.84 ppm.[73] The efficiency of the compound 2-fluorine-4-nitrophenyl β-D-galactopyranoside was studied on prostate and breast tumor cells, overexpressing β-gal under the CMV promoter. The major weakness of these substrates is the relative toxicity of the products. In view of the cytotoxic side effect of the product, attempts were made to synthesize new substrates, conjugated with trifluoromethyl (CF3–), an addition that enhances MR signal, allowing reduction of substrate concentration.[74] This series of compounds had relatively low aqueous solubility with a varied rate of cleavage. In contrast to the earlier cleavage products, these aglycones revealed a rather moderate chemical shift that did not exceed 1.14 ppm, but was sufficient for acquisition of resolved images by chemical shift imaging.

11.8 THE SUICIDE GENES THYMIDINE KINASE AND CYTOSINE DEAMINASE AS REPORTERS OF GENE EXPRESSION

An elegant approach for chemotherapeutic treatments is the design of nontoxic prodrugs that can be enzymatically converted to form cytotoxic agents at the target location, thus reducing systemic toxicity and side effects. The prodrug activating enzyme can be targeted to specific antigens in the target tissue. This concept is referred to as antibody-directed prodrug therapy (ADEPT) and was first introduced in 1973.[75]

In a different context, the prodrug activating enzymes can be used as reporter genes, allowing the detection of transcriptional activation through detection of prodrugs that act as smart contrast agents. The herpes simplex virus 1 thymidine kinase (TK) and cytosine deaminase (CD) catalyze the conversion of the nontoxic prodrugs ganciclovir and 5-fluoro-cytosine, respectively, to form the cytotoxic chemotherapeutic agents. Radiotracer-labeled paramagnetic or fluorinated prodrugs can allow the use of these enzymes as reporter genes for detection by SPECT, PET, MRI, or MRS.

Although these reporter genes are not inert and direct biological effects of the reporter probe should be evaluated, clinical use of these enzymes for cancer-targeted gene therapy makes the reporter probes important for molecular imaging of gene expression. Broader utilization of these enzymes as general reporter genes is limited by the toxicity of the reaction product, but could be overcome through development of alternative prodrug analogs generating nontoxic products.

The underlying principle for use of HSV1-TK in both therapeutic and imaging applications is manifested in the broad substrate specificity of HSV1-TK activity relative to endogenous TK, enabling robust phosphorylation of nucleoside analogs, such as ganciclovir, compared with endogenous TK normally present in mammalian cells. Approximately 50% of cell-associated nucleoside is converted to the triphosphate form and incorporated into DNA within 1 h,[76] where it acts to terminate DNA synthesis. These metabolic steps result in trapping of nucleoside analogs like ganciclovir selectively within cells that express HSV1-TK.

To introduce HSV1-TK into animals for imaging, investigators have either transfected tumor cells with HSV1-TK or delivered the reporter gene into liver via adenoviral vectors.[76–80] TK was applied for PET and SPECT imaging of constitutive human promoters, such as elongation factor 1α,[81] or promoters regulated by endogenous transcription factors, such as p53 or nuclear factor of activated T cells.[82,83] Protein–protein interactions were imaged *in vivo* by microPET and fluorescence imaging using an engineered fusion reporter gene comprising a mutant HSV1-TK and GFP for readout of a tetracycline-inducible two-hybrid system *in vivo*.[81]

MRI was applied during thymidine kinase-based gene therapy, primarily for functional evaluation of therapy success.[84–86] Therapeutic intervention using thymidine kinase leads to cell death and associated changes in MRI contrast.[87] Such changes should be taken into account when evaluating contrast changes for analysis of gene expression. New ganciclovir analogs that can serve as substrates of herpes simplex virus types 1 and 2 (HSV-1 and -2) were developed to allow direct tracking of prodrug conversion by ^{19}F NMR. These compounds showed reduced antiviral effects but cytostatic activity against human osteosarcoma tumor cells expressing HSV1-TK.[88]

Cytosine deaminase is a microbial enzyme that converts cytosine to uracyl. This catalytic activity made it attractive for gene therapy based on the conversion of the relatively nontoxic prodrug 5-fluorocytosine (5-FC) to the chemotherapeutic cytotoxic agent 5-fluorouracyl (5-FU), which is incorporated into fluorinated nucleotides/nuclesides (F-nuc).[89] All three species, including 5-FC, 5-FU, and F-nuc, show distinct, well-resolved ^{19}F NMR resonances, allowing the kinetic analysis of the utilization and clearance of 5-FC along with the generation of 5-FU and F-nuc.[89–93] The conversion of 5-FC to 5-FU generates a potent cytotoxic chemotherapy. Toxicity was followed by diffusion MRI and bioluminescence imaging using tumors expressing both the yeast cytosine deaminase and luciferase.[92]

11.9 CREATINE AND ARGININE KINASES AS REPORTER GENES

Creatine and arginine kinases are a family of enzymes that catalyze the bidirectional exchange of phosphate between ATP and creatine (Cr) or arginine (Arg), respectively. By maintaining a high-energy phosphate in a readily accessible phosphocreatine (PCr) or phosphoarginine (PArg) pool, these enzymatic reactions assist in buffering the intracellular level of ATP and allow shuttling of high-energy phosphate from one part of the cell to another (CK/PCr shuttle). Phosphorylation of Cr releases H+, and thus the PCr/Cr ratio tends to reflect changes in intracellular pH.

^{31}P MRS provides a unique tool for multiparametric measurement of the creatine kinase reaction, allowing direct detection of PCr^{2-}, ADP (if levels are high enough), and $MgATP^{2-}$. In addition, intracellular pH can be measured from the chemical shift of inorganic phosphate (Pi), while Mg content can be determined from the chemical shifts of ATP. The rate of exchange can be determined by magnetization transfer for the creatine kinase reaction alone and as part of a network of coupled metabolic kinases.[94,95]

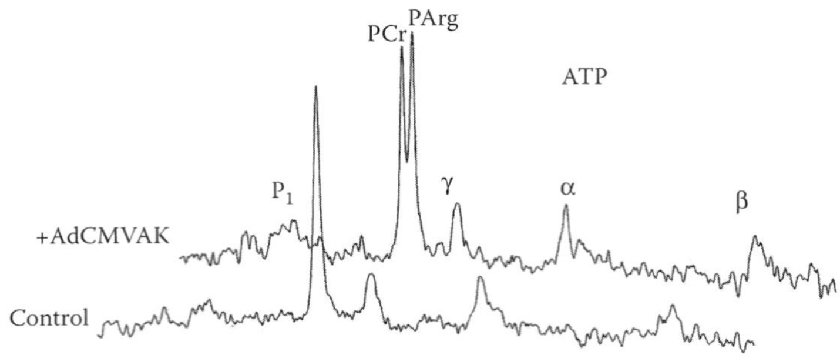

FIGURE 11.1 *In vivo* basal ^{31}P spectra from the hind limbs of a 6-month-old mouse. ^{31}P MRS spectra from the rAdCMVAK-injected limb (upper spectrum) reveal a ^{31}P resonance at the chemical shift for PArg that is not present in the contralateral control limb (lower spectrum). (Reproduced from Walter, G. et al., *Proc. Natl. Acad. Sci. U.S.A.*, 97, 5151–5155, 2000. Copyright (2000) National Academy of Sciences, U.S.A. With permission.)

A number of transgenic mice were generated with an altered pattern of creatine kinase expression.[96] Overexpression of creatine kinase results in improved tissue maintenance under hypoxic conditions with no apparent ill effects. Mice deficient for the cytosolic or mitochondrial CK are viable and fertile but show altered cardiac function[97,98] and have reduced capacity for endurance exercise.[99] Due to its endogenous abundance in many organs, the use of creatine kinase as a reporter gene is limited to the liver, in which the enzyme is not expressed, resulting in lack of PCr signal.

Creatine kinase-β-gal fusion protein was applied as a reporter gene for studying regulation of gene expression by ^{31}P NMR in *E. coli* in 1989.[100] Subsequently, overexpression of creatine kinase in the liver was demonstrated in transgenic mice.[101] Viral gene transfer to the liver was found to be detectable through NMR analysis of creatine kinase,[102] providing the foundation for ^{31}P NMR detection of internal ribosomal entry site (IRES)-expressed creatine kinase, as a reporter for expression of low-density lipoprotein receptor in viral gene therapy delivered to a transgenic mouse model for hypercholesterolemia.[103]

Arginine kinase, the invertebrate analog of creatine kinase, was also applied as a potential reporter gene for NMR spectroscopy. Analysis of the activity of arginine kinase and comparison of creatine and arginine kinases by NMR spectroscopy demonstrated the feasibility of following both reactions and the similarity in coformation of the enzyme-bound substrates.[104,105]

Arginine kinase catalyzes the reversible phosphorylation of arginine, and when expressed in mammalian cells, this enzyme leads to the accumulation of PArg, which can be resolved by ^{31}P spectroscopy even in the presence of PCr[106] (Figure 11.1). The function of this enzyme is to provide a buffer for ATP, and thus overexpression could lead to changes in cellular energetics. Despite the expected effects, adenoviral infection of animals in the hind limb muscle led to no apparent biological effects. Specific signal due to PArg was persistant and visible many months after infection. The major advantage of arginine kinase is that the reaction product PArg can be found in mammalian tissues only when the transgenic enzyme is produced, and thus there is no background, while the substrates of the reaction are all ubiquitous, not requiring exogenous administration of any reporter probe.

11.10 APPLICATION OF BIOTIN LIGASE FOR CONSTRUCTING REPORTER GENES

Enzymatic biotinylation of a target cell surface receptor can be used to yield a reporter gene that is detectable via an avidin-conjugated reporter probe. In mammalian cells, the endogenous enzyme biotin ligase can link biotin molecules to specific recognition sequences on biotin-accepting proteins (BAP/biotinylation sequence). Hence, the substance is a natural resident and neither is immunogenic nor requires an exogenously administered linker. The biotinylated sequence can then be imaged with contrast agents that are attached to avidin; however, one hindrance of extensive *in vivo* use of avidin is its immunogenicity.

Proof of principle was demonstrated using recombinant biotinylated PDGF-R and EGF-R as reporter genes.[107–110] The fusion receptors were detected in cell cultures by a fluorescently tagged antibiotin antibody and using a fused reporter with a strepavidin-CLIO T_2-weighted contrast agent. *In vivo* detection of subcutaneous tumors initiated from transfected cells was carried out by optical imaging with fluorescence molecular tomography (FMT).

11.11 CEST AGENTS AS REPORTER GENES

Another contrast mechanism that was applied for reporter gene imaging with MR is the chemical exchange saturation transfer (CEST). CEST relies on the magnetization transfer that occurs between the bulk water protons and macromolecular protons. A saturation RF pulse is applied at the macromolecular proton frequency, while exchange of the labile protons will attenuate the signal of the bulk water protons. The decrease in signal amplitude depends on the rate of exchange between the two proton pools. Motivated with this notion, Gilad and coworkers constructed cells that express an artificial sequence coding for lysine-rich protein (LRP), and therefore exhibit an endogenous CEST agent.[111] Extracts originating from LRP-expressing cells demonstrated a significant increase in MR contrast relative to extracts of control cells.[111] A pivotal benefit of CEST is the ability to turn the contrast on and off at will, by selecting the irradiation pulse. Furthermore, the frequency dependence of CEST may provide the possibility for multiple spectrally resolved reporters, each designed to yield a different offset for the CEST effect.

11.12 TYROSINASE AS A REPORTER GENE

Unlike most tumors, which have low signal intensities on T_1-weighted MR images, melanotic melanomas commonly have high signal intensities. This paramagnetic effect characteristic to melanotic melanomas is attributed to the high affinity and binding capacity of melanin, a biopolymeric pigment abundant in these tumors, for metal ions.[112] Found in animals and plants, melanin is mainly known for its coloration, photoprotection, and metal ion scavenging abilities.[113] The high affinity of melanin for a large variety of metal ions was exploited for use as a reporter gene.[113] Melanin synthesis is mainly regulated by tyrosinase, which catalyzes the oxidation of tyrosine to dopaquinone.

The tyrosinase–melanin system, in which tyrosinase is overexpressed and leads to melanin synthesis, was suggested as a reporter gene for MRI.[113] Mouse fibroblasts and human embryonic kidney cells transfected with a vector encoding for constitutive expression of human tyrosinase showed elevated levels of tyrosinase mRNA and melanin production in transfected cells, as well as significantly higher metal binding capacity, resulting in enhanced MR signal intensity. These results paved the way for the development of tetracycline-inducible expression of tyrosinase for detection of gene expression by MRI.[17]

MCF-7 breast cancer cells were transfected with a plasmid containing the tetracycline-controlled transactivator and a second construct that coded for the human tyrosinase under the control of the tetracycline response element.[17] This co-transfection created a tetracycline switched-off system that allowed the suppression of human tyrosinase expression upon tetracycline addition to the medium. MRI showed significantly higher signal intensity upon tetracycline withdrawal, along with enhanced human tyrosinase mRNA expression and induced tyrosinase and melanin synthesis. Since melanin is a highly stable molecule, even upon switching off tyrosinase expression, the MRI contrast is maintained for a considerable time interval. Development of melanin homologs of reduced stability would create reporter genes with improved temporal resolution.

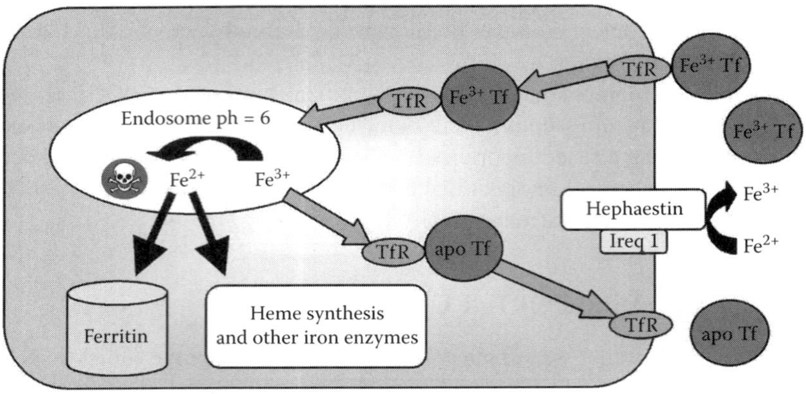

FIGURE 11.2 Regulation of iron uptake and homeostasis. Fe^{3+}-Tf is internalized as a complex with TfR by endocytosis into an acidic (pH 6.0) endosome where iron is released. Fe^{3+} is reduced by an endosomal ferric reductase. Fe^{2+} is transported to the cytoplasm by the Fe transporter, where it can be used for heme synthesis or as a catalizator/cofactor for numerous enzymes or stored in ferritin molecule as an oxidized trivalent ferric iron (Fe^{3+}).

11.13 IRON UPTAKE AND STORAGE PROTEINS AS REPORTER GENES

Iron is an essential nutrient for the functionality and viability of cells. Due to its ability to mediate one-electron exchange reactions, iron participates in many metabolic pathways and is required for the proper function of numerous essential proteins, such as the heme-containing proteins, electron transport chain, and microsomal electron transport protein.[114–116] However, the reactivity of iron may also be detrimental for living cells, as free hydroxyl radicals are generated through an iron-catalyzed Fenton reaction. Thus, maintenance of iron homeostasis is essential for the survival of animals, plants, and microorganisms.[114] In the course of evolution, specialized iron-binding proteins have been evolved, allowing iron to be maintained in a thermodynamically stable form, but also kinetically available for biological processes;[114] among these proteins are ferritin, transferrin, and the transferrin receptor (Figure 11.2).

11.14 TRANSFERRIN RECEPTOR AS A REPORTER GENE FOR MRI

In vertebrates, iron is transported through the bloodstream by the plasma protein transferrin (Tf). Internalization of iron into the cytosol occurs by binding of iron-loaded transferrin to the transferrin receptor (TfR) and its consequent internalization.[117] The first attempt to use TfR as a reporter gene for MRI was reported in 1996.[118] A fibroblast cell line transfected with a DNA construct encoding for the human transferrin receptor (hTfR) was used to produce tumors in mice. Overexpression of hTfR resulted in elevated iron levels both *in vitro* and *in vivo* and induction of ferritin expression. Tumors expressing hTfR had a decreased signal intensity on T_2-weighted images, and accordingly had a 20% shorter T_2 than control tumors.

The ability to detect TfR expression by MRI was subsequently amplified by the use of transferrin covalently conjugated to Mion (Tf-MION[119]). This approach allowed MRI detection of TfR activity in proliferating tumor cells with an endogenously elevated expression of TfR. Tf-MION was applied as a reporter probe for detection of hTfR overexpressed as a reporter gene in 9L gliosarcoma cell lines stably expressing several clones of hTfR.[32,120,121] A significant transgene-dependent loss of MR signal intensity was observed upon Tf-MION supplementation. Nude mice implanted with hTfR and control 9L gliosarcoma tumors showed significant differences in MR signal intensity 24 h after administration of Tf-MION.[120,121] Improvement of the affinity of the Tf-based reporter probe to TfR was achieved using a disulfide cross-linker (N-succimidyl 3-(2-pyridyl thio) propionate, or SPDP) between transferrin and the iron oxide particles. The new probe, Tf-S-S-CLIO, shows a higher transferrin proteins-to-iron ratio and a higher affinity for the transferrin receptor.[122]

The application of the transferrin receptor as an MR reporter gene using an HSV-based amplicon vector system for the delivery of several transgenes demonstrated the ability to monitor therapeutic gene expression by MRI reporter genes.[123] The gene therapy vector contained several transgenes: an engineered human transferrin receptor as an MR reporter gene, CYP2B1, a prodrug therapy gene, and LacZ as a second reporter. This vector was inoculated into human glioma cells, and the correlation in expression levels of each transgene was examined by Western blot as well as immunohistochemical staining, while TfR expression was also analyzed by MRI.

11.15 FERRITIN AS A REPORTER GENE

Ferritin is a ubiquitous, highly conserved iron storage protein.[115] The spherical apoferritin shell is composed of 24 heavy and light chains of ferritin in variable proportions.[124] The heavy chain has ferroxidase activity, i.e., catalytic oxidation of ferrous ions or ferrous complexes to the ferric state that promotes iron oxidation and incorporation.[125] Overexpression of the H-chain ferritin resulted in the upregulation of the transferrin receptor and increased iron uptake by the transfected cells.[18,126] The MRI properties of ferritin were the focus of extensive research and showed an anomaly at very low iron loading and a peculiar linear dependence on the magnetic field.[127–129] The heavy chain of ferritin (h-ferritin) was applied as a reporter gene for MRI for monitoring tetracycline-inducible expression in C6 glioma tumors and transgenic mice,[18,130] while adenoviral infection was demonstrated for the combination of both light (l-ferritin) and heavy chains of ferritin,[131] as well as for the combination of h-ferritin with TfR.[132]

Tetracycline-regulated expression of h-ferritin was demonstrated in C6 rat glioma cells co-transfected with an expression vector encoding for murine ferritin H-chain and enhanced green fluorescent protein (EGFP) under tetracycline control and an expression vector encoding for tetracycline transactivator (tTA; Figure 11.3). Overexpression of the ferritin H-chain in these cells generated an *in vitro* increase in cellular iron content that was manifested in a significant increase in MR relaxation rates (R_1 and R_2).[18] MRI analysis of tumors inoculated in nude mice showed tetracycline-regulated expression of the ferritin H-chain and EGFP fluorescence. R_1 and R_2 were significantly elevated in transgene-expressing tumors relative to mice in which transgene expression was suppressed by treatment with tetracycline.[18]

Recently, transgenic mice that expressed the ferritin H-chain in a tissue-specific tetracycline-regulated manner were generated.[130] R_2 maps of these mice demonstrated a significant change in liver relaxation rate relative to single transgenic siblings (that do not express the transgene). Unexpectedly, R_2 values were lower in ferritin-expressing mice; a similar change in R_2 was previously reported for ferritin iron overload in IRP-2 knockout mice.[133]

Ferritin was also applied as a reporter gene for *in vivo* infections using a replication defective adenovirus that encoded for the human H- and L-chains of ferritin.[131] *In vitro* studies of infected A549 — human lung carcinoma cells — showed elevated ferritin levels that were 60 times higher than background ferritin levels, and a higher iron uptake upon ^{59}Fe-enriched transferrin supplementation that did not influence cell viability. A 2.5-fold elevation of R_2 was observed in infected cells that were incubated with a Fe supplementation. *In vivo* infection of AdV-FT into brain parenchyma of mice showed significant MRI contrast that was only visible at the inoculation site, while no significant contrast was detected at inoculation sites of mice infected with control viruses.

Recently, the combined transfection of a mouse neural stem cell line with both h-ferritin and TfR led to an increased accumulation of iron from the iron-supplemented environment, which resulted in an increased signal loss on T_2- and T_2*-weighted MRI.[132]

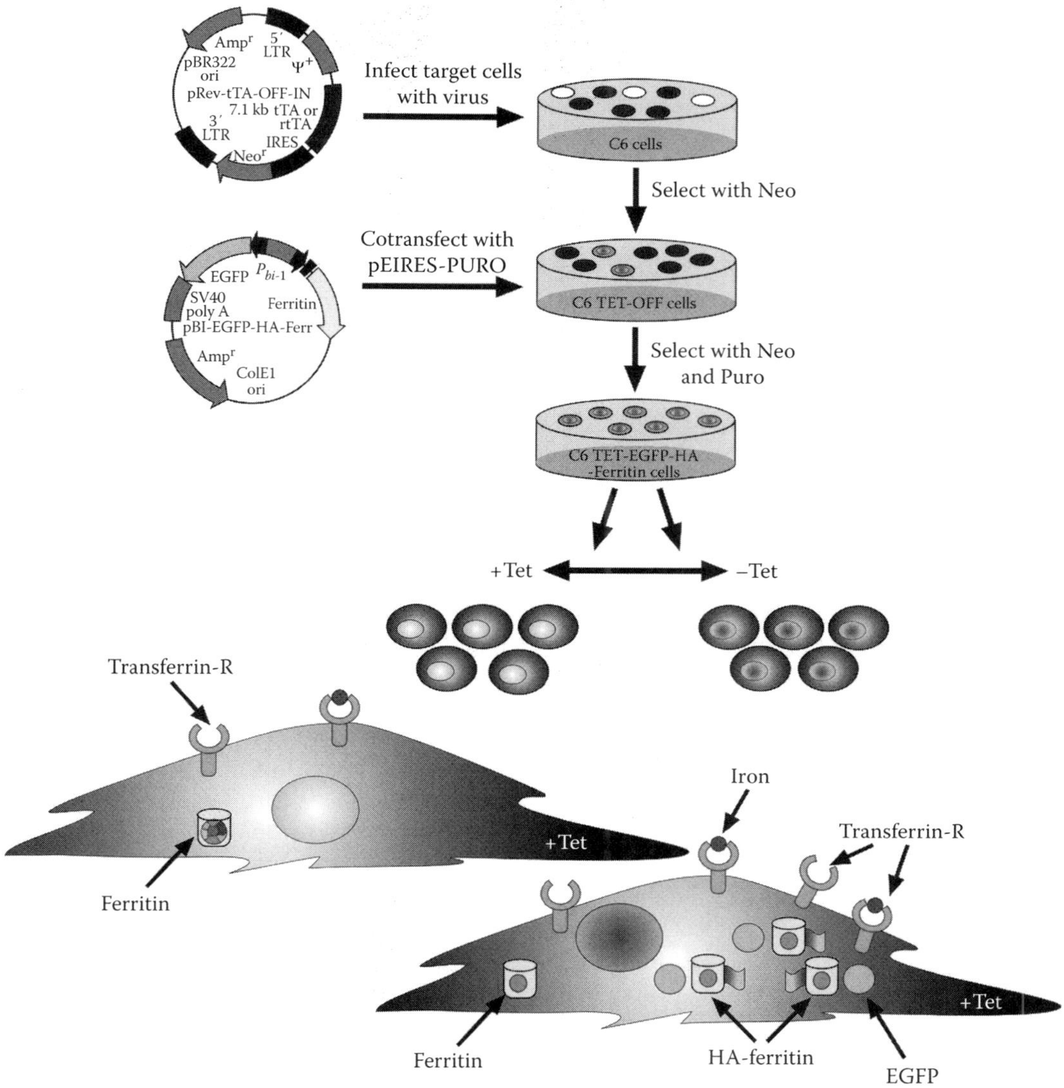

FIGURE 11.3 (please see color insert following page 210) Tetracycline-regulated expression of EGFP-HA-ferritin as a multimodality endogenous reporter of gene expression for MRI and optical imaging. C6-TET-EGFP-ferritin was generated by infection of C6 cells with viruses carrying the tetracycline transactivator (tTA) under a constitutive promoter (pRev-tTA-OFF-IN). The cells were then transfected to express TET-EGFP-HA-ferritin using a bidirectional vector (pBI-EGFP-HA-Ferr). Selected clones showed overexpression of EGFP and HA-tagged ferritin, both of which could be suppressed by administration of tetracycline (+Tet). In the absence of tetracycline (Tet), overexpression of ferritin leads to redistribution of intracellular ferritin iron and chelation of intracellular labile iron, thereby generating MR contrast by increasing R_1 and R_2 relaxation rates. Iron homeostasis is restored by the compensatory expression of transferrin receptor and increased iron uptake, providing further gain in MR contrast. (Reproduced from Cohen, B. et al., *Neoplasia*, 7, 109–117, 2005. With permission.)

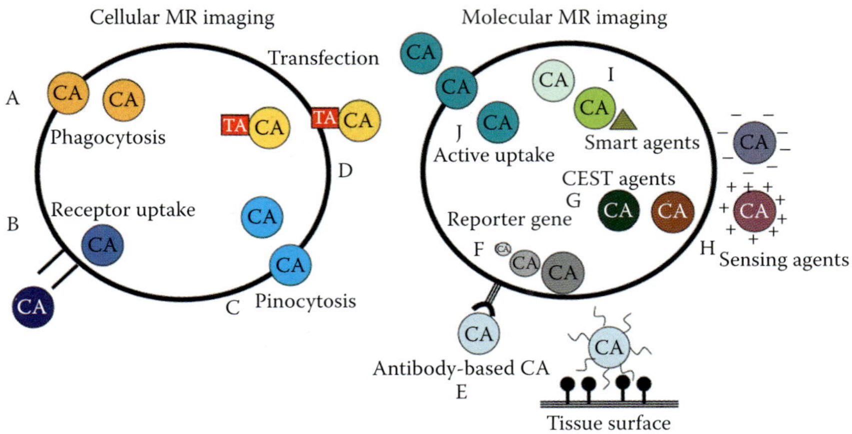

FIGURE 1.3

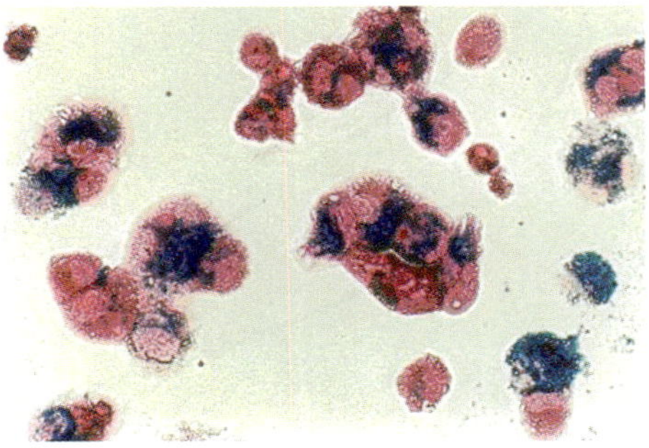

FIGURE 4.3

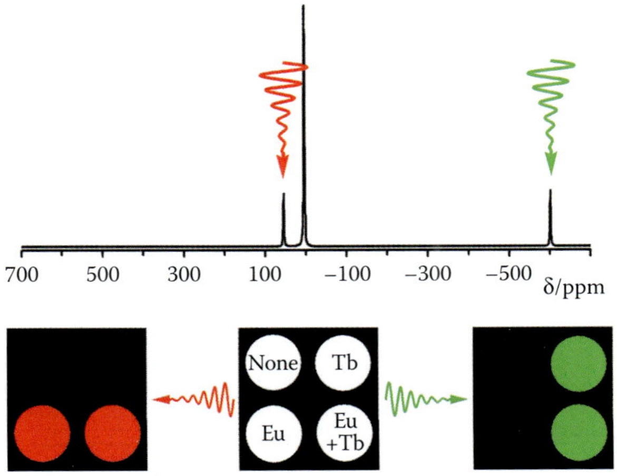

FIGURE 6.5

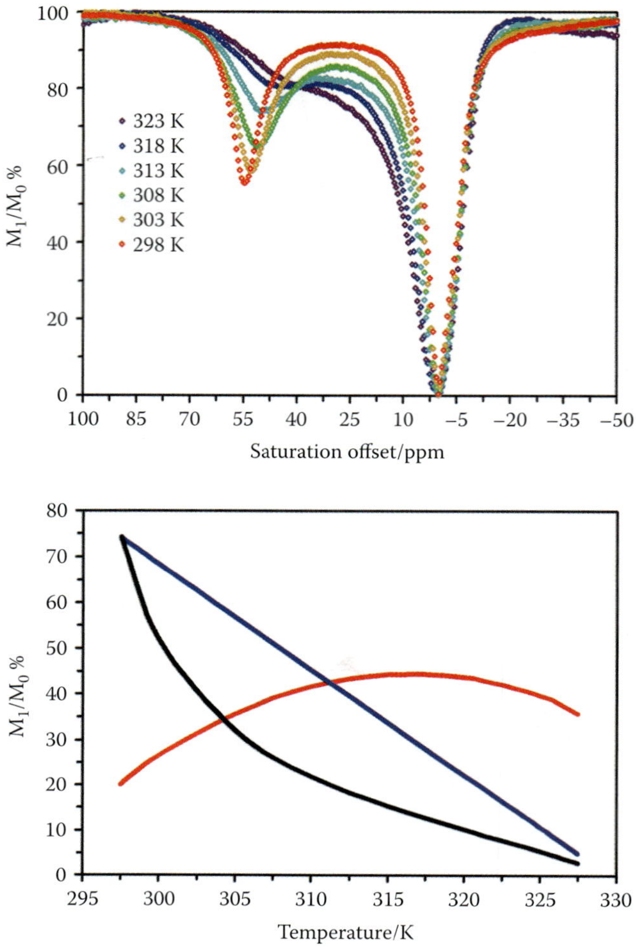

FIGURE 6.7

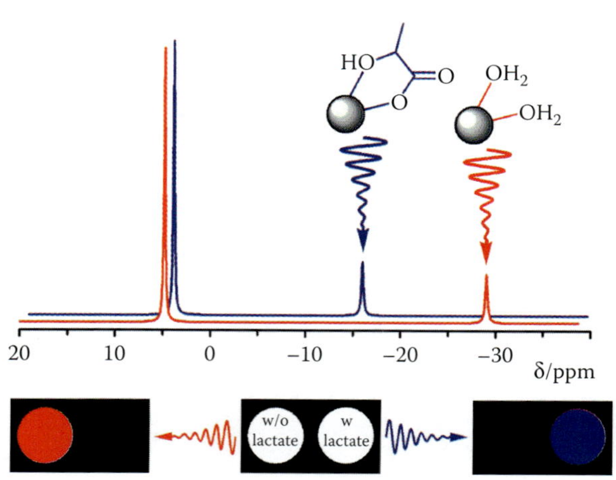

FIGURE 6.8

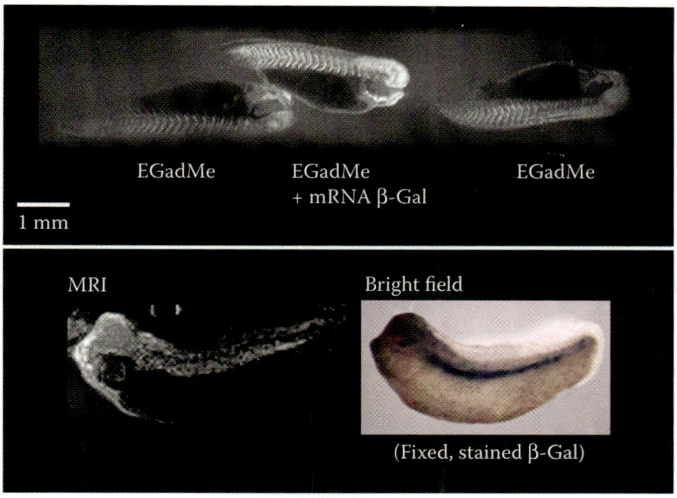

FIGURE 7.1

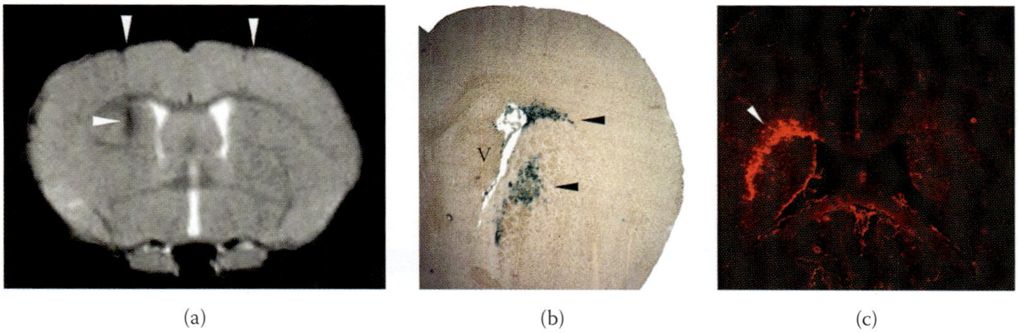

FIGURE 7.2

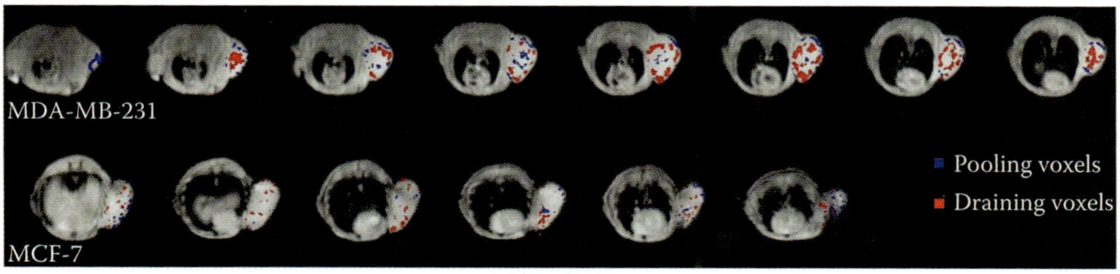

FIGURE 8.5

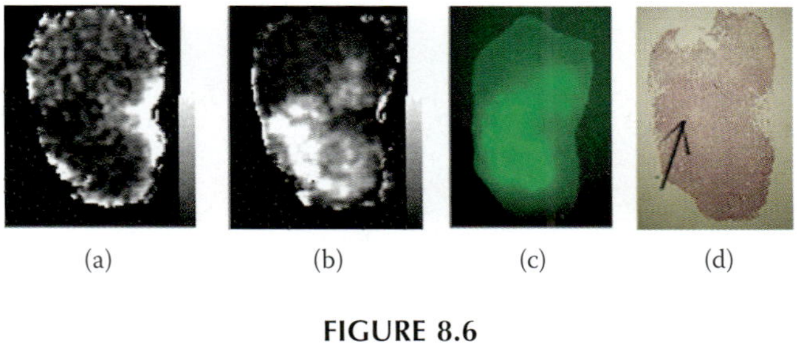

FIGURE 8.6

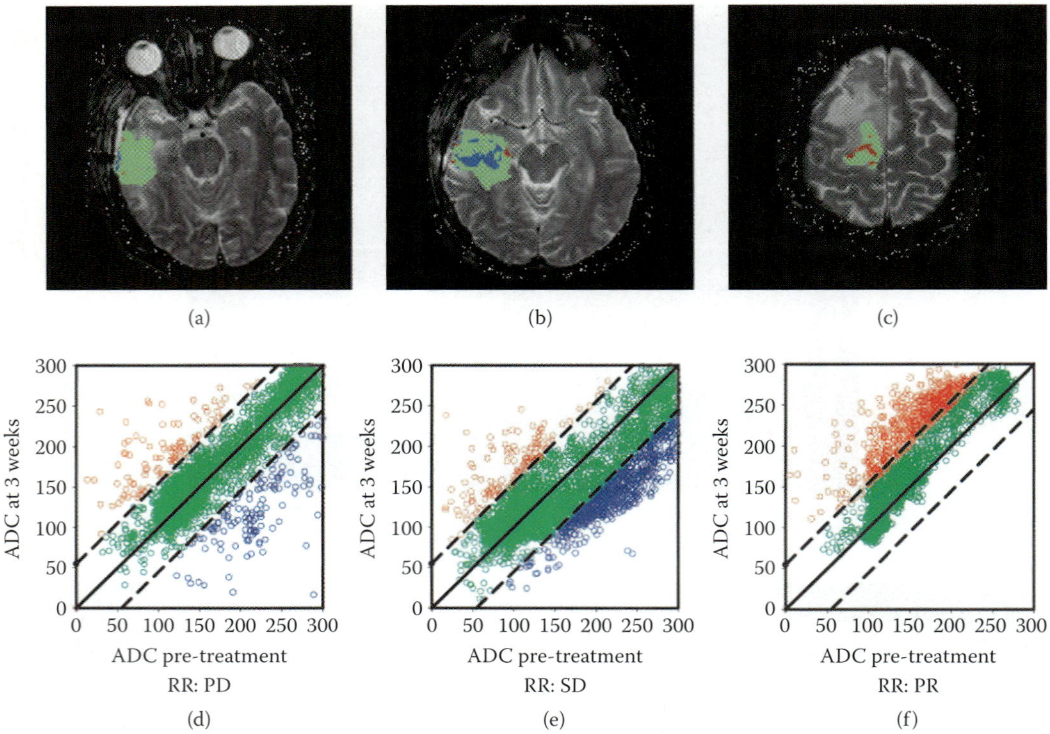

FIGURE 8.7

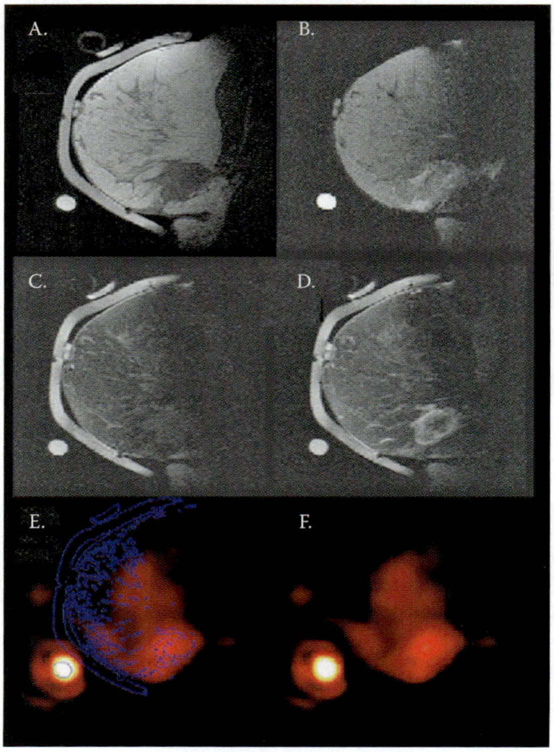

FIGURE 8.8

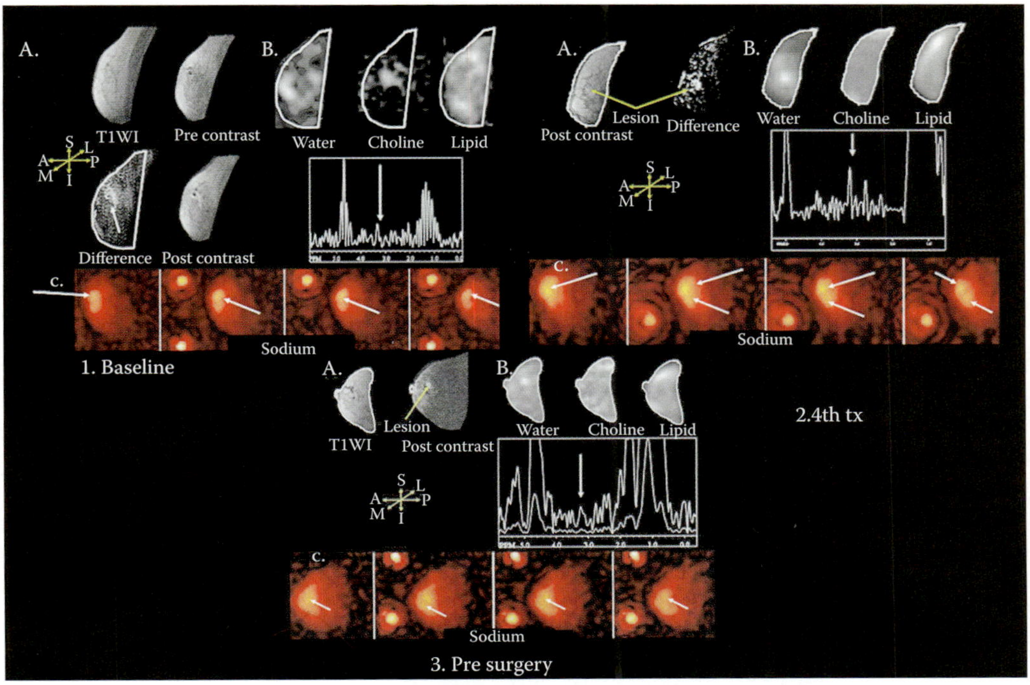

FIGURE 8.10

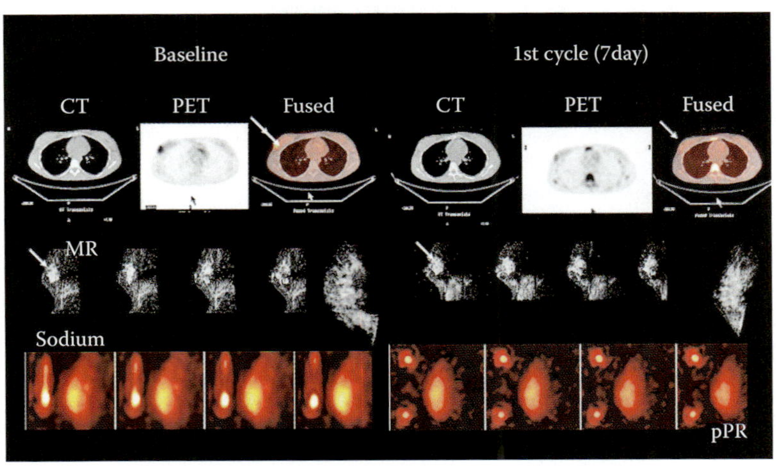

FIGURE 8.12

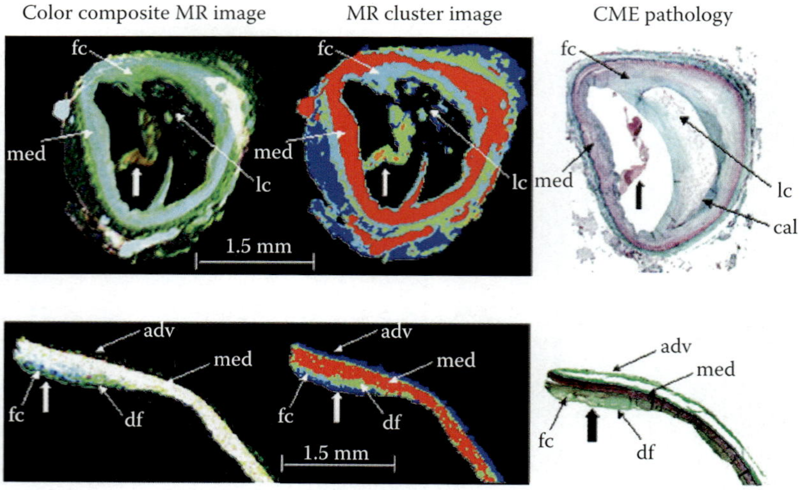

FIGURE 9.1

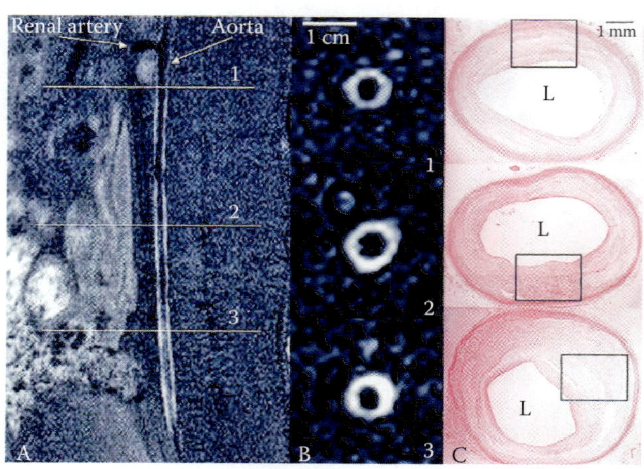

FIGURE 9.5

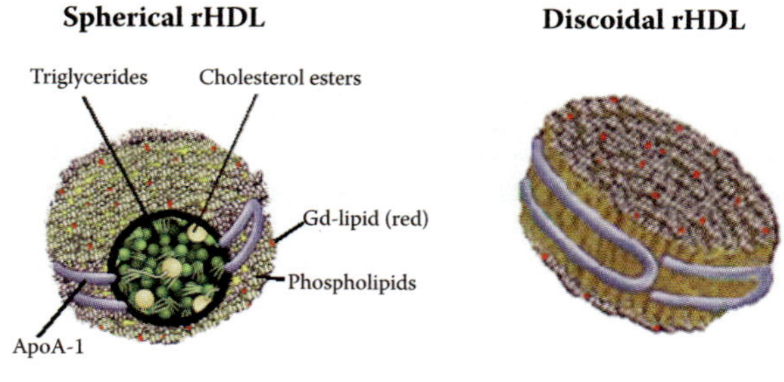

FIGURE 9.7

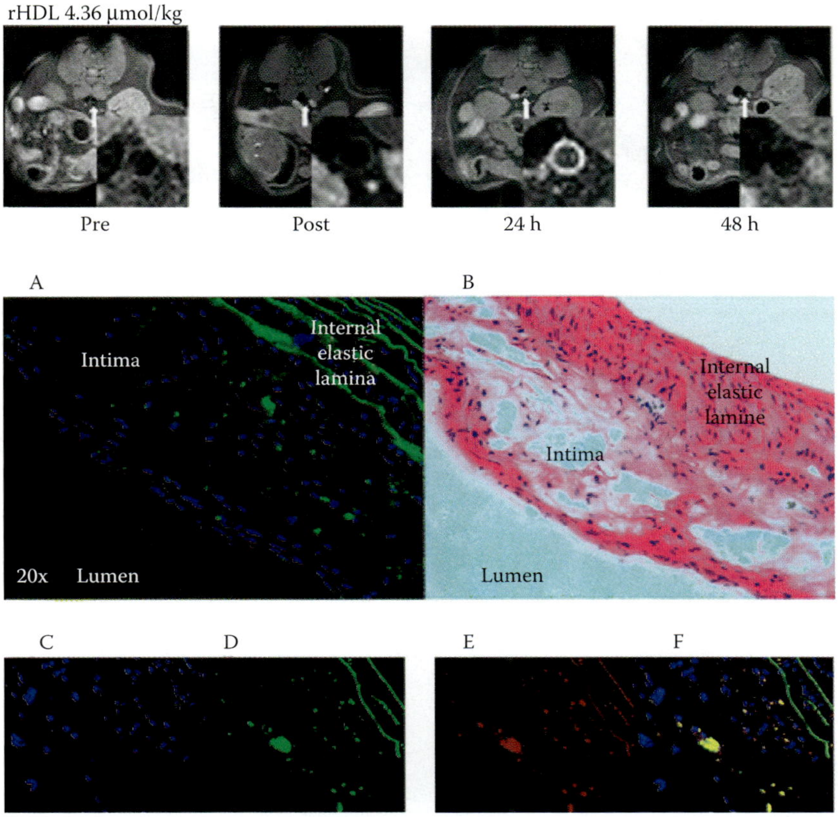

FIGURE 9.8

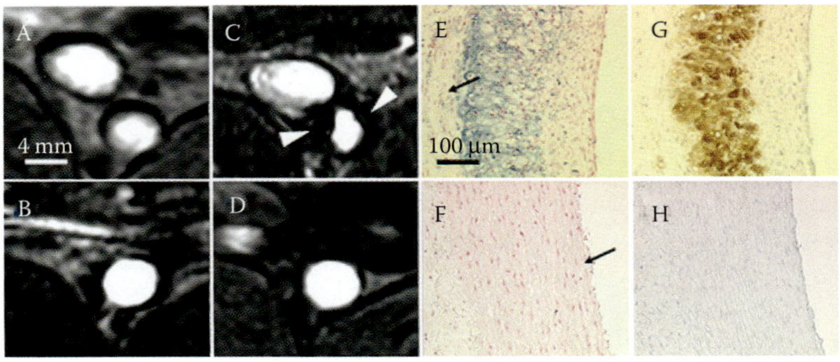

FIGURE 9.9

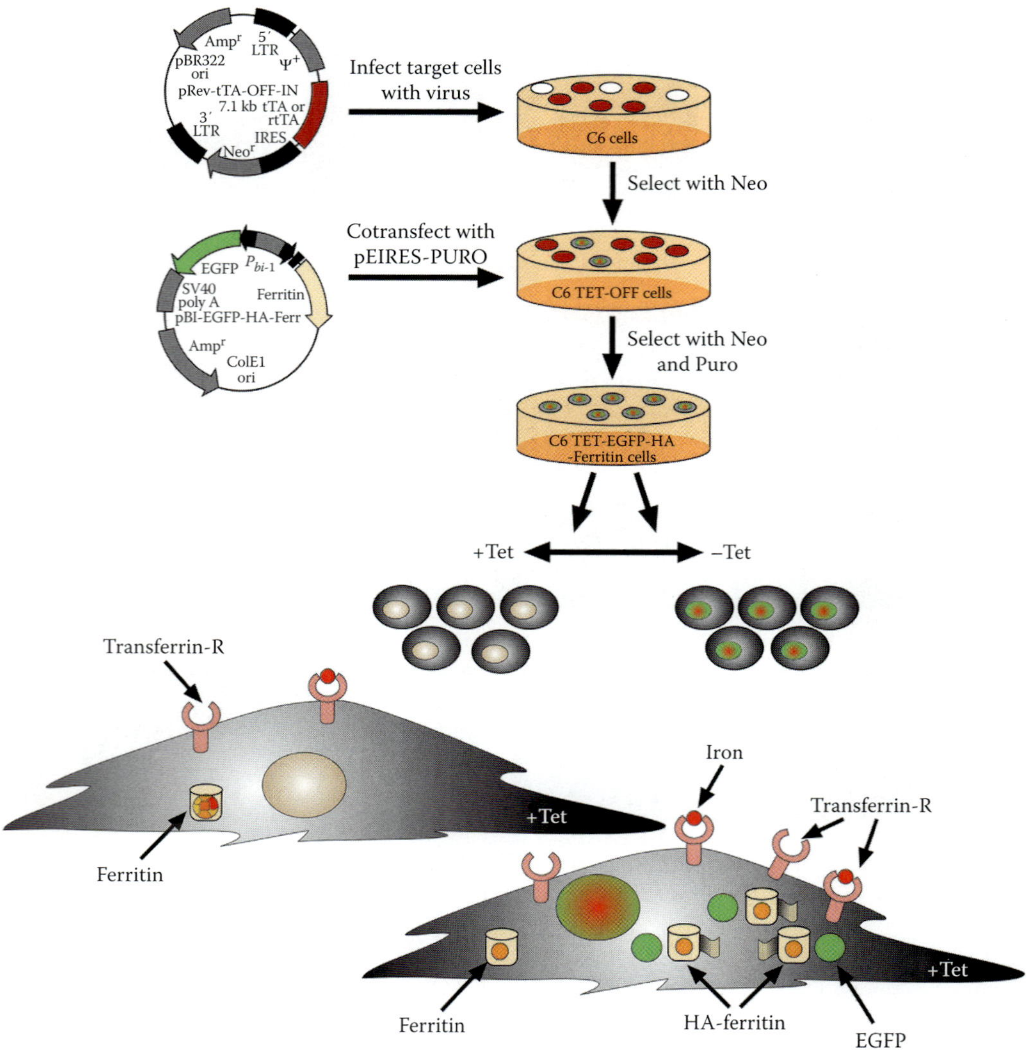

FIGURE 11.3

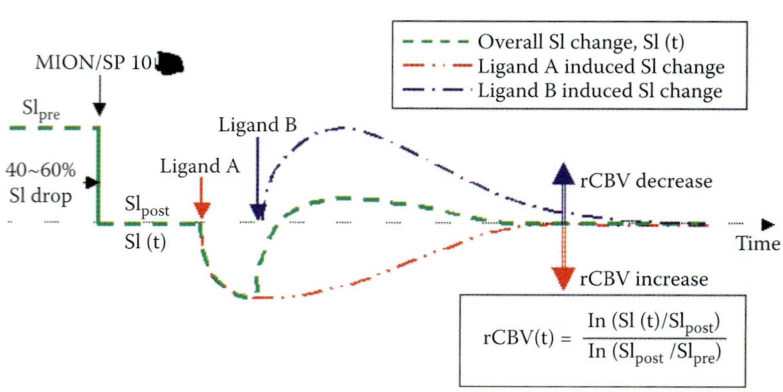

FIGURE 12.1

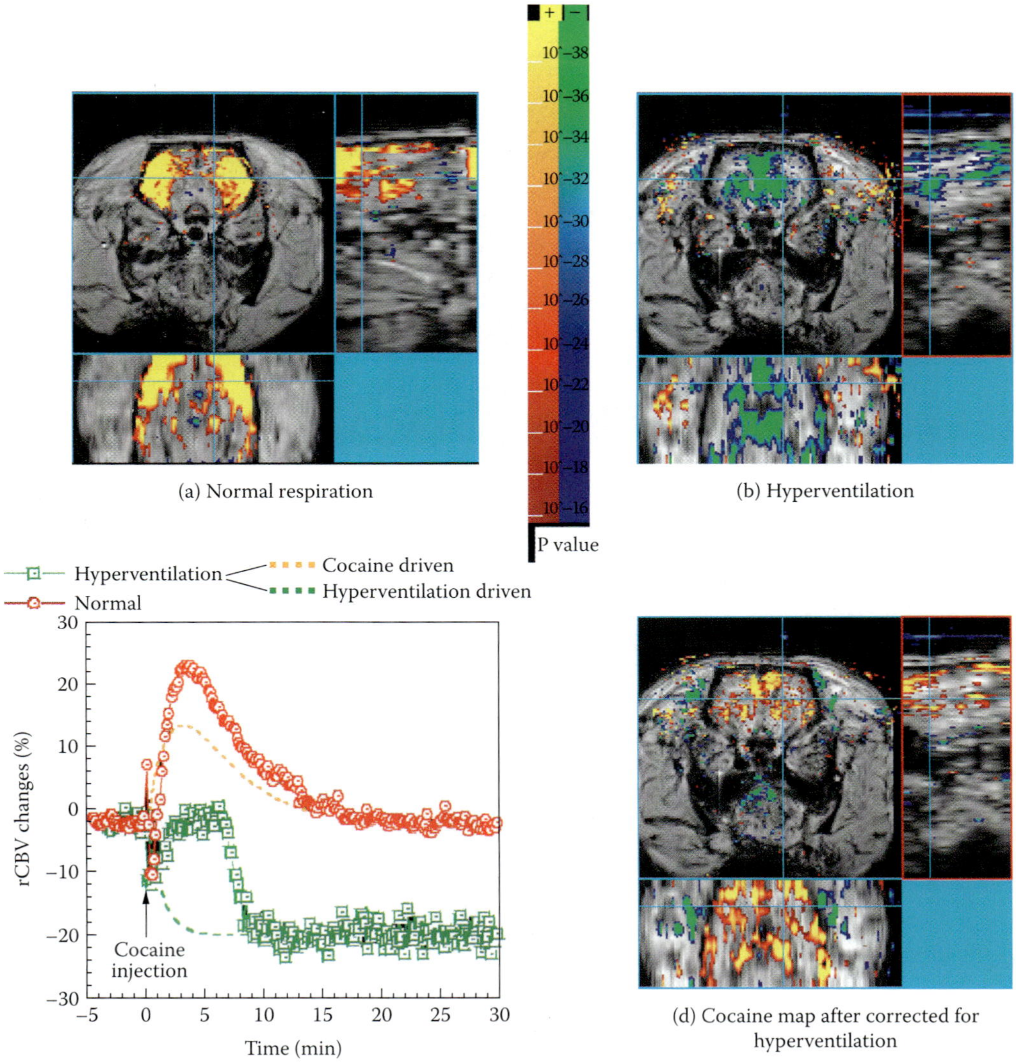

(a) Normal respiration

(b) Hyperventilation

(c) rCBV time course

(d) Cocaine map after corrected for hyperventilation

FIGURE 12.4

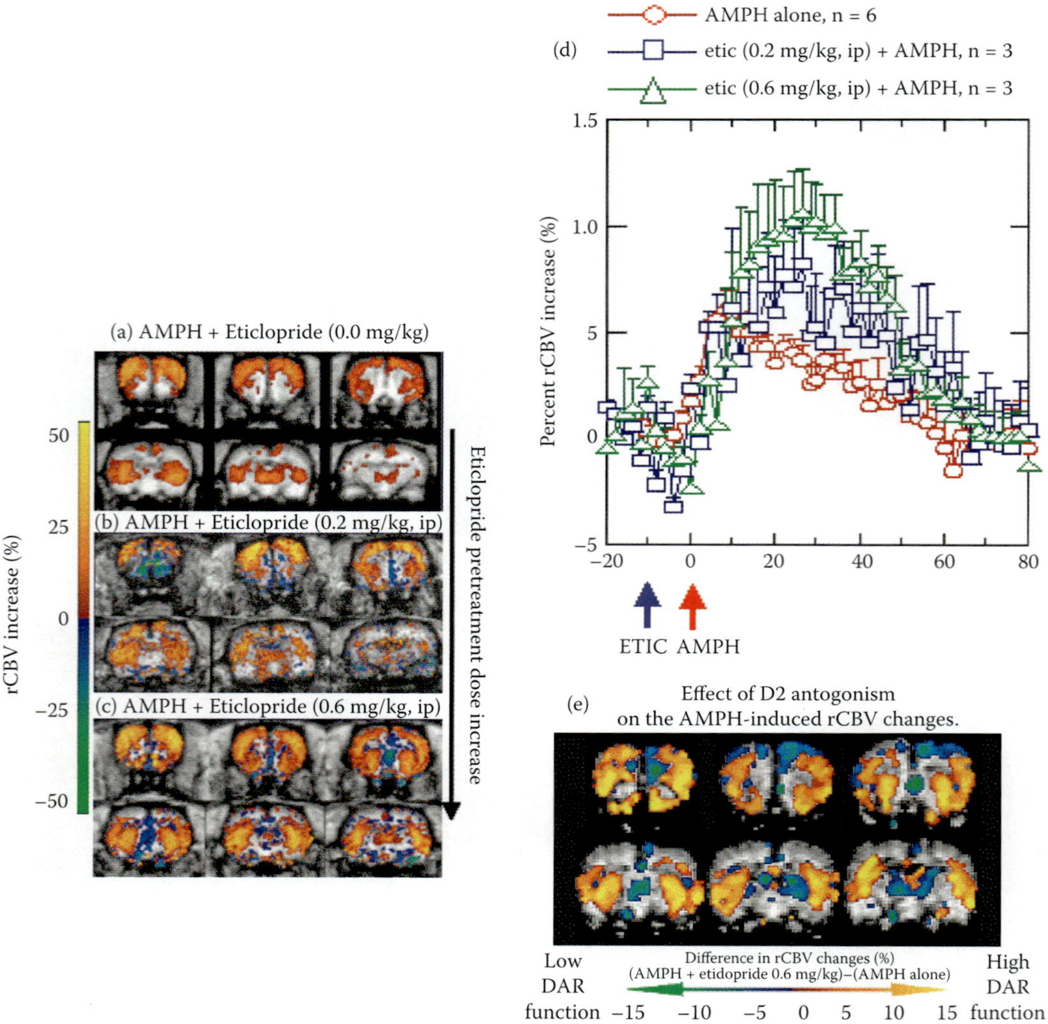
(a) AMPH + Eticlopride (0.0 mg/kg)
(b) AMPH + Eticlopride (0.2 mg/kg, ip)
(c) AMPH + Eticlopride (0.6 mg/kg, ip)
rCBV increase (%)
50
25
0
−25
−50
Eticlopride pretreatment dose increase
(d)
AMPH alone, n = 6
etic (0.2 mg/kg, ip) + AMPH, n = 3
etic (0.6 mg/kg, ip) + AMPH, n = 3
Percent rCBV increase (%)
1.5
1.0
5
0
−5
−20
0
20
40
60
80
ETIC
AMPH
(e)
Effect of D2 antogonism
on the AMPH-induced rCBV changes.
Low
DAR
function
Difference in rCBV changes (%)
(AMPH + etidopride 0.6 mg/kg)−(AMPH alone)
−15
−10
−5
0
5
10
15
High
DAR
function

FIGURE 12.6

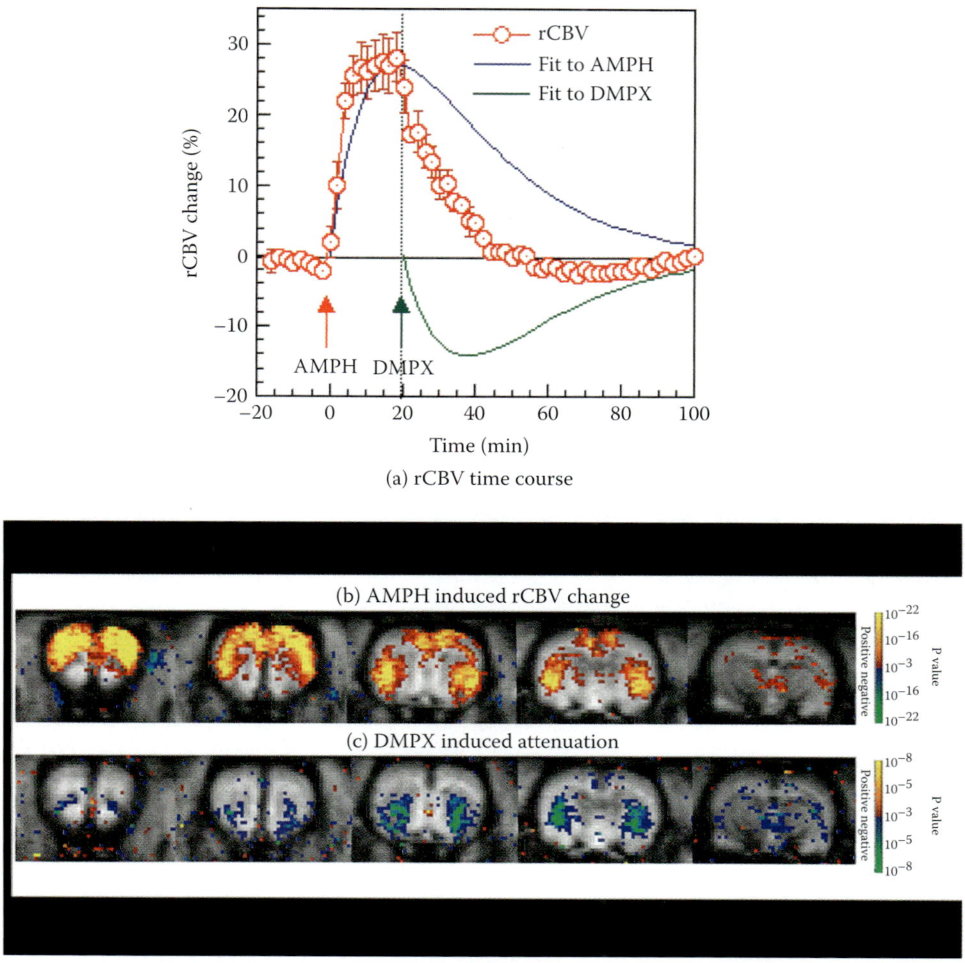

FIGURE 12.7

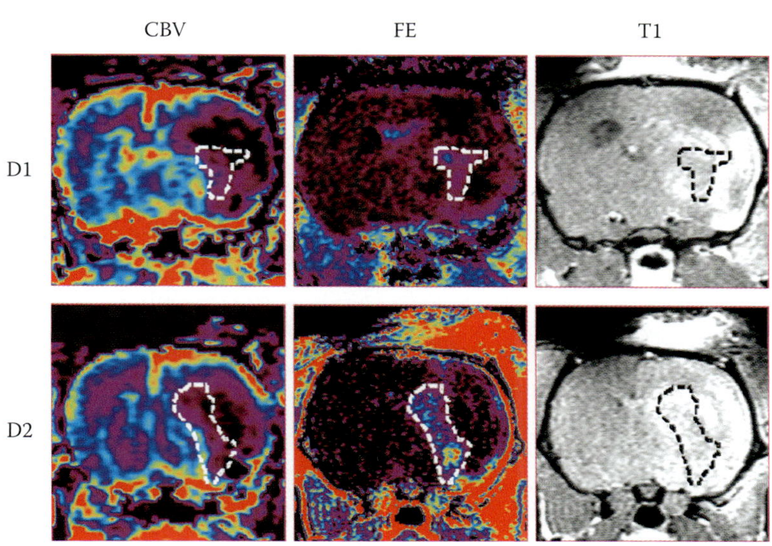

FIGURE 14.1

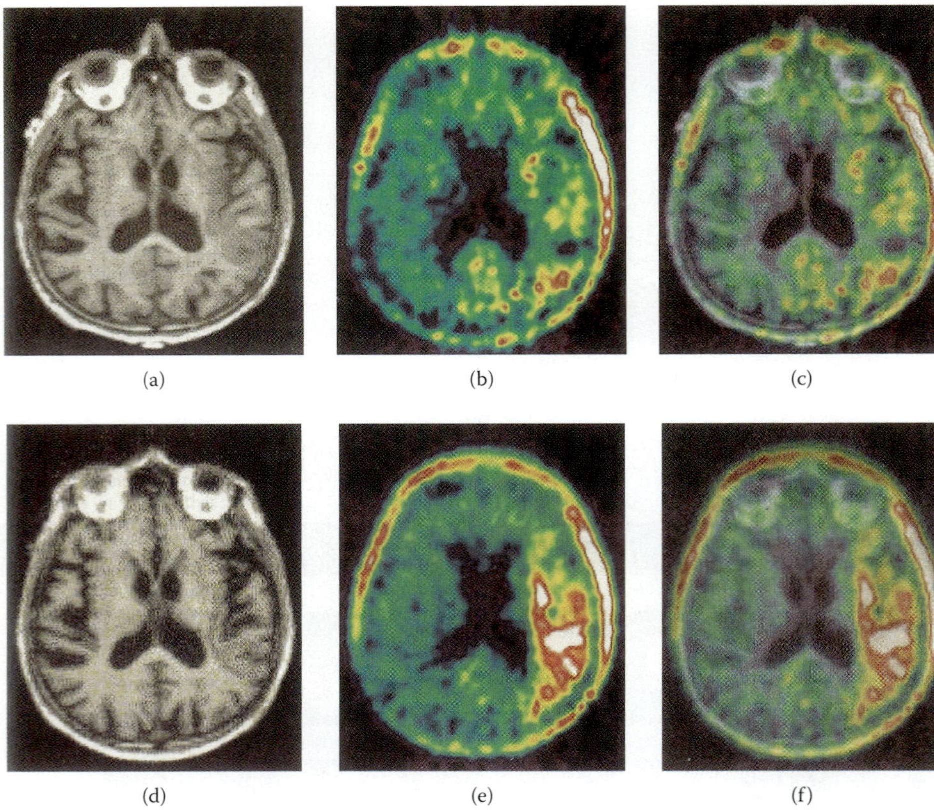

FIGURE 15.2

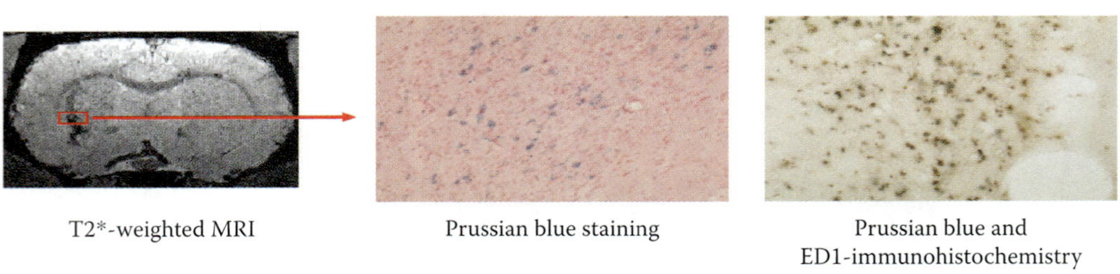

FIGURE 15.5

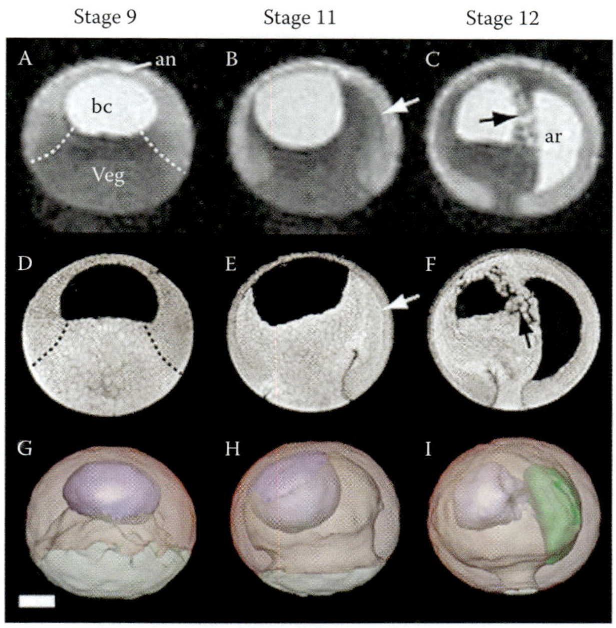

FIGURE 16.2

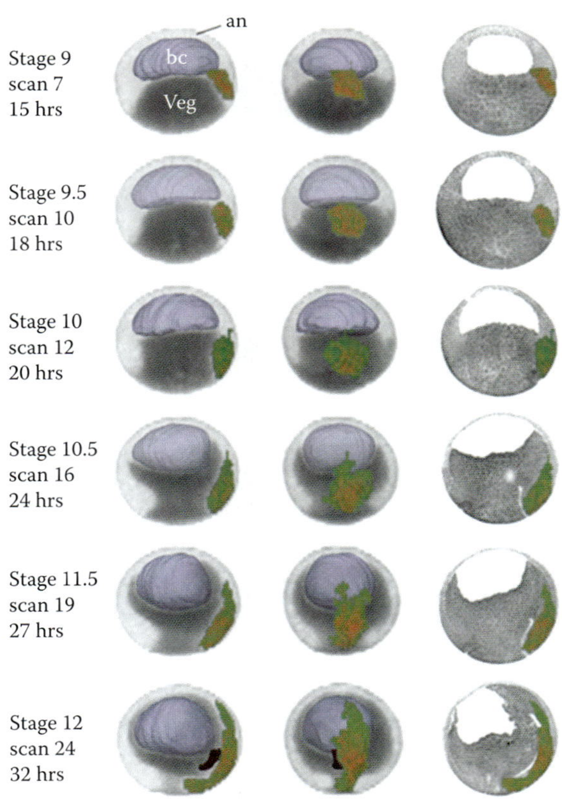

FIGURE 16.4

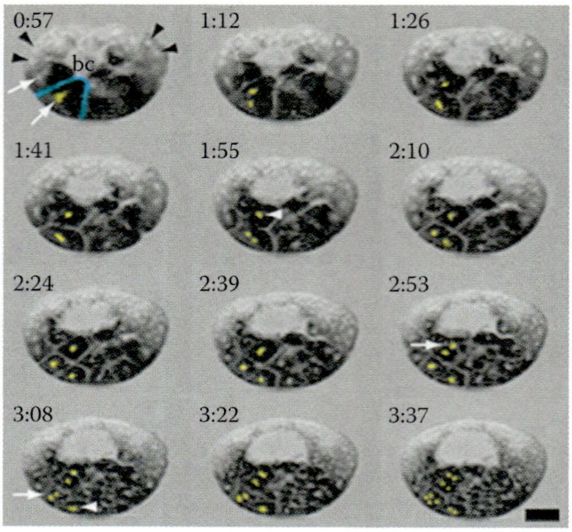

FIGURE 16.5

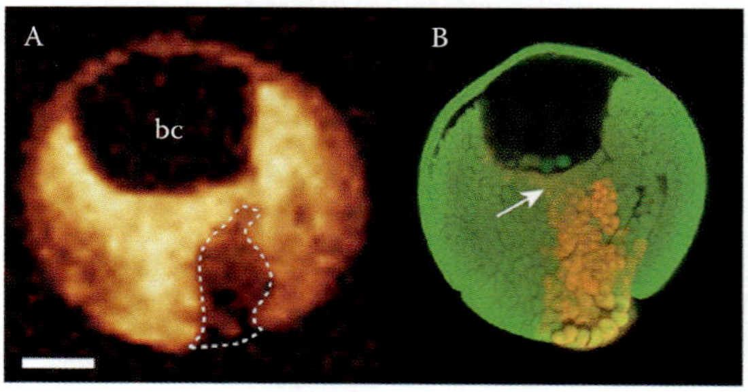

FIGURE 16.7

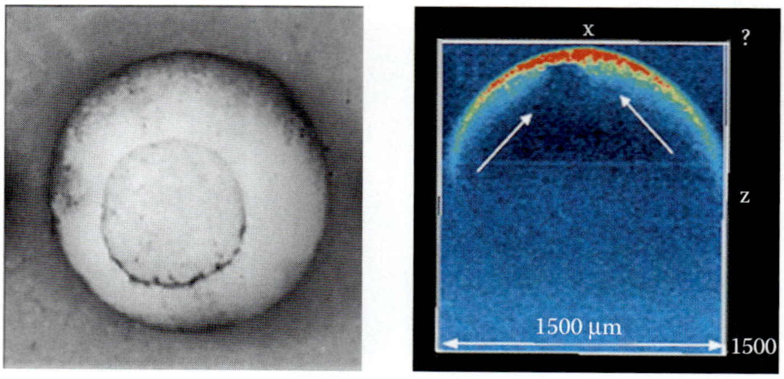

(a) (b)

FIGURE 16.8

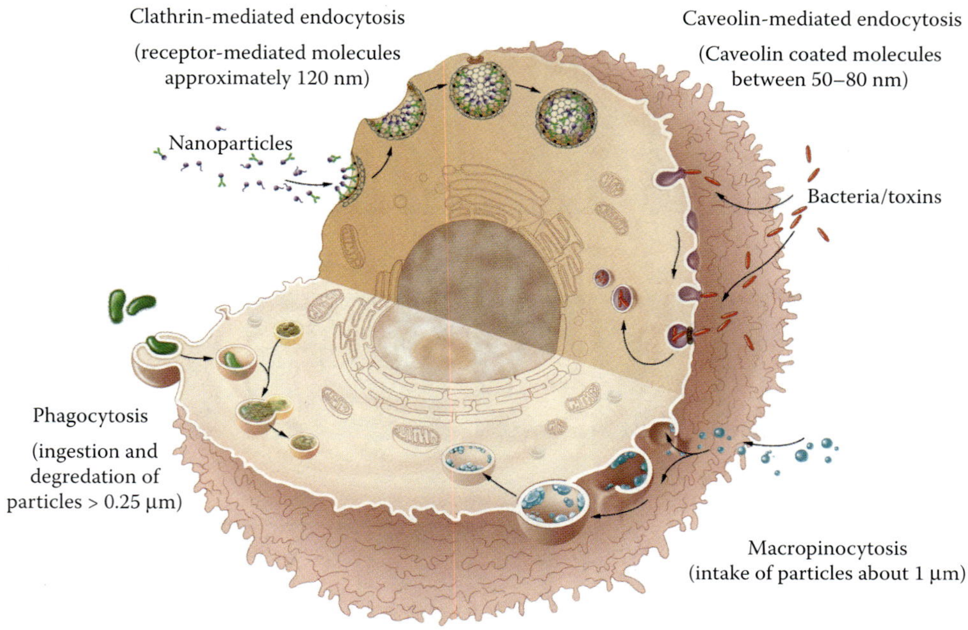

FIGURE 17.1

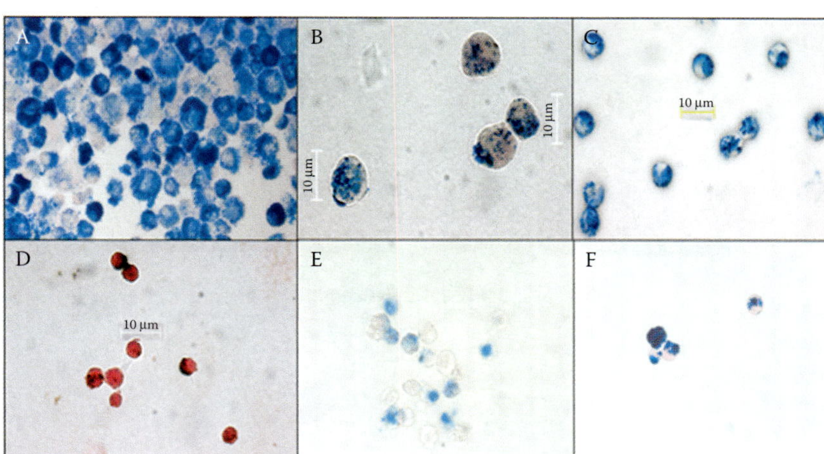

FIGURE 17.5

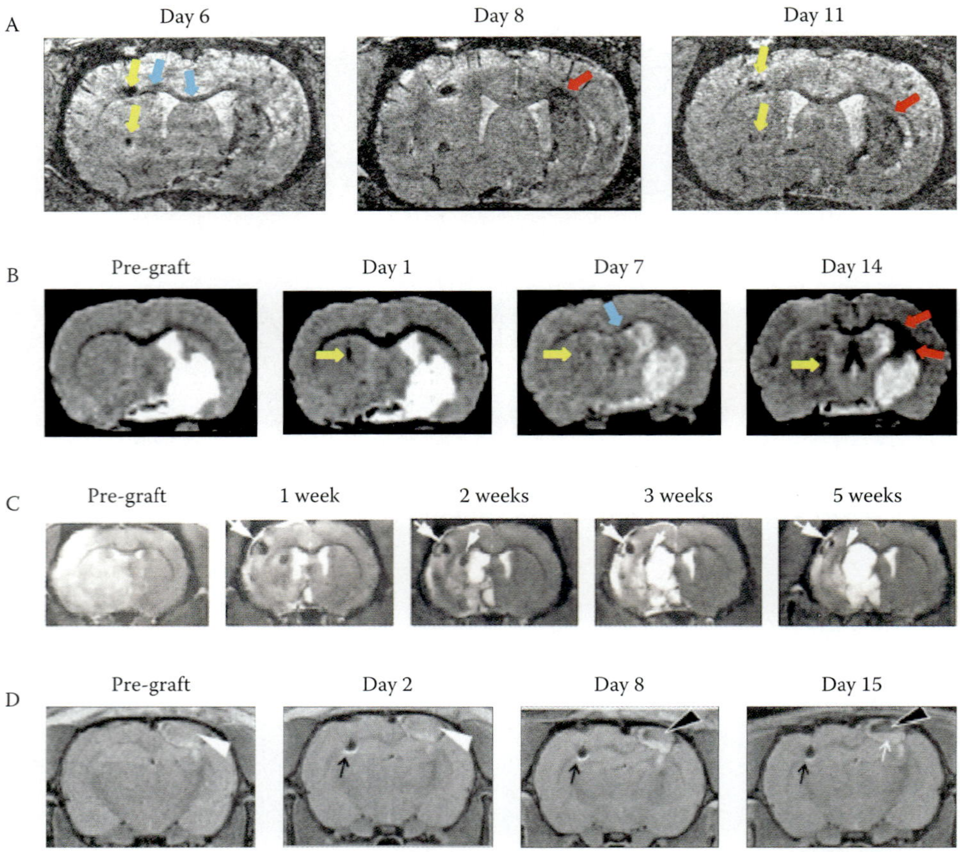

FIGURE 18.2

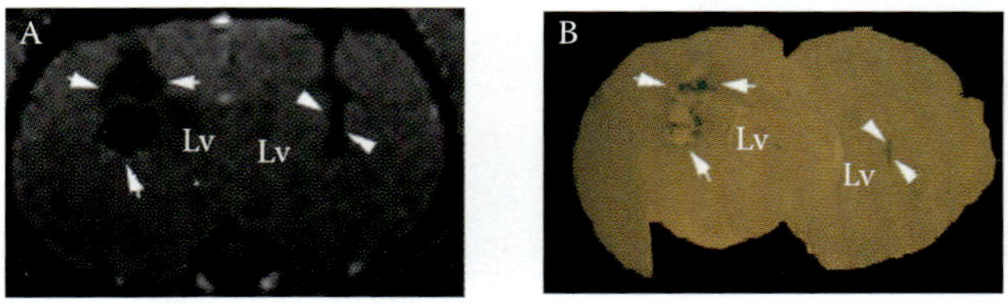

FIGURE 18.3

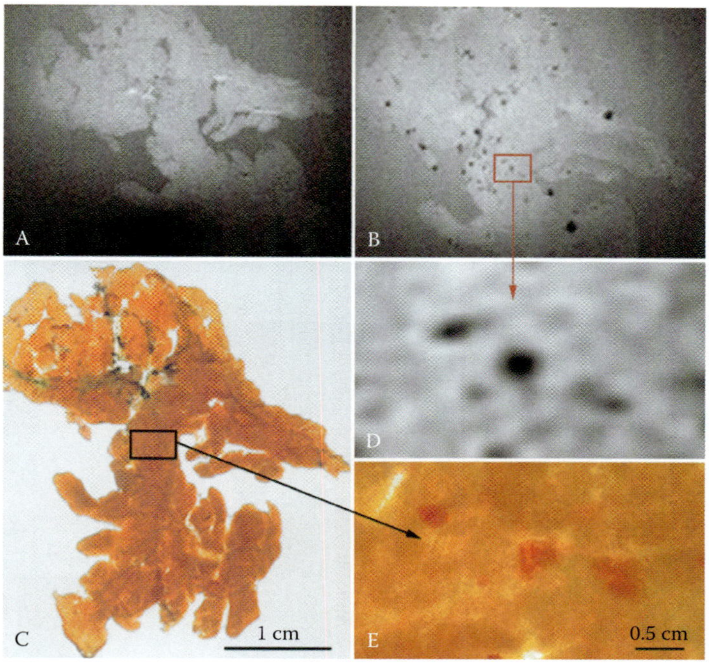

FIGURE 19.4

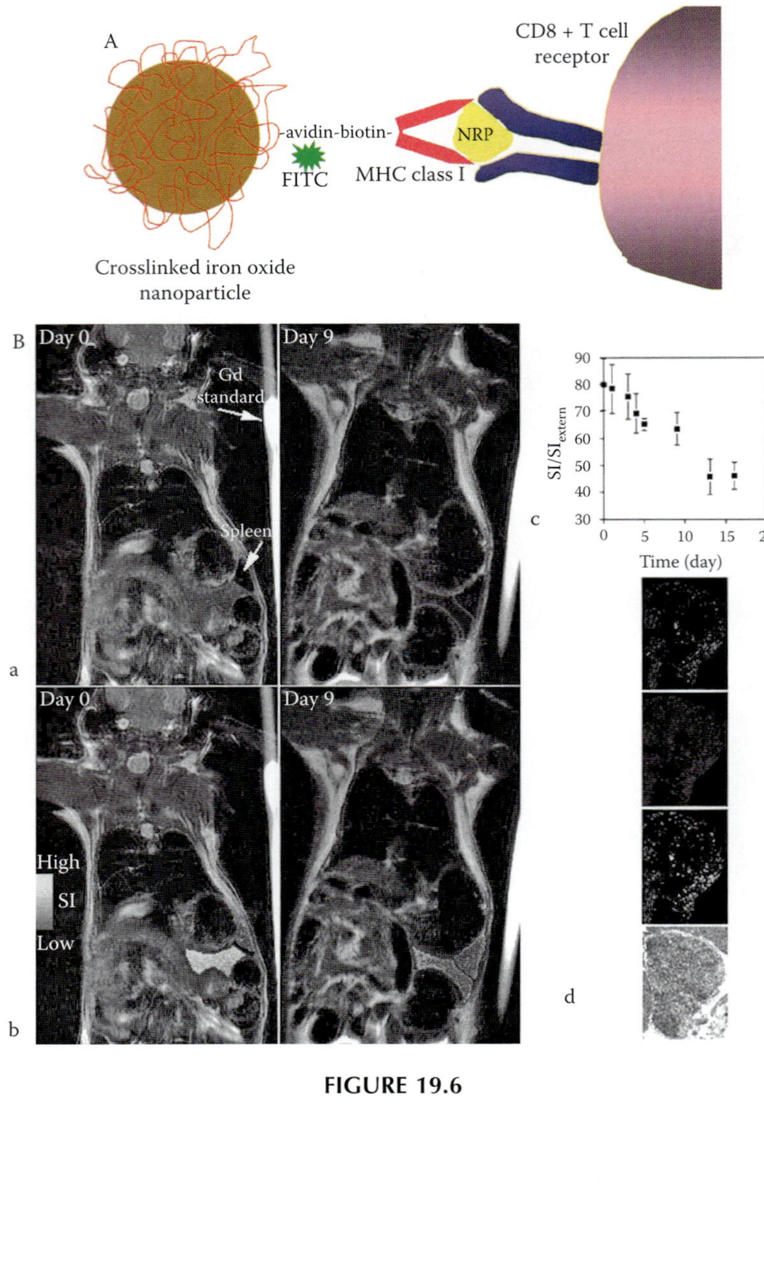
A
CD8 + T cell
receptor
-avidin-biotin-
NRP
FITC
MHC class I
Crosslinked iron oxide
nanoparticle
B
Day 0
Day 9
Gd
standard
Spleen
a
Day 0
Day 9
High
SI
Low
b
90
80
70
60
50
40
30
0
5
10
15
20
SI/SI$_{extern}$
Time (day)
c
d

FIGURE 19.6

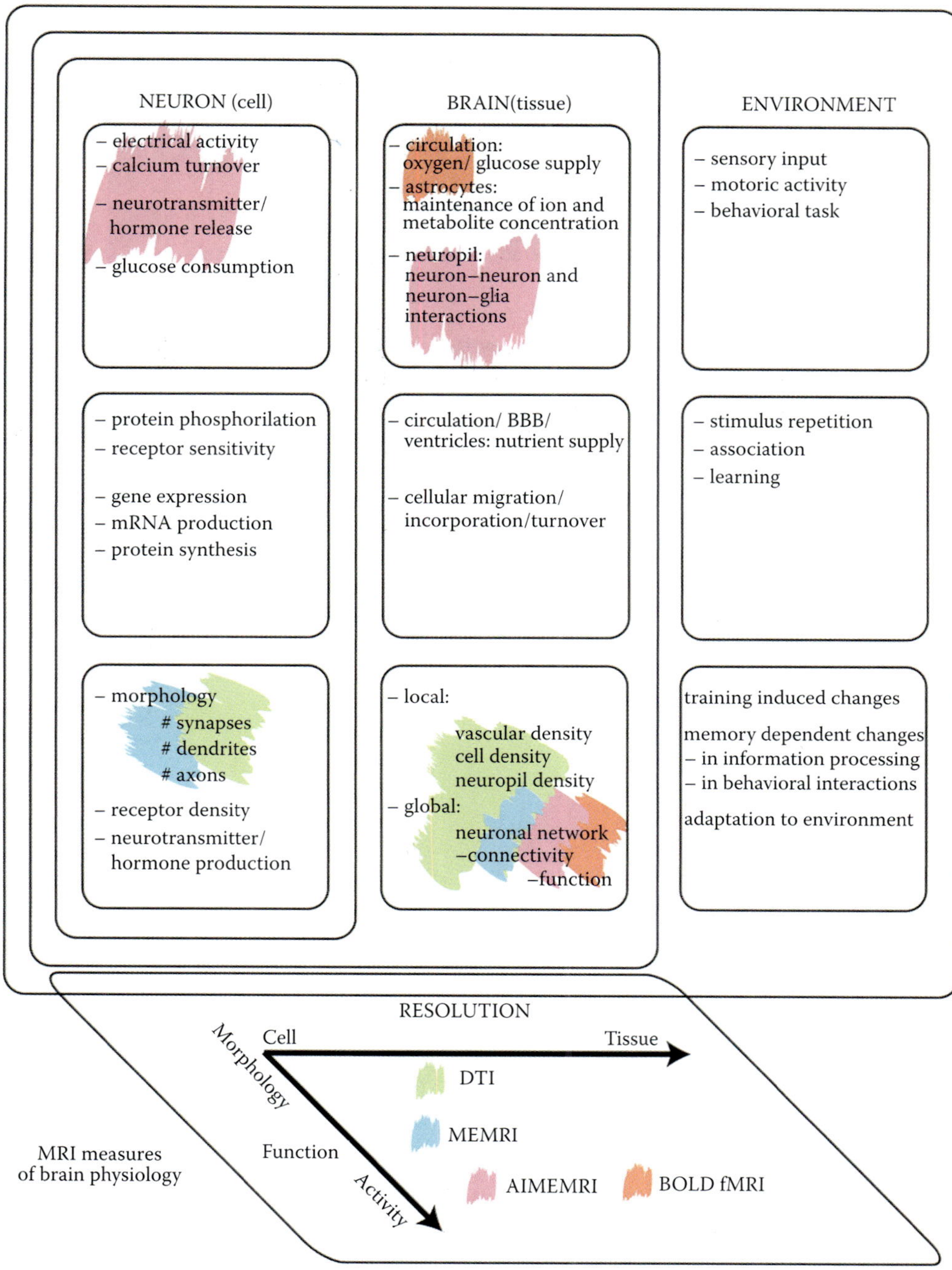
NEURON (cell)
BRAIN(tissue)
ENVIRONMENT
Short term (after single activation)
– electrical activity
– calcium turnover
– neurotransmitter/ hormone release
– glucose consumption
– circulation: oxygen/ glucose supply
– astrocytes: maintenance of ion and metabolite concentration
– neuropil: neuron–neuron and neuron–glia interactions
– sensory input
– motoric activity
– behavioral task
Intermediate term (upon repeated activations)
– protein phosphorilation
– receptor sensitivity
– gene expression
– mRNA production
– protein synthesis
– circulation/ BBB/ ventricles: nutrient supply
– cellular migration/ incorporation/turnover
– stimulus repetition
– association
– learning
Long term (prolongued effects)
– morphology
synapses
dendrites
axons
– receptor density
– neurotransmitter/ hormone production
– local:
vascular density
cell density
neuropil density
– global:
neuronal network
–connectivity
–function
training induced changes
memory dependent changes
– in information processing
– in behavioral interactions
adaptation to environment
RESOLUTION
Cell
Tissue
Morphology
Function
Activity
DTI
MEMRI
AIMEMRI
BOLD fMRI
MRI measures of brain physiology

FIGURE 20.1

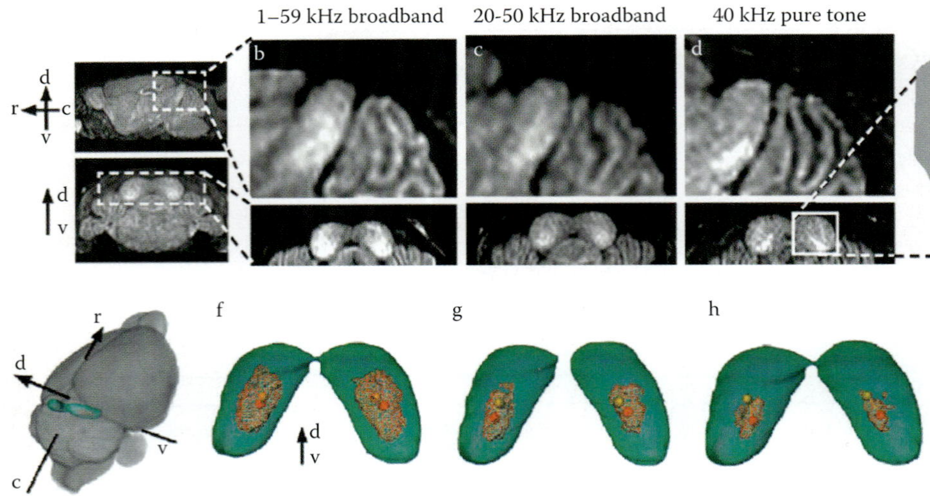
1–59 kHz broadband
20-50 kHz broadband
40 kHz pure tone
b
c
d
d
r
c
v
d
v
10
20
40
50
60
f
g
h
r
d
v
c
d
v

FIGURE 20.4

11.16 SUMMARY AND FUTURE PERSPECTIVE

MRI shows exquisite endogenous contrast and high spatial resolution for deep tissue imaging, along with a rapidly expanding arsenal of contrast mechanisms, which can be utilized for extraction of multiple structural, functional, and molecular parameters simultaneously and noninvasively. Over the past decades, exciting progress led to the generation of a range of reporter genes that can be utilized for the detection of the transcriptional regulation of gene expression by MRI and MRS. These reporter genes supplement the progress of *in vivo* detection of gene expression available by optical imaging, SPECT, and PET.

In addition to the complementary anatomical and functional information, an important advantage of MR-based reporter genes is the unique sensitivity of MR to the chemical and molecular environment. Thus, while reporter gene approaches of PET and SPECT are limited to detection of accumulation of the reporter probe, albeit at very high sensitivity, MRI and MRS can be applied for detection of changes in chemical shift or relaxation that can be used for reporting molecular events such as binding, phosphorylation, or degradation.

The high sensitivity of MR to the molecular environment implies that a thorough understanding of the system is required to account for the multiple effects that can alter signal intensity, including follow-up of reporter probe delivery and confounding effects of the microenvironment. These largely ignored parameters affect reporter gene readout for other modalities as well, but MRI offers the possibility for multiparametric imaging, and thus the possibly to account for such effects so as to improve specificity and accuracy, allowing quantitative determination of transgene expression or activity.

So far, the development of MR reporter genes has centered on providing proof-of-principle demonstrations of contrast with relatively little quantitative follow-up studies and applications of these tools for biological, preclinical, and clinical studies. Translation of these reporter genes for all three levels will require evaluation of the dynamics of activation and stability of the various reporter genes, along with the determination of toxicity and the biological consequence of an acute and chronic expression. Successful applications will provide temporal information that is not available by current biological tools and will be important in drug discovery and development. Clinical utilization can be envisaged for monitoring delivery and differentiation of cells in stem cell therapy and for noninvasive monitoring of gene therapy.

ACKNOWLEDGMENTS

Supported by the Israel Science Foundation, the Minerva Foundation, and NIH RO1 CA075334.

REFERENCES

1. Herschman, H.R., Non-invasive imaging of reporter genes, *J Cell Biochem Suppl*, 39, 36–44, 2002.
2. Ray, P., De, A., Min, J.J., Tsien, R.Y., and Gambhir, S.S., Imaging tri-fusion multimodality reporter gene expression in living subjects, *Cancer Res*, 64, 1323–1330, 2004.
3. Graham, F.L. and Prevec, L., Methods for construction of adenovirus vectors, *Mol Biotechnol*, 3, 207–220, 1995.
4. Wilson, J.M., Adenoviruses as gene-delivery vehicles, *N Engl J Med*, 334, 1185–1187, 1996.
5. Samulski, R.J., Adeno-associated virus: integration at a specific chromosomal locus, *Curr Opin Genet Dev*, 3, 74–80, 1993.
6. Fink, D.J. and Glorioso, J.C., Herpes simplex virus-based vectors: problems and some solutions, *Adv Neurol*, 72, 149–156, 1997.
7. Quinonez, R. and Sutton, R.E., Lentiviral vectors for gene delivery into cells, *DNA Cell Biol*, 21, 937–951, 2002.

8. Miyatake, S., Iyer, A., Martuza, R.L., and Rabkin, S.D., Transcriptional targeting of herpes simplex virus for cell-specific replication, *J Virol*, 71, 5124–5132, 1997.
9. Talbot, D., Descombes, P., and Schibler, U., The 5' flanking region of the rat LAP (C/EBP beta) gene can direct high-level, position-independent, copy number-dependent expression in multiple tissues in transgenic mice, *Nucleic Acids Res*, 22, 756–766, 1994.
10. Shi, Q., Wang, Y., and Worton, R., Modulation of the specificity and activity of a cellular promoter in an adenoviral vector, *Hum Gene Ther*, 8, 403–410, 1997.
11. Schlaeger, T.M., Bartunkova, S., Lawitts, J.A., Teichmann, G., Risau, W., Deutsch, U., and Sato, T.N., Uniform vascular-endothelial-cell-specific gene expression in both embryonic and adult transgenic mice, *Proc Natl Acad Sci USA*, 94, 3058–3063, 1997.
12. Sarao, R. and Dumont, D.J., Conditional transgene expression in endothelial cells, *Transgenic Res*, 7, 421–427, 1998.
13. Motoike, T., Loughna, S., Perens, E., Roman, B.L., Liao, W., Chau, T.C., Richardson, C.D., Kawate, T., Kuno, J., Weinstein, B.M., Stainier, D.Y., and Sato, T.N., Universal GFP reporter for the study of vascular development, *Genesis*, 28, 75–81, 2000.
14. Gory, S., Vernet, M., Laurent, M., Dejana, E., Dalmon, J., and Huber, P., The vascular endothelial-cadherin promoter directs endothelial-specific expression in transgenic mice, *Blood*, 93, 184–192, 1999.
15. Staggers, W.R., Paterson, A.J., and Kudlow, J.E., Sequence of the functional human keratin K14 promoter, *Gene*, 153, 297–298, 1995.
16. Yarranton, G.T., Inducible vectors for expression in mammalian cells, *Curr Opin Biotechnol*, 3, 506–511, 1992.
17. Alfke, H., Stoppler, H., Nocken, F., Heverhagen, J.T., Kleb, B., Czubayko, F., and Klose, K.J., *In vitro* MR imaging of regulated gene expression, *Radiology*, 228, 488–492, 2003.
18. Cohen, B., Dafni, H., Meir, G., Harmelin, A., and Neeman, M., Ferritin as an endogenous MRI reporter for noninvasive imaging of gene expression in C6 glioma tumors, *Neoplasia*, 7, 109–117, 2005.
19. Herr, W., The herpes simplex virus VP16-induced complex: mechanisms of combinatorial transcriptional regulation, *Cold Spring Harbor Symp Quant Biol*, 63, 599–607, 1998.
20. No, D., Yao, T.P., and Evans, R.M., Ecdysone-inducible gene expression in mammalian cells and transgenic mice, *Proc Natl Acad Sci USA*, 93, 3346–3351, 1996.
21. Braselmann, S., Graninger, P., and Busslinger, M., A selective transcriptional induction system for mammalian cells based on Gal4-estrogen receptor fusion proteins, *Proc Natl Acad Sci USA*, 90, 1657–1661, 1993.
22. Dymecki, S.M. and Tomasiewicz, H., Using Flp-recombinase to characterize expansion of Wnt1-expressing neural progenitors in the mouse, *Dev Biol*, 201, 57–65, 1998.
23. Branda, C.S. and Dymecki, S.M., Talking about a revolution: the impact of site-specific recombinases on genetic analyses in mice, *Dev Cell*, 6, 7–28, 2004.
24. Xu, T. and Rubin, G.M., Analysis of genetic mosaics in developing and adult Drosophila tissues, *Development*, 117, 1223–1237, 1993.
25. O'Gorman, S., Fox, D.T., and Wahl, G.M., Recombinase-mediated gene activation and site-specific integration in mammalian cells, *Science*, 251, 1351–1355, 1991.
26. Orban, P.C., Chui, D., and Marth, J.D., Tissue- and site-specific DNA recombination in transgenic mice, *Proc Natl Acad Sci USA*, 89, 6861–6865, 1992.
27. Hoess, R.H., Ziese, M., and Sternberg, N., P1 site-specific recombination: nucleotide sequence of the recombining sites, *Proc Natl Acad Sci USA*, 79, 3398–3402, 1982.
28. McLeod, M., Craft, S., and Broach, J.R., Identification of the crossover site during FLP-mediated recombination in the *Saccharomyces cerevisiae* plasmid 2 microns circle, *Mol Cell Biol*, 6, 3357–3367, 1986.
29. Weissleder, R. and Ntziachristos, V., Shedding light onto live molecular targets, *Nat Med*, 9, 123–128, 2003.
30. Hoffman, R.M., Visualization of GFP-expressing tumors and metastasis *in vivo*, *Biotechniques*, 30, 1016–1022, 1024–1026, 2001.
31. Yang, M., Baranov, E., Moossa, A.R., Penman, S., and Hoffman, R.M., Visualizing gene expression by whole-body fluorescence imaging, *Proc Natl Acad Sci USA*, 97, 12278–12282, 2000.

32. Moore, A., Basilion, J.P., Chiocca, E.A., and Weissleder, R., Measuring transferrin receptor gene expression by NMR imaging, *Biochim Biophys Acta*, 1402, 239–249, 1998.
33. Wunderbaldinger, P., Josephson, L., Bremer, C., Moore, A., and Weissleder, R., Detection of lymph node metastases by contrast-enhanced MRI in an experimental model, *Magn Reson Med*, 47, 292–297, 2002.
34. Yang, M., Baranov, E., Jiang, P., Sun, F.X., Li, X.M., Li, L., Hasegawa, S., Bouvet, M., Al-Tuwaijri, M., Chishima, T., Shimada, H., Moossa, A.R., Penman, S., and Hoffman, R.M., Whole-body optical imaging of green fluorescent protein-expressing tumors and metastases, *Proc Natl Acad Sci USA*, 97, 1206–1211, 2000.
35. Fukumura, D., Xavier, R., Sugiura, T., Chen, Y., Park, E.C., Lu, N., Selig, M., Nielsen, G., Taksir, T., Jain, R.K., and Seed, B., Tumor induction of VEGF promoter activity in stromal cells, *Cell*, 94, 715–725, 1998.
36. Bumann, D., Examination of *Salmonella* gene expression in an infected mammalian host using the green fluorescent protein and two-colour flow cytometry, *Mol Microbiol*, 43, 1269–1283, 2002.
37. Marra, A., Asundi, J., Bartilson, M., Lawson, S., Fang, F., Christine, J., Wiesner, C., Brigham, D., Schneider, W.P., and Hromockyj, A.E., Differential fluorescence induction analysis of *Streptococcus pneumoniae* identifies genes involved in pathogenesis, *Infect Immun*, 70, 1422–1433, 2002.
38. Labow, D., Lee, S., Ginsberg, R.J., Crystal, R.G., and Korst, R.J., Adenovirus vector-mediated gene transfer to regional lymph nodes, *Hum Gene Ther*, 11, 759–769, 2000.
39. Van Den Pol, A.N., Vieira, J., Spencer, D.D., and Santarelli, J.G., Mouse cytomegalovirus in developing brain tissue: analysis of 11 species with GFP-expressing recombinant virus, *J Comp Neurol*, 427, 559–580, 2000.
40. Rubin, A.D., Hogikyan, N.D., Sullivan, K., Boulis, N., and Feldman, E.L., Remote delivery of rAAV-GFP to the rat brainstem through the recurrent laryngeal nerve, *Laryngoscope*, 111 (Pt 1), 2041–2045, 2001.
41. Prasher, D.C., Eckenrode, V.K., Ward, W.W., Prendergast, F.G., and Cormier, M.J., Primary structure of the *Aequorea victoria* green-fluorescent protein, *Gene*, 111, 229–233, 1992.
42. Shaner, N.C., Steinbach, P.A., and Tsien, R.Y., A guide to choosing fluorescent proteins, *Nat Methods*, 2, 905–909, 2005.
43. Lippincott-Schwartz, J. and Patterson, G.H., Development and use of fluorescent protein markers in living cells, *Science*, 300, 87–91, 2003.
44. Verkhusha, V.V. and Lukyanov, K.A., The molecular properties and applications of Anthozoa fluorescent proteins and chromoproteins, *Nat Biotechnol*, 22, 289–296, 2004.
45. Chudakov, D.M., Verkhusha, V.V., Staroverov, D.B., Souslova, E.A., Lukyanov, S., and Lukyanov, K.A., Photoswitchable cyan fluorescent protein for protein tracking, *Nat Biotechnol*, 22, 1435–1439, 2004.
46. Habuchi, S., Ando, R., Dedecker, P., Verheijen, W., Mizuno, H., Miyawaki, A., and Hofkens, J., Reversible single-molecule photoswitching in the GFP-like fluorescent protein Dronpa, *Proc Natl Acad Sci USA*, 102, 9511–9516, 2005.
47. Contag, C.H. and Bachmann, M.H., Advances in *in vivo* bioluminescence imaging of gene expression, *Annu Rev Biomed Eng*, 4, 235–260, 2002.
48. Contag, C.H., Jenkins, D., Contag, P.R., and Negrin, R.S., Use of reporter genes for optical measurements of neoplastic disease *in vivo*, *Neoplasia*, 2, 41–52, 2000.
49. Wetterwald, A., van der Pluijm, G., Que, I., Sijmons, B., Buijs, J., Karperien, M., Lowik, C.W., Gautschi, E., Thalmann, G.N., and Cecchini, M.G., Optical imaging of cancer metastasis to bone marrow: a mouse model of minimal residual disease, *Am J Pathol*, 160, 1143–1153, 2002.
50. Costa, G.L., Sandora, M.R., Nakajima, A., Nguyen, E.V., Taylor-Edwards, C., Slavin, A.J., Contag, C.H., Fathman, C.G., and Benson, J.M., Adoptive immunotherapy of experimental autoimmune encephalomyelitis via T cell delivery of the IL-12 p40 subunit, *J Immunol*, 167, 2379–2387, 2001.
51. Burns, S.M., Joh, D., Francis, K.P., Shortliffe, L.D., Gruber, C.A., Contag, P.R., and Contag, C.H., Revealing the spatiotemporal patterns of bacterial infectious diseases using bioluminescent pathogens and whole body imaging, *Contrib Microbiol*, 9, 71–88, 2001.
52. Contag, C.H., Spilman, S.D., Contag, P.R., Oshiro, M., Eames, B., Dennery, P., Stevenson, D.K., and Benaron, D.A., Visualizing gene expression in living mammals using a bioluminescent reporter, *Photochem Photobiol*, 66, 523–531, 1997.

53. Contag, P.R., Olomu, I.N., Stevenson, D.K., and Contag, C.H., Bioluminescent indicators in living mammals, *Nat Med*, 4, 245–247, 1998.
54. Contag, C.H. and Stevenson, D.K., *In vivo* patterns of heme oxygenase-1 transcription, *J Perinatol*, 21 (Suppl 1), S119–S124, 2001; discussion, S125–S127.
55. Bhaumik, S. and Gambhir, S.S., Optical imaging of Renilla luciferase reporter gene expression in living mice, *Proc Natl Acad Sci USA*, 99, 377–382, 2002.
56. Honigman, A., Zeira, E., Ohana, P., Abramovitz, R., Tavor, E., Bar, I., Zilberman, Y., Rabinovsky, R., Gazit, D., Joseph, A., Panet, A., Shai, E., Palmon, A., Laster, M., and Galun, E., Imaging transgene expression in live animals, *Mol Ther*, 4, 239–249, 2001.
57. Weng, Y.H., Tatarov, A., Bartos, B.P., Contag, C.H., and Dennery, P.A., HO-1 expression in type II pneumocytes after transpulmonary gene delivery, *Am J Physiol Lung Cell Mol Physiol*, 278, L1273–L1279, 2000.
58. Wu, J.C., Sundaresan, G., Iyer, M., and Gambhir, S.S., Noninvasive optical imaging of firefly luciferase reporter gene expression in skeletal muscles of living mice, *Mol Ther*, 4, 297–306, 2001.
59. Zhang, W., Feng, J.Q., Harris, S.E., Contag, P.R., Stevenson, D.K., and Contag, C.H., Rapid *in vivo* functional analysis of transgenes in mice using whole body imaging of luciferase expression, *Transgenic Res*, 10, 423–434, 2001.
60. Vooijs, M., Jonkers, J., Lyons, S., and Berns, A., Noninvasive imaging of spontaneous retinoblastoma pathway-dependent tumors in mice, *Cancer Res*, 62, 1862–1867, 2002.
61. Ray, P., Pimenta, H., Paulmurugan, R., Berger, F., Phelps, M.E., Iyer, M., and Gambhir, S.S., Noninvasive quantitative imaging of protein-protein interactions in living subjects, *Proc Natl Acad Sci USA*, 99, 3105–3110, 2002.
62. Carlsen, H., Moskaug, J.O., Fromm, S.H., and Blomhoff, R., *In vivo* imaging of NF-kappa B activity, *J Immunol*, 168, 1441–1446, 2002.
63. Gross, S. and Piwnica-Worms, D., Real-time imaging of ligand-induced IKK activation in intact cells and in living mice, *Nat Methods*, 2, 607–614, 2005.
64. Tung, C.H., Zeng, Q., Shah, K., Kim, D.E., Schellingerhout, D., and Weissleder, R., *In vivo* imaging of beta-galactosidase activity using far red fluorescent switch, *Cancer Res*, 64, 1579–1583, 2004.
65. Moats, R.A., Fraser, S.E., and Meade, T.J., A "smart" magnetic resonance imaging agent that reports on specific enzymatic activity, *Angew Chem Int Ed Engl*, 726–728, 1997.
66. Horrocks, W.D. and Sundick, D.R., Lanthanide ion probes of structure in biology. Laser-induced luminescence decay constants provide a direct measure of the number of metal-coordinated water molecules, *J Am Chem Soc*, 101, 334–340, 1979.
67. Zhang, X., Chang, C.A., Brittain, H.G., Garrison, J.M., Telser, J., and Tweedle, M.F., pH dependence of relaxivities and hydration numbers of gadolinium(III) complexes of macrocyclic amino carboxylates, *Inorg Chem*, 31, 5597–5600, 1992.
68. Louie, A.Y., Huber, M.M., Ahrens, E.T., Rothbacher, U., Moats, R., Jacobs, R.E., Fraser, S.E., and Meade, T.J., *In vivo* visualization of gene expression using magnetic resonance imaging, *Nat Biotechnol*, 18, 321–325, 2000.
69. Kroll, K.L. and Amaya, E., Transgenic *Xenopus* embryos from sperm nuclear transplantations reveal FGF signaling requirements during gastrulation, *Development*, 122, 3173–3183, 1996.
70. Wetts, R. and Fraser, S.E., Slow intermixing of cells during *Xenopus* embryogenesis contributes to the consistency of the blastomere fate map, *Development*, 105, 9–15, 1989.
71. Cui, W., Otten, P., Li, Y., Koeneman, K.S., Yu, J., and Mason, R.P., Novel NMR approach to assessing gene transfection: 4-fluoro-2-nitrophenyl-beta-D-galactopyranoside as a prototype reporter molecule for beta-galactosidase, *Magn Reson Med*, 51, 616–620, 2004.
72. Cui, W., Otten, P., Merritt, M., and Mason, R.P., A novel NMR reporter molecule for transmembrane pH gradients: para-fluoro-ortho-nitrophenol, in *Proceedings of the 10th Annual Meeting of ISMRM*, Hononulu, 2002, p. 385.
73. Yu, J., Otten, P., Ma, Z., Cui, W., Liu, L., and Mason, R.P., Novel NMR platform for detecting gene transfection: synthesis and evaluation of fluorinated phenyl beta-D-galactosides with potential application for assessing LacZ gene expression, *Bioconjug Chem*, 15, 1334–1341, 2004.
74. Yu, J., Liu, L., Kodibagkar, V.D., Cui, W., and Mason, R.P., Synthesis and evaluation of novel enhanced gene reporter molecules: detection of beta-galactosidase activity using 19F NMR of trifluoromethylated aryl beta-D-galactopyranosides, *Bioorg Med Chem*, 14, 326–333, 2006.

75. Philpott, G.W., Shearer, W.T., Bower, R.J., and Parker, C.W., Selective cytotoxicity of hapten-substituted cells with an antibody-enzyme conjugate, *J Immunol*, 111, 921–929, 1973.
76. Haubner, R., Avril, N., Hantzopoulos, P.A., Gansbacher, B., and Schwaiger, M., *In vivo* imaging of herpes simplex virus type 1 thymidine kinase gene expression: early kinetics of radiolabelled FIAU, *Eur J Nucl Med*, 27, 283–291, 2000.
77. Hospers, G.A., Calogero, A., van Waarde, A., Doze, P., Vaalburg, W., Mulder, N.H., and de Vries, E.F., Monitoring of herpes simplex virus thymidine kinase enzyme activity using positron emission tomography, *Cancer Res*, 60, 1488–1491, 2000.
78. Gambhir, S.S., Bauer, E., Black, M.E., Liang, Q., Kokoris, M.S., Barrio, J.R., Iyer, M., Namavari, M., Phelps, M.E., and Herschman, H.R., A mutant herpes simplex virus type 1 thymidine kinase reporter gene shows improved sensitivity for imaging reporter gene expression with positron emission tomography, *Proc Natl Acad Sci USA*, 97, 2785–2790, 2000.
79. Tjuvajev, J.G., Chen, S.H., Joshi, A., Joshi, R., Guo, Z.S., Balatoni, J., Ballon, D., Koutcher, J., Finn, R., Woo, S.L., and Blasberg, R.G., Imaging adenoviral-mediated herpes virus thymidine kinase gene transfer and expression *in vivo*, *Cancer Res*, 59, 5186–5193, 1999.
80. Sangro, B., Qian, C., Ruiz, J., and Prieto, J., Tracing transgene expression in cancer gene therapy: a requirement for rational progress in the field, *Mol Imaging Biol*, 4, 27–33, 2002.
81. Luker, G.D., Sharma, V., Pica, C.M., Dahlheimer, J.L., Li, W., Ochesky, J., Ryan, C.E., Piwnica-Worms, H., and Piwnica-Worms, D., Noninvasive imaging of protein-protein interactions in living animals, *Proc Natl Acad Sci USA*, 99, 6961–6966, 2002.
82. Ponomarev, V., Doubrovin, M., Lyddane, C., Beresten, T., Balatoni, J., Bornman, W., Finn, R., Akhurst, T., Larson, S., Blasberg, R., Sadelain, M., and Tjuvajev, J.G., Imaging TCR-dependent NFAT-mediated T-cell activation with positron emission tomography *in vivo*, *Neoplasia*, 3, 480–488, 2001.
83. Doubrovin, M., Ponomarev, V., Beresten, T., Balatoni, J., Bornmann, W., Finn, R., Humm, J., Larson, S., Sadelain, M., Blasberg, R., and Gelovani Tjuvajev, J., Imaging transcriptional regulation of p53-dependent genes with positron emission tomography *in vivo*, *Proc Natl Acad Sci USA*, 98, 9300–9305, 2001.
84. Oshiro, E.M., Viola, J.J., Oldfield, E.H., Walbridge, S., Bacher, J., Frank, J.A., Blaese, R.M., and Ram, Z., Toxicity studies and distribution dynamics of retroviral vectors following intrathecal administration of retroviral vector-producer cells, *Cancer Gene Ther*, 2, 87–95, 1995.
85. Ram, Z., Culver, K.W., Walbridge, S., Frank, J.A., Blaese, R.M., and Oldfield, E.H., Toxicity studies of retroviral-mediated gene transfer for the treatment of brain tumors, *J Neurosurg*, 79, 400–407, 1993.
86. Haberkorn, U. and Altmann, A., Imaging methods in gene therapy of cancer, *Curr Gene Ther*, 1, 163–182, 2001.
87. Grohn, O.H., Valonen, P.K., Lehtimaki, K.K., Vaisanen, T.H., Kettunen, M.I., Yla-Herttuala, S., Kauppinen, R.A., and Garwood, M., Novel magnetic resonance imaging contrasts for monitoring response to gene therapy in rat glioma, *Cancer Res*, 63, 7571–7574, 2003.
88. Ostrowski, T., Golankiewicz, B., De Clercq, E., and Balzarini, J., Fluorosubstitution and 7-alkylation as prospective modifications of biologically active 6-aryl derivatives of tricyclic acyclovir and ganciclovir analogues, *Bioorg Med Chem*, 13, 2089–2096, 2005.
89. Stegman, L.D., Rehemtulla, A., Beattie, B., Kievit, E., Lawrence, T.S., Blasberg, R.G., Tjuvajev, J.G., and Ross, B.D., Noninvasive quantitation of cytosine deaminase transgene expression in human tumor xenografts with *in vivo* magnetic resonance spectroscopy, *Proc Natl Acad Sci USA*, 96, 9821–9826, 1999.
90. Hamstra, D.A., Lee, K.C., Tychewicz, J.M., Schepkin, V.D., Moffat, B.A., Chen, M., Dornfeld, K.J., Lawrence, T.S., Chenevert, T.L., Ross, B.D., Gelovani, J.T., and Rehemtulla, A., The use of 19F spectroscopy and diffusion-weighted MRI to evaluate differences in gene-dependent enzyme prodrug therapies, *Mol Ther*, 10, 916–928, 2004.
91. Dresselaers, T., Theys, J., Nuyts, S., Wouters, B., de Bruijn, E., Anne, J., Lambin, P., Van Hecke, P., and Landuyt, W., Non-invasive 19F MR spectroscopy of 5-fluorocytosine to 5-fluorouracil conversion by recombinant *Salmonella* in tumours, *Br J Cancer*, 89, 1796–1801, 2003.
92. Rehemtulla, A., Hall, D.E., Stegman, L.D., Prasad, U., Chen, G., Bhojani, M.S., Chenevert, T.L., and Ross, B.D., Molecular imaging of gene expression and efficacy following adenoviral-mediated brain tumor gene therapy, *Mol Imaging*, 1, 43–55, 2002.

93. Corban-Wilhelm, H., Hull, W.E., Becker, G., Bauder-Wust, U., Greulich, D., and Debus, J., Cytosine deaminase and thymidine kinase gene therapy in a Dunning rat prostate tumour model: absence of bystander effects and characterisation of 5-fluorocytosine metabolism with 19F-NMR spectroscopy, *Gene Ther*, 9, 1564–1575, 2002.
94. Brindle, K.M. and Radda, G.K., Measurements of exchange in the reaction catalysed by creatine kinase using 14C and 15N isotope labels and the NMR technique of saturation transfer, *Biochim Biophys Acta*, 829, 188–201, 1985.
95. Neeman, M., Rushkin, E., Kaye, A.M., and Degani, H., 31P-NMR studies of phosphate transfer rates in T47D human breast cancer cells, *Biochim Biophys Acta*, 930, 179–192, 1987.
96. Nicolay, K., van Dorsten, F.A., Reese, T., Kruiskamp, M.J., Gellerich, J.F., and van Echteld, C.J., *In situ* measurements of creatine kinase flux by NMR. The lessons from bioengineered mice, *Mol Cell Biochem*, 184, 195–208, 1998.
97. Nahrendorf, M., Spindler, M., Hu, K., Bauer, L., Ritter, O., Nordbeck, P., Quaschning, T., Hiller, K.H., Wallis, J., Ertl, G., Bauer, W.R., and Neubauer, S., Creatine kinase knockout mice show left ventricular hypertrophy and dilatation, but unaltered remodeling post-myocardial infarction, *Cardiovasc Res*, 65, 419–427, 2005.
98. ten Hove, M., Lygate, C.A., Fischer, A., Schneider, J.E., Sang, A.E., Hulbert, K., Sebag-Montefiore, L., Watkins, H., Clarke, K., Isbrandt, D., Wallis, J., and Neubauer, S., Reduced inotropic reserve and increased susceptibility to cardiac ischemia/reperfusion injury in phosphocreatine-deficient guanidinoacetate-N-methyltransferase-knockout mice, *Circulation*, 111, 2477–2485, 2005.
99. Momken, I., Lechene, P., Koulmann, N., Fortin, D., Mateo, P., Doan, B.T., Hoerter, J., Bigard, X., Veksler, V., and Ventura-Clapier, R., Impaired voluntary running capacity of creatine kinase-deficient mice, *J Physiol*, 565 (Pt 3), 951–964, 2005.
100. Koretsky, A.P. and Traxler, B.A., The B isozyme of creatine kinase is active as a fusion protein in *Escherichia coli*: *in vivo* detection by 31P NMR, *FEBS Lett*, 243, 8–12, 1989.
101. Koretsky, A.P., Brosnan, M.J., Chen, L.H., Chen, J.D., and Van Dyke, T., NMR detection of creatine kinase expressed in liver of transgenic mice: determination of free ADP levels, *Proc Natl Acad Sci USA*, 87, 3112–3116, 1990.
102. Auricchio, A., Zhou, R., Wilson, J.M., and Glickson, J.D., *In vivo* detection of gene expression in liver by 31P nuclear magnetic resonance spectroscopy employing creatine kinase as a marker gene, *Proc Natl Acad Sci USA*, 98, 5205–5210, 2001.
103. Li, Z., Qiao, H., Lebherz, C., Choi, S.R., Zhou, X., Gao, G., Kung, H.F., Rader, D.J., Wilson, J.M., Glickson, J.D., and Zhou, R., Creatine kinase, a magnetic resonance-detectable marker gene for quantification of liver-directed gene transfer, *Hum Gene Ther*, 16, 1429–1438, 2005.
104. Rao, B.D., Buttlaire, D.H., and Cohn, M., 31P NMR studies of the arginine kinase reaction. Equilibrium constants and exchange rates at stoichiometric enzyme concentration, *J Biol Chem*, 251, 6981–6986, 1976.
105. Nageswara Rao, B.D. and Cohn, M., 31P NMR of enzyme-bound substrates of rabbit muscle creatine kinase. Equilibrium constants, interconversion rates, and NMR parameters of enzyme-bound complexes, *J Biol Chem*, 256, 1716–1721, 1981.
106. Walter, G., Barton, E.R., and Sweeney, H.L., Noninvasive measurement of gene expression in skeletal muscle, *Proc Natl Acad Sci USA*, 97, 5151–5155, 2000.
107. Howarth, M., Takao, K., Hayashi, Y., and Ting, A.Y., Targeting quantum dots to surface proteins in living cells with biotin ligase, *Proc Natl Acad Sci USA*, 102, 7583–7588, 2005.
108. Chen, I., Howarth, M., Lin, W., and Ting, A.Y., Site-specific labeling of cell surface proteins with biophysical probes using biotin ligase, *Nat Methods*, 2, 99–104, 2005.
109. Dafni, H., Najjar, A., Gelovani, J., and Ronen, S., Metabolic biotinylation as a molecular imaging method: imaging the PDGFR-β cell surface receptor, *Mol Imaging*, 4, 329, 2005.
110. Tannous, B.A., Grimm, J., Perry, K., Weissleder, R., and Breakefield, X.O., Metabolic biotinylation of cell surface receptors for tumor imaging *in vivo*, *Mol Imaging*, 4, 219, 2005.
111. Gilad, A.A., McMahon, M.T., Winnard, P.T., Jr., Raman, V., Bulte, J.W.M., and van Zijl, P.C.M., Artificial reporter gene for MRI providing frequency-selective contrast, *Mol Imaging*, 4, 213, 2005.
112. Enochs, W.S., Petherick, P., Bogdanova, A., Mohr, U., and Weissleder, R., Paramagnetic metal scavenging by melanin: MR imaging, *Radiology*, 204, 417–423, 1997.

113. Weissleder, R., Simonova, M., Bogdanova, A., Bredow, S., Enochs, W.S., and Bogdanov, A., Jr., MR imaging and scintigraphy of gene expression through melanin induction, *Radiology*, 204, 425–429, 1997.
114. Arredondo, M. and Nunez, M.T., Iron and copper metabolism, *Mol Aspects Med*, 26, 313–327, 2005.
115. Chiancone, E., Ceci, P., Ilari, A., Ribacchi, F., and Stefanini, S., Iron and proteins for iron storage and detoxification, *Biometals*, 17, 197–202, 2004.
116. Siah, C.W., Trinder, D., and Olynyk, J.K., Iron overload, *Clin Chim Acta*, 358, 24–36, 2005.
117. Egyed, A., Carrier mediated iron transport through erythroid cell membrane, *Br J Haematol*, 68, 483–486, 1988.
118. Koretsky, A.P., Lin, Y.J., Schorle, H., and Jaenisch, R., Genetic control of MRI contrast by expression of the transferrin receptor, *Proc Int Soc Magn Reson Med*, 1, 69, 1996.
119. Kresse, M., Wagner, S., Pfefferer, D., Lawaczeck, R., Elste, V., and Semmler, W., Targeting of ultrasmall superparamagnetic iron oxide (USPIO) particles to tumor cells *in vivo* by using transferrin receptor pathways, *Magn Reson Med*, 40, 236–242, 1998.
120. Weissleder, R., Moore, A., Mahmood, U., Bhorade, R., Benveniste, H., Chiocca, E.A., and Basilion, J.P., *In vivo* magnetic resonance imaging of transgene expression, *Nat Med*, 6, 351–355, 2000.
121. Moore, A., Josephson, L., Bhorade, R.M., Basilion, J.P., and Weissleder, R., Human transferrin receptor gene as a marker gene for MR imaging, *Radiology*, 221, 244–250, 2001.
122. Hogemann, D., Josephson, L., Weissleder, R., and Basilion, J.P., Improvement of MRI probes to allow efficient detection of gene expression, *Bioconjug Chem*, 11, 941–946, 2000.
123. Ichikawa, T., Hogemann, D., Saeki, Y., Tyminski, E., Terada, K., Weissleder, R., Chiocca, E.A., and Basilion, J.P., MRI of transgene expression: correlation to therapeutic gene expression, *Neoplasia*, 4, 523–530, 2002.
124. Harrison, P.M. and Arosio, P., The ferritins: molecular properties, iron storage function and cellular regulation, *Biochim Biophys Acta*, 1275, 161–203, 1996.
125. Treffry, A., Zhao, Z., Quail, M.A., Guest, J.R., and Harrison, P.M., Dinuclear center of ferritin: studies of iron binding and oxidation show differences in the two iron sites, *Biochemistry*, 36, 432–441, 1997.
126. Cozzi, A., Corsi, B., Levi, S., Santambrogio, P., Albertini, A., and Arosio, P., Overexpression of wild type and mutated human ferritin H-chain in HeLa cells: *in vivo* role of ferritin ferroxidase activity, *J Biol Chem*, 275, 25122–25129, 2000.
127. Gottesfeld, Z. and Neeman, M., Ferritin effect on the transverse relaxation of water: NMR microscopy at 9.4 T, *Magn Reson Med*, 35, 514–520, 1996.
128. Vymazal, J., Brooks, R.A., Bulte, J.W., Gordon, D., and Aisen, P., Iron uptake by ferritin: NMR relaxometry studies at low iron loads, *J Inorg Biochem*, 71, 153–157, 1998.
129. Gossuin, Y., Muller, R.N., and Gillis, P., Relaxation induced by ferritin: a better understanding for an improved MRI iron quantification, *NMR Biomed*, 17, 427–432, 2004.
130. Ziv, K., Cohen, B., Kalchenko, V., Harmelin, A., and Neeman, M., Ferritin as a tissue specific MRI reporter of inducible gene expression in transgenic mice, *Proc Int Soc Magn Reson Med*, 14, 1839, 2006.
131. Genove, G., DeMarco, U., Xu, H., Goins, W.F., and Ahrens, E.T., A new transgene reporter for *in vivo* magnetic resonance imaging, *Nat Med*, 11, 450–454, 2005.
132. Deans, A. E., Wadghiri, Y. Z., Bernas, L. H., Yu, X., Rutt, B. K., and Turnbull, D. H., Cellular MRI contrast via coexpression of transferrin receptor and ferritin. *Magn Reson Med*, 56, 51–59, 2006.
133. Grabill, C., Silva, A.C., Smith, S.S., Koretsky, A.P., and Rouault, T.A., MRI detection of ferritin iron overload and associated neuronal pathology in iron regulatory protein-2 knockout mice, *Brain Res*, 971, 95–106, 2003.

12 Pharmacological MRI as a Molecular Imaging Technique

Y. Iris Chen and Bruce G. Jenkins

CONTENTS

12.1 Introduction 219
12.2 Multimodal Experiment Integration..221
12.3 Physiological Contributors to Pharmacologically Induced Hemodynamic Changes..........223
12.3.1 Ligand–Vascular Coupling 223
12.3.2 Ligand–Neuronal Coupling: The Dopaminergic Example 224
12.3.2.1 Presynaptic Autoreceptor Regulation 226
12.3.2.2 Postsynaptic Function 226
12.3.2.3 Receptor-Mediated Vascular Regulation 227
12.4 Using Pharmacological MRI to Probe the Dopaminergic Function 228
12.4.1 D-Amphetamine (AMPH)..228
12.4.2 Using Receptor Ligand to Monitor Changes in Receptor Density 229
12.4.3 Mapping Dopamine Autoreceptor (DAR) Function 229
12.4.4 Interaction of the Function between the Dopamine D2 Receptor and the Adenosine A2A Receptor..231
12.4.5 Alteration of the Functional Circuitry: Drug Addiction................................231
12.5 Which Is the Better Model for Studying the Neurotransmitter Function—Awake vs. Anesthetized Protocol?...233
12.5.1 Awake Animal Model...234
12.5.2 Anesthetized Animal Model..234
12.6 Conclusion 235
References 236

12.1 INTRODUCTION

Brain function represents an integrated outcome from series of complicated tasks such as dynamic adjustment of neuronal metabolism, signal communication, and transmission. Macroscopically, it appears as clusters of neurons firing simultaneously or sequentially at different brain locations. This kind of macroscopic representation of neuronal activity is typically observed at the level of parcellated cerebral structures (such as subunits in the thalamus) using event-detecting techniques such as functional magnetic resonance imaging (fMRI) for hemodynamic responses, electroencephalography (EEG) and magnetoencephalography (MEG) for the electrical magnetic responses, and positron emission tomography (PET) for the hemodynamic and metabolite responses. These techniques help to realize the concept that the brain functions through a circuitry-like network with information (or neuronal signals) flowing and communicating among major functional structures. The neuronal signal gets relayed between neurons by releasing neurotransmitters and neuropeptides into synapses to modulate the activity on postsynaptic neurons. The synaptic function can be

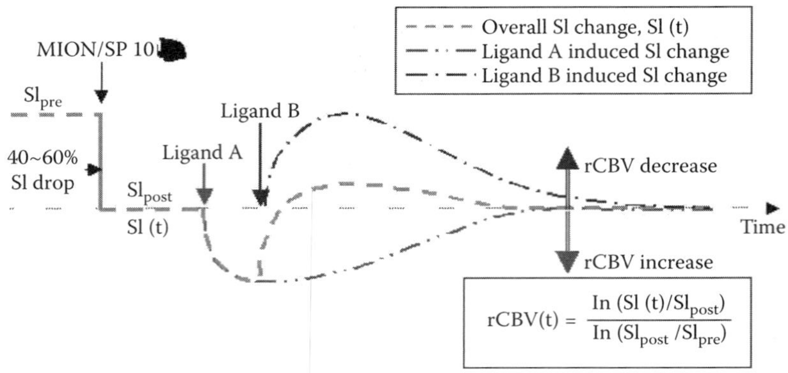

FIGURE 12.1 (please see color insert following page 210) A typical phMRI protocol using the IRON method. MR images are acquired repeatedly without interruption throughout the whole study. Two series of baselines (before and after MION/SPIO injection) are acquired. In order to reach the optimal sensitivity, the average post-MION baseline signal intensity (SI_{post}) is about 40 ~ 60% smaller than the average pre-MION signal intensity (SI_{pre}). After the post-MION baseline is established, single or multiple pharmacological ligands are given via bolus injections or slow infusion, simultaneously or sequentially. Ideally, the repetitive acquisition shall last until the post-ligand signal intensity is back to the SI_{post} level. With pre- and post-MION baselines, MR signal (SI(t)) can be converted to relative cerebral blood volume, rCBV(t), using the formula shown in the box. In some cases, when SI_{pre} is not available (in some awake monkey studies, monkeys were injected with MION before being sent into the magnet), the IRON method is used to enhance the contrast-to-noise ratio but not to convert the MR signal to rCBV changes. Note that the signs for MR signals and rCBV changes are opposite, i.e., MR signal decreases indicate rCBV increases, and signal increases indicate rCBV decreases. The drug-induced hemodynamic time course can be characterized by gamma function ($t \times \exp(-t/t)$) using the general linear model (GLM).[157] SI, MR signal intensity.

provoked or interfered with by pharmacological ligands that are specific to a designate neurotransmitter system. The drug-induced alteration in the synaptic activity often leads to hemodynamic change that is measurable by fMRI. We thus dub such an fMRI task with pharmacological challenge as pharmacological MRI (phMRI). Different than the traditional task-relative fMRI, phMRI usually has a prolonged time course that often associates with the pharmacokinetic property of the ligand. In this chapter, we will discuss the concept of using phMRI to probe the synaptic function. Specifically, we will take the dopaminergic system as an example to discuss how to use phMRI to probe different dopamine-relevant functions. We will address the importance of combining phMRI with other molecular techniques to get a fuller picture of the molecular mechanisms involved in brain function. We will discuss factors contributing to the phMRI-detected signal.

Similar to a task-related fMRI study, a typical phMRI protocol acquires signals repeatedly without interruption to cover time points, including baseline, drug administrations, and some portion of the pharmacodynamic and pharmacologic time course. Single or multiple pharmacological agents can be administered (simultaneously or sequentially), depending on the neuronal mechanism one wishes to investigate. Blood oxygen level-dependent (BOLD) and increased relaxation with iron oxide nanoparticles (IRON) techniques are the most popular methods used in phMRI, although the IRON method is currently used in animals only due to the concern of excess body iron accumulation. The IRON technique uses superparamagnetic iron oxide particles such as monocrystalline iron oxide nanocompound (MION)[1] or superparamagnetic iron oxide (SPIO) nanoparticles to sensitize the MR signal to regional cerebral blood volume (rCBV) changes.[2,3] A schematic illustration of a typical phMRI protocol using the IRON method is shown in Figure 12.1. The advantage of the IRON technique includes:

1. The MR signal can be converted to a physiologically meaningful parameter, i.e., rCBV.
2. rCBV change is field strength independent.[4]

3. The contrast-to-noise ratio (CNR) is dramatically enhanced, especially at lower magnetic field strength; thus, less averaging across studies are needed.[5,6]
4. The high sensitivity of BOLD to veins leading to spatial shifts in the activation maxima is not encountered.[5–8]

12.2 MULTIMODAL EXPERIMENT INTEGRATION

Performing a simple sensory, motor, or cognitive task requires complicated neuronal interactions involving evoking and modulating activity from the molecular to the structural level. For example, finger tapping is the result of coordinating neuronal activity in the ascending sensory and descending motor pathways involving the thalamus, the basal ganglia feedback loop, and sensory, premotor, and motor cortices. The cascading neuronal activity traveling through those functional units reflects a series of neurotransmitter-mediating synaptic transmission and modulation. When a neuronal signal is transmitted at the synapse, it can either evoke or suppress the postsynaptic neuron. For example, glutamate, which induces excitatory postsynaptic potentials (EPSPs), is a major excitatory neurotransmitter in the brain. GABA, which induces inhibitory postsynaptic potentials (IPSPs), is a major inhibitory neurotransmitter. Both glutamate and GABA can be released into the same synapse and alter the function of the postsynaptic neuron competitively. They can also innervate each other to boost (glutamate on GABAergic neuron) or attenuate (GABA on glutamatergic neuron) their ability of releasing the GABA and glutamate neurotransmitters, respectively.

Some other neurotransmitters act as modulators and can have influence both ways. For example, depending on which receptor subtype it binds to, dopamine (DA) can be either excitatory (when acting on D1-like receptors) or inhibitory (when acting on D2-like receptors). However, the neurotransmitter–receptor-mediated modulation of the postsynaptic potential only describes a superficial part of the neuronal function. The neurotransmitter–receptor innervation often leads to a chain modulation in molecular and gene expression levels. As reviewed in Greengard,[9] many neurotransmitter activities are coupled to second messengers such as cAMP, cGMP, calcium, and diacylglycerol. The second messenger thus serves as a functional converging center for multiple neurotransmitters and consequently modulates gene transcription and function of membrane-bound receptors through a series of reactions at the molecular level. In the case of drug addiction or plasticity remodeling, a modulation at the molecular genetic profile can be a major contributor to the overall alteration of the neuronal function. In order to understand brain function thoroughly, it is necessary to link functions from genomolecular to neuroreceptoral through to functional circuitry levels. Fortunately, as science and technology advance, tools have been developed to allow the investigation of brain activity at different functional levels. There are imaging methods, including fMRI and PET, for mapping the activity across the whole brain in response to a particular task. *In vivo* neurotransmitter activity can be assessed via invasive procedures such as microdialysis and cyclic voltametry, and noninvasively via pharmacological MRI (phMRI), as we will discuss below. Changes in the molecular expression are usually done *in vitro* or *ex vivo* via immunohistology or through gene expression techniques such as polymerase chain reaction (PCR). Furthermore, recent developments show that *in vivo* access of molecular expression can be done via MRI using superparamagnetic iron oxide (SPIO) nanoparticle labeled precursors.[10–14] In spite of all the state-of-the-art tools, linkage of knowledge is often lost or misguided by the segmented and isolated data.

A great approach is to use multimodal experiments to address the questions of *where*, *when*, and *how* the brain reacts to a particular task at different neuronal activity levels. fMRI or phMRI provides a macroscopic view of the whole brain and serves as a functional pinpointer (or identifier) With the advantages of noninvasive high spatial and temporal resolutions, fMRI and phMRI provide information of where in the brain and when along the time axis, in both acute and chronic conditions, the neuronal adaptations take place.

In the case that a pharmacological ligand is used to provoke neuronal activity, the MRI-measured hemodynamic change (either BOLD or rCBV changes) can serve as an indirect index

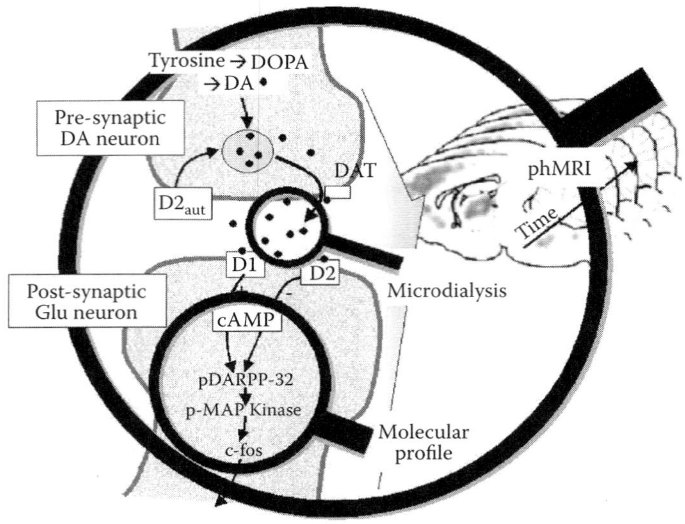

FIGURE 12.2 Integrated multimodal approach to probe an AMPH-induced neuronal response in the brain. phMRI provides a macroscopic picture of where in the brain and when along the time axis the cascading neuronal function (how) is taking place. The "where" feature also serves as a pointer to identify brain areas involved in a particular task. Based on a proposed theoretical model and acquired phMRI findings, a microscopic study is performed to evaluate the modulation of the responding neurotransmitter and molecular profile — this is a direct answer to how the brain reacts to a given task. In the example here, the activity of the presynaptic DA neurons is assessed by microdialysis measurement of DA release. Knowing that DA release is altered, the expected change in the postsynaptic neuron is assessed by probing the MAP kinase phosphorylation (p-MAPK), a marker of neuronal plasticity and survival, and Fos expression, a marker of neuronal activity. Combining integrated information obtained at the molecular and neurotransmitter levels with the gross neuroanatomical functional map will provide a fuller picture of the brain function.

to a specific neurotransmission function and provide us with the knowledge of how the brain works from focal neuronal innervation to an extended circuitry. As we will discuss in depth later, studying D-amphetamine (AMPH; a dopamine releaser) challenge with phMRI not only allows us to investigate the neuronal response at the site of dopaminergic innervation (such as caudate/putamen and nucleus accumbens), but also allows us to follow activation along the basal ganglia feedback loop, which includes downstream structures such as the thalamus and the neocortex.[6,15–17] Whole-brain mapping by imaging methods thus provides us with the opportunity to identify the responsible cerebral organs and circuits for conditions or disorders that are less understood. For other conditions or disorders that may already have a sound hypothesis regarding which circuitry is involved, phMRI can also help to deepen our understanding of when the neuroadaptation happens and what is the impact on the downstream structures. Once the active centers are identified by the imaging techniques, more fundamental mechanistic changes (how) have to be probed using a direct measurement. Take the AMPH challenge as an example again: phMRI first identifies the sites of interests to be the striatum, thalamus, and nucleus accumbens (NAc). Afterwards, the real neurotransmitter dynamics have to be assessed using well-established traditional standards, such as measuring the alteration in dopamine (DA) release using microdialysis or voltametry, measuring consequent postsynaptic changes by probing the mitogen-activated protein (MAP) kinase pathway, c-Fos and ΔfosB expression, and probing the modulation of the mRNA profile via the expression of the D1/D2 receptor. An example of such an integrated multimodal design is schematically illustrated in Figure 12.2. Combining integrated information obtained at the molecular and neurotransmitter levels with the gross neuroanatomical functional map provides a fuller picture of a complicated cerebral task.

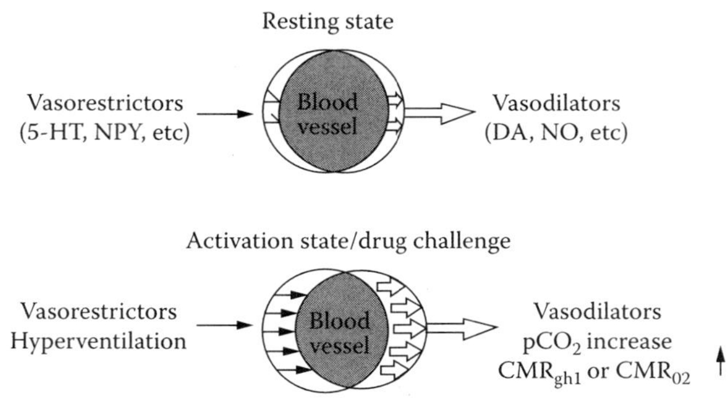

FIGURE 12.3 Schematic illustration of factors modulating the hemodynamic responses. Vasorestrictors such as serotonin (5-HT) and neuropeptide Y (NPY) act directly on vessels to decrease the vessel diameter, and vasodilators such as dopamine (DA) and NO dilate the blood vessel. At resting state, tension created between the vasorestrictors and vasodilators set the basal vessel tone. During an activation state, such as post-drug challenge or a functional task, the vessel tone or hemodynamic response is modulated by (1) increased availability of vasorestrictors or vasodilators, (2) change in respiration, such as hyperventilation and blood acidity, (3) neuronal activity such as glucose utilization and oxygen consumption, and (4) more complicated cardiac function impact, since the brain is capable of regulating blood flow in response to changes in blood pressure (blood flow autoregulation).

12.3 PHYSIOLOGICAL CONTRIBUTORS TO PHARMACOLOGICALLY INDUCED HEMODYNAMIC CHANGES

The pharmacologically induced hemodynamic responses (both BOLD and rCBV changes) can be coupled to vascular and neuronal activities. Precaution thus has to be taken in an attempt to interpret their physiological meanings. Figure 12.3 illustrates factors contributing to hemodynamic tone at resting and activation states.

12.3.1 LIGAND–VASCULAR COUPLING

Major contributors to the ligand–vascular couplings are (1) direct coupling on the endothelium, (2) global cardiac modulation, and (3) global respiratory modulation. The first class of ligands includes drugs that are capable of dilating or constricting endothelium directly. For example, nitric oxide (NO) and acetylcholine are classical endothelium vasodilators, while serotonin (5-HT), neuropeptide Y (NPY), antihistamines, and epinephrine are vasoconstrictors. This class of ligands usually generates a hemodynamic response that overestimates the neuronal activity if the direct vascular contribution is not accounted for. It is a great challenge to separate the direct vascular contribution from the neuronal contribution using BOLD or rCBV techniques. Unlike positron emission tomography (PET), which is sensitive to the radioactive substance with its site-/receptor-specific binding potential (after correction of the blood residue and signals from known nonspecific binding sites, such as the cerebellum), the phMRI technique is sensitive not only at the sites of ligand action, but also at its downstream structures. In other words, it is hard to isolate a brain area that is a pure nonspecific acting site in phMRI as the nonspecific binding site in PET. For example, PET imaging with ^{11}C-labeled 2-beta-carbomethoxy-3-beta-(4-fluorophenyl)tropane (^{11}C-CFT), a dopamine transporter protein blocker, shows a focal uptake in the striatum. The phMRI detects hemodynamic changes with CFT challenge covering broader brain areas.[6,16] Although CFT induces local synaptic dopamine increases at the dopaminergic terminals (i.e., striatum), the increased dopamine concentration triggers cascading neuronal responses in the downstream circuitry, including the hypothalamus, thalamus, and neocortex. The neuronal activation at the downstream structures involves function of many other neurotransmitters, such as glutamate and GABA. In brain regions where there may not be a large release, or concentration of the neurotransmitter of interest, it becomes more difficult to determine which of the many potential substances lead to the vascular response.

The second class of ligands includes drugs that alter cardiac function and induce changes in heart rate and blood pressure. The cardiac influence is global, and its hemodynamic contribution can be corrected by deconvolving the hemodynamic time course from the blood pressure curve or by independent component analysis (ICA). However, a prolonged blood pressure depression or elevation may compromise the neuronal function, and an estimate of the receptor function under this condition may no longer be reliable, even after a correction for the cardiac modulation.

The third class of ligands includes drugs that induce hyper- or hypoventilation. Hyper- or hypoventilation changes the blood acidity and thus alters blood flow and blood volume globally. Recordings of respiratory rate and end tidal or blood CO_2 level help to determine the blood acidity contribution to the overall hemodynamic changes. If a ligand typically induces hyperventilation, it is likely to induce significant rCBV decreases in most of the brain. This decreased rCBV time course correlates to end tidal CO_2 and blood CO_2 curves and will not be a real representation for the neuronal function induced by the ligand. The hemodynamic effect induced by hyper- or hypoventilation is usually very strong and masks the neuronal effect. However, a focal rCBV increase under the condition of hyperventilation likely represents an underestimation of a true neuronal activity that is buried under the hyperventilation-induced global rCBV decrease. A similar consideration could be made for hypoventilation. The respiratory side effect can be minimized by mechanically ventilating the animals. For example, cocaine challenge typically induces rCBV increases in free-breathing Sprague-Dawley rats. However, some Sprague-Dawley rats hyperventilate after cocaine challenge, and thus a global rCBV decrease in the brain is observed (Figure 12.4). If hyperventilation is not taken into account, the rCBV response curve can be mistakenly interpreted as inhibitory neuronal action.

12.3.2 Ligand–Neuronal Coupling: The Dopaminergic Example

After taking the vascular contributions into consideration, the pharmacologically induced hemodynamic responses can be linked to multiple neuronal events. Major neuronal contributors include (1) presynaptic neuronal activity, (2) postsynaptic neuronal activity, (3) coupling to other neurotransmitters, and (4) neuronal activity-driven vasodilation or vasoconstriction. Taking the AMPH challenge as an example, there is a massive increase of the synaptic dopamine concentration at the dopaminergic terminals. However, the massive dopamine release is actually the consequence and precursor of a series of neuronal events. AMPH binds to a Na^+/Cl^--dependent dopamine transporter protein (DAT) on the presynaptic dopaminergic neuron and subsequently internalizes DAT[18,19] and stimulates a DAT-mediated DA efflux[18] (detailed review in Kitayama and Sogawa[20]). Up to this stage, the amount of dopamine released represents the presynaptic neuronal activity indirectly. The synaptic dopamine then innervates various targets, leading to different consequences:

1. Binding to the dopamine autoreceptor (DAR) on the presynaptic DA terminals to regulate dopamine release
2. Binding to the dopamine receptor on the postsynaptic neurons to regulate signal transduction and interaction with other neurotransmitters
3. Binding to DAT and recycling back to the dopaminergic neuron
4. Diffusion to neighboring blood vessels or glial cells, leading to direct impact on the vessel regulation

Figure 12.5 illustrates neuronal events contributing to hemodynamic changes. Since similar mechanisms can be applied to other neurotransmitters, we will discuss the dopamine-induced neuronal consequences in detail.

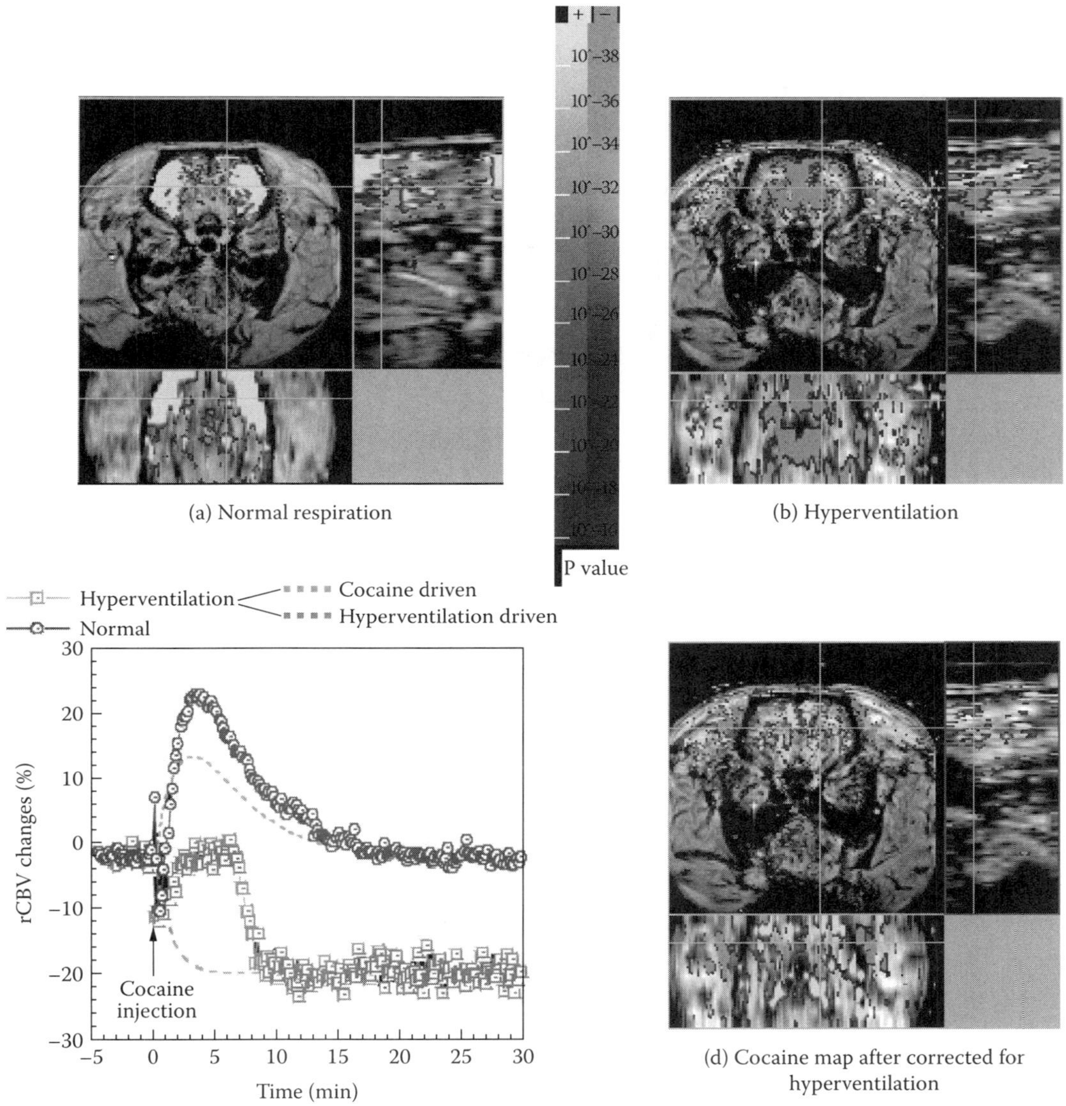

FIGURE 12.4 (please see color insert) The rCBV response to cocaine under normal respiration and hyperventilation. (a) When the animal breathed normally, the cocaine challenge induced rCBV increases in selected brain areas. (b) When the animal was hyperventilated, a global rCBV decrease over the whole brain was expected. The cocaine-induced rCBV increase was counteracted by the hyperventilation-induced rCBV decrease, thus giving the appearance of negative or no response. (c) Striatal rCBV time courses from normal breathing and hyperventilated rats. The rCBV time course from a hyperventilated rat can be decomposed into a cocaine-driven component and a hyperventilation-driven component. (d) After correction for hyperventilation, a relatively normal cocaine map was restored, although it is still not 100% when compared to a normal cocaine map, as shown in (a).

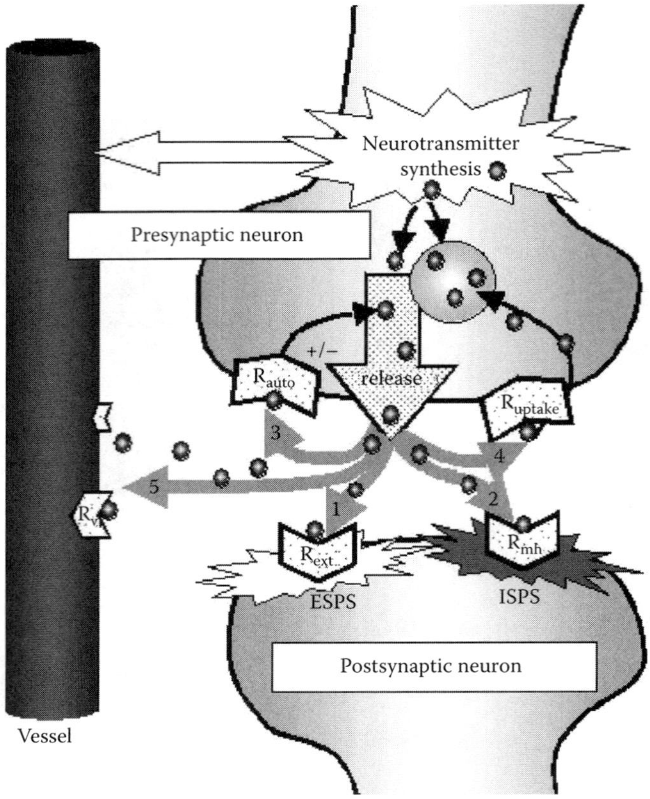

FIGURE 12.5 A schematic illustration of neuronal events contributing to hemodynamic changes. Presynaptic activities such as neurotransmitter synthesis and release are directly coupled to glucose utilization and oxygen consumption, and thus the hemodynamic response. Once the neurotransmitter is released into the synapse, it can (1) bind to an excitatory receptor (R_{ext}) and induce (such as glutamate) or facilitate (such as the DA D1 receptor) an ESPS, (2) bind to an inhibitory receptor (R_{inh}) to induce (such as GABA) or facilitate (such as the DA D2 receptor) an ISPS, (3) bind to an autoreceptor (R_{auto}) to regulate release, (4) bind to an uptake transporter to be recycled back to the presynaptic neuron, or (5) diffuse to a neighboring vessel and dilate or restrict the vessel via glia cells or vessel-bound receptors.

12.3.2.1 Presynaptic Autoreceptor Regulation

Dopamine release is locally modulated by the autoreceptor (a receptor that is sensitive to the neurotransmitter released by its own neuron) located on the presynaptic membrane. The DAR belongs to the D2 receptor family,[21] which is part of the superfamily of G-protein-coupled receptors, and functions through regulating potassium[22,23] and calcium[24–26] currents. Agonizing the axonal dopamine autoreceptor (DAR) suppresses dopamine release, while antagonizing the DAR potentiates dopamine release. We will later discuss how to map the DAR function *in vivo* (see Section 12.4.3).

12.3.2.2 Postsynaptic Function

Dopaminergic receptors can be subdivided into two families: (1) the D1-like family consists of D1 and D5 receptors, which can modulate excitatory postsynaptic potentials (ESPSs), and (2) the D2-like family consists of D2, D3, and D4 receptors, which can modulate inhibitory postsynaptic potentials (ISPSs). Both D1 and D2 receptors are coupled to calcium currents and the function of cyclic adenosine monophosphate (cAMP), which subsequently leads to the phosphorylation/dephosphorylation of

DARPP-32 (dopamine and cAMP-regulated phosphoprotein [see detailed review in Greengard[9]]). The phosphorylative status of DARPP-32 is an important signal transduction factor regulating activity of many receptors, enzymes, ion channels, and transcription factors,[9] including the MAP kinase pathway[27] — an important mediator of neuronal plasticity,[28–30] growth,[31,32] and survival.[33,34] Increase of synaptic DA concentrations induced by AMPH or cocaine can activate phosphorylation of MAP kinases[35–38] and subsequently lead to activation of proteins such as CREB (cAMP response element-binding protein), c-Fos, and ΔfosB — all are important mediators of psychostimulant-induced plasticity.[39–41] Thus, the regulation of MAP kinase phosphorylation serves as an ideal marker for neuronal activation and dendritic branching and morphology.

The c-*fos* gene is an immediate early gene (IEG) that is activated by multiple stimuli, including psychostimulants.[42–46] It is a transcription factor that regulates expression of other neuronal genes that lead to alterations in neuronal function. Acute AMPH-induced activation of c-Fos has been studied extensively in the nucleus accumbens (NAc), medial prefrontal cortex (mPFC),[47–51] and striatum (str) in the brain. Even though c-Fos expression is transient and its function in plasticity is unknown, it serves as an ideal marker of psychostimulant-mediated changes in dopamine signaling.[42–46] ΔFosB belongs to the *fos* family of proteins that has been found to be activated by recurrent psychostimulants but, unlike the IEG c-Fos, remains upregulated for long periods and has been found to correlate with sensitized behavior.[41,52,53] Examining the activation of the MAP kinase pathway, c-Fos and ΔfosB expression allow us to correlate the AMPH-induced dopamine release and functional adaptations of brain circuitry. In addition, these molecular markers (MAP kinase, c-Fos, and ΔfosB) allow us to follow the progression of functional neuroanatomical and neurochemical changes during the course of AMPH-induced sensitization and plasticity in the brain. Regulation of the DARPP-32 phosphorylation status and its subsequent MAP kinase pathway is not unique to dopamine innervation, but is also carried out by various neuotransmitters, including glutamate, adenosine, serotonin, and GABA. As illustrated in Figure 2 of the review article from Greengard,[9] activity of cAMP, DARPP-32, and calcium current serves as a local funnel point, with various neuroreceptors acting synergetically to fine-tune the phosphorylation status of DARPP-32, and thus its downstream signal transduction. Thus, similar mechanisms can be applied and considered when studying the neuronal consequences induced by neurotransmitters other than dopamine.

As we discussed earlier in Section 12.2, fMRI and phMRI can be used as location pointers to identify brain regions responsible for performing certain tasks or challenges; phMRI can also reverse the process to investigate the overall function after a molecular mechanism is identified. For example, it is possible to selectively inhibit or excite a particular molecular reaction via a specific pharmacological agent, and its impact on the overall function can be probed by phMRI.

12.3.2.3 Receptor-Mediated Vascular Regulation

Instead of being recycled back to the dopaminergic neuron through the dopamine transporter protein (DAT), a majority of DA diffuses to neighboring glial cells and vessels.[54–57] Electron microscopic studies show that a subset of cortical interneurons are surrounded by astrocytic processes that strongly express D2 receptors.[58,59] Agonizing the astrocyte-bound D2 receptors elevates Ca^{2+} flow,[59] which in turn can regulate the blood vessel via the end feet of the astrocyte onto blood vessels.[60–63] On the other hand, immunocytochemistry and Western blots on isolated cerebral microvessels show expression of D1 and D5 receptor proteins, while the D2 receptor family is expressed in the adventia and adventia–media border.[64,65] The location segregating protein expression of the D1 and D2 receptor families suggests that dopamine can lead to vasodilation via the vessel-bound D1 receptor family and vasoconstriction via the glia-bound D2 receptor family, and thus rCBV change.[60–62,65] Although this kind of neurotransmitter–vessel coupling does not appear to be a true neuronal activity, such coupling is actually driven by the availability of free dopamine in the extracellular space and is indirectly linked to its presynaptic neuronal activity.

12.4 USING PHARMACOLOGICAL MRI TO PROBE THE DOPAMINERGIC FUNCTION

Dopamine is an important neuromodulator in the brain. The fine balance between its excitatory (D1-like receptors) and inhibitory (D2-like receptors) innervation on the postsynaptic neurons controls the initiation and fine-tuning of many brain functions, such as cognitive, emotional, and motor controls. In the brain, dopaminergic cells are located in the substantia nigra pars compacta, ventral tegmental areas, and arcuate nucleus of the mediobasal hypothalamus. There are three major dopamine pathways with the dopaminergic axons projecting to the striatum (the nigralstrial pathway), the prefrontal and basal forebrain (mesolimbic pathway), and the median eminence (the tuberoinfundibular pathway). Many diseases, including cocaine addiction, Parkinson's disease, and schizophrenia, are linked to the disruption of the dopaminergic function along those three pathways. In this section, we will discuss how to use phMRI to probe the function of the dopaminergic pathways in the brain.

12.4.1 D-Amphetamine (AMPH)

The ability to release a massive amount of DA in the brain makes D-amphetamine (AMPH) a good agent to probe the functional status of the dopaminergic neurons (see a more detailed review of mechanism in Section 12.3). In naïve rats, a behavioral dose of AMPH (1 ~ 3 mg/kg iv or ip) increases the synaptic DA concentration two- to tenfold in the areas of dopaminergic innervation (i.e., caudate/putamen (CPu) and nucleus accumbens (NAc)). The dramatic increase of synaptic DA concentration by AMPH drives the pharmacological-neurochemical-neurovascular consequences into a large-scale response, providing a much greater contrast to the resting state, and thus gives us a magnified means to evaluate the DA function. In naïve rats and monkeys, behavioral doses of AMPH lead to BOLD/rCBV increases in the CPu, NAc, thalamus, and various cortical areas — including both the DA terminals (CPu and NAc) and downstream structures (thalamus and cortices) along the basal ganglia circuit.[15,16,65] Simultaneous recordings of blood pCO_2 and blood pressure do not show a good correlation with the BOLD/rCBV curves, indicating that the cardiac/respiratory components are not major contributors to the hemodynamic changes. In addition, the BOLD/rCBV time courses in CPu and NAc showed a good correlation to the microdialysis measurements of DA release. This suggests that the BOLD/rCBV response largely reflects the synaptic DA concentration — a consequence from the activation of the dopaminergic neurons.

To further validate the coupling between the hemodynamic response and DA activity, striatal DA function was first eliminated by lesioning the nigrostriatal DA projections and was later restored by transplanting dopaminergic fetal or stem cells back to the nonfunctional CPu. The DA function can be abolished unilaterally by injecting 6-hydroxy-dopamine (6-OHDA or MPTP (1-methyl-4-phenyl-1,2,3,6-tetrahydropyridine)). Animals pretreated with desipramine were injected with 6-OHDA unilaterally into the ipsilateral medial forebrain bundle or with MPTP globally by systematical injection —models routinely used for Parkinson's disease in rodents (6-OHDA) and primates (MPTP). AMPH challenge of the 6-OHDA lesioned rats showed a normal BOLD/rCBV response in the contralateral hemisphere but a near absence of hemodynamic response in the hemisphere ipsilateral to the lesioned side.[15,16] The imbalance of DA innervation in the two hemispheres upon AMPH challenge also showed up behaviorally as animals kept turning toward the ipsilateral side. The behavioral and DA function deficits were later restored by transplanting dopaminergic fetal or stem cells into the ipsilateral striatum. In addition, the AMPH-induced hemodynamic response was restored at the graft site and also in the cortex.[15,66] The successful grafting of DA cells was validated by tyrosine–hydroxylase (TH) staining with evidence of neuronal sprouting toward cortical areas in postmortem brains.[15,66] Although TH staining proved the existence of DA neurons, the return of the phMRI signal suggested that these DA neurons were capable of releasing DA and reconnecting the cortical linkage, which was either lost or interrupted due to the lesioning. This kind of functional evaluation is hard to measure by other techniques that only provide static information. Similar experiments to measure DA functional loss were observed in Parkinsonian monkeys with MPTP lesioning and amphetamine challenge.

12.4.2 Using Receptor Ligand to Monitor Changes in Receptor Density

The expression of receptors in the brain is an innervation-dependent dynamic process. Overstimulating a receptor often leads to a downregulation, while understimulation leads to an upregulation of receptor expression. The receptor upregulation and downregulation are often observed in neuronal disorders relative to long-term modulation of the receptor innervation. For example, Parkinson's disease (PD) is due to a progressive loss of dopaminergic cells. At the early stages of PD, although the dopamine neuron population is reduced, the dopaminergic function is largely preserved because the remaining dopaminergic neurons are able to compensate for the loss by releasing more DA. However, at later stages of PD, the remaining dopamine neurons are not able to release enough DA to maintain a normal dopaminergic function and the DA receptors will not be able to get sufficient innervation. The inefficient innervation prompts the neuron to generate more DA receptors — this phenomenon is called receptor supersensitivity. The receptor supersensitivity can be easily detected by phMRI. For example, apomorphine induced a stronger rCBV increase in the CPu ipsilateral to the 6-OHDA lesioning in PD rats.[17] Although supersensitivity can also be detected by PET, the phMRI approach of not involving radioactivity may make it a useful diagnostic tool. Long-term modulation of receptor density may also be the cause or consequence of many other disorders, such as drug addiction and schizophrenia. The ability to detect receptor expression will help in understanding better the natural course of a disease and in evaluating the therapeutic efficacy. For example, it is proposed that the D3 receptor is associated with dyskinesia, although it is not clear that changes in D3 receptor expression are the consequence or cause for the disorder. Nonetheless, the hypothesis of D3 to link to dyskinesia can be tested *in vivo* by performing phMRI with D3 ligands. Furthermore, if dyskinesia is therapeutically reversible, the hemodynamic changes resulting from the D3 ligand challenge can be used as an indicator for the therapeutic efficacy.

12.4.3 Mapping Dopamine Autoreceptor (DAR) Function

The dopamine autoreceptor (DAR) is a major modulator regulating dopamine release from the dopaminergic dendrite and terminals on presynaptic neurons. Antagonism of the DAR enhances DA release, while agonism of the DAR attenuates DA release. So far, determination of autoreceptor density on presynaptic neurons has not been attained. However, even if one were able to determine the DAR density on the presynaptic neuron using autoradiography or immunohistological techniques, this static information would not be able to address the question of how the DAR dynamically regulates dopamine release. With the emerging *in vivo* imaging techniques, mapping of the DAR function can be realized using PET and phMRI. There are major issues to be resolved using ligand labeling techniques such as PET to investigate DAR function. Since the protein structure of the DAR is indifferent or similar to the postsynaptic D2/D3 receptors, a DAR-specific ligand is not available so far. And a general D2/D3 ligand will not be able to distinguish the receptor binding pre- or postsynaptically.

Second, even if a DAR-specific PET ligand is available, the receptor-binding technique again only provides static information, such as the availability or receptor density of DAR. However, the degree of dopamine release can be estimated using a displacement approach[67–70] in which the endogenous dopamine competes with the radiolabeled ligands — greater reduction of the measured binding potential indicates a greater degree of dopamine release (more radiolabeled ligands have been displaced by the endogenous dopamine). The DAR function can also be probed by using phMRI.[71] Although there is no DAR-specific ligand, most of the D2 ligands have greater affinity

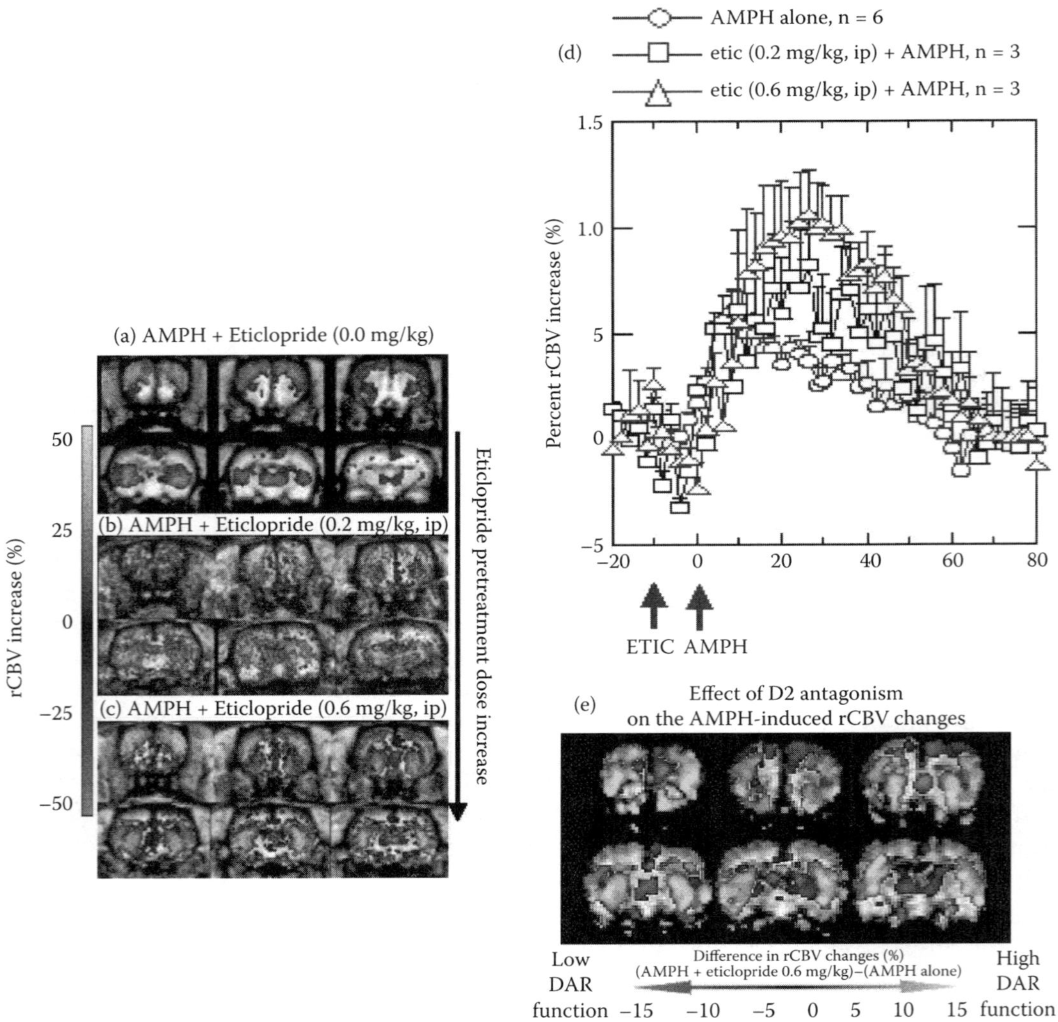

FIGURE 12.6 (please see color insert) Probe function of dopamine autoreceptor (DAR) function by combining AMPH with different doses of the D2 antagonist eticlopride operating at a low-dose regimen. Eticlopride lead to a dose-dependent increase of rCBV in the caudate/putamen and nucleus accumbens (a–d), which is parallel to the microdialysis measurement in NAc, as shown in Chen et al.[71] The effect of DAR function was demonstrated by subtracting the rCBV map of AMPH plus eticlopride from the rCBV map of AMPH alone, as shown in (e). A stronger enhancement of rCBV increase (toward yellow-red tone) indicated a stronger DA release and stronger antagonism of the DAR function.

to DAR than to the postsynaptic D2 receptor.[72–77] In other words, a low behavioral dose of a D2 ligand acts mostly on the presynaptic DAR rather than on the postsynaptic D2 receptor. Evidence of the dose-dependent hemodynamic changes of the D2 ligands was shown in Chen et al.,[71] reporting that a higher dose of eticlopride decreased rCBV in the caudate/putamen (CPu) and nucleus accumbens (NAc), while a lower dose of eticlopride induced rCBV increases in those areas (Figure 12.6). When operating at low dose range, a D2 antagonist such as eticlopride will enhance the amphetamine-induced dopamine release in a dose-dependent manner. Indeed, at a low behavioral dose range, eticlopride potentiated the AMPH-induced dopamine release, and hence we observed a potentiation of the corresponding rCBV increase in a dose-dependent manner (see Figure 12.6 and Chen et al.[71]). Since the DAR function may be altered in various disorders, including cocaine addiction and schizophrenia, the phMRI measured dose–response curve of the DAR function can help to evaluate the status of the dopaminergic system, and thus the efficacy to a specific therapy.

12.4.4 Interaction of the Function between the Dopamine D2 Receptor and the Adenosine A2A Receptor

The brain functions through highly complicated neuronal networks mediated synergetically or disergically by multiple neurotransmitters. As reviewed in the Section 12.3.2.2 and in Greengard,[9] almost no single neurotransmitter function works alone, as they are typically coupled to the function of many other neurotransmitters. Traditionally, the neurotransmitter coupling effect is examined *in vitro* on cell culture or *in vivo* using invasive techniques such as microdialysis probing. *In vitro* culture is under an artificial environment that often lacks the complicated neuronal innervations of an *in vivo* setting. An *in vivo* invasive procedure such as inserting the intracranial microdialysis probe usually causes damage to the tissue. The damage to the tissue may not be 100% recoverable even after a long recovery time is allowed, and the results of the measurements are often offset by such damage and disruption. Thus, the *in vivo* and noninvasive approach using phMRI is an ideal setting for studying the neuronal coupling between two selected neurotransmitters.

We take the example of using phMRI to study the interaction between the adenosinergic and dopaminergic systems. It has been shown that adenosine alters the dopaminergic release and function in the brain.[78–81] Adenosine A2A receptor works synergetically with the D1 receptor and disergically with the D2 receptor on the regulation of DARPP-32,[82–84] expression of enkephalin and immediate early genes (IEGs),[84,85] and behavioral activity.[86,87] Specifically, A2A receptors and dopamine D2 receptors are densely co-localized in the caudate/putamen (CPu), nucleus accumbens (NAc), and olfactory tubercule.[79–81] Thus, agonizing or antagonizing the A2A receptor produces effects similar to those via antagonizing or agonizing the D2 receptor, respectively.

Indeed, phMRI with an A2A antagonist 3,7-dimethyl-1-propargylaxanthine (DMPX) challenge induced rCBV decrease in CPu and NAc, the areas with a dense population of A2A receptor (Figure 12.7),[88] similar to the rCBV changes induced by a D2 agonist.[71] The effect of the A2A antagonist on the dopaminergic function can be investigated by DMPX post-treatment on AMPH pretreatment animals (20-min gap between the two treatments). Post-treatment of DMPX attenuated the AMPH-induced rCBV increase in the A2A receptor populated areas and also in thalamus, cingulate cortex, and frontal cortex — areas not populated with A2A receptors, but downstream structures along the basal ganglia circuitry.[88] The DMPX-induced rCBV modulation in CPu was parallel to the microdialysis measurement of DA release but disassociated with the global cardiac and respiratory function.[88]

This example shows that phMRI can be used to monitor the interaction between neurotransmitter systems at the molecular (i.e., as a dopamine release indicator) and the neurocircuitry (i.e., detecting functional changes in downstream structures) levels. Similar *in vivo* approaches can be used to study functional coupling between many neurotransmitter systems, such as GABA, DA, glutamate, serotonin, etc.

12.4.5 Alteration of the Functional Circuitry: Drug Addiction

The brain does not work focally at a single node but through a complicated network. To perform a task, the neuronal signal is rectified and relayed from one neuron to another, forming a neuronal circuit that allows information to flow and facilitates communication among many structures. Functioning of the brain circuitry is mainly controlled by the synaptic innervation that converges, gates, justifies, and initiates the relay of the neuronal signals. The alteration of innervation at a local synapse usually leads to an alteration that propagates throughout the entire circuit. In some cases, long-term alteration of synaptic activity leads to functional adaptation, sometimes including changes of structures and routes involved in the circuitry. This phenomenon is observed in many

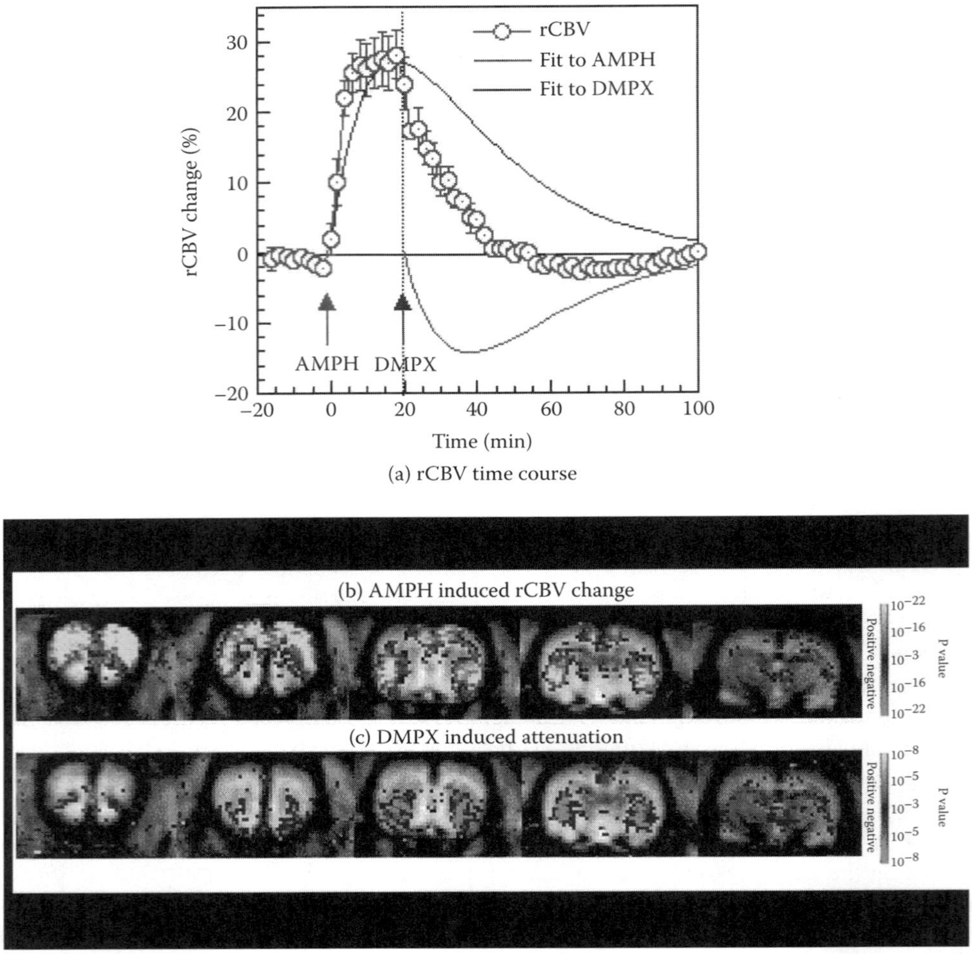

FIGURE 12.7 (please see color insert) Effect of A2A antagonist DMPX on the AMPH-induced rCBV changes. The raw rCBV time course can be fit to AMPH and DMPX components using the gamma function (a). Maps for the AMPH and DMPX components are shown in (b) and (c). (c) Post-treatment of DMPX attenuated the AMPH-induced rCBV increase in CPu and NAc, the areas with a dense population of the A2A receptor.

disorders, such as drug addiction. For example, the dopaminergic and glutamatergic projections to the striatum have contacts mainly on medium-sized GABAergic projection neurons and the aspiny dentrites of the striatal interneurons. The close proximity of the glutamatergic and dopaminergic inputs to the striatal interneurons suggests that the functional interaction and modulation of these two neurotransmitters occur both pre- and postsynaptically. In other words, normal striatal function is based and maintained by the tension and balance between these two neurotransmitters. Consequently, the striatal output has major influences on the function of other structures along the basal ganglia loop. Such tension and balance created by the influence of Glu and DA in the striatum is disturbed when one component is repeatedly and excessively stimulated or suppressed. For example, after being repeatedly exposed to psychostimulants such as cocaine or AMPH, a dose of psychostimulant at a later time tends to induce a greater DA release and locomotor activity — a phenomenon called sensitization. The sensitization is thought to be partly due to changes in the glutamatergic regulation on the DA release.[89–91] Genomolecular studies showed that cortico-striatal glutamate input modulates the ability of AMPH to engage striatal neurons through the

MAP kinase pathway,[92,93] which subsequently changes the mRNA expression for c-FOS, ΔfosB, and dopaminergic and glutamatergic receptors.[36,89,94–99] In other words, repeated exposure to psychostimulants leads to alteration of neurotransmitter transmission along the functional circuitry and alteration of the neuronal status (receptor expression and neurotransmitter synthesis, etc.) via changes in gene expression.

In the case of disorders linked to the dopaminergic system, phMRI has the potential to detect the changes in the neurotransmitter transmission. For example, phMRI with acute AMPH challenge on previously AMPH treated rats showed significant rCBV increases in CPu, NAc, mPFC, and thalamus.[100] The high-spatial-resolution phMRI map allows us to divide the major structures into subunits — NAc into shell and core portions and mPFC into cingulate 1, cingulate 2, infralimbic cortex (IL), and prelimbic cortex (PrL) portions.[101] The rCBV responses from those subunits can be evaluated independently and provide further information on how neuroadaptation occurs along this VTA-mPFC-NAc circuitry. As mentioned previously in Section 12.2 and in Figure 12.2, the detailed spatial mapping using fMRI helps to determine the sites in which to perform microdialysis and gene profile measurements. Indeed, we found a significant increase of ΔfosB expression in mPFC and NAc (core > shell). In this manner, phMRI indirectly reflects a view of the molecular changes within the neurons. In addition to the spatial information, the phMRI also adds information in the time domain. For instance, the rCBV increase in CPu was much shorter in time in the AMPH-pretreated rats than in the saline-pretreated rats. Such dynamic adjustments may not be capable of being captured by static techniques such as immunohistology or techniques requiring long averaging times, such as PET and microdialysis.

12.5 WHICH IS THE BETTER MODEL FOR STUDYING THE NEUROTRANSMITTER FUNCTION — AWAKE VS. ANESTHETIZED PROTOCOL?

Traditionally, phMRI studies are carried out on anesthetized animals. The advantage of using an anesthetized animal is the ease of controlling motion and physiological parameters, such as blood CO_2, cardiac cycles, and stress. The drawback is the unknown coupling and compromise of the neuronal function due to the anesthetics. A study performed on awake animals simulates the conscious condition in human studies by avoiding the interference of anesthetics in the brain. However, neuropsychological factors such as motivation and stress are harder to control and can present a different set of confounds — especially in the context of a pharmacologic stimulus. Table 12.1 lists observations encountered when performing phMRI in awake and anesthetized animals.

TABLE 12.1
Comparison of ph/fMRI Studies in Awake and Anesthetized Animals

Observations	Awake		Anesthetized	
Alteration of selected neurotransmitter function	Presumably normal[a]	✓[b]	Yes; choice of anesthetics to use is critical	
Baseline CBF/CBV	Presumably normal	✓	Usually altered	
Baseline CMRglu	Presumably normal	✓	Usually depressed	
Motion movement	Hard to control and correct		Well controlled	✓
Stress level	Potentially high		Low to none	✓
Training	Requires extensive training sessions		None	✓
Respiratory and blood gas regulation	Hard to control without paralyzation		Controlled by ventilation	✓

[a] Many awake protocols utilize an initial anesthesia to place the animal in the magnet. It is hard to evaluate if the animal has fully recovered from anesthesia.[106,109,112] Stress can also alter neurotransmitter function.

[b] Indicates an advantage to this approach.

12.5.1 Awake Animal Model

fMRI on awake behaving monkeys[102–105] and rats[106–112] has been demonstrated in the past several years. Two of the most important issues to be addressed in awake animal studies are the control of motion and stress. Worse than in human studies, the fMRI setting induces tremendous stress on awake behaving animals and often generates severe body and head movement. Thus, a successful awake animal study calls for a good motion restriction system and an extensive training period. A highly intelligent nonhuman primate such as a rhesus macaque can be trained for eye fixation using a reward to minimize the stress and head motion.[102–105] Such a task is more challenging in cynomologous macaques, squirrel monkeys, and rodents. In rats, stress induced by acute or chronic immobilization increases multiple gene encodings, including the epinephrine-synthesizing enzyme phenylethanolamine N-methyltransferase (PNMT), which stimulates epinephrine production.[113] In addition, stress is known to induce sustained high levels of dopamine,[114–121] cortisol,[122] vasopressin,[123–127] endorphins,[128] and neurotensin.[129–132] The complicated neurological effects induced by stress are a problem that cannot be ignored, particularly when the targeted neurotransmitter system for the designated phMRI study is greatly modulated by stress. For example, cocaine and immobilization stress have similar effects on the levels of heat shock proteins and stress-activated protein kinases in the rat hippocampus.[121] Furthermore, cocaine-seeking behavior can be reinstated by foot shock stress in rats that previously self-administered cocaine,[133] and the stress-induced reinstatement can be blocked by antagonizing the dopamine D3 receptor.[134]

In addition to stress, emotion and expectancy are important neuropsychological factors to be considered. As mentioned earlier, awake animal studies are usually achieved via a reward training paradigm in which the animal receives a food reward when it performs a designated task correctly. Dopaminergic neurons are responsible for motivation, expectation for reward, and correction for reward expectation errors.[135–139] The buildup of striatal dopamine activity during the period of reward expectancy[135,140] indicates changes in baseline conditions of the dopaminergic system and may potentially compromise the dopaminergic function one would like to investigate using an awake model.

12.5.2 Anesthetized Animal Model

The advantages of using anesthetized animals are the easiness to control movement and physiological parameters. In addition, the anesthetized animals allow the study of specific neuroreceptor function without the interference of the emotional and stress components. For example, when animals have been previously exposed to psychostimulants such as cocaine and AMPH, changes in DA receptor densities and DA release capacity often occur due to the drug-mediated chronic excess in the innervation of the dopaminergic system. In addition, the emotional components, including euphoria and craving, while possibly a consequence of the altered receptor density, can change the degree of neurotransmitter release, and thus the hemodynamic response. The complicated emotional factors and contributions can be excluded using anesthetized animals and allow us to study the static neuroadaptation of drug addiction.

Although the neuronal mechanisms for most of the generally used anesthetics are not well understood, the choice of anesthetics in fMRI studies usually depends on the field experience based on the particular task to be investigated. For example, halothane is typically used for the monitoring of the dopaminergic function,[4,6,15–17,71,141,142] while α-chloralose is used for detecting the somatosensory response, such as forepaw stimulation.[7,8,143–147] The choice of anesthetics can have a great and unexpected impact on the hemodynamics observed using MRI. Dopaminergic ligands such as cocaine or AMPH induced robust BOLD and rCBV changes under halothane anesthesia,[4,6,15–17,71,141,142] and this hemodynamic response was gradually abolished when halothane was switched to α-chloralose anesthesia (Mandeville, unpublished data). This phenomenon is partly due to the fact that α-chloralose blocks dopamine release, as shown in a 3H-dopamine labeling study.[148]

On the other hand, forepaw stimulation generated robust BOLD and rCBV responses[5,7,142,143,149] in the sensorimotor cortex only with the animals under α-chloralose, and not those under halothane anesthesia (Mandeville, unpublished data). Although α-chloralose anesthesia preserved the hemodynamic response in the cortical area, fewer BOLD/rCBV responses were observed in subcortical areas, including the thalamus[5,7,142,143,149] and basal ganglia. Although the choice of α-chloralose or halothane seems to be largely experimental, there is some scientific basis for the choice over particular anesthetics. α-Chloralose is widely used as a laboratory anesthetic due to its minimal effects on autonomic and cardiovascular systems. However, the mechanism of α-chloralose is less clear, with limited data showing that α-chloralose acts through the GABA(A) receptor.[150]

Volatile anesthetics such as halothane and isoflurane in general act by decreasing the extent of gap junction-mediated cell–cell coupling and by altering the action potential activity-associated channels.[151] Halothane inhibits the synaptic transmission partly via prolonging the GABA(A) receptor-mediated IPSP[152,153] and blocking the presynaptic glutamate release and the postsynaptic AMPA receptors.[154] Deep electrode recordings of the EEG waves in rats showed that halothane decreased the median frequency wave but increased the low-frequency delta wave,[155] which indicates a decrease in basal neuronal activity. Compared to halothane anesthesia, EEG waves under α-chloralose anesthesia had increased median frequency waves.[155] Note that isoflurane may not be a better choice since it suppressed EEG waves to a greater degree than halothane did in rats.[156] And deep isoflurane anesthesia induced an EEG burst suppression that was absent from halothane anesthesia.[156] The inhibition of the synaptic transmission by halothane may be the responsible factor for blocking the relays of sensorimotor signal from forepaw to the central nerve system. Conversely, under halothane anesthesia, the degree of hind paw twitches induced by direct stimulation on the hind paw motor cortex was progressively weakened as the dose of halothane was increased.[155]

Evidence from both the forepaw and direct motor cortex stimulation suggests that both the ascending and descending synaptic transmissions were blocked by halothane anesthesia. When the animal was under α-chloralose anesthesia, signal transmissions along both the ascending and descending sensorimotor pathways were preserved.[155] Interestingly, direct cortical stimulation under halothane anesthesia generated a consistent BOLD signal in the ipsilateral motor cortex, and the degree of BOLD response was independent of halothane dose,[155] indicating that the observed BOLD signal was not synaptic transmission dependent. Similarly, when an animal was challenged with a dopaminergic ligand under halothane anesthesia, the local hemodynamic change was sensitive to local dopamine release and neuroreceptor innervation, and thus the synaptic transmission might play a lesser role in the hemodynamic contribution (see discussion in Section 12.3).

In conclusion, although there are many anesthetics available for laboratory use, the impacts of anesthetics on the neurological function are far from known. The decision of which anesthetics to use in an fMRI study should be based on the known physiological parameters, and one must consult similar protocols previously used for successful outcomes.

12.6 CONCLUSION

fMRI has turned out to be a great tool for investigation of brain function. However, its hemodynamic basis most accurately reflects neuronal activity at the level of changes in metabolism due to alterations in blood flow/volume and oxygen consumption. However, phMRI lacks detailed information on molecular events at the neuronal level. Under favorable circumstances, the indirect hemodynamic response can be utilized to reflect specific neurotransmitter function by using pharmacological ligands that selectively agonize or inhibit specific neurotransmitter systems. We have presented data above, specific for the dopamine system, demonstrating that there are numerous molecular events (such as dopamine cell loss, receptor upregulation and modulation) amenable to analysis using phMRI techniques. Clearly this is a field in its infancy, and future experiments can be expected to help ascertain the possibilities and limitations of this methodology. The existing literature shows that even now, although it still relies on hemodynamic measurements, phMRI can be a powerful molecular imaging tool for examination of brain function at the neurotransmitter level.

REFERENCES

1. Shen, T., Weissleder, R., Papisov, M., Bogdanov, A., Jr., and Brady, T.J., Monocrystalline iron oxide nanocompounds (MION): physicochemical properties, *Magn Reson Med*, 29, 599–604, 1993.
2. Belliveau, J.W., Kennedy, D.N., McKinstry, R.C., Buchbinder, B.R., Weisskoff, R.M., Cohen, M.S., Vevea, J.M., Brady, T.J., and Rosen, B.R., Functional mapping of the human visual cortex by magnetic resonance imaging, *Science*, 254, 716–719, 1991.
3. Villringer, A., Rosen, B.R., Belliveau, J.W., Ackerman, J.L., Lauffer, R.B., Buxton, R.B., Chao, Y.S., Wedeen, V.J., and Brady, T.J., Dynamic imaging with lanthanide chelates in normal brain: contrast due to magnetic susceptibility effects, *Magn Reson Med*, 6, 164–174, 1988.
4. Mandeville, J.B., Jenkins, B.G., Chen, Y.I., Choi, J.K., Kim, Y.R., Belen, D., Liu, C., Kosofsky, B.E., and Marota, J.J., Exogenous contrast agent improves sensitivity of gradient-echo functional magnetic resonance imaging at 9.4 T, *Magn Reson Med*, 52, 1272–1281, 2004.
5. Mandeville, J.B., Marota, J.J., Kosofsky, B.E., Keltner, J.R., Weissleder, R., Rosen, B.R., and Weisskoff, R.M., Dynamic functional imaging of relative cerebral blood volume during rat forepaw stimulation, *Magn Reson Med*, 39, 615–624, 1998.
6. Chen, Y.I., Mandeville, J.B., Nguyen, T.V., Talele, A., Cavagna, F., and Jenkins, B.G., Improved mapping of pharmacologically induced neuronal activation using the IRON technique with superparamagnetic blood pool agents, *J Magn Reson Imaging*, 14, 517–524, 2001.
7. Mandeville, J.B. and Marota, J.J., Vascular filters of functional MRI: spatial localization using BOLD and CBV contrast, *Magn Reson Med*, 42, 591–598, 1999.
8. Marota, J.J., Ayata, C., Moskowitz, M.A., Weisskoff, R.M., Rosen, B.R., and Mandeville, J.B., Investigation of the early response to rat forepaw stimulation, *Magn Reson Med*, 41, 247–252, 1999.
9. Greengard, P., The neurobiology of slow synaptic transmission, *Science*, 294, 1024–1030, 2001.
10. Bulte, J.W., Hoekstra, Y., Kamman, R.L., Magin, R.L., Webb, A.G., Briggs, R.W., Go, K.G., Hulstaert, C.E., Miltenyi, S., The, T.H., et al., Specific MR imaging of human lymphocytes by monoclonal antibody-guided dextran-magnetite particles, *Magn Reson Med*, 25, 148–157, 1992.
11. Kolbel, C.B., Holtmann, G., McRoberts, J.A., Scholer, S., Aengenvoordt, P., Singer, M.V., and Mayer, E.A., Involvement of chloride channels in the receptor-mediated activation of longitudinal colonic muscle, *Neurogastroenterol Motil*, 10, 489–498, 1998.
12. Moore, A., Medarova, Z., Potthast, A., and Dai, G., *In vivo* targeting of underglycosylated MUC-1 tumor antigen using a multimodal imaging probe, *Cancer Res*, 64, 1821–1827, 2004.
13. Pirko, I., Johnson, A., Ciric, B., Gamez, J., Macura, S.I., Pease, L.R., and Rodriguez, M., *In vivo* magnetic resonance imaging of immune cells in the central nervous system with superparamagnetic antibodies, *FASEB J*, 18, 179–182, 2004.
14. So, P.W., Hotee, S., Herlihy, A.H., and Bell, J.D., Generic method for imaging transgene expression, *Magn Reson Med*, 54, 218–221, 2005.
15. Chen, Y.I., Galpern, W.R., Brownell, A.L., Matthews, R.T., Bogdanov, M., Isacson, O., Keltner, J.R., Beal, M.F., Rosen, B.R., and Jenkins, B.G., Detection of dopaminergic neurotransmitter activity using pharmacologic MRI: correlation with PET, microdialysis, and behavioral data, *Magn Reson Med*, 38, 389–398, 1997.
16. Chen, Y.I., Brownell, A.L., Galpern, W., Isacson, O., Bogdanov, M., Beal, M.F., Livni, E., Rosen, B.R., and Jenkins, B.G., Detection of dopaminergic cell loss and neural transplantation using pharmacological MRI, PET and behavioral assessment, *Neuroreport*, 10, 2881–2886, 1999.
17. Nguyen, T.V., Brownell, A.L., Chen, Y.I., Livni, E., Coyle, J.T., Rosen, B.R., Cavagna, F., and Jenkins, B.G., Detection of the effects of dopamine receptor supersensitivity using pharmacological MRI and correlations with PET, *Synapse*, 36, 57–65, 2000.
18. Sulzer, D., Maidment, N.T., and Rayport, S., Amphetamine and other weak bases act to promote reverse transport of dopamine in ventral midbrain neurons, *J Neurochem*, 60, 527–535, 1993.
19. Sorkina, T., Doolen, S., Galperin, E., Zahniser, N.R., and Sorkin, A., Oligomerization of dopamine transporters visualized in living cells by fluorescence resonance energy transfer microscopy, *J Biol Chem*, 278, 28274–28283, 2003.
20. Kitayama, S. and Sogawa, C., Regulated expression and function of the somatodendritic catecholamine neurotransmitter transporters, *J Pharmacol Sci*, 99, 121–127, 2005.

21. Usiello, A., Baik, J.H., Rouge-Pont, F., Picetti, R., Dierich, A., LeMeur, M., Piazza, P.V., and Borrelli, E., Distinct functions of the two isoforms of dopamine D2 receptors, *Nature*, 408, 199–203, 2000.
22. Liu, L., Shen, R.Y., Kapatos, G., and Chiodo, L.A., Dopamine neuron membrane physiology: characterization of the transient outward current (IA) and demonstration of a common signal transduction pathway for IA and IK, *Synapse*, 17, 230–240, 1994.
23. Freedman, J.E. and Weight, F.F., Quinine potently blocks single K+ channels activated by dopamine D-2 receptors in rat corpus striatum neurons, *Eur J Pharmacol*, 164, 341–346, 1989.
24. Grace, A.A., The regulation of dopamine neurons as determined by *in vivo* and *in vitro* intracellular recordings, in *Neurophysiology of Dopamine Neuronal Systems: Current Status and Clinical Perspectives*, Chiodo, L. and Freeman, A., Eds., Lakeshore Publishing Co., Grosse Pointe, MI, 1987, pp. 1–66.
25. Grace, A.A. and Onn, S.P., Morphology and electrophysiological properties of immunocytochemically identified rat dopamine neurons recorded *in vitro*, *J Neurosci*, 9, 3463–3481, 1989.
26. Chiodo, L. and Kapatos, G., Mesencephalic dopamine-containing neurons in culture: morphological and electrophysiological characterization, in *Neurophysiology of Dopamine Neuronal Systems: Current Status and Clinical Perspectives*, Chiodo, L. and Freeman, A., Eds., Lakeshore Publishing Co., Grosse Pointe, MI, 1987, pp. 67–91.
27. Valjent, E., Pascoli, V., Svenningsson, P., Paul, S., Enslen, H., Corvol, J.C., Stipanovich, A., Caboche, J., Lombroso, P.J., Nairn, A.C., Greengard, P., Herve, D., and Girault, J.A., Regulation of a protein phosphatase cascade allows convergent dopamine and glutamate signals to activate ERK in the striatum, *Proc Natl Acad Sci USA*, 102, 491–496, 2005.
28. Kornhauser, J.M. and Greenberg, M.E., A kinase to remember: dual roles for MAP kinase in long-term memory, *Neuron*, 18, 839–842, 1997.
29. Sweatt, J.D., The neuronal MAP kinase cascade: a biochemical signal integration system subserving synaptic plasticity and memory, *J Neurochem*, 76, 1–10, 2001.
30. Impey, S., Obrietan, K., and Storm, D.R., Making new connections: role of ERK/MAP kinase signaling in neuronal plasticity, *Neuron*, 23, 11–14, 1999.
31. Huang, E.J. and Reichardt, L.F., Neurotrophins: roles in neuronal development and function, *Annu Rev Neurosci*, 24, 677–736, 2001.
32. Patapoutian, A. and Reichardt, L.F., Trk receptors: mediators of neurotrophin action, *Curr Opin Neurobiol*, 11, 272–280, 2001.
33. Hetman, M. and Gozdz, A., Role of extracellular signal regulated kinases 1 and 2 in neuronal survival, *Eur J Biochem*, 271, 2050–2055, 2004.
34. Grewal, S.S., York, R.D., and Stork, P.J., Extracellular-signal-regulated kinase signalling in neurons, *Curr Opin Neurobiol*, 9, 544–553, 1999.
35. Rajadhyaksha, A., Husson, I., Satpute, S.S., Kuppenbender, K.D., Ren, J.Q., Guerriero, R.M., Standaert, D.G., and Kosofsky, B.E., L-type Ca2+ channels mediate adaptation of extracellular signal-regulated kinase 1/2 phosphorylation in the ventral tegmental area after chronic amphetamine treatment, *J Neurosci*, 24, 7464–7476, 2004.
36. Valjent, E., Corvol, J.C., Pages, C., Besson, M.J., Maldonado, R., and Caboche, J., Involvement of the extracellular signal-regulated kinase cascade for cocaine-rewarding properties, *J Neurosci*, 20, 8701–8709, 2000.
37. Choe, E.S., Chung, K.T., Mao, L., and Wang, J.Q., Amphetamine increases phosphorylation of extracellular signal-regulated kinase and transcription factors in the rat striatum via group I metabotropic glutamate receptors, *Neuropsychopharmacology*, 27, 565–575, 2002.
38. Choe, E.S. and Wang, J.Q., CaMKII regulates amphetamine-induced ERK1/2 phosphorylation in striatal neurons, *Neuroreport*, 13, 1013–1016, 2002.
39. Berke, J.D. and Hyman, S.E., Addiction, dopamine, and the molecular mechanisms of memory, *Neuron*, 25, 515–532, 2000.
40. Nestler, E.J., Molecular basis of long-term plasticity underlying addiction, *Nat Rev Neurosci*, 2, 119–128, 2001.
41. Nestler, E.J., Molecular neurobiology of addiction, *Am J Addict*, 10, 201–217, 2001.
42. Konradi, C., Leveque, J.C., and Hyman, S.E., Amphetamine and dopamine-induced immediate early gene expression in striatal neurons depends on postsynaptic NMDA receptors and calcium, *J Neurosci*, 16, 4231–4239, 1996.

43. Turgeon, S.M., Pollack, A.E., and Fink, J.S., Enhanced CREB phosphorylation and changes in c-Fos and FRA expression in striatum accompany amphetamine sensitization, *Brain Res*, 749, 120–126, 1997.
44. Ostrander, M.M., Richtand, N.M., and Herman, J.P., Stress and amphetamine induce Fos expression in medial prefrontal cortex neurons containing glucocorticoid receptors, *Brain Res*, 990, 209–214, 2003.
45. Graybiel, A.M., Moratalla, R., and Robertson, H.A., Amphetamine and cocaine induce drug-specific activation of the c-fos gene in striosome-matrix compartments and limbic subdivisions of the striatum, *Proc Natl Acad Sci USA*, 87, 6912–6916, 1990.
46. Cole, R.L., Konradi, C., Douglass, J., and Hyman, S.E., Neuronal adaptation to amphetamine and dopamine: molecular mechanisms of prodynorphin gene regulation in rat striatum, *Neuron*, 14, 813–823, 1995.
47. Feldpausch, D.L., Needham, L.M., Stone, M.P., Althaus, J.S., Yamamoto, B.K., Svensson, K.A., and Merchant, K.M., The role of dopamine D4 receptor in the induction of behavioral sensitization to amphetamine and accompanying biochemical and molecular adaptations, *J Pharmacol Exp Ther*, 286, 497–508, 1998.
48. Meng, Z.H., Feldpaush, D.L., and Merchant, K.M., Clozapine and haloperidol block the induction of behavioral sensitization to amphetamine and associated genomic responses in rats, *Brain Res Mol Brain Res*, 61, 39–50, 1998.
49. Uslaner, J., Badiani, A., Day, H.E., Watson, S.J., Akil, H., and Robinson, T.E., Environmental context modulates the ability of cocaine and amphetamine to induce c-fos mRNA expression in the neocortex, caudate nucleus, and nucleus accumbens, *Brain Res*, 920, 106–116, 2001.
50. Hedou, G., Jongen-Relo, A.L., Murphy, C.A., Heidbreder, C.A., and Feldon, J., Sensitized Fos expression in subterritories of the rat medial prefrontal cortex and nucleus accumbens following amphetamine sensitization as revealed by stereology, *Brain Res*, 950, 165–179, 2002.
51. Mead, A.N., Vasilaki, A., Spyraki, C., Duka, T., and Stephens, D.N., AMPA-receptor involvement in c-fos expression in the medial prefrontal cortex and amygdala dissociates neural substrates of conditioned activity and conditioned reward, *Eur J Neurosci*, 11, 4089–4098, 1999.
52. Nestler, E.J., Barrot, M., and Self, D.W., DeltaFosB: a sustained molecular switch for addiction, *Proc Natl Acad Sci USA*, 98, 11042–11046, 2001.
53. Nestler, E.J., Kelz, M.B., and Chen, J., DeltaFosB: a molecular mediator of long-term neural and behavioral plasticity, *Brain Res*, 835, 10–17, 1999.
54. Cragg, S.J., Nicholson, C., Kume-Kick, J., Tao, L., and Rice, M.E., Dopamine-mediated volume transmission in midbrain is regulated by distinct extracellular geometry and uptake, *J Neurophysiol*, 85, 1761–1771, 2001.
55. van Horne, C., Hoffer, B.J., Stromberg, I., and Gerhardt, G.A., Clearance and diffusion of locally applied dopamine in normal and 6-hydroxydopamine-lesioned rat striatum, *J Pharmacol Exp Ther*, 263, 1285–1292, 1992.
56. Rice, M.E., Distinct regional differences in dopamine-mediated volume transmission, *Prog Brain Res*, 125, 277–290, 2000.
57. Zoli, M., Torri, C., Ferrari, R., Jansson, A., Zini, I., Fuxe, K., and Agnati, L.F., The emergence of the volume transmission concept, *Brain Res Brain Res Rev*, 26, 136–147, 1998.
58. Vincent, S.L., Khan, Y., and Benes, F.M., Cellular distribution of dopamine D1 and D2 receptors in rat medial prefrontal cortex, *J Neurosci*, 13, 2551–2564, 1993.
59. Khan, Z.U., Koulen, P., Rubinstein, M., Grandy, D.K., and Goldman-Rakic, P.S., An astroglia-linked dopamine D2-receptor action in prefrontal cortex, *Proc Natl Acad Sci USA*, 98, 1964–1969, 2001.
60. Paulson, O.B. and Newman, E.A., Does the release of potassium from astrocyte endfeet regulate cerebral blood flow? *Science*, 237, 896–898, 1987.
61. Zonta, M., Angulo, M.C., Gobbo, S., Rosengarten, B., Hossmann, K.A., Pozzan, T., and Carmignoto, G., Neuron-to-astrocyte signaling is central to the dynamic control of brain microcirculation, *Nat Neurosci*, 6, 43–50, 2003.
62. Takano, T., Tian, G.F., Peng, W., Lou, N., Libionka, W., Han, X., and Nedergaard, M., Astrocyte-mediated control of cerebral blood flow, *Nat Neurosci*, 9, 260–267, 2006.
63. Wahl, M. and Schilling, L., Regulation of cerebral blood flow: a brief review, *Acta Neurochir Suppl*, 59, 3–10, 1993.

64. Amenta, F., Barili, P., Bronzetti, E., Felici, L., Mignini, F., and Ricci, A., Localization of dopamine receptor subtypes in systemic arteries, *Clin Exp Hypertens*, 22, 277–288, 2000.
65. Choi, J.K., Chen, Y.I., Hamel, E., and Jenkins, B.G., Brain hemodynamic changes mediated by dopamine receptors: role of the cerebral microvasculature in dopamine-mediated neurovascular coupling, *Neuroimage*, 30, 700–712, 2006.
66. Bjorklund, L.M., Sanchez-Pernaute, R., Chung, S., Andersson, T., Chen, Y.I., McNaught, K.S., Brownell, A.L., Jenkins, B.G., Wahlestedt, C., Kim, K.S., and Isacson, O., Embryonic stem cells develop into functional dopaminergic neurons after transplantation in a Parkinson rat model, *Proc Natl Acad Sci USA*, 99, 2344–2349, 2002.
67. Doudet, D.J. and Holden, J.E., Sequential versus nonsequential measurement of density and affinity of dopamine D2 receptors with [11C]raclopride: effect of methamphetamine, *J Cereb Blood Flow Metab*, 23, 1489–1494, 2003.
68. Minuzzi, L., Nomikos, G.G., Wade, M.R., Jensen, S.B., Olsen, A.K., and Cumming, P., Interaction between LSD and dopamine D2/3 binding sites in pig brain, *Synapse*, 56, 198–204, 2005.
69. Wilson, A.A., McCormick, P., Kapur, S., Willeit, M., Garcia, A., Hussey, D., Houle, S., Seeman, P., and Ginovart, N., Radiosynthesis and evaluation of [11C]-(+)-4-propyl-3,4,4a,5,6,10b-hexahydro-2H-naphtho[1,2-b][1,4]oxazin-9-ol as a potential radiotracer for *in vivo* imaging of the dopamine D2 high-affinity state with positron emission tomography, *J Med Chem*, 48, 4153–4160, 2005.
70. Seneca, N., Finnema, S.J., Farde, L., Gulyas, B., Wikstrom, H.V., Halldin, C., and Innis, R.B., Effect of amphetamine on dopamine D2 receptor binding in nonhuman primate brain: a comparison of the agonist radioligand [(11)C]MNPA and antagonist [(11)C]raclopride, *Synapse*, 59, 260–269, 2006.
71. Chen, Y.I., Choi, J.K., Andersen, S.L., Rosen, B.R., and Jenkins, B.G., Mapping dopamine D2/D3 receptor function using pharmacological magnetic resonance imaging, *Psychopharmacology*, 180, 705–715, 2004.
72. Tidey, J.W. and Bergman, J., Drug discrimination in methamphetamine-trained monkeys: agonist and antagonist effects of dopaminergic drugs, *J Pharmacol Exp Ther*, 285, 1163–1174, 1998.
73. Cory-Slechta, D.A., Widzowski, D.V., and Pokora, M.J., Functional alterations in dopamine systems assessed using drug discrimination procedures, *Neurotoxicology*, 14, 105–114, 1993.
74. Schoemaker, H., Claustre, Y., Fage, D., Rouquier, L., Chergui, K., Curet, O., Oblin, A., Gonon, F., Carter, C., Benavides, J., and Scatton, B., Neurochemical characteristics of amisulpride, an atypical dopamine D2/D3 receptor antagonist with both presynaptic and limbic selectivity, *J Pharmacol Exp Ther*, 280, 83–97, 1997.
75. Furmidge, L., Tong, Z.Y., Petry, N., and Clark, D., Effects of low, autoreceptor selective doses of dopamine agonists on the discriminative cue and locomotor hyperactivity produced by d-amphetamine, *J Neural Transm Gen Sect*, 86, 61–70, 1991.
76. Widzowski, D.V. and Cory-Slechta, D.A., Apparent mediation of the stimulus properties of a low dose of quinpirole by dopaminergic autoreceptors, *J Pharmacol Exp Ther*, 266, 526–534, 1993.
77. Semba, J., Functional role of dopamine D3 receptor in schizophrenia, *Nihon Shinkei Seishin Yakurigaku Zasshi*, 24, 3–11, 2004.
78. Ferre, S., O'Connor, W.T., Snaprud, P., Ungerstedt, U., and Fuxe, K., Antagonistic interaction between adenosine A2A receptors and dopamine D2 receptors in the ventral striopallidal system. Implications for the treatment of schizophrenia, *Neuroscience*, 63, 765–773, 1994.
79. Rosin, D.L., Robeva, A., Woodard, R.L., Guyenet, P.G., and Linden, J., Immunohistochemical localization of adenosine A2A receptors in the rat central nervous system, *J Comp Neurol*, 401, 163–186, 1998.
80. Sebastiao, A.M. and Ribeiro, J.A., Adenosine A2 receptor-mediated excitatory actions on the nervous system, *Prog Neurobiol*, 48, 167–189, 1996.
81. Svenningsson, P., Le Moine, C., Aubert, I., Burbaud, P., Fredholm, B.B., and Bloch, B., Cellular distribution of adenosine A2A receptor mRNA in the primate striatum, *J Comp Neurol*, 399, 229–240, 1998.
82. Svenningsson, P., Lindskog, M., Ledent, C., Parmentier, M., Greengard, P., Fredholm, B.B., and Fisone, G., Regulation of the phosphorylation of the dopamine- and cAMP-regulated phosphoprotein of 32 kDa *in vivo* by dopamine D1, dopamine D2, and adenosine A2A receptors, *Proc Natl Acad Sci USA*, 97, 1856–1860, 2000.

83. Uematsu, K., Futter, M., Hsieh-Wilson, L.C., Higashi, H., Maeda, H., Nairn, A.C., Greengard, P., and Nishi, A., Regulation of spinophilin Ser94 phosphorylation in neostriatal neurons involves both DARPP-32-dependent and independent pathways, *J Neurochem*, 95, 1642–1652, 2005.
84. Svenningsson, P., Le Moine, C., Fisone, G., and Fredholm, B.B., Distribution, biochemistry and function of striatal adenosine A2A receptors, *Prog Neurobiol*, 59, 355–396, 1999.
85. Svenningsson, P., Fourreau, L., Bloch, B., Fredholm, B.B., Gonon, F., and Le Moine, C., Opposite tonic modulation of dopamine and adenosine on c-fos gene expression in striatopallidal neurons, *Neuroscience*, 89, 827–837, 1999.
86. Tanganelli, S., Sandager Nielsen, K., Ferraro, L., Antonelli, T., Kehr, J., Franco, R., Ferre, S., Agnati, L.F., Fuxe, K., and Scheel-Kruger, J., Striatal plasticity at the network level. Focus on adenosine A2A and D2 interactions in models of Parkinson's disease, *Parkinsonism Relat Disord*, 10, 273–280, 2004.
87. Prediger, R.D., Da Cunha, C., and Takahashi, R.N., Antagonistic interaction between adenosine A2A and dopamine D2 receptors modulates the social recognition memory in reserpine-treated rats, *Behav Pharmacol*, 16, 209–218, 2005.
88. Chen, Y.I., Choi, J.K., and Jenkins, B.G., Mapping interactions between dopamine and adenosine A2a receptors using pharmacologic MRI, *Synapse*, 55, 80–88, 2005.
89. Mao, L. and Wang, J.Q., Differentially altered mGluR1 and mGluR5 mRNA expression in rat caudate nucleus and nucleus accumbens in the development and expression of behavioral sensitization to repeated amphetamine administration, *Synapse*, 41, 230–240, 2001.
90. Giorgetti, M., Hotsenpiller, G., Ward, P., Teppen, T., and Wolf, M.E., Amphetamine-induced plasticity of AMPA receptors in the ventral tegmental area: effects on extracellular levels of dopamine and glutamate in freely moving rats, *J Neurosci*, 21, 6362–6369, 2001.
91. West, A.R., Floresco, S.B., Charara, A., Rosenkranz, J.A., and Grace, A.A., Electrophysiological interactions between striatal glutamatergic and dopaminergic systems, *Ann NY Acad Sci*, 1003, 53–74, 2003.
92. Ujike, H., Takaki, M., Kodama, M., and Kuroda, S., Gene expression related to synaptogenesis, neuritogenesis, and MAP kinase in behavioral sensitization to psychostimulants, *Ann NY Acad Sci*, 965, 55–67, 2002.
93. Ferguson, S.M. and Robinson, T.E., Amphetamine-evoked gene expression in striatopallidal neurons: regulation by corticostriatal afferents and the ERK/MAPK signaling cascade, *J Neurochem*, 91, 337–348, 2004.
94. Radwanska, K., Caboche, J., and Kaczmarek, L., Extracellular signal-regulated kinases (ERKs) modulate cocaine-induced gene expression in the mouse amygdala, *Eur J Neurosci*, 22, 939–948, 2005.
95. Miller, C.A. and Marshall, J.F., Molecular substrates for retrieval and reconsolidation of cocaine-associated contextual memory, *Neuron*, 47, 873–884, 2005.
96. Mattson, B.J., Bossert, J.M., Simmons, D.E., Nozaki, N., Nagarkar, D., Kreuter, J.D., and Hope, B.T., Cocaine-induced CREB phosphorylation in nucleus accumbens of cocaine-sensitized rats is enabled by enhanced activation of extracellular signal-related kinase, but not protein kinase A, *J Neurochem*, 95, 1481–1494, 2005.
97. Freeman, W.M., Nader, M.A., Nader, S.H., Robertson, D.J., Gioia, L., Mitchell, S.M., Daunais, J.B., Porrino, L.J., Friedman, D.P., and Vrana, K.E., Chronic cocaine-mediated changes in non-human primate nucleus accumbens gene expression, *J Neurochem*, 77, 542–549, 2001.
98. Hope, B.T., Cocaine and the AP-1 transcription factor complex, *Ann NY Acad Sci*, 844, 1–6, 1998.
99. Lu, W. and Wolf, M.E., Repeated amphetamine administration alters AMPA receptor subunit expression in rat nucleus accumbens and medial prefrontal cortex, *Synapse*, 32, 119–131, 1999.
100. Chen, Y.I., Ren, J.Q., Satpute, S.S., Kosofsky, B.E., Rosen, B.R., Jenkins, B.G., and Rajadhyaksha, A.M., Mapping the modulation of dopaminergic circuitry in amphetamine-sensitized rats using pharmacological MRI, Annual meeting, Society for Neuroscience, Washington, DC, 2004, p. 688.2.
101. Paxinos, G. and Watson, C., *The Rat Brain in Stereotaxic Coordinates*, Academic Press, San Diego, 1997.
102. Orban, G.A., Claeys, K., Nelissen, K., Smans, R., Sunaert, S., Todd, J.T., Wardak, C., Durand, J.B., and Vanduffel, W., Mapping the parietal cortex of human and non-human primates, *Neuropsychologia*, 44(13): 2647–2667, 2006.
103. Orban, G.A., Fize, D., Peuskens, H., Denys, K., Nelissen, K., Sunaert, S., Todd, J., and Vanduffel, W., Similarities and differences in motion processing between the human and macaque brain: evidence from fMRI, *Neuropsychologia*, 41, 1757–1768, 2003.

104. Vanduffel, W., Fize, D., Mandeville, J.B., Nelissen, K., Van Hecke, P., Rosen, B.R., Tootell, R.B., and Orban, G.A., Visual motion processing investigated using contrast agent-enhanced fMRI in awake behaving monkeys, *Neuron*, 32, 565–577, 2001.
105. Vanduffel, W., Fize, D., Peuskens, H., Denys, K., Sunaert, S., Todd, J.T., and Orban, G.A., Extracting 3D from motion: differences in human and monkey intraparietal cortex, *Science*, 298, 413–415, 2002.
106. Febo, M., Segarra, A.C., Nair, G., Schmidt, K., Duong, T.Q., and Ferris, C.F., The neural consequences of repeated cocaine exposure revealed by functional MRI in awake rats, *Neuropsychopharmacology*, 30, 936–943, 2005.
107. Lahti, K.M., Ferris, C.F., Li, F., Sotak, C.H., and King, J.A., Comparison of evoked cortical activity in conscious and propofol-anesthetized rats using functional MRI, *Magn Reson Med*, 41, 412–416, 1999.
108. Peeters, R.R., Tindemans, I., De Schutter, E., and Van der Linden, A., Comparing BOLD fMRI signal changes in the awake and anesthetized rat during electrical forepaw stimulation, *Magn Reson Imaging*, 19, 821–826, 2001.
109. Sicard, K., Shen, Q., Brevard, M.E., Sullivan, R., Ferris, C.F., King, J.A., and Duong, T.Q., Regional cerebral blood flow and BOLD responses in conscious and anesthetized rats under basal and hypercapnic conditions: implications for functional MRI studies, *J Cereb Blood Flow Metab*, 23, 472–481, 2003.
110. Skoubis, P.D., Hradil, V., Chin, C.L., Luo, Y., Fox, G.B., and McGaraughty, S., Mapping brain activity following administration of a nicotinic acetylcholine receptor agonist, ABT-594, using functional magnetic resonance imaging in awake rats, *Neuroscience*, 137, 583–591, 2006.
111. Tenney, J.R., Duong, T.Q., King, J.A., and Ferris, C.F., fMRI of brain activation in a genetic rat model of absence seizures, *Epilepsia*, 45, 576–582, 2004.
112. Tenney, J.R., Duong, T.Q., King, J.A., Ludwig, R., and Ferris, C.F., Corticothalamic modulation during absence seizures in rats: a functional MRI assessment, *Epilepsia*, 44, 1133–1140, 2003.
113. Wong, D.L., Tai, T.C., Wong-Faull, D.C., Claycomb, R., and Kvetnansky, R., Genetic mechanisms for adrenergic control during stress, *Ann NY Acad Sci*, 1018, 387–397, 2004.
114. Cooper, D.O. and Stolk, J.M., Differences in the response of superior cervical ganglion dopamine-beta-hydroxylase activity to immobilization stress between inbred rat strains, *Commun Psychopharmacol*, 1, 291–299, 1977.
115. Watanabe, H., Activation of dopamine synthesis in mesolimbic dopamine neurons by immobilization stress in the rat, *Neuropharmacology*, 23, 1335–1338, 1984.
116. Ekker, M. and Sourkes, T.L., Decreased activity of adrenal S-adenosylmethionine decarboxylase in rats subjected to dopamine agonists, metabolic stress, or bodily immobilization, *Endocrinology*, 120, 1299–1307, 1987.
117. Desole, M.S., Miele, M., Esposito, G., Enrico, P., De Natale, G., and Miele, E., Analysis of immobilization stress-induced changes of ascorbic acid, noradrenaline, and dopamine metabolism in discrete brain areas of the rat, *Pharmacol Res*, 22 (Suppl 3), 43–44, 1990.
118. Harada, K., Noguchi, K., and Wakusawa, R., Effects of immobilization stress and of a benzodiazepine derivative on rat central dopamine system, *J Anesth*, 6, 167–171, 1992.
119. McMahon, A., Kvetnansky, R., Fukuhara, K., Weise, V.K., Kopin, I.J., and Sabban, E.L., Regulation of tyrosine hydroxylase and dopamine beta-hydroxylase mRNA levels in rat adrenals by a single and repeated immobilization stress, *J Neurochem*, 58, 2124–2030, 1992.
120. Nankova, B., Devlin, D., Kvetnansky, R., Kopin, I.J., and Sabban, E.L., Repeated immobilization stress increases the binding of c-Fos-like proteins to a rat dopamine beta-hydroxylase promoter enhancer sequence, *J Neurochem*, 61, 776–779, 1993.
121. Hayase, T., Yamamoto, Y., Yamamoto, K., Muso, E., Shiota, K., and Hayashi, T., Similar effects of cocaine and immobilization stress on the levels of heat-shock proteins and stress-activated protein kinases in the rat hippocampus, and on swimming behaviors: the contribution of dopamine and benzodiazepine receptors, *Behav Pharmacol*, 14, 551–562, 2003.
122. Zimmerman, R.S. and Frohlich, E.D., Stress and hypertension, *J Hypertens Suppl*, 8, S103–S107, 1990.
123. Whitnall, M.H., Stress selectively activates the vasopressin-containing subset of corticotropin-releasing hormone neurons, *Neuroendocrinology*, 50, 702–707, 1989.
124. de Goeij, D.C., Kvetnansky, R., Whitnall, M.H., Jezova, D., Berkenbosch, F., and Tilders, F.J., Repeated stress-induced activation of corticotropin-releasing factor neurons enhances vasopressin stores and colocalization with corticotropin-releasing factor in the median eminence of rats, *Neuroendocrinology*, 53, 150–159, 1991.

125. Bartanusz, V., Aubry, J.M., Jezova, D., Baffi, J., and Kiss, J.Z., Up-regulation of vasopressin mRNA in paraventricular hypophysiotrophic neurons after acute immobilization stress, *Neuroendocrinology*, 58, 625–629, 1993.
126. Rabadan-Diehl, C., Lolait, S.J., and Aguilera, G., Regulation of pituitary vasopressin V1b receptor mRNA during stress in the rat, *J Neuroendocrinol*, 7, 903–910, 1995.
127. Hatakeyama, S., Kawai, Y., Ueyama, T., and Senba, E., Nitric oxide synthase-containing magnocellular neurons of the rat hypothalamus synthesize oxytocin and vasopressin and express Fos following stress stimuli, *J Chem Neuroanat*, 11, 243–256, 1996.
128. Forman, L.J. and Estilow, S., Estrogen influences the effect of immobilization stress on immunoreactive beta-endorphin levels in the female rat pituitary, *Proc Soc Exp Biol Med*, 187, 190–196, 1988.
129. Honkaniemi, J., Colocalization of peptide- and tyrosine hydroxylase-like immunoreactivities with Fos-immunoreactive neurons in rat central amygdaloid nucleus after immobilization stress, *Brain Res*, 598, 107–113, 1992.
130. Castagliuolo, I., Leeman, S.E., Bartolak-Suki, E., Nikulasson, S., Qiu, B., Carraway, R.E., and Pothoulakis, C., A neurotensin antagonist, SR 48692, inhibits colonic responses to immobilization stress in rats, *Proc Natl Acad Sci USA*, 93, 12611–12615, 1996.
131. Pang, X., Alexacos, N., Letourneau, R., Seretakis, D., Gao, W., Boucher, W., Cochrane, D.E., and Theoharides, T.C., A neurotensin receptor antagonist inhibits acute immobilization stress-induced cardiac mast cell degranulation, a corticotropin-releasing hormone-dependent process, *J Pharmacol Exp Ther*, 287, 307–314, 1998.
132. Singh, L.K., Pang, X., Alexacos, N., Letourneau, R., and Theoharides, T.C., Acute immobilization stress triggers skin mast cell degranulation via corticotropin releasing hormone, neurotensin, and substance P: a link to neurogenic skin disorders, *Brain Behav Immun*, 13, 225–239, 1999.
133. Wang, B., Shaham, Y., Zitzman, D., Azari, S., Wise, R.A., and You, Z.B., Cocaine experience establishes control of midbrain glutamate and dopamine by corticotropin-releasing factor: a role in stress-induced relapse to drug seeking, *J Neurosci*, 25, 5389–5396, 2005.
134. Xi, Z.X., Gilbert, J., Campos, A.C., Kline, N., Ashby, C.R., Jr., Hagan, J.J., Heidbreder, C.A., and Gardner, E.L., Blockade of mesolimbic dopamine D3 receptors inhibits stress-induced reinstatement of cocaine-seeking in rats, *Psychopharmacology*, 176, 57–65, 2004.
135. Schultz, W., Getting formal with dopamine and reward, *Neuron*, 36, 241–263, 2002.
136. Martin-Soelch, C., Leenders, K.L., Chevalley, A.F., Missimer, J., Kunig, G., Magyar, S., Mino, A., and Schultz, W., Reward mechanisms in the brain and their role in dependence: evidence from neurophysiological and neuroimaging studies, *Brain Res Brain Res Rev*, 36, 139–149, 2001.
137. Satoh, T., Nakai, S., Sato, T., and Kimura, M., Correlated coding of motivation and outcome of decision by dopamine neurons, *J Neurosci*, 23, 9913–9923, 2003.
138. Schultz, W., Neural coding of basic reward terms of animal learning theory, game theory, microeconomics and behavioural ecology, *Curr Opin Neurobiol*, 14, 139–147, 2004.
139. Abler, B., Walter, H., Erk, S., Kammerer, H., and Spitzer, M., Prediction error as a linear function of reward probability is coded in human nucleus accumbens, *Neuroimage*, 31(2): 790–795, 2006.
140. de la Fuente-Fernandez, R., Phillips, A.G., Zamburlini, M., Sossi, V., Calne, D.B., Ruth, T.J., and Stoessl, A.J., Dopamine release in human ventral striatum and expectation of reward, *Behav Brain Res*, 136, 359–363, 2002.
141. Marota, J.J., Mandeville, J.B., Weisskoff, R.M., Moskowitz, M.A., Rosen, B.R., and Kosofsky, B.E., Cocaine activation discriminates dopaminergic projections by temporal response: an fMRI study in rat, *Neuroimage*, 11, 13–23, 2000.
142. Mandeville, J.B., Jenkins, B.G., Kosofsky, B.E., Moskowitz, M.A., Rosen, B.R., and Marota, J.J., Regional sensitivity and coupling of BOLD and CBV changes during stimulation of rat brain, *Magn Reson Med*, 45, 443–447, 2001.
143. Mandeville, J.B., Marota, J.J., Ayata, C., Moskowitz, M.A., Weisskoff, R.M., and Rosen, B.R., MRI measurement of the temporal evolution of relative CMRO(2) during rat forepaw stimulation, *Magn Reson Med*, 42, 944–951, 1999.
144. Kim, Y.R., Huang, I.J., Lee, S.R., Tejima, E., Mandeville, J.B., van Meer, M.P., Dai, G., Choi, Y.W., Dijkhuizen, R.M., Lo, E.H., and Rosen, B.R., Measurements of BOLD/CBV ratio show altered fMRI hemodynamics during stroke recovery in rats, *J Cereb Blood Flow Metab*, 25, 820–829, 2005.

145. Kerskens, C.M., Hoehn-Berlage, M., Schmitz, B., Busch, E., Bock, C., Gyngell, M.L., and Hossmann, K.A., Ultrafast perfusion-weighted MRI of functional brain activation in rats during forepaw stimulation: comparison with T2-weighted MRI, *NMR Biomed*, 9, 20–23, 1996.
146. Duong, T.Q., Silva, A.C., Lee, S.P., and Kim, S.G., Functional MRI of calcium-dependent synaptic activity: cross correlation with CBF and BOLD measurements, *Magn Reson Med*, 43, 383–392, 2000.
147. Dijkhuizen, R.M., Singhal, A.B., Mandeville, J.B., Wu, O., Halpern, E.F., Finklestein, S.P., Rosen, B.R., and Lo, E.H., Correlation between brain reorganization, ischemic damage, and neurologic status after transient focal cerebral ischemia in rats: a functional magnetic resonance imaging study, *J Neurosci*, 23, 510–517, 2003.
148. Nieoullon, A. and Dusticier, N., Effects of alpha-chloralose on the activity of the nigrostriatal dopaminergic system in the cat, *Eur J Pharmacol*, 65, 403–410, 1980.
149. Keilholz, S.D., Silva, A.C., Raman, M., Merkle, H., and Koretsky, A.P., Functional MRI of the rodent somatosensory pathway using multislice echo planar imaging, *Magn Reson Med*, 52, 89–99, 2004.
150. Garrett, K.M. and Gan, J., Enhancement of gamma-aminobutyric acid A receptor activity by alpha-chloralose, *J Pharmacol Exp Ther*, 285, 680–686, 1998.
151. He, D.S. and Burt, J.M., Mechanism and selectivity of the effects of halothane on gap junction channel function, *Circ Res*, 86, E104–E109, 2000.
152. Lecharny, J.B., Salord, F., Henzel, D., Desmonts, J.M., and Mantz, J., Effects of thiopental, halothane and isoflurane on the calcium-dependent and -independent release of GABA from striatal synaptosomes in the rat, *Brain Res*, 670, 308–312, 1995.
153. Li, X. and Pearce, R.A., Effects of halothane on GABA(A) receptor kinetics: evidence for slowed agonist unbinding, *J Neurosci*, 20, 899–907, 2000.
154. Perouansky, M., Kirson, E.D., and Yaari, Y., Mechanism of action of volatile anesthetics: effects of halothane on glutamate receptors *in vitro*, *Toxicol Lett*, 100/101, 65–69, 1998.
155. Austin, V.C., Blamire, A.M., Allers, K.A., Sharp, T., Styles, P., Matthews, P.M., and Sibson, N.R., Confounding effects of anesthesia on functional activation in rodent brain: a study of halothane and alpha-chloralose anesthesia, *Neuroimage*, 24, 92–100, 2005.
156. Antunes, L.M., Golledge, H.D., Roughan, J.V., and Flecknell, P.A., Comparison of electroencephalogram activity and auditory evoked responses during isoflurane and halothane anaesthesia in the rat, *Vet Anaesth Analg*, 30, 15–23, 2003.
157. Mandeville, J.B., Liu, C., Kosofsky, B.E., and Marota, J.J., Transient signal changes in pharmacological fMRI: effects of no interest? in *International Society of Magnetic Resonance Medicine*, Miami, 2005, p. 1512.

Part III

Cellular MR Imaging

13 Cellular MR Imaging of the Liver Using Contrast Agents

N. Cem Balci and Sukru Mehmet Erturk

CONTENTS

13.1 Introduction 247
13.2 Liver Anatomy and Physiology 247
13.3 Liver Disease 248
13.3.1 Benign Lesions of the Liver 248
13.3.2 Malignant Lesions of the Liver 248
13.4 MR Contrast Agents for the Liver 249
13.4.1 Nonspecific Extracellular Gadolinium Chelates 249
13.4.2 Liver-Specific Contrast Agents 250
13.4.2.1 Hepatocyte-Selective Agents 250
13.4.2.2 Agents with Combined Perfusion and Hepatocyte-Selective Properties 251
13.4.2.3 RES-Specific Contrast Agents 253
13.5 Future Directions 254
References 255

13.1 INTRODUCTION

Magnetic resonance imaging (MRI) is considered a state-of-the-art technique for liver imaging. It is superior to other imaging modalities because the anatomic resolution provided by MR images is as low as 1 to 2 mm in-plane at clinical field strengths of 1.5 tesla and tissue-specific techniques can aid to further characterize pathologic conditions in the liver. The liver contains a vascular and biliary network within different types of cells. Extracellular gadolinium chelates have been used for liver imaging. With the use of extracellular contrast agents, focal liver lesions and diffuse parenchymal pathologies in the liver can be evaluated based on contrast agent dynamics within the vascular tree of the liver. Cellular contrast agents have been developed in the past 10 to 15 years for liver MRI, which are distinctively taken up by liver cells. The use of cellular contrast agents is helpful for further evaluation of liver disease.

13.2 LIVER ANATOMY AND PHYSIOLOGY

The liver is the largest and most metabolically complex organ in our body. **Blood supply** to the liver is from both the portal vein and the hepatic artery; the former supplies about 75% of the total flow. Small branches of the terminal portal venule and terminal hepatic arteriole enter each acinus at the portal triad. The pooled blood then flows through sinusoids between plates of hepatocytes. Hepatocytes are responsible for the liver's central role in metabolism, including formation and excretion of bile; regulation of carbohydrate homeostasis; lipid synthesis and secretion of plasma lipoproteins; control of cholesterol metabolism; formation of urea, serum albumin, clotting factors, enzymes, and numerous other proteins; and metabolism or detoxification of drugs and other foreign substances. The spaces of Disse separate hepatocytes from the porous sinusoidal lining, where the nutrients are exchanged. Flow from sinusoids merges at terminal hepatic venules. The hepatic venules coalesce and eventually form the hepatic vein, which carries all efferent blood into the inferior vena cava. A rich supply of lymphatic vessels also drains the liver. The **intrahepatic biliary network begins** as tiny bile canaliculi formed by adjacent hepatocytes. These microvilli-lined structures progressively coalesce into ductules, interlobular bile ducts, and larger hepatic ducts. Outside the porta hepatis, the main hepatic duct joins the cystic duct from the gallbladder to form the common bile duct, which drains into the duodenum. **Sinusoidal lining cells** consist of mainly four types: endothelial, Kuppfer, perisinusoidal fat-storing, and pit cells. Endothelial cells are responsible for the exchange of nutrients and macromolecules with nearby hepatocytes across the spaces of Disse. Endothelial cells also endocytose various molecules and particles, synthesize proteins that influence the extracellular matrix, and play a role in lipoprotein metabolism. Spindle-shaped Kuppfer cells line the sinusoids and form an important part of the reticuloendothelial system; they derive from bone marrow precursors and serve as tissue macrophages. Major functions include phagocytosis of foreign particles, removal of endotoxins and other noxious substances, and modulation of the immune response. Perisinusoidal fat-storing cells (Ito cells) store vitamin A, synthesize various matrix proteins, and can transform into fibroblasts in response to hepatic injury. The uncommon pit cells are believed to be tissue lymphocytes with natural killer cell functions. The **extracellular matrix** of the liver includes the organ's reticulin framework, consisting of several molecular forms of collagen, laminin, fibronectin, and other extracellular glycoproteins. Matrix interactions and functions are not fully understood.[1–4]

13.3 LIVER DISEASE

The liver is one of the most common targets for metastatic disease, due to its dual blood supply. Because of its rich cellular and vascular structure, benign lesions are also not uncommon. Imaging of the liver is performed mainly for the detection of focal liver lesions and differentiation of benign from malignant lesions.[5] To understand the need for extracellular or hepatocellular targeted contrast agents in liver imaging, we will review liver diseases and their cellular nature.

13.3.1 Benign Lesions of the Liver

Benign hepatic cystic lesions of the liver have a high fluid content with well-defined margins; histopathologically, they can be simple hepatic cysts, biliary cystadenomas, or hamartomas. Hepatic hemangiomas are sponge-like blood-filled mesenchymal tumors and are the most common benign hepatic lesions. Hepatic adenomas contain liver cells including mainly hepatocytes and few Kuppfer cells. Focal nodular hyperplasia (FNH) of the liver is defined by a localized region of hyperplasia within otherwise normal liver. FNH contains bile ductules and moderate to significant amounts of Kuppfer cells. Regenerative nodules arise in the cirrhotic liver as a result of heterogenous regeneration in the grossly distorted liver architecture.[6–12]

13.3.2 Malignant Lesions of the Liver

Primary malignant tumors of liver cells are called hepatocellular carcinoma (HCC). HCC arises in cirrhotic livers and patients with hepatitis. A malignant transformation of the hepatocytes follows the stages of mildly to severely dysplastic nodules, ending up with HCC. HCC is histopathologically graded as moderately and poorly differentiated tumors according to their non-neoplastic hepatocellular content. Poor differentiation of HCC is associated with increased arterial neovascularization of the lesion. Cholangiocarcinoma is the neoplasia of the biliary tree, can arise from the main intrahepatic bile ducts, or can have peropheric intrahepatic location. Cholangiocarcinomas have poor arterial vascularization and are known as hypovascular tumors. Secondary tumors of the liver are metastatic lesions to the liver.[13–16]

Focal liver lesions are classified according to their vascularity in three groups: hypovascular, isovascular with the liver parenchyma, and hypervascular lesions. On imaging, the degree of contrast enhancement in sequential imaging techniques can determine the nature of the lesions as described below.[13] In overlapping cases of hypervascular benign and malignant tumors, hepatocellular-specific contrast agents can determine the nature of the lesions.

13.4 MR CONTRAST AGENTS FOR THE LIVER

Two general types of contrast agents for MR imaging are commonly used; this also applies to liver imaging, those that primarily decrease T1 time, leading to increased signal on T1-weighted images, and contrast agents that predominantly decrease T2, leading to a reduction of signal on T2-weighted images. Contrast agents affecting T1 are gadolinium-based paramagnetic chelates, whereas T2 agents consist of superparamagnetic complexes, such as iron oxide nanoparticles. The level of induced contrast by a paramagnetic contrast agent depends on its relaxivity, which is defined by the expression $1/T_{1,2obs} = 1/T_{1,2d} + r_1\,[CA]$. $1/T_{obs}$ is the observed T1 or T2 in an MR imaging experiment, $1/T_{1,2d}$ is the diamagnetic contribution to the T1 or T2, and [CA] is the relaxivity and is equal to $1/T_{1,2p}$, the paramagnetic contribution to the relaxation rate. Relaxivity is defined as the relaxation rate of water protons in 1 mmol $l^{-1}sec^{-1}$.[17]

In the case of superparamagnetic iron oxide (SPIO) particles, the increased relaxivity leading to hypointense changes observed on T2-and T2*-weighted images is induced primarily by local magnetic susceptibility effects that lead to dephasing of nearby protons.[17–19]

13.4.1 Nonspecific Extracellular Gadolinium Chelates

The use of these agents has been considered essential for evaluation of the full complexity of abdominal disease in patients evaluated with MR imaging for a diverse range of indications. After intravenous application, the extracellular gadolinium chelates distribute into the intravascular and interstitial spaces. In liver imaging, the time-related distribution of the extracellular agents aids to depict and characterize focal hepatic lesions that have arterial or portal venous blood supply. Diffuse parenchymal pathologies such as hepatitis and cirrhosis can be evaluated by the degree of enhancement. In order to achieve an enhancement of diagnostic value, contrast timing is crucial, after application. Hepatic imaging with nonspecific extracellular contrast agents is performed mainly in three consecutive temporal phases related to the location of the bulk of the contrast agent in the abdomen and the liver. The central phase, encoding steps of the hepatic arterial-dominant phase, is approximately 28 sec after the initiation of the bolus injection of the contrast agent. In this phase, contrast enhancement is observed in the hepatic artery and portal vein, and not in the hepatic veins. Hepatic parenchyma is mildly enhanced. During this phase, other abdominal organs with rich capillary blood supply, such as the kidneys and the pancreas, reveal contrast enhancement. During the hepatic arterial phase hypervascular liver pathologies are depicted and characterized. The portal venous phase (early hepatic venous phase) is approximately 1 min after the initiation of the contrast injection. During this phase contrast enhancement in the hepatic veins is observed and the hepatic parenchyma is nearly maximally enhanced. Maximal vascular enhancement is observed in this phase. Assessment of hypovascular liver pathologies and washout of hypervascular pathologies also occur in this phase. The late hepatic venous or interstitial phase is the time interval, approximately 90 sec to 5 min, after initiation of contrast injection. During this phase, some of the hepatic lesions reveal their imaging characteristics. Hemangiomas reveal progressive enhancement, persistent enhancement is observed in small-sized hemangiomas, and washout of hypervascular metastases and HCC is also apparent during this phase.[20–23]

Serial contrast-enhanced imaging with the use of nonspecific extracellular contrast agents has high accuracy in detecting and characterizing focal liver lesions. In some focal liver lesions such as HCC, FNH, and adenomas, the enhancement pattern may not be confident for the differential diagnosis between lesions of liver origin and malignant lesions. Targeting the liver cells, such as the hepatocytes and macrophages of the reticuloendothelial system (RES), is an effective approach for liver-specific agents.[20]

13.4.2 Liver-Specific Contrast Agents

Liver-specific contrast agents make up a group of contrast agents that reveal intracellular uptake by cells located in the liver. According to the hepatic cells that show uptake, they are divided into two groups: hepatocyte-selective and Kuppfer cell-specific contrast agents.

13.4.2.1 Hepatocyte-Selective Agents

Hepatocyte-selective agents either are only hepatocyte selective, such as Mn-DPDP, or are gadolinium chelates (Gd-BOPTA and Gd-EOB) that are both distributed in the extracellular space and hepatocyte selective. Kuppfer cell-specific contrast agents are iron-containing compounds. The hepatocyte-specific Gd chelates and Mn-DPDP are T1 agents that shorten T1 time and result in increased signal on T1-weighted images. Kuppfer cell-specific agents shorten T2 and T1 times, with a predominant effect of decreasing the T2 signal, but increasing the T1 signal in some settings.[20]

Mn-DPDP (Mangafodipir, Amersham Health) was the first manganese complex to be used as a liver contrast agent in clinical trials. It is an anionic manganese chelate salified with four molecules of meglumine. The Mn^{++} ion is a powerful T1 relaxation agent because its five unpaired electrons exhibit optimal correlation times.

Mn-DPDP dissociates rapidly following administration, yielding free Mn^{++} ion. Free Mn^{++} is taken up by the hepatocytes and eliminated by the hepatobiliary pathway.[20,24–26] Its chemical similarity to vitamin B6 is cited as the reason for the hepatocyte uptake, but some of the liver accumulation of paramagnetic manganese is probably caused by metabolism of the parent compound. Mn-DPDP also shows uptake by the renal cortex, pancreas, and gastric mucosa. Free Mn^{++} may cause an increased neurological risk in patients with hepatic impairment. Mn-DPDP shortens the T1 time and causes increased signal in the liver on T1-weighted images.[26–29] Mn-DPDP is administered as a slow intravenous infusion over 1 to 2 min with a dose of 5 to 10 mmol/kg. Maximum liver enhancement is observed within 10 to 15 min of the infusion. After slow-drip infusion, facial flushing and perception of increased body temperature may occur as a reported side effect. Serious side effects have not been described.[20,24–29] On postcontrast images, lesions without hepatocyte content remain unenhanced, including metastases, benign liver cysts, and hemangiomas. Most tumors of nonhepatocellular origin typically are hypointense relative to enhanced liver parenchyma on T1-weighted images and are more conspicuous than on unenhanced images (Figure 13.1).

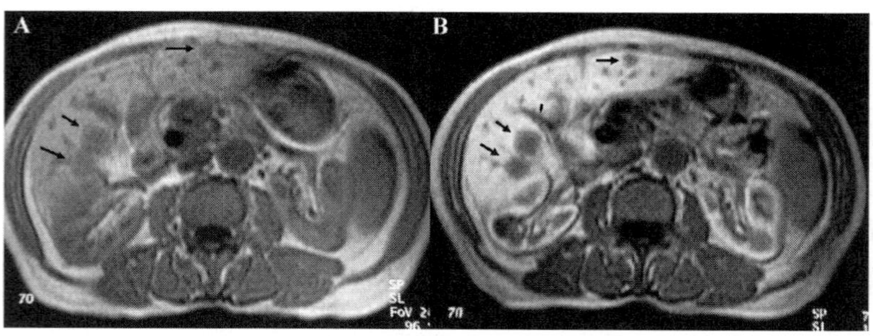

FIGURE 13.1 T1-weighted spoiled gradient echo image (A) reveals three focal liver lesions that are hypo- to isointense (arrows). After Mn-DPDP application (B), the liver becomes hyperintense and the lesions do not show contrast uptake (arrows) and become obvious; they are consistent with metastases.

Tumors of hepatocellular origin such as focal nodular hyperplasia (FNH), adenoma, and well-differentiated hepatocellular carcinomas (HCCs) have been shown to accumulate Mn-DPDP, providing characterization information to discriminate hepatocellular from nonhepatocellular tumors. Although Mn-DPDP can differentiate between hepatocyte- and non-hepatocyte-containing lesions, it may not be that effective in the differentiation between benign and malignant lesions. Regenerative nodules, well-differentiated hepatocellular carcinomas, and metastases from endocrine tumors reveal contrast uptake and increased enhancement. Benign and malignant hepatocellular tumors reveal varying degrees of enhancement that can be observed up to 24 h after administration.[20,24–29]

13.4.2.2 Agents with Combined Perfusion and Hepatocyte-Selective Properties

13.4.2.2.1 Gd-BOPTA (Gadobenate Dimeglumine)

Gadobenate dimeglumine (Gd-BOPTA; Multihance; Bracco, Milan, Italy) combines the properties of a conventional nonspecific gadolinium agent with those of a hepatocyte-selective agent. In Gd-BOPTA, the Gd ion forms a complex with 4-carboxy-5,8,11-tris (carboxymethyl)-1-phenyl-2-oxa-5,8,11-triazatridecan-3-oic acid, forming a highly stable octadenate coordination spherem, which is subsequently salified with two molecules of meglumine. This agent differs from other available gadolinium chelates in that it not only distributes to the extracellular fluid space, but is selectively taken up by functioning hepatocytes and excreted into the bile by the canalicular multispecific organic anion transporter, which is used to eliminate bilirubin.[20,23,29–32] Gd-BOPTA is mainly eliminated by the kidneys. While the biliary excretion rate is 55% in rats and 25% in rabbits, respectively, it is only 3 to 5% in humans.[20] This agent results in prolonged enhancement of the liver parenchyma combined with the plasma kinetics of an extracellular agent. The hepatobiliary contrast enhancement is most prominent 60 to 120 min after intravenous injection. The liver parenchyma enhancement obtained with gadobenate dimeglumine is comparable to the enhancement level of purely liver specific contrast media.[29–32] Gd-BOPTA has a higher relaxivity than equimolar formulations of other approved extracellular contrast agents, such as gadopentetate dimeglumine (Magnevist; Schering AG, Berlin, Germany), gadodiamide (Omniscan; Amersham-Health, Oslo, Norway), and gadoterate meglumine (Dotarem; Guerbet, Aulnay-sous-Bois, France), due to its more lipophilic structure and its capacity for weak and transient interaction with serum albumin.[20,30–38] In the liver, the estimated relaxivity is about 30 $mmol^{-1}sec^{-1}$, compared with calculated values of 16.6 $mmol^{-1}sec^{-1}$ for Gd-EOB-DTPA (Schering AG, Berlin, Germany) and 21.7 $mmol^{-1}sec^{-1}$ for mangafodipir trisodium.[38] This effect is thought to be due more to increased intracellular microviscosity within the hepatocytes than to transient interactions with intracellular proteins.[37,38] Serial contrast-enhanced liver imaging can be performed with the use of gadobenate dimeglumine after bolus injection, in the same fashion as with other nonspecific extracellular contrast agents.[20,30] Serial contrast-enhanced images exploit the differences in blood supply between lesions and normal liver parenchyma. The results are comparable to other conventional extracellular contrast agents, particularly for the improved visualization of hypervascular lesions.[20] Furthermore, gadobenate dimeglumine, similar to other extracellular contrast agents, allows improved assessment of lesion hemodynamics.[20,23] Improvement in the detection of hypovascular lesions has also been reported with Gd-BOPTA, compared to standard extracellular agents.[41] This reflects the fact that an increased fraction of gadobenate dimeglumine is taken up by the hepatocytes, which translates into increased detection and delineation of hypovascular lesions on delayed (40 to 120 min postinjection) or static hepatobiliary liver imaging. Uptake of the agent into regenerative nodules and well-differentiated HCCs is observed. Kuwatsuru et al. compared gadobenate dimeglumine with gadopentetate dimeglumine in 257 patients suspected of having malignant liver tumors,[39] and they observed that the contrast efficacy on early dynamic postcontrast images was comparable, while on delayed images gadobenate dimeglumine was significantly superior to gadopentetate dimeglumine in terms of improvement over the nonenhanced scans (44.5 vs. 19.0% on breath-hold gradient echo sequences).

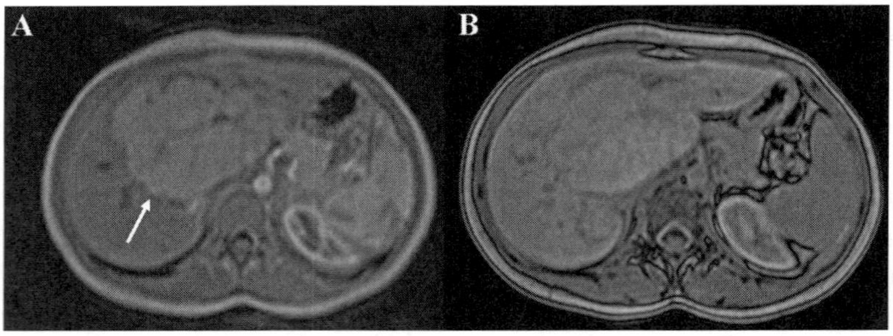

FIGURE 13.2 T1-weighted spoiled gradient echo image of the liver in arterial phase (A) with the use of Gd-BOPTA shows a lesion with increased enhancement (arrow). In the hepatocellular phase (B), the liver reveals enhancement and the lesion retains contrast due to the hepatocellular and biliary contrast uptake of FNH.

In a multicenter, multireader study involving 214 patients, Petersein et al. reported a significantly increased number of lesions detected on delayed postcontrast images.[40] Furthermore, the average size of detected lesions was smaller, reflecting improved depiction of small lesions, while the conspicuity of all lesions improved. All on-site readers and two of three off-site readers reported an increase in overall diagnostic confidence. In addition, further information on lesion characterization was provided in up to 25% of dynamic phase images and 59% of delayed phase postcontrast images, compared to the noncontrast scans.[20] Schneider et al. similarly reported similiar results after a half dose of gadobenate dimeglumine, compared to the standard dose of gadopentetate dimeglumine in a study of 43 patients.[41] Morana et al.[42] examined 249 patients with a variety of primary and secondary hypervascular tumors on both dynamic and delayed imaging. They found that delayed imaging gave additional information for lesion characterization with high accuracy in distinguishing benign lesions like FNH and regenerative hyperplasia from other lesion types (sensitivity, 79.7%; specificity, 96.1%). Differentiation between hepatocellular adenomas and FNH is possible with the use of Gd-BOPTA during the hepatobiliary phase. FNH contains biliary ducts, whereas hepatocellular adenoma does not have biliary ducts; thus, in the hepatobiliary phase FNH reveals increased enhancement, compared to hepatocellular adenomas[43] (Figure 13.2). Gd-BOPTA is an effective extracellular contrast agent. The greater T1 relaxivity allows better visualization of hypervascular lesions than conventional extracellular agents and equivalent lesion detection when used at half the dose. The hepatocyte selectivity further helps to distinguish hepatocyte-containing lesions from other focal liver lesions that do not contain hepatocytes. However, long waiting times for hepatobiliary phase imaging may be a disadvantage in daily practice.

13.4.2.2.2 Gd-EOB-DTPA

Gd-EOB-DTPA (gadolinium [4S]-4-[4-ethoxybenzyl]-3,6,9-tris[carboxylatomethyl]-3,69-triazaundecandioic acid-disodium salt; SH L 569 B; Primovist, Schering, Germany) is a paramagnetic hepatobiliary contrast agent with hepatocellular uptake via the anionic transporter protein.[20,23,29,44,45] Gd-EOB-DTPA has a higher T1-relaxivity in human plasma ($R1 = 8.2\ mmol^{-1}sec^{-1}$) than gadopentetate dimeglumine ($R1 = 5.0\ mmol^{-1}sec^{-1}$). This may be explained by the greater degree of protein binding compared to gadopentetate dimeglumine. Gd-EOB-DTPA provides a triphasic pharmacokinetic profile similar to that of Gd-BOPTA. The lipophilic side-chain EOB produces a high affinity to the organic anion transporter system, which is also responsible for the uptake of Gd-BOPTA. After intravenous bolus injection, Gd-EOB-DTPA is rapidly cleared from the intravascular space to the extracellular space; from here the compound is both taken up by hepatocytes and eliminated by glomerular filtration. In contrast to Gd-BOPTA, urinary filtration and fecal excretion by way of bile fluid account for approximately equal portions of the administered dose. Although the degree of renal elimination rises with increasing doses, its hepatic clearance reveals a moderate saturation phenomenon in higher doses.

Hepatobiliary contrast enhancement with Gd-EOB-DTPA reaches the maximum level at about 10 to 20 min postinjection and is followed by a plateau phase that has a duration of 2 h. The highest liver-to-lesion contrast is observed during the imaging window 20 to 45 min after injection of Gd-EOB-DTPA, compared to a 60- to 120-min postinjection period for delayed phase imaging with Gd-BOPTA.

Without serious adverse effects in phases I and II for dosages between 12.5 and 100 μmol/kg body weight, Gd-EOB-DTPA-enhanced imaging is noted to be superior to Gd-DTPA-enhanced imaging. The diagnostic performance of Gd-EOB-DTPA-enhanced MRI for detection of liver lesions was evaluated in a prospective, open-label, within-patient-comparison phase 3 study with the use of a 25 μmol/kg dose.[46] Acquired images were assessed during early phase serial contrast-enhanced imaging and hepatobiliary late phase imaging. A total of 302 histopathologically and intraoperative ultrasound-verified lesions in 131 patients were evaluated. Among the lesions, 215 were malignant, 80 were benign, and 7 were not assessable. The malignant lesions were metastases (n = 172), HCC (n = 31), and cholangiocellular carcinomas (CCCs) (n = 12). The benign lesions included 41 liver cysts, 18 hemangiomas, 7 FNHs, and 14 other benign lesions (adenomas, hydatid cysts, abcesses, etc.). The percentage of correctly matched lesions increased from 80.8% on precontrast MRI to 87.4% on postcontrast MRI. The correct classification of lesions also improved significantly. In the off-site reading, as in the clinical on-site study, more small lesions were detected on postcontrast than on precontrast images.

Hypervascular metastatic lesions revealed their most prominent enhancement in the early arterial phase, while hypovascular metastases showed highest enhancement 90 to 120 sec after IV injection of Gd-EOB-DTPA, then gradually decreased and stabilized at >10 min after contrast injection. Hepatocellular carcinomas demonstrated increased enhancement in the initial distribution phase 60 sec after IV injection of Gd-EOB-DTPA, similar to liver parenchyma, with more prolonged enhancement than metastases and liver parenchyma, and enhancement of HCC was similar during the complete observation period. It has also been demonstrated that benign hepatocyte-containing solid liver tumors, such as liver adenoma and FNH, exhibit prolonged tumor enhancement because of specific intracellular uptake of Gd-EOB-DTPA.[20,23,29]

13.4.2.3 RES-Specific Contrast Agents

Reticuloendothelial system (RES)-specific contrast agents contain iron oxide particles that are selectively taken up by Kuppfer cells in the liver, spleen, and bone marrow. Iron oxide particles have been developed in two different sizes. Superparamagnetic iron oxide (SPIO) particles have a mean iron oxide particulate size of 50 nm, while ultrasmall superparamagnetic iron oxide (USPIO) particles have a mean size of less than 50 nm.

Iron oxide formulations that are currently available are: (1) SPIO ferumoxides — Endorem (Guerbet, Aulneysous-Bois, France) and Feridex (Advanced Magnetics, Cambridge, MA); and (2) USPIO ferucarbotran (Resovist, Schering AG, Berlin, Germany). RES-specific contrast agents are superparamagnetic, causing shortening of both T2 and T1 relaxivity. T1- and T2*-weighted gradient echo and T2-weighted echo train spin echo imaging sequences are used for image acquisition. Ferumoxides are administered by intravenous infusion. According to one standard technique, a 15 μmol/kg SPIO concentrate is mixed in 100 ml of 5% dextrose solution and administered in drip infusion over 30 min. Before the infusion the patient undergoes a noncontrast MRI of the liver with the use of T1-weighted gradient echo and T2-weighted sequences. Approximately 30 min after completion of contrast administration, the patient undergoes a repeat MRI with the same imaging sequences. The particles are cleared from the plasma by the RES of the liver (80%) and spleen (12%). Minimal uptake occurs in the lymph nodes and bone marrow. Ferucarbotran is administered by direct bolus injection of a small volume (<2 ml) of contrast. Serial contrast-enhanced liver imaging is performed with the use of T1-weighted gradient echo images.[47–49] During arterial and portal venous phases there is no uptake in the Kuppffer cells, and increased signal in

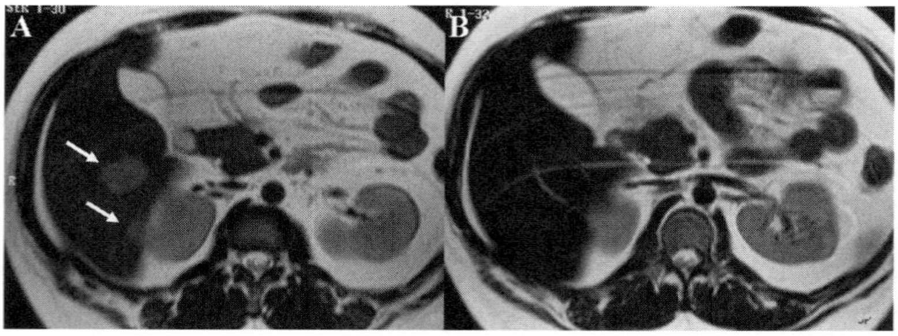

FIGURE 13.3 T2-weighted fast spin echo image of the liver with two hepatocellular adenomas. The adenomas are hyperintense (A) (arrows). After SPIO application (B), the lesions and the liver become hypointense because of contrast uptake of Kuppfer cells.

the vascular space and the liver parenchyma is observed due to the presence of ferucarbotran in the liver sinisoids. In the late venous phase (90 sec) there is increased intravascular signal and low signal in the liver parenchyma because of minimal uptake in the Kuppfer cells and the intravascular blood pool effect of the agent. During the late phase imaging (10 min after injection) there is increased uptake of the iron particles in the Kuppfer cells, and liver parenchyma is rendered hypointense on both T1- and T2-weighted images. Focal liver lesions with negligible Kuppfer cell content reveal higher signal relative to the contrast-mediated lower signal of the liver parenchyma. FNH, hepatocellular adenomas, regenerative nodules, and dysplastic nodules contain Kuppfer cells to a varying degree; therefore, their relative signal loss may parallel the signal loss in normal hepatic parenchyma after iron oxide administration (Figure 13.3).[20,47–49]

Hemangiomas reveal prolonged pooling of iron oxide in the enlarged venous channels, causing a mildly increased signal on T1-weighted images because of the blood pool effect of the agent. Limitations of SPIO-enhanced MRI are the relatively increased signal of cross-sectioned vessels in the low-signal-background liver parenchyma and well-differentiated HCC may contain Kuppfer cells and reveal contrast uptake. In patients with cirrhosis, heterogenous and diminished uptake of iron oxide by fibrotic tissue can mimic HCC.[47–49]

Furthermore, liposomal-encapsulated gadolinium DTPA or DOTA complexes were investigated as liver-specific agents.[29] Depending on the preparation, the vesicles range from 50 to 500 nm in diameter. The vesicles are recognized by the RES, permitting delivery of the gadolinium label to the liver and spleen. Decomplexation with release of free gadolinium is a clinical concern.[50–52] The tolerance of such preparations and the stability and synthesis of liposomal-encapsulated gadolinium complexes are still critical issues.

13.5 FUTURE DIRECTIONS

Numerous synthetic molecules such as polylysine, dextran, polyethylene glycol, a brush-like copolymer of polylysine and dextran, and a synthetic pegylated linear copolymer with polylysine have been investigated for their suitability as a core of the contrast agent rather than albumin. Thus, the vast majority of polymers, which were employed as cores, have been molecules of linear, branched linear, or circular chains. While convenient and reasonably available, limitations to the use of linear polyamines include difficulties in the synthesis, characterization, and purification of the agent to produce a final product of a single physical size with consistent chemical characteristics. Therefore, since polyamidoamine (PAMAM) or diaminobutane (DAB) dendrimers and gadomers, which are based on a different type of synthetic polymers made by Schering (Berlin, Germany), are spherical molecules, the advantage of their use as cores or platforms for forming contrast agents is the ability to better control the physical sizes of the molecules. Dendrimers are a class of highly branched synthetic spherical polymers consisting of a vast array of types, chemical structures, and functional groups. Two types of dendrimers, PAMAM and DAB, are commercially available.[53] They are highly soluble in aqueous solution, and both have a unique surface topology of primary amino groups. DAB dendrimers have a pure aliphatic polyamine core in contrast to the PAMAM dendrimers, which have an amide functional group core component. Based on the same general principle that hydrophobicity targets the liver, DAB-Am64-(1B4M-Gd)64 (DABG5), containing a generation 5 DAB dendrimer with a 64-surface-amine core unit, was synthesized and evaluated as a liver MR contrast agent. DABG5 homogeneously enhanced the liver parenchyma and was excreted more rapidly through both the liver and kidney than the analogous PAMAM dendrimer of similar molecular size.[53] Furthermore, dynamic micro-MRIs employing DABG5 were able to visualize liver metastatic tumors of colon cancer cells (LS174T) as small as 0.3 mm in diameter in living mice via a reverse contrast image.[54]

In conclusion, MR contrast agents increase the diagnostic accuracy in liver MRI. Contrast agents with combined extracellular and hepatocellular enhancement patterns are used in clinical daily practice. Future directions include development of ultrasmall particles to be used for cellular imaging in the liver.

REFERENCES

1. Liver anatomy and physiology, in *Merck Manual*, 17th ed., Beers, M.H. and Berkow, R., Eds., Wiley, New York, 1999, chap. 36.
2. Knell, A.J., Liver function and failure: the evolution of liver physiology, *J R Coll Physicians Lond*, 14, 205–208, 1980.
3. Read, A.E., Clinical physiology of the liver, *Br J Anaesth*, 44, 910–917, 1972.
4. Corless, J.K. and Middleton, H.M., 3rd, Normal liver function. A basis for understanding hepatic disease, *Arch Intern Med*, 143, 2291–2294, 1983.
5. Hyslop, W.B. and Semelka, R.C., Future directions in body magnetic resonance imaging, *Top Magn Reson Imaging*, 16, 3–14, 2005.
6. Shoenut, J.P., Semelka, R.C., Levi, C., and Greenberg, H., Ciliated hepatic foregut cysts: US, CT, and contrast-enhanced MR imaging, *Abdom Imaging*, 19, 150–152, 1994.
7. Mosetti, M.A., Leonardou, P., Motohara, T., Kanematsu, M., Armao, D., and Semelka, R.C., Autosomal dominant polycystic kidney disease: MR imaging evaluation using current techniques, *J Magn Reson Imaging*, 18, 210–215, 2003.
8. Semelka, R.C., Hussain, S.M., Marcos, H.B., and Woosley, J.T., Biliary hamartomas: solitary and multiple lesions shown on current MR techniques including gadolinium enhancement, *J Magn Reson Imaging*, 10, 196–201, 1999.
9. Palacios, E., Shannon, M., Solomon, C., and Guzman, M., Biliary cystadenoma: ultrasound, CT, and MRI, *Gastrointest Radiol*, 15, 313–316, 1990.
10. Semelka, R.C. and Sofka, C.M., Hepatic hemangiomas, *Magn Reson Imaging Clin North Am*, 5, 241–253, 1997.
11. Paulson, E.K., McClellan, J.S., Washington, K., Spritzer, C.E., Meyers, W.C., and Baker, M.E., Hepatic adenoma: MR characteristics and correlation with pathologic findings, *Am J Roentgenol*, 163, 113, 1994.
12. Mortelè, K.J., Praet, M., Van Vlierberghe, H., Kunnen, M., and Ros, P.R., CT and MR imaging findings in focal nodular hyperplasia of the liver: radiologic-pathologic correlation, *Am J Roentgenol*, 175, 687, 2000.
13. Semelka, R.C., Shoenut, J.P., Ascher, S.M., Kroeker, M.A., Greenberg, H.M., Yaffe, C.S., and Micflikier, A.B., Solitary hepatic metastasis: comparison of dynamic contrast-enhanced CT and MR imaging with fat-suppressed T2-weighted, breath-hold T1-weighted FLASH, and dynamic gadolinium-enhanced FLASH sequences, *J Magn Reson Imaging*, 4, 319–323, 1994.
14. Kelekis, N.L., Semelka, R.C., Worawattanakul, S., de Lange, E.E., Ascher, S.M., Ahn, I.O., Reinhold, C., Remer, E.M., Brown, J.J., Bis, K.G., Woosley, J.T., and Mitchell, D.G., Hepatocellular carcinoma in North America: a multiinstitutional study of appearance on T1-weighted, T2-weighted, and serial gadolinium-enhanced gradient-echo images, *Am J Roentgenol*, 170, 1005–1013, 1998.

15. Hussain, S.M., Semelka, R.C., and Mitchell, D.G., MR imaging of hepatocellular carcinoma, *Magn Reson Imaging Clin North Am*, 10, 31–52, 2002.
16. Hamrick-Turner, J., Abbitt, P.L., and Ros, P.R., Intrahepatic cholangiocarcinoma: MR appearance, *Am J Roentgenol*, 158, 77–79, 1992.
17. Delikatny, E.J. and Poptani, H., MR techniques for *in vivo* molecular and cellular imaging, *Radiol Clin North Am*, 43, 205–220, 2005.
18. Modo, M., Hoehn, M., and Bulte, J.W., Cellular MR imaging, *Mol Imaging*, 4, 143–164, 2005.
19. Bulte, J.W. and Kraitchman, D.L., Iron oxide MR contrast agents for molecular and cellular imaging, *NMR Biomed*, 17, 484–499, 2004.
20. Balci, N.C. and Semelka, R.C., Contrast agents for MR imaging of the liver, *Radiol Clin North Am*, 43, 887–898, 2005.
21. Semelka, R.C. and Helmberger, T.K., Contrast agents for MR imaging of the liver, *Radiology*, 218, 27–38, 2001.
22. Hamm, B., Mahfouz, A.E., Taupitz, M., Mitchell, D.G., Nelson, R., Halpern, E., Speidel, A., Wolf, K.J., and Saini, S., Liver metastases: improved detection with dynamic gadolinium-enhanced MR imaging? *Radiology*, 202, 677–682, 1997.
23. Reimer, P., Schneider, G., and Schima, W., Hepatobiliary contrast agents for contrast-enhanced MRI of the liver: properties, clinical development and applications, *Eur Radiol*, 14, 559–578, 2004.
24. Wang, C., Mangafodipir trisodium (MnDPDP)-enhanced magnetic resonance imaging of the liver and pancreas, *Acta Radiol Suppl*, 415, 1–31, 1998.
25. Wang, C., Ahlstrom, H., and Ekholm, S., Diagnostic efficacy of MnDPDP in MR imaging of the liver. A phase III multicentre study, *Acta Radiol*, 38, 643, 1997.
26. Aicher, K.P., Laniado, M., Kopp, A.F., Gronewaller, E., Duda, S.H., and Claussen, C.D., Mn-DPDP-enhanced MR imaging of malignant liver lesions: efficacy and safety in 20 patients, *J Magn Reson Imaging*, 3, 731–737, 1993.
27. Bartolozzi, C., Donati, F., Cioni, D., Crocetti, L., and Lencioni, R., MnDPDP-enhanced MRI vs. dual-phase spiral CT in the detection of hepatocellular carcinoma in cirrhosis, *Eur Radiol*, 10, 1697–1702, 2000.
28. Marti-Bonmati, L., Fog, A.F., de Beeck, B.O., Kane, P., and Fagertun, H., Safety and efficacy of Mangafodipir trisodium in patients with liver lesions and cirrhosis, *Eur Radiol*, 13, 1685–1692, 2003.
29. Weinmann, H.J., Tissue-specific MR contrast agents, *Eur J Radiol*, 46, 33, 2003.
30. de Haen, C., Lorusso, V., and Tirone, P., Hepatic transport of gadobenate dimeglumine in TR-rats, *Acad Radiol*, 3 (Suppl 2), S452–S454, 1996.
31. Kirchin, M.A., Pirovano, G.P., and Spinazzi, A., Gadobenate dimeglumine (Gd-BOPTA). An overview, *Invest Radiol*, 33, 798–809, 1998.
32. Spinazzi, A., Lorusso, V., Pirovano, G., and Kirchin, M., Safety, tolerance, biodistribution, and MR imaging enhancement of the liver with gadobenate dimeglumine: results of clinical pharmacologic and pilot imaging studies in nonpatient and patient volunteers, *Acad Radiol*, 6, 282–291, 1999.
33. de Haen, C. and Gozzini, L., Soluble-type hepatobiliary contrast agents for MR imaging, *J Magn Reson Imaging*, 3, 179–186, 1993.
34. Schima, W., Petersein, J., Hahn, P.F., Harisinghani, M., Halpern, E., and Saini, S., Contrast-enhanced MR imaging of the liver: comparison between Gd-BOPTA and Mangafodipir, *J Magn Reson Imaging*, 7, 130, 1997.
35. de Haen, C., La Ferla, R., and Maggioni, F., Gadobenate dimeglumine 0.5 *M* solution for injection (MultiHance) as contrast agent for magnetic resonance imaging of the liver: mechanistic studies in animals, *J Comput Assist Tomogr*, 23 (Suppl 1), S169–S179, 1999.
36. Cavagna, F.M., Maggioni, F., Castelli, P.M., Dapra, M., Imperatori, L.G., Lorusso, V., and Jenkins, B.G., Gadolinium chelates with weak binding to serum proteins. A new class of high-efficiency, general purpose contrast agents for magnetic resonance imaging, *Invest Radiol*, 32, 780–796, 1997.
37. Schuhmann-Giampieri, G., Liver contrast media for magnetic resonance imaging. Interrelations between pharmacokinetics and imaging, *Invest Radiol*, 28, 753–761, 1993.
38. Spinazzi, A., Lorusso, V., Pirovano, G., Taroni, P., Kirchin, M., and Davies, A., Multihance clinical pharmacology: biodistribution and MR enhancement of the liver, *Acad Radiol*, 5 (Suppl 1), S86–S89, 1998; discussion, S93–S94.

39. Kuwatsuru, R., Kadoya, M., Ohtomo, K., Tanimoto, A., Hirohashi, S., Murakami, T., Tanaka, Y., Yoshikawa, K., and Katayama, H., Comparison of gadobenate dimeglumine with gadopentetate dimeglumine for magnetic resonance imaging of liver tumors, *Invest Radiol*, 36, 632–641, 2001.
40. Petersein, J., Spinazzi, A., Giovagnoni, A., Soyer, P., Terrier, F., Lencioni, R., Bartolozzi, C., Grazioli, L., Chiesa, A., Manfredi, R., Marano, P., Van Persijn Van Meerten, E.L., Bloem, J.L., Petre, C., Marchal, G., Greco, A., McNamara, M.T., Heuck, A., Reiser, M., Laniado, M., Claussen, C., Daldrup, H.E., Rummeny, E., Kirchin, M.A., Pirovano, G., and Hamm, B., Focal liver lesions: evaluation of the efficacy of gadobenate dimeglumine in MR imaging — a multicenter phase III clinical study, *Radiology*, 215, 727–736, 2000.
41. Schneider, G., Maas, R., Schultze Kool, L., Rummeny, E., Gehl, H.B., Lodemann, K.P., and Kirchin, M.A., Low-dose gadobenate dimeglumine versus standard dose gadopentetate dimeglumine for contrast-enhanced magnetic resonance imaging of the liver: an intra-individual crossover comparison, *Invest Radiol*, 38, 85–94, 2003.
42. Morana, G., Grazioli, L., Schneider, G., Testoni, M., Menni, K., Chiesa, A., and Procacci, C., Hypervascular hepatic lesions: dynamic and late enhancement pattern with Gd-BOPTA, *Acad Radiol*, 9 (Suppl 2), S476–S479, 2002.
43. Grazioli, L., Morana, G., Kirchin, M.A., Caccia, P., Romanini, L., Bondioni, M.P., Procacci, C., and Chiesa, A., MRI of focal nodular hyperplasia (FNH) with gadobenate dimeglumine (Gd-BOPTA) and SPIO (ferumoxides): an intra-individual comparison, *J Magn Reson Imaging*, 17, 593–602, 2003.
44. Weinmann, H.J., Schuhmann-Giampieri, G., Schmitt-Willich, H., Vogler, H., Frenzel, T., and Gries, H., A new lipophilic gadolinium chelate as a tissue-specific contrast medium for MRI, *Magn Reson Med*, 22, 233–237, 1991; discussion, 242.
45. Schuhmann-Giampieri, G., Schmitt-Willich, H., Press, W.R., Negishi, C., Weinmann, H.J., and Speck, U., Preclinical evaluation of Gd-EOB-DTPA as a contrast agent in MR imaging of the hepatobiliary system, *Radiology*, 183, 59–64, 1992.
46. Huppertz, A., Balzer, T., Blakeborough, A., Breuer, J., Giovagnoni, A., Heinz-Peer, G., Laniado, M., Manfredi, R.M., Mathieu, D.G., Mueller, D., Reimer, P., Robinson, P.J., Strotzer, M., Taupitz, M., and Vogl, T.J., Improved detection of focal liver lesions at MR imaging: multicenter comparison of gadoxetic acid-enhanced MR images with intraoperative findings, *Radiology*, 230, 266–275, 2004.
47. Bellin, M.F., Zaim, S., Auberton, E., Sarfati, G., Duron, J.J., Khayat, D., and Grellet, J., Liver metastases: safety and efficacy of detection with superparamagnetic iron oxide in MR imaging, *Radiology*, 193, 657–663, 1994.
48. Winter, T.C., III, Freeny, P.C., Nghiem, H.V., Mack, L.A., Patten, R.M., Thomas, C.R., Jr., and Elliott, S., MR imaging with i.v. superparamagnetic iron oxide: efficacy in the detection of focal hepatic lesions, *Am J Roentgenol*, 161, 1191–1198, 1993.
49. Weissleder, R., Elizondo, G., Wittenberg, J., Rabito, C.A., Bengele, H.H., and Josephson, L., Ultrasmall superparamagnetic iron oxide: characterization of a new class of contrast agents for MR imaging, *Radiology*, 175, 489–493, 1990.
50. Unger, E., Cardenas, D., Zerella, A., Fajardo, L.L., and Tilcock, C., Biodistribution and clearance of liposomal gadolinium-DTPA, *Invest Radiol*, 25, 638–644, 1990.
51. Tilcock, C., Ahkong, Q.F., Koenig, S.H., Brown, R.D., 3rd, Davis, M., and Kabalka, G., The design of liposomal paramagnetic MR agents: effect of vesicle size upon the relaxivity of surface-incorporated lipophilic chelates, *Magn Reson Med*, 27, 44–51, 1992.
52. Kim, S.K., Pohost, G.M., and Elgavish, G.A., Gadolinium complexes of [(myristoyloxy)propyl]diethylenetriaminetetraacetate: new lipophilic, fatty acyl conjugated NMR contrast agents, *Bioconjug Chem*, 3, 20–26, 1992.
53. Kobayashi, H. and Brechbiel, M.W., Nano-sized MRI contrast agents with dendrimer cores, *Adv Drug Deliv Rev*, 57, 2271–2286, 2005.
54. Kobayashi, H., Saga, T., Kawamoto, S., Sato, N., Hiraga, A., Ishimori, T., Konishi, J., Togashi, K., and Brechbiel, M.W., Dynamic micro-magnetic resonance imaging of liver micrometastasis in mice with a novel liver macromolecular magnetic resonance contrast agent DAB-Am64-(1B4M-Gd)(64), *Cancer Res*, 61, 4966–4970, 2001.

14 Cellular Imaging of Macrophage Activity in Infection and Inflammation

Martin Rausch, Markus Rudin, Peter R. Allegrini, and Nicolau Beckmann

CONTENTS

14.1 The Function of Immune Cells in Inflammation and Infection 259
14.2 Cell Labeling for Studying Inflammatory Diseases in the Brain 260
  14.2.1 The Role of Immune Cells in CNS Diseases 260
  14.2.2 Labeling Immune Cells with Iron Oxide-Based MR Contrast Agents 260
  14.2.3 Imaging Techniques and Potential Problems to Detect USPIO-Labeled Macrophages by MRI 261
  14.2.4 Tracking Macrophages in Animal Models of MS 261
  14.2.5 Tracking of Nonphagocytotic Immune Cells 263
14.3 MRI and NIRF in Animal Models of Arthritis 263
14.4 Using Macrophage Tracking to Study Graft Status 264
  14.4.1 Macrophage Infiltration into Kidney Transplants 264
  14.4.2 Macrophage Infiltration into Cardiac and Lung Grafts 266
  14.4.3 Macrophage Tracking by MRI in Infected Tissue 267
14.5 Conclusions 269
References 269

14.1 THE FUNCTION OF IMMUNE CELLS IN INFLAMMATION AND INFECTION

Inflammatory cells are derived from pluripotent stem cells of the bone marrow that have the potential to differentiate into any blood or lymphatic cell depending on environmental stimuli. Blood-borne stem cells (myeloid stem cells) further differentiate into erythrocytes, platelets, and inflammatory cells such as macrophages, neutrophils, eosinophils, and basophils. Lymphatic (lymphoid) stem cells develop into the various types of lymphocytes, the B- and T-lymphocytes, and the natural killer cells.

Macrophages are bone marrow-derived cells of the immune system that are characterized by a high phagocytotic activity. Macrophages internalize and digest antigen–antibody complexes, foreign pathogens such as viruses and bacteria, inorganic substances, and even entire cells. They belong to the *mononuclear phagocytotic system* (MPS), which includes only cells that are involved in defense-related phagocytosis. Macrophages in tissue originate from blood-borne monocytes, which possess migratory, chemotactic, pinocytic, and phagocytic activities. They differentiate further upon migration to the target tissue to become multifunctional tissue macrophages.[1]

Macrophages are involved in essentially all stages of the immune response of an organism. They play a key role in host defense against intracellular pathogens such as bacteria or protozoa, as well as against tumor cells. Macrophages are important killer cells (K-cells) via antibody-dependent cell-mediated cytotoxicity. Macrophages are also effector and regulatory cells of the inflammatory response by secretion of inflammatory mediators. Finally, they are important mediators of tissue repair by releasing substances participating in tissue reorganization.[1,2]

Due to its key role in host defense, the MPS plays a significant part in many disease processes. Therefore, visualization of the migration of cells from the MPS, in particular that of macrophages, is of high relevance for both diagnostic purposes and the evaluation of therapeutic interventions. This requires the development of suitable cell labeling strategies, which in the case of macrophages is straightforward, as it is their biological purpose to internalize any foreign particles. Intravenous administration of iron-labeled nanoparticles, which are removed from the circulation by blood-borne cells of the MPS, in particular by monocytes, is a strategy routinely applied for tracking macrophages by magnetic resonance imaging (MRI). Uptake efficiency depends on the size of the nanoparticles. Competitive uptake by other cells of the immune system has been found to be negligible; these cells would have to be harvested and labeled *in vitro* in order to achieve effective incorporation of marker molecules. The labeled monocytes then migrate to sites of tissue inflammation attracted by chemotactic signals. Activated endothelial cells enable adhesion, rolling, sticking, and finally extravasation of the monocytes into affected tissue.

14.2 CELL LABELING FOR STUDYING INFLAMMATORY DISEASES IN THE BRAIN

14.2.1 The Role of Immune Cells in CNS Diseases

Immune cells play a crucial role in neuroinflammation and degeneration, and their distribution and activity have been studied in great detail for stroke, Alzheimer's disease (AD), and multiple sclerosis (MS), or the related animal model experimental autoimmune encephalomyelitis (EAE). Lymphocytes (T-cells/B-cells) are involved in inflammatory processes in several ways. They are able to recognize their specific antigen (memory T-cells), release cytotoxic agents (cytotoxic T-cells), regulate the immune response and the differentiation of naïve T-cells into different T-helper cells, and release antibodies (B-cells). Brain mononuclear phagocytes (brain-residing microglia, bone marrow monocyte-derived and perivascular macrophages) can either protect the nervous system by acting as debris scavengers or as regulators of the immune responses, or induce cell damage by releasing cytotoxic agents.

14.2.2 Labeling Immune Cells with Iron Oxide-Based MR Contrast Agents

The majority of *in vivo* labeling studies focused on macrophages, taking advantage of their phagocytotic activity. Macrophages can take up nanoparticles and store them in their vesicles. This process is facilitated when plasma proteins bind to the nanoparticles, which depends on, e.g., their surface charge or particle size. This mechanism has already been used for a couple of years to label Kupffer cells of the liver with superparamagnetic iron oxide (SPIO)-based nanoparticles. SPIO particles usually have a size of around 200 nm (see Chapter 13). They are rapidly opsonized and phagocytosed. Tracking of different types of macrophages (e.g., lymph node or blood-borne monocytes) requires the use of nanoparticles, which are not rapidly captured by the liver, and hence have a longer circulation time. USPIO particles were initially designed for demarcation of benign from malignant lymph node lesions.[3] Due to their high T2 relaxivity, USPIO particles are mostly used as a negative contrast agent. To label macrophages, USPIO particles are administered systemically about 24 h before the MRI measurements. This long delay ensures that free USPIO particles are cleared from the circulation and all signal alterations arise from USPIO particles captured by phagocytotic cells.[4]

14.2.3 Imaging Techniques and Potential Problems to Detect USPIO-Labeled Macrophages by MRI

There are several disease-relevant applications for macrophage tracking in the brain. In stroke, macrophages start to infiltrate brain tissue during the first day.[5] This process might be linked to additional brain damage and was the subject of many histological studies in the past.[5,6] Their infiltration can also be observed when USPIO particles are administered intravenously some hours after ischemia. In this case, circulating USPIO particles are captured by macrophages and can be visualized by MRI when entering the brain due to their strong transversal and longitudinal relaxation enhancement. This process has been demonstrated for both permanent and transient occlusion of the middle cerebral artery, where vascular occlusion was induced unilaterally for 30 min. In both cases, a clear contrast change was observed in the affected brain tissue.[7,8] Interestingly, USPIO accumulation was also observed in the cerebral cortex, while vascular edema was only found in the basal ganglia. This result suggests that early infiltration of macrophages can lead to a delayed increase of the area of tissue damage and might render USPIO-enhanced MRI as an interesting tool for clinical stroke studies. Since USPIO particles increase transversal and longitudinal proton relaxation, both T1- and T2-weighted MRI could be used for detection. Although transversal relaxivity is much stronger than longitudinal relaxivity, in particular at high field strengths, T2-weighted imaging has an inherent drawback because the hyperintense edema has the opposite effect on the signal intensity as the USPIO-induced signal attenuation. Therefore, both effects might cancel out or the amplitude of USPIO enhancement might be underestimated. Therefore, T1 imaging, although less sensitive, might be advantageous because the T1 relaxation of ischemic tissue remains rather constant.

In longitudinal studies, the visibility of USPIO enhancement was studied over several days. Interestingly, USPIO remained visible for about 48 h before contrast loss became apparent. In neuroinflammatory processes, macrophages can potentially leave the site of inflammation. However, physicochemical processes might also contribute to this effect: the core of USPIO particles like AMI-227 is covered by dextran. If this layer is enzymatically cleaved off within the vesicles of macrophages, USPIO particles tend to agglomerate, leading to a reduced longitudinal relaxivity.

Disruption of the blood-brain barrier (BBB) is frequently observed in neuroinflammatory diseases. In stroke, BBB damage can occur as a direct response of endothelial cells to hypoxia.[9] In EAE or MS, it is assumed that transcytosis of leukocytes might be linked to a transient loss of gap junction tightness.[10] This damage of the endothelial lineage and basement membrane could allow particle efflux from the vascular lumen to brain tissue. Bulte et al. used a freezing lesion for inducing BBB disruption.[11] In this model, systemically injected dextran–magnetite particles were taken up by endothelial cells within the lesion, but they did not pass the basement membrane. In contrast, opening of the BBB by mannitol allowed monocrystalline iron oxide nanoparticles (MIONs) having a mean particle size of 20 nm to diffuse into the interstitial space.[12,13] MIONs were captured by pericytes or were found in the extracellular space around capillaries. Hence, extravasation and free diffusion of nanoparticles depend on the particle size and the severity of the BBB disruption.

In order to use USPIO as a selective marker for macrophages, particles should not diffuse freely into brain tissue across the impaired BBB. Mannitol and freezing lesions seem to induce structural BBB damage, which is sufficient to allow USPIO to enter the interstitium. It is questionable, however, whether this structural damage resembles the situation of ischemia or inflammation-related BBB damage. Observations made in an experiment on rats with permanent middle cerebral artery occlusion (MCAO) suggest that BBB damage (indicated by Gd-DOTA enhancement) is not necessarily linked to USPIO efflux (Figure 14.1).

14.2.4 Tracking Macrophages in Animal Models of MS

Several EAE models are in use to study neuroinflammatory or degenerative processes and to test new drugs. Most models are based on rats (Lewis rat, DA rat) or mice (SJL, C57B6), in which

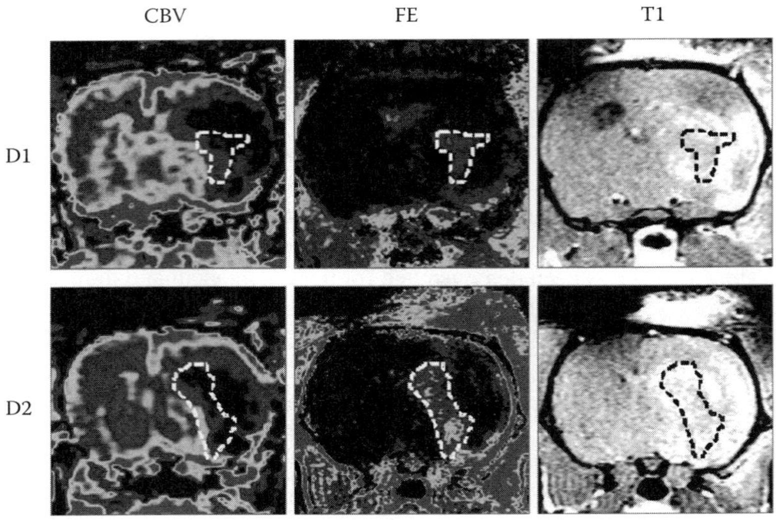

FIGURE 14.1 (please see color insert following page 210) Comparison of maps indicating cerebral blood volume (CBV), final enhancement (FE), and USPIO distribution. Areas of increased vascular permeability are observed on days 1 and 2. They are located in the transition zone between normal and severely compromised blood flow (white dashed line). USPIO accumulation can be observed in the cortex and the outer rim of the putamen by the signal increase in T1-weighted images. In this stroke model, BBB damage is not sufficient to allow USPIO particles to accumulate in brain tissue, but areas without BBB demonstrate USPIO accumulation.

EAE can be induced by active immunization with spinal cord homogenates, or with whole myelin-oligodendrocyte glycoproteins (MOGs), myelin basic proteins (MBPs), proteolipid proteins (PLPs), or fragments thereof. Although there are certain differences in disease evolution, all animal models are characterized by an acute phase with strong neurological symptoms, including paresis of the tail and hind paws. In some models, a relapse or a chronic phase can be observed. Macrophages play a pivotal role in the acute inflammatory phase, where they can be tracked by systemic administration of USPIO.[14,15] Macrophage infiltration can usually be observed in the spinal cord, the cerebellum, and the medulla. Inflammation during the acute phase is also accompanied by BBB damage, which might even precede macrophage infiltration.[16] The state of the BBB reflects several inflammatory processes, including early interaction of leukocytes and endothelial cells, opening of tight junctions due to transcytosis, and repair mechanisms. Repeated administration of USPIO allows visualization of acute infiltrating macrophages, and therefore USPIO-enhanced MRI can help to selectively demarcate phases of active inflammation from remission.[17]

A detailed analysis of Gd-DOTA- and USPIO-enhancing lesions in the EAE model has demonstrated a clear mismatch between these two parameters.[14] Qualitatively, Gd enhancement is more diffuse than the distribution of USPIO-labeled cells. However, there was also a clear mismatch during the acute phase of the disease in that some areas showed enhancement of either USPIO or Gd-DOTA. During the chronic phase, only USPIO enhancement could be observed. Although the EAE model cannot be transferred in a one-to-one fashion to human MS, these findings raise the question of whether BBB damage always reflects acute inflammation and if acute inflammation could also occur without BBB damage. This reservation is even more justified since BBB damage as measured by Gd extravasation is often poorly correlated with the disease outcome.

In vivo tracking of macrophages has already been used to test various anti-inflammatory or neuroprotective drugs. Prophylactic and therapeutic treatment of Lewis rats with FTY720, a sphingosin-1 phosphate receptor agonist, completely abolished infiltration of macrophages.[18] In contrast, anti-VLA-4 antibody treatment did not affect macrophage infiltration in the same model, although it reduced lymphocyte infiltration.[19] Floris et al. have observed abolishment of macrophage infiltration in EAE rats, which were treated with the immunomodulator Lovastatin.[16]

14.2.5 Tracking of Nonphagocytotic Immune Cells

Nonphagocytotic immune cells cannot be tracked *in vivo* by MRI as easily as macrophages because they will not take up nanoparticles from the circulation. An alternative approach is to harvest cells from a donor animal and label these cells *ex vivo*. It has been demonstrated that T-cells can be labeled with iron oxide-based nanoparticles by incubating these cells for several hours in a medium containing the contrast agent.[20] However, this approach is not very efficient since incorporation of the contrast agent by pinocytosis occurs at a rather low rate. Improved methods are based on conjugating nanoparticles with transfection agents (TAs) such as TAT[21] and poly-L-lysine[22] and lipid-based TAs such as FuGENE,[23] which serve as a translocation signal and enhance the uptake of the particle, or by using magnetodendrimers.[24] Up to now, *in vivo* tracking of labeled T-cells by MRI has not been used extensively. The reason for that might be the relatively low sensitivity of the technique and the complex preparation of the cell cultures. Several factors have to be considered in this respect: First, T-cells have to be myelin specific. For observation of bystanders, tremendous amounts of labeled cells must be administered because only a low fraction of these cells will enter the brain. Second, the label will be diluted after clonal expansion. And finally, the specificity of T-cells may be altered by the labeling procedure.

14.3 MRI AND NIRF IN ANIMAL MODELS OF ARTHRITIS

Macrophages possess widespread pro-inflammatory, destructive, and remodeling capabilities that critically contribute to the acute and chronic phases of rheumatoid arthritis (RA).[25] Therefore, monitoring noninvasively the macrophage infiltration into sites of inflammation may play an important role in RA models.

In vivo labeling of macrophages by SPIO particles administered intravenously or intra-articularly has been exploited in murine,[26] rat,[27] and rabbit[28] models of rheumatoid arthritis to follow the infiltration of macrophages into inflamed areas in the living animal.

In a rat antigen-induced arthritis (AIA) model, in which previous systemic immunization with antigen in Freund's complete adjuvant was followed by intra-articular injection of the same soluble antigen (methylated bovine serum albumin), a significant negative correlation was found between the MRI signal intensity in the knee and the histologically determined iron content in macrophages located in the same region of animals that had received SPIO (Endorem®) 24 h before image acquisition.[27] Starting 2 days postantigen injection, images from arthritic knees exhibited distinctive signal attenuation in the synovium. This signal attenuation was significantly smaller in knees from animals treated with dexamethasone (0.3 mg/kg/day p.o.) and completely absent in contralateral knees that had been challenged with vehicle. These results suggest the feasibility of detecting macrophage infiltration into the knee synovium in this AIA model by labeling the cells with SPIO. This readout may have an impact in preclinical studies by shortening the duration of the experimental period and by facilitating the investigation of novel immunomodulatory therapies acting on macrophages. The great advantage of this model is that the contralateral, unchallenged knee serves as a reference in the same animal.

Using a cathepsin B-activatable near-infrared fluorescence (NIRF) probe, early signs of osteoarthritis were detected in a murine model involving intra-articular collagenase injection.[29] Early signs of experimental arthritis have also been measured in a murine model of AIA, in which the target of NIRF was the F4/80 antigen present on the surface of macrophages infiltrating the inflamed synovial membrane.[30] Imaging performed using anti-F4/80 monoclonal antibodies (mAbs) labeled with Cy5.5 fluorochromes administered intravenously revealed an accumulation of fluorochrome probes in inflamed knee joints and, to a lesser extent, in contralateral (nonarthritic) knee joints.

NIRF in conjunction with protease-activated probes has been shown as well to be a sensitive means for imaging the presence of target enzymes in arthritic joints, thereby providing the potential for early monitoring of treatment response to antirheumatic drugs.[31] The rationale for the approach lies in the fact that different proteases are highly upregulated in RA and contribute significantly to joint destruction. Proteases that target the Lys-Lys cleavage site, including cathepsin B, activate the fluorescence of the probe administered intravenously.

14.4 USING MACROPHAGE TRACKING TO STUDY GRAFT STATUS

Organ transplantation is nowadays the preferred and accepted treatment option in end-stage organ disease. The patient and transplant survival time following organ transplantation have improved substantially over the last decades. Several factors contributed to this development: progress in organ procurement, more stringent criteria for the selection of suitable donors and recipients, a better understanding of the biology, treatment, and prevention of acute graft rejection, improved diagnosis and treatment of infectious complications, and better monitoring of transplanted patients.

Since the introduction of calcineurin inhibitors as immunosuppressive therapy, acute rejection can be well managed. However, chronic rejection remains the main complication, and its treatment represents a major challenge.

Many methods for minimizing the immunosuppressive requirements following allotransplantation have been proposed based on a growing understanding of physiological and allospecific immunity. Major reasons for these efforts are the clinical risks associated with immunosuppression and the specific safety concerns of the molecules used (e.g. nephrotoxicity, diabetogenicity, etc.). Before these regimens can be developed for clinical application, they require validation in models that are reasonably predictive of their performance in humans. Several animal models are available to develop organ transplant protocols and to test the effects of immunomodulatory compounds.[32] These models are also essential for identifying and validating possible biomarkers of acute and chronic rejection in view of translational research.

14.4.1 Macrophage Infiltration into Kidney Transplants

Orthotopic kidney transplantation models are adopted in rodents to study renal allograft rejection.[33] One of the recipient's kidneys is nephrectomized and replaced with a kidney from a syngeneic animal or from an allogeneic animal. The graft may have a life-supporting function, in which case the contralateral kidney from the recipient animal is also removed. However, the recipient's contralateral kidney may as well remain in place, in which case the entire rejection process can be followed with minimal physiological alterations.

Leukocyte infiltration, predominantly recipient-derived lymphocytes and macrophages, is a prominent feature of allograft rejection.[34] Detection of macrophage infiltration into allografts by MRI in combination with the administration of SPIO preparations has been proposed as a possible strategy to noninvasively characterize the graft rejection status in several rat models of transplantation.

Macrophage infiltration during the acute rejection process of kidney grafts has been demonstrated in the Dark Agouti (DA)-to-Brown Norway (BN) model.[35,36] Administration of USPIO particles, of mean diameter of the order of 30 nm, at day 4 posttransplantation led one day later to distinct signal attenuation in the cortex of allogeneic kidneys. Immunohistological staining for ED1+ macrophages and CD4+ and CD8+ T-cells in allogeneic transplanted kidneys indicated the accumulation of these immune cells as acute rejection occurred. Morphological studies by electron microscopy confirmed the existence of iron particles inside the lysosomes of macrophages of rejecting kidneys, while Prussian blue staining detected the presence of iron plaques in macrophages. However, no signal reduction was observed in isografts and allografts of recipients receiving triple immunosuppressant treatment with daily subcutaneous injections of methylprednisolone (2 mg/kg), rapamycin (1 mg/kg), and cyclosporine A (CsA) (5 mg/kg).[35]

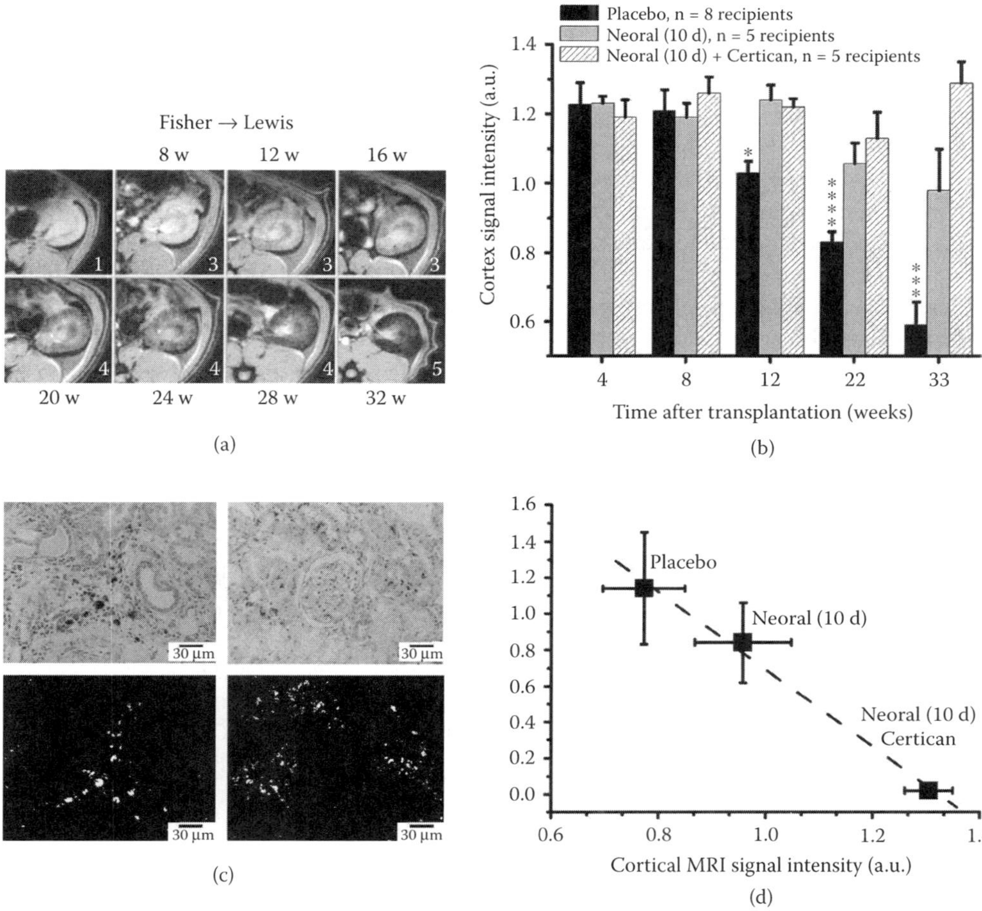

FIGURE 14.2 (a) Gradient echo images of a Fisher kidney transplanted into a Lewis rat, acquired at several time points with respect to transplantation. The recipient received SPIO (500 μl/kg body weight i.v.) 4 weeks prior to each measurement. (b) Cortical MRI signal intensity in grafts from untreated recipients, from recipients treated with Sandimmun Neoral (1.5 mg/kg/day p.o. for 10 days), and from recipients treated with Neoral (1.5 mg/kg/day p.o. for 10 days) followed by Certican (1.25 mg/kg/day) for the remainder of the study. (c) Histological sections stained with Perls' blue at the end of the study demonstrating iron-labeled macrophages. (d) Mean (±s.e.m. for five animals in each group) percentage of iron in the kidney cortex vs. mean (±s.e.m. for five animals in each group) cortical MRI signal intensity (in arbitrary units). A significant negative correlation ($r = -0.86$, $p < 0.0001$) was found between the iron content determined histologically and the MRI signal intensity. (See Beckmann et al.[45] for more details.)

Although numerous experimental and human studies have emphasized their presence during acute renal allograft rejection,[35,37] early infiltration by macrophages is considered to be a poor prognostic sign for allograft survival.[38] Nevertheless, several studies support the hypothesis that macrophage-derived inflammation is a cofactor for chronic allograft nephropathy,[39–41] monocytes/macrophages and T-cells being the predominant graft-invading cells of rat renal allografts with chronic rejection.[42–44] In view of the fact that chronic graft dysfunction represents the leading cause for the still unsatisfactory long-term results after organ transplantation, investigating macrophage infiltration into allografts during chronic rejection constitutes an important paradigm.

In the less stringent Fisher 344-to-Lewis model, starting 12 weeks posttransplantation, MR images from grafts of untreated recipients that had received SPIO contrast agent (mean size of particles = 150 nm) exhibited distinctive signal attenuation in the cortex (Figure 14.2).[45] Animals treated with CsA (Neoral, 1.5 mg/kg/day p.o. by gavage for 10 days after transplantation) to prevent acute rejection showed a signal attenuation in the cortex at 33 weeks posttransplantation, while kidneys from rats treated additionally with everolimus (Certican, 1.25 mg/kg/day p.o. by gavage until the end of the study), a rapamycin derivative, had no changes in anatomical appearance. A significant negative correlation was found between the MRI cortical signal intensity and the histologically determined iron content in macrophages located in the cortex. Moreover, a very strong and highly significant negative correlation was found between the MRI signal in the cortex and the rejection scores according to the Banff classification, suggesting that this method might be considered an alternative to histologic evaluation. Renography revealed a significantly reduced functionality of the kidneys of untreated controls 33 weeks after transplantation, while no significant changes in perfusion were observed in any group of rats. Thus, graft nephropathy was detected significantly earlier than changes in graft function brought about by labeling macrophages with SPIO. Analysis of biochemical parameters in the urine and the blood revealed proteinuria starting at 16 weeks in CsA-treated recipients. However, creatinine and the glomerular filtration rate calculated from the amount of urine collected in a 24-h period remained unchanged up to week 28.

14.4.2 Macrophage Infiltration into Cardiac and Lung Grafts

Similarly to kidney transplantation, gradient echo MRI in combination with the administration of USPIO 24 h before image acquisition was shown to offer promise as a noninvasive method for detecting acute cardiac transplant allograft rejection. In a DA-to-BN rat model of working-heart and lung transplantation,[46] at postoperative day 7 a significant reduction in MR signal intensity was observed in allotransplanted hearts, which developed moderate rejection. Syngeneic transplants showed no differences in MR signal intensities before and after USPIO injections. Following administration of USPIO particles at postoperative day 6, a group of allotransplanted rats was treated with CsA (3 mg/kg). Recipients treated with CsA for 7 days showed no reduction in MR signal intensity after USPIO reinjection at day 14, whereas animals treated for 4 days only showed a significant decrease in MR signal intensity in the transplanted hearts, indicative of acute allograft rejection. Pathological analysis of these animals revealed that dextran-coated USPIO particles were taken up by the infiltrating macrophages that accumulated within the rejecting cardiac graft. The MRI signal attenuation due to USPIO labeling correlated with rejection.[47]

It has been reported recently that with much larger, micrometer-sized spheres containing iron oxide,[48,49] single particles can be visualized by MRI. Referred to as micrometer-sized paramagnetic iron oxide (MPIO) particles, they are highly effective T_2* contrast agents composed of polystyrene/divinyl benzene polymer microspheres containing a magnetite (iron oxide) core. With the smaller USPIO particles, a cell must ingest millions of particles to cause enough local field distortion to be detectable by MRI.[49] MPIO particles are approximately 35 times larger in diameter and about 42,875 times larger in volume than USPIO particles. Potentially, each MPIO particle can produce up to 40,000 times larger background attenuation. Thus, the loading of only a few MPIO particles per cell is sufficient to produce detectable signal attenuation within one imaging voxel.

Wu et al. explored the possibility of noninvasively monitoring individual immune cells, primarily macrophages, *in vivo* in the acute DA-to-BN rat model of working-heart and lung transplantation by intravenously administering MPIO.[50] The distinctive contrast patterns of punctate dots caused by *in vivo* MPIO labeling in the rejecting heart was comparable, yet with some peculiar differences, to the contrast pattern observed by using USPIO particles. On postoperational day 6, while USPIO labeling manifested itself over a large, continuous area of contrast, MPIO labeling had a punctate pattern. Isograft controls did not show any detectable contrast after either MPIO or USPIO administration.

To avoid interference of abdominal gases and bowel and respiratory movements during acquisition of MR images, Kanno et al. developed a new model of heterotopic heart and lung transplantation into the inguinal region for the strain combination of DA rats as donors and BN rats as recipients.[51] For this model, at day 5 after transplantation, lung allografts in animals without CsA treatment were severely rejected as revealed by histology. Correspondingly, USPIO led to a significant signal attenuation in images of allografts 24 h after being administered. Syngeneic transplants showed no evidence of rejection and no differences in MRI signals between the images before and after injection of USPIO. Allotransplants in animals treated with CsA showed a mild rejection. USPIO-induced MRI signal changes in CsA-treated animals were smaller than those observed in untreated allografts. Immunohistochemistry and iron staining of lung allografts indicated that USPIO particles were taken up by the infiltrating macrophages that accumulated at the rejection site.[51]

14.4.3 Macrophage Tracking by MRI in Infected Tissue

Early attempts by MRI to visualize macrophages involved in inflammatory events after infections were made without specific labeling of macrophages. Demonstration of presence and activity of macrophages was the primary goal of these studies rather than investigation of the homing of these cells. Multiple concentric rims in *Nocardia asteroides* abscesses were hypothesized to be due to the organization of the necrotic debris and phagocytosis by macrophages in the capsule.[52] The multiple-ring structure was probably due to the slow action of the bacterium *N. asteriodes*, which allowed for modification of the host defense and was claimed to be specific for infection.

The rather typical feature of abscess is a dark rim demarcating the capsule on T_2-weighted MR images as mentioned above. In an equine study of a brain abscess, the occurrence of the dark ring was found to be due to the presence of numerous hemosiderin-laden macrophages.[53]

The presence and activity of macrophages was also demonstrated by ^{1}H MRS in a patient with hemophagocytic lymphhistiocytosis-associated encephalopathy that developed after rotavirus infection. The spectra revealed elevated lactate at the site of the lesion, indicating that macrophages not exhibiting aerobic metabolism had infiltrated into the central nervous system (CNS).[54]

The above-mentioned approaches were rather unspecific. First efforts to develop a specific readout were by labeling inflammatory cells.[55] The authors experimented with iron oxide particles with various coatings in order to load macrophages. They could successfully observe hypointensities, i.e., macrophage invasion in the rim of turbentine-induced sterile abscesses, only with lipid-coated SPIO particles. The longest plasma half-life of the SPIO particles with this coating was key for phagocytosis by macrophages. Macrophages in the abscess could then be visualized from 12 h up to 5 days after SPIO injection.

Later macrophage tracking into soft tissue infection by phagocytosis of USPIO particles was carried out in a similar way, as originally proposed for an EAE model[15] and extensively discussed in the earlier sections. Activated macrophages are powerful phagocytes that ingest all kinds of small particulate material, as well as dextran-coated USPIO particles with a long residence time in the bloodstream.[56] This renders activated macrophages MRI detectable by their T_2* effect. This technique also indicated increased high metabolic activity, as phagocytosis is energy consuming. The largest contrast, i.e., signal reduction in T_2*-weighted images in the abscess of calf muscle infected with *Staphylococcus aureus*, was observed 3 h after USPIO infusion. Thereafter, until 72 h, dynamic rearrangement of the macrophages could be followed. Histological analysis and electron microscopy of the abscess confirmed intracellular accumulation of the iron oxide particles within macrophages.[56] Functional-morphologic cellular MRI in experimental soft tissue infection[57] increased the specificity of MR findings in chronic infection. In the chronic phase of infection, the cellular infiltrate is dominated by macrophages and lymphocytes. Granulation tissue in septic and aseptic disease cannot be reliably differentiated from active inflammation with conventional MRI, whereas tracking of USPIO-loaded macrophages allows differentiation between areas with active inflammation and areas of reparative granulation (Figure 14.3). One limitation of this experimental study was the high amount of contrast agent; i.e. fourfold the approved clinical dose applied to the animals. However, due to the twice-as-long half-life of the USPIO in humans, compared to in rats, similar effects may be expected in patients, even with a decreased dose.

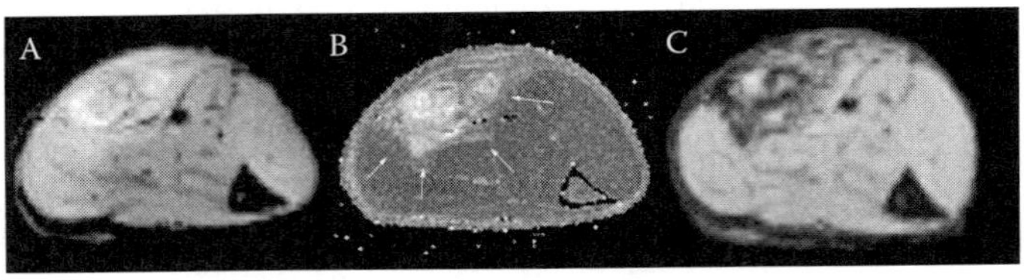

FIGURE 14.3 MR images of late chronic soft tissue infection. (A) High signal intensity is seen within infected muscle on a transverse T_2*-weighted gradient echo MR image obtained before intravenous USPIO administration. (B) Areas of high signal intensity (arrows) are delineated on a transverse T_2 map obtained before USPIO administration. The rather nonspecific term *edema pattern* is used to describe the extent of pathomorphologic changes. (C) On a transverse T_2*-weighted gradient echo MR image obtained 24 h after USPIO administration, macrophage invasion appears circumscribed in the abscess wall around the area of central necrosis (*) and is surrounded by a less intense and relatively blurred hypointense pattern.

Macrophage tracking in the same animal model and USPIO, but in a clinical imager at 1.5 T, were performed to investigate dose–response.[58] Distinct differences between high and low doses were observed. Although the number of ED1+ cells was not different between the groups, dark rims of susceptibility effect around the abscess could only be observed for the high dose. This effect was not visible at the low USPIO dose, but the entire abscess became brighter on T_1-weighted images. Histological investigation demonstrated that the macrophages had a completely different, i.e., much lower, iron content at the lower dose.

Short-term signal intensity alterations within bacterially induced abscesses were studied up to 60 min after USPIO administration in a rat model.[59] Staphylogenic abscesses of the right hind leg were induced in six Sprague-Dawley rats. This study demonstrated that different mechanisms exist for the accumulation of USPIO particles in soft tissue abscesses. One of the early ones is transcapillary transport. Several parameters, such as dose, size of the USPIO particles (here 26 nm), and impairment of the capillary integrity, affected the iron oxide-induced alteration of signal intensity. After the very short distribution period of 60 min, intracellular accumulation of USPIO could be mainly seen in the abscess wall, but extracellular deposits predominated. At later time points, i.e., 24 h or later after contrast agent administration, iron oxide particles could just be found inside macrophages.[56,57]

Alternatively, Pirko et al. developed an MRI technique to image immune cell location, including macrophages, by attaching superparamagnetic antibodies specific for cell surface markers.[60] Instead of loading immune cells such as CD4+ and CD8+ T-cells or macrophages with (U)SPIO particles, they intravenously injected superparamagnetic anti-CD4, anti-CD8, or anti-Mac1 antibodies into chronically infected mice. Theiler's murine encephalomyelitis virus-infected IFN-$R^{-/-}$ mice or C57BL/7 mice were used as test models. The authors verified by histological methods that immune cell brain infiltration could be followed over time. This technique does not require *in vitro* labeling of the cells of interest, provided that monoclonal antibodies are available against unique surface antigens and the cells are accessible from the bloodstream. Interestingly, with this method cells can be visualized on T_1-weighted images beside conventional T_2* ones. Then labeled cells appear bright rather than dark on the images, which is advantageous for image analysis. Therefore, these agents would be ideal for accurate imaging in clinical practice.

Positron emission tomography (PET) imaging of brain macrophages was performed in a macaque model of neuroAIDS.[61] Peripheral benzodiazepine receptors are abundant on activated macrophages and are expressed in low levels in the uninfected brain. Nevertheless, the authors could demonstrate the presence of macrophages in simian immunodeficiency virus (SIV) encephalitis *in vivo* by using the ^{11}C-labeled R-enantiomer of the specific ligand [^{11}C](R)-PK11195. Postmortem binding studies of [^{3}H](R)-PK11195 in brains with SIV encephalitis showed overlapping regions rich in activated macrophages and the radioactive ligand. They suggested that the technique could be useful to predict the development of HIV encephalitis.

Despite the fact that the PET approach was demonstrated to be practical, up to now clinical applications of noninvasive leukocyte tracking have only been shown with scintigraphic methods.[62] The high intrinsic sensitivity of scintigraphy is the main cause for its use in the clinical environment. However, like PET, scintigraphic imaging lacks an anatomical reference, which renders the precise localization of the cells difficult. Furthermore, leukocytes have to be labeled *in vitro* and then reapplied to patients, which is in contrast to macrophage labeling with iron oxide nanoparticles, as described above.

14.5 CONCLUSIONS

Many examples from preclinical research illustrate the feasibility of tracking phagocytotic cells *in vivo* with MRI. This cell type can directly be labeled by systemic administration of SPIO, USPIO or cross-linked iron oxide (CLIO) particles. The technology was successfully used for tracking macrophages entering the brain in neuroinflammatory diseases, allograft rejection, arthritis, and soft tissue infection. Moreover, some interesting examples also demonstrate the usefulness of this approach for testing drugs in pharmaceutical research. Tracking of macrophages could also enhance the clinical imaging portfolio provided that these contrast agents will be approved for use in patients. Tracking of nonphagocytotic cells such as lymphocytes is still difficult, and only a few studies have shown sufficient labeling efficacy to monitor cell migration *in vivo*. It remains to be demonstrated whether MRI can be routinely applied for tracking nonphagocytotic cells as well.

REFERENCES

1. Gordon, S., The macrophage, *Bioessays*, 17, 977, 1995.
2. Barron, K.D., The microglial cell. A historical review, *J. Neurol. Sci.*, 134, 57, 1995.
3. Bellin, M.-F., Roy, C., Kinkel, K., Thoumas, D., Zaim, S., Vanel, D., Tuchmann, C., Richard, F., Jacqmin, D., Delcourt, A., Challier, E., Lebret, T., and Cluzel, P., Lymph node metastases: safety and effectiveness of MR imaging with ultrasmall superparamagnetic iron oxide particles — initial clinical experience, *Radiology*, 207, 799, 1998.
4. Dousset, V., Gomez, C., Petry, K.G., Delalande, C., and Caille, J.M., Dose and scanning delay using USPIO for central nervous system macrophage imaging, *MAGMA*, 8, 185, 1999.
5. Schroeter, M., Jander, S., Huitinga, I., Witte, O., and Stoll, G., Phagocytic response in photochemically induced infarction of rat cerebral cortex, *Stroke*, 28, 382, 1997.
6. Schroeter, M., Jander, S., Witte, O., and Stoll, A.L., Local immune responses in the rat cerebral cortex after middle cerebral artery occlusion, *J. Neuroimmunol.*, 55, 195, 1994.
7. Rausch, M., Sauter, A., Frohlich, J., Neubacher, U., Radu, E.W., and Rudin, M., Dynamic patterns of USPIO enhancement can be observed in macrophages after ischemic brain damage, *Magn. Reson. Med.*, 46, 1018, 2001.
8. Rausch, M., Baumann, D., Neubacher, U., and Rudin, M., *In-vivo* visualization of phagocytotic cells in rat brains after transient ischemia by USPIO, *NMR Biomed.*, 15, 278, 2002.
9. Neumann-Haefelin, T., Kastrup, A., de Crespigny, A., Yenari, M.A., Ringer, T., Sun, G.H., and Moseley, M.E., Serial MRI after transient focal cerebral ischemia in rats: dynamics of tissue injury, blood-brain barrier damage, and edema formation, *Stroke*, 31, 1965, 2000.
10. de Vries, H.E., Kuiper, J., de Boer, A.G., Van Berkel, T.J., and Breimer, D.D., The blood-brain barrier in neuroinflammatory diseases, *Pharmacol. Rev.*, 49, 143, 1997.
11. Bulte, J.W., de Jonge, M.W., de Leij, L., The, T.H., Kamman, R.L., Blaauw, B., Zuiderveen, F., and Go, K.G., Passage of DMP across a disrupted BBB in the context of antibody-mediated MR imaging of brain metastases, *Acta Neurochir. Suppl.*, 51, 43, 1990.
12. Muldoon, L.L., Pagel, M.A., Kroll, R.A., Roman-Goldstein, S., Jones, R.S., and Neuwelt, E.A., A physiological barrier distal to the anatomic blood-brain barrier in a model of transvascular delivery, *Am. J. Neuroradiol.*, 20, 217, 1999.

13. Muldoon, L.L., Sandor, M., Pinkston, K.E., and Neuwelt, E.A., Imaging, distribution, and toxicity of superparamagnetic iron oxide magnetic resonance nanoparticles in the rat brain and intracerebral tumor, *Neurosurgery*, 57, 785, 2005.
14. Rausch, M., Hiestand, P., Baumann, D., Cannet, C., and Rudin, M., MRI-based monitoring of inflammation and tissue damage in acute and chronic relapsing EAE, *Magn. Reson. Med.*, 50, 309, 2003.
15. Dousset, V., Delalande, C., Ballarino, L., Quesson, B., Seilhan, D., Coussemacq, M., Thiaudière, E., Brochet, B., Canioni, P., and Caille, J.M., *In vivo* macrophage activity imaging in the central nervous system detected by magnetic resonance, *Magn. Reson. Med.*, 41, 329, 1999.
16. Floris, S, Blezer, E.L.A., Schreibelt, G., Doepp, E., van der Pol, S.M.A., Schadee-Eesterman, I.L., Nicoley, K., Dijkstra, C.D., and de Vries, H.E., Blood-brain barrier permeability and monocyte infiltration in experimental allergic encephalomyelitis: a quantitative MRI study, *Brain*, 127, 1, 2004.
17. Berger, C., Hiestand, P., Kindler-Baumann, D., Rudin, M., and Rausch, M., Analysis of lesion development during acute inflammation and remission in a rat model of experimental autoimmune encephalomyelitis by visualization of macrophage infiltration, demyelination and blood-brain barrier damage, *NMR Biomed.*, 19, 101, 2006.
18. Rausch, M., Hiestand, P., Foster, C.A., Baumann, D., Cannet, C., and Rudin, M., Predicability of FTY720 efficacy in experimental autoimmune encephalomyelitis by *in vivo* macrophage tracking: clinical implications for ultrasmall superparamagnetic iron oxide-enhanced magnetic resonance imaging, *J. Magn. Reson. Imaging*, 20, 16, 2004.
19. Deloire, M.S.A., Touil, T., Brochet, B., Dousset, V., Caille, J.M., and Petry, K.G., Macrophage brain infiltration in experimental autoimmune encephalomyelitis is not completely compromised by suppressed T-cell invasion: *in vivo* magnetic resonance imaging illustration in effective anti-VLA-4 antibody treatment, *Multiple Sclerosis*, 10, 540, 2004.
20. Yeh, T., Zhang, W., Ildstad, S.T., and Ho, C., *In vivo* dynamic MRI tracking of rat T-cells labeled with superparamagnetic iron-oxide particles, *Magn. Reson. Med.*, 33, 200, 1995.
21. Dodd, C.H., Hsu, H.C., Chu, W.J., Yang, P., Zhang, H.G., Mountz, J.D., Jr., Zinn, K., Forder, J., Josephson, L., Weissleder, R., Mountz, J.M., and Mountz, J.D., Normal T-cell response and *in vivo* magnetic resonance imaging of T cells loaded with HIV transactivator-peptide-derived superparamagnetic nanoparticles, *J. Immunol. Methods*, 256, 89, 2001.
22. Frank, J.A., Zywicke, H., Jordan, E.K., Mitchell, J., Lewis, B.K., Miller, B., Bryant, L.H., Jr., and Bulte, J.W., Magnetic intracellular labeling of mammalian cells by combining (FDA-approved) superparamagnetic iron oxide MR contrast agents and commonly used transfection agents, *Acad. Radiol.*, 9, (Suppl. 2), 484, 2002.
23. Arbab, A.S., Yocum, G.T., Wilson, L.B., Parwana, A., Jordan, E.K., Kalish, H., and Frank, J.A., Comparison of transfection agents in forming complexes with ferumoxides, cell labeling efficiency, and cellular viability, *Mol. Imaging*, 3, 24, 2004.
24. Bulte, J.W., Douglas, T., Witwer, B., Zhang, S.C., Lewis, B.K., van Gelderen, P., Zywicke, H., Duncan, I.D., and Frank, J.A., Monitoring stem cell therapy *in vivo* using magnetodendrimers as a new class of cellular MR contrast agents, *Acad. Radiol.*, 9, S332, 2002.
25. Cutolo, M., Macrophages as effectors of the immunoendocrinologic interactions in autoimmune rheumatic diseases, *Neuroendocrine Immune Basis Rheum. Dis.*, 876, 32, 1999.
26. Dardzinski, B.J., Schmithorst, V.J., Holland, S.K., Boivin, G.P., Imagawa, T., Watanabe, S., Lewis, J.M., and Hirsch, R., MR imaging of murine arthritis using ultrasmall superparamagnetic iron oxide particles, *Magn. Reson. Imaging*, 19, 1209, 2001.
27. Beckmann, N., Falk, R., Zurbrugg, S., Dawson, J., and Engelhardt, P., Macrophage infiltration into the rat knee detected by MRI in a model of antigen-induced arthritis, *Magn. Reson. Med.*, 49, 1047, 2003.
28. Lutz, A.M., Seemayer, C., Corot, C., Gay, R.E., Goepfert, K., Michel, B.A., Marincek, B., Gay, S., and Weishaupt, D., Detection of synovial macrophages in an experimental rabbit model of antigen-induced arthritis: ultrasmall superparamagnetic iron oxide-enhanced MR imaging, *Radiology*, 233, 149, 2004.
29. Lai, W.F., Chang, C.H., Tang, Y., Bronson, R., and Tung, C.H., Early diagnosis of osteoarthritis using cathepsin B sensitive near-infrared fluorescent probes, *Osteoarthritis Cartilage*, 12, 239, 2004.

30. Hansch, A., Frey, O., Sauner, D., Hilger, I., Haas, M., Malich, A., Brauer, R., and Kaiser, W.A., *In vivo* imaging of experimental arthritis with near-infrared fluorescence, *Arthritis Rheum.*, 50, 961, 2004.
31. Wunder, A., Tung, C.H., Muller-Ladner, U., Weissleder, R., and Mahmood, U., *In vivo* imaging of protease activity in arthritis: a novel approach for monitoring treatment response, *Arthritis Rheum.*, 50, 2459, 2004.
32. Schuurman, H.J., Weckbecker, G., Bruns, C., Beckman, N., Rudin, M., and Cook, N.S., Preclinical models of chronic rejection: promises and pitfalls, *Transplant Proc.*, 29, 2624, 1997.
33. Engelbrecht, G., Kahn, D., Duminy, F., and Hickman, R., New rapid technique for renal-transplantation in the rat, *Microsurgery*, 13, 340, 1992.
34. Rocha, P.N., Plumb, T.J., Crowley, S.D., and Coffman, T.M., Effector mechanisms in transplant rejection, *Immunol. Rev.*, 196, 51, 2003.
35. Zhang, Y., Dodd, S.J., Hendrich, K.S., Williams, M., and Ho, C., Magnetic resonance imaging detection of rat renal transplant rejection by monitoring macrophage infiltration, *Kidney Int.*, 58, 1300, 2000.
36. Ye, Q., Yang, D., Williams, M., Williams, D.S., Pluempitiwiriyawej, C., Moura, J.M., and Ho, C., *In vivo* detection of acute rat renal allograft rejection by MRI with USPIO particles, *Kidney Int.*, 61, 1124, 2002.
37. Grau, V., Herbst, B., and Steiniger, B., Dynamics of monocytes/macrophages and T lymphocytes in acutely rejecting rat renal allografts, *Cell Tissue Res.*, 291, 117, 1998.
38. Grimm, P.C., McKenna, R., Nickerson, P., Russell, M.E., Gough, J., Gospodarek, E., Liu, B., Jeffery, J., and Rush, D.N., Clinical rejection is distinguished from subclinical rejection by increased infiltration by a population of activated macrophages, *J. Am. Soc. Nephrol.*, 10, 1582, 1999.
39. Croker, B.P., Clapp, W.L., bu Shamat, A.R., Kone, B.C., and Peterson, J.C., Macrophages and chronic renal allograft nephropathy, *Kidney Int. Suppl.*, 57, S42, 1996.
40. Herrero-Fresneda, I., Torras, J., Cruzado, J.M., Condom, E., Vidal, A., Riera, M., Lloberas, N., Alsina, J., and Grinyo, J.M., Do alloreactivity and prolonged cold ischemia cause different elementary lesions in chronic allograft nephropathy? *Am. J. Pathol.*, 162, 127, 2003.
41. Azuma, H., Takahara, S., Matsumoto, K., Ichimaru, N., Wang, J.D., Moriyama, T., Waaga, A.M., Kitamura, M., Otsuki, Y., Okuyama, A., Katsuoka, Y., Chandraker, A., Sayegh, M.H., and Nakamura, T., Hepatocyte growth factor prevents the development of chronic allograft nephropathy in rats, *J. Am. Soc. Nephrol.*, 12, 1280, 2001.
42. Hamar, P., Szabo, A., Muller, V., and Heemann, U., Involvement of interleukin-2 and growth factors in chronic kidney allograft rejection in rats, *Transplant Proc.*, 33, 2160, 2001.
43. Ziai, F., Nagano, H., Kusaka, M., Coito, A.J., Troy, J.L., Nadeau, K.C., Rennke, H.G., Tilney, N.L., Brenner, B.M., and MacKenzie, H.S., Renal allograft protection with losartan in FisherLewis rats: hemodynamics, macrophages, and cytokines, *Kidney Int.*, 57, 2618, 2000.
44. Yang, J., Reutzel-Selke, A., Steier, C., Jurisch, A., Tullius, S.G., Sawitzki, B., Kolls, J., Volk, H.D., and Ritter, T., Targeting of macrophage activity by adenovirus-mediated intragraft overexpression of TNFRp55-Ig, IL-12p40, and vIL-10 ameliorates adenovirus-mediated chronic graft injury, whereas stimulation of macrophages by overexpression of IFN-gamma accelerates chronic graft injury in a rat renal allograft model, *J. Am. Soc. Nephrol.*, 14, 214, 2003.
45. Beckmann, N., Cannet, C., Fringeli-Tanner, M., Baumann, D., Pally, C., Bruns, C., Zerwes, H.G., Andriambeloson, E., and Bigaud, M., Macrophage labeling by SPIO as an early marker of allograft chronic rejection in a rat model of kidney transplantation, *Magn. Reson. Med.*, 49, 459, 2003.
46. Wu, Y.J., Sato, K., Ye, Q., and Ho, C., MRI investigations of graft rejection following organ transplantation using rodent models, *Methods Enzymol.*, 386, 73, 2004.
47. Kanno, S., Wu, Y.J., Lee, P.C., Dodd, S.J., Williams, M., Griffith, B.P., and Ho, C., Macrophage accumulation associated with rat cardiac allograft rejection detected by magnetic resonance imaging with ultrasmall superparamagnetic iron oxide particles, *Circulation*, 104, 934, 2001.
48. Hinds, K.A., Hill, J.M., Shapiro, E.M., Laukkanen, M.O., Silva, A.C., Combs, C.A., Varney, T.R., Balaban, R.S., Koretsky, A.P., and Dunbar, C.E., Highly efficient endosomal labeling of progenitor and stem cells with large magnetic particles allows magnetic resonance imaging of single cells, *Blood*, 102, 867, 2003.
49. Shapiro, E.M., Skrtic, S., Sharer, K., Hill, J.M., Dunbar, C.E., and Koretsky, A.P., MRI detection of single particles for cellular imaging, *Proc. Natl. Acad. Sci. U.S.A.*, 101, 10901, 2004.

50. Wu, Y.L., Ye, Q., Foley, L.M., Hitchens, T.K., Sato, K., Williams, J.B., and Ho, C., *In situ* labeling of immune cells with iron oxide particles: an approach to detect organ rejection by cellular MRI, *Proc. Natl. Acad. Sci. U.S.A.*, 103, 1852, 2006.
51. Kanno, S., Lee, P.C., Dodd, S.J., Williams, M., Griffith, B.P., and Ho, C., A novel approach with magnetic resonance imaging used for the detection of lung allograft rejection, *J. Thorac. Cardiovasc. Surg.*, 120, 923, 2000.
52. Pyhtinen, J., Paakko, E., and Jartti, P., Cerebral abscess with multiple rims of MRI, *Neuroradiology*, 39, 857, 1997.
53. Audigie, F., Tapprest, J., George, C., Didierlaurent, D., Foucher, N., Faurie, F., Houssin, M., and Denoix, J.M., Magnetic resonance imaging of a brain abscess in a 10-month-old filly, *Vet. Radiol. Ultrasound*, 45, 210, 2004.
54. Takahashi, S., Oki, J., Miyamoto, A., Koyano, S., Ito, K., Azuma, H., and Okuno, A., Encephalopathy associated with haemophagocytic lymphohistiocytosis following rotavirus infection, *Eur. J. Pediatr.*, 158, 133, 1999 (review).
55. Chan, T.W., Eley, C., Liberti, P., So, A., and Kressel, H.Y., Magnetic resonance imaging of abscesses using lipid-coated iron oxide particles, *Invest. Radiol.*, 27, 443, 1992.
56. Kaim, A.H., Wischer, T., O'Reilly, T., Jundt, G., Frohlich, J., von Schulthess, G.K., and Allegrini, P.R., MR imaging with ultrasmall superparamagnetic iron oxide particles in experimental soft-tissue infections in rats, *Radiology*, 225, 808, 2002.
57. Kaim, A.H., Jundt, G., Wischer, T., O'Reilly, T., Frohlich, J., von Schulthess, G.K., and Allegrini, P.R., Functional-morphologic MR imaging with ultrasmall superparamagnetic particles of iron oxide in acute and chronic soft-tissue infection: study in rats, *Radiology*, 227, 169, 2003.
58. Lutz, A.M., Weishaupt, D., Persohn, E., Goepfert, K., Froehlich, J., Sasse, B., Gottschalk, J., Marincek, B., and Kaim, A.H., Imaging of macrophages in soft-tissue infection in rats: relationship between ultrasmall superparamagnetic iron oxide dose and MR signal characteristics, *Radiology*, 234, 765, 2005.
59. Gellissen, J., Axmann, Ch., Prescher, A., Bohndorf, K., and Lodemann, K.-P., Extra- and intracellular accumulation of ultrasmall superparamagnetic iron oxides (USPIO) in experimentally induced abscesses of the peripheral soft tissues and their effects on magnetic resonance imaging, *Magn. Reson. Imaging*, 17, 557, 1999.
60. Pirko, I., Johnson, A., Ciric, B., Gamez, J., Macura, S.I., Pease, L.R., and Rodriguez, M., *In vivo* magnetic resonance imaging of immune cells in the central nervous system with superparamagnetic antibodies, *FASEB J.*, 17, 2003.
61. Venneti, S., Lopresti, B.J., Wang, G., Bissel, S.J., Mathis, C.A., Meltzer, C.C., Boada, F., Capuano S., III, Kress, G.J., Davis, D.K., Ruszkiewicz, J., Reynolds, I.J., Murphey-Corb, M., Trichel, A.M., Wisniewski, S.R., and Wiley, C.A., PET imaging of brain macrophages using the peripheral benzodiazepine receptor in a macaque model of neuroAIDS, *J. Clin. Invest.*, 113, 981, 2004.
62. Kiessling, F. and Semmler, W., Status and perspectives of noninvasive cell tracking, *Z. Med. Phys.*, 15, 169, 2005 (German).

15 Cellular Imaging of Macrophage Activity in Stroke

Ralph Weber and Mathias Hoehn

CONTENTS

15.1 Inflammatory Mechanisms after Ischemic Stroke: A Potential Approach for Neuroprotective Therapy 273
15.2 Detection of Macrophage Activity by Positron Emission Tomography 275
15.3 Detection of Macrophage Activity by Magnetic Resonance Imaging 276
15.4 Outlook 280
References 280

15.1 INFLAMMATORY MECHANISMS AFTER ISCHEMIC STROKE: A POTENTIAL APPROACH FOR NEUROPROTECTIVE THERAPY

Ischemic brain injury first causes direct cell damage through energy failure, resulting in necrosis and delayed apoptosis of brain cells. Yet, the postischemic inflammatory response after focal cerebral ischemia also plays a crucial role in the pathophysiological cascade of stroke evolution in the chronic phase (Figure 15.1).[1–3] Inflammatory cells and cytotoxic mediators produced by these cells contribute to the final extent and severity of the ischemic brain injury.[4,5] In particular, the potentially salvagable tissue surrounding the ischemic core, the so-called penumbra, is affected by this postischemic inflammatory reaction,[6] which can still contribute to a delayed, yet substantial, expansion of the infarction and a subsequent clinical deterioration.[7]

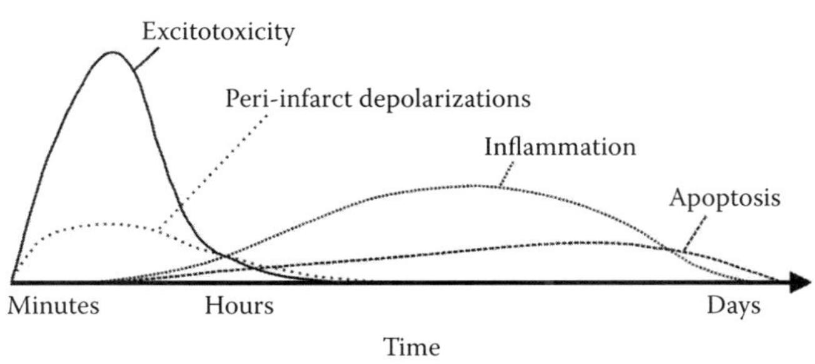

FIGURE 15.1 Putative cascade of damaging events in focal cerebral ischemia. Very early after the onset of the focal perfusion deficit, excitotoxic mechanisms can damage neurons and glia cells lethally. In addition, excitotoxicity triggers a number of events that can further contribute to the demise of the tissue. Such events include peri-infarct depolarizations and the more delayed mechanisms of inflammation and programmed cell death (apoptosis). The x-axis reflects the evolution of the cascade over time, while the y-axis aims to illustrate the impact of each element of the cascade on final outcome. (Reproduced from Dirnagl, U. et al., *Trends Neurosci.*, 22, 391–397, 1999.[1] With permission.)

Hematogenous granulocytes are the first cells entering the ischemic brain parenchyma within hours of stroke onset.[8] The infiltration of granulocytes peaks at 24 h after infarction, while only a few granulocytes are still present in the ischemic tissue after 7 days.[2] Macrophages become the predominant cells 4 to 7 days after ischemia and are present in the boundary zone of the infarction for many weeks. The macrophage response in the ischemic brain consists of two cellular populations, both contributing to the inflammation process. Blood-borne monocytes (hematogenous macrophages) enter the injured brain area via cerebral blood vessels, while microglia, the resident macrophages of the brain,[2] are activated by the ischemia. After activation, blood-borne monocytes and cerebral microglia cannot be distinguished by cellular markers anymore. Recently, two groups used bone marrow chimera mice expressing green fluorescent protein (GFP) in hematogenous monocytes to monitor the temporal pattern of monocyte invasion and microglia activation after focal brain ischemia in mice.[9,10] These studies showed an earlier activation of resident microglia within 1 day of ischemia, compared to a delayed invasion of hematogenous monocytes.

Therapeutic interventions after stroke are still scarce, and only recombinant tissue plasminogen activator (rtPA) has been shown to be effective in human stroke patients, when given in the first 6 h after stroke onset.[11] Therefore, other therapeutic strategies aim at the postischemic phase, hoping to reduce the harmful effects of macrophage infiltration, macrophage activation, and production of toxic mediators, and to extend the therapeutic window (see Barone and Feuerstein[12] and Becker[13] for reviews). Experiments in rodent models of focal brain ischemia have demonstrated that pro-inflammatory chemokines that attract hematogenous monocytes to the site of ischemia, like monocyte chemoattractant protein-1 (MCP-1),[14] or chemokines that are predominantly produced by macrophages, like tumor necrosis factor (TNF)[15,16] or matrix metalloproteinases,[17] are possible targets for an anti-inflammatorry strategy. Animals treated with neuroregulin-1 showed an attenuated expression of several inflammatory and stress genes, including heat shock protein-70, interleukin-1beta, and MCP-1 in activated macrophages.[18] An anti-MCP-1 gene therapy was effective in reducing infarct volume and infiltration of inflammatory cells in hypertensive rats subjected to focal brain ischemia.[19]

The anti-inflammatory properties of a number of already approved drugs by the Food and Drug Adminstration (FDA) have also recently been tested in experimental focal brain ischemia.[20,21] The angiotensin-1 receptor antagonist irbesartan showed improved motor functions and an inhibition of microglia activation and macrophage invasion on days 3 and 7 after focal cerebral ischemia when continuously infused intracerebroventricularly.[20] Sundararajan and coworkers tested the anti-inflammatory potential of a new class of insulin-sensitizing agents, which are already used as antidiabetic drugs.[21] The peroxisome proliferator-activated receptor-gamma (PPAR) agonists troglitazone and pioglitazone reduced infarction volume, improved neurological function, reduced macrophage activity, and decreased expression of inflammatory mediators after focal brain ischemia in rats.

Despite promising results using experimental anti-inflammatory therapies after focal brain ischemia in animal studies, no such therapy has been proven effective in human stroke studies to date. A randomized placebo-controlled trial of a murine intercellular adhesion molecule-1 (ICAM-1) antibody to reduce leukocyte adhesion in 625 stroke patients even showed a worse stroke outcome in patients treated with the antibody.[22]

Since invasive histological or immunohistochemical techniques to detect and monitor the postischemic response after focal brain ischemia are not feasible in human stroke studies, attempts have been made to study cellular macrophage activity after brain ischemia with molecular imaging techniques, like positron emission tomography (PET) or magnetic resonance imaging (MRI). Such a noninvasive approach of inflammation monitoring is of particular importance, as serial measurements can be performed, a prerequisite for studying both time-dependent inflammatory activity after stroke onset and therapeutical anti-inflammatory strategies in human stroke patients.

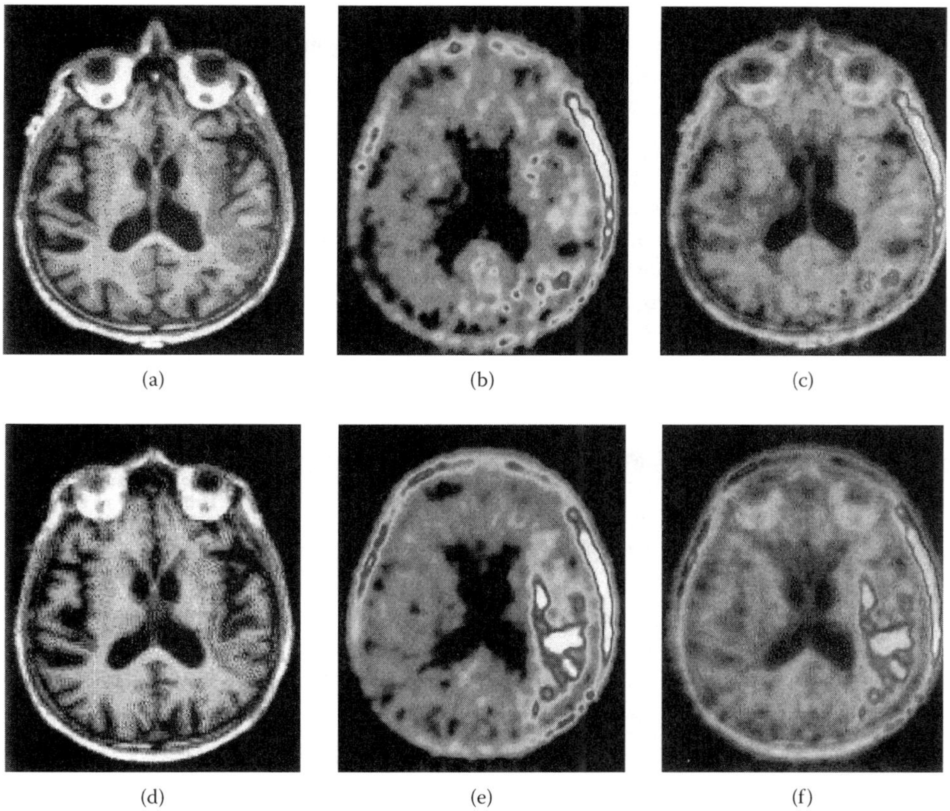

FIGURE 15.2 (please see color insert following page 210) [^{11}C]-PK11195 PET and T_1-weighted MRI of a patient with left hemispheric MCA infarct. (a) MRI at 5 days after the infarct, (b) PET at same time, and (c) PET co-registered with MRI. (d–f) Corresponding images 13 days after infarct. At the early time points, slight PK binding is visualized in left MCA territory, especially in the temporal lobe. At 13 days, this binding is more pronounced and the affected area exceeds that of signal change in the MRI. (Reproduced from Gerhard, A. et al., *Neuroreport*, 11, 2957–2960, 2000.[31] With permission.)

15.2 DETECTION OF MACROPHAGE ACTIVITY BY POSITRON EMISSION TOMOGRAPHY

The first attempts to detect macrophage accumulation *in vivo* after focal brain ischemia were based on positron emission tomography (PET) using the ligand PK11195. This ligand has been shown to bind specifically and selectively to peripheral-type benzodiazepine receptors.[23] These binding sites are normally only found at very low levels in the healthy brains of rat and human, but increase dramatically in lesioned brain areas.[24,25] Autoradiographic studies showed that the binding of PK11195 correlated, in both space and time, with the appearance of macrophages (both activated microglia and invading blood-borne monocytes) in the border zone of the ischemic lesion in photothrombotically induced cortical infarcts in rats.[26] Brain regions labeled with PK11195 correlated with regions of positive ED-1 staining, a monoclonal antibody against macrophages. These regions became detectable as soon as 2 days after stroke induction as small cell clusters; they were present around the whole infarct area after 7 days.

Labeling with the positron-emitting carbon isotope [^{11}C]-PK11195 has been used to visualize the accumulation of macrophages after focal brain ischemia, both in animal models[27] and in humans[28–31] (Figure 15.2).

Sette and coworkers evaluated the time patterns of macrophage accumulation by sequential PET imaging in anesthetized baboons subjected to focal brain ischemia.[27] Extensive PET studies were performed between 1 and 91 days after stroke induction, consisting of assessments of [^{11}C]-PK11195 binding, [^{11}C]-flumazenil binding, cerebral blood flow, and oxygen consumption. Additionally, the lesion morphology was assessed by computer tomography (CT) scanning. The authors found a time-dependent increase in [^{11}C]-PK11195 uptake in the peri-infarct area and, to a lesser extent, in the infarct core, with a maximum binding between 20 and 40 days after occlusion of the middle cerebral artery. Unfortunately, no histological analysis and co-registration with the *in vivo* PET data was performed in this study to evaluate the specificity and sensitivity of the [^{11}C]-PK11195 binding sites with the macrophage accumulation.

In a human case study using longitudinal PET recordings, Ramsay and coworkers found a quantitatively increased [^{11}C]-PK11195 uptake at both 13 and 20 days after the patient had experienced a cardioembolic stroke.[29] In the scans at 6 days after stroke onset, only a small rim of increased binding around the outer borders of the lesion was visible. Gerhard and coworkers examined five patients with ischemic stroke 5 to 53 days after infarction onset.[31] Regions of increased [^{11}C]-PK11195 binding extended beyond the lesioned area as identified by T_1-weighted MRI.

[^{11}C]-PK11195 has recently also been used as a marker for the detection of remote macrophage activation of resting microglia in the ipsilateral thalamus in ischemic stroke patients.[30]

All human and animal PET studies using [^{11}C]-PK11195 labeling showed a delayed [^{11}C]-PK11195 uptake compared with the histological time pattern of microglial activation and macrophage infiltration after ischemic stroke, which may be attributable to the poor spatial resolution of PET imaging at earlier time points.

15.3 DETECTION OF MACROPHAGE ACTIVITY BY MAGNETIC RESONANCE IMAGING

Compared to PET imaging, magnetic resonance imaging (MRI) has a much higher spatial resolution, allowing visualization with more detail of cellular processes like macrophage infiltration and activation, especially at higher magnetic field strengths, now available for experimental animal and human investigations.

To date, two different approaches have been used to delineate macrophage accumulation after stroke by MRI: high-resolution multiparametric MRI, including gadolinium (Gd)-enhanced T_1-weighted images, as used in other central nervous system diseases with inflammatory lesion activity, such as multiple sclerosis,[32] and contrast-enhanced MRI using injectable ultrasmall superparamagnetic iron oxide (USPIO) nanoparticles[33] for labeling of blood-borne macrophages. USPIO particles have a hydrodynamic diameter of 20 to 30 nm, are taken up by macrophages, and produce a signal loss on T_2- and T_2*-weighted MRI due to susceptibility effects caused by the dextran-coated iron oxide.

Sbarbati and coworkers described a layered pattern in the lesioned ischemic cortex 1 week after stroke induction on T_2-weighted images and quantitative T_2 maps, at a field strength of 4.7 T, in approximately 25% of the rats subjected to permanent occlusion of the middle cerebral artery (MCAO).[34] The superficial layer and deepest layer were both hyperintense, while the intermediate layer was iso-only slightly hyperintense with respect to healthy cortex. This layered pattern disappeared again between 2 and 4 weeks after MCAO. Upon histological and ultrastructural examination, the intermediate layer was found to be composed of degenerating brain tissue and macrophages accumulating around large capillaries. The authors speculated about the presence of iron in the cytoplasm of macrophages as an explanation for the reduction of the signal intensity on T_2-weighted MRI. However, no Prussian blue staining was performed in this study to verify the hypothesis.

However, using the photothrombotic model of focal brain ischemia, Schroeter and coworkers were not able to delineate distinct areas of macrophage accumulation between 3 and 14 days after stroke induction.[35] Those authors used an extensive multiparametric MRI protocol, comprising quantitative apparent diffusion coefficient (ADC) and T_2 mapping, perfusion-weighted imaging (PWI), and Gd-DTPA-enhanced T_1-weighted MRI at a field strength of 7 T. As the iron-induced susceptibility change becomes more sensitive and more pronounced with increasing field strength, it is noteworthy that Schroeter et al. were not able to confirm the observations by Sbarbati and colleagues.

Since imaging of macrophage activity and discrimination of inflamed from noninflamed brain tissue were not feasible with standard, even though quantitative, MRI protocols, and without administration of an MRI-compatible, macrophage-specific contrast agent, several groups used ultrasmall superparamagnetic iron oxide (USPIO) nanoparticles[33] to monitor macrophage infiltration *in vivo* in various rat models of experimental focal brain ischemia.[36–40] When administered intravenously, dextran-coated USPIO particles are phagocytosed by blood-borne monocytes, and then are stored in lysosomes of these macrophages. Iron-loaded macrophages do not penetrate the blood–brain barrier of healthy brain tissue. Histologically, iron-labeled macrophages are visualized by a combined Prussian blue staining for iron and ED-1 antibody staining for macrophages.[38]

Rausch and coworkers were the first to demonstrate the dynamic infiltration of blood-borne macrophages into the ischemic brain tissue of rats subjected to permanent MCAO.[36] Due to susceptibility effects produced by the incorporated iron oxide, USPIO-labeled blood-borne macrophages could be visualized in the boundary zone of the ischemic lesion as hypointense signal changes on T_2-weighted MRI, as early as 24 h after systemic USPIO administration. The USPIO accumulation was still detectable 7 days after administration, although its concentration was markedly reduced at this time. Surprisingly, the same authors were not able to reproduce these findings in a model of transient MCAO on T_2-weighted images, but depicted hyperintense signal changes on T_1-weighted images 48 h after USPIO administration, which disappeared again between days 4 and 7.[37] To better describe the spatiotemporal pattern of macrophage infiltration, Kleinschnitz and coworkers injected USPIO particles at different time points after photothrombotic infarction and performed MRI 24 h after each injection.[38] In this project, USPIO-loaded macrophages, identified again as hypointense signal changes on T_2-weighted and three-dimensional constructed interference in steady state (CISS) sequences, accumulated in a rim-like manner at the outer margin of the ischemic cortical lesion only from day 5.5 on, independent of a disruption of the blood–brain barrier. No iron-loaded macrophage accumulation was visible at time points later than 9 days after stroke induction, although iron-negative, ED-1-positive macrophages were still found histologically after this time. Another important finding was the fact that once USPIO-loaded macrophages invaded the ischemic boundary zone, they did not further migrate from the site of their primary infiltration (Figure 15.3).

With the help of an improved protocol for the histochemical detection of USPIO-mediated iron[40] and high-resolution three-dimensional T_2*-weighted MR images, Saleh and coworkers were able to detect USPIO-loaded macrophages with an in-plane resolution of 48 μm. They observed perfect spatial agreement of the hypointense MRI signal changes with histochemically detected paramagnetic iron in macrophages at 6 days after photothrombosis (Figure 15.4).[39]

At the high magnetic field strength of 7 T, a faint hypointense signal loss in the ischemic boundary zone was also described by these authors in animals subjected to photothrombosis but without an intravenous injection of USPIO particles.[39] The origin of these endogenous susceptibility effects was later further investigated by us in rats subjected to transient MCAO and not receiving USPIO particles.[41] A time-dependent accumulation of iron-rich macrophages, both blood-borne monocytes and microglia transformed into macrophages, was detected in the chronic phase (between 2 and 10 weeks after MCAO) on T_2*-weighted MRI, in excellent spatial congruence with the corresponding histological detection of ED-1-positive, iron-loaded macrophages. Most of the iron-loaded macrophages were found in the neighborhood of cerebral blood vessels in the ischemic striatum, indicating delayed microhemorrhages due to late-onset vascular degradation (Figure 15.5).

This finding has implications for the use of iron-containing contrast agents as biomarkers for cellular imaging of macrophage invasion or for therapeutic cell implantation in stroke animals[42,43] or patients, since the endogenous susceptibility effects, caused by the uptake of iron from erythrocytes

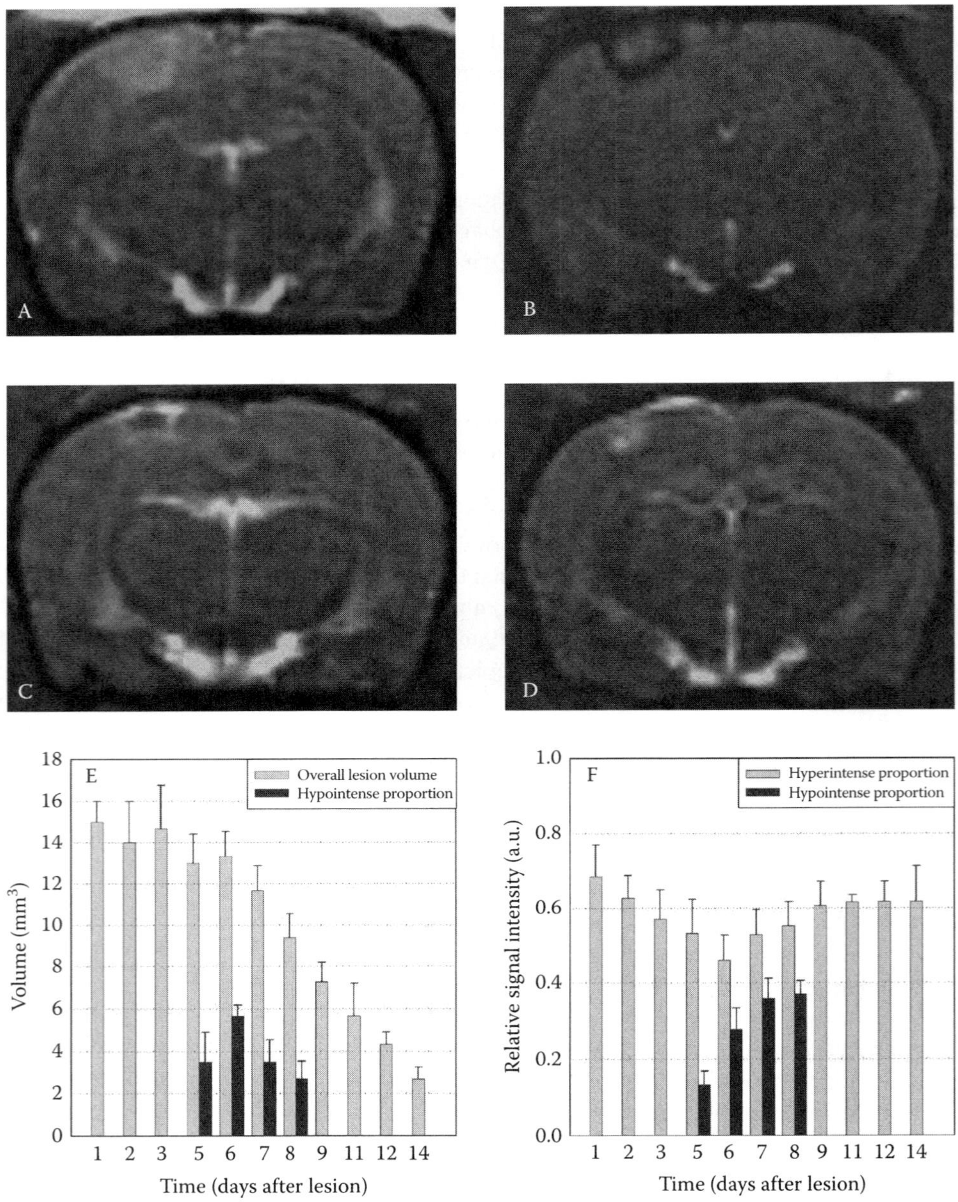

FIGURE 15.3 *In vivo* assessment of macrophage infiltration by magnetic resonance imaging as indicated by focal signal loss: coronal CISS-weighted images of photothrombotic infarcts at days 3 (A), 6 (B), 8 (C), and 14 (D). Animals always received SPIO particles systemically 24 h before MRI. Infarcts appeared as hyperintense lesions at early stages of infarct development because of tissue damage and edema formation (A). At day 6, signal loss indicative of local iron accumulation was apparent at the outer margins of infarcts (B); at day 8, the core of the lesions exhibited signal loss, whereas the periphery appeared hyperintense again (C). At day 14, lesions were entirely hyperintense and had shrunk because of tissue remodeling and scar formation (D). (E) Quantification of the overall lesion volume based upon the area of hyperintensity on CISS images (grey bars) at various stages of infarct development. (F) Mean signal intensities in arbitrary units (a.u.) of the hyper- and hypointense areas after normalization to cerebrospinal fluid. Note the lowest signal values indicating maximum SPIO accumulation at day 5. (Reproduced from Kleinschnitz, C. et al., *J. Cereb. Blood Flow Metab.*, 23, 1356–1361, 2003.[38] With permission.)

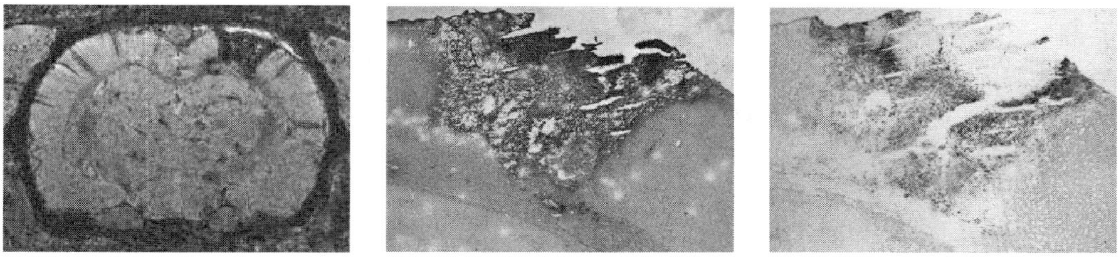

FIGURE 15.4 Coronal T_2*-weighted MRI (left) through the edge of a photothrombotic cortical lesion in the rat brain, 6 days after lesion induction. The hypointense region indicates the massive accumulation of USPIO-labeled macrophages in the infarct border zone. Directly following the *in vivo* MRI, Prussian blue staining for iron content (center) and ED-1 immunohistochemistry (right) were performed, confirming the interpretation of iron-rich macrophages in the region of hypointensity on the MR image.

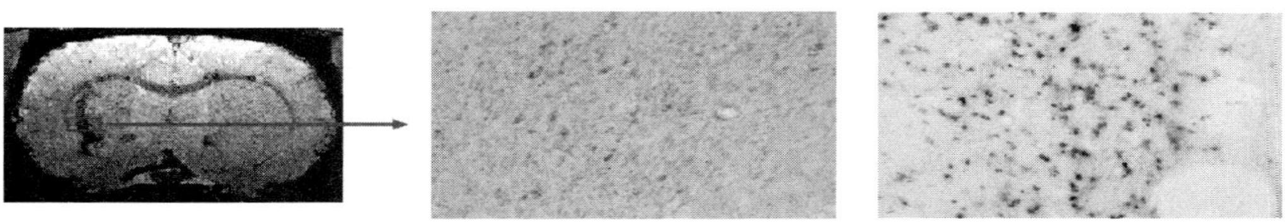

FIGURE 15.5 (please see color insert) Coronal T_2*-weighted MRI (left) of rat brain, 10 weeks after transient middle cerebral artery occlusion for infarct induction. In the ischemic territory on the left hemisphere, a diffuse dark region is discernible. (At an earlier time point still, a clear, discrete line stemming from an intrastriatal vessel is visible.) Prussian blue staining (center) for iron indicates many iron-rich cells, perferentially in the close neighborhood of vessels. Double staining for Prussion blue and ED-1 immunohistochemistry for macrophages (right) confirms the interpretation of iron-rich macrophages close to degrading vessels in the ischemic territory.

leaking out of degraded vessels, cannot be distinguished from deliberately USPIO-labeled macrophages or stem cells, neither by MRI signal changes nor by Prussian blue staining. Ways around this problem have recently been discussed by the same authors. Endogenous, vascular-origin macrophage labeling can be unequivocally interpreted if neither systemic USPIO application nor labeled cell implantation has been performed. In the case of labeled (stem) cell implantation, it is noteworthy to point out the different temporal profile of contrast generation. While the iron-positive macrophages, generated by vascular degradation, appear only at about 10 weeks after stroke, labeled stem cells are usually implanted at much earlier time points (approximately 2 weeks after stroke). Because the major migrational dynamics of these stem cells take place within 2 weeks of implantation,[42] the time window for monitoring the implanted cells is distinctly separate from the onset of vascular degradation.

An important aspect to be considered in this general strategy using a systemic contrast agent application is the selective labeling of blood-borne macrophages. Microglia, on the other hand, resident in the brain parenchyma and only transforming into macrophages under pathological conditions like focal brain ischemia, will remain unlabeled and, consequently, are not detected by MRI.

Recently, Saleh and coworkers conducted a first clinical phase II study in 10 patients suffering from ischemic stroke to visualize postischemic macrophage infiltration by USPIO-enhanced MRI.[44] They performed multimodal MRI including T_1-, T_2- and T_2*-weighted imaging sequences at 24 to 36 and 48 to 72 h after a single intravenous injection of USPIO particles, approximately 1 week after stroke onset. Hyperintense signal changes that increased over time after USPIO administration were visible on T_1-weighted images in the ischemic brain parenchyma. These hyperintense areas did not match with brain areas displaying a disrupted blood–brain barrier on gadolinium-enhanced MRI. Interestingly, and not yet understood, hypointense signal changes on T_2/T_2*-weighted images were not detected in the ischemic territory of the brain (in contrast to animal experiments), but transiently in the direct neighborhood of cerebral blood vessels.

15.4 OUTLOOK

Selective studies of the contribution and temporal activity profile of activated microglia in ischemic brain areas under *in vivo* conditions are still a challenge for the future. Generation of a transgenic mouse expressing an imaging reporter gene coupled to the promotor responsible for microglia in its activated state may be an exciting approach. As has already been shown for optical imaging techniques, coupling the corresponding promotor to GFP, other fluorescence proteins, or near-infrared imaging, or to luciferase expression for bioluminescence imaging,[45,46] will allow observation of various cell dynamic processes. Recently, in proof-of-concept fashion, it was demonstrated that engineered cells may even produce their own MRI contrast mechanism, under the control of specific promotors, thus allowing, in principle, the *in vivo* observation of cell function changes.[47] These activatable smart contrast agents open new frontiers to true highly resolved three-dimensional imaging by MRI, not only registering the whereabouts of cells like macrophages, but also following their functional changes, with the help of molecular MRI technology.[48] The future may see *in vivo* monitoring of phagocytic activity, release of cytokines or tissue-remodeling enzymes like metalloproteinases, or specific drug effects on macrophage function, all depictable by such smart contrast agents.

REFERENCES

1. Dirnagl U, Iadecola C, Moskowitz MA. Pathobiology of ischaemic stroke: an integrated view. *Trends Neurosci* 1999; 22: 391–397.
2. Stoll G, Jander S, Schroeter M. Inflammation and glial response in ischemic brain lesions. *Prog Neurobiol* 1998; 56: 149–171.
3. Danton HG, Dietrich WD. Inflammatory mechanisms after ischemia and stroke. *J Neuropathol Exp Neurol* 2003; 62: 127–136.
4. Arvin B, Neville LF, Barone FC, Feuerstein GZ. The role of inflammation and cytokines in brain injury. *Neurosci Biobehav Rev* 1996; 20: 445–452.
5. del Zoppo G, Ginis I, Hallenbeck JM, Iadecola C, Wang X, Feuerstein GZ. Inflammation and stroke: putative role for cytokines, adhesion molecules and iNOS in brain response to ischemia. *Brain Pathol* 2000; 10: 95–112.
6. Lehrmann E, Christensen T, Zimmer J, Diemer NH, Finsen B. Microglial and macrophage reactions mark progressive changes and define the penumbra in the rat neocortex and striatum after transient middle cerebral artery occlusion. *J Comp Neurol* 1997; 386: 461–476.
7. Mabuchi T, Kitagawa K, Ohtsuki T, Kuwabara K, Yagita Y, Yanagihara T, Hori M, Matsumuto M. Contribution of microglia/macrophages to expansion of infarction and response of oligodendrocytes after focal cerebral ischemia in rats. *Stroke* 2000; 31: 1735–1743.
8. Garcia JH, Liu KF, Yoshida Y, Lian J, Chen S, del Zoppo G. Influx of leukocytes and platelets in an evolving brain infarct (Wistar rat). *Am J Pathol* 1994; 144: 188–189.
9. Tanaka R, Komine-Kobayashi M, Mochizuki H, Yamada M, Furuya T, Migita M, Shimada T, Mizuno Y, Urabe T. Migration of enhanced green fluorescent protein expressing bone-marrow-derived microglia/macrophage into the mouse brain following permanent focal ischemia. *Neuroscience* 2003; 117: 531–539.
10. Schilling M, Besselmann M, Muller M, Strecker JK, Ringelstein EB, Kiefer R. Predominant phagocytic activity of resident microglia over hematogenous macrophages following transient focal cerebral ischemia: an investigation using green fluorescent protein transgenic bone marrow chimeric mice. *Exp Neurol* 2005; 196: 290–297.

11. Hacke W, Donnan G, Fieschi C, Kaste M, von Kummer R, Broderick JP, Brott T, Frankel M, Grotta JC, Haley EC, Jr., Kwiatkowski T, Levine SR, Lewandowski C, Lu M, Lyden P, Marler JR, Patel S, Tilley BC, Albers G, Bluhmki E, Wilhelm M, Hamilton S. Association of outcome with early stroke treatment: pooled analysis of ATLANTIS, ECASS, and NINDS rt-PA stroke trials. *Lancet* 2004; 363: 768–774.
12. Barone FC, Feuerstein GZ. Inflammatory mediators and stroke: new opportunities for novel therapeutics. *J Cereb Blood Flow Metab* 1999; 10: 819–834.
13. Becker KJ. Targeting the central nervous system inflammatory response in ischemic stroke. *Curr Opin Neurol* 2001; 14: 349–353.
14. Chen YI, Hallenbeck JM, Ruetzler C, Bol D, Thoams K, Berman NE, Vogel SN. Overexpression of monocyte chemoattractant protein 1 in the brain exacerbates ischemic brain injury and is associated with recruitment of inflammatory cells. *J Cereb Blood Flow Metab* 2003; 23: 748–755.
15. Gregersen R, Lambertsen K, Finsen B. Microglia and macrophages are the major source of tumor necrosis factor in permanent middle cerebral artery occlusion in mice. *J Cereb Blood Flow Metab* 2000; 20: 53–65.
16. Lambertsen K, Meldgaard M, Ladeby R, Finsen B. A quantitative study of microglial-macrophage synthesis of tumor necrosis factor during acute and late focal cerebral ischemia in mice. *J Cereb Blood Flow Metab* 2005; 25: 119–135.
17. Romanic AM, White RF, Arleth AJ, Ohlstein EH, Barone FC. Matrix metalloproteinase expression increases after cerebral focal ischemia in rats: inhibition of matrix metalloproteinase-9 reduces infarct size. *Stroke* 1998; 29: 1020–1030.
18. Xu Z, Ford GD, Croslan DR, Jiang J, Gates A, Allen R, Ford BD. Neuroprotection by neuroregulin-1 following focal stroke is associated with the attenuation of ischemia-induced pro-inflammatory and stress gene expression. *Neurobiol Dis* 2005; 19: 461–470.
19. Kumai Y, Ooboshi H, Takada J, Kamouchi M, Kitazono T, Egashira K, Ibayashi S, Iida M. Anti-monoctye chemoattractant protein-1 gene therapy protects against focal brain ischemia in hypertensive rats. *J Cereb Blood Flow Metab* 2004; 24: 1359–1368.
20. Lou M, Blume A, Zhao Y, Gohlke P, Deuschl G, Herdegen T, Culman J. Sustained blockade of brain AT1 receptors before and after focal cerebral ischemia alleviates neurologic deficits and reduces neuronal injury, apoptosis, and inflammatory responses in the rat. *J Cereb Blood Flow Metab* 2004; 24: 536–547.
21. Sundararajan S, Gamboa JL, Victor NA, Wanderi EW, Lust WD, Landreth GE. Peroxisome proliferator-actived receptor-gamma ligands reduce inflammation and infarction size in transient focal ischemia. *Neuroscience* 2005; 130: 685–696.
22. Investigators EAST. Use of anti-ICAM-1 therapy in ischemic stroke: results of the Enlimomab Acute Stroke Trial. *Neurology* 2001; 57: 1428–1434.
23. Benavides J, Savaki HE, Malgouris C, Laplace C, Daniel M, Begassat M, Desban M, Uzat A, Dubroeuccq MC, Renault C, Gueremy C, Le Fur G. Autoradiographic localization of peripheral benzodiazepine binding sites in the cat brain with (^{3}H)-PK 11195. *Brain Res Bull* 1984; 13: 69–77.
24. Benavides J, Dubois A, Dennis T, Hamel E, Scatton B. Imaging of human brain lesions with an ω_3 site radioligand. *Ann Neurol* 1988; 24: 708–712.
25. Dubois A, Benavides J, Peny B, Duverger D, Fage D, Gotii B, MacKenzie ET, Scatton B. Imaging of primary and remote ischaemic and excitotoxic brain lesions: an autoradiographic study of peripheral type benzodiazepine binding sites in the rat and cat. *Brain Res* 1988; 445: 77–90.
26. Myers R, Manjil LG, Cullen BM, Price GW, Frackowiak RS, Cremer JE. Macrophage and astrocyte populations in relation to (3H)PK 11195 binding in rat cerebral cortex following a local ischaemic lesion. *J Cereb Blood Flow Metab* 1991; 11: 314–322.
27. Sette G, Baron JC, Young AR, Miyazawa H, Tillet I, Barre L, Travere J-M, Derlon J-M, MacKenzie ET. *In vivo* mapping of brain benzodiazepine receptor changes by positron emission tomography after focal ischemia in the anesthetized baboon. *Stroke* 1993; 24: 2046–2058.
28. Junck L, Jewett DM, Kilbourn MR, Young AB, Kuhl DE. PET imaging of cerebral infarcts using a ligand for the periperal benzodiazepine binding site. *Neurology* 1990; 40(Suppl 1): 265.
29. Ramsay SC, Weiller C, Myers R, Cremer JE, Luthra SK, Lammertsma AA, Frackowiak RS. Monitoring by PET of macrophage accumulation in brain after ischaemic stroke. *Lancet* 1992; 25: 1054–1055.
30. Pappata S, Levasseur M, Gunn RN, Myers S, Crouzel C, Syrota A, Jones T, Kreutzberg GW, Banati RB. Thalamic microglial activation in ischemic stroke detected *in vivo* by PET and [11C]-PK 11195. *Neurology* 2000; 55: 1052–1054.

31. Gerhard A, Neumaier B, Elitok E, Glatting G, Ries V, Tomczak R, Ludolph AC, Reske SN. *In vivo* imaging of activated microglia using [11C]PK11195 and positron emission tomography in patients after ischemic stroke. *Neuroreport* 2000; 11: 2957–2960.
32. Barkhof F, Rocca M, Francis G, Van Waesberghe JH, Uitdehaag BM, Hommes OR. Validation of diagnostic magnetic resonance imaging criteria for multiple sclerosis and response to interferon beta1a. *Ann Neurol* 2003; 53: 718–724.
33. Weissleder R, Elizondo G, Wittenberg J, Rabito C, Bengele H, Josephson L. Ultrasmall superparamagnetic iron oxide: characterization of a new class of contrast agent for MR imaging. *Radiology* 1990; 175: 489–494.
34. Sbarbati A, Reggiani A, Nicolato E, Arban R, Berardi P, Lunati E, Asperio RM, Marzola P, Osculati F. Correlation MRI/ultrastructure in cerebral ischemic lesions: application to the interpretation of cortical layered areas. *Magn Reson Imaging* 2002; 20: 479–486.
35. Schroeter M, Franke C, Stoll G, Hoehn M. Dynamic changes of magnetic resonance imaging abnormalities in relation to inflammation and glial responses after photothrombotic cerebral infarction in the rat brain. *Acta Neuropathol* 2001; 101: 114–122.
36. Rausch M, Sauter A, Fröhlich J, Neubacher U, Radü EW, Rudin M. Dynamic patterns of USPIO enhancement can be observed in macrophages after ischemic brain damage. *Magn Reson Med* 2001; 46: 1018–1022.
37. Rausch M, Baumann D, Neubacher U, Rudin M. *In-vivo* visualization of phagocytic cells in rat brains after transient ischemia by USPIO. *NMR Biomed* 2002; 15: 278–283.
38. Kleinschnitz C, Bendzus M, Frank M, Solymosi L, Toyka KV, Stoll G. *In vivo* monitoring of macrophage infiltration in experimental ischemic brain lesions by magnetic resonance imaging. *J Cereb Blood Flow Metab* 2003; 1356–1361, 23.
39. Saleh A, Wiedermann D, Schroeter M, Jonkmanns C, Jander S, Hoehn M. Central nervous system inflammatory response after cerebral infarction as detected by magnetic resonance imaging. *NMR Biomed* 2004; 17: 163–169.
40. Schroeter M, Saleh A, Wiedermann D, Hoehn M, Jander S. Histochemical detection of ultrasmall superparamagnetic iron oxide (USPIO) contrast medium uptake in experimental brain ischemia. *Magn Reson Med* 2004; 52: 403–406.
41. Weber R, Wegener S, Ramos-Cabrer P, Wiedermann D, Hoehn M. MRI detection of macrophage activity after experimental stroke in rats: new indicators for late appearance of vascular degradation? *Magn Reson Med* 2005; 54: 59–66.
42. Hoehn M, Kustermann E, Blunk J, Wiedermann D, Trapp T, Wecker S, Focking M, Arnold H, Hescheler J, Fleischmann BK, Schwindt W, Buhrle C. Monitoring of implanted stem cell migration *in vivo*: a highly resolved *in vivo* magnetic resonance imaging investigation of experimental stroke in rat. *Proc Natl Acad Sci USA* 2002; 99: 16267–16272.
43. Jendelova P, Herynek V, Urdzikova L, Glogarova K, Kroupova J, Andersson B, Bryja V, Burian M, Hajek M, Sykova E. Magnetic resonance tracking of transplanted bone marrow and embryonic stem cells labeled by iron oxide nanoparticles in rat brain and spinal cord. *J Neurosci Res* 2004; 76: 232–243.
44. Saleh A, Schroeter M, Jonkmanns C, Hartung HP, Mödder U, Jander S. *In vivo* MRI of brain inflammation in human ischemic stroke. *Brain* 127, 1670–1677, 2004.
45. Ciana P, Raviscioni M, Mussi P, Vegeto E, Que I, Parker MG, Lowik C, Maggi A. *In vivo* imaging of transcriptionally active estrogen receptors. *Nat Med* 2003; 9: 82–86.
46. de Boer J, van Blitterswijk C, Löwik C. Bioluminescent imaging: emerging technology for non-invasive imaging of bone tissue engineering. *Biomaterials* 2006; 27: 1851–1858.
47. Cohen B, Dafni H, Meir G, Harmelin A, Neeman M. Ferritin as an endogenous MRI reporter for noninvasive imaging of gene expression in C6 glioma tumors. *Neoplasia* 2005; 7: 109–117.
48. Himmelreich U, Aime S, Hieronymus T, Justicia C, Uggeri F, Zenke M, Hoehn M. A responsive MRI contrast agent to monitor functional cell status. *Neuroimage* 32, 1142–1149, 2006.

16 Magnetic Resonance Imaging in Developmental Biology

Cyrus Papan, J. Michael Tyszka, and Russell E. Jacobs

CONTENTS

16.1 Advantages of MRI for Studying Embryonic Development 283
16.2 Applications of MRI to Developmental Biology 283
16.2.1 Rodent Embryo MRI 284
16.2.2 Avian Embryo MRI 284
16.2.3 Human Embryo MRI 284
16.3 Magnetic Resonance Microscopy of Frog Development 284
16.4 Volumetric Imaging of Development in *Xenopus laevis* Embryos 285
16.5 High-Spatiotemporal-Resolution Two-Dimensional Imaging of Development 287
16.6 Cell Tracking by Magnetic Resonance Microscopy 287
16.7 Cell Tracking with T2 Contrast Agents 289
16.8 Comparing Magnetic Resonance Microscopy with Optical Methods 291
16.9 Magnetic Resonance Histology in Developmental Biology 291
16.10 Conclusions 292
Acknowledgments 292
References 292

16.1 ADVANTAGES OF MRI FOR STUDYING EMBRYONIC DEVELOPMENT

Magnetic resonance imaging (MRI) has two major strengths for *in vivo* imaging of developing embryos: First, the modality is noninvasive and nondestructive. MRI does not use ionizing radiation, and the bioeffects (including static field exposure and radio frequency (RF) heating) are well understood and can be reduced to negligible levels for most MRI studies. Second, the radiation wavelength is on the order of tens of centimeters to meters, so scattering and attenuation effects in embryos are negligible. MRI can therefore visualize the interior of an optically opaque embryo with essential unlimited penetration. MRI is, however, an intrinsically insensitive technique, even at very high field strengths. 1H nuclear polarization fractions of less than 100 ppm (0.01%) are typical field strengths in the 7- to 12-T range, leading to severe constraints on spatial and temporal resolution. For small samples, such as *Xenopus laevis* embryos, spatial and temporal resolutions in an 11.7-T microimaging system are limited to about 20 microns and 10 min for coverage of the whole embryo (about 1 mm in diameter). For larger embryos, the best achievable spatial resolution scales with the sample size, so for quail and chicken embryos *in ovo*, voxel sizes on the order of 100 to 200 microns are typical.

16.2 APPLICATIONS OF MRI TO DEVELOPMENTAL BIOLOGY

MRI has been applied to developmental biology questions in several key model systems, including chicken, quail, rat, mouse, and frog embryos. Avian and amphibian embryos are popular subjects for MR microscopy (MRM) studies, since they develop externally and can be incubated within the MR imager with careful preparation and experimental design. The embryo also occupies a much larger fraction of the imaged volume than can be achieved with mammalian embryos, further improving image acquisition efficiency and sensitivity.

16.2.1 Rodent Embryo MRI

MRI of rodent embryos *in utero* is a technically challenging task, since the embryos are small relative to the size of the mother, placing limits on the filling fraction, and therefore sensitivity, of the imaging experiment. Anesthesia of the mother helps to minimize embryonic motion with no gross effect on embryonic development reported in rat pups, even following repeated time-course imaging during the pregnancy (Smith et al. 1998). Early work in this area focused on contrast-enhanced localization of embryonic implantation in the rat uterus (Hamilton et al. 1993, 1994). The utility of MRI for time-course imaging of developing rat embryos was later demonstrated by Smith et al. (1998). More recently, two groups have described imaging mouse embryos *in utero* (Chapon et al. 2002; Hogers et al. 2000), but no follow-up studies or applications of the methodology have been reported, suggesting that the more widespread adoption of MRI of rodent embryo imaging requires a more convincing demonstration of utility.

16.2.2 Avian Embryo MRI

Chicken and quail are popular development models that have traditionally been studied using optical microscopy via shell windows *in ovo* or in culture. MRI of intact embryos has more appeal at later stages of development, once the embryo has turned and its size precludes useful optical imaging. Bone et al. (1986) reported *in ovo* MRI of a live chick embryo as early as 1986, with subsequent demonstrations by Effmann et al. (1988) and (in conjunction with 31P spectroscopy) Moseley et al. (1989). The chick embryo has also been studied for bioeffects of MRI, demonstrating no significant effects from prolonged exposure to a 1.5-T magnetic field with associated RF and gradient pulsing (Yip et al. 1994a, 1994b, 1995). Hypoxia and aerobic requirements in chicken embryos have been studied with both MR spectroscopy and diffusion imaging (Peebles et al. 2003), and 1H MR spectroscopy has also been reported in quail embryos (Tyszka et al. 2002).

16.2.3 Human Embryo MRI

In humans, ultrafast MRI of the fetus has become more commonplace in clinical practice (Sandrasegaran et al. 2005; Hubbard 2003; Hubbard and States 2001), and MRM of fixed human embryos (Figure 16.1) has been reported by several groups (Matsuda et al. 2003; Smith 1999; Smith et al. 1999). For *in utero* imaging, high-speed imaging, typically single-shot rapid acquisition with relaxation enhancement (RARE) (Hennig 1986), is required to “freeze” fetal motion, just as for unanesthetized avian embryos *in ovo*. The construction of atlases from fixed specimens at high sensitivity and spatial resolution is likely to be the primary application of MRM in the field of human embryonic development in the near future.

16.3 MAGNETIC RESONANCE MICROSCOPY OF FROG DEVELOPMENT

A notably successful application of magnetic resonance microscopy in developmental biology has been the study of early-stage amphibian embryos, particularly of gastrulation in frogs. *Xenopus laevis* (the African clawed frog) is a classical model of early vertebrate development, particularly gastrulation and neurulation. Progress in understanding the developmental mechanisms operating in the *Xenopus* embryo has often been hampered by the inability to visualize cells within the interior of the embryo due to light scattering by the intracellular yolk inclusions. Experiments on morphogenetic movements and their molecular control have therefore relied upon analyzing the morphology

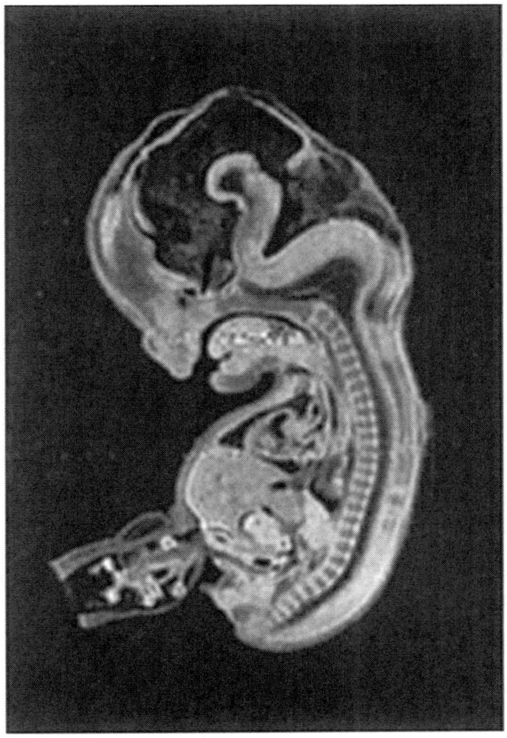

FIGURE 16.1 Magnetic resonance microscopy of a chemically fixed human embryo with an isotropic spatial resolution of 120 microns. (From Matsuda, Y. et al., *Magn. Reson. Med.*, 50, 183–189, 2003. Copyright © Wiley-Liss, Inc. With permission.) Magnetic resonance histology has the advantage of being nondestructive, allowing detailed volumetric imaging of rare specimens from repositories such as, for example, the Carnegie Collection of Human Embryos (Smith et al. 1999).

of fixed preparations or the motions of cells in tissue explants. Although experiments on explants have offered tremendous insights, they leave many open questions concerning the nature and timing of morphogenetic movements in their normal context. Furthermore, cellular and molecular studies of the signals that pattern the early embryo have been hampered by imperfect knowledge of the positions of the signaling and responding cells at the stages of presumed interactions. Thus, both cellular and molecular studies would benefit greatly from an imaging technique that can directly observe *in vivo* the internal features and movements within normal and perturbed *Xenopus* embryos.

Because MRI does not interfere with the normal development of early *Xenopus* embryos (Kay et al. 1988), dynamic events directly follow with MRM by repeatedly imaging the same specimen rather than inferring dynamics from different specimens fixed and sectioned at different times. Earlier work (Jacobs and Fraser 1994) employed extrinsic contrast from microinjecting a dextran-linked MRI contrast agent to characterize the labeled cell movements over time. More recently, combinations of both intrinsic and extrinsic contrast have been used to improve spatio-temporal resolution and to follow labeled clones *in vivo* in the context of the whole *Xenopus* embryo.

16.4 VOLUMETRIC IMAGING OF DEVELOPMENT IN *XENOPUS LAEVIS* EMBRYOS

MRM applications with small specimens, such as *Xenopus* embryos, often take advantage of the added sensitivity afforded by working at high magnetic field strengths with RF coils customized to the sample. In a typical experiment, a live embryo will be placed within a narrow-diameter

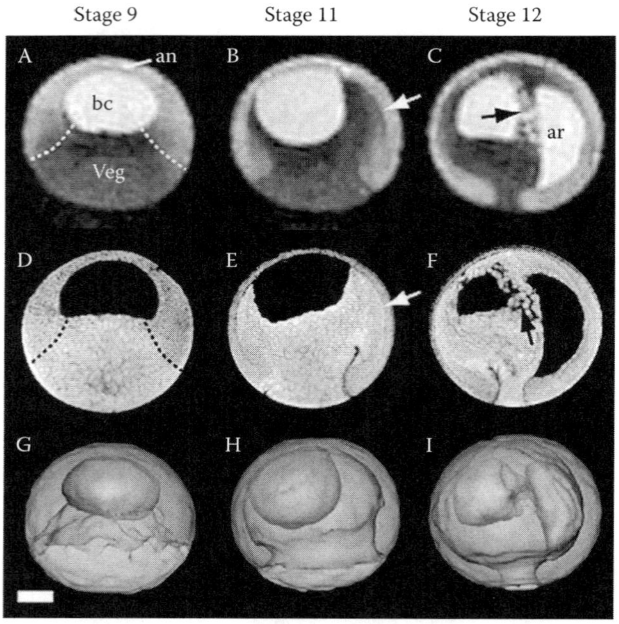

FIGURE 16.2 (please see color insert following page 210) Intrinsic contrast of MR images correlates with tissue type in the early *Xenopus* embryo. View is lateral with animal oriented to the top and dorsal to the right. The liquid-filled blastocoel (bc) and the archenteron (ar) exhibit the highest signal intensity, the vegetal cell mass (veg) the lowest, and the animal cap (an) an intermediate signal intensity. At stage 9 (A, D, G), the embryo is rotationally symmetric along the animal–vegetal axis. Note the sharp boundary between the animal cap and the vegetal cell mass in the MR images (highlighted with a dashed line in A), which correlates with differences in cell size seen in the histological image (highlighted with a dashed line in D). At stage 11 (B, E, H), gastrulation is under way. The mesendodermal mantle is moving animal-ward, and the animal cap tissue is extending vegetal-ward by epiboly. The cleft of Brachet routinely observed in histological images (white arrow in E) can also be clearly recognized in the MR image as a sharp and distinctive boundary (white arrow in B) on the dorsal as well as on the ventral side. At stage 12 (C, F, I), the archenteron (ar) has formed, displacing the blastocoel cavity to the ventral side. The histological image shows loosely packed cells between the archenteron and the blastocoel (black arrow in F), which matches with intermixed bright and dark intensities in the MR image (white arrow in C). Scale bar, 300 μm.

glass tube containing 10 to 50 μl of rearing buffer. The tube is in turn placed within a small solenoid coil (approximately 3 mm in diameter) and imaged using simple, but robust spin echo and gradient echo pulse sequences. Temporal resolution ranges from 10 to 60 min per volume with a nominal isotropic spatial resolution of 15 to 50 μm (Figure 16.2). Development of embryos in the scanner is indistinguishable from control sibling embryos developing in a Petri dish to at least stage 24, when the surrounding drop size becomes too small to accommodate the extending embryo. The low spatiotemporal resolution of MRI can in part be compensated by cooling the embryo to 15°C, at which temperature gastrulation takes about 12 h. The 1H NMR spectrum of a live frog embryo is dominated by the singlet resonance of water at 4.7 ppm and multiple resonances from small mobile lipids between 0.5 and 2 ppm. Consequently, MR images can be generated from each resonance or resonance group separately by either spectral selective saturation of the unwanted region or selective excitation of the required resonances. Water and fat imaging of *Xenopus* embryos allows independent study of the animal and vegetal cells that have significantly different lipid compositions. Water-only images highlight embryonic fluid spaces such as the blastocoel and archenteron, whereas fat-only images emphasize the vegetal cell mass, the internal dynamics of which have remained relatively mysterious to optical microscopists.

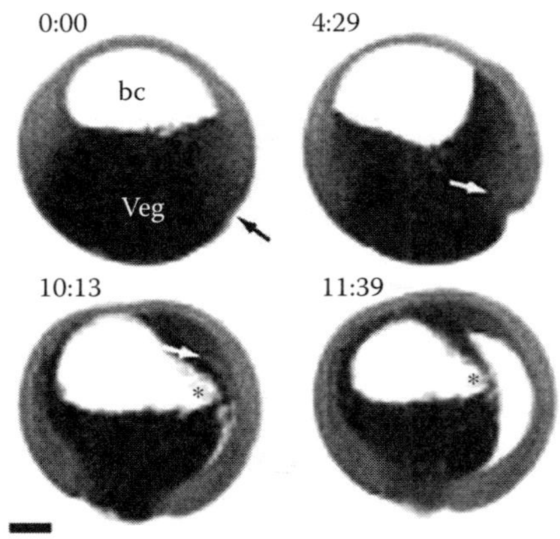

FIGURE 16.3 Archenteron formation visualized by two-dimensional high-resolution MR time-lapse imaging. A sagittal midline slice is shown with the animal pole at the top and dorsal to the right. Times are in hours with respect to the first image. At time 0:00 (approximately stage 10) the dorsal blastopore lip (black arrow) has formed. At time 4:29 (approximately stage 11), the archenteron has invaginated about 200 µm (white arrow). Invagination continues over several hours. At times 8:04 and 8:47, the tip of the archenteron is difficult to discern in the images. By time 10:13 (approximately stage 12), the archenteron becomes clearly visible (white arrow). The archenteron inflates further at the expense of the shrinking blastocoel, and the blastocoel forms a characteristic protrusion (asterisk) on the dorsal side.

16.5 HIGH-SPATIOTEMPORAL-RESOLUTION TWO-DIMENSIONAL IMAGING OF DEVELOPMENT

While three-dimensional imaging provides important information about the spatial arrangement of structures within the embryo, it often requires extensive data collection times. Fortunately, two-dimensional imaging of the mid-sagittal plane of early embryos is an acceptable substitute for many purposes. The disadvantage of losing the three-dimensional information is offset by significantly higher resolution and image quality in a much shorter imaging time. For reliable embryonic axis identification in two-dimensional experiments, the early-blastula-stage embryos can be oriented using the asymmetric pigmentation either perpendicular or parallel to the long axis of the sample tube. The resulting two-dimensional images exploit the fact that the embryo has a strong symmetry across the midline, so the thickness of the slice has less impact on the resulting image quality than might be expected.

Figure 16.3 demonstrates one specific application of two-dimensional midline imaging: *in vivo* measurement of archenteron extension during gastrulation. The position of the archenteron tip is visible in these images, and data derived from them suggest that the motors that drive archenteron invagination or elongation might operate somewhat independently of the motors that drive vegetal cell mass rotation. This conclusion would be very difficult to draw from a study of fixed embryos or from optical microscopy.

16.6 CELL TRACKING BY MAGNETIC RESONANCE MICROSCOPY

To further define the cell motions that create the cleft of Brachet, three-dimensional time-series images can be used to track the descendants of one or more blastomeres labeled with a macromolecular T_1 contrast agent such as GRID (Huber et al. 1998). Figure 16.4 follows the descendants of a GRID-labeled C1 blastomere showing that both mesendoderm and neurectoderm derive from C1. The clone extends radially into the depth of the embryo, with the deep part of the clone becoming mes/endoderm and the superficial part contributing to posterior neurectoderm.

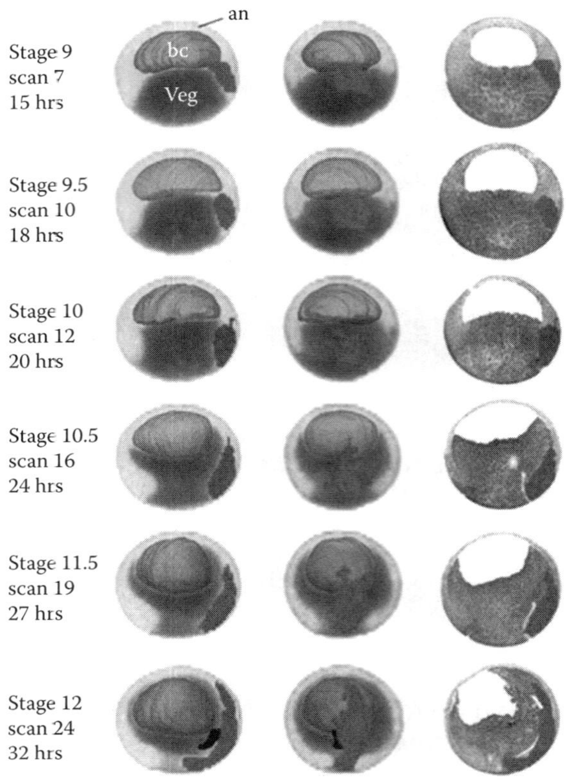

FIGURE 16.4 (please see color insert) Selected stages of a time series of three-dimensional MR images in lateral (left column) and dorsal (middle column) views. The descendants of the labeled C1 blastomere are shown in a green/orange color scale and overlaid on an equivalent sagittal histological section (right column). At stage 9, the C1 clone resides in the upper marginal zone of the embryo, extending radially from the blastocoel to the embryo surface. By stage 9.5, the C1 clone has begun to move toward the vegetal pole, and extends to the blastopore by stage 10. By stage 12, the C1 clone has extended vegetal-ward to span nearly the entire dorsal marginal zone of the embryo before any external signs of gastrulation are normally observed.

The cleft of Brachet is a well-defined landmark in the embryo, marking the boundary between the interior endo-mesodermal cells and exterior ectodermal cells (Brachet 1921; Keller 1991) and serving as a useful morphological landmark in defining the germ layer identity of the cells within labeled clones. As can be seen in the histological images, the cleft becomes apparent at approximately the onset of gastrulation and is located at the intersection of the blastocoel roof and floor. It subsequently extends vegetally parallel to the surface of the embryo toward the blastopore lip (Figure 16.4, stages 10 to 12, red line). Its initial formation is a result of vegetal rotation (Winklbauer et al. 1999) when cells of the vegetal cell mass appose the blastocoel roof.

Although the resolution of the three-dimensional MRM does not permit us to visualize thin structures such as Brachet's cleft directly, its location and time of appearance can be deduced in the MRM renderings by discontinuities in the intrinsic contrast between the vegetal cell mass and animal cap and the extrinsic contrast between the inner subclone and the unlabeled ectodermal tissue. In our MRM images, the formation of the cleft first becomes evident at stage 10, when the outwardly rotating vegetal cell mass apposes the ectoderm (stage 10, large open arrow). The cleft runs through the labeled C1 clone, separating it into inner mesendodermal and outer ectodermal subclones (stages 10.5 to 12). The inner and outer subclones then behave differently. The vegetal limits of the two parts move together ventral-ward with the blastopore (asterisk). However, the animal limit of the outer subclone remains stationary (solid arrow), while the animal limit of the

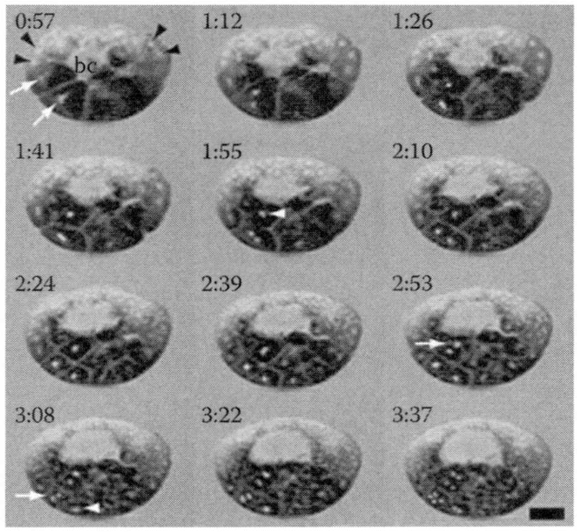

FIGURE 16.5 (please see color insert) Example lineage tracing of a vegetal blastomere. The figure shows selected stages of a two-dimensional time-lapse sequence. Image slices are sagittally oriented, with animal toward the top and dorsal to the right. Scale bar, 300 μm. Many nuclei can be identified in a 300-μm-thick image slice (black arrowheads at 0:57). By taking consecutive two-dimensional slice images, the cleavages of the nuclei can be traced. In the first frame, both nuclei of the D4 daughters are indicated with a white arrow. One nucleus is highlighted in yellow, and the cell boundary is colored in blue. In the second frame (time 1:12), the highlighted nucleus has cleaved in the radial direction and daughter nuclei move apart in the third and fourth frames (1:26 and 1:41). At 1:55, only the nucleus of the outer blastomere can be seen to divide, and it changes the cleavage plane by 90° to divide in the planar direction. The nuclei of the inner sibling blastomeres do not appear to cleave (white arrow) due to the cleavage plane being oriented in the image plane. At 2:53 and 3:03, two sibling cells are dividing, albeit not synchronous (white arrows). Again, one sibling does not appear to cleave. At 3:37, all highlighted nuclei of the clone are cleaving in the planar direction, resulting in 10 clonally related sibling cells. After this, nuclei cannot be reliably traced.

inner subclone moves anteriorly, following the leading edge of gastrulation (open arrow). The final configuration of the labeled clone confirms the tissue identity of the inner and outer subclones as mesendodermal and ectodermal, respectively. At the end of gastrulation (stage 12), the inner subgroup of the C1 clone lies within the mesendoderm, dorsal to the developing archenteron. The dorsal view of the embryo (Figure 16.4, middle column) shows how the clone changes its morphology from a somewhat squat appearance to an elongated one, consistent with the normal convergent-extension behavior of the axial mesoderm during gastrulation.

These observations show that MRM can be used for qualitative and quantitative morphological analysis of developing systems. Although spatial resolution is often insufficient to visualize individual cells, information about rearrangements and movements of sheets of cells and clonal populations is readily available. Using intrinsic contrast, we can observe the development of anatomic structures as they are forming (e.g., the archenteron). The introduction of contrast agent into a specific cell allows the tracking of that cell's progeny and identification of its fate within the context of the whole embryo. Moreover, the ability of MR to image different chemical species present naturally in the sample (i.e., fat vs. water) provides a uniquely MR view into the specimen.

16.7 CELL TRACKING WITH T_2 CONTRAST AGENTS

In early amphibian embryos, T_1 agents provide no significant enhancement of vegetal cells, most probably due to the lower fraction of free water in these cells (Sehy et al. 2001). The lack of enhancement is compounded by the shorter intrinsic T_2* of these cells, which reduces available signal at even short echo times (see, for example, Figure 16.3 and Figure 16.5). As an alternative,

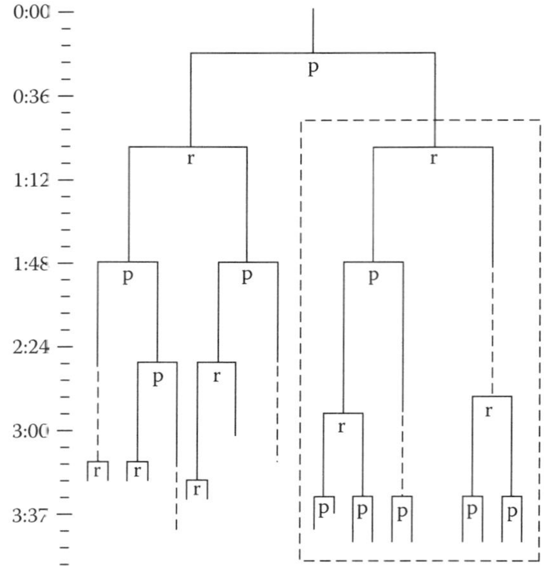

FIGURE 16.6 Lineage diagram of a D4 blastomere noted in MR images shown in Figure 16.5. The timescale on the left of the diagram corresponds to the times shown in Figure 16.6. The part of the lineage tree observed in Figure 16.5 is enclosed in the dashed box. The first three cleavages occur relatively synchronously. Then the right part of the lineage, which is the vegetal daughter cell of the D4 blastomere shown in Figure 16.5, begins to slow. This is expected, because the blastomeres at the vegetal pole tend to divide more slowly than the animal blastomeres (Niewkoop and Faber 1994). The cleavage orientation of the nuclei alternates in most cases by 90°, as has been observed at the animal pole in *Xenopus* (Chalmers et al. 2003). However, early divisions do not always change orientation by 90°. In one case, two successive planar cleavages were observed, and in another case, two successive radial cleavages were observed.

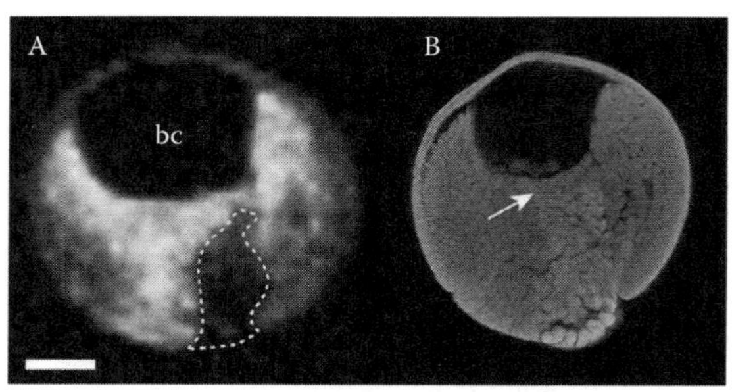

FIGURE 16.7 (please see color insert) Cell labeling of a vegetal blastomere with a T_2 contrast agent visualized by three-dimensional fat-only MR microscopy. The embryo was injected with a mix of Feridex and TexasRed-Dextran at the 64- to 128-cell stage into a vegetal blastomere. (A) Volume rendering of the three-dimensional fat-only MR image with the front half cut away, showing the labeled region. The cells labeled with Feridex can be identified as a dark mass (white dotted outline). (B) Fluorescence microscopy of the same embryo in section following chemical fixation, demonstrating the location of the TexasRed colabeling.

a T_2 contrast agent such as Feridex® (Berlex) might be used to label a vegetal blastomere, resulting in a strong negative contrast between the labeled region and surrounding vegetal cells. This approach is particularly effective when used in conjunction with water-suppressed or fat-only imaging, as described previously. The negative contrast of the labeled clones allows cell tracking throughout gastrulation, where questions of vegetal cell mass movements have been difficult to answer in the past due to the opacity of the embryo (Figure 16.7).

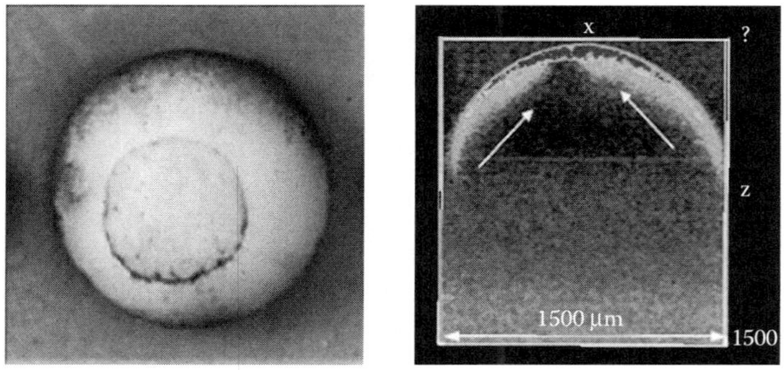

(a) (b)

FIGURE 16.8 (please see color insert) Example optical images of *Xenopus* gastrulae. (a) Wide-field optical microscopy demonstrates the opacity of the embryo, shown inverted with the blastopore at center. (b) OCT (from Haskell et al. 2004. With permission.) of a gastrulating *Xenopus* embryo demonstrates high resolution (on the order of 15 μm) and sensitivity in the superficial animal cell layers, but little to no information content at depth.

16.8 COMPARING MAGNETIC RESONANCE MICROSCOPY WITH OPTICAL METHODS

Many externally developing embryos, such as those of amphibians, are optically opaque at early stages (Figure 16.8). Tremendous progress has been made by many groups in increasing the image information content obtained from optical methods at depth in biological tissues. Recently, Haskell et al. (2004) and Hoeling et al. (2000) demonstrated the application of optical coherence tomography (OCT) to developmental biology. Comparison of the OCT results with current MRM results reveals the key differences in spatial resolution, tissue contrast, and penetration between the two modalities. State-of-the-art OCT at wavelengths from 850 to 1300 nm is capable of spatial resolutions on the order of 1 to 15 microns. However, the exponential attenuation coefficient in specimens such as frog embryos at these wavelengths is on the order of 20 to 30 μm, ensuring that the side of the embryo farthest from the light source cannot be observed.

16.9 MAGNETIC RESONANCE HISTOLOGY IN DEVELOPMENTAL BIOLOGY

The use of MRM to study fixed embryos at high spatial resolution deserves mention here, since it provides a convenient bridge between *in vivo* imaging and conventional histology. In scanning fixed specimens, problems associated with sample motion are completely eliminated and scan times are only limited by instrument stability issues. This enables lengthy signal averaging to overcome the poor signal-to-noise ratio encountered when imaging at very high spatial resolution. In addition, fixed specimens may be manipulated with contrast agents or other moieties that selectively alter the physiochemical properties of the sample in ways not possible *in vivo* (Smith et al. 1992). This allows the researcher to key in on a particular organ system, particular cell types, or areas of interest. Imaging of fixed samples is also a reasonable end point for a time-course series of *in vivo* imaging experiments, allowing the acquisition of high-fidelity, high-resolution images to compare with the final *in vivo* image. MRM cannot approach the spatial resolution typical of optical histological imaging (on the order of microns), but the same strengths that make it attractive for *in vivo* imaging also apply to the study of fixed specimens. A number of MRM studies of fixed specimens of animal and human embryos have been reported (Smith et al. 1994, 1996; Smith and Huff 1994; Johnson et al. 1993; Chong et al. 1997; Jacobs et al. 1999; Price et al. 1999; Dhenain et al. 2001; Mori et al. 2001; Puerta-Fonolla et al. 2001; Smith 1999, 2001; Schneider et al. 2003; Zhang et al. 2003a, 2003b) and reviewed elsewhere (Narasimhan and Jacobs 2005; Tyszka et al. 2005).

16.10 CONCLUSIONS

The notion that microscopic resolution magnetic resonance imaging might be feasible, and would certainly be exciting, was first proposed by Lauterbur (1973) in his initial description of MRI. The challenge in MR microscopy has been to optimize the experimental setup (hardware and software) to overcome the poor intrinsic signal-to-noise ratio (SNR) to obtain a respectable image in a reasonable amount of time. Molecular imaging (i.e., the judicious use of contrast agents) offers the ability to highlight particular aspects of a changing system, while doing little to augment the SNR of the MRM experiment. Spatial resolution, quality, and applicability of MRM have all increased dramatically in the last decade. It is now straightforward to obtain 50-μm resolution in *in vivo* three-dimensional images of small animals in reasonable amounts of time, and higher resolution is obtainable in embryonic-sized specimens (millimeter dimensions). Longitudinal *in vivo* imaging of the same specimen developing over time offers obvious advantages over serial imaging of different *ex vivo* specimens sampled at different times; imaging of morphological transformations in near real time is a prerequisite for investigating mechanisms underlying developmental changes. In the spirit of using the appropriate tool for the job at hand, there are several efforts toward combining MRM with other imaging modalities, such as optical microscopy, confocal microscopy, micro-CT, and positron emission tomography (PET) (Barillot et al. 1993; Slates et al. 1999; Jacobs and Cherry 2001; Townsend and Cherry 2001; Kahn et al. 2002, 2003; Allen and Meade 2003; Humm et al. 2003; MacKenzie-Graham et al. 2004). We fully expect that multimodal images utilizing agents visible in more than one modality will reveal details of internal organization and function in living things that are not apparent in any single type of image.

ACKNOWLEDGMENTS

The authors acknowledge support from NIH through NIBIB research grant R01 Eb000993, NCRR grant U24 RR021760, and additional support from the Moore Foundation and Beckman Institute.

REFERENCES

Allen MJ, Meade TJ. 2003. Synthesis and visualization of a membrane-permeable MRI contrast agent. *J Biol Inorg Chem* 8, 746–750.

Barillot C, Lemoine D, Le Briquer L, Lachmann F, Gibaud B. 1993. Data fusion in medical imaging: merging multimodal and multipatient images, identification of structures and 3D display aspects. *Eur J Radiol* 17, 22–27.

Bone SN, Johnson GA, Thompson MB. 1986. Three-dimensional magnetic resonance microscopy of the developing chick embryo. *Invest Radiol* 21, 782–787.

Chalmers AD, Strauss B, Papalopulu N. 2003. Oriented cell divisions asymmetrically segregate aPKC and generate cell fate diversity in the early *Xenopus* embryo. *Development* 130, 2657–2668.

Chapon C, Franconi F, Roux J, Marescaux L, Le Jeune JJ, Lemaire L. 2002. *In utero* time-course assessment of mouse embryo development using high resolution magnetic resonance imaging. *Anat Embryol* 206, 131–137.

Chong BW, Babcook CJ, Pang D, Ellis WG. 1997. A magnetic resonance template for normal cerebellar development in the human fetus. *Neurosurgery* 41, 924.

Dhenain M, Ruffins S, Jacobs RE. 2001. Three dimensional digital mouse atlas using high resolution MRI. *Dev Biol* 232, 458–470.

Effmann EL, Johnson GA, Smith BR, Talbott GA, Cofer G. 1988. Magnetic resonance microscopy of chick embryos *in ovo*. *Teratology* 38, 59–65.

Hamilton GS, Kennedy TG, Karlik SJ. 1994. Early identification of sites of embryo implantation in rats by means of gadolinium-enhanced MR imaging. *J Magn Reson Imaging* 4, 481–484.

Hamilton GS, Kennedy TG, Norley CJ, Karlik SJ. 1993. Gadolinium-DTPA enhanced MRI demonstrates uterine vascular changes associated with artificially induced decidualization and ovoimplantation in rats. *Magn Reson Med* 29, 817–821.

Haskell RC, Williams ME, Petersen DC, Hoeling BM, Schile AJ, Pennington JD, Seetin MG, Castelaz JM, Fraser SE, Papan C, Ren H, de Boer JF, Chen Z. 2004. Visualizing early frog development with motion-sensitive 3-D optical coherence microscopy. In Proceedings of the IEEE Engineering in Medicine and Biology Society, San Francisco, pp. 5296–5299.

Hennig J, Nauerth A, Friedburg H. 1986. RARE imaging: a fast imaging method for clinical MR, *Magn Reson Med* 3(6), 823–833.

Hoeling BM, Fernandez AD, Haskell RC, Huang E, Myers WR, Petersen DC, Ungersma SE, Wang RY, Williams ME, Fraser SE. 2000. An optical coherence microscope for 3-dimensional imaging in developmental biology. *Opt Express* 6, 136–146.

Hogers B, Gross D, Lehmann V, Zick K, De Groot HJ, Gittenberger-De Groot AC, Poelmann RE. 2000. Magnetic resonance microscopy of mouse embryos *in utero*. *Anat Rec* 260, 373–377.

Hubbard AM. 2003. Ultrafast fetal MRI and prenatal diagnosis. *Semin Pediatr Surg* 12, 143–153.

Hubbard AM, States LJ. 2001. Fetal magnetic resonance imaging. *Top Magn Reson Imaging* 12, 93–103.

Huber MM, Staubli A B, Kustedjo K, Gray MH, Shih J, Fraser SE, Jacobs RE, Meade TJ. 1998. Fluorescently detectable magnetic resonance imaging agents, *Bioconjug Chem* 9(2), 242–249.

Humm JL, Ballon D, Hu YC, Ruan S, Chui C, Tulipano PK, Erdi A, Koutcher J, Zakian K, Urano M, Zanzonico P, Mattis C, Dyke J, Chen Y, Harrington P, O'Donoghue JA, Ling CC. 2003. A stereotactic method for the three-dimensional registration of multi-modality biologic images in animals: NMR, PET, histology, and autoradiography. *Med Phys* 30, 2303–2314.

Jacobs RE, Ahrens ET, Dickinson ME, Laidlaw D. 1999. Towards a microMRI atlas of mouse development. *Comput Med Imaging Graph* 23, 15–24.

Jacobs RE, Cherry SR. 2001. Complementary emerging techniques: high-resolution PET and MRI. *Curr Opin Neurobiol* 11, 621–629.

Jacobs RE, Fraser SE. 1994. Magnetic resonance microscopy of embryonic cell lineages and movements. *Science* 263, 681–684.

Johnson GA, Benveniste H, Black RD, Hedlund LW, Maronpot RR, Smith BR. 1993. Histology by magnetic resonance microscopy. *Magn Reson Q* 9, 1–30.

Kahn E, Tessier C, Lizard G, Petiet A, Bernengo JC, Coulaud D, Fourre C, Frouin F, Clement O, Jourdain JR, Delain E, Guiraud-Vitaux F, Colas-Linhart N, Siauve N, Cuenod CA, Frija G, Todd-Pokropek A. 2003. Analysis of the distribution of MRI contrast agents in the livers of small animals by means of complementary microscopies. *Cytometry* 51A, 97–106.

Kahn E, Tessier C, Lizard G, Petiet A, Brau F, Clement O, Frouin F, Jourdain JR, Guiraud-Vitaux F, Colas-Linhart N, Siauve N, Cuenod CA, Frija G, Todd-Pokropek A. 2002. Distribution of injected MRI contrast agents in mouse livers studied by confocal and SIMS microscopy. *Anal Quant Cytol Histol* 24, 295–302.

Kay HH, Herfkens RJ, Kay BK. 1988. Effect of magnetic resonance imaging on *Xenopus laevis* embryogenesis. *Magn Reson Imaging* 6, 501–506.

Keller R. 1991. Early embryonic Development in Xenopus laevis. *Methods in Cell Biology.* BK Kay and HB Peng. San Diego, Academic Press. 36, 62–113.

Lauterbur P. 1973. Measurements of local nuclear magnetic-resonance relaxation-times. *Bull Am Phys Soc* 18, 86–86.

MacKenzie-Graham A, Lee EF, Dinov ID, Bota M, Shattuck DW, Ruffins S, Yuan H, Konstantinidis F, Pitiot A, Ding Y, Hu G, Jacobs RE, Toga AW. 2004. A multimodal, multidimensional atlas of the C57BL/6J mouse brain. *J Anat* 204, 93–102.

Matsuda Y, Utsuzawa S, Kurimoto T, Haishi T, Yamazaki Y, Kose K, Anno I, Marutani M. 2003. Super-parallel MR microscope. *Magn Reson Med* 50, 183–189.

Mori S, Itoh R, Zhang J, Kaufmann WE, van Zijl PC, Solaiyappan M, Yarowsky P. 2001. Diffusion tensor imaging of the developing mouse brain. *Magn Reson Med* 46, 18–23.

Moseley ME, Wendland MF, Darnell DK, Gooding CA. 1989. Metabolic and anatomic development of the chick embryo as studied by phosphorus-31 magnetic resonance spectroscopy and proton MRI. *Pediatr Radiol* 19, 400–405.

Narasimhan PT, Jacobs RE. 2005. Microscopy in magnetic resonance imaging. *Annu Rep NMR Spectrosc* 55, 262–299.

Niewkoop PD, Faber J. 1994. *Normal Table of Xenopus laevis (Daudin): A Systematical and Chronological Survey of the Development from the Fertilized Egg till the End of Metamorphosis*. New York: Garland.

Peebles DM, Dixon JC, Thornton JS, Cady EB, Priest A, Miller SL, Blanco CE, Mulder TL, Ordidge RJ, Rodeck CH. 2003. Magnetic resonance proton spectroscopy and diffusion weighted imaging of chick embryo brain *in ovo*. *Brain Res Dev Brain Res* 141, 101–107.

Price WS, Kobayashi A, Ide H, Natori S, Arata Y. 1999. Visualizing the postembryonic development of *Sarcophaga peregrina* (flesh fly) by NMR microscopy. *Physiol Entomol* 24, 386–390.

Puerta-Fonolla J, Ruiz-Cabello J, Vazquez-Osorio T, Murillo-Gonzalez J, Pena-Melian A. 2001. The human embryo development through MMR. *Ital J Anat Embryol* 106 (Suppl 2), 155–160.

Sandrasegaran K, Lall C, Aisen AA, Rajesh A, Cohen MD. 2005. Fast fetal magnetic resonance imaging. *J Comput Assist Tomogr* 29, 487–498.

Schneider JE, Bamforth SD, Farthing CR, Clarke K, Neubauer S, Bhattacharya S. 2003. High-resolution imaging of normal anatomy, and neural and adrenal malformations in mouse embryos using magnetic resonance microscopy. *J Anat* 202, 239–247.

Sehy JV, Ackerman JJH, Neil JJ. 2001. Water and lipid MRI of the *Xenopus* oocyte. *Magn Reson Med* 46, 900–906.

Slates RB, Farahani K, Shao YP, Marsden PK, Taylor J, Summers PE, Williams S, Beech J, Cherry SR. 1999. A study of artefacts in simultaneous PET and MR imaging using a prototype MR compatible PET scanner. *Phys Med Biol* 44, 2015–2027.

Smith BR. 1999. Visualizing human embryos. *Sci Am* 280, 76–81.

Smith BR. 2001. Magnetic resonance microscopy in cardiac development. *Microsc Res Tech* 52, 323–330.

Smith BR, Effmann EL, Johnson GA. 1992. MR microscopy of chick embryo vasculature. *J Magn Reson Imaging* 2, 237–240.

Smith B, Huff D. 1994. Magnetic-resonance microscopy of human embryos. *Lab Invest* 70, P9–P9.

Smith BR, Huff DS, Johnson GA. 1999. Magnetic resonance imaging of embryos: an Internet resource for the study of embryonic development. *Comput Med Imaging Graph* 23, 33–40.

Smith BR, Johnson GA, Groman EV, Linney E. 1994. Magnetic-resonance microscopy of mouse embryos. *Proc Natl Acad Sci USA* 91, 3530–3533.

Smith BR, Linney E, Huff DS, Johnson GA. 1996. Magnetic resonance microscopy of embryos. *Comput Med Imaging Graph* 20, 483–490.

Smith BR, Shattuck MD, Hedlund LW, Johnson GA. 1998. Time-course imaging of rat embryos *in utero* with magnetic resonance microscopy. *Magn Reson Med* 39, 673–677.

Townsend DW, Cherry SR. 2001. Combining anatomy and function: the path to true image fusion. *Eur Radiol* 11, 1968–1974.

Tyszka JM, Fraser SE, Jacobs RE. 2005. Magnetic resonance microscopy: recent advances and applications. *Curr Opin Biotechnol* 16, 93–99.

Tyszka JM, Pautler RG, Lansford R, Wood J, Jacobs RE. 2002. Near real-time T2-weighted MR microscopy of Japanese quail embryos *in ovo*. In *Proc Intl Soc Magn Reson in Med*, Honolulu, Hawaii, p. 2532.

Winklbauer R, Schürfeld M. 1999. Vegetal rotation, a new gastrulation movement involved in the internalization of the mesoderm and endoderm in Xenopus. *Development* 126, 3703–3713.

Yip YP, Capriotti C, Norbash SG, Talagala SL, Yip JW. 1994a. Effects of MR exposure on cell proliferation and migration of chick motoneurons. *J Magn Reson Imaging* 4, 799–804.

Yip YP, Capriotti C, Talagala SL, Yip JW. 1994b. Effects of MR exposure at 1.5 T on early embryonic development of the chick. *J Magn Reson Imaging* 4, 742–748.

Yip YP, Capriotti C, Yip JW. 1995. Effects of MR exposure on axonal outgrowth in the sympathetic nervous system of the chick. *J Magn Reson Imaging* 5, 457–462.

Zhang JY, Richards LJ, Yarowsky P, Huang H, van Zijl PCM, Mori S. 2003b. Three-dimensional anatomical characterization of the developing mouse brain by diffusion tensor microimaging. *Neuroimage* 20, 1639–1648.

Zhang X, Yelbuz TM, Cofer GP, Choma MA, Kirby ML, Johnson GA. 2003a. Improved preparation of chick embryonic samples for magnetic resonance microscopy. *Magn Reson Med* 49, 1192–1195.

17 Methods for Labeling Nonphagocytic Cells with MR Contrast Agents

Joseph A. Frank, Stasia A. Anderson, and Ali S. Arbab

CONTENTS

17.1 Introduction 296
17.2 Why Magnetically Label Cells for *In Vivo* MRI? 296
17.3 Entryways into Cells: Endocytosis 296
17.4 Physiochemical Characterization of SPIO MRI Contrast Agents Used in Cell Labeling 298
17.5 Labeling Nonphagocytic Cells with MR Contrast Agents 300
17.5.1 Labeling with Paramagnetic Contrast Agents 300
17.5.2 Incubation with Unmodified (U)SPIO Nanoparticles 302
17.5.3 Mechanical Methods 303
17.5.4 Micron-Sized Superparamagnetic Iron Oxide Particles 304
17.5.5 Virus, Peptides, or Antibody-Coated (U)SPIO Nanoparticles 304
17.5.6 Other (Nondextran) Coated (U)SPIO Nanoparticles 305
17.5.7 Transfection Agents and Dextran-Coated (U)SPIO Nanoparticles 306
17.6 Evaluation of (U)SPIO-Labeled Nonphagocytic Cells 310
17.7 Validation That Cells Are Labeled with (U)SPIO 310
17.7.1 Prussian Blue Staining with or without DAB Enhancement 310
17.7.2 Electron Microscopy 310
17.7.3 MRI of Cells 310
17.8 Determination of Iron Content 311
17.8.1 NMR Relaxometry 311
17.8.2 Spectrophotometric Methods 312
17.8.3 ICP Mass Spectrometry 312
17.9 Determination of Cell Viability, Function, and Differentiation 312
17.9.1 Viability 312
17.9.2 Proliferative Capacity 313
17.9.3 Apoptosis in Labeled Cells 313
17.9.4 ROS in Labeled Cells 313
17.9.5 Functional and Phenotypical Analysis of Labeled Cells 313
17.9.6 Differential Capacity of Labeled Cells 314
17.10 Translation to Clinic 314
Acknowledgment 316
References 316

17.1 INTRODUCTION

The ability to monitor the migration and trafficking of cells in target tissues using noninvasive imaging techniques may contribute to the understanding of the role that cellular therapy will play in repair, replacement, and treatment of diseases. There are a variety of imaging approaches and methods that use contrast agents to label cells, including single-photon emission tomography with labeling done by indium^{111} oxine[1–4] and technetium^{99m} chelates,[5,6] positron emission tomography by incubating cells with copper 64 pyruvaldehyde-bis (N4-methylthiosemicarbazone),[7] optical or bioluminescent imaging by incorporating luciferase or green fluorescent protein-like molecules into the cell genome,[8–10] and magnetic resonance imaging (MRI) using paramagnetic contrast agents (e.g., gadolinium chelates, gadolinium nanoparticles, manganese chloride, etc.) or superparamagnetic iron oxide (SPIO) nanoparticle contrast agents for *ex vivo* labeling and *in vivo* cell tracking.[11–14] This chapter will focus on the methods for intracellular uptake of MRI contrast agents, the methods used to label *ex vivo* nonphagocytic cells (i.e., primary cells, including stem cells, progenitor cells, or terminally differentiated cells such as lymphocytes, B-cells, neurons, and beta cells in islets) with paramagnetic contrast agents and SPIO nanoparticles, the validation steps used to determine if the agents are toxic or alter cell function proliferation or differential capacity, and the steps needed to translate cell labeling techniques from bench to bedside.

17.2 WHY MAGNETICALLY LABEL CELLS FOR *IN VIVO* MRI?

Cellular therapy is a personalized medicine approach for the treatment of diseases. Cellular therapies have been used for more than 50 years in the form of autologous or allogenic intravenous bone marrow stem cell transplantation following ablative chemotherapy or radiation therapy to repopulate and reconstitute the bone marrow for hematopoiesis. Repopulation studies following bone marrow transplantation can be easily monitored by complete blood counts with differential cell counts or bone marrow biopsy and aspirate. However, stem cell or other cell-based therapies under investigation in preclinical studies or clinical studies involving not readily accessible tissue will require noninvasive imaging techniques coupled with a nontoxic method for labeling cells to monitor the temporal-spatial migration into target tissue.

There are presently no techniques to determine the location of therapeutic cells infused or transplanted into the body over extended periods in humans. This can potentially be addressed by noninvasive techniques such as MRI in order to translate cell-based therapy into the clinic. Labeling stem cells and other mammalian cells with clinically approved or experimental MR contrast agents has the potential of monitoring cell trafficking and answers questions regarding the exogenous cells' contact with or entry into the targeted pathology. Magnetic labeling of cells may provide researchers with a tool to understand the role or contribution of a specific cell population in normal and abnormal development or pathologic processes. Magnetic labeling may also allow for the development of innovative experimental and clinical cellular therapy trials using stem cells or genetically engineered cells or adoptive transfer of immune cells that are used as part of repair, replacement, or treatment strategies for various disease processes. Although magnetically labeling cells with existing MRI contrast agents cannot be used to interrogate the transplanted labeled cells' viability, function, or ability to differentiate toward desired lineage, there is a possibility that a multimodality imaging approach combining MRI with nuclear medicine approaches (e.g., positron emission tomography or single-photon emission tomography) or optical and bioluminescent imaging may be useful in addressing these issues.

17.3 ENTRYWAYS INTO CELLS: ENDOCYTOSIS

There are a variety of methods that cells use to regulate their internal environment and allow for the uptake of nutrients, cell-to-cell communication, and the internalization of macromolecules and

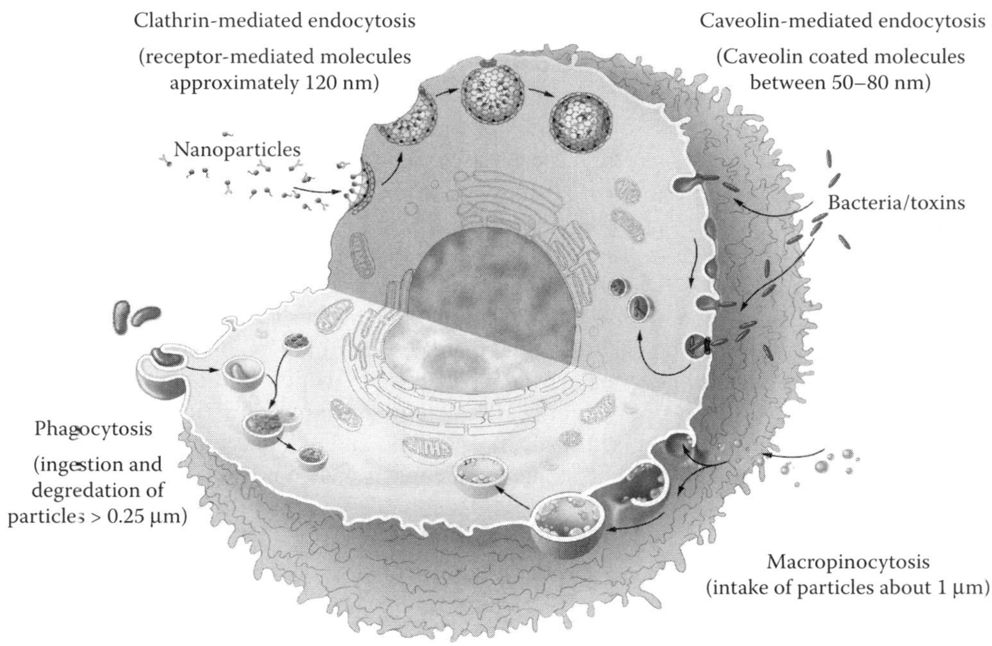

FIGURE 17.1 (please see color insert following page 210) Cartoon of a cell that depicts the various methods cells use to take up macromolecules and particles from the environment, including phagocytosis, pinocytosis, clathrin-mediated endocytosis, caveolin-mediated endocytosis, and clathrin- and caveolin-independent endocytosis. (See text for more details on endocytosis of SPIO nanoparticles.)

particles across the plasma membrane. Endocytosis is a global term to describe the various mechanisms that are used by cells to internalize macromolecules and particles within vesicles derived from the cell membrane and located in the cytoplasm. Endocytosis by definition includes phagocytosis (cell eating), pinocytosis (cell drinking), clathrin-mediated endocytosis, caveolin-mediated endocytosis, and clathrin- and caveolin-independent endocytosis.[15–17] Figure 17.1 is a cartoon of the different ways cells may incorporate SPIO nanoparticles inside cells. Specialized or professional phagocytic cells such as neutrophils, monocytes, macrophages, and microglia use phagocytosis to clear large particulate debris and bacteria. Phagocytosis is an active process involving specific cell surface receptors and signaling cascades, energy metabolites, and regulation of actin as part of the cell cytoskeleton.

In contrast to phagocytosis, all cells exhibit pinocytosis or fluid phase uptake from the surrounding environment, and the amount of uptake of a macromolecular MR contrast agent is directly proportional to the concentration in the media and the volume of the intracellular vesicles. In general, nonspecific binding of an agent to cell surfaces results in pinocytosis. A highly efficient method of intracellular incorporation of contrast agents into cells is through clathrin-mediated endocytosis (also referred to as receptor-mediated endocytosis), which concentrates the contrast agent within clathirin-coated vesicles. The receptor binding typically triggers the local accumulation of clathrin structures on the cytoplasmic surface of the plasma membrane as it begins to invaginate, forming clathrin-coated pits. Clathrin-mediated endocytosis requires specific receptor–ligand interactions to stimulate pinocytosis, such as binding of transferrin to its receptor or the high- or low-density lipoprotein (HDL or LDL) particles to their respective cell surface receptors.[13,19] MR contrast agents with ligands or site-specific antibodies covalently bound to the nanoparticles attach to the specific receptor on the cell surface and start the formation of the complex into a coated pit on the plasma membrane.[20–26] The clathrin-coated pits invaginate into the membrane and ultimately are pinched off to form endosomes. For example, the transferrin receptor has been primarily used to shuttle superparamagnetic iron oxide nanoparticles into cells either by attaching transferrin to the surface coat or by covalently attaching a monoclonal antibody directed to the rat transferrin receptor present on rat progenitor oligodendrocytes.[26] By genetically engineering tumor cells to express high levels of transferrin receptor, uptake of the transferrin-USPIO (ultrasmall SPIO) contrast agent has been used to monitor gene expression in flank tumors in experimental models, leading to the possibility of using this labeling approach to tag cells *in vivo* with MR contrast agents in preclinical studies.[20–22,24,25]

Macropinocytosis is a method of endocytosis that allows for the uptake of larger particles from the extracellular space.[16,17] It is an active process starting as an actin-driven membrane ruffle or extension from the plasma membrane. After encompassing the particles or ligands, the extension collapses into the plasma membrane, forming irregular-sized macropinosomes. There is recent evidence indicating that macropinocytosis is the mechanism by which arginine-containing protein transduction domains, including polyarginine, HIV transactivator transcription (Tat) proteins, peptides, polycationic macromolecular transfection agents, and cationic liposomes, get incorporated into the cell.[17,27,28] Depending on its contents, the macropinosome may or may not fuse with deep lysosomes in the cells, and the contents would then undergo digestion by lysosomal enzymes and buffers.

Caveolea-mediated endocytosis is primarily used by endothelial cells for extensive transcellular shuttling of serum proteins from the bloodstream into tissue; however, other cells in culture may also use this method of endocytosis.[16,17] Caveolea are flask-shaped invagination in the plasma membrane enriched in cholesterol and sphingolipods and marked by an integral membrane protein, caveolin. Caveolea internalize large molecular complexes, toxins, and some viruses. For most cells, caveolea formation is a slow process and the vesicles are relatively small in size (50 to 60 nm) and volume compared to clathrin-coated vesicles, which can rapidly form and are about 120 nm in diameter.[17] Further experiments need to be performed to determine what role caveolea-mediated endocytosis plays in the uptake of viral-coated SPIO contrast agents[29] or other SPIO complexes in cell culture.

Clathrin- and caveolea-independent endocytosis tends to occur in neurons and neuroendocrine cells and is primarily the mechanism of the reuptake following postsynaptic release of vesicles. This method of endocytosis is probably not involved in the uptake of SPIO nanoparticles into neuronal stem cells or neuronal progenitor cells.[16,17]

17.4 PHYSIOCHEMICAL CHARACTERIZATION OF SPIO MRI CONTRAST AGENTS USED IN CELL LABELING

Superparamagnetic iron oxide nanoparticles are a family of MRI contrast agents that are presently used to efficiently label cells for cellular imaging. There are various methods used to prepare SPIO nanoparticles, resulting in a wide range of physiochemical differences, including core size (e.g., ultrasmall (U)SPIO), shape, mono- or oligocrystalline composition, and outer coating, that may alter the ability to use these agents to label cells. Corot and coauthors present an extensive discussion of the physiochemical properties of SPIO nanoparticles in Chapter 4.

SPIO nanoparticles shorten the T1, T2, and T2* relaxation time properties of water or tissue when present in high enough concentrations.[31,36] Table 17.1 summarizes the various physiochemical characteristics of (U)SPIO nanoparticles that have been used for labeling cells. In general, (U)SPIO nanoparticles will alter the T2/T2* of the surrounding tissue compared to the T1 relaxation times in part due to field gradients surrounding the nanoparticles resulting in a rapid dephasing of the protons in the environment. (U)SPIO nanoparticles effect on MRI signal intensities depend on various factors including particle size, hydrodynamic radius, concentration of particles within the voxel, image acquisition parameters, and whether the MR contrast agent is in solution or compartmentalized within a cell.[12,14,37] Long echo time T2 weighted spin echo pulse sequences or T2*

TABLE 17.1
Summary of Physiochemical Characteristics of Superparamagnetic Iron Oxide Nanoparticles Used to Label Cells

Name (Trade Names)	Coating	Crystal Size	Particle Size	R1 (m*M*-sec)–1	R2 (m*M*-sec)–1	Reference
Ferumoxides (Feridex, Endorem)	Dextran, incomplete	5 nm	160 nm	23.7	107	30
Ferumoxtran (MION, Combidex, Sinerem)	Dextran	6 nm	35 nm	24.5	51.9	30
Ferucarbotran (Resovist)	Carboxydextran	4 nm	60 nm	19.4	185.8	49
CLIO-Tat	Cross-linked dextran	5 nm	37–41 nm	22.4	74.2	114
VSOP	Citrate	5 nm	31.3 nm	34.1	61.4	38
WSIO (water-soluble iron oxide)	(3-Carboxypropyl) trimethyl ammonium	12 nm	Not reported	Not reported	40.3	42
HEVJ particle	Viral coat	5 nm (Feridex)	Not reported	Not reported	Not reported	29
AMNPs	Meso-2,3-dimercaptosuccinic acid	8.7 nm	Not reported	11.6	363.2	48
Magnetodendrimers	PAMM dendrimer	7–8 nm	>60 nm	7	200–406	34
BANG particles	POLYVINYL benzene		300–5800 nm	Not reported	Not reported	53

weighted gradient echo MR pulse sequences are usually used to detect the presence of (U)SPIO nanoparticles within tissues and these agents usually appear as hypointensities with or without associated susceptibility artifacts on the images. For MR cellular imaging, (U)SPIO nanoparticles are usually compartmentalized within endosomes or macropinosomes within the cytoplasma of cells and cause a decrease in the signal intensity of the target tissue on T2 and T2* weighted images because of rapid dephasing of the water proton spins set up by the magnetic field gradients that develop around the magnetically labeled cells.

The size of the (U)SPIO nanoparticles depends on the surface coating used and will determine if the particle is monocrystalline (ferumoxtran) or consists of multiple or oligocrystalline, such as ferumoxides.[30] Surface coatings on (U)SPIO nanoparticles may be of various sizes and have surface-charged molecules, including dextran and modified cross-linked dextran, dendrimers, starches, citrate, or viral particles.[24,29–49] The coating is usually added during formation of the Fe_3O_4 to Fe_2O_3 crystals and allows the SPIO nanoparticles to exist in a colloidal suspension in aqueous solutions. For several clinically approved SPIO nanoparticles (e.g., ferumoxides, ferumoxtran, and ferucarbotran), the coating is dextran, that is, attached through electrostatic interaction to the iron core by hydrogen bonds between some of the dextran hydroxyl groups and the surface oxide groups of the iron core.[30] The unattached dextran tail covers the rest of the iron crystals and contributes to most of the hydrodynamic diameter of the (U)SPIO nanoparticles.[30] Figure 17.2 is a scanning electron micrograph of ferumoxide with its oligocrystal structure and dextran coating. For SPIO nanoparticles (e.g., ferumoxides or ferucarbotran) the dextran coating links multiple iron oxide crystals together, and they have a hydrodynamic diameter of about 60 to 200 nm.[30]

The coating molecules will contribute to the surface charge or zeta potential of the (U)SPIO in water. The zeta potential is the average potential difference in millivolts existing between the surface of the (U)SPIO nanoparticles immersed in a conduction liquid (water) and the bulk of the liquid. Dextran-coated ferumoxides have a zeta potential of –32 mV, while ferumoxtran that is

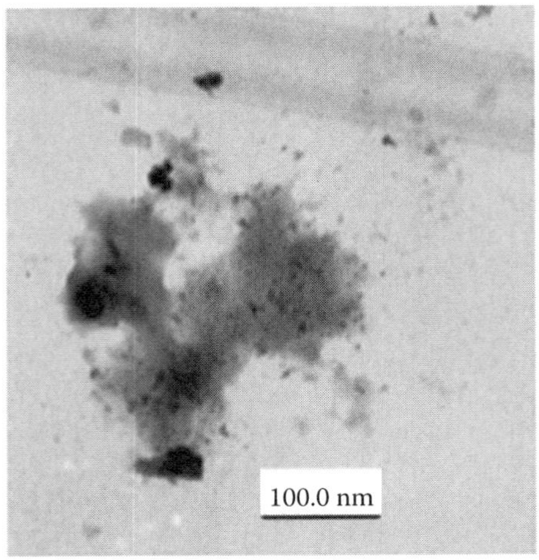

FIGURE 17.2 Transmission electron micrograph of a ferumoxide nanoparticle obtained at a magnification of 9300. Multiple iron oxide crystals of about 4 to 5 nm can be detected in the dextran coat.

coated with a shorter-chain dextran has a measured zeta potential of –2 to 0 mV,[50] and the near-neutral surface potential of ferumoxtran possibly contributes to the long blood half-life, compared to the larger SPIO nanoparticles. Zeta potentials have not been reported for other (U)SPIO nanoparticles characterized in the literature as anionic or cationic because of the coating. Cationic USPIO nanoparticles have been used to label primary and cloned cell population in culture because of the negative surface charge of cell membrane, such as HeLa cells for *in vitro* studies. HeLa cells have a negative zeta potential,[51] and the (U)SPIO agent initially interacts with the plasma membrane through electrostatic interactions, followed by endocytosis.

Larger micron-sized particles of iron oxide (MPIOs), commercially available particles or beads, are also used to label cells for cellular MRI studies in experimental models.[52–56] These agents are from 0.3 to >5 microns in size and contain greater than 60% magnetite in a polymer coating, which can include a fluorescent marker that allows for dual detection of labeled cells by MRI and fluorescent microscopy. MPIOs can be purchased with either terminal amines or carboxyl groups on the surface, thus allowing chemical modification to attach peptides, ligands, or monoclonal antibodies to specific targets and receptors on cells.

17.5 LABELING NONPHAGOCYTIC CELLS WITH MR CONTRAST AGENTS

17.5.1 Labeling with Paramagnetic Contrast Agents

In comparison to labeling cells with iron oxide-based nanoparticles, relatively few studies have used paramagnetic MR contrast agents such as gadolinium chelates (e.g., DTPA, HPDO3A) or its derivatives, gadolinium nanoparticles, or manganese chloride to label cells for cell trafficking studies. The major goal of labeling cells with a paramagnetic contrast agent is to shorten the T1 relaxation properties of the intra- and extracellular water undergoing exchange across the cell membrane, resulting in an increase in signal intensity due to the labeled cells on T1-weighted MRI. The T1 relaxation shortening effects of typical paramagnetic agents such as gadolinium chelates are significantly weaker on a molar basis compared to iron oxides nanoparticles because the relaxation mechanism is based on a dipole–dipole interaction of water protons with the gadolinium vs. a through-space or dephasing of the water proton spins due to the alteration in the local magnetic field gradients with the SPIO nanoparticles. However, compartmentalization of the paramagnetic contrast agent in the cell is concentration dependent and can result in a T2 and T2* shortening at high concentration, thereby making it difficult to detect labeled cells on T1-weighted MRI.[57,58]

Therefore, the labeling of cells with paramagnetic agents needs to be optimized to obtain proton relaxation enhancement and increased signal intensity on T1-weighted imaging.

Several approaches have been reported for cell labeling using gadolinium chelates. By directly injecting gadolinium chelates into the early stages of division of a *Xenopus laevis* egg, it was possible to track the proliferation of the labeled cell population and migration into different organs during development using MRI and optical imaging.[59] Relatively high concentrations of gadolinium-based contrast agents were introduced into a cell at the 16-cell stage and MR imaging was performed at 11.7 T. The direct injection approach is not practical or efficient for labeling mammalian cells with MR contrast agents.

Crich et al.[60] incubated endothelial progenitor cells with GdHPDO3A at various concentrations and were able to demonstrate uptake into cells, presumably by micropinocytosis. These authors demonstrated a linear correlation between the amount of GdHPDO3A in the media and uptake of gadolinium per gram of protein in lysate of endothelial precursor cells. The authors directly implanted 10^6 labeled cells in matrigel under the capsule of mouse kidneys and were able to identify the location of these cells as high signal intensity on T1-weighted MRI. By covalently attaching spermidine to GdDOTA, Crich et al.[61] used *in vivo* electroporation to introduce the MRI contrast agent and DNA plasmid into cells and were able to identify the location of labeled cells by MRI.

Daldrup-Link et al.[62] labeled human hematopoietic stem cells by combining gadophorin-2, a gadolinium-based paramagnetic and fluorescent metalloporphyrin, with cationic liposomes using standard transfection techniques (see Section 17.5.6). These authors were able to detect 10^6 gadophorin-2-labeled cells by fluorescent imaging in normal mouse; however, in order to detect a 20% increase in signal intensity on T1-weighted images at 1.5 tesla in the liver, spleen, or bone marrow, approximately 10^8 labeled cells needed to be injected intravenously into the mouse. Daldrup-Link et al.[25] also labeled cells by combining liposomes with GdDTPA and were able to demonstrate an increase in signal intensity on T1-weighted images compared with unlabeled stem cells. A dual-labeled nanoparticles approach based on lipid monomers combining gadolinium chelates and rhodamine has recently been shown to incorporate into tumor cells and could be detected by MRI and optical imaging.[63] Modo et al.[57,58] labeled neural stem cells with a bifunctional gadolinium chelate rhodamine dextran (GRID) complex by simple incubation; however, following intracerebral injection of 50×10^4 cells in rats, labeled cells could only be detected as hypointense regions on T2*-weighted images.

Liposomal "artificial virus-like envelopes" have been used to label cells and showed enhanced T1 relaxivity with degradation of the envelope.[64] GdDOTA covalently attached to D-Tat facilitated, via macropinocytosis, the entry of the agent into leukemia cells and demonstrated T1 contrast enhancement at 4.7 tesla.[65] By targeting the amino acid active transport in cancer, Lattuada et al.[66] developed gadolinium complexes conjugated to agmatine, arginine, and glutamine and labeled tumor cells in culture. In this study, glutamine (gln)–GdDTPA-labeled rat hepatocellular carcinoma cell pellets had a greater than 50% decrease in T1 than unlabeled control cells at 7 tesla. Comparison of membrane-bound to internalized forms of the contrast agent was made and demonstrated a moderate reduction of T1 when the gadolinium chelate was internalized in the hepatocellular carcinoma cell, compared to bound to its surface.

By inserting a macrocyclic gadolinium chelate that embeds two hydrophobic alkyl chains in the cell membrane with the gadolinium complex moiety outside the cell, Zheng et al.[67] have been able to label cells. In this study, diffusion-enhanced fluorescence resonance energy transfer was used to determine the location of the gadolinium chelate. By placing the macrocylic gadolinium chelate outside the cell membrane, there is a greater chance of water protons interacting with the last coordination site on the gadolinium, and the movement of the cell would potentially dominate the increased T1 relaxivity of this agent. The ability of such a label to remain attached to the membrane would be essential for cell labeling and tracking studies. Whether the attached label is later lost or internalized would remain to be determined; however, testing of transplanted labeled cells in this study showed a persistent label effect over 2 weeks.

MR contrast agents with very high T1 relaxivity would be valuable for labeling cells, as they have the potential to retain significant detectability on MRI even in low concentrations, as well as to retain sufficiently high T1 relaxivity even at high magnetic field strengths. Gadolinium fullerene compounds, composed of a Gd^{3+} ion trapped in a 60- to 82-carbon fullerene cage, have been shown to have much higher relaxivites than conventional GdDTPA.[68,69] Although gadolinium fullerenes are highly stable, they must be made water soluble by the addition of hydroxyl groups to the surface of the nanoparticles so that they can be used as MR contrast agents.[70] Gadolinium fullerenol (Gd@C82) has a high T1 relaxivity, presumably due to the unique distribution of the unpaired electrons to the surface of the fullerenol. Water does not have direct access to the gadolinium ion through the cage; the presumed relaxation mechanism is through contact with the outer cage. Recently, Anderson et al.[71] used gadolinium fullerenol nanoparticles consisting of a carbon fullerene cage with external hydroxyl functionality with protamine sulfate to form a complex to label mesenchymal stem cells. Gadolinium fullerenol-labeled mesenchymal stem cell pellets in gel demonstrated an increase in signal intensity on T1-weighted images at 1.5 and 7.0 tesla. Moreover, the Gd@C82-labeled cells could be detected with 7-T MRI following direct injection of 10^6 cells into the rat thigh. Anderson et al. noted that gadolinium fullerenol initially decreased the stem cell's ability to proliferate following labeling, suggesting that the agent may be altering mitochondrial function, which normalized by day 3 after labeling.[71] Electron microscopy of labeled cells showed aggregates of the Gd@C82 nanoparticles in endosomes. This is the first report of a transient negative effect of a gadolinium-based agent on cell proliferation, and because more gadolinium-based nanoparticles are used for cellular and molecular imaging, further evaluation will be needed to ensure there is no long-term toxicity to the cells.

Manganese (Mn^{2+}) chloride was the first paramagnetic contrast agent used in MRI, and it has been shown that it can be taken up by cells *in vivo* through calcium channels in the cell membrane.[72] T1-weighted MRI has been used to monitor antegrade movement of accumulated Mn^{2+} in neurons through axons following inhalation or direct injection of manganese chloride into the brain, thus serving as a functional MRI agent.[73–75] Aoki et al.[76] have recently reported that lymphocytes could be labeled following incubation with manganese chloride without altering function or toxicity at concentrations of less than 0.5 m*M* in media. MRI of Mn^{2+}-labeled lymphocytes in gelatin demonstrated increased signal intensity on T1-weighted images. However, it is not clear if there would be sufficient contrast enhancement to detect Mn-labeled cells *in vivo* by MRI. The major drawbacks to the use of Mn^{2+} as an MR contrast agent are its narrow therapeutic window and significant cardiotoxicity.

17.5.2 Incubation with Unmodified (U)SPIO Nanoparticles

Various approaches have been developed that use both experimental and clinically approved (U)SPIO nanoparticles to label cells. There are dextran-coated SPIO nanoparticles that are Food and Drug Administration (FDA)-approved MR contrast agents for use in hepatic imaging (SPIO), and USPIO particles have been used in clinical trials as blood pool agents or for lymphangiography.[77–80] Initial studies that labeled nonphagocytic cells such as lymphocytes, C6 glioma cells, and progenitor stem cells[41,43,81–83] were based on the principles used to label phagocytic cells such as neutrophils, macrophages, and monocytes.[38,82,84–87] The native unaltered (U)SPIO nanoparticles would simply be incubated at high concentrations (>2 mg iron/ml), with the adherent or cells grown in suspension for 48 to 72 h.[35,36,81,88,89] No effect on viability was reported as a result of labeling cells with high concentrations of iron oxide nanoparticles; however, detailed analysis of cell proliferation, reactive oxygen species formation, and differential capacity was not performed (see section 17.9 for details on validating the safety of labeling cells with SPIO nanoparticles). Cells will incorporate dextran-coated (U)SPIO nanoparticles through endocytosis pathways; however, incubation times are long and the labeling efficiency using this approach is not as high as with other techniques (see below). Therefore, this is not the preferred method for labeling nonphagocytic cells *in vitro*.[90–92] Jirak et al.[90] have shown that beta cells from harvested pancreatic islets will take up SPIO, and clearly demonstrated that the beta cells continue to produce insulin after infusion of cells into the portal vein of a diabetic mouse model. Islets labeled with SPIO were detected 18 weeks after infusion into the mice and the animals remained normoglycemic; therefore, iron labeling did not alter the beta cells' function. However, Frank et al. have reported that incubating mesenchymal stem cells or HeLa cells with low levels of ferumoxides or MION-46L for 24 to 48 h did not result in labeling of the cells, compared to labeling cells using transfection agents complexed with dextran-coated (U)SPIO.[93,94]

Although simple incubation is not the preferred method for labeling nonphagocytic cells with dextran-coated (U)SPIO nanoparticles, a recent clinical trial used ferumoxides to label dendritic cells derived from peripheral blood monocytes to track cells *in vivo*. Monocytes were incubated with ferumoxides for 2 days in appropriate conditioned media, and the labeled dendritic cells were transplanted directly into lymph nodes of patients with melanoma as part of an early-phase clinical trial.[95] Magnetically labeled dendritic cells were monitored at 3 tesla, migrating through lymphatic vessels into adjacent lymph nodes, demonstrating the clinical utility of this approach for monitoring cellular therapy.

17.5.3 Mechanical Methods

Mechanical approaches that have been used primarily to introduce DNA or plasmids into the nucleus have been modified to directly introduce MRI contrast agents into cells. The gene gun fires DNA, plasmids, or DNA-coated nanoparticles directly into cells in culture, driving the particles through the cell membrane or directly into the nucleus,[96,97] using a ballistic gas charge. SPIO nanoparticles and magnetic beads have been introduced into stem cells using a gene gun with high labeling efficiency,[98] and the cells were implanted into rats, but no information was provided about the effects of the method on long-term cell viability, proliferation, differentiation capabilities, or reactive oxygen species. It should be noted that the gene gun approach is indiscriminant and will introduce MRI SPIO nanoparticles directly into the cell nucleus. Although it is beneficial to deliver DNA directly into the nucleus, the presence of SPIO nanoparticles in the nucleus could initiate a Fenton reaction and, through Haber–Weiss chemistry,[99] result in the development of free radicals that could cause damage to DNA.

Electroporation is used to introduce plasmids and DNA into cells without the use of transfection agents.[100] This technique has recently been applied to labeling cells with gadolinium chelates and SPIO nanoparticles.[61,101] The cell membrane acts as an electrical capacitor that is unable to pass current except through ion channels. Applying a short electric field pulse to cells results in an elevation of the transmembrane potential, causing a reorganization of the membrane, and the development of electropores. The opening of the electropores allows for the diffusion of DNA or MR contrast agents into the cells. After the pulsed electric field, the electropores close quickly, trapping the agents in the cytoplasma or nucleus. There is relatively little experience with using this approach with MRI contrast agents to label cells, and it is unclear what the short- and long-term effects on reactive oxygen species or cell viability will be. In addition, there is presently no available FDA-approved electroporation device operating in good manufacturing practice (GMP) blood product facilities that could be used to introduce DNA or MRI contrast agents into cells.

Magnetofection is another mechanical approach that combines gene vectors with coated SPIO nanoparticles and uses an external magnetic field to target the DNA–SPIO complex into cells.[102,103] Cells in culture plates are positioned on top of a Nd-Fe-B magnet strong enough to quickly sediment the applied SPIO gene vectors onto or into the cells, resulting in an increased concentration of agent on the surface of the cells, with rapid incorporation of the gene presumably with the carrier SPIO nanoparticles into the cell. The DNA gene vector is complexed to the SPIO nanoparticle with the use of transfection agents, and this approach appears to be superior to incorporating genes into adherent cells by retroviral transfection. This method of labeling cells can also be used to transfect cells *in vivo*; however, it is not clear if cells in suspension can be magnetically labeled, and there is no indication of the concentration of SPIO in the cells.[104]

17.5.4 Micron-Sized Superparamagnetic Iron Oxide Particles

Recently, investigators labeled cells with micron-sized particles of iron oxide (MPIOs) such as BANG particles (www.bangslabs.com) or magnetic beads used in positive or negative magnetic cell separation.[52–56,105,106] Although these agents have not been used to effectively label cells with a high nuclear-to-cytoplasmic ratio, such as lymphocytes, phagocytic and adherent cells incorporate these large particles into the cytoplasma via endocytosis. The MPIOs have certain advantages in that they are synthesized with a fluorochrome embedded in the polyvinyl benzene coating, allowing for dual detection using fluorescent microscopy and MRI of labeled cells within tissues. Hinds et al.[53] have shown that labeling mesenchymal stem cells (MSCs) or CD34 hematopoietic stem cells (HSCs) with MPIO did not alter cell proliferation or differentiation capacity; however, excess MPIO in appropriate media did suppress proliferation of MSCs. The MPIO-labeled MSCs injected into the hearts of swine with myocardial infarctions could be visualized on 1.5-tesla MRI scans. Shapiro et al.[52] demonstrated uptake of very large MPIOs (5.8 microns) in cultured hepatocytes and have been able to visualize single cells at 7 tesla. Although the work with MPIO allows for the detection of single cells *in vivo* at high magnetic field strength, there are some aspects of this agent that have not been addressed. No studies have been performed to specifically address how the micron-sized particles are eliminated from the cells or whether the endosome/ lysosome can digest the polymer coating of the micron-sized particles. If these agents are expelled from the labeled cell *in vivo*, it is unclear if the agents will be eliminated from the body or if they will localize to the extracellular spaces in the neighborhood of the previously labeled cells or travel along lymphatic pathways and be phagocytosed by macrophages in lymph nodes. Since MPIOs are not approved for clinical studies, these agents can only be used in preclinical or experimental studies.

17.5.5 Virus, Peptides, or Antibody-Coated (U)SPIO Nanoparticles

Several modifications of the dextran coat of (U)SPIO nanoparticles have been shown to increase the efficiency and facilitate the incorporation of the MR contrast agents in cells. After reconstituting Sendai virus envelopes in the presence of dextran-coated iron oxide nanoparticles, Hawrylak et al.[33] incubated this novel iron-loaded virus in rat fetal neuronal cells in suspension and reported a moderate degree of uptake by cells; however, cell viability was between 45 and 65%. Labeled cells were implanted directly into the brains of rats, and the migration of labeled cells was demonstrated on MRI and correlated to Prussian blue and immunohistochemical staining of brains. Recently, a nonviral vector system was developed — the hemagglutinating virus of Japan (HVJ) envelope — encapsulating the SPIO, ferucarbotran, and labeled microglial cells.[29,107,108] MRI could detect clusters of HVJ SPIO-labeled cells within 1 day of transplantation in the brains of mice after intracardiac injection. The HVJ SPIO particles were reportedly more efficient at labeling cells than was dextran-coated SPIO combined with the transfection agent lipofectamine;[29] however, since HVJ envelopes are not commercially available, the use of this agent is limited.

By encapsulating SPIO nanoparticles in unilamellar vesicles known as magnetoliposomes and incubating them with human peripheral blood mononuclear cells, Bulte et al.[109,110] demonstrated uptake of the iron oxide particles in phagosomes and secondary lysosomes by electron microscopy. Cationic magnetoliposomes were also developed to label cancer cells and, in conjunction with alternating magnetic field hyperthermia, have been used to effectively treat tumors in experimental models.[111,112] Lectin, a protein derived from lentils that is also used as an antigen to stimulate T-cells in culture, when attached to USPIO, was used to label lymphocytes *ex vivo*, and T2 relaxivity of cells in gels suggested uptake of the nanoparticles into the cells.[113] No further work was performed with the complex.

Josephson et al.[114] increased the efficiency and versatility of dextran-coated monodispersed superparamagnetic iron oxide nanoparticles (MIONs) in labeling nonphagocytic cells by first cross-linking the dextran strands (CLIO, or cross-linked iron oxide) with epichlohydrin, and then covalently attaching HIV-1 Tat proteins to the surface. CLIO-Tat nanoparticles are efficiently taken up by a variety of nonphagocytic cells, presumably because of the arginine-rich Tat peptide that stimulates macropinocytosis and incorporation of this agent into micropinosomes.[27,28] CLIO-Tat has been used to label lymphocytes and hematopoietic stem cells with high labeling efficiencies in these nonphagocytic cells.[40,115–119] There are no reported differences in CLIO-Tat-labeled and unlabeled T-cells' reactivity to antigen, viability, or ability to home to target tissues.[119,120] CLIO-Tat-labeled T-cells have been used for adoptive transfer of autoimmune diabetes in a mouse model, and labeled cells have been shown to selectively home to specific antigens in B16 melanoma in a mouse model by *in vivo* MRI.[117,118,120] CLIO-Tat-labeled hematopoietic stem cells expressed the same surface markers over a period of 3 weeks, had the same number of colony-forming units, and were able to repopulate the bone marrow of lethally irradiated mice at the same level as unlabeled stem cells.[119] Although CLIO-Tat and its derivatives, which include optical imaging agents (i.e., FITC, CY5.5) to its surface, are very efficient at labeling cells *ex vivo*,[116] their use is relatively limited since CLIO-Tat is a custom synthesized agent and not commercially available.

Functionalizing (U)SPIO nanoparticles by conjugating antigen-specific internalizing monoclonal antibodies (mAbs) to the surface dextran coating has been used to label progenitor oligodendrocytes, neural precursor cells, and dendritic cells by clathrin-mediated endocytosis.[20,21,26,121–123] The monoclonal antibody to the rat transferrin receptor was covalently bound to the MION-46L and used to label rat progenitor oligodendrocytes that were then directly implanted into spinal cords of myelin-deficient rats.[26] *Ex vivo* MR images 10 to 14 days after implantation of labeled cells demonstrated excellent correlation between the blooming susceptibility artifact and the degree of myelination in the spinal cord detected on immunohistochemistry to the proteolipid protein of myelin. The results demonstrated that magnetically labeled cells would not interfere with the cell differentiation, function (formation of myelin wraps around axons), and margination along areas of pathology in the spinal cord. Ahrens et al. labeled dendritic cells by biotinylating anti-CD11 mAb in conjunction with strepavidin attached to dextran-coated iron oxide nanoparticles.[121] The advantage of this two-stage labeling approach is that the CD11 mAb is commercially available, and the biotin–strepavidin interaction is a commonly used molecular biology procedure resulting in the SPIO being incorporated in endosomes. Since dendritic cells are phagocytic and can be derived from stimulated monocytes, simple incubation will result in uptake of ferumoxides into endosomes.[95]

17.5.6 Other (Nondextran) Coated (U)SPIO Nanoparticles

For most cellular MRI studies that involve monitoring of the migration of cells *in vivo*, the nonphagocytic cells are labeled with dextran-coated (U)SPIO nanoparticles because these agents are commercially available from pharmaceutical companies or synthesized in laboratories. Instead of precipitating the magnetite or maghemite in the presence of dextran, various other molecules have been used to stabilize the iron oxide crystals in solution, including citrate (VSOP), dimercaptosuccinic acid (AMNP), carboxypropyl trimethyl ammonium (WSIO), and dendrimers (MD 100).[34,38,42,44–48,85,124,125] Anionic maghemite nanoparticles (AMNPs) have surface charges, mainly due to carboxyl groups on the surface, and these agents are reported to be taken up by HeLa cells and macrophages in a greater amount than dextran-coated iron oxides.[85] Cationic-coated USPIO nanoparticles such as WSIO and VSOP were designed with coating that would attach to the negative surface charge of plasma membranes through electrostatic interactions.[38,42,125–127] VSOP C125 reportedly labeled macrophages with iron oxide nanoparticles to a greater concentration and more quickly than a carboxydextran-coated SPIO.[38] VSOP is presently in early-phase clinical trials as an intravascular MRI contrast agent for angiography in Europe.[124] It can be assumed that if appropriate preclinical studies are preformed using VSOP, these nanoparticles may be available for *ex vivo* labeling of stem cells for clinical trials. Cationic WSIO nanoparticles have been used to efficiently label fibroblasts, neural stem cells, and cancer cell lines. The migration of WSIO-labeled neural stem cells implanted into the rat spinal cord was detected on *in vivo* MRI, and WSIO-labeled cells were identified on histology by Prussian blue stain.[42]

Dendrimers are branched synthetic polymers with a layered architecture that can be of various sizes or generation and have multiple biologic and drug applications, including serving as the backbone for macromolecular gadolinium chelated MR contrast agents.[128–133] Dendrimers are highly charged, commercially available macromolecules used to compact DNA and transfect oligonucleotides into cells by binding to the plasma membrane and stimulating endocytosis. By adding dendrimers during the synthesis of iron oxide nanoparticles, the carboxyl groups on the surface of a generation 4.5 polyamidoamine (PAMAM) dendrimer served as a nidus for crystal formation, resulting in the creation of magnetodendrimers (MD 100). MD 100 were shown to efficiently and effectively label a variety of mammalian cells from the mouse, rat, and human at relatively low concentrations of iron oxide (25 μg/ml) in media when cells were incubated for 24 to 48 h. There was no difference between MD 100-labeled cells and unlabeled cells in terms of proliferation and viability, migration, differentiation potential, and cell function. Although MD 100 has been shown to label a wide range of cell types in culture, including mouse embryonic stem cells, fibroblasts, C2C12 muscle progenitors, rat and human neural stem cells, mesenchymal stem cells, and various cancer cell lines,[34,44,134–137] it has not been shown to effectively label hematopoietic stem cells or lymphocytes in cell culture suspension, possibly due to the surface charge of these cells.

In vivo MRI studies of MD 100-labeled rat progenitor oligodendrocytes implanted in the lateral ventricles of Long Evans Shaker rats demonstrated migration of cells into white matter, and on histology showed areas of myelination in this animal model of dysmyelination.[34] This was the first study to demonstrate that magnetically labeled cells could be detected and tracked over 44 days at 1.5 tesla using standard gradient echo techniques, indicating that cellular MR imaging could translate from bench to bedside for ultimate use in clinical trials. A disadvantage of MD 100 is its involved synthesis, and that it is not commercially available. It is unlikely to be developed for clinical studies because of alternative methods using FDA-approved agents to label cells.

Recently, Kohler et al.[138,139] developed a novel MRI and controlled drug delivery by coating the surface of SPIO nanoparticles with 3-aminopropyl-trimethoxysilane and covalently attaching methotrexate (MTX), a chemotherapeutic agent that enters the cell through binding to the folate receptor on the plasma membrane. Because the folate receptor is upregulated in cancer cells such as HeLa cells, MCF-7 breast cancer cells, or 9L glioma cells, compared to cardiac myocytes, the SPIO nanoparticles with MTX are quickly taken up by clathrin-mediated endocytosis and the cancer cells containing high concentrations of iron per cell can be detected by MRI and electron microscopy.[138,139] Methotrexate is released in the endosomes and will kill cells over time; cell death can be prevented by the addition of a folate analogue, leucovorin, to the media.[138]

17.5.7 Transfection Agents and Dextran-Coated (U)SPIO Nanoparticles

Although several approaches for labeling nonphagocytic cells with iron oxide nanoparticles have been explored, most of the agents used were proprietary compounds, involving unique or complex synthesis or biochemical modification of the dextran coat of the SPIO to stimulate endocytosis by cells. The combination of transfection agents such as cationic liposomes encapsulating large crystal-sized SPIO nanoparticles was initially used to label phagocytic cancer cells in culture as part of experiments to develop the use of alternating magnetic field hyperthermia as a cancer therapy.[111,112] As an alternative to retroviral gene transduction, de Marco et al.[140] combined poly-L-lysine conjugated to dextran chains anchored to a superparamagnetic iron oxide core and complexed this particle to DNA encoding for humanized green fluorescent protein (GFP). The constructs resulted in low transfection efficiency (4.1%) for GFP, and no determination was made of the presence of intracellular iron.

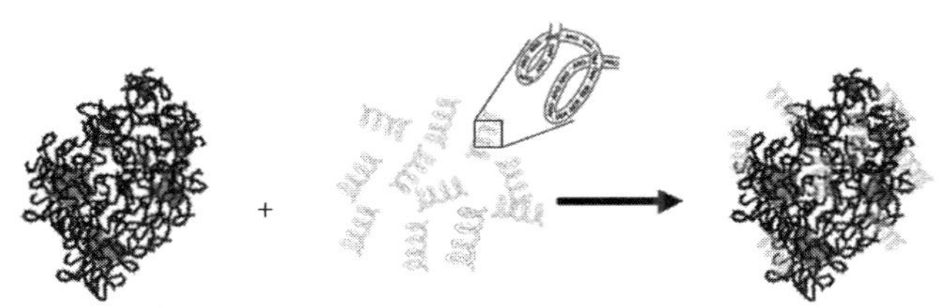

FIGURE 17.3 A cartoon depicting the formation of the nanoparticle complex when ferumoxides are incubated in the presence of protamine sulfate. Protamine sulfate is approximately 60% arginine (enlargement), has a helical structure in solution, and binds to ferumoxides through electrostatic interactions. The ferumoxides–protamine sulfate complex used to label cells has a negative zeta potential, and therefore the transfection agent does not completely coat the dextran coating of ferumoxides.

Based on the success of using magnetodendrimers[34] to label cells in which dendrimers, a type of transfection agent (TA), were used as a coating, a straightforward approach was developed in our laboratory by combining (U)SPIO nanoparticles (e.g., ferumoxides, ferumoxtran-10, and MION-46L) with commonly available polycationic transfection agents to label cells.[93,94,141,142] Ferumoxides are FDA-approved colloidal dextran-coated SPIO nanoparticles used as MR contrast agents for hepatic imaging and have a zeta potential of –32 mV because of the carboxyl groups on their surface.[50] Transfection agents are polydisperse macromolecules that are either positively or negatively charged and include classes of commercially available agents such as dextrans, phosphates, artificial lipids (FuGene™, Lipofectin™, Lipofectamine™, Lipovec™), proteins (poly-L-lysine, poly-L-arginine), and dendrimers (Superfect™, Polyfect™).[143–145] Polycationic transfection agents have zeta potentials that vary from 2 to >48 mV and, when mixed with dextran-coated (U)SPIO nanoparticles, form complexes through electrostatic interactions. Kalish et al.[50] characterized the physical–chemical interactions of TA with dextran-coated SPIO nanoparticles and observed that the relative concentration and class of transfection agent had an effect on the formation of complexes and the ability of the complexes to alter the relaxation properties of the water interacting with the paramagnetic sites on the iron oxide crystals. Serial dilutions of a ferumoxides–poly-L-lysine (PLL) complex demonstrated effective switching from a negative to positive zeta potential, depending on the relative ratio of the SPIO to PLL, with a decrease in the T1 and T2 relaxation rates of the solutions when the measured zeta potential of complex was neutral.[50] Increasing the ratio of cationic transfection agent to ferumoxides results in the clumping of multiple nanoparticles together, and the T2 relaxivity increases to greater than the MRI contrast alone (Figure 17.3). It has been suggested that the dextran-coated SPIO nanoparticles are completely coated by the TA to facilitate binding to the plasma membrane of cells; however, the concentration ratios of ferumoxides to TA commonly used to label cells have negative zeta potentials, suggesting that both negative and positive charges would have to be present on the nanoparticle complexes to initiate adherence to the cell membrane and endocytosis[50,146] (Figure 17.4).

Although synthesized transfection agents are commercially available for DNA transfection, none are FDA approved.[144] Protamine sulfate (Pro) USP is a low molecular weight (~4200 daltons), FDA-approved, naturally occurring peptide containing about 60 to 70% arginine that has been in use as an antidote to heparin-induced anticoagulation as well as in clinical gene therapy protocols to introduce genes into cells *ex vivo*.[147,148] Protamine sulfate is well tolerated by cells with a high lethal dose at 50%, is greater than 50 μg/ml, and has a greater efficiency of transfecting DNA into cells than does PLL.[147] Table 17.2 is the protocol presently used in our laboratory to magnetically label adherent cells and cells grown in suspension with ferumoxides complexed to protamine sulfate (FePro).[146,149] The most effective ratio of ferumoxides to protamine sulfate for

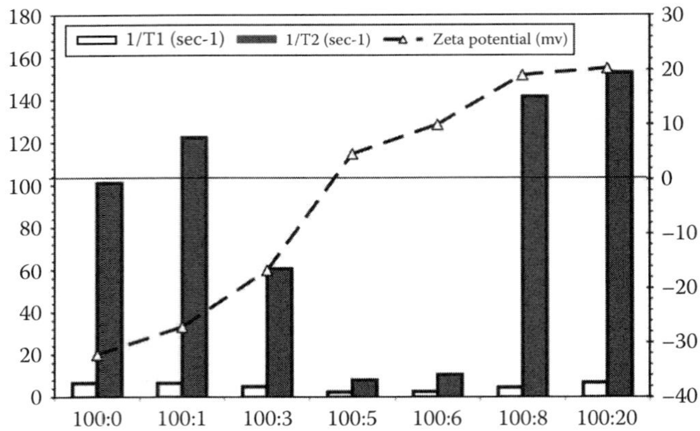

FIGURE 17.4 Graph of R1 and R2 (sec^{-1}) relaxation rates and zeta potentials (millivolts) for concentration ratios of ferumoxides to protamine sulfate in solution. The R1 and R2 decreases for the complex around ferumoxides–protamine sulfate ratios of 100 μg/ml to 4 to 6 μg/ml as the zeta potential goes from positive to negative, which is most likely due to nanoparticles partially precipitating out of suspension. At higher concentrations of protamine sulfate, the zeta potential is positive and R2 increases past that value of ferumoxides alone, presumably due to the formation of larger soluble particles.

TABLE 17.2
Protocol Used to Label Adherent Cells and Cells Grown in Suspension Using Ferumoxides Complexed to Protamine Sulfate (FePro)[146]

Media

1. Serum-free media
   a. RPMI-1640 (with 25 m*M* HEPES, without glutamate; biofluids) plus l-glutamine at 1× or 4 m*M*, sodium pyruvate at 1×, and MEM nonessential amino acids — serum-free media.
2. Serum-free FePro labeling media
   a. Add ferumoxides (Fe) (100 μg/ml of media) and protamine sulfate (Pro) (4 to 6 μg/ml) to serum-free media in a tube (protamine and ferumoxides must be sterile).
   b. Shake/invert tube for 1 to 2 min to allow for complex formation.
3. Adherent cell labeling (e.g., mesenchymal stem cells, monocytes, macrophages, HeLa)
   a. Culture cells to 80% confluence.
   b. Remove culture media and add serum-free labeling media (5 ml for T75 and 15 ml for T175 flasks; FePro concentration of 100:4–6 μg/ml) and culture for 2 h at 37°C and 5% CO_2.
   c. Add an equal amount of appropriate complete media for cell type being labeled to bring FePro to a final concentration of 50:2–3 μg/ml and culture overnight.
   d. Remove media and wash cells two or three times with PBS (add 2 to 10 units/ml of heparin to three PBS washes to improve washing if desired).
   e. Trypsinize, filter with 40-micron mesh.
   f. Collect labeled cells.
4. Suspension cell labeling (e.g., T-cells, CD34+ cells)
   a. Add serum-free labeling media (FePro at 100:4–6 μg/ml) to cells such that cell concentration is 4×10^6/ml and culture at 37°C and 5% CO_2 for 2 h.
   b. Add an equal amount of appropriate complete media for cell type being labeled to bring FePro to a final concentration of 50:2–3 μg/ml and the cell concentration to 2×10^6/ml and culture overnight.
   c. Remove media and wash cells two or three times with PBS (add 7 to 10 units/ml of heparin to PBS during washes to improve washing if desired) by centrifuging and decanting/aspirating.
   d. Filter with 40-micron mesh and repeat pippetting.
   e. Collect labeled cells

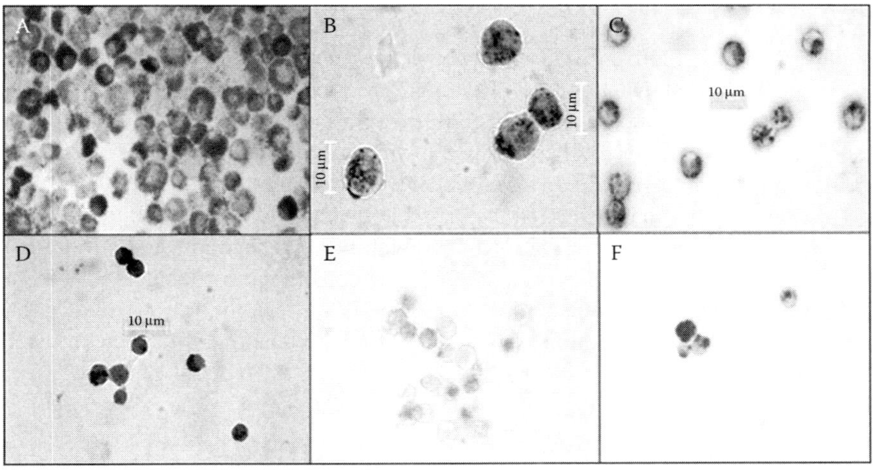

FIGURE 17.5 (please see color insert) Examples of different types of magnetically labeled cells using ferumoxides–protamine sulfate complexes at a ratio of 100 μg to 5 μg. (A) Mouse macrophage. (B) Human mesenchymal stem cells. (C) Mouse lymphoblast. (D) Mouse splenocytes. (E) Human CD34+/AC133+ stem cells collected from cord blood. (F) Human glioma cells (U251).

efficiently labeling cells is 100 μg/ml of ferumoxides to 4 to 6 μg/ml of protamine sulfate. Initial procedures for making the FePro complex should be in serum-free media, because serum or albumin in the media would interfere with the electrostatic interaction between the negative zeta potential for ferumoxides and the positive zeta potential for protamine sulfate (or other transfection agents) (Figure 17.4). Moreover, making an appropriately sized complex is also important for the effective internalization of FePro by nonphagocytic cells. This protocol has been used to efficiently and effectively label mouse, rat, human, or nonhuman primate hematopoietic stem cells, embryonic stem cells, mesenchymal stem cells, neural stem cells, nonstimulated and proliferating lymphocytes, B-cells, fibroblasts, beta cells, dendritic cells, and various cancer cell lines.[146,149] Of note, adding 2 to 10 units of heparin sulfate to the cell wash after labeling with FePro will result in a breakdown of the complexes, since the heparin molecules will bind more avidly to the protamine sulfate than the dextran coat on ferumoxides. Figure 17.5 contains photomicrographs of a variety of Prussian blue-stained stem cells and other mammalian cells labeled with a ferumoxides–protamine sulfate complex.

Labeled cells with clinically approved dextran-coated (U)SPIO nanoparticles complexed to protamine sulfate, PLL, or other TAs have demonstrated no changes in short- or long-term toxicity to the cell. Cell viability, proliferation, increased production of reactive oxygen species, increased rates of apoptosis, or the ability to differentiate have also not been altered as a result of labeling cells with clinically approved (U)SPIO and TA.[13,25,94,141,145,146,149–153] In addition, infusing ferumoxides–PLL-labeled human mesenchymal stem cells into nude rats did not alter biochemical or hematologic measures of organ function in a clinically relevant or preclusive manner.[154] Although there is a report suggesting that intracellular ferumoxides–PLL inhibited chondrogenesis of mesenchymal stem cells in culture, likely representing an interference by ferumoxides–PLL complexes and the iron loading of the cells,[155] there are now several reports demonstrating that labeling MSCs with the ferumoxides–protamine sulfate complex does not inhibit differentiation under appropriate conditions of mesenchymal stem cells along chondrocytic lines, nor does the ferumoxides–protamine sulfate complex inhibit differentiation of hematopoietic or cord blood stem cells.[146,149,152] Section 17.9, on monitoring the effects of (U)SPIO labeling, will discuss the tests used for the *in vitro* evaluation of cell function, metabolism, proliferation, differentiation, and iron content of labeled stem cells and terminally differentiated cells.

17.6 EVALUATION OF (U)SPIO-LABELED NONPHAGOCYTIC CELLS

Since commercially available iron oxides are not fluorescently tagged, magnetic labeling of cells cannot be assessed without using proper staining for iron oxide nanoparticles or detection of electron-dense material in endosomes by electron microscopy, magnetic resonance imaging, or other methods that indirectly detect the presence of iron in cells. There are also different methods to determine the amount of iron taken up by the cells. Determination of intracellular iron is necessary to validate findings on MRI and allow for the possible quantitation of the number of labeled cells in a region of interest. Moreover, preservation of cell viability, cell function, and differential capacity of stem cells is an important parameter that needs to be considered for labeled cells because ionic iron (intracellular) is toxic to the cells and may damage DNA.[94] In the following sections, we will present some of the commonly used methods to determine the labeling efficiency, intracellular iron concentration, and preservation of cell viability, function, and differential capacity.

17.7 VALIDATION THAT CELLS ARE LABELED WITH (U)SPIO

17.7.1 Prussian Blue Staining with or without DAB Enhancement

Prussian blue (PB) staining is the easiest method for determining the presence of intracellular iron using Perl's reagent. Perl's reagent is made of 2% potassium ferrocyanide and 3.7% hydrochloric acid. Iron oxide-labeled cells are fixed with 3 to 4% paraformaldehyde (or any other cell fixatives), washed, incubated for 20 to 30 min with Perl's reagent for PB staining, washed again, and counterstained with nuclear fast red (for making contrast between blue iron particles and red nucleus). Sometimes it is necessary to enhance PB staining with diaminobenzide (DAB), oxidizing the reduced iron and inducing a blue-to-brown color transformation. DAB enhancement is needed to determine iron labeling efficiency in cells with small cytoplasm, such as lymphocytes.[150] For DAB enhancement, slides after PB staining are put into hydrogen peroxide-activated DAB solution for 5 to 10 min, washed with phosphate-buffered saline (PBS), and counterstained with either nuclear fast red or hematoxillin. Iron labeling efficiency can be qualitatively determined by manual counting of PB-stained and unstained cells using any microscope. Cells are considered PB positive if intracytoplasmic blue or brown (DAB-enhanced) granules are detected (Figure 17.5). Schroeter et al.[156] recently reported on a method to improve the sensitivity of PB staining to detect the presence of USPIO-labeled cells as they migrate into photochemically induced strokes in rat brains. These authors suggested that standard PB staining might not be sensitive enough to detect the presence of USPIO nanoparticles in cells in the brain even after diaminobenzidine (DAB) enhancement, and demonstrated improved sensitivity with the addition of gold and silver impregnation of the histological section.[157]

17.7.2 Electron Microscopy

Electron microscopy determines not only the intracellular iron, but also the exact intracellular distribution of iron particles. We have demonstrated the incorporation of iron particles in cells from the initiation of the labeling process to the lysosomal disintegration of iron oxide nanoparticles by electron microcopy.[158] On transmission electron microscopy, the (U)SPIO nanoparticles may be seen as compact, electron-dense (black) granules in the cytoplasm in endosomes (Figure 17.6). Electron microscope examination can also be used to determine whether (U)SPIO nanoparticles have entered into the nucleus. It is possible that (U)SPIO nanoparticles may be located in the nucleus after labeling with CLIO-Tat.[159]

17.7.3 MRI of Cells

MRI can easily differentiate iron-labeled cells from unlabeled cells using either T2-weighted or T2*-weighted images. SPIO-labeled cells in pellets or suspended in gelatin are usually hypointense

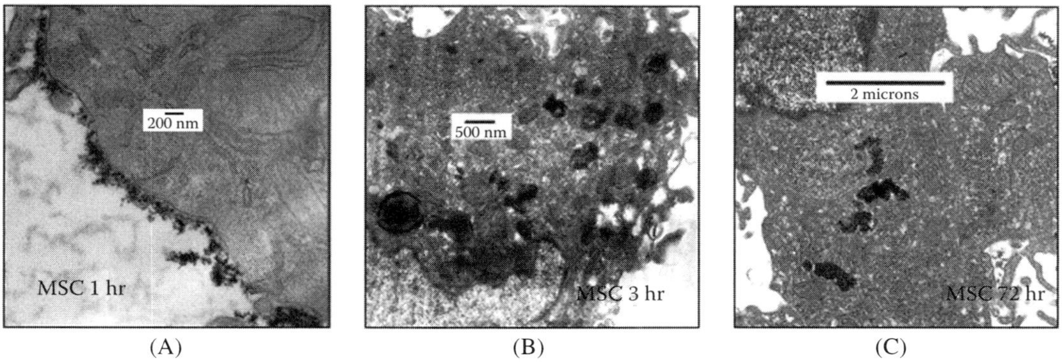

FIGURE 17.6 Uptake and internalization of ferumoxides–transfection agent complexes into the endosomes of human mesenchymal stem cells (MSCs) depicted by electron microscopic examination: (A) 1 h, (B) 3 h, and (C) 72 h after incubation. Note the attachment of the complexes onto the cell membrane by 1 h, endosomal incorporation by 3 h, and the complexes become relatively compact inside endosomes by 72 h. Scale bars: A, 200 nm; B, 500 nm; C, 2 μm.

or have a low signal intensity and blooming artifact, due to their magnetic susceptibility effect on T2*-weighted gradient echo imaging, that is not observed in control unlabeled cells. Recently, several novel pulse sequences have been developed that are sensitive to the off-resonance effects that occur due to the presence of SPIO in cells, and these positive-contrast (hyperintensity) or "white marker" sequences can also be used to determine the presence of labeled cells.[160–162] Investigators have used MRI over a wide range of magnetic field strengths to demonstrate the labeling efficiency of SPIO nanoparticles for various labeling protocols and agents.[29,38,55,81,93,94,114,119,153,163] By using multispin echo or gradient echo MR pulse sequences, one can estimate T2 or T2* maps to calculate the known number of SPIO-labeled cells *ex vivo* and to approximate the number of labeled cells observed on *in vivo* MRI.

17.8 DETERMINATION OF IRON CONTENT

A variety of methods are used to determine the intracellular iron content of magnetically labeled cells.[26,93,94,150,164] Commercial kits (e.g., BioAssay Systems QuantiChrom™ iron assay kit) are also available to determine iron in solution, although the detection range may be of limited use for determining the amount of iron in labeled cells. To determine the concentration of iron in cells, intracellular (U)SPIO nanoparticles need to be completely dissolved in solution. Below are some methods commonly used to determine the concentration of intracellular iron.

17.8.1 NMR Relaxometry

Our group has routinely used T1 and T2 relaxation rates for determining the iron content in labeled cells or in culture media.[34,94,150,151] Cell suspensions are first dried at 110°C overnight, and then completely digested in a mixture (500 μl) of perchloric and nitric acid at a 3:1 ratio. The samples are digested for at least 3 h at 60°C by using a heating block. For these 500-μl samples, 1/T1 and 1/T2 are measured at room temperature and at 0.1 and 1.0 T with an MR relaxometer. Determination of iron content in cells can also be performed on a clinical MRI scanner with appropriate pulse sequences and analysis.[165,166] Iron concentration in the cell sample is then determined from a standard curve that is derived from calibration standards of ferrous chloride containing 0.01 to 10 mmol/l iron mix in acid and using identical nuclear magnetic resonance (NMR) pulse sequences at room temperature. Ideally, a standard curve should be obtained each time a new set of iron determinations in labeled cells is performed.

17.8.2 Spectrophotometric Methods

Various ultraviolet spectrophotometric methods have been developed to determine the iron content in acid solution. It is important to dissolve the intracellular iron completely overnight using a mixture of nitric acid and perchloric acids at a 3:1 ratio, and the resultant solution can be analyzed using the Ferrozine assay for iron.[34] Moreover, 10 and 5 mol/l hydrochloric acid solutions can also be used to dissolve the intracellular iron. Currently, we are validating the use of citrate buffer in determining intracellular iron nanoparticles. Citrate buffer has been used previously to determine the reduced iron in any solution.[158,167] Heating the mixture of any acid and cell for 1 to 3 h at 60 to 70°C will dissolve iron nanoparticles in cells. Potassium ferrocyanide (5%) can also be used with hydrochloric acid (5 mol/l) to determine the iron concentration. Various groups have used a wide range of absorption wavelengths for UV spectrophotometric reading.[25,168,169] We have observed effective absorption wavelengths for hydrochloric acid, hydrochloric acid plus ferrocyanide, and citrate buffer: 351 to 360 nm, 690 to 710 nm, and 535 nm, respectively. It is recommended that when determining the intracellular iron concentration by UV spectrophotometric methods, results should be compared to a known standard containing iron.

17.8.3 ICP Mass Spectrometry

Inductively coupled plasma (ICP) mass spectrometry is a rapid, sensitive way of measuring the elemental concentrations of solutions.[170] More than 75 elements can be determined, most of them at detection limits of less than 1 ppb. The first step in the procedure is conversion of the molecules in the sample to individual atoms and ions using a high-temperature, radio frequency-induced argon plasma. The sample is introduced into the plasma as a solution. It is then pumped using a peristaltic pump to a nebulizer, where it is converted to a fine spray and mixed with argon in a spray chamber. The purpose of the spray chamber is to ensure that only droplets of the appropriate size enter the plasma and are instantly excited by the high temperatures (5000 to 10,000 K). Atoms become ionized with 99% efficiency (arsenic and selenium are a couple of exceptions, ionizing only at 52 and 33%). Either ICP optical emission spectrometry (ICP-OES) or ICP mass spectrometry (ICP-MS) can be used to analyze samples. ICP mass spectrometry is one of the popular methods for determining iron or any other elements in cells; however, samples usually are sent out to a core facility or analytic laboratory for analysis.

17.9 DETERMINATION OF CELL VIABILITY, FUNCTION, AND DIFFERENTIATION

Ferrous ions in high concentrations and synthetic transfection agents alone have been shown to be toxic to cells.[145] Under appropriate conditions, ferrous ions through a Haber–Weiss chemical reaction can stimulate the increased production of H_2O_2 and increase the amount of nascent hydroxyl free radicals, to cause a denaturation of DNA.[99,171,172] Therefore, it is essential to determine the short- and long-term toxicities of the intracellular iron by assessing the viability, ability to proliferate, metabolic activity, rate of apoptosis, reactive oxygen species (ROS) formation, and functional and differential capacities of labeled cells, and comparing the results to those of control unlabeled cells under the same conditions.

17.9.1 Viability

Cell viability can be assessed soon and at different time points after labeling of cells. The easiest and most cost effective technique is the Trypan blue dye exclusion test. Propidium iodide (PI) can also be used to detect dead cells by a fluorescent microscope or by a flow cytometer. Both Trypan blue and PI will penetrate the cell membrane of dead cells and stain the nucleus blue or red, respectively.

17.9.2 Proliferative Capacity

The short- and long-term proliferation capacities of the labeled cells should be determined and compared to those of control cells. The proliferation of cells can be assessed by thymidine (radioactive) uptake, Bromodeoxyuridine (BRDU) incorporation, or MTT assay.[34,150] For our studies, cellular proliferation is determined by the MTT (3-[4,5-dimethylthiazol-2-yl]-2,5-diphenyl tetrazolium bromide) assay (Roche Molecular Biochemicals, Indianapolis, IN) at different time points. After labeling and washing, specific numbers of unlabeled (control) and labeled cells were allowed to grow in 96-well plates and the MTT assay was performed at specific time intervals. The number of cells in each well should be optimized such that on later days the absorption of formazen product will not exceed the upper threshold of the UV spectrometer. The absorbance of the formazen product is usually measured at a wavelength of 570 nm, with 750 nm (subtracted) as reference. MTT assay values for labeled cells can be expressed as the percentage of corresponding average values in control cells. Cell proliferation can also be assessed by counting cell numbers at regular intervals after incubation in their respective complete medium and comparing with those of control cells grown side by side.

17.9.3 Apoptosis in Labeled Cells

Because of the possible release of free iron in the cytoplasm, it is important to determine the long-term induction of apoptosis in labeled cells.[150] To determine if iron labeling results in changes in the rate of apoptosis, both unlabeled and labeled cells are collected and washed twice with ice-cold PBS and resuspended in 1 ml of annexin media (Vybrant apoptosis assay kit 2, Molecular Probes, Inc., Eugene, OR) at 1×10^6 cells/ml. Ten microliters of fluorescent-labeled annexin V and 2 μl of propidium iodide solution are added to 100 μl of cell suspension, which is then kept at room temperature for 15 to 20 min, and flow cytometry is performed using a fluorescent-activated cell sorter (FACScalibur, Becton Dickinson, Mountain View, CA).

17.9.4 ROS in Labeled Cells

Determination of reactive oxygen species (ROS) should be the most important part of the toxicity assay for any cells labeled with iron.[94,150] For the ROS assay, both (U)SPIO-labeled and unlabeled cells are collected and resuspended in respective complete media at 1×10^6 cells/ml. The intracellular formation of ROS is detected by using the fluorescent probe CM-H_2DCFDA (Molecular Probes, Inc., Eugene, OR). CM-H_2DCFDA is added at a final concentration of 10 μ*M* and cells are incubated for 60 min at 37°C. CM-H_2DCFDA is a nonfluorescent agent that forms fluorescent esters when reacted with ROS inside cells. For ROS assays, fluorescence is analyzed using a fluorescent plate reader set at a 490- to 500-nm wavelength for excitation and 525 nm for emission.[150] Values were normalized to those obtained from corresponding unlabeled cells. The samples can also be analyzed by flow cytometry to determine the intensity of fluorescence in labeled and unlabeled cells.

17.9.5 Functional and Phenotypical Analysis of Labeled Cells

It is important that (U)SPIO-labeled cells not deviate from their original function or be stimulated, resulting in a change in phenotype. Cellular function assays should be performed after labeling and compared to control unlabeled cells. Cell activation and surface markers should be assessed in primary cells following labeling with (U)SPIO nanoparticles to ensure that the cells have not changed phenotype. It is therefore recommended to perform a battery of phenotypic markers (e.g., CD markers, CD4, CD8, CD11a, CD19, CD25) of T-cells and production of cytokines before and after labeling. In our previous experiments, we have analyzed different CD markers of T-cells as well as markers for activated T-cells. T-cells labeled with ferumoxides–transfection agents were not altered in phenotype or immunoreactivity in response to secondary antigen stimuli as unlabeled T-cells.[173]

Cytokine production by enzyme-linked immunosorbent assay (ELISA) is also valuable to determine cell function. We have shown that labeling cells with ferumoxides did not alter the production of cytokines by T-cells (i.e., INF-gamma, TNF-alpha, and interleukins), compared to control cells.[173]

For hematopoietic stem cells (HSCs), it is important to analyze different lineages before and after labeling with iron oxides. It is well known that HSCs lose stem cell markers such as CD34 and CD133 over time; however, determination of different surface markers should run side by side for both labeled and unlabeled control cells. Of note, it is worth determining markers for T-cell (CD3), B-cell (CD20), and macrophage-monocytic (CD14) lineages along with the markers of stem cells (CD34, CD133, CD31, CXCR4, etc.) following (U)SPIO labeling of hematopoietic stem cells. To determine the preservation of functional capacity of labeled HSCs (CD34+), a stromal derived factor 1 (SDF-1)-mediated cell migration study should be performed[146,149] to demonstrate response to this cytokine. For neural and embryonic stem cells, respective markers should be assessed before and after labeling.

Of note, clinically established biomarkers or surrogate markers such as serum glucose can be used to assess beta cells in islet function following labeling with (U)SPIO nanoparticles.[90,174]

17.9.6 Differential Capacity of Labeled Cells

The determination of differential capacity of (U)SPIO-labeled cells is especially important for nonhematopoietic stem cells such as mesenchymal (MSCs) and neural (NSCs) stem cells. We have shown preservation of the differential capacity of ferumoxides–protamine sulfate-labeled MSCs into adipogenic, osteogenic, and chondrogenic lineages;[146,149] however, it has been reported that ferumoxides–poly-L-lysine-labeled MSC's ability to differentiate along the chondrogenic lineage[155] is impaired. The difference in results might be due to the iron loading of the mesenchymal stem cells with ferumoxides–PLL vs. ferumoxides–protamine sulfate complexes, the length of time cells were incubated with the agents, and the characteristics of ferumoxides–PLL (PLL has a molecular weight of 300 K vs. protamine sulfate's 4.2 K) that resulted in inhibiting the differentiation along chondrocytic lineages. For hematopoietic stem cells, differentiation to colony-forming units should be determined for labeled and unlabeled cells.[53,119,149] Endothelial progenitor cells (EPCs) can be differentiated to colony-forming units, endothelial cells, and cord-like structures in a matrigel system. Investigators that have labeled neural stem cells with pharmaceutical-grade (U)SPIO nanoparticles have reported unaltered differentiation compared to unlabeled control cells. SPIO-labeled NSCs can be differentiated along neuron and glial cell lineages.[98]

17.10 TRANSLATION TO CLINIC

Translation of the various labeling approaches from bench to bedside for clinical trials will require that agents used are of pharmaceutical grade (USP) or clinical-grade ancillary products or made with good manufacturing practices.[11] Ferumoxides alone have been used to label dendritic cells as part of early-phase clinicals trial in Europe;[95] however, in the United States, the FDA will require preclinical evaluation and testing of ferumoxides alone or complexed to protamine sulfate before filing an investigational new drug as part of an investigator-driven early-phase clinical trial using magnetically labeled stem cells or other mammalian cells. Investigators planning on using (U)SPIO or ferumoxides complexed to transfection agents to label cells will need to demonstrate the following:

1. All products — cytokines, growth factors, media, antibiotics, plastics, and instrumentation — are approved for clinical use, or an exemption from the FDA will be required to use these products for clinical use in cell labeling.
2. The scale-up technique is reproducible and can be performed using good manufacturing practices (GMPs) in an approved facility.

3. The GMP facility would have approved standard operating procedures for handling, processing, and evaluating stem cells or other cells and be able to perform labeling on a large scale.
4. The GMP facility must be able to identify and track cells obtained from a subject that will be used for transplantation.
5. Labeled cells have similar viability as unlabeled cells without significant loss during the labeling process.
6. The phenotype of the stem cells or other mammalian cells is unaltered as a result of the labeling with (U)SPIO.
7. Magnetically labeled cells are able to differentiate and retain the same cell functional capacity as unlabeled cells.
8. There are no toxins or infection agents present in the resulting product.

Preclinical studies performed in experimental disease models will probably need to include the infusion of magnetically labeled and unlabeled cells along with sham controls to determine if there is any toxicity in the animal. Serum chemistries and complete blood count evaluations will probably be required. The preclinical studies will require that cells be labeled using the same process as in scale-up procedures planned for the clinical trials, and that the labeled cells are safe, do not stimulate immune reaction, and can be detected in target tissue by MRI and on histological examination. At this time, disease-specific indications for certain cell types that will be used for transplantation will need to be developed along with documentation submitted to the FDA with institution review board-approved clinical protocols.

Although most pharmacotoxicology studies of new drugs are usually performed in at least two species (i.e., mouse or rat, canine or swine or nonhuman primate), the safety and toxicity studies for ferumoxides–protamine sulfate complex labeling of autologous human stem cells will not require testing in a large-animal species and will go directly from mouse into humans. Large-animal (i.e., swine, canine, nonhuman primate) models may better present the types of challenges or have a similar disease course as humans; however, preclinical cell labeling studies using ferumoxides–protamine sulfate complexes for labeling human stem cells for inclusion in investigative new drug (IND) submission are performed in immune-compromised mice and rats. Cell escalation (e.g., 10^5, 10^6, or 10^7 cells per animal) experimental studies using human stem cells labeled in a clinical GMP facility are presently required in the animal models, with the results included as part of the IND submission. The investigator-driven IND for using human cells (i.e., stem cells or other mammalian cells) labeled with ferumoxides–protamine sulfate complexes, along with institution review board (IRB) phase I clinical trials, will probably be evaluated by the Office of Cellular, Tissue and Gene Therapies in the Center for Biologics Evaluation and Research (CBER) at the FDA. CBER has indicated that the ferumoxide-labeled human stem cells will be considered a biologic or cellular contrast agent for phase I/II clinical trials using MRI to monitor cell trafficking.

If magnetically labeled cell trafficking studies with human stem cells or immune cells were to be performed in a large-animal model (e.g., swine), immunosuppression would be required to prevent transplant rejection or graft vs. host disease, and would probably alter results on safety and toxicity of the labeled cells needed for the IND. Moreover, human stem cells may not traffic or migrate to target tissue in the immune-competent large animal treated with immunosuppressive agents. Alternatively, researchers could consider using, for example, swine stem cells labeled with (U)SPIO nanoparticles and infuse or directly transplant these autologous or allogenic magnetically labeled cells into the target organ of the pig and monitor the migration of the labeled cells with MRI. However, unless the swine stem cells were labeled and processed using clinical-grade materials, in a clinical GMP facility using FDA-approved equipment, and in the same rigorous manner as the human cells would be used in a clinical trial, the results, although providing valuable scientific information, may not be considered or evaluated in the IND submission to the FDA. It is also unlikely that a clinical GMP facility in academic centers will have a duplicate setup or separate equipment to process nonhuman stem cells. In addition, the investigators would be required to demonstrate that the (U)SPIO-labeled swine stem cells would have the same properties as the labeled human cells. It is recommended that researchers have a discussion and obtain guidance from the appropriate regulatory agencies before starting to obtain data that will be used for submission of an IND of magnetically labeled stem cells for a clinical trial.

It is likely that labeling cells with (U)SPIO nanoparticles will be performed in a cell processing section of transfusion medicine departments in collaboration with treating physicians and radiologists as part of a multidisciplinary approach in the evaluation of cellular therapies.

ACKNOWLEDGMENT

This work was supported in part by the intramural research program of the Clinical Center at the National Institutes of Health.

REFERENCES

1. Botti, C., Negri, D.R., Seregni, E., Ramakrishna, V., Arienti, F., Maffioli, L., Lombardo, C., Bogni, A., Pascali, C., Crippa, F., Massaron, S., Remonti, F., Nerini-Molteni, S., Canevari, S., and Bombardieri, E., Comparison of three different methods for radiolabelling human activated T lymphocytes, *Eur J Nucl Med*, 24, 497–504, 1997.
2. Read, E.J., Keenan, A.M., Carter, C.S., Yolles, P.S., and Davey, R.J., *In vivo* traffic of indium-111-oxine labeled human lymphocytes collected by automated apheresis, *J Nucl Med*, 31, 999–1006, 1990.
3. Griffith, K.D., Read, E.J., Carrasquillo, J.A., Carter, C.S., Yang, J.C., Fisher, B., Aebersold, P., Packard, B.S., Yu, M.Y., and Rosenberg, S.A., *In vivo* distribution of adoptively transferred indium-111-labeled tumor infiltrating lymphocytes and peripheral blood lymphocytes in patients with metastatic melanoma, *J Natl Cancer Inst*, 81, 1709–1717, 1989.
4. Fisher, B., Packard, B.S., Read, E.J., Carrasquillo, J.A., Carter, C.S., Topalian, S.L., Yang, J.C., Yolles, P., Larson, S.M., and Rosenberg, S.A., Tumor localization of adoptively transferred indium-111 labeled tumor infiltrating lymphocytes in patients with metastatic melanoma, *J Clin Oncol*, 7, 250–261, 1989.
5. Zhang, Z., van den Bos, E.J., Wielopolski, P.A., de Jong-Popijus, M., Duncker, D.J., and Krestin, G.P., High-resolution magnetic resonance imaging of iron-labeled myoblasts using a standard 1.5-T clinical scanner, *MAGMA*, 17, 201–209, 2004.
6. Love, C. and Palestro, C.J., Radionuclide imaging of infection, *J Nucl Med Technol*, 32, 47–57, 2004; quiz, 58–59.
7. Adonai, N., Nguyen, K.N., Walsh, J., Iyer, M., Toyokuni, T., Phelps, M.E., McCarthy, T., McCarthy, D.W., and Gambhir, S.S., *Ex vivo* cell labeling with 64Cu-pyruvaldehyde-bis(N4-methylthiosemicarbazone) for imaging cell trafficking in mice with positron-emission tomography, *Proc Natl Acad Sci USA*, 99, 3030–3035, 2002.
8. Doyle, T.C., Burns, S.M., and Contag, C.H., *In vivo* bioluminescence imaging for integrated studies of infection, *Cell Microbiol*, 6, 303–317, 2004.
9. Michalet, X., Pinaud, F.F., Bentolila, L.A., Tsay, J.M., Doose, S., Li, J.J., Sundaresan, G., Wu, A.M., Gambhir, S.S., and Weiss, S., Quantum dots for live cells, *in vivo* imaging, and diagnostics, *Science*, 307, 538–544, 2005.
10. Hildebrandt, I.J. and Gambhir, S.S., Molecular imaging applications for immunology, *Clin Immunol*, 111, 210–224, 2004.
11. Frank, J.A., Anderson, S.A., Kalsih, H., Jordan, E.K., Lewis, B.K., Yocum, G.T., and Arbab, A.S., Methods for magnetically labeling stem and other cells for detection by *in vivo* magnetic resonance imaging, *Cytotherapy*, 6, 621–625, 2004.
12. Modo, M., Hoehn, M., and Bulte, J.W., Cellular MR imaging, *Mol Imaging*, 4, 143–164, 2005.
13. Bulte, J.W. and Kraitchman, D.L., Iron oxide MR contrast agents for molecular and cellular imaging, *NMR Biomed*, 17, 484–499, 2004.

14. Bulte, J.W., Arbab, A.S., Douglas, T., and Frank, J.A., Preparation of magnetically labeled cells for cell tracking by magnetic resonance imaging, *Methods Enzymol*, 386, 275–299, 2004.
15. Mukherjee, S., Ghosh, R.N., and Maxfield, F.R., Endocytosis, *Physiol Rev*, 77, 759–803, 1997.
16. Conner, S.D. and Schmid, S.L., Regulated portals of entry into the cell, *Nature*, 422, 37–44, 2003.
17. Medina-Kauwe, L.K., Xie, J., and Hamm-Alvarez, S., Intracellular trafficking of nonviral vectors, *Gene Ther*, 12, 1734–1751, 2005.
18. Brodsky, F.M., Chen, C.Y., Knuehl, C., Towler, M.C., and Wakeham, D.E., Biological basket weaving: formation and function of clathrin-coated vesicles, *Annu Rev Cell Dev Biol*, 17, 517–568, 2001.
19. Schmid, S.L., Clathrin-coated vesicle formation and protein sorting: an integrated process, *Annu Rev Biochem*, 66, 511–548, 1997.
20. Moore, A., Josephson, L., Bhorade, R.M., Basilion, J.P., and Weissleder, R., Human transferrin receptor gene as a marker gene for MR imaging, *Radiology*, 221, 244–250, 2001.
21. Moore, A., Basilion, J.P., Chiocca, E.A., and Weissleder, R., Measuring transferrin receptor gene expression by NMR imaging, *Biochim Biophys Acta*, 1402, 239–249, 1998.
22. Kresse, M., Wagner, S., Pfefferer, D., Lawaczeck, R., Elste, V., and Semmler, W., Targeting of ultrasmall superparamagnetic iron oxide (USPIO) particles to tumor cells *in vivo* by using transferrin receptor pathways, *Magn Reson Med*, 40, 236–242, 1998.
23. Ichikawa, T., Hogemann, D., Saeki, Y., Tyminski, E., Terada, K., Weissleder, R., Chiocca, E.A., and Basilion, J.P., MRI of transgene expression: correlation to therapeutic gene expression, *Neoplasia*, 4, 523–530, 2002.
24. Hogemann, D., Josephson, L., Weissleder, R., and Basilion, J.P., Improvement of MRI probes to allow efficient detection of gene expression, *Bioconjug Chem*, 11, 941–946, 2000.
25. Daldrup-Link, H.E., Rudelius, M., Oostendorp, R.A., Settles, M., Piontek, G., Metz, S., Rosenbrock, H., Keller, U., Heinzmann, U., Rummeny, E.J., Schlegel, J., and Link, T.M., Targeting of hematopoietic progenitor cells with MR contrast agents, *Radiology*, 228, 760–767, 2003.
26. Bulte, J.W., Zhang, S., van Gelderen, P., Herynek, V., Jordan, E.K., Duncan, I.D., and Frank, J.A., Neurotransplantation of magnetically labeled oligodendrocyte progenitors: magnetic resonance tracking of cell migration and myelination, *Proc Natl Acad Sci USA*, 96, 15256–15261, 1999.
27. Nakase, I., Niwa, M., Takeuchi, T., Sonomura, K., Kawabata, N., Koike, Y., Takehashi, M., Tanaka, S., Ueda, K., Simpson, J.C., Jones, A.T., Sugiura, Y., and Futaki, S., Cellular uptake of arginine-rich peptides: roles for macropinocytosis and actin rearrangement, *Mol Ther*, 10, 1011–1022, 2004.
28. Kaplan, I.M., Wadia, J.S., and Dowdy, S.F., Cationic TAT peptide transduction domain enters cells by macropinocytosis, *J Control Release*, 102, 247–253, 2005.
29. Toyoda, K., Tooyama, I., Kato, M., Sato, H., Morikawa, S., Hisa, Y., and Inubushi, T., Effective magnetic labeling of transplanted cells with HVJ-E for magnetic resonance imaging, *Neuroreport*, 15, 589–593, 2004.
30. Jung, C.W. and Jacobs, P., Physical and chemical properties of superparamagnetic iron oxide MR contrast agents: ferumoxides, ferumoxtran, ferumoxsil, *Magn Reson Imaging*, 13, 661–674, 1995.
31. Ittrich, H., Lange, C., Dahnke, H., Zander, A.R., Adam, G., and Nolte-Ernsting, C., Labeling of mesenchymal stem cells with different superparamagnetic particles of iron oxide and detectability with MRI at 3T, *Rofo*, 177, 1151–1163, 2005.
32. Mikhaylova, M., Kim do, K., Bobrysheva, N., Osmolowsky, M., Semenov, V., Tsakalakos, T., and Muhammed, M., Superparamagnetism of magnetite nanoparticles: dependence on surface modification, *Langmuir*, 20, 2472–2477, 2004.
33. Hawrylak, N., Ghosh, P., Broadus, J., Schlueter, C., Greenough, W.T., and Lauterbur, P.C., Nuclear magnetic resonance (NMR) imaging of iron oxide-labeled neural transplants, *Exp Neurol*, 121, 181–192, 1993.
34. Bulte, J.W., Douglas, T., Witwer, B., Zhang, S.C., Strable, E., Lewis, B.K., Zywicke, H., Miller, B., van Gelderen, P., Moskowitz, B.M., Duncan, I.D., and Frank, J.A., Magnetodendrimers allow endosomal magnetic labeling and *in vivo* tracking of stem cells, *Nat Biotechnol*, 19, 1141–1147, 2001.
35. Yeh, T.C., Zhang, W., Ildstad, S.T., and Ho, C., *In vivo* dynamic MRI tracking of rat T-cells labeled with superparamagnetic iron-oxide particles, *Magn Reson Med*, 33, 200–208, 1995.
36. Yeh, T.C., Zhang, W., Ildstad, S.T., and Ho, C., Intracellular labeling of T-cells with superparamagnetic contrast agents, *Magn Reson Med*, 30, 617–625, 1993.
37. Shen, T.T., Bogdanov, A., Jr., Bogdanova, A., Poss, K., Brady, T.J., and Weissleder, R., Magnetically labeled secretin retains receptor affinity to pancreas acinar cells, *Bioconjug Chem*, 7, 311–316, 1996.

38. Fleige, G., Seeberger, F., Laux, D., Kresse, M., Taupitz, M., Pilgrimm, H., and Zimmer, C., *In vitro* characterization of two different ultrasmall iron oxide particles for magnetic resonance cell tracking, *Invest Radiol*, 37, 482–488, 2002.
39. Kaufman, C.L., Williams, M., Ryle, L.M., Smith, T.L., Tanner, M., and Ho, C., Superparamagnetic iron oxide particles transactivator protein-fluorescein isothiocyanate particle labeling for *in vivo* magnetic resonance imaging detection of cell migration: uptake and durability, *Transplantation*, 76, 1043–1046, 2003.
40. Koch, A.M., Reynolds, F., Kircher, M.F., Merkle, H.P., Weissleder, R., and Josephson, L., Uptake and metabolism of a dual fluorochrome Tat-nanoparticle in HeLa cells, *Bioconjug Chem*, 14, 1115–1121, 2003.
41. Ho, C. and Hitchens, T.K., A non-invasive approach to detecting organ rejection by MRI: monitoring the accumulation of immune cells at the transplanted organ, *Curr Pharm Biotechnol*, 5, 551–566, 2004.
42. Song, H., Choi, J.S., Huh, Y.M., Kim, S., Jun, Y.W., Suh, J.S., and Cheon, J., Surface modulation of magnetic nanocrystals in the development of highly efficient magnetic resonance probes for intracellular labeling, *J Am Chem Soc*, 127, 9992–9993, 2005.
43. Schulze, E., Ferrucci, J.T., Jr., Poss, K., Lapointe, L., Bogdanova, A., and Weissleder, R., Cellular uptake and trafficking of a prototypical magnetic iron oxide label *in vitro*, *Invest Radiol*, 30, 604–610, 1995.
44. Bulte, J.W., Douglas, T., Witwer, B., Zhang, S.C., Lewis, B.K., van Gelderen, P., Zywicke, H., Duncan, I.D., and Frank, J.A., Monitoring stem cell therapy *in vivo* using magnetodendrimers as a new class of cellular MR contrast agents, *Acad Radiol*, 9 (Suppl 2), S332–S335, 2002.
45. Smirnov, P., Gazeau, F., Lewin, M., Bacri, J.C., Siauve, N., Vayssettes, C., Cuenod, C.A., and Clement, O., *In vivo* cellular imaging of magnetically labeled hybridomas in the spleen with a 1.5-T clinical MRI system, *Magn Reson Med*, 52, 73–79, 2004.
46. Riviere, C., Boudghene, F.P., Gazeau, F., Roger, J., Pons, J.N., Laissy, J.P., Allaire, E., Michel, J.B., Letourneur, D., and Deux, J.F., Iron oxide nanoparticle-labeled rat smooth muscle cells: cardiac MR imaging for cell graft monitoring and quantitation, *Radiology*, 235, 959–967, 2005.
47. Brillet, P.Y., Gazeau, F., Luciani, A., Bessoud, B., Cuenod, C.A., Siauve, N., Pons, J.N., Poupon, J., and Clement, O., Evaluation of tumoral enhancement by superparamagnetic iron oxide particles: comparative studies with ferumoxtran and anionic iron oxide nanoparticles, *Eur Radiol*, 15, 1369–1377, 2005.
48. Wilhelm, C., Billotey, C., Roger, J., Pons, J.N., Bacri, J.C., and Gazeau, F., Intracellular uptake of anionic superparamagnetic nanoparticles as a function of their surface coating, *Biomaterials*, 24, 1001–1011, 2003.
49. Wang, Y.X., Hussain, S.M., and Krestin, G.P., Superparamagnetic iron oxide contrast agents: physicochemical characteristics and applications in MR imaging, *Eur Radiol*, 11, 2319–2331, 2001.
50. Kalish, H., Arbab, A.S., Miller, B.R., Lewis, B.K., Zywicke, H.A., Bulte, J.W., Bryant, L.H., Jr., and Frank, J.A., Combination of transfection agents and magnetic resonance contrast agents for cellular imaging: relationship between relaxivities, electrostatic forces, and chemical composition, *Magn Reson Med*, 50, 275–282, 2003.
51. Walliser, S. and Redmann, K., Effect of 5-fluorouracil and thymidine on the transmembrane potential and zeta potential of HeLa cells, *Cancer Res*, 38, 3555–3559, 1978.
52. Shapiro, E.M., Skrtic, S., and Koretsky, A.P., Sizing it up: cellular MRI using micron-sized iron oxide particles, *Magn Reson Med*, 53, 329–338, 2005.
53. Hinds, K.A., Hill, J.M., Shapiro, E.M., Laukkanen, M.O., Silva, A.C., Combs, C.A., Varney, T.R., Balaban, R.S., Koretsky, A.P., and Dunbar, C.E., Highly efficient endosomal labeling of progenitor and stem cells with large magnetic particles allows magnetic resonance imaging of single cells, *Blood*, 102, 867–872, 2003.
54. Rodriguez, O., Fricke, S., Chien, C., Dettin, L., Vanmeter, J., Shapiro, E., Dai, H.N., Casimiro, M., Ileva, L., Dagata, J., Johnson, M.D., Lisanti, M.P., Koretsky, A., and Albanese, C., Contrast-enhanced *in vivo* imaging of breast and prostate cancer cells by MRI, *Cell Cycle*, 5, 113–119, 2006.
55. Shapiro, E.M., Skrtic, S., Sharer, K., Hill, J.M., Dunbar, C.E., and Koretsky, A.P., MRI detection of single particles for cellular imaging, *Proc Natl Acad Sci USA*, 101, 10901–10906, 2004.
56. Hill, J.M., Dick, A.J., Raman, V.K., Thompson, R.B., Yu, Z.X., Hinds, K.A., Pessanha, B.S., Guttman, M.A., Varney, T.R., Martin, B.J., Dunbar, C.E., McVeigh, E.R., and Lederman, R.J., Serial cardiac magnetic resonance imaging of injected mesenchymal stem cells, *Circulation*, 108, 1009–1014, 2003.

57. Modo, M., Cash, D., Mellodew, K., Williams, S.C., Fraser, S.E., Meade, T.J., Price, J., and Hodges, H., Tracking transplanted stem cell migration using bifunctional, contrast agent-enhanced, magnetic resonance imaging, *Neuroimage*, 17, 803–811, 2002.
58. Modo, M., Mellodew, K., Cash, D., Fraser, S.E., Meade, T.J., Price, J., and Williams, S.C., Mapping transplanted stem cell migration after a stroke: a serial, *in vivo* magnetic resonance imaging study, *Neuroimage*, 21, 311–317, 2004.
59. Jacobs, R.E. and Fraser, S.E., Magnetic resonance microscopy of embryonic cell lineages and movements, *Science*, 263, 681–684, 1994.
60. Crich, S.G., Biancone, L., Cantaluppi, V., Duo, D., Esposito, G., Russo, S., Camussi, G., and Aime, S., Improved route for the visualization of stem cells labeled with a Gd-/Eu-chelate as dual (MRI and fluorescence) agent, *Magn Reson Med*, 51, 938–944, 2004.
61. Crich, S.G., Lanzardo, S., Barge, A., Esposito, G., Tei, L., Forni, G., and Aime, S., Visualization through magnetic resonance imaging of DNA internalized following "*in vivo*" electroporation, *Mol Imaging*, 4, 7–17, 2005.
62. Daldrup-Link, H.E., Rudelius, M., Metz, S., Piontek, G., Pichler, B., Settles, M., Heinzmann, U., Schlegel, J., Oostendorp, R.A., and Rummeny, E.J., Cell tracking with gadophrin-2: a bifunctional contrast agent for MR imaging, optical imaging, and fluorescence microscopy, *Eur J Nucl Med Mol Imaging*, 31, 1312–1321, 2004.
63. Vuu, K., Xie, J., McDonald, M.A., Bernardo, M., Hunter, F., Zhang, Y., Li, K., Bednarski, M., and Guccione, S., Gadolinium-rhodamine nanoparticles for cell labeling and tracking via magnetic resonance and optical imaging, *Bioconjug Chem*, 16, 995–999, 2005.
64. Heverhagen, J.T., Graser, A., Fahr, A., Muller, R., and Alfke, H., Encapsulation of gadobutrol in AVE-based liposomal carriers for MR detectability, *Magn Reson Imaging*, 22, 483–487, 2004.
65. Prantner, A.M., Sharma, V., Garbow, J.R., and Piwnica-Worms, D., Synthesis and characterization of a Gd-DOTA-D-permeation peptide for magnetic resonance relaxation enhancement of intracellular targets, *Mol Imaging*, 2, 333–341, 2003.
66. Lattuada, L., Demattio, S., Vincenzi, V., Cabella, C., Visigalli, M., Aime, S., Crich, S.G., and Gianolio, E., Magnetic resonance imaging of tumor cells by targeting the amino acid transport system, *Bioorg Med Chem Lett*, 16, 4111–4114, 2006.
67. Zheng, Q., Dai, H., Merritt, M.E., Malloy, C., Pan, C.Y., and Li, W.H., A new class of macrocyclic lanthanide complexes for cell labeling and magnetic resonance imaging applications, *J Am Chem Soc*, 127, 16178–16188, 2005.
68. Mikawa, M., Kato, H., Okumura, M., Narazaki, M., Kanazawa, Y., Miwa, N., and Shinohara, H., Paramagnetic water-soluble metallofullerenes having the highest relaxivity for MRI contrast agents, *Bioconjug Chem*, 12, 510–514, 2001.
69. Kato, H., Kanazawa, Y., Okumura, M., Taninaka, A., Yokawa, T., and Shinohara, H., Lanthanoid endohedral metallofullerenols for MRI contrast agents, *J Am Chem Soc*, 125, 4391–4397, 2003.
70. Cagle, D.W., Kennel, S.J., Mirzadeh, S., Alford, J.M., and Wilson, L.J., *In vivo* studies of fullerene-based materials using endohedral metallofullerene radiotracers, *Proc Natl Acad Sci USA*, 96, 5182–5187, 1999.
71. Anderson, S.A., Lee, K.K., and Frank, J.A., Gadolinium-fullerenol as a paramagnetic contrast agent for cellular imaging, *Invest Radiol*, 41, 332–338, 2006.
72. Aoki, I., Wu, Y.J., Silva, A.C., Lynch, R.M., and Koretsky, A.P., *In vivo* detection of neuroarchitecture in the rodent brain using manganese-enhanced MRI, *Neuroimage*, 22, 1046–1059, 2004.
73. Lee, J.H. and Koretsky, A.P., Manganese enhanced magnetic resonance imaging, *Curr Pharm Biotechnol*, 5, 529–537, 2004.
74. Saleem, K.S., Pauls, J.M., Augath, M., Trinath, T., Prause, B.A., Hashikawa, T., and Logothetis, N.K., Magnetic resonance imaging of neuronal connections in the macaque monkey, *Neuron*, 34, 685–700, 2002.
75. Thuen, M., Singstad, T.E., Pedersen, T.B., Haraldseth, O., Berry, M., Sandvig, A., and Brekken, C., Manganese-enhanced MRI of the optic visual pathway and optic nerve injury in adult rats, *J Magn Reson Imaging*, 22, 492–500, 2005.
76. Aoki, I., Takahashi, Y., Chuang, K.H., Silva, A.C., Igarashi, T., Tanaka, C., Childs, R.W., and Koretsky, A.P., Cell labeling for magnetic resonance imaging with the T(1) agent manganese chloride, *NMR Biomed*, 19, 50–59, 2006.

77. Bluemke, D., Weber, T.M., Rubin, D., de Lange, E.E., Semelka, R., Redvanly, R.D., Chezmar, J., Outwater, E., Carlos, R., Saini, S., Holland, G.A., Mammone, J.F., Brown, J.J., Milestone, B., Javitt, M.C., and Jacobs, P., Hepatic MR imaging with ferumoxides: multicenter study of safety and effectiveness of direct injection protocol, *Radiology*, 228, 457–464, 2003.
78. Mack, M., Balzer, J.O., Straub, R., Eichler, K., and Vogl, T.J., Superparamagnetic iron oxide-enhanced MR imaging of head and neck lymph nodes, *Radiology*, 222, 239–244, 2002.
79. Harisinghani, M., Barentsz, J., Hahn, P.F., Deserno, W.M., Tabatabaei, S., van de Kaa, C.H., de la Rosette, J., and Weissleder, R., Noninvasive detection of clinically occult lymph-node metastases in prostate cancer, *N Engl J Med*, 343, 2491–2499, 2003.
80. Harisinghani, M., Saini, S., Weissleder, R., Hahn, P.F., Yantiss, R.K., Tempany, C., Wood, B.J., and Mueller, P.R., MR lymphangiography using ultrasmall superparamagnetic iron oxide in patients with primary abdominal and pelvic malignancies: radiographic-pathologic correlation, *Am J Roentgenol*, 172, 1347–1351, 1999.
81. Sipe, J.C., Filippi, M., Martino, G., Furlan, R., Rocca, M.A., Rovaris, M., Bergami, A., Zyroff, J., Scotti, G., and Comi, G., Method for intracellular magnetic labeling of human mononuclear cells using approved iron contrast agents, *Magn Reson Imaging*, 17, 1521–1523, 1999.
82. Moore, A., Weissleder, R., and Bogdanov, A., Jr., Uptake of dextran-coated monocrystalline iron oxides in tumor cells and macrophages, *J Magn Reson Imaging*, 7, 1140–1145, 1997.
83. Franklin, R., Blaschuk, K.L., Bearchell, M.C., Prestoz, L.L., Setzu, A., Brindle, K.M., and ffrench-Constant, C., Magnetic resonance imaging of transplanted oligodendrocyte precursors in the rat brain, *Neuroreport*, 10, 3961–3965, 1999.
84. Chan, T., Eley, C., Liberti, P., So, A., and Kressel, H.Y., Magnetic resonance imaging of abscesses using lipid-coated iron oxide particles, *Invest Radiol*, 27, 443–449, 1992.
85. Billotey, C., Wilhelm, C., Devaud, M., Bacri, J.C., Bittoun, J., and Gazeau, F., Cell internalization of anionic maghemite nanoparticles: quantitative effect on magnetic resonance imaging, *Magn Reson Med*, 49, 646–654, 2003.
86. Weissleder, R., Cheng, H.C., Bogdanova, A., and Bogdanov, A., Jr., Magnetically labeled cells can be detected by MR imaging, *J Magn Reson Imaging*, 7, 258–263, 1997.
87. Fleige, G., Nolte, C., Synowitz, M., Seeberger, F., Kettenmann, H., and Zimmer, C., Magnetic labeling of activated microglia in experimental gliomas, *Neoplasia*, 3, 489–499, 2001.
88. Metz, S., Bonaterra, G., Rudelius, M., Settles, M., Rummeny, E.J., and Daldrup-Link, H.E., Capacity of human monocytes to phagocytose approved iron oxide MR contrast agents *in vitro*, *Eur Radiol*, 14, 1851–1858, 2004.
89. Sun, R., Dittrich, J., Le-Huu, M., Mueller, M.M., Bedke, J., Kartenbeck, J., Lehmann, W.D., Krueger, R., Bock, M., Huss, R., Seliger, C., Grone, H.J., Misselwitz, B., Semmler, W., and Kiessling, F., Physical and biological characterization of superparamagnetic iron oxide- and ultrasmall superparamagnetic iron oxide-labeled cells: a comparison, *Invest Radiol*, 40, 504–513, 2005.
90. Jirak, D., Kriz, J., Herynek, V., Andersson, B., Girman, P., Burian, M., Saudek, F., and Hajek, M., MRI of transplanted pancreatic islets, *Magn Reson Med*, 52, 1228–1233, 2004.
91. Jendelova, P., Herynek, V., Urdzikova, L., Glogarova, K., Kroupova, J., Andersson, B., Bryja, V., Burian, M., Hajek, M., and Sykova, E., Magnetic resonance tracking of transplanted bone marrow and embryonic stem cells labeled by iron oxide nanoparticles in rat brain and spinal cord, *J Neurosci Res*, 76, 232–243, 2004.
92. Jendelova, P., Herynek, V., DeCroos, J., Glogarova, K., Andersson, B., Hajek, M., and Sykova, E., Imaging the fate of implanted bone marrow stromal cells labeled with superparamagnetic nanoparticles, *Magn Reson Med*, 50, 767–776, 2003.
93. Frank, J.A., Zywicke, H., Jordan, E.K., Mitchell, J., Lewis, B.K., Miller, B., Bryant, L.H., Jr., and Bulte, J.W., Magnetic intracellular labeling of mammalian cells by combining (FDA-approved) superparamagnetic iron oxide MR contrast agents and commonly used transfection agents, *Acad Radiol*, 9 (Suppl 2), S484–S487, 2002.
94. Frank, J.A., Miller, B.R., Arbab, A.S., Zywicke, H.A., Jordan, E.K., Lewis, B.K., Bryant, L.H., Jr., and Bulte, J.W., Clinically applicable labeling of mammalian and stem cells by combining superparamagnetic iron oxides and transfection agents, *Radiology*, 228, 480–487, 2003.

95. de Vries, I.J., Lesterhuis, W.J., Barentsz, J.O., Verdijk, P., van Krieken, J.H., Boerman, O.C., Oyen, W.J., Bonenkamp, J.J., Boezeman, J.B., Adema, G.J., Bulte, J.W., Scheenen, T.W., Punt, C.J., Heerschap, A., and Figdor, C.G., Magnetic resonance tracking of dendritic cells in melanoma patients for monitoring of cellular therapy, *Nat Biotechnol*, 23, 1407–1413, 2005.
96. Chen, D., Maa, Y.F., and Haynes, J.R., Needle-free epidermal powder immunization, *Expert Rev Vaccines*, 1, 265–276, 2002.
97. Gan, W.B., Grutzendler, J., Wong, W.T., Wong, R.O., and Lichtman, J.W., Multicolor "DiOlistic" labeling of the nervous system using lipophilic dye combinations, *Neuron*, 27, 219–225, 2000.
98. Zhang, Z.G., Jiang, Q., Zhang, R., Zhang, L., Wang, L., Zhang, L., Arniego, P., Ho, K.L., and Chopp, M., Magnetic resonance imaging and neurosphere therapy of stroke in rat, *Ann Neurol*, 53, 259–263, 2003.
99. Emerit, J., Beaumont, C., and Trivin, F., Iron metabolism, free radicals, and oxidative injury, *Biomed Pharmacother*, 55, 333–339, 2001.
100. Gehl, J., Electroporation: theory and methods, perspectives for drug delivery, gene therapy and research, *Acta Physiol Scand*, 177, 437–447, 2003.
101. Walczak, P., Kedziorek, D.A., Gilad, A.A., Lin, S., and Bulte, J.W., Instant MR labeling of stem cells using magnetoelectroporation, *Magn Reson Med*, 54, 769–774, 2005.
102. Scherer, F., Anton, M., Schillinger, U., Henke, J., Bergemann, C., Kruger, A., Gansbacher, B., and Plank, C., Magnetofection: enhancing and targeting gene delivery by magnetic force *in vitro* and *in vivo*, *Gene Ther*, 9, 102–109, 2002.
103. Plank, C., Scherer, F., Schillinger, U., Bergemann, C., and Anton, M., Magnetofection: enhancing and targeting gene delivery with superparamagnetic nanoparticles and magnetic fields, *J Liposome Res*, 13, 29–32, 2003.
104. Martina, M.S., Fortin, J.P., Menager, C., Clement, O., Barratt, G., Grabielle-Madelmont, C., Gazeau, F., Cabuil, V., and Lesieur, S., Generation of superparamagnetic liposomes revealed as highly efficient MRI contrast agents for *in vivo* imaging, *J Am Chem Soc*, 127, 10676–10685, 2005.
105. Jendelova, P., Herynek, V., Urdzikova, L., Glogarova, K., Rahmatova, S., Fales, I., Andersson, B., Prochazka, P., Zamecnik, J., Eckschlager, T., Kobylka, P., Hajek, M., and Sykova, E., Magnetic resonance tracking of human CD34+ progenitor cells separated by means of immunomagnetic selection and transplanted into injured rat brain, *Cell Transplant*, 14, 173–182, 2005.
106. Zhang, Z., Jiang, Q., Jiang, F., Ding, G., Zhang, R., Wang, L., Zhang, L., Robin, A.M., Katakowski, M., and Chopp, M., *In vivo* magnetic resonance imaging tracks adult neural progenitor cell targeting of brain tumor, *Neuroimage*, 23, 281–287, 2004.
107. Miyoshi, S., Flexman, J.A., Cross, D.J., Maravilla, K.R., Kim, Y., Anzai, Y., Oshima, J., and Minoshima, S., Transfection of neuroprogenitor cells with iron nanoparticles for magnetic resonance imaging tracking: cell viability, differentiation, and intracellular localization, *Mol Imaging*, 4, 1–10, 2005.
108. Song, Y., Morikawa, S., Morita, M., Inubushi, T., Takada, T., Torii, R., and Tooyama, I., Magnetic resonance imaging using hemagglutinating virus of Japan-envelope vector successfully detects localization of intra-cardially administered microglia in normal mouse brain, *Neurosci Lett*, 395, 42–45, 2006.
109. Bulte, J.W., Ma, L.D., Magin, R.L., Kamman, R.L., Hulstaert, C.E., Go, K.G., The, T.H., and de Leij, L., Selective MR imaging of labeled human peripheral blood mononuclear cells by liposome mediated incorporation of dextran-magnetite particles, *Magn Reson Med*, 29, 32–37, 1993.
110. Bulte, J.W. and De Cuyper, M., Magnetoliposomes as contrast agents, *Methods Enzymol*, 373, 175–198, 2003.
111. Ito, A., Matsuoka, F., Honda, H., and Kobayashi, T., Antitumor effects of combined therapy of recombinant heat shock protein 70 and hyperthermia using magnetic nanoparticles in an experimental subcutaneous murine melanoma, *Cancer Immunol Immunother*, 53, 26–32, 2004.
112. Shinkai, M., Le, B., Honda, H., Yoshikawa, K., Shimizu, K., Saga, S., Wakabayashi, T., Yoshida, J., and Kobayashi, T., Targeting hyperthermia for renal cell carcinoma using human MN antigen-specific magnetoliposomes, *Jpn J Cancer Res*, 92, 1138–1145, 2001.
113. Bulte, J., Laughlin, P.G., Jordan, E.K., Tran, V.A., Vymazal, J., and Frank, J.A., Tagging of T cells with superparamagnetic iron oxide: uptake kinetics and relaxometry, *Acad Radiol*, 3 (Suppl 2), S301–S303, 1996.
114. Josephson, L., Tung, C.H., Moore, A., and Weissleder, R., High-efficiency intracellular magnetic labeling with novel superparamagnetic-Tat peptide conjugates, *Bioconjug Chem*, 10, 186–191, 1999.

115. Zhao, M., Kircher, M.F., Josephson, L., and Weissleder, R., Differential conjugation of tat peptide to superparamagnetic nanoparticles and its effect on cellular uptake, *Bioconjug Chem*, 13, 840–844, 2002.
116. Josephson, L., Kircher, M.F., Mahmood, U., Tang, Y., and Weissleder, R., Near-infrared fluorescent nanoparticles as combined MR/optical imaging probes, *Bioconjug Chem*, 13, 554–560, 2002.
117. Kircher, M.F., Allport, J.R., Graves, E.E., Love, V., Josephson, L., Lichtman, A.H., and Weissleder, R., *In vivo* high resolution three-dimensional imaging of antigen-specific cytotoxic T-lymphocyte trafficking to tumors, *Cancer Res*, 63, 6838–6846, 2003.
118. Moore, A., Sun, P.Z., Cory, D., Hogemann, D., Weissleder, R., and Lipes, M.A., MRI of insulitis in autoimmune diabetes, *Magn Reson Med*, 47, 751–758, 2002.
119. Lewin, M., Carlesso, N., Tung, C.H., Tang, X.W., Cory, D., Scadden, D.T., and Weissleder, R., Tat peptide-derivatized magnetic nanoparticles allow *in vivo* tracking and recovery of progenitor cells, *Nat Biotechnol*, 18, 410–414, 2000.
120. Moore, A., Grimm, J., Han, B., and Santamaria, P., Tracking the recruitment of diabetogenic CD8+ T-cells to the pancreas in real time, *Diabetes*, 53, 1459–1466, 2004.
121. Ahrens, E.T., Feili-Hariri, M., Xu, H., Genove, G., and Morel, P.A., Receptor-mediated endocytosis of iron-oxide particles provides efficient labeling of dendritic cells for *in vivo* MR imaging, *Magn Reson Med*, 49, 1006–1013, 2003.
122. Bulte, J.W., Hoekstra, Y., Kamman, R.L., Magin, R.L., Webb, A.G., Briggs, R.W., Go, K.G., Hulstaert, C.E., Miltenyi, S., The, T.H., et al., Specific MR imaging of human lymphocytes by monoclonal antibody-guided dextran-magnetite particles, *Magn Reson Med*, 25, 148–157, 1992.
123. Berry, C.C., Charles, S., Wells, S., Dalby, M.J., and Curtis, A.S., The influence of transferrin stabilised magnetic nanoparticles on human dermal fibroblasts in culture, *Int J Pharm*, 269, 211–225, 2004.
124. Taupitz, M., Wagner, S., Schnorr, J., Kravec, I., Pilgrimm, H., Bergmann-Fritsch, H., and Hamm, B., Phase I clinical evaluation of citrate-coated monocrystalline very small superparamagnetic iron oxide particles as a new contrast medium for magnetic resonance imaging, *Invest Radiol*, 39, 394–405, 2004.
125. Taupitz, M., Schnorr, J., Abramjuk, C., Wagner, S., Pilgrimm, H., Hunigen, H., and Hamm, B., New generation of monomer-stabilized very small superparamagnetic iron oxide particles (VSOP) as contrast medium for MR angiography: preclinical results in rats and rabbits, *J Magn Reson Imaging*, 12, 905–911, 2000.
126. Stroh, A., Zimmer, C., Gutzeit, C., Jakstadt, M., Marschinke, F., Jung, T., Pilgrimm, H., and Grune, T., Iron oxide particles for molecular magnetic resonance imaging cause transient oxidative stress in rat macrophages, *Free Radic Biol Med*, 36, 976–984, 2004.
127. Taupitz, M., Schmitz, S., and Hamm, B., Superparamagnetic iron oxide particles: current state and future development, *Rofo*, 175, 752–765, 2003.
128. Wiener, E.C., Brechbiel, M.W., Brothers, H., Magin, R.L., Gansow, O.A., Tomalia, D.A., and Lauterbur, P.C., Dendrimer-based metal chelates: a new class of magnetic resonance imaging contrast agents, *Magn Reson Med*, 31, 1–8, 1994.
129. Sato, N., Kobayashi, H., Hiraga, A., Saga, T., Togashi, K., Konishi, J., and Brechbiel, M.W., Pharmacokinetics and enhancement patterns of macromolecular MR contrast agents with various sizes of polyamidoamine dendrimer cores, *Magn Reson Med*, 46, 1169–1173, 2001.
130. Yan, G.P., Hu, B., Liu, M.L., and Li, L.Y., Synthesis and evaluation of gadolinium complexes based on PAMAM as MRI contrast agents, *J Pharm Pharmacol*, 57, 351–357, 2005.
131. Kobayashi, H., Sato, N., Kawamoto, S., Saga, T., Hiraga, A., Haque, T.L., Ishimori, T., Konishi, J., Togashi, K., and Brechbiel, M.W., Comparison of the macromolecular MR contrast agents with ethylenediamine-core versus ammonia-core generation-6 polyamidoamine dendrimer, *Bioconjug Chem*, 12, 100–107, 2001.
132. Kobayashi, H., Kawamoto, S., Jo, S.K., Bryant, H.L., Jr., Brechbiel, M.W., and Star, R.A., Macromolecular MRI contrast agents with small dendrimers: pharmacokinetic differences between sizes and cores, *Bioconjug Chem*, 14, 388–394, 2003.
133. Bryant, L.H., Jr., Brechbiel, M.W., Wu, C., Bulte, J.W., Herynek, V., and Frank, J.A., Synthesis and relaxometry of high-generation (G = 5, 7, 9, and 10) PAMAM dendrimer-DOTA-gadolinium chelates, *J Magn Reson Imaging*, 9, 348–352, 1999.
134. Tunici, P., Bulte, J.W., Bruzzone, M.G., Poliani, P.L., Cajola, L., Grisoli, M., Douglas, T., and Finocchiaro, G., Brain engraftment and therapeutic potential of stem/progenitor cells derived from mouse skin, *J Gene Med*, 8, 506–513, 2006.

135. Walter, G.A., Cahill, K.S., Huard, J., Feng, H., Douglas, T., Sweeney, H.L., and Bulte, J.W., Noninvasive monitoring of stem cell transfer for muscle disorders, *Magn Reson Med*, 51, 273–277, 2004.
136. Lee, I.H., Bulte, J.W., Schweinhardt, P., Douglas, T., Trifunovski, A., Hofstetter, C., Olson, L., and Spenger, C., *In vivo* magnetic resonance tracking of olfactory ensheathing glia grafted into the rat spinal cord, *Exp Neurol*, 187, 509–516, 2004.
137. Bulte, J.W., Ben-Hur, T., Miller, B.R., Mizrachi-Kol, R., Einstein, O., Reinhartz, E., Zywicke, H.A., Douglas, T., and Frank, J.A., MR microscopy of magnetically labeled neurospheres transplanted into the Lewis EAE rat brain, *Magn Reson Med*, 50, 201–205, 2003.
138. Kohler, N., Sun, C., Fichtenholtz, A., Gunn, J., Fang, C., and Zhang, M., Methotrexate-immobilized poly(ethylene glycol) magnetic nanoparticles for MR imaging and drug delivery, *Small*, 2, 785–792, 2006.
139. Kohler, N., Sun, C., Wang, J., and Zhang, M., Methotrexate-modified superparamagnetic nanoparticles and their intracellular uptake into human cancer cells, *Langmuir*, 21, 8858–8864, 2005.
140. de Marco, G., Bogdanov, A., Marecos, E., Moore, A., Simonova, M., and Weissleder, R., MR imaging of gene delivery to the central nervous system with an artificial vector, *Radiology*, 208, 65–71, 1998.
141. Hoehn, M., Kustermann, E., Blunk, J., Wiedermann, D., Trapp, T., Wecker, S., Focking, M., Arnold, H., Hescheler, J., Fleischmann, B.K., Schwindt, W., and Buhrle, C., Monitoring of implanted stem cell migration *in vivo*: a highly resolved *in vivo* magnetic resonance imaging investigation of experimental stroke in rat, *Proc Natl Acad Sci USA*, 99, 16267–16272, 2002.
142. Xiang, J.J., Tang, J.Q., Zhu, S.G., Nie, X.M., Lu, H.B., Shen, S.R., Li, X.L., Tang, K., Zhou, M., and Li, G.Y., IONP-PLL: a novel non-viral vector for efficient gene delivery, *J Gene Med*, 5, 803–817, 2003.
143. Bielinska, A.U., Chen, C., Johnson, J., and Baker, J.R., Jr., DNA complexing with polyamidoamine dendrimers: implications for transfection, *Bioconjug Chem*, 10, 843–850, 1999.
144. Luo, D. and Saltzman, W.M., Synthetic DNA delivery systems, *Nat Biotechnol*, 18, 33–37, 2000.
145. Arbab, A.S., Yocum, G.T., Wilson, L.B., Parwana, A., Jordan, E.K., Kalish, H., and Frank, J.A., Comparison of transfection agents in forming complexes with ferumoxides, cell labeling efficiency, and cellular viability, *Mol Imaging*, 3, 24–32, 2004.
146. Arbab, A.S., Yocum, G.T., Kalish, H., Jordan, E.K., Anderson, S.A., Khakoo, A.Y., Read, E.J., and Frank, J.A., Efficient magnetic cell labeling with protamine sulfate complexed to ferumoxides for cellular MRI, *Blood*, 104, 1217–1223, 2004.
147. Sorgi, F.L., Bhattacharya, S., and Huang, L., Protamine sulfate enhances lipid-mediated gene transfer, *Gene Ther*, 4, 961–968, 1997.
148. Cornetta, K. and Anderson, W.F., Protamine sulfate as an effective alternative to polybrene in retroviral-mediated gene-transfer: implications for human gene therapy, *J Virol Methods*, 23, 187–194, 1989.
149. Arbab, A.S., Yocum, G.T., Rad, A.M., Khakoo, A.Y., Fellowes, V., Read, E.J., and Frank, J.A., Labeling of cells with ferumoxides-protamine sulfate complexes does not inhibit function or differentiation capacity of hematopoietic or mesenchymal stem cells, *NMR Biomed*, 18, 553–559, 2005.
150. Arbab, A.S., Bashaw, L.A., Miller, B.R., Jordan, E.K., Lewis, B.K., Kalish, H., and Frank, J.A., Characterization of biophysical and metabolic properties of cells labeled with superparamagnetic iron oxide nanoparticles and transfection agent for cellular MR imaging, *Radiology*, 229, 838–846, 2003.
151. Arbab, A.S., Bashaw, L.A., Miller, B.R., Jordan, E.K., Bulte, J.W., and Frank, J.A., Intracytoplasmic tagging of cells with ferumoxides and transfection agent for cellular magnetic resonance imaging after cell transplantation: methods and techniques, *Transplantation*, 76, 1123–1130, 2003.
152. Arbab, A., Yocum, G.T., Anderson, S.A., Kalish, H., Read, E.J., and Frank, J.A., Response to letter to the editor, re ferumoxides-protamine sulfate labeling does not alter differentiation of mysenchymal stem cells, *Blood*, 104, 3412–3413, 2004.
153. Daldrup-Link, H.E., Rudelius, M., Oostendorp, R.A., Jacobs, V.R., Simon, G. H., Gooding, C., and Rummeny, E.J., Comparison of iron oxide labeling properties of hematopoietic progenitor cells from umbilical cord blood and from peripheral blood for subsequent *in vivo* tracking in a xenotransplant mouse model XXX, *Acad Radiol*, 12, 502–510, 2005.
154. Yocum, G.T., Wilson, L.B., Ashari, P., Jordan, E.K., Frank, J.A., and Arbab, A.S., Effect of human stem cells labeled with ferumoxides-poly-L-lysine on hematologic and biochemical measurements in rats, *Radiology*, 235, 547–552, 2005.

155. Kostura, L., Kraitchman, D.L., Mackay, A.M., Pittenger, M.F., and Bulte, J.W., Feridex labeling of mesenchymal stem cells inhibits chondrogenesis but not adipogenesis or osteogenesis, *NMR Biomed*, 17, 513–517, 2004.
156. Schroeter, M., Saleh, A., Wiedermann, D., Hoehn, M., and Jander, S., Histochemical detection of ultrasmall superparamagnetic iron oxide (USPIO) contrast medium uptake in experimental brain ischemia, *Magn Reson Med*, 52, 403–406, 2004.
157. Moos, T. and Mollgard, K., A sensitive post-DAB enhancement technique for demonstration of iron in the central nervous system, *Histochemistry*, 99, 471–475, 1993.
158. Arbab, A.S., Wilson, L.B., Ashari, P., Jordan, E.K., Lewis, B.K., and Frank, J.A., A model of lysosomal metabolism of dextran coated superparamagnetic iron oxide (SPIO) nanoparticles: implications for cellular magnetic resonance imaging, *NMR Biomed*, 18, 383–389, 2005.
159. Dodd, C.H., Hsu, H.C., Chu, W.J., Yang, P., Zhang, H.G., Mountz, J.D., Jr., Zinn, K., Forder, J., Josephson, L., Weissleder, R., Mountz, J.M., and Mountz, J.D., Normal T-cell response and *in vivo* magnetic resonance imaging of T cells loaded with HIV transactivator-peptide-derived superparamagnetic nanoparticles, *J Immunol Methods*, 256, 89–105, 2001.
160. Mani, V., Briley-Saebo, K.C., Itskovich, V.V., Samber, D.D., and Fayad, Z.A., Gradient echo acquisition for superparamagnetic particles with positive contrast (GRASP): sequence characterization in membrane and glass superparamagnetic iron oxide phantoms at 1.5T and 3T, *Magn Reson Med*, 55, 126–135, 2006.
161. Seppenwoolde, J.H., Viergever, M.A., and Bakker, C.J., Passive tracking exploiting local signal conservation: the white marker phenomenon, *Magn Reson Med*, 50, 784–790, 2003.
162. Cunningham, C.H., Arai, T., Yang, P.C., McConnell, M.V., Pauly, J.M., and Conolly, S.M., Positive contrast magnetic resonance imaging of cells labeled with magnetic nanoparticles, *Magn Reson Med*, 53, 999–1005, 2005.
163. de Laquintane, B.D., Dousset, V., Solanilla, A., Petry, K.G., and Ripoche, J., Iron particle labeling of haematopoietic progenitor cells: an *in vitro* study, *Biosci Rep*, 22, 549–554, 2002.
164. Heyn, C., Bowen, C.V., Rutt, B.K., and Foster, P.J., Detection threshold of single SPIO-labeled cells with FIESTA, *Magn Reson Med*, 53, 312–320, 2005.
165. Bowen, C.V., Zhang, X., Saab, G., Gareau, P.J., and Rutt, B.K., Application of the static dephasing regime theory to superparamagnetic iron-oxide loaded cells, *Magn Reson Med*, 48, 52–61, 2002.
166. Bulte, J.W., Miller, G.F., Vymazal, J., Brooks, R.A., and Frank, J.A., Hepatic hemosiderosis in non-human primates: quantification of liver iron using different field strengths, *Magn Reson Med*, 37, 530–536, 1997.
167. Skotland, T., Sontum, P.C., and Oulie, I., *In vitro* stability analyses as a model for metabolism of ferromagnetic particles (Clariscan), a contrast agent for magnetic resonance imaging, *J Pharm Biomed Anal*, 28, 323–329, 2002.
168. Hoppe, M., Hulthen, L., and Hallberg, L., Serum iron concentration as a tool to measure relative iron absorption from elemental iron powders in man, *Scand J Clin Lab Invest*, 63, 489–496, 2003.
169. Pieroni, L., Khalil, L., Charlotte, F., Poynard, T., Piton, A., Hainque, B., and Imbert-Bismut, F., Comparison of bathophenanthroline sulfonate and ferene as chromogens in colorimetric measurement of low hepatic iron concentration, *Clin Chem*, 47, 2059–2061, 2001.
170. Marshall, P., Leavens, B., Heudi, O., and Ramirez-Molina, C., Liquid chromatography coupled with inductively coupled plasma mass spectrometry in the pharmaceutical industry: selected example, *J Chromatogr A*, 1056, 3–12, 2004.
171. Gutteridge, J.M. and Halliwell, B., The role of the superoxide and hydroxyl radicals in the degradation of DNA and deoxyribose induced by a copper-phenanthroline complex, *Biochem Pharmacol*, 31, 2801–2805, 1982.
172. Gutteridge, J.M. and Toeg, D., Iron-dependent free radical damage to DNA and deoxyribose. Separation of TBA-reactive intermediates, *Int J Biochem*, 14, 891–893, 1982.
173. Anderson, S.A., Shukaliak-Quandt, J., Jordan, E.K., Arbab, A.S., Martin, R., McFarland, H., and Frank, J.A., Magnetic resonance imaging of labeled T-cells in a mouse model of multiple sclerosis, *Ann Neurol*, 55, 654–659, 2004.
174. Evgenov, N.V., Medarova, Z., Dai, G., Bonner-Weir, S., and Moore, A., *In vivo* imaging of islet transplantation, *Nat Med*, 12, 144–148, 2006.

18 Cellular Imaging of Cell Transplants

Michel M.J. Modo and Jeff W.M. Bulte

CONTENTS

18.1 Introduction 325
18.2 Imaging Cell Transplants 327
18.2.1 Demyelinating Disease 327
18.2.2 Spinal Cord Injury 328
18.2.3 Cerebral Ischemia 328
18.2.4 Neurodegenerative Disorders 330
18.2.5 Brain Tumors 331
18.2.6 Myocardial Infarction 332
18.3 Imaging of Endogenous Stem Cells 333
18.4 Clinical Implementation of Cellular MR Imaging 334
18.5 Conclusion 336
References 336

18.1 INTRODUCTION

Visualizing the presence and migration of transplanted cells has, until recently, almost exclusively been the domain of postmortem immunohistochemical studies. The use of magnetic resonance (MR) contrast agents to label cells prior to transplantation, however, has provided new vistas for the noninvasive serial *in vivo* study of transplanted cells.[1] The incorporation of MR contrast agents into the cell renders the grafted cells visible on MR scans by changing the signal intensity of the area where transplanted cells can be found.

Without the prior incubation of cells with an MR contrast agent, these cells cannot be detected on MR scans.[2] Different types of contrast agents using different metal ions or particles (Fe, Gd, Mn) and fluorine have been used for this purpose. These contrast agents possess different characteristics in terms of cellular incorporation and relaxivity. Depending on the application of cell transplantation, the choice of agent can be determined based on these basic characteristics. For instance, detection of very small clusters of cells or even individual cells would require a contrast agent with a very high relaxivity, such as micron-sized particles of iron oxide (MPIOs).[3,4] In contrast, if the aim is to visualize the distribution of several millions of cells, a contrast agent with less sensitivity and detectable in a separate channel, such as fluorine-based agents,[5] might be advantageous.

Apart from relaxivity and cell incorporation, other aspects need to be considered to successfully visualize transplanted cells *in vivo* (Figure 18.1). As contrast agents are generally only poorly incorporated into nonphagocytic cells, the route of incorporation might need to be modified (see Chapter 17 for a detailed discussion). For instance, excellent results have been reported by using transfection agents[6] and magnetoelectroporation[7] to improve cellular uptake of iron oxide particles. It is essential to evaluate the effect this cell labeling exerts on cellular functions and to

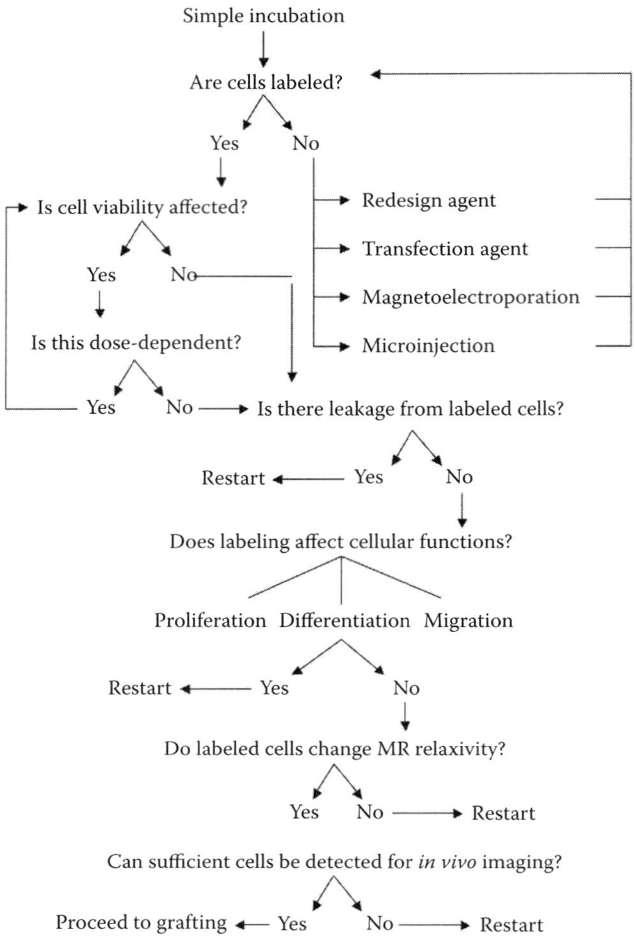

FIGURE 18.1 Overview of evaluating cell labeling prior to *in vivo* tracking of cells by MR imaging. The most basic approach to cell labeling with contrast agents is to test if adding the MR contrast agent to the incubation media results in a sufficient cellular uptake that would allow detection. If this is not the case, various approaches, such as the use of transfection agents, can be used to optimize the cellular uptake process. Once a sufficient cell labeling has been achieved, it is important to evaluate if this labeling has any detrimental effects on cellular functions that could compromise the therapeutic efficacy of the cells. To avoid potential false positives, the contrast agent cannot leak from the labeled cell and be taken up by host cells. To ensure *in vivo* detection, sufficient contrast agent needs to be taken up by cells to allow measurement of how many cells could be detected on an MR image based on relaxivity measurements. These *in vitro* studies will ensure that *in vivo* imaging can study the survival, migration, and differentiation of transplanted cells.

accordingly adjust the concentration or type of contrast agent used for cell imaging. To ensure a robust *in vivo* detection, it is also very commendable to investigate the relaxivity of labeled cells and determine if sufficient signal can be detected upon *in vivo* implantation. *In vivo*, the injection pressure and migration of cells will distribute the amount of contrast agent over many voxels, and it is therefore important to measure how many cells will be needed per voxel to warrant detection. An iterative process between these various aspects will be required to allow the *in vivo* tracking of implanted cells.

For the *in vivo* visualization of cells by MR imaging, special emphasis also needs to be given to reduce artifacts that could mistakenly be taken for transplanted cells. For instance, the site of injection can often be seen on T_2*-weighted images due to small bleeds causing hypointensities or a slight hyperintensity on T_2-weighted images as the result of an inflammatory responses (i.e., edema). The hypointense injection tract filled with transplanted cells can also have the same appearance as some poorly oxygenated blood vessels on T_2*-weighted images. Increasing the oxygen content in the anaesthetic regime can overcome this potential source of artifact.[8] However, hypointensities due to, for instance, macrophage infiltration[9] can only be distinguished from iron oxide-labeled stem cells based on their pattern of migration and time frame over which these hypointensities are apparent (see Chapter 15). Although many of these artifacts can be seen on T_2-based images, T_1-weighted images of transplanted cells with paramagnetic agents, such as Gd and Mn, often suffer from the inherent low relaxivity of these agents.

Once the initial challenges of labeling cells with contrast agents have been overcome, this approach can be used to study serially *in vivo* the survival, location, and migration of transplanted cells. This *in vivo* assessment of transplanted cells is needed to allow a thorough investigation of how transplanted cells respond to damaged tissue and promote functional recovery.[10] As damaged tissue generally evolves, a dynamic assessment of these changes in relation to transplantation is needed to make firm conclusions regarding the effects of the transplanted cells.[11] In some cases, such as neurodegenerative disease, for instance, it is possible that the transplanted cells will not reverse tissue damage, but might merely slow down the degenerative processes. The temporal change therefore could be crucial to assess if this therapeutic intervention is beneficial. Histopathological assessments that require postmortem tissue collection can only provide analyses at a single time point, and it is therefore difficult to infer the evolution to this final time point. The development of cellular MR imaging is likely to overcome this challenge and provide researchers and clinicians with a new tool to assess damage and cell transplantation.[10]

18.2 IMAGING CELL TRANSPLANTS

18.2.1 Demyelinating Disease

Autoimmune conditions, such as multiple sclerosis, are characterized by a specific loss of cells, such as oligodendrocytes, which myelinate axons. This demyelination leads to severe motor impairments in patients that are not easily addressed by pharmacological means. Transplantation of cells differentiating, for instance, into oligodendrocytes can reverse the pathology and may provide persistent relief of these impairments. The diffuse nature and widespread migration of cells after transplantation, however, make it difficult to relate cell infiltration/migration and functional recovery. Prelabeling of (stem-cell derived) oligodendrocyte progenitors with MR contrast agents affords the visualization and tracking of the movement of the cells in the spinal cord and brain.[12–15] Other cells able to myelinate damaged myelin sheets, such as Schwann or olfactory ensheathing cells, have also been shown to incorporate superparamagnetic iron oxide (SPIO) particles by using dendrimers[16] or fluid phase pinocytosis while preserving their functional ability to remyelinate damaged fibers *in vivo*.[17]

Investigation of the autoimmune nature of demyelinating disease could greatly benefit from the *in vivo* tracking of lymphocytes as they induce the disease process. *In vitro* labeling of mouse lymphocytes, such as T-cells, and their tracking *in vivo* after injection into mice by MR imaging[18–20] enable the early visualization of pathogenesis and evolution while potentially allowing a direct study of how interventions interfere with these processes. In contrast, *in vivo* labeling of lymphocytes will be complicated by the uptake of contrast agents into phagocytic cells that are not thought to be involved in the early pathogenetic events, but rather in the response to inflicted cell loss. A specific uptake mechanism for T-cells[21,22] and inhibitory signal to macrophages would be desirable to allow a distinct and specific *in vivo* labeling of T-cell and macrophage activity in demyelinating disease. Macrophage imaging by MR imaging has already been shown to be adept to illustrate its contribution to understanding the disease progress and assessing the efficacy of promising pharmaceutical interventions.[23,24] Being able to visualize the trafficking and contribution of particular cell fractions to the disease process and recovery can open new avenues for therapeutic interventions.

18.2.2 Spinal Cord Injury

As with demyelination of the spinal cord, it is also possible to track the migration of transplanted cells in the spinal cord aiming to restore lost function due to transection or damage of the cord. Early studies in the spinal cord assessing fetal graft survival by MR imaging indicated that T_1-weighted images only provided useful information on the graft border and lesion boundary, but that T_2-weighted MR scans allowed a distinction of viable and failed grafts based on a medium signal intensity reflecting viable grafts.[25,26] Similar approaches have been developed in the clinic,[27] but largely depend on subtle changes on MR scans that are nonspecific to the transplanted cells, but can be used to assess the effect of transplanted cells on the evolution of pathology.[28–30]

Although this approach might be sufficient to monitor nonmigrating fetal grafts, cell transplants with a high migratory potential would not allow an acute assessment of distribution or overgrowth. At present, the use of contrast agents to prelabel transplanted cells is overcoming this issue. For instance, recently Lee et al.[16] were able to detect SPIO-labeled olfactory ensheathing cells (OECs) for at least 2 months postinjection. OECs migrated in the intact spinal cord, but only showed limited migration in the damaged cord. The physical barrier of the transection severely limits the OECs' ability to populate both sides of the stump. No behavioral improvement was observed in transplanted animals in comparison to lesioned animals. Transplantation of cells into both sides of the lesion might be needed to increase the functional repair mediated by cell transplantation. It was also observed that in lesioned animals, small hemorrhages resulted in hypointensities on the MR images. These small bleeds stained positive for Prussian Blue, a histochemical marker of iron often used to indicate the presence of iron oxide particles in transplanted cells. The use of an antidextran antibody[7] that allows the immunohistochemical detection of dextran-chelated contrast agents provides a very selective approach to visualize only incorporated particles rather than an endogenous iron pool that can be found in macrophages. Similar issues arise for transplanted cells in the brain, where the injection of cells can easily damage small blood vessels and provide a source of false positives. The potential to develop specific imaging sequences, such as FIESTA, that are insensitive to hemorrhages but detect iron oxide particles might overcome these limitations.[31]

18.2.3 Cerebral Ischemia

Stroke is the most common cause of adult disability in the Western world. Not much progress has been achieved in reducing the incidence of stroke or the impact of impairment in survivors. Transplantation of stem cells that might recover/replace lost cells is a promising strategy that might improve the chronic disability observed after an infarct. Cellular MR imaging in stroke allows monitoring of the survival and migration of transplanted stem cells aiming to improve behavioral impairments. The prelabeling of cells with iron oxide-[32–35] or gadolinium-based[36] contrast agents has allowed different groups to visualize the presence and migration of transplanted stem cells to the site of cerebral ischaemia (Figure 18.2). Zhang et al.[34,35] calculated that transplanted subventricular zone cells migrate at an average speed of about 65 μm/h to the site of lesion, approximately half the speed of cells migrating along the rostral migratory stream in intact brains. It remains unclear, though, if speed of migration or infiltration of cells to the site of lesion is related to functional recovery. A common observation in these studies, irrespective of the type of transplanted cells (embryonic stem cell, neural stem cell, or bone marrow stromal cell), is that grafted cells start migration from remote implantation sites within the first 2 days of transplantation and typically arrive at the site of injury between 10 and 14 days. Some of the above-mentioned studies, combined with immunohistochemical staining using cell-specific antibodies, could show a substantial differentiation of the implanted cells into mature neurons and glia in the target zone, but the relevance

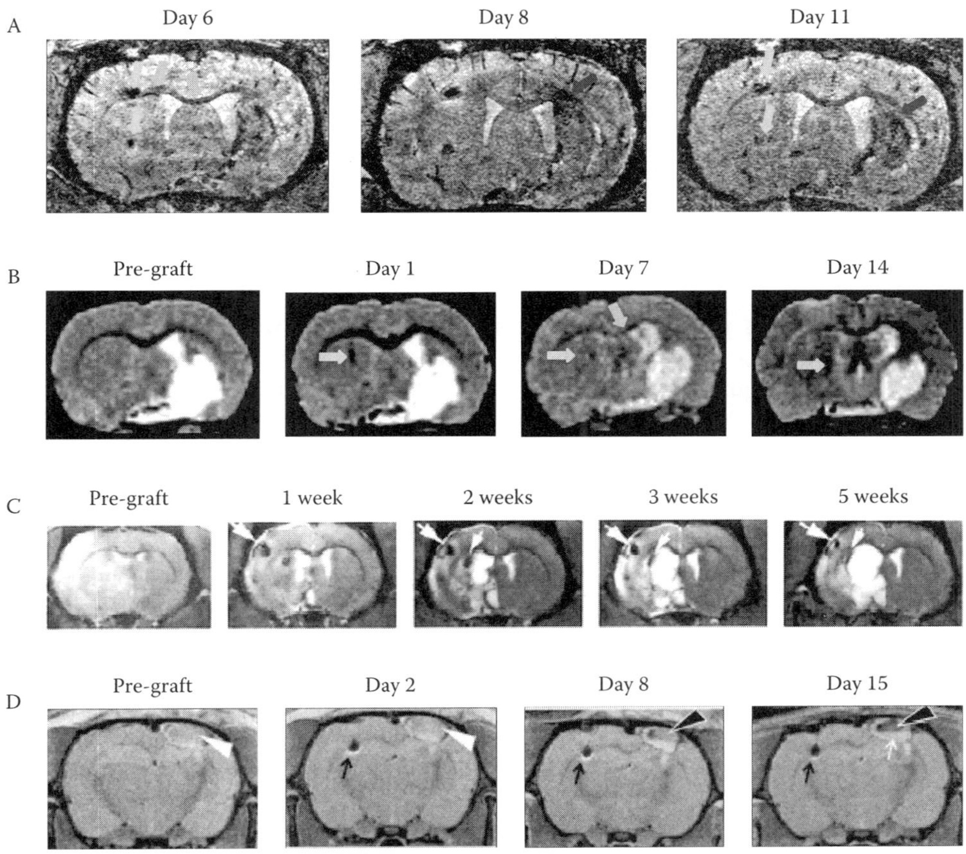

FIGURE 18.2 (please see color insert following page 210) Visualizing the migration of transplanted cells to the site of injury. The migration of transplanted embryonic (A) and neural (B) stem cells from the contralateral hemisphere to stroke damage over 14 days has been demonstrated by Hoehn et al.[32] and Modo et al.,[36] respectively. Migrating cells appear as hypointense areas on the T_2*- (A) and T_2- (B) weighted images. Migration to the site of stroke has also been demonstrated with subventricular zone cells injected intracisternally by Zhang et al.[34] (C), allowing a detection of transplanted cells over 5 weeks by MR imaging. However, this pathology-seeking ability is not specific to neural cells and stroke. For instance, Jendelova et al.[33] demonstrated that bone marrow stromal cells migrate to a stab wound in the contralateral cortex (D), highlighting that many different types of stem cells seek out pathology.

of migration to functional recovery remains to be established. Although migration and survival of stem cells will be important to *in vivo* monitoring, this assessment also needs to encompass the effect of transplanted cells on the evolution of the lesion.[37,38] Clinical studies so far only implanted cells into the infarcted region and used MR imaging or positron emission tomography (PET) to measure the effect of transplanted cells on the host brain.[39–42] The further development of cellular MR imaging would be very valuable to human studies trying to ascertain survival and functionality of transplanted cells over many years. Being able to establish how the cells interact with the damaged brain parenchyma might provide a tool to understand how transplanted cells promote recovery. Visualizing the mechanisms of recovery *in vivo* would bring more experimental control to clinical trials and allow a refinement of interventions.

As stroke is a cerebrovascular disease, special attention must also be given to angiogenesis in the peri-infarct area. Being able to detect an infiltration of endothelial cells into this area could be very instrumental in understanding the mechanisms involved in spontaneous recovery after stroke. To this end, different MR measurements can provide an *in vivo* analysis of various processes involved in angiogenesis in the peri-infarct area.[43] However, the evolving lesion, the need for serial assessment, and the consequent realignment of images complicate this type of analysis. To increase the therapeutic effect of angiogenesis, in some cases it might be opportune to transplant CD34+ (identifying endothelial progenitor) cells and investigate their contribution to lesion repair by tracking their migration and integration *in vivo* by MR imaging.[44] The ability to trace the pathways along which these cells migrate permits study of their dynamic behavior *in vivo*. Nonetheless, more extensive and clinically oriented studies will be needed to guide future cell therapy in patients.[45]

18.2.4 Neurodegenerative Disorders

Transplantation of fetal cells in neurodegenerative disease is showing promising results in both preclinical and clinical settings. Although in clinical settings cellular therapy for Parkinson's and Huntington's disease is under way, MR imaging has been mainly used to ensure the safety of the grafting surgery,[46] the guidance of implantation,[47] or to determine if transplanted cells are responsive to particular pharmacological challenges in preclinical models.[48,49] The importance of imaging to clinical studies is illustrated by establishing that a restoration of dopamine in the caudate putamen by PET relates to behavioral improvements.[50] However, little research has been devoted to the cellular MR imaging of transplanted cells in these neurodegenerative disorders.

Preliminary studies by Sandhu et al.[51] reported that uni- or bilateral transplantation of GRID-labeled neural stem cells in a rat model of Huntington's disease allowed visualization of the grafted cells for at least 28 days after injection. Upon unilateral implantation, cells not only populated the transplanted hemisphere, but also migrated to the damage in the nontransplanted hemisphere, suggesting that cells respond to distant pathology even if they are injected into a site of damage. The functional significance of these observations, however, remains to be elucidated. Similar to clinical studies, the use of functional MR imaging to assess graft-mediated activity[52] could reveal the functional link between anatomy and behavior. The possibility to administer stem cells into one particular site in widespread damage would minimize iatrogenic damage caused by many injection tracts and lessen the surgical impact of the procedure. Still, the evolution of degenerative disease requires the integration of cellular MR imaging with other approaches that allow a monitoring of the disease progression.[11] For instance, in addition to the tracking of grafted cells, the measurement of anatomical changes induced by transplanted cells by means of serial MR imaging,[53] or the downstream effects on brain activity by functional or pharmacological MR imaging, will provide a more holistic assessment of cell transplants.

Of all neurological diseases, the clinical development of cell therapy is the most advanced for Parkinson's disease. To date, mainly PET has been used to elucidate the pharmacological effects of fetal transplants in these patients. The small size of the sustantia nigra affected by cell loss in Parkinson's disease and its difficulty to define on MR imaging have so far not provided much interest in the use of MR imaging. Recently, the development of pharmacological MRI (phMRI) and diffusion tensor imaging (DTI), which allow the detection of dopaminergic cell and fiber loss, has drawn interest to MR imaging as a technique to monitor the progression of Parkinson's disease. Cellular MR imaging integrated with these approaches can address many important issues of cell transplantation in Parkinson's disease. For instance, it will be possible to define in preclinical models to what degree of accuracy cell delivery and graft survival are important in promoting functional recovery. The quantity of cells to be transplanted could also be calculated in relation to the degree of dopaminergic cell loss, which can be assessed by phMRI. MR imaging can be considered a tool for the integration of various aspects of cell transplantation, as it can be used to select suitable patients, to balance experimental groups, to provide a baseline measure of pathology, and to evaluate the functional effects of the transplant.[45] The potential to assess graft survival and migration *in vivo* by MR imaging will help to further refine cell therapy for neurodegenerative disease.

It is important, though, not to compromise the detection of pathology while visualizing grafted cells. Some neurodegenerative disease, such as Alzheimer's, which might benefit from widespread

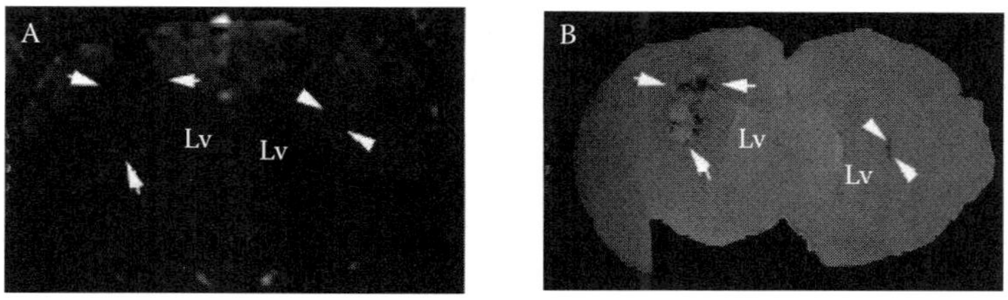

FIGURE 18.3 (please see color insert) *In vivo* tracking of neural progenitors migrating toward implanted tumor cells. (A) An MR image of iron oxide-labeled neural progenitor cells migrating toward a needle tract containing tumor cells (arrows), as indicated by an expansion of the signal loss around the needle tract. However, no expansion of a sham needle tract (arrowheads) was observed, as no iron oxide-labeled cells migrated to this site. Zhang et al.[55] further corroborated these results by Prussian Blue staining for iron oxide-labeled cells (B).

migration and early intervention, will benefit from cellular imaging of transplanted cells by providing correlations among cell survival, stagnation of neurodegeneration, and cognitive decline. However, visualizing transplanted cells based on a T_2 or T_2* hypointensity could limit efforts to visualize iron-containing plaques.[54] Being able to visualize amyloid plaques, a major pathological feature of this condition, could help to diagnose and monitor the progression of Alzheimer's disease. It is therefore important to maintain the ability to detect the plaques. Alternative strategies, such as fluorine-based agents, might be very attractive to preserve the diagnostic potential of ^{1}H-based MR imaging and allow the selective detection of widespread stem cell migration in a separate channel. Visualizing stem cells in a different channel, however, requires a sufficient signal being present to allow reliable detection.

18.2.5 Brain Tumors

A lack of sufficient signal can also be a problem in detecting tumor cells implanted into animals. As tumor cells proliferate extensively, they will halve the amount of contrast agent per cell with each cell division. Then as these cells proliferate, each cell will become more difficult to detect.

In contrast, visualizing the migration of transplanted cells to a brain tumor is similar to visualizing the migration of stem cells to any other pathology. For instance, Zhang et al.[55] have followed neural progenitors and bone marrow stromal cells migrating toward previously established 9L-gliosarcomas in rats (Figure 18.3). Within 5 days of contralateral implantation, cells migrated to the hemisphere implanted with the tumor and caused a hypointensity surrounding the neoplastic growth. Verification of the contrast agent by histological means verified that the signal loss on MR images was indeed due to implanted cells. In any case, it will be important to support *in vivo* MR imaging observations by postmortem histology, as tumors are known to result in small bleeds and a macrophage infiltration that can also cause hypointense signals on MR images.

Transplanted cells lining the outer rim of the tumor mass spread over several cubic millimeters that can easily be detected by MR imaging,[55] but the ability to visualize only a few or even a single stem cell that tracks down infiltrating tumor cells[56] poses an entirely different challenge. This can only be achieved by using agents with a very high superparamagentic effect, such as MPIOs.[57] Being able to trace a few stem cells *in vivo* possibly responding to an invading tumor cell would still require the ability to detect the tumor cell *in vivo* using a mechanism that is distinct from the one used to detect the stem cell.

One potential approach to track two populations of cells is based on chemical exchange saturation transfer (CEST; see Chapters 5 and 6), which, when used with paramagnetic ions (PARACEST), can selectively visualize separate contrast agents.[58] As PARACEST agents respond to different resonance frequencies, these can be selectively excited at very narrow bands and produce a very selective signal. Especially at high field strengths, which are ideal for detecting small quantities of cells, these agents show their potential. Being able to trace two independent cell populations could also be a very informative approach to study how macrophages interact with tumor or stem cells.

Especially in animal models, being able to study *in vivo* how the host's immune system attacks the implanted tumor[59–61] might provide insights as to how this native response could be harnessed as a therapeutic intervention. Antigen-specific cytotoxic T cells have been found to specifically home to OVA+[62] and HER2/neu+[63] tumors. As cellular MR imaging allows serial monitoring of the same animal/tumor over time, the biodynamics (site of cell infiltration and its consequent distribution over time) can be determined in a robust manner. SPIO-labeled hematopoietic bone marrow cells have a different distribution pattern compared to free SPIO particles. Cells rapidly home in to bone marrow following intravenous injection, whereas free SPIO particles do not show this directional migration.[64] Being able to visualize several populations of cells would dramatically enhance the information that can be derived from cellular MR imaging.

Tracking the evolution of various components of tumors *in vivo* would allow the determination of how distinct types of cells contribute to tumor survival and expansion. For instance, the neovasculature supports tumor expansion, but the interplay between tumor cell proliferation and endothelial cells constructing the new vasculature remains to be assessed serially *in vivo*. Although the infusion of iron oxide-labeled bone marrow-derived endothelial precursors illustrates the integration of endothelial progenitor cells to the neovasculature supporting brain tumors,[65] it fails to track the expansion of tumor cells. Only 3 days after implantation, endothelial progenitor cells were observed as signal hypointensity encompassing the preimplanted tumors.[66] If injected at the same time as the tumors, the prelabeled endothelial cells contributing to the neovasculature could only be observed when the tumor mass reached 1 cm, after 12 to 14 days of growth. Being able to visualize *in vivo* neovascularization will be an important step in assessing therapies geared toward suppressing nutrient support of the tumor.

18.2.6 Myocardial Infarction

Visualizing the formation of newly formed blood vessels and the incorporating of cells into the heart muscle will be essential to assess cell therapy for myocardial infarction. However, imaging of transplanted cells after myocardial infarction poses a further technical challenge, as the heart is constantly moving, and therefore requires a fast image acquisition protocol. Various strategies, such as cardiac gating of image acquisition, have been devised to allow cardiac imaging. This approach could also alleviate potential partial volume effects that would limit the detection of small clusters of cells that contribute to blood vessels. However, the time constraint and potential movement artifact considerably complicate the image acquisition of implanted cells by MR imaging.[67]

These technical complications can be overcome by implantation of iron oxide-labeled transplanted cells that can be detected using MR scanners.[68–74] Injection of iron oxide particles alone (without cells) into the area of damage in the heart only results in a transient (<12 h) signal attenuation on T_2-weighted scans; unlabeled control cells do not affect the MR signal *in vivo*, suggesting that the long-term identification of iron oxide-labeled cells is not a mere artifact of the surgical procedure or the incorporation of free iron oxide particles into host cells.[75] A further strategy to visualize the engraftment and survival of transplanted cells in the damaged myocardium will consist of the infusion of Mn-DPDP, which is incorporated into viable cardiocytes via Ca^{2+} channels[76] and should restore the signal hyperintensity on T_1-weighted MR scans.

Prelabeling of endothelial progenitor cells with CD34 antibody-coated magnetic beads also allows their tracking *in vivo* and provides a means to study the revascularization of areas of damage, which ensues repopulation by cardiocytes and could provide a surrogate marker of treatment efficacy without being directly related to graft survival.[77] Special care needs to be taken with antibody-based MR contrast agents for labeling cells, as these particles can detach from the cell surface and nonselectively attach to other elements in the system, providing a potential source of false positives.

Although transplantation of stem cells for cardiovascular disease is progressing rapidly to clinical applications, at present the common use of implanted devices, such as pacemakers, in this patient group still remains largely prohibitive for MR imaging.[78] However, significant advances have been achieved in designing catheters that allow MR-guided delivery and monitoring of transplanted cells. MR fluoroscopy, for instance, allows real-time monitoring of the injection and ensures accurate placement of cells.[79–81] A similar approach could be used to ensure that a sufficient number of cells are implanted in different damaged regions. Clinical translation of these approaches is on the horizon and promises to revolutionize the treatment of patients with heart attacks.[78]

Aside from imaging of migration or trafficking, the ability of cellular MR imaging to verify accurate cell delivery in the targeted organ (region) will become very important in itself. This is the case for mesenchymal stem cell (MSC) delivery in myocardial infarct, as well as MSC delivery to organs (liver, kidney) through their feeding arteries.[82,83] Systemic administration of MSCs through intravenous administration would be highly desirable, as it is less invasive. However, the number of MSCs that home to injured areas, e.g., the heart or kidney, may become below the threshold of detection for *in vivo* MR imaging, as many cells become initially entrapped in the liver.[84]

The use of imaging in cell-based therapies will be essential to ensure the clinical efficacy of this therapeutic intervention.[85] Although cell therapy for myocardial therapy has advanced to clinical trials,[86] there remain questions as to how the injected cells mediate therapy. For instance, Limbourg et al.[87] have shown that cardiac function can improve without survival of transplanted cells, raising the question of whether continued cell survival is necessary for a continued effect, and casting doubt on the idea that transplanted cells will integrate and replace damaged host cells. The use of serial *in vivo* cellular and molecular MR imaging of the survival, integration, and differentiation of infused cells will be necessary to address this issue.

18.3 IMAGING OF ENDOGENOUS STEM CELLS

Most organ systems regenerate perpetually from endogenous stem cells. For instance, all cells found in the hematopoeitic system regenerate from bone marrow stem cells. However, only recently have endogenous stem cells been discovered in the central nervous system. The endogeous neural stem cells are thought to mainly reside in the subventricular zone (SVZ) and the dentate gyrus. Although the extent and time frame over which cells are replaced in the brain is less than in other organs, it is a necessary condition for the brain's normal functioning. A decrease of endogenous stem cells in the brain has been linked to neurodegenerative[88] and psychiatric disease.[89] Being able to visualize the presence, migration, and integration of this endogenous stem cell pool can therefore impart insights into mechanisms of disease and recovery.

The *in vivo* visualization of this endogenous stem cell fraction requires *in vivo* labeling of cells with contrast agents. Although in general iron oxide particles have been found to be inefficient to label stem cells by themselves, with sufficient time a few particles will be incorporated into cells. To be able to visualize migrating endogenous stem cells, it is important to provide sufficient signal attenuation on an MR scan that would allow the detection of a single cell with potentially only a single particle of iron oxide being incorporated. Micron-sized particles of iron oxide (MPIOs) have a sufficient superparamagentic effect to achieve this.[57] Similar to the injection of fluorescent dyes, injection of MPIOs into the cerebroventricular system or the SVZ has been shown to label adult endogenous stem cells and allow their visualization by MR imaging.[90] MPIOs are taken up through the ventricular wall and incorporated into stem cells. Upon migration out of the SVZ, these cells carry the MPIOs and allow visualization by MR imaging.[90] However, to ensure that this approach truly selectively labels neural stem cells, it is important to conduct immunohistological analyses confirming that the label is indeed contained within cells and not merely moving in the extracellular matrix. Moreover, as nonphagocytic stem cells generally do not easily take up large particles, it is important to confirm that the cells containing MPIOs are indeed neural stem cells, rather than resident microglia or macrophages. This allowed Shapiro et al.[57] to detect a single cell at embryonic day 11.5 after injection of the MPIO into a single-cell-stage mouse embryo.

Detection of single cells requires that cells are sufficiently loaded with a superparamagnetic agent without affecting cell function. Labeling of cells with MPIOs, for instance, results in a cellular loading of 100 pg of an iron oxide concentration about three times higher than with nanometer-sized agents.[91] It is therefore important to be able to calculate how much paramagnetic agent needs to be incorporated to ensure *in vivo* detection. Heyn et al.[92] recently developed an approach to calculate the minimum required detectable mass of SPIO against a uniform background using the FIESTA sequence, with the main determining factors being signal-to-noise ratio and voxel volume. Under typical conditions, the required cellular load of ferumoxides was in the femtomolar range, which is similar to the quantity of radiotracer needed for detection by PET imaging. Based on this work, single cells could be detected *in vivo* after injection into the circulation of mice.[4] However, further work needs to be undertaken to provide methods to optimize cell detection based on the concentration of contrast agent uptake. Ideally, quantitative methods will be developed that will allow the tracing of single or multiple cells. These developments will facilitate the study of the influence of endogenous stem cells on brain repair. However, it is unclear whether the studying of only a few endogenous stem cells will allow scientists to gain an understanding of how these cells respond to damage.

18.4 CLINICAL IMPLEMENTATION OF CELLULAR MR IMAGING

The many developments cellular MR imaging has achieved over the past few years have opened new vistas for the study of cell transplants in preclinical settings. However, as cell therapy is increasingly finding its translation into clinical settings, it will be important to also implement cellular MR imaging that will allow the monitoring of this new therapeutic approach in patients (see Chapter 21).

To ensure that cellular MR imaging will find its clinical implementation, certain safeguards need to be observed to guarantee the safety of patients. Although clinically approved MR contrast agents have a solid safety record,[93] the application of these agents to cellular MR imaging might need further scrutiny. Typically, clinically approved contrast agents are contained within the extracellular matrix and are cleared fairly rapidly by the reticuloendothelial system (RES), which clears them through the kidney or liver. Labeling the intracellular compartments of cells with these agents for continued *in vivo* detection demands a more thorough *in vitro* assessment of the agents to ensure that they do not affect normal cellular functions (see Chapter 17 for detailed discussion).

There have been reports of iron oxide agents affecting normal cell differentiation of mesenchymal stem cells into chondrocytes.[94] In certain cases, a limited differentiation might not pose a problem, but if differentiation is needed for therapeutic efficacy, it will be an essential function that should not be affected by cell labeling. Although many of these effects might be detected *in vitro*, some functions can only be tested *in vivo*. For instance, neural stem cells might differentiate *in vitro* into appropriate phenotypes, but differentiation of neural stem cells might only be one way how these cells promote repair. It is therefore essential to test the labeled cells' ability to promote functional repair *in vivo* by testing behavioral recovery.

The continued presence of contrast agents within cells also needs to be thoroughly investigated, as deleterious effects might only start to appear after protracted time points. Many MR contrast agents use dextran as a chelating agent that prevents exposure of the cell to the metal ions used to induce an MR signal attenuation.[1] If dextran is decomposed inside the cells, the cell will be exposed to the potentially toxic metal particle. Although this process might take months to occur, cell transplants in human patients would be required to survive many years. It will therefore be an essential test to evaluate the survival and function of contrast agent-labeled cells over many months in preclinical settings.

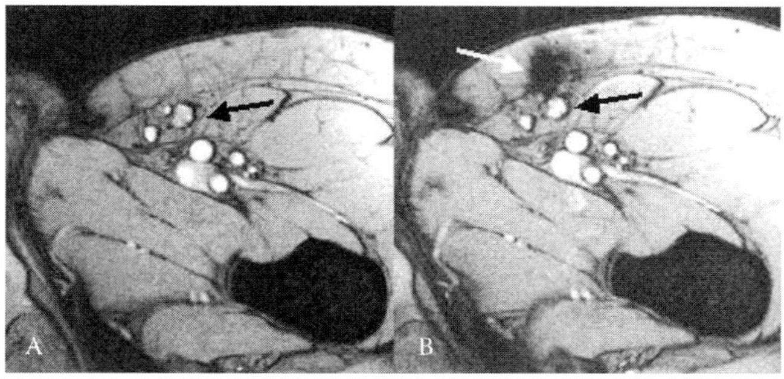

FIGURE 18.4 Monitoring of the accuracy of delivery of SPIO-labeled cells using cellular MR imaging. (A) MR scan before grafting; the inguinal lymph node to be injected is indicated by a black arrow. (B) MR scan after grafting, demonstrating that the dendritic cells were not accurately delivered into the draining inguinal lymph node (black arrow), but instead into the perinodal, subcutaneous fat (white arrow). Accidental misinjection was found to have occurred in four of eight patients.[96]

Another essential element of cellular MR imaging in patients will be that cellular tracking cannot interfere with the routine assessment of the disease. MR imaging is routinely used to assess various diseases and provide important differential information to the physician who will adjust therapies based on these assessments. If cellular MR imaging interferes with this process, it could endanger patient management. For instance, many diseases produce hyperintense signals on T_2-weighted MR images. This often reflects an inflammatory response to ongoing pathological processes. If these patients are implanted with iron oxide-labeled cells that are detected as hypointense signals, this could interfere with the accurate assessment of inflammatory responses to damage, as the signal output on the MR imaging is an average for each voxel. The use of contrast agents for cellular MR imaging therefore needs to be planned in accordance with the patient's routine assessment protocol.

There are circumstances where these considerations are less of a concern, for instance, imaging of immune cells that are generally involved in the disposal of contrast agents from the bloodstream or liver/kidney cells that are involved in the clearing of these particles from the body. Upon *in vivo* injection of iron oxide particles, blood-borne macrophages will take up some of the particles and allow the visualization of macrophage trafficking to sites of damage[95] (see Chapter 14). Prelabeling of cells *in vitro* gives more control over how much contrast agent is incorporated into cells and also allows for a more selective uptake in specific cells. The advantage of using phagocytic immune cells (i.e., immature dendritic cells) as a first target for translation is that these cells can readily incorporate unmodified MR contrast agents, and therefore do not require additional materials, such as transfection agents, which are used to shuttle iron oxide particles into the cells. Using as few compounds as possible will ensure that it is easier to trace any problem to its source.

Using this approach, de Vries et al.[96] prelabeled immature dendritic cells with either iron oxide particles or ^{111}In-oxine destined for injection into the same lymph nodes of patients with stage III melanoma. The use of two distinct imaging modalities permitted a direct comparison of the accuracy of scintigraphic and MR imaging to deliver cells under ultrasound guidance into the destined local draining lymph nodes (Figure 18.4). The excellent soft tissue contrast and high anatomical resolution provided a greater accuracy for MR imaging in determining correct dendritic cell delivery than for scintigraphy. In only 50% of cases were cells correctly injected. Only accurately placed cells migrated, which is a mandatory requirement for an efficient therapeutic effect. Thus, real-time MR imaging with its excellent anatomical information could help to ensure more accurate delivery of cells, and hence improve therapeutic outcome. On the other hand, scintigraphy allowed a quantitative assessment of the number of cells injected, whereas the inherent difficulty of quantifying contrast agent-induced MR signal changes did not allow for this information to be derived from the acquired MR images. This study clearly exemplifies the potential of using cellular MR imaging as a diagnostic and monitoring technique for pursuing novel experimental cell therapies, which may be further enhanced when combined with other imaging modalities, such as scintigraphy or PET imaging.

18.5 CONCLUSION

Cell therapy is currently finding its translation from the bench to the bedside. As part of this process, it will be important to monitor the survival, distribution, and differentiation of implanted cells. Being able to visualize stem and immune cell activity is crucial. Cellular MR imaging is in the process of developing the tools to make this happen.

REFERENCES

1. Modo, M., Hoehn, M., and Bulte, J.W., Cellular MR imaging, *Mol Imaging*, 4, 143–164, 2005.
2. Modo, M., Cash, D., Mellodew, K., Williams, S.C., Fraser, S.E., Meade, T.J., Price, J., and Hodges, H., Tracking transplanted stem cell migration using bifunctional, contrast agent-enhanced, magnetic resonance imaging, *Neuroimage*, 17, 803–811, 2002.
3. Shapiro, E.M., Sharer, K., Skrtic, S., and Koretsky, A.P., *In vivo* detection of single cells by MRI, *Magn Reson Med*, 55, 242–249, 2006.
4. Heyn, C., Ronald, J.A., Mackenzie, L.T., MacDonald, I.C., Chambers, A.F., Rutt, B.K., and Foster, P.J., *In vivo* magnetic resonance imaging of single cells in mouse brain with optical validation, *Magn Reson Med*, 55, 23–29, 2006.
5. Ahrens, E.T., Flores, R., Xu, H., and Morel, P.A., *In vivo* imaging platform for tracking immunotherapeutic cells, *Nat Biotechnol*, 23, 983–987, 2005.
6. Frank, J.A., Miller, B.R., Arbab, A.S., Zywicke, H.A., Jordan, E.K., Lewis, B.K., Bryant, L.H., Jr., and Bulte, J.W., Clinically applicable labeling of mammalian and stem cells by combining superparamagnetic iron oxides and transfection agents, *Radiology*, 228, 480–487, 2003.
7. Walczak, P., Kedziorek, D.A., Gilad, A.A., Lin, S., and Bulte, J.W., Instant MR labeling of stem cells using magnetoelectroporation, *Magn Reson Med*, 54, 769–774, 2005.
8. Himmelreich, U., Weber, R., Ramos-Cabrer, P., Wegener, S., Kandal, K., Shapiro, E.M., Koretsky, A.P., and Hoehn, M., Improved stem cell MR detectability in animal models by modification of the inhalation gases, *Mol Imaging*, 4, 1–6, 2005.
9. Weber, R., Wegener, S., Ramos-Cabrer, P., Wiedermann, D., and Hoehn, M., MRI detection of macrophage activity after experimental stroke in rats: new indicators for late appearance of vascular degradation? *Magn Reson Med*, 54, 59–66, 2005.
10. Modo, M., Understanding stem cell-mediated brain repair through neuroimaging, *Curr Stem Cell Res Ther*, 1, 55–64, 2006.
11. Kirik, D., Breysse, N., Bjorklund, T., Besret, L., and Hantraye, P., Imaging in cell-based therapy for neurodegenerative diseases, *Eur J Nucl Med Mol Imaging*, 32 (Suppl 2), S417–S434, 2005.
12. Bulte, J.W.M., Zhang, S., van Gelderen, P., Herynek, V., Jordan, E.K., Duncan, I.D., and Frank, J.A., Neurotransplantation of magnetically labeled oligodendrocyte progenitors: magnetic resonance tracking of cell migration and myelination, *Proc Natl Acad Sci USA*, 96, 15256–15261, 1999.
13. Bulte, J.W.M., Ben-Hur, T., Miller, B.R., Mizrachi-Kol, R., Einstein, O., Reinhartz, E., Zywicke, H.A., Douglas, T., and Frank, J.A., MR microscopy of magnetically labeled neurospheres transplanted into the Lewis EAE rat brain, *Magn Reson Med*, 50, 201–205, 2003.
14. Franklin, R.J., Blaschuk, K.L., Bearchell, M.C., Prestoz, L.L., Setzu, A., Brindle, K.M., and ffrench-Constant, C., Magnetic resonance imaging of transplanted oligodendrocyte precursors in the rat brain, *Neuroreport*, 10, 3961–3965, 1999.
15. Bulte, J.W.M., Douglas, T., Witwer, B., Zhang, S.C., Strable, E., Lewis, B.K., Zywicke, H., Miller, B., van Gelderen, P., Moskowitz, B.M., Duncan, I.D., and Frank, J.A., Magnetodendrimers allow endosomal magnetic labeling and *in vivo* tracking of stem cells, *Nat Biotechnol*, 19, 1141–1147, 2001.

16. Lee, I.H., Bulte, J.W., Schweinhardt, P., Douglas, T., Trifunovski, A., Hofstetter, C., Olson, L., and Spenger, C., *In vivo* magnetic resonance tracking of olfactory ensheathing glia grafted into the rat spinal cord, *Exp Neurol*, 187, 509–516, 2004.
17. Dunning, M.D., Lakatos, A., Loizou, L., Kettunen, M., ffrench-Constant, C., Brindle, K.M., and Franklin, R.J., Superparamagnetic iron oxide-labeled Schwann cells and olfactory ensheathing cells can be traced *in vivo* by magnetic resonance imaging and retain functional properties after transplantation into the CNS, *J Neurosci*, 24, 9799–9810, 2004.
18. Anderson, S.A., Shukaliak-Quandt, J., Jordan, E.K., Arbab, A.S., Martin, R., McFarland, H., and Frank, J.A., Magnetic resonance imaging of labeled T-cells in a mouse model of multiple sclerosis, *Ann Neurol*, 55, 654–659, 2004.
19. Stoll, G., Wessig, C., Gold, R., and Bendszus, M., Assessment of lesion evolution in experimental autoimmune neuritis by gadofluorine M-enhanced MR neurography, *Exp Neurol*, 197, 150–156, 2006.
20. Oweida, A.J., Dunn, E.A., and Foster, P.J., Cellular imaging at 1.5 T: detecting cells in neuroinflammation using active labeling with superparamagnetic iron oxide, *Mol Imaging*, 3, 85–95, 2004.
21. Pirko, I., Johnson, A., Ciric, B., Gamez, J., Macura, S.I., Pease, L.R., and Rodriguez, M., *In vivo* magnetic resonance imaging of immune cells in the central nervous system with superparamagnetic antibodies, *FASEB J*, 18, 179–182, 2004.
22. Pirko, I., Ciric, B., Johnson, A.J., Gamez, J., Rodriguez, M., and Macura, S., Magnetic resonance imaging of immune cells in inflammation of central nervous system, *Croat Med J*, 44, 463–468, 2003.
23. Deloire, M.S., Touil, T., Brochet, B., Dousset, V., Caille, J.M., and Petry, K.G., Macrophage brain infiltration in experimental autoimmune encephalomyelitis is not completely compromised by suppressed T-cell invasion: *in vivo* magnetic resonance imaging illustration in effective anti-VLA-4 antibody treatment, *Mult Scler*, 10, 540–548, 2004.
24. Kleinschnitz, C., Bendszus, M., Frank, M., Solymosi, L., Toyka, K.V., and Stoll, G., *In vivo* monitoring of macrophage infiltration in experimental ischemic brain lesions by magnetic resonance imaging, *J Cereb Blood Flow Metab*, 23, 1356–1361, 2003.
25. Wirth, E.D., 3rd, Theele, D.P., Mareci, T.H., Anderson, D.K., Brown, S.A., and Reier, P.J., *In vivo* magnetic resonance imaging of fetal cat neural tissue transplants in the adult cat spinal cord, *J Neurosurg*, 76, 261–274, 1992.
26. Wirth, E.D., 3rd, Theele, D.P., Mareci, T.H., Anderson, D.K., and Reier, P.J., Dynamic assessment of intraspinal neural graft survival using magnetic resonance imaging, *Exp Neurol*, 136, 64–72, 1995.
27. Wirth, E.D.R., Reier, P.J., Fessler, R.G., Thompson, F.J., Uthman, B., Behrman, A., Beard, J., Vierck, C.J., and Anderson, D.K., Feasibility and safety of neural tissue transplantation in patients with syringomyelia, *J Neurotrauma*, 18, 911–929, 2001.
28. Iannotti, C., Li, H., Stemmler, M., Perman, W.H., and Xu, X.M., Identification of regenerative tissue cables using *in vivo* MRI after spinal cord hemisection and Schwann cell bridging transplantation, *J Neurotrauma*, 19, 1543–1554, 2002.
29. Schwartz, E.D., Chin, C.L., Shumsky, J.S., Jawad, A.F., Brown, B.K., Wehrli, S., Tessler, A., Murray, M., and Hackney, D.B., Apparent diffusion coefficients in spinal cord transplants and surrounding white matter correlate with degree of axonal dieback after injury in rats, *Am J Neuroradiol*, 26, 7–18, 2005.
30. Schwartz, E.D., Shumsky, J.S., Wehrli, S., Tessler, A., Murray, M., and Hackney, D.B., *Ex vivo* MR determined apparent diffusion coefficients correlate with motor recovery mediated by intraspinal transplants of fibroblasts genetically modified to express BDNF, *Exp Neurol*, 182, 49–63, 2003.
31. Dunn, E.A., Weaver, L.C., Dekaban, G.A., and Foster, P.J., Cellular imaging of inflammation after experimental spinal cord injury, *Mol Imaging*, 4, 53–62, 2005.
32. Hoehn, M., Kustermann, E., Blunk, J., Wiedermann, D., Trapp, T., Wecker, S., Focking, M., Arnold, H., Hescheler, J., Fleischmann, B.K., Schwindt, W., and Buhrle, C., Monitoring of implanted stem cell migration *in vivo*: a highly resolved *in vivo* magnetic resonance imaging investigation of experimental stroke in rat, *Proc Natl Acad Sci USA*, 99, 16267–16272, 2002.
33. Jendelova, P., Herynek, V., deCroos, J., Glogarova, K., Andersson, B., Hajek, M., and Sykova, E., Imaging the fate of implanted bone marrow stromal cells labeled with superparamagnetic nanoparticles, *Magn Reson Med*, 50, 767–776, 2003.
34. Zhang, Z.G., Jiang, Q., Zhang, R., Zhang, L., Wang, L., Arniego, P., Ho, K.L., and Chopp, M., Magnetic resonance imaging and neurosphere therapy of stroke in rat, *Ann Neurol*, 53, 259–263, 2003.

35. Zhang, R.L., Zhang, L., Zhang, Z.G., Morris, D., Jiang, Q., Wang, L., Zhang, L.J., and Chopp, M., Migration and differentiation of adult rat subventricular zone progenitor cells transplanted into the adult rat striatum, *Neuroscience*, 116, 373–382, 2003.
36. Modo, M., Mellodew, K., Cash, D., Fraser, S.E., Meade, T.J., Price, J., and Williams, S.C.R., Mapping transplanted stem cell migration after a stroke: a serial, *in vivo* magnetic resonance imaging study, *Neuroimage*, 21, 311–317, 2004.
37. Nomura, T., Honmou, O., Harada, K., Houkin, K., Hamada, H., and Kocsis, J.D., I.V. infusion of brain-derived neurotrophic factor gene-modified human mesenchymal stem cells protects against injury in a cerebral ischemia model in adult rat, *Neuroscience*, 136, 161–169, 2005.
38. Kurozumi, K., Nakamura, K., Tamiya, T., Kawano, Y., Kobune, M., Hirai, S., Uchida, H., Sasaki, K., Ito, Y., Kato, K., Honmou, O., Houkin, K., Date, I., and Hamada, H., BDNF gene-modified mesenchymal stem cells promote functional recovery and reduce infarct size in the rat middle cerebral artery occlusion model, *Mol Ther*, 9, 189–197, 2004.
39. Kondziolka, D., Wechsler, L., Goldstein, S., Meltzer, C., Thulborn, K.R., Gebel, J., Jannetta, P., DeCesare, S., Elder, E.M., McGrogan, M., Reitman, M.A., and Bynum, L., Transplantation of cultured human neuronal cells for patients with stroke, *Neurology*, 55, 565–569, 2000.
40. Savitz, S.I., Dinsmore, J., Wu, J., Henderson, G.V., Stieg, P., and Caplan, L.R., Neurotransplantation of fetal porcine cells in patients with basal ganglia infarcts: a preliminary safety and feasibility study, *Cerebrovasc Dis*, 20, 101–107, 2005.
41. Meltzer, C.C., Kondziolka, D., Villemagne, V.L., Wechsler, L., Goldstein, S., Thulborn, K.R., Gebel, J., Elder, E.M., DeCesare, S., and Jacobs, A., Serial [18F] fluorodeoxyglucose positron emission tomography after human neuronal implantation for stroke, *Neurosurgery*, 49, 586–591, 2001; discussion, 591–592.
42. Bang, O.Y., Lee, J.S., Lee, P.H., and Lee, G., Autologous mesenchymal stem cell transplantation in stroke patients, *Ann Neurol*, 57, 874–882, 2005.
43. Jiang, Q., Zhang, Z.G., Ding, G.L., Zhang, L., Ewing, J.R., Wang, L., Zhang, R., Li, L., Lu, M., Meng, H., Arbab, A.S., Hu, J., Li, Q.J., Pourabdollah Nejad, D.S., Athiraman, H., and Chopp, M., Investigation of neural progenitor cell induced angiogenesis after embolic stroke in rat using MRI, *Neuroimage*, 28, 698–707, 2005.
44. Jendelova, P., Herynek, V., Urdzikova, L., Glogarova, K., Rahmatova, S., Fales, I., Andersson, B., Prochazka, P., Zamecnik, J., Eckschlager, T., Kobylka, P., Hajek, M., and Sykova, E., Magnetic resonance tracking of human CD34+ progenitor cells separated by means of immunomagnetic selection and transplanted into injured rat brain, *Cell Transplant*, 14, 173–182, 2005.
45. Modo, M., Roberts, T.J., Sandhu, J.K., and Williams, S.C.R., *In vivo* monitoring of cellular transplants by magnetic resonance imaging and positron emission tomography, *Exp Opin Biol Ther*, 4, 145–155, 2004.
46. Rosser, A.E., Barker, R.A., Harrower, T., Watts, C., Farrington, M., Ho, A.K., Burnstein, R.M., Menon, D.K., Gillard, J.H., Pickard, J., and Dunnett, S.B., Unilateral transplantation of human primary fetal tissue in four patients with Huntington's disease: NEST-UK safety report ISRCTN no. 36485475, *J Neurol Neurosurg Psychiatry*, 73, 678–685, 2002.
47. Donovan, T., Fryer, T.D., Pena, A., Watts, C., Carpenter, T.A., and Pickard, J.D., Stereotactic MR imaging for planning neural transplantation: a reliable technique at 3 tesla? *Br J Neurosurg*, 17, 443–449, 2003.
48. Chen, Y.C., Galpern, W.R., Brownell, A.L., Matthews, R.T., Bogdanov, M., Isacson, O., Keltner, J.R., Beal, M.F., Rosen, B.R., and Jenkins, B.G., Detection of dopaminergic neurotransmitter activity using pharmacologic MRI: correlation with PET, microdialysis, and behavioral data, *Magn Reson Med*, 38, 389–398, 1997.
49. Chen, Y.I., Brownell, A.-L., Galpern, W., Isacson, O., Bogdanov, M., Beal, M.F., Livni, E., Rosen, B.R., and Jenkins, B.G., Detection of dopaminergic cell loss and neural transplantation using pharmacological MRI, PET, and behavioral assessment, *Neuroreport*, 10, 2881–2886, 1999.
50. Piccini, P., Brooks, D.J., Bjorklund, A., Gunn, R.N., Grasby, P.M., Rimoldi, O., Brundin, P., Hagell, P., Rehncrona, S., Widner, H., and Lindvall, O., Dopamine release from nigral transplants visualized *in vivo* in a Parkinson's patient, *Nat Neurosci*, 2, 1137–1140, 1999.
51. Sandhu, J.K., Roberts, T.J., Price, J., Meade, T.J., Williams, S.C.R., and Modo, M., A quantitative comparison of unilateral versus bilateral neural stem cell transplantation in the 3-nitroproprionic acid model of Huntington's disease by contrast agent-enhanced MRI, *Proc ISMRM*, 11, 12, 2420, 2004.

52. Bluml, S., Kopyov, O., Jacques, S., and Ross, B.D., Activation of neurotransplants in humans, *Exp Neurol*, 158, 121–125, 1999.
53. Roberts, T.J., Price, J., Williams, S.C., and Modo, M., Preservation of striatal tissue and behavioral function after neural stem cell transplantation in a rat model of Huntington's disease, *Neuroscience*, 144, 1, 100–109, 2007.
54. Jack, C.R., Jr., Wengenack, T.M., Reyes, D.A., Garwood, M., Curran, G.L., Borowski, B.J., Lin, J., Preboske, G.M., Holasek, S.S., Adriany, G. and Poduslo, J.F., *In vivo* magnetic resonance microimaging of individual amyloid plaques in Alzheimer's transgenic mice, *J Neurosci*, 25, 10041–10048, 2005.
55. Zhang, Z., Jiang, Q., Jiang, F., Ding, G., Zhang, R., Wang, L., Zhang, L., Robin, A.M., Katakowski, M., and Chopp, M., *In vivo* magnetic resonance imaging tracks adult neural progenitor cell targeting of brain tumor, *Neuroimage*, 23, 281–287, 2004.
56. Aboody, K.S., Brown, A., Rainov, N.G., Bower, K.A., Liu, S., Yang, W., Small, J.E., Herrlinger, U., Ourednik, V., Black, P.M., Breakefield, X.O., and Snyder, E.Y., Neural stem cells display extensive tropism for pathology in adult brain: evidence from intracranial gliomas, *Proc Natl Acad Sci USA*, 97, 12846–12851, 2000.
57. Shapiro, E.M., Skrtic, S., Sharer, K., Hill, J.M., Dunbar, C.E., and Koretsky, A.P., MRI detection of single particles for cellular imaging, *Proc Natl Acad Sci USA*, 101, 10901–10906, 2004.
58. Aime, S., Carrera, C., Delli Castelli, D., Geninatti Crich, S., and Terreno, E., Tunable imaging of cells labeled with MRI-PARACEST agents, *Angew Chem Int Ed Engl*, 44(12), 1813–1815, 2005.
59. Hu, D.E., Kettunen, M.I., and Brindle, K.M., Monitoring T-lymphocyte trafficking in tumors undergoing immune rejection, *Magn Reson Med*, 54, 1473–1479, 2005.
60. Hu, D.E., Beauregard, D.A., Bearchell, M.C., Thomsen, L.L., and Brindle, K.M., Early detection of tumour immune-rejection using magnetic resonance imaging, *Br J Cancer*, 88, 1135–1142, 2003.
61. Nolte, I., Vince, G.H., Maurer, M., Herbold, C., Goldbrunner, R., Solymosi, L., Stoll, G., and Bendszus, M., Iron particles enhance visualization of experimental gliomas with high-resolution sonography, *Am J Neuroradiol*, 26, 1469–1474, 2005.
62. Kircher, M.F., Allport, J.R., Graves, E.E., Love, V., Josephson, L., Lichtman, A.H., and Weissleder, R., *In vivo* high resolution three-dimensional imaging of antigen-specific cytotoxic T-lymphocyte trafficking to tumors, *Cancer Res*, 63, 6838–6846, 2003.
63. Daldrup-Link, H.E., Meier, R., Rudelius, M., Piontek, G., Piert, M., Metz, S., Settles, M., Uherek, C., Wels, W., Schlegel, J., and Rummeny, E.J., *In vivo* tracking of genetically engineered, anti-HER2/neu directed natural killer cells to HER2/neu positive mammary tumors with magnetic resonance imaging, *Eur Radiol*, 15, 4–13, 2005.
64. Daldrup-Link, H.E., Rudelius, M., Piontek, G., Metz, S., Brauer, R., Debus, G., Corot, C., Schlegel, J., Link, T.M., Peschel, C., Rummeny, E.J. and Oostendorp, R.A.J., Migration of iron-oxide labeled human hematopoietic progenitor cells in a mouse model: *in vivo* monitoring with 1.5-T MR imaging equipment, *Radiology*, 234, 197–205, 2005.
65. Anderson, S.A., Glod, J., Arbab, A.S., Noel, M., Ashari, P., Fine, H.A., and Frank, J.A., Noninvasive MR imaging of magnetically labeled stem cells to directly identify neovasculature in a glioma model, *Blood*, 105, 420–425, 2005.
66. Arbab, A.S., Frenkel, V., Pandit, S.D., Anderson, S.A., Yocum, G.T., Bur, M., Khuu, H.M., Read, E.J., and Frank, J.A., Magnetic resonance imaging and confocal microscopy studies of magnetically labeled endothelial progenitor cells trafficking to sites of tumor angiogenesis, *Stem Cells*, 24, 671–678, 2006.
67. Fuster, V., Sanz, J., Viles-Gonzalez, J.F., and Rajagopalan, S., The utility of magnetic resonance imaging in cardiac tissue regeneration trials, *Nat Clin Pract Cardiovasc Med*, 3 (Suppl 1), S2–S7, 2006.
68. Garot, J., Unterseeh, T., Teiger, E., Champagne, S., Chazaud, B., Gherardi, R., Hittinger, L., Gueret, P., and Rahmouni, A., Magnetic resonance imaging of targeted catheter-based implantation of myogenic precursor cells into infarcted left ventricular myocardium, *J Am Coll Cardiol*, 41, 1841–1846, 2003.
69. Hill, J.M., Dick, A.J., Raman, V.K., Thompson, R.B., Yu, Z.X., Hinds, K.A., Pessanha, B.S., Guttman, M.A., Varney, T.R., Martin, B.J., Dunbar, C.E., McVeigh, E.R., and Lederman, R.J., Serial cardiac magnetic resonance imaging of injected mesenchymal stem cells, *Circulation*, 108, 1009–1014, 2003.
70. Kraitchman, D.L., Heldman, A.W., Atalar, E., Amado, L.C., Martin, B.J., Pittenger, M.F., Hare, J.M., and Bulte, J.W.M., *In vivo* magnetic resonance imaging of mesenchymal stem cells in myocardial infarction, *Circulation*, 107, 2290–2293, 2003.

71. Stuckey, D.J., Carr, C.A., Martin-Rendon, E., Tyler, D.J., Willmott, C., Cassidy, P.J., Hale, S.J., Schneider, J.E., Tatton, L., Harding, S.E., Radda, G.K., Watt, S., and Clarke, K., Iron particles for non-invasive monitoring of bone marrow stromal cell engraftment into, and isolation of viable engrafted donor cells from the heart, *Stem Cells*, 24(8), 1968–1975, 2006.
72. Cahill, K.S., Germain, S., Byrne, B.J., and Walter, G.A., Non-invasive analysis of myoblast transplants in rodent cardiac muscle, *Int J Cardiovasc Imaging*, 20, 593–598, 2004.
73. Kustermann, E., Roell, W., Breitbach, M., Wecker, S., Wiedermann, D., Buehrle, C., Welz, A., Hescheler, J., Fleischmann, B.K., and Hoehn, M., Stem cell implantation in ischemic mouse heart: a high-resolution magnetic resonance imaging investigation, *NMR Biomed*, 18, 362–370, 2005.
74. Weber, A., Pedrosa, I., Kawamoto, A., Himes, N., Munasinghe, J., Asahara, T., Rofsky, N.M., and Losordo, D.W., Magnetic resonance mapping of transplanted endothelial progenitor cells for therapeutic neovascularization in ischemic heart disease, *Eur J Cardiothorac Surg*, 26, 137–143, 2004.
75. Himes, N., Min, J.Y., Lee, R., Brown, C., Shea, J., Huang, X., Xiao, Y.F., Morgan, J.P., Burstein, D., and Oettgen, P., *In vivo* MRI of embryonic stem cells in a mouse model of myocardial infarction, *Magn Reson Med*, 52, 1214–1219, 2004.
76. Nordhoy, W., Anthonsen, H.W., Bruvold, M., Brurok, H., Skarra, S., Krane, J., and Jynge, P., Intracellular manganese ions provide strong T1 relaxation in rat myocardium, *Magn Reson Med*, 52, 506–514, 2004.
77. Weber, A., Pedrosa, I., Kawamoto, A., Himes, N., Munasinghe, J., Asahara, T., Rofsky, N.M., and Losordo, D.W., Magnetic resonance mapping of transplanted endothelial progenitor cells for therapeutic neovascularization in ischemic heart disease, *Eur J Cardiothorac Surg*, 26, 137–143, 2004.
78. Frangioni, J.V. and Hajjar, R.J., *In vivo* tracking of stem cells for clinical trials in cardiovascular disease, *Circulation*, 110, 3378–3383, 2004.
79. Karmarkar, P.V., Kraitchman, D.L., Izbudak, I., Hofmann, L.V., Amado, L.C., Fritzges, D., Young, R., Pittenger, M., Bulte, J.W., and Atalar, E., MR-trackable intramyocardial injection catheter, *Magn Reson Med*, 51, 1163–1172, 2004.
80. Dick, A.J., Guttman, M.A., Raman, V.K., Peters, D.C., Hill, J.M., Pessanha, B.S., Scott, G., Smith, S., McVeigh, E.R., and Lederman, R.J., Real-time MRI enables targeted injection of labeled stem cells to the border of recent porcine myocardial infarction based on functional and tissue characteristics, *Proc ISMRM*, 11, 365, 2003.
81. Dick, A.J., Guttman, M.A., Raman, V.K., Peters, D.C., Pessanha, B.S., Hill, J.M., Smith, S., Scott, G., McVeigh, E.R., and Lederman, R.J., Magnetic resonance fluoroscopy allows targeted delivery of mesenchymal stem cells to infarct borders in swine, *Circulation*, 108, 2899–2904, 2003.
82. Bos, C., Delmas, Y., Desmouliere, A., Solanilla, A., Hauger, O., Grosset, C., Dubus, I., Ivanovic, Z., Rosenbaum, J., Charbord, P., Combe, C., Bulte, J.W., Moonen, C.T., Ripoche, J., and Grenier, N., *In vivo* MR imaging of intravascularly injected magnetically labeled mesenchymal stem cells in rat kidney and liver, *Radiology*, 233, 781–789, 2004.
83. Mahmood, U., Can MR imaging be used to track delivery of intravascularly administered stem cells? *Radiology*, 233, 625–626, 2004.
84. Hauger, O., Frost, E.E., van Heeswijk, R., Deminiere, C., Xue, R., Delmas, Y., Combe, C., Moonen, C.T., Grenier, N., and Bulte, J.W., MR evaluation of the glomerular homing of magnetically labeled mesenchymal stem cells in a rat model of nephropathy, *Radiology*, 238, 200–210, 2006.
85. Bengel, F.M., Schachinger, V., and Dimmeler, S., Cell-based therapies and imaging in cardiology, *Eur J Nucl Med Mol Imaging*, 32 (Suppl 2), S404–S416, 2005.
86. Dib, N., Michler, R.E., Pagani, F.D., Wright, S., Kereiakes, D.J., Lengerich, R., Binkley, P., Buchele, D., Anand, I., Swingen, C., Di Carli, M.F., Thomas, J.D., Jaber, W.A., Opie, S.R., Campbell, A., McCarthy, P., Yeager, M., Dilsizian, V., Griffith, B.P., Korn, R., Kreuger, S.K., Ghazoul, M., MacLellan, W.R., Fonarow, G., Eisen, H.J., Dinsmore, J., and Diethrich, E., Safety and feasibility of autologous myoblast transplantation in patients with ischemic cardiomyopathy: four-year follow-up, *Circulation*, 112, 1748–1755, 2005.
87. Limbourg, F.P., Ringes-Lichtenberg, S., Schaefer, A., Jacoby, C., Mehraein, Y., Jager, M.D., Limbourg, A., Fuchs, M., Klein, G., Ballmaier, M., Schlitt, H.J., Schrader, J., Hilfiker-Kleiner, D., and Drexler, H., Haematopoietic stem cells improve cardiac function after infarction without permanent cardiac engraftment, *Eur J Heart Fail*, 7, 722–729, 2005.

88. Armstrong, R.J. and Barker, R.A., Neurodegeneration: a failure of neuroregeneration? *Lancet*, 358, 1174–1176, 2001.
89. Kempermann, G., Regulation of adult hippocampal neurogenesis: implications for novel theories of major depression, *Bipolar Disord*, 4, 17–33, 2002.
90. Shapiro, E.M., Skrtic, S., and Koretsky, A., Long term cellular MR imaging using micron sized iron oxide particles, *Proc ISMRM*, 11, 166, 2004.
91. Shapiro, E.M., Skrtic, S., and Koretsky, A.P., Sizing it up: cellular MRI using micron-sized iron oxide particles, *Magn Reson Med*, 53, 329–338, 2005.
92. Heyn, C., Bowen, C.V., Rutt, B.K., and Foster, P.J., Detection threshold of single SPIO-labeled cells with FIESTA, *Magn Reson Med*, 53, 312–320, 2005.
93. Runge, V.M., Safety of approved MR contrast media for intravenous injection, *J Magn Reson Imaging*, 12, 205–213, 2000.
94. Kostura, L., Kraitchman, D.L., Mackay, A.M., Pittenger, M.F., and Bulte, J.W., Feridex labeling of mesenchymal stem cells inhibits chondrogenesis but not adipogenesis or osteogenesis, *NMR Biomed*, 17, 513–517, 2004.
95. Saleh, A., Schroeter, M., Jonkmanns, C., Hartung, H.P., Modder, U., and Jander, S., *In vivo* MRI of brain inflammation in human ischaemic stroke, *Brain*, 127 (Pt 7), 1670–1677, 2004.
96. de Vries, I.J., Lesterhuis, W.J., Barentsz, J.O., Verdijk, P., van Krieken, J.H., Boerman, O.C., Oyen, W.J., Bonenkamp, J.J., Boezeman, J.B., Adema, G.J., Bulte, J.W., Scheenen, T.W., Punt, C.J., Heerschap, A., and Figdor, C.G., Magnetic resonance tracking of dendritic cells in melanoma patients for monitoring of cellular therapy, *Nat Biotechnol*, 23, 1407–1413, 2005.

19 Cellular and Molecular Imaging of the Diabetic Pancreas

Zdravka Medarova and Anna Moore

CONTENTS

19.1 Introduction 343
19.2 Imaging of Endogenous Beta-Cell Mass 345
19.2.1 Current Strategies for Assessment of Beta-Cell Mass (BCM) 345
19.2.2 Challenges Associated with the Measurement of Beta-Cell Mass 345
19.2.3 Potential Strategies for Imaging Beta-Cell Mass 345
19.2.3.1 Nuclear Imaging of Beta-Cell Mass 345
19.2.3.2 Magnetic Resonance Imaging of Beta-Cell Mass 347
19.3 Imaging of Immune Cell Infiltration 351
19.3.1 The Role of Immune Cell Infiltration in Diabetes 351
19.3.2 Potential Strategies for Imaging Immune Cell Infiltration in Diabetes 351
19.3.2.1 Non-Antigen-Specific Immune Cell Tracking 351
19.3.2.2 Antigen-Specific Strategies for Imaging Immune Cell Infiltration in Diabetes 353
19.4 Imaging of Apoptosis 354
19.4.1 The Role of Apoptosis in Diabetes 354
19.4.2 Potential Strategies for Imaging Diabetes-Associated Apoptosis 356
19.4.2.1 The Annexin V Experience 356
19.4.2.2 Imaging Diffusion as a Marker of Cell Viability 356
19.4.2.3 Imaging Oxygen Consumption as a Marker of Cell Viability 357
19.5 Imaging of Islet Transplantation 357
19.5.1 The Promise of Islet Transplantation: Current Status and Challenges 357
19.5.2 Current Strategies for Islet Graft Assessment 358
19.5.3 Potential Strategies for Imaging Transplanted Islets 358
19.6 Future Outlook 360
References 363

19.1 INTRODUCTION

Diabetes is a group of metabolic disorders that result from defects in insulin secretion or action. Normally, insulin is secreted by the beta-cells in the pancreatic islets of Langerhans in response to a rise in blood glucose levels (for example, after a meal) and serves as a signal for glucose uptake and assimilation by peripheral tissues. As diabetes develops, however, the body loses the capacity for insulin production/assimilation, resulting in elevated blood glucose (hyperglycemia).

Diabetes is a chronic disease for which there is no cure. Over time, diabetes can lead to serious cardiovascular, retinal, kidney, and neural complications, increasing the risk of death two-fold. It is a staggering epidemic affecting 20.8 million people (about 7% of the population) in the U.S., with 1.5 million new cases diagnosed in people over 20 years of age in 2005. It is believed that one in three people born in the U.S. in the year 2000, will develop diabetes in their lifetime. The estimated cost to the U.S. healthcare system in 2002 was $132 billion, of which $23.2 billion was for direct care, $24.6 billion was for treatment of complications from diabetes, and $44.1 billion was due to increased incidence of other diabetes-associated diseases.[1]

Considering the remarkable toll diabetes is having in terms of human life, it is clear that the development of strategies for the noninvasive assessment of molecular events associated with this disease constitutes an important healthcare priority. The ability to image the pathology on that scale would be instrumental in understanding the time course of the disease, identifying the key initiating events, and possibly designing novel therapeutic approaches and monitoring their efficacy.

Accomplishing the goal of cellular imaging in diabetes, however, presents a tremendous challenge. The underlying reasons extend from both the unique structure and distribution of pancreatic islets and the metabolic complexity of the disease. With respect to the first challenge, pancreatic islets are small organ-like entities (about 100 microns in diameter) dispersed throughout the pancreas at a low density and comprising only about 1.7% of the pancreatic volume.[2] The islet itself is a complex structure, consisting of insulin-producing beta-cells (which constitute approximately 50% of the islet), glucagon-secreting alpha-cells (15 to 20%), delta-cells involved in somatostatin production (3 to 10%), and cells that release pancreatic polypeptide (1%). Hormone production and secretion by all of these cells is a tightly regulated dynamic process driven by the need to respond to ever-changing energy demands and influenced by metabolic and environmental factors continually throughout the life of an organism. In diabetes, this delicate functional balance is disrupted. Identifying the key cellular events that define the pathology of diabetes and become manifest at early enough stages of the disease to allow intervention is a demanding process critical for the success of imaging.

Diabetes is a collection of diseases, which apparently originate from different causes and progress along seemingly different paths. The unifying event that defines these pathologies as diabetes is the development of hyperglycemia. Therefore, the study of diabetes has focused mainly on the insulin-secreting pancreatic beta-cell, since beta-cell failure has been implicated as a central event in the progression to hyperglycemia. There are two major types of diabetes, type 1 and type 2. Whereas type 1 diabetes is characterized by the specific autoimmune recognition and destruction of insulin-generating beta-cells, the signals mediating beta-cell destruction in type 2 diabetes seem to be of a metabolic nature.[3,4] Considering the central role of the beta-cell, strategies for cellular imaging of diabetes would focus on the detection of beta-cells via a variety of markers. Beta-cells secrete insulin, zinc, and C-peptide, which represent promising molecular screening targets. Autoimmune beta-cell destruction in type 1 diabetes is accompanied by a progressive and specific immune cell infiltration of pancreatic islets, which defines an alternative pathway for tracking the natural course of the disease. In addition, the pancreatic islet is a highly vascularized structure, receiving 10 to 20% of the blood flow to the pancreas.[5] Imaging methods can be developed to detect blood flow, volume, or vessel permeability, as indicators of islet viability and function. Furthermore, molecular imaging based on markers of apoptosis and necrosis would be valuable for the detection of the earliest stages of beta-cell damage in diabetes.[6] Finally, recent years have seen encouraging results in the area of islet transplantation as a means to achieve insulin independence and near normal glucose control. Therefore, the ability to follow the fate of transplanted islets noninvasively is urgently needed.

Molecular imaging can provide answers to many of the questions related to diabetes. It offers the unprecedented potential to unravel the complex natural history of the disease and to permit diagnosis at the earliest causative stages, characterized by the first signs of metabolic or molecular disturbance. Furthermore, by combining the global anatomical/physiologic scale of currently available *in vivo* imaging modalities with the detailed molecular/cellular scale of biochemistry and cell and molecular biology, molecular imaging allows the noninvasive real-time monitoring of diabetes progression as well as response to therapy noninvasively and in authentic physiologic environments.

The current goals of diabetes imaging can be assigned to four core areas: assessment of beta-cell mass, imaging of autoimmune attack in type 1 diabetes, evaluation of diabetes-associated beta-cell death, and imaging of islet transplantation. Although research in these categories is not advancing at an equal pace, and for some of them there are limited or no reports related to MRI, we are going to dedicate attention to all of them, in an attempt to give the reader a comprehensive view of the current accomplishments, as well as the future directions of this rapidly growing field.

19.2 IMAGING OF ENDOGENOUS BETA-CELL MASS

19.2.1 Current Strategies for Assessment of Beta-Cell Mass (BCM)

Since beta-cell mass (BCM) is tightly linked to the pathogenesis of diabetes, the ability to evaluate this property of the pancreas quantitatively and temporally is one of the central goals of diabetes research. In type 1 diabetes, interventional strategies depend on a clear understanding of whether there is a slow, gradual, or sudden loss of beta-cells, whereas with respect to type 2 diabetes, there is still an active debate whether the observed functional failure is due to altered rates of beta-cell formation or loss. Assessment of beta-cell mass in human diabetes is limited to autopsy studies, which usually provide inadequate clinical information.[7] Indirect determination of beta-cell mass and function is derived through blood sampling for measurement of glucose, glucagon, tolbutamide, or insulin in response to various secretagogues.[8] However, the precise association of these serological parameters with beta-cell mass is unclear, particularly since the threshold of beta-cell loss beyond which these symptoms become apparent is unknown and is likely influenced by multiple factors.

19.2.2 Challenges Associated with the Measurement of Beta-Cell Mass

Considering the therapeutic implications of being able to measure beta-cell mass in intact subjects and in real time, it becomes important to devise strategies toward overcoming the issues associated with beta-cell imaging. As was mentioned earlier, the endocrine pancreas represents 1 to 2% of the volume of the adult pancreas and is scattered as islets of Langerhans throughout the organ.[2] Though beta-cells constitute the majority of the cells in the islets of Langerhans, their proportion might change as the individual ages. Pancreatic islets are organized in a nonrandom fashion and are scattered throughout the acinar tissue of the pancreas.[9] Furthermore, in rodents the pancreas is a diffuse organ difficult to isolate in one plane for planar imaging. From a functional perspective, measuring beta-cell mass represents a challenge since BCM is dynamic with compensatory changes (both expansion and involution) to maintain glucose homeostasis. These changes can be in the number or volume of beta-cells in the islet. For example, beta-cell mass is directly proportional to body weight/body mass index.[10] In the context of diabetes, the dynamic response of the beta-cell to changes in its environment is a major concern, having in mind its exposure to chronic hyperglycemia. With active secretion, granule membrane molecules may be inserted into the plasma membrane to a greater degree, some surface receptors such as GLUT2 and GLP-1 may become downregulated during hyperglycemia, and the cell itself may become degranulated.[11]

19.2.3 Potential Strategies for Imaging Beta-Cell Mass

19.2.3.1 Nuclear Imaging of Beta-Cell Mass

Much has been written about the difficulties of imaging the pancreas, but there are at least four aspects of the pancreas that make it admirably suited to clinical imaging as well as molecular imaging: vasculature, innervation, physiology, and rare metal concentration.

Pancreatic islets are highly vascularized and have a direct arteriolar blood flow. Islet capillaries are fenestrated, and hence are highly permeable. These capillaries, resembling a glomerulus, course through the islet in a tortuous fashion that is ideal for cell–blood and blood–cell interactions. In addition, the blood flow to the islets has been found to be disproportionally large (10 to 20% of the pancreatic blood flow) for the 1 to 2% of pancreatic volume. These features create a favorable environment for delivery of pharmaceutical or imaging agents.[11]

Another distinctive feature of the pancreas is its heavy innervation. Parasympathetic nerves reach the pancreas through the vagus nerve and its branches, which interact with ganglia dispersed throughout the organ. In turn, these ganglia control the exocrine and endocrine pancreas through unmyelinated nerve fibers that secrete neurotransmitters such as acetylcholine and other polypeptides. Pancreatic islets are associated with higher levels of choline acetyltransferase and acetylcholinesterase than the remainder of the pancreas, and the majority of cholinergic activity appears focused on modulating beta-cell insulin secretion, glucose tolerance, and beta-cell proliferation. Consequently, the development of contrast agents that exploit these differential patterns of innervation and their relationship to insulin homeostasis represents a clear avenue to pursue. One example of such an agent is [^{18}F]4-fluorobenzyltrozamicol (FBT), which is a radioactive tracer that binds to the vesicular acetylcholine transporter on presynaptic cholinergic neurons.[12] Nuclear imaging with this tracer demonstrated a remarkably good accumulation in the pancreas of murine and primate models, higher than in any other organ. Furthermore, it was suggested that FBT activity may correlate closely with insulin-producing beta-cell innervation, mass, and function. This is particularly important, since it is believed that cholinergic innervation is reduced in type 1 diabetes.[12]

Another set of markers that could potentially be explored for imaging of beta-cell mass relies on the unique physiology of the beta-cell. For example, the zinc content in the pancreatic beta-cell is among the highest of the body. Zinc appears to be an important metal for insulin-secreting cells, as insulin is stored inside secretory vesicles as a solid hexamer bound with two Zn^{2+} ions per hexamer. Zinc is also an important component of the mechanisms behind insulin secretion. Therefore, beta-cells express specialized zinc transporters, which can be targeted.[13,14]

A logical approach for imaging the endocrine pancreas would involve labeled sugars, since beta-cells are sensitive to changes in blood glucose. Different tracers based on radiolabeled glucose have been explored[15–18] with variable success. Particularly encouraging, however, have been studies with the ketoheptose D-mannoheptulose. D-mannoheptulose is taken up into cells mainly through the GLUT-2 receptor, which is exclusively present on hepatocytes and insulin-producing cells, and therefore displays discriminating capacity between pancreatic islets and exocrine pancreas. In addition, its accumulation in pancreatic islets is proportional to beta-cell mass and can be utilized to assess metabolic status in diabetic animals. Therefore, despite its relatively high accumulation in the liver, mannoheptulose has been proposed as one of the more promising candidate tracers.[19–22]

Numerous attempts have been made to utilize pharmacologic agents for beta-cell imaging. The vast majority of them have met with either partial or no success. Extensive attention has been dedicated to the investigation of sulfonylurea receptor ligands as imaging agents. Insulin secretion is regulated by the membrane potential of the beta-cell, which depends on the activity of ATP-sensitive K^+ channels in the plasma membrane. These channels are composed of a small, inwardly rectifying K^+ channel subunit (Kir6.1 or Kir6.2) plus a sulfonylurea receptor (SUR).[23] SURs represent the target for hypoglycemic sulfonylureas (e.g., glyburide, tolbutamide), a group of well-known antidiabetic agents that have been in clinical use for years.[24] So far, three sulfonylurea receptors have been cloned: SUR1, SUR2A, and SUR2B. SUR1 is expressed at a very high density at the internal face of the plasma membrane[25] of the pancreatic islet cells, but not in the exocrine part of the pancreas. Therefore, a radiolabeled glyburide or tolbutamide derivative could be a feasible tracer for visualizing pancreatic islet cell mass. Different labeled glyburide/tolbutamide derivatives have been tested.[14,26–28] For several reasons, however, including overall radiochemical yield, binding affinity, deposition time at the receptor, uptake in the liver, plasma protein binding, etc., a lot of these agents have been found unsuitable for *in vivo* evaluations.

Alternative tracers have been designed based on a nonsulfonylurea hypoglycemic drug, named repaglinide, which functions similarly to glyburide by inhibiting potassium efflux from beta-cells. This drug is considered superior to glyburide by virtue of its enhanced absorption and elimination profile, as well as its high affinity for SUR1. Initial biodistribution experiments and *in vitro* experiments with ^{18}F- and ^{11}C-labeled repaglinide and its metoxy analog suggest that the improved uptake of repaglinide by the pancreas makes repaglinide and its derivatives promising candidates for *in vivo* imaging by nuclear methods.[29,30]

A strategy that has met with some success involves targeting beta-cell-specific markers with labeled antibodies. One is the monoclonal antibody R2D6 directed against gangliosides on the plasma membranes of pancreatic beta-cells, but not found on other pancreatic cells.[31–35] Nuclear imaging with ^{128}I-labeled R2D6 antibody demonstrated a significantly higher binding to isolated islets than to acinar tissue. However, no differences were detected when comparing pancreatic pieces or isolated islets from diabetic and control animals.[36] Another candidate is a protein on the surface of islet beta-cells recognized by the IC2 antibody.[37] It was shown to be beta-cell specific among islet cells.[37–39] Electron microscopy of IC2-labeled islet beta- cells showed exclusive binding to the surface membrane.[39] The IC2 antibody was labeled with the radioisotope ^{111}In and injected intravenously into healthy and streptozotocin (STZ)-induced diabetic mice. Animals with STZ-induced diabetes showed approximately 50% loss of BCM, with significantly lower accumulation of the probe than normal mice. Furthermore, nuclear imaging of excised pancreas from these animals showed a clear difference in signal intensity between normal and diabetic pancreases, though the weights of these organs were approximately the same.[40] These results demonstrated proof of principle that it is feasible to estimate BCM using the accumulation of antibody-based radiolabeled probes.[11]

19.2.3.2 Magnetic Resonance Imaging of Beta-Cell Mass

Considering the difficulties that nuclear imaging has encountered in trying to image the endogenous beta-cell, questions arise as to the potential of MRI, which is comparatively less sensitive, to accomplish the same goal. A possible way to enhance sensitivity involves binding Gd^{3+} to a protein or other beta-cell-targeting molecule to act as a tissue-specific contrast agent. An interesting approach toward addressing the potential of such complexes for measurement of beta-cell mass on a theoretical basis has been made with respect to employing agents that target the GLUT-2 receptor. By taking the relaxivity values obtained for a particular contrast module isolated by phage display, GdG80BP (R_b = 8.3 mM^{-1}sec^{-1}, R_s + 44.8 mM^{-1}sec^{-1}), and assuming that a binding constant of 10^{-7} could be attained, a theoretical tissue contrast model (TCM) could then be applied to estimate the minimum amount of targeted agent that could be detected by MRI.[41]

Assuming good clearance from adjacent sites, with a concentration of the contrast agent lower than 10 μ*M*, a detection limit of 3.4 μ*M* for GdG80BP is estimated by the TCM.

In addition, assigning a diameter of about 300 μm to pancreatic islets and a 60% beta-cell content by volume, with a diameter of just 8 μm, there would be around 31,000 beta-cells per islet. So if one further assumes that a typical voxel resolution of 1 mm^3 can be achieved, one could expect to find about 71 islets, or 2.23×10^6 beta-cells, per voxel. An estimated detection limit of 3.4 μ*M* in the voxel would require 2×10^{12} molecules of the contrast agent, or 9.2×10^5 binding sites per beta-cell. Although the number of GLUT2 receptors per beta-cell is unknown, this estimate is within the upper concentration range for receptors in other cell types. However, these detection limits could be improved considerably by using low molecular weight multimers of targeted Gd agents (100-fold should be easily achievable) and by further increasing the relaxivity of the targeting moiety (~2-fold). Based on this model, it appears that the application of MRI for the measurement of endogenous beta-cell mass is feasible.[41]

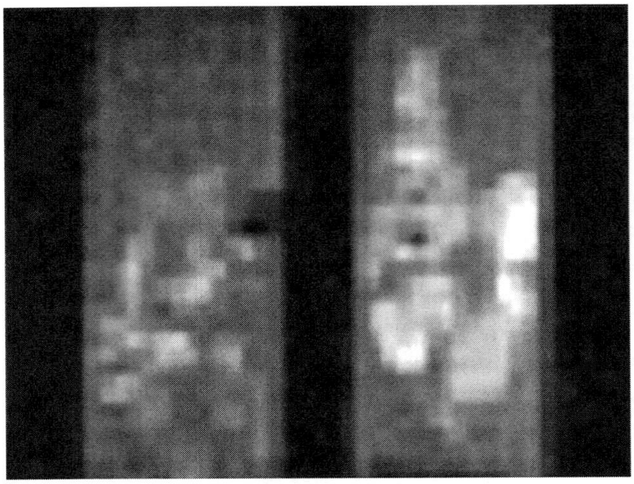

FIGURE 19.1 Mn^{2+}-enhanced T1-weighted contrast of rat pancreatic islets. Two capillary tubes containing pancreatic islets were incubated for 30 min in the presence of 25 m*M* $MnCl_2$. The tubes were incubated in 5 m*M* glucose (left) and 20 m*M* glucose (right). The image was acquired at 500 MHz with TR = 400 msec, TE = 7.2 msec, slice thickness = 100 mm, field of view = 4.8 × 2.4 mm, acquisition matrix = 128 × 64, and number of averages = 64. (Reprinted from Gimi, B. et al., *Magn. Res. Eng.*, 18B, 1, 2003. With kind permission from Wiley InterScience.)

Nevertheless, to date there have been no reports of the *in vivo* detection of beta-cells in the pancreas by MRI. Some preliminary results were reported at the 2003 NIH Workshop on Imaging Pancreatic Beta-Cells. These include *in vitro* studies using ^{1}H nuclear magnetic resonance (NMR) spectroscopy, which has been suggested for measuring choline levels[42] and C-13 spectroscopy for studying glucose-stimulated insulin secretion,[43] as well as the development of novel lipophilic lanthanide contrast agents.[44]

One of the most interesting approaches utilizes Ca^{2+} metabolism by the beta-cell as a tool to deliver MR contrast agent. The influx of Ca^{2+} into beta-cells precedes insulin secretion. Therefore, surrogate paramagnetic ions can be used as Ca^{2+} mimics. One such candidate is Mn^{2+}. Mn^{2+} accumulates in beta-cells in proportion to the glucose concentration presented to the cells, presumably via Ca^{2+} channels. Images collected of isolated pancreatic islets in the presence of an ultrahigh concentration of $MnCl_2$ (25 m*M*) at two different glucose concentrations (5 and 20 m*M*) showed that a clear contrast enhancement is obtained when the stimulatory levels of glucose are higher (Figure 19.1;[45] reviewed in Woods et al.[41]). Another method utilizes conformational changes in the chelates. This method was used to design a Ca^{2+}-sensitive MR agent. This agent uses a heptadentate DO3A substructure to chelate Gd^{3+}, thereby leaving two coordination sites vacant for ligation by water. Two of these units were attached to a 1,2-bis(o-aminophenoxy)ethane-N,N,N',N'-tetraacetic acid (BAPTA) derivative via propyloxy linkers to allow the carboxylates of the BAPTA to coordinate the vacant sites on Gd^{3+}.[41] In the absence of Ca^{2+}, the carboxylates of the central BAPTA unit bind to the water sites on the two appended GdDO3A moieties, thereby excluding water molecules from each Gd^{3+} center. Upon exposure to Ca^{2+}, these same carboxylates are required to bind Ca^{2+} into the central BOPTA moiety and are released from each GdDO3A unit, thereby exposing Gd^{3+} to water and facilitating inner-sphere relaxation (Figure 19.2). Ca^{2+} binding to GdDOPTA occurs in the micromolar range and gives rise to an increase in relaxivity from 3.26 to 5.76 mM^{-1}sec^{-1} (500 MHz, 25°C), an enhancement of some 80%.[41] This agent has been proposed as a natural sensor of Ca^{2+} metabolism and, indirectly, insulin secretion and beta-cell mass and function. An analogous strategy was applied for the design of a Zn^{2+}-sensitive T1 agent.[41,46] In view of the role that Ca^{2+} and Zn^{2+} play in glucose responsiveness and insulin secretion, the development of contrast agents

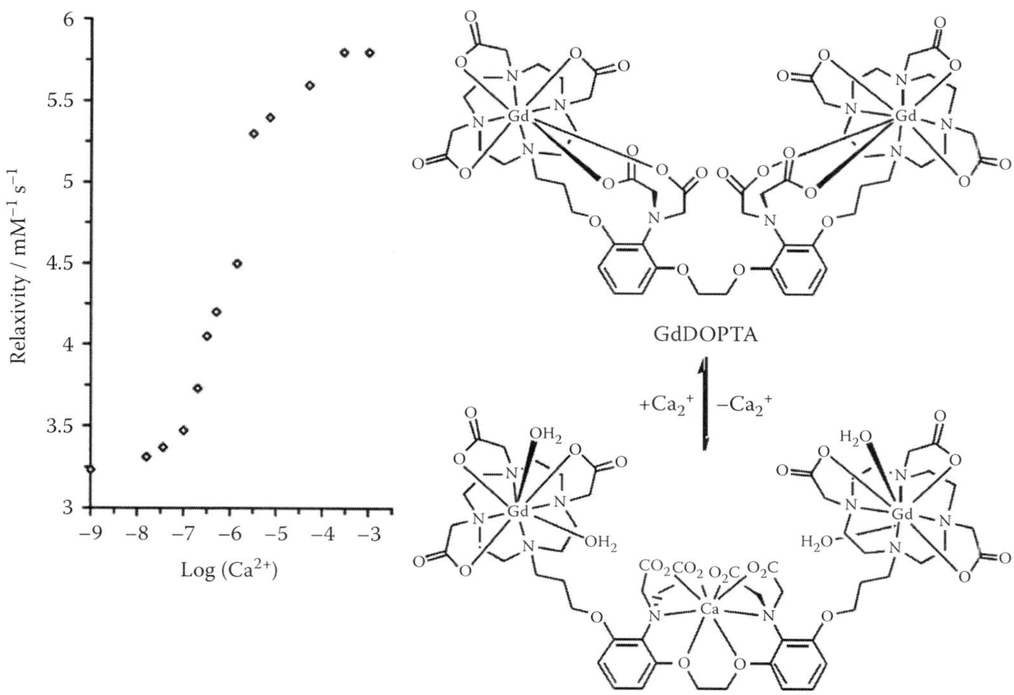

FIGURE 19.2 The binding of calcium(II) by GdDOPTA results in a change of hydration state at the gadolinium(III) centers, "switching on" the inner-sphere relaxivity. (Reprinted from Woods, M. et al., *Curr. Med. Chem. Immun. Endoc. Metab. Agents*, 4, 349, 2004. With kind permission from Bentham Science Publishers, Ltd.)

sensitive to the concentration of these ions would provide invaluable information about the metabolic and functional competence of beta-cells.

An exciting new development in the field of MRI with potential implications for the imaging of endogenous beta-cells is chemical exchange saturation transfer (CEST). CEST is based on the slow exchange of magnetization between low molecular weight diamagnetic compounds, which introduces proton density contrast that can be turned on or off. After irradiating a millimolar pool of metabolite, for instance, the energy is transferred to a nearby pool of water, generating a strong increase in enhancement by means of chemical exchange. An alternative technique, paramagnetic CEST (PARACEST), is based on the unusually slow water exchange of paramagnetic lanthanide complexes that increases the utility of CEST and has the potential to be used as a biologically responsive agent capable of registering molecular exchange phenomena in tissue with superior sensitivity (<1 μ*M*).[47] With CEST and PARACEST, specific metabolites can be detected. Using this method, the concentration of molecules such as glucose can be evaluated noninvasively *in vivo* in all organs. In addition, multiparametric molecular MRI might become feasible, as CEST agents with different absorption frequencies of the exchanging protons can be designed. In the context of beta-cell mass, recent studies have reported on the development of a novel class of PARACEST agents that contain lanthanides, europium, or dysprosium for labeling beta-cells. Very encouraging results were obtained with the glucose-sensitive PARACEST agent, EuDOTA-4AmBBA. Binding of glucose to EuDOTA-4AmBBA led to a slower water exchange in this system. MR images (4.7 T) of phantoms containing different amounts of glucose showed clear differences in signal intensity between the four samples (Figure 19.3). Furthermore, the sensitivity of this system for differences in glucose concentration was within the physiologically relevant range (5 m*M*), opening up new possibilities for the quantification of metabolites in tissue.[41,47]

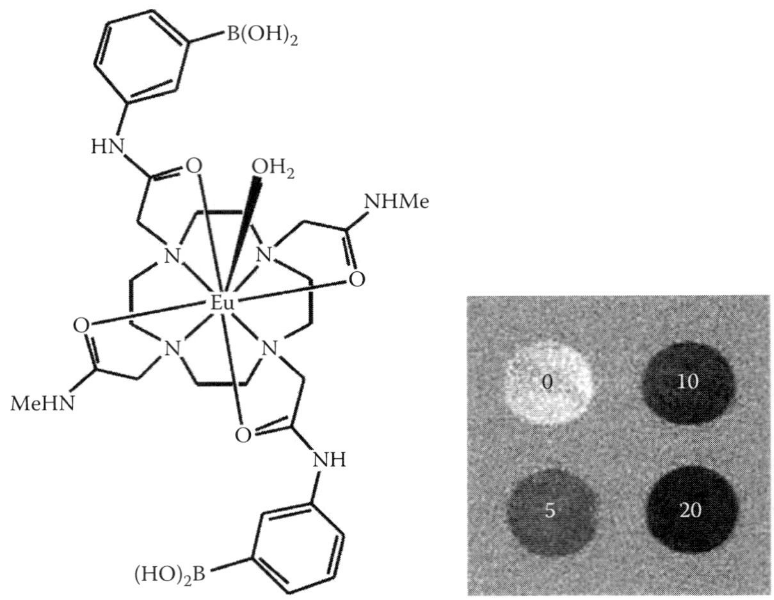

EuDOTA-4AmBBA^{3+}

FIGURE 19.3 CEST images of phantoms containing 10 m*M* EuDOTA-4AmBBA plus either 0, 5, 10, or 20 m*M* glucose, which was obtained by subtracting the image obtained by saturating at 50 ppm from that at 30 ppm. (Reprinted from Zhang, S. et al., *J. Am. Chem. Soc.*, 125, 15288, 2003. With kind permission from the American Chemical Society.)

A novel approach for the measurement of beta-cell mass may include the introduction of a hyperpolarized contrast agent based on an endogenous substance. The function of the injected substance is no longer to modify the amplitude of the signal generated by the protons. Instead, the hyperpolarized nuclei, contained in each molecule, are the source of the NMR signal.[48,49] Nuclear polarization contrast agents greatly increase (hyperpolarize) the weak nuclear polarization of atomic nuclei from parts per million (ppm) toward unity. The inherently weak nuclear polarization at room temperature is responsible for the low sensitivity of MRI, compared with other imaging methods. Hyperpolarization can be achieved using a variety of methods, but recent advances in dynamic nuclear polarization (DNP) have shown tremendous potential for molecular imaging. DNP is based on polarizing the nuclear spins of a contrast medium in the solid state by partly transferring the high electron spin polarization to the nuclear spins by microwave irradiation, and quickly returning the contrast medium to the liquid state. The contrast medium is then administered and MRI acquisitions performed within a short time frame (typically 6-sec transfer, MRI within 1 min), before the hyperpolarization enhancement is lost owing to thermal equilibrium, i.e., T1 relaxation. DNP provides novel contrast to MRI by massively increasing the signal that can be derived from a given number of spins (>100,000×). Therefore, nuclei other than hydrogen can be imaged, e.g., ^{129}Xe, not present in the body, and ^{13}C, which is found endogenously at levels below the detection threshold. Consequently, images of the distribution of the contrast agent can be generated without background. To comply with medical diagnostic needs, however, the T1 of the injected contrast agent needs to be sufficiently long to permit distribution of the substance to target organs before the hyperpolarization has been lost. ^{13}C in small molecular weight substances meets this requirement, since it can have T1 relaxation times longer than 10 sec. Thus, this method could potentially address questions dealing with glucose metabolism, as well as the distribution of small molecular weight endogenous or exogenous contrast agents through the circulation and their homing to the pancreas with high spatial and temporal resolution.[48,49]

19.3 IMAGING OF IMMUNE CELL INFILTRATION

19.3.1 The Role of Immune Cell Infiltration in Diabetes

Autoimmune destruction of beta-cells is primarily a T-cell-mediated process. The most likely sequence of events leading to beta-cell death involves initial acquisition of beta-cell antigens by antigen presenting cells (APCs) in the islets of Langerhans, followed by activation and migration of APCs to pancreatic lymph nodes. Interaction between activated APCs and naïve T-lymphocytes present in the immediately draining pancreatic lymph nodes leads to activation of beta-cell-reactive T-lymphocytes, which subsequently migrate to the islets, where they reencounter beta-cell-derived antigen and are retained.[3] Therefore, overt diabetes, characterized by hyperglycemia, is preceded by an occult phase, termed insulitis, during which leukocyte invasion of pancreatic islets occurs. This process effects the eventual specific destruction of the insulin-producing beta-cells. It is postulated that a loss of more than 90% of the beta-cells takes place before insulin production is no longer sufficient to regulate blood glucose levels, resulting in hyperglycemia.[27] Importantly, this initial stage of insulitis begins a long time before the manifestation of overt symptoms, persists for many years, and progressively decreases after diabetes onset, as beta-cell mass declines.[50,51] Therefore, the early detection and continuous monitoring of immune cell infiltration of the pancreas in real time would represent a significant step toward identifying the initial insult leading to beta-cell destruction and permit effective curative and not just palliative intervention. Furthermore, progress in this specific area would provide a better understanding of the time course of the pathology and possibly help unravel the mechanisms behind its progress.

19.3.2 Potential Strategies for Imaging Immune Cell Infiltration in Diabetes

19.3.2.1 Non-Antigen-Specific Immune Cell Tracking

Limited attempts have been made to diagnose insulitis by pancreatic biopsy.[52] However, the invasiveness of this procedure makes it unattractive in a clinical setting. A logical strategy applicable to this scenario of disease progression would naturally involve immune cell tracking. Cell tracking is a well-established method that, among other scenarios, can be employed for direct *in vivo* visualization of lymphocyte migration and involves cell labeling with contrast agents suitable for detection using a variety of imaging modalities.

Magnetic resonance imaging of cells labeled with iron oxides is probably the most promising modality for cell tracking. Iron oxides are biocompatible and have strong effects on T2-star relaxation. Furthermore, MRI has excellent spatial resolution and permits tracking of labeled cells over prolonged periods, due to the persistence of the label inside cells and the stability of the resultant signal.[53] In particular, superparamagnetic iron oxides have been used for cell labeling by fluid phase endocytosis[54–57] and receptor-mediated endocytosis.[58,59] Improved labeling efficiency has been achieved via derivatization of iron oxide nanoparticles, followed, for example, by the introduction of a membrane translocation signal (HIV-Tat peptide).[60] A variety of cells have been labeled and detected with this compound (CLIO-Tat; cross-linked iron oxide nanoparticles modified with Tat), including normal splenocytes, neuroprogenitor cells, and CD34+ stem cells.[61] More recently, superparamagnetic iron oxide nanoparticles have been used with remarkable success to label purified cells such as neural progenitor cells,[62,63] embryonic stem cells,[64] dendritic cells,[65] bone marrow-derived endothelial precursors,[66] and mesenchymal stem cells,[67–69] to name a few. In addition, high-resolution imaging has been applied to visualize single cells *in vivo*.[70–72] The clinical relevance of these findings is indisputable, considering that in order to develop effective cell therapies, the location and distribution of these cells must be known.

In addition to monitoring cell therapies, cell tracking by MRI has also been employed to reveal disease-related patterns of lymphocyte homing. In particular, MRI has been used to detect the migration of autoreactive lymphocytes in models of multiple sclerosis,[73] AIDS,[74] and cancer.[75]

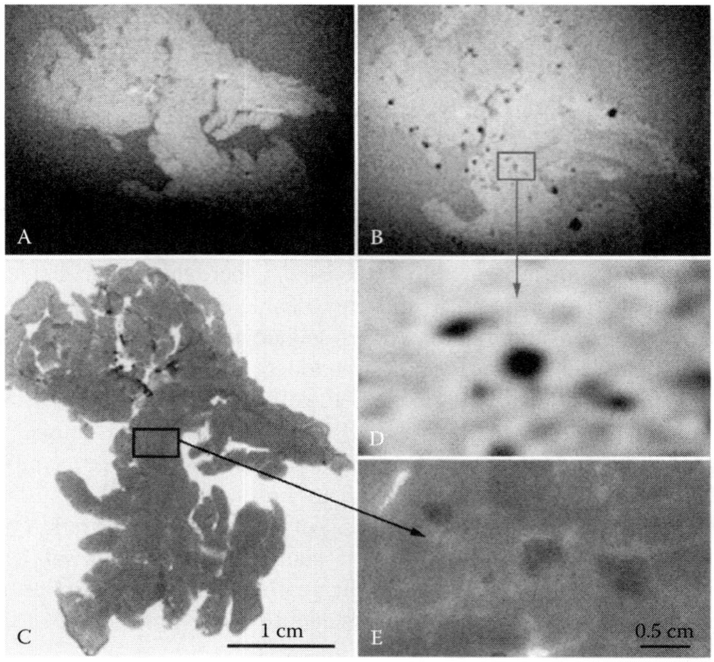

FIGURE 19.4 (please see color insert following page 210) MRI of the pancreas after adoptive transfer of diabetic splenocytes labeled with CLIO-Tat into NOD.scid mice. (A) T1-weighted image showing physiological details of the pancreas. (B) T2-weighted image showing infiltration of labeled cells into pancreatic islets. (C) Dithizone-stained pancreas. Correlation between MR images (D) and staining for DTZ (E) shows the presence of infiltrated cells labeled with CLIO-Tat (dark dots on (D)) in the pancreatic islets (purple DTZ staining on (E)). Magnification bars: A to C, 1 cm; D and E, 0.5 mm. (Reprinted from Moore, A. et al., *Magn. Reson. Med.*, 47, 751, 2002. With kind permission from Wiley-Liss, Inc.)

In the context of diabetes, there are examples of lymphocyte tracking using different modalities. Kaldany et al.[76] demonstrated accumulation of radiolabeled autologous lymphocytes in pancreata of patients with diabetes. Indirect labeling methods have also shown some success. Radiolabeled cytokines, such as IL-2,[77–79] have been used for tracking of autoimmune attack in diabetes by scintigraphy. However, nuclear imaging methods suffer some significant drawbacks, including the short half-life of the signal and potential toxic effects of radiation if long-half-life radiotracers are used. A novel application of optical imaging for lymphocyte detection and monitoring involved fluorescent labeling of T-lymphocytes and the visualization of their homing to islet grafts through a body window device.[80] This investigation will be discussed in more detail with respect to imaging of islet transplantation.

Our studies have focused on lymphocyte labeling for the purposes of their detection in the diabetic pancreas by MRI. In our initial investigation,[81] we labeled splenocytes derived from nonobese diabetic (NOD) mice, as a model of type 1 diabetes, with CLIO-Tat and adoptively transferred them into immunocompromised mice, lacking mature lymphocytes. Our hypothesis was that we could visualize the accumulation of these autoreactive clones in the pancreata of recipient mice. MR imaging was performed on isolated islets using high-field 14-T MR microscopy and on excised pancreas using a 1.5-T clinical MR imaging system. Both methods demonstrated T2 changes consistent with accumulation of magnetically labeled cells in recipient islets (Figure 19.4).

More recently, Billotey et al.[82] corroborated these findings by labeling CD3+ T-lymphocytes with a novel class of anionic magnetic nanoparticles (AMNPs) and monitoring their homing to the pancreas of NOD mice *in vivo* for up to 20 days after adoptive transfer of labeled cells (Figure 19.5).

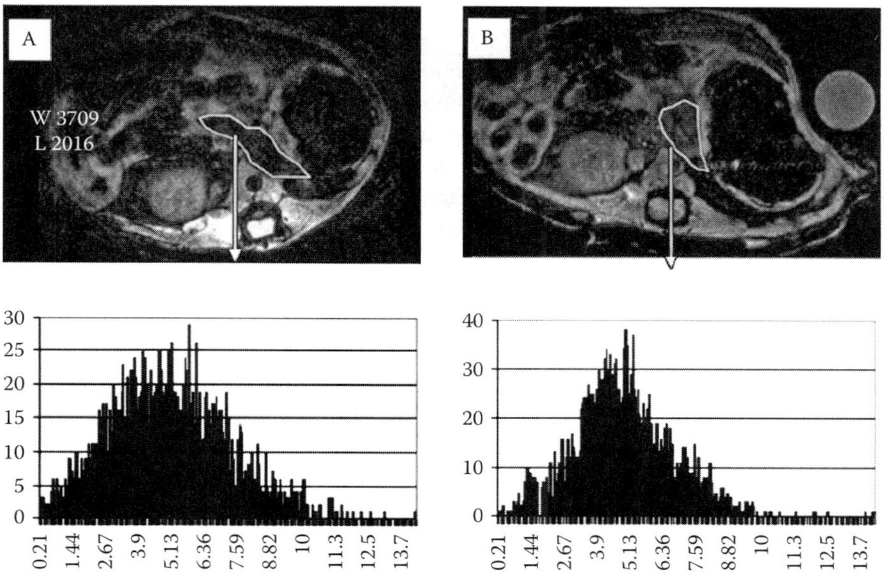

FIGURE 19.5 *In vivo* MR images and histograms obtained in a mouse 20 days after transfer. Transverse views (25/10) centered on pancreas in a mouse transferred with labeled (A) and unlabeled (B) T-cells. The pancreas is obviously much darker in A than in B, which is confirmed by the discrepancies on the histograms. Water phantoms are visible alongside B. (Reprinted from Billotey, C. et al., *Radiology*, 236, 579, 2005. With kind permission from the Radiological Society of North America, Inc.)

19.3.2.2 Antigen-Specific Strategies for Imaging Immune Cell Infiltration in Diabetes

These initial studies confirmed the feasibility of tracking the migration of autoreactive T-cell clones at the earliest stages of insulitis in an adoptive transfer model. However, a fundamental limitation of this and other labeling techniques in the context of autoimmunity is that they are not cell specific and do not allow us to discriminate between autoreactive and irrelevant lymphocyte specificities.

In our search for markers that characterize the autoreactive lymphocyte population, we derived knowledge relevant to the identification of the specific T-cell subpopulations effecting the incipient beta-cell assault in type 1 diabetes. It has been proposed that autoimmune destruction of beta-cells is triggered by an initial insult by CD8+ T-cells, which causes the shedding of beta-cell autoantigens, their subsequent loading onto antigen-presenting cells, and the activation of an autoreactive cascade of events leading to overt diabetes.[83–85] Although the nature of the CD8+ T-cell subpopulation that contributes to the initiation of autoimmune diabetes remains largely unknown, several lines of evidence suggest that this subpopulation is dominated by clonotypes expressing Vα17-Jα42 T-cell receptor α (TCR-α) chains, specific for a major histocompatibility complex (MHC) I-restricted family of peptides. These clonotypes are present at a very high frequency in the peripheral blood of young NOD mice.[86] They also constitute a large fraction of the CD8+ cells that can be propagated from the earliest insulitic lesions of NOD mice.[85] Moreover, they are highly diabetogenic in T-cell receptor (TCR) transgenic mice[83,84] and undergo avidity maturation during the progression of insulitis to overt disease in wild-type NOD mice.[87,88] Promising data have been obtained recently in identifying the beta-cell antigen (islet-specific glucose-6-phosphatase catalytic subunit-related protein, or IGRP) targeted by the Vα17-Jα42 TCR-α lymphocyte population.[89] The relevance of these findings to diabetes in humans is underscored by the fact that the human IGRP gene maps to a diabetes susceptibility locus, raising the possibility that IGRP may also be the target of the human diabetogenic response.

In an attempt to devise a target-selective approach for tracking immune cells to the diabetic pancreas, we engineered a new generation of magnetic imaging probes capable of labeling the

Vα17-Jα42 TCR-α autoantigen-specific population of lymphocytes. The basis of the recognition was the presence of the so-called NRP-V7 peptide, similar to $IGRP_{206-214}$, which is bound by the autoreactive CD8+ T-cell population with high avidity, when presented in the context of the relevant MHC I.[87] These probes were designed by conjugating NRP-V7 peptides, complexed with the appropriate MHC I, to superparamagnetic iron oxide nanoparticles[88] (Figure 19.6A). The resulting agent, termed CLIO-NRP-V7, specifically reacted with CD8+ T-cells from transgenic NOD mice (8.3-NOD mice),[84] expressing a highly diabetogenic TCR, and did not react with CD8+ T-cells from nontransgenic healthy NOD mice. Imaging experiments performed in an adoptive transfer model using the CLIO-NRP-V7 probe allowed for visualization of the accumulation of labeled cells up to 16 days after their transfer into a live animal (Figure 19.6B). Currently, studies are under way in our laboratory to adapt our T-cell-specific probe for direct labeling of endogenous diabetogenic immune cells after systemic administration.

These studies represent the first step in the developing field of *in vivo* cell tracking of diabetogenic immune cells to the pancreas. Providing that similar mechanisms of autoreactive T-lymphocyte selection and homing to pancreatic islets operate in humans, a natural direction of future efforts should be the development of systemic target-specific agents that would allow for the visualization of prevalent populations of autoreactive T-cells in humans noninvasively. The clinical repercussions of such an approach are significant, not only by virtue of its imaging implications, but also its therapeutic potential. Administration of autoantigenic proteins or peptides in solution can blunt the initiation or progression of autoimmunity in experimental models of autoimmune disease.[90–94] In particular, repeated treatment of prediabetic NOD mice with soluble NRP-A7, which is a lower-affinity peptide counterpart to NRP-V7, blunted avidity maturation of the $IGRP_{206-214}$-reactive CD8+ subset and afforded near complete protection from diabetes.[87] Therefore, it is feasible that delivery of tolerogenic peptide/MHC complexes as part of multivalent structures, such as CLIO-NRP-V7, would allow not only the imaging of pancreatic inflammation in real time, but also the delivery of tolerogenic doses of many epitopes simultaneously, enabling the depletion of different autoreactive T-cell pools (such as those recognizing most IGRP epitopes) below the threshold required for diabetogenesis (P. Santamaria, unpublished data).

19.4 IMAGING OF APOPTOSIS

19.4.1 The Role of Apoptosis in Diabetes

Evidence of beta-cell death exists in cases of both type 1 and type 2 diabetes. However, whereas in type 1 diabetes the loss of beta-cells is initiated by an autoimmune attack, the underlying causes behind beta-cell destruction in type 2 diabetes seem to be metabolic. Importantly, current theories about the origin of type 1 diabetes implicate a neonatal wave of beta-cell apoptosis in the process of early beta-cell remodeling as the main trigger of autoimmunity,[95–97] whereas in type 2 diabetes, it appears that beta-cell apoptosis can be triggered by the buildup of toxic levels of metabolic products, such as glucose[98–100] and free fatty acids.[101–104]

Current evidence of diabetes-associated beta-cell loss is exclusively collected *ex vivo*, based on morphologic changes associated with apoptosis, nuclear fragmentation, and activation of the caspase apoptotic cascade. All of these methods, however, represent frame-by-frame *in situ* approaches, and therefore have been ineffective at quantifying the rate of beta-cell loss due to the rapid clearing of dying beta-cells, as well as the long time course of the process. A promising alternative to these strategies would be the development of *in vivo*-delivered probes for the detection of apoptosis in intact animals and in real time. Such probes would permit the tracking of the time course of diabetes-associated beta-cell loss and would expand the range of tools available for study and ultimately for early detection of the disease.

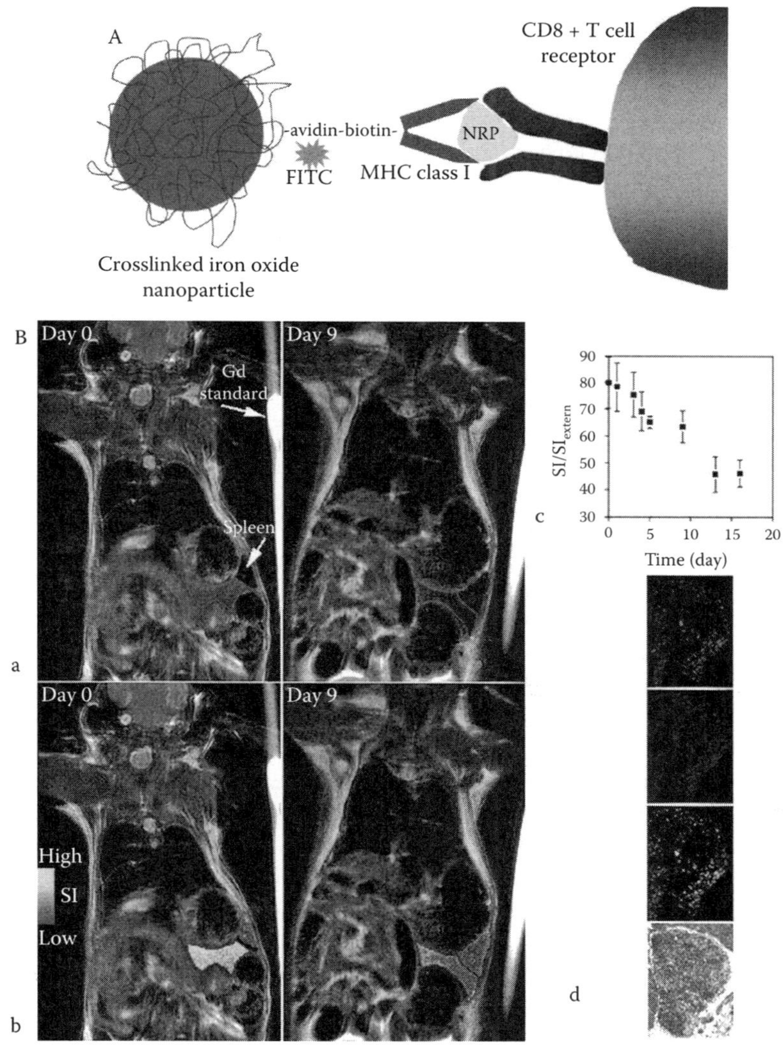

FIGURE 19.6 (please see color insert) (A) Schematic representation of CLIO-NRP probe. A biotinilated NRP-V7/H-2Kd complex was coupled to CLIO particles modified with FITC-avidin. The CLIO-NRP-V7 probe is recognized by the T-cell receptor on NRP-V7-reactive CD8+ T-cells. Note that there are four binding sites for biotin on avidin, and hence for the biotinylated peptide–MHC complex. (B) *In vivo* imaging of pancreatic infiltration of CD8+ T-cells labeled with CLIO-NRP-V7 and adoptively transferred into 5-week-old NOD.scid mouse (a). Color-encoded signal intensity of the spleen and the tail of the pancreas (b). The decrease of signal intensity from day 0 to day 16 in the pancreas (outlined in red) is due to the accumulation of CLIO-NRP-V7-labeled CD8+ T-cells (c). Dual-channel fluorescence microscopy (d) of the excised pancreas after the 16th day of MR imaging. Green channel shows the presence of CLIO-NRP-V7-labeled cells in the islet; staining with anti-CD8-PE antibodies (red channel) confirmed the presence of transferred cells. Superimposition of two images shows that CLIO-NRP-V7-labeled cells were also positive for the CD8+ T-cell marker. Histological hematoxylin and eosin (H&E) stain of the same islet (consecutive section) shows the infiltration of mononuclear cells in the islet (20× magnification). (Reproduced from Moore et al., *Diabetes*, 53, 1459, 2004. With kind permission from the American Diabetes Association.)

19.4.2 Potential Strategies for Imaging Diabetes-Associated Apoptosis

19.4.2.1 The Annexin V Experience

Recently, we have described the application of an apoptosis-specific imaging probe, annexin V-Cy5.5, in murine models of type 1 and type 2 diabetes.[105] This probe takes advantage of the externalization of phosphatidylserine (PS), an aminophospholipid normally found exclusively on the cytoplasmic leaflet of the plasma membrane. The externalization of PS, which occurs early in the process of apoptosis, makes it available for recognition by exogenous annexin V, a naturally occurring calcium-dependent 35-kDa protein that binds with high avidity (Kd = 1 to 10 M^{-1}) to membrane-associated aminophospholipids.[106,107] Annexin V labeled with a fluorescent tag is routinely used for histological and cell-sorting studies to identify apoptotic cells.[108] *In vivo*, annexin has been conjugated to technetium-99m for scintigraphy in a variety of disease models, including cancer,[109,110] heart disease,[111] and immune rejection.[112] A probe identical to the one employed by us has been applied for near-infrared optical imaging of chemotherapy-induced tumor cell death.[113,114]

The success of imaging apoptosis using labeled annexin V and the fact that PS externalization is an early apoptotic event prompted us to investigate the potential of a Cy5.5-conjugated annexin V probe for the detection of beta-cell death in diabetes. In our studies, we demonstrated the successful localization of annexin V-Cy5.5 to apoptotic cells in diabetic islets, following intravenous administration, using several model systems of type 1 and type 2 diabetes. Our results established the feasibility of utilizing fluorescently labeled annexin V for detection of the earliest stages of diabetes-associated beta-cell death.

The relevance of our findings to magnetic resonance imaging becomes apparent when considering the fact that annexin V labeled with superparamagnetic iron oxides (annexin V-CLIO) has shown promise as an apoptosis imaging agent in tumor cells.[115] In that study, tumor cells treated with the pro-apoptotic agent camptothecin were incubated with annexin V-CLIO, resulting in a significant signal decrease in phantom MRI experiments even at very low concentrations of magnetic substrate. More recently, annexin V-CLIO-Cy5.5 has been applied *in vivo* to a model of cardiomyocyte apoptosis.[116] Intravenous injection of this nanoparticle into mice subjected to transient coronary artery occlusion resulted in reduction in T2* relaxivity values. In another study, superparamagnetic iron oxides were conjugated to the C2 domain of synaptotagmin I, which binds to anionic phospholipids in cell membranes.[117] This conjugate successfully detected cell death *in vitro* in isolated apoptotic tumor cells, and *in vivo* in a tumor treated with chemotherapeutic drugs. Having in mind the demonstrated applicability of fluorescently labeled annexin V to diabetes imaging and the encouraging results using iron oxide conjugates for the molecular imaging of apoptosis by MRI, it appears that the development of similar strategies in the context of diabetes-associated cell death represents a promising avenue to pursue in the future, particularly since levels of apoptosis are comparable in these models.[105,116] In fact, the feasibility of this approach is underscored by the earlier observation that an increase of as much as 30% in T2* relaxation (1/T2*) can be effected at levels of apoptosis around 3%.[116]

19.4.2.2 Imaging Diffusion as a Marker of Cell Viability

As indicated earlier, the pancreatic islet is a highly vascularized structure, receiving 10 to 20% of the blood flow to the pancreas, despite the fact that it only represents 1 to 2% of the pancreatic volume.[5] Imaging methods can be developed to detect blood flow, volume, or vessel permeability, as indicators of islet viability and function.[6] Again, valuable lessons can be derived from tumor imaging, where another MRI category, namely, dynamic contrast-enhanced MRI (DCE-MRI), has been used extensively to evaluate vascular volume and permeability.[118–120]

Furthermore, in tumor models, it has been demonstrated that MRI can provide information about cell viability in the absence of contrast agents. MRI is sensitive to changes in tissue water. Water bound to macromolecules has a very short T2, whereas free water has a long T2. Therefore, changes in the proportion of free and bound water can be detected on T2-weighted images. As a result, cell swelling is associated with an increase in local T2, whereas cell shrinkage linked to apoptosis appears to mediate a drop in T2.[121] Alternatively, diffusion-weighted MRI (DW-MRI) can be used to monitor apoptosis *in vivo*. There are convincing reports of an increase in the apparent diffusion coefficient (ADC) in models of chemotherapy-induced cell death.[122–124] This effect correlates with a decrease in cellular density and an expansion of the extracellular compartment and is consistent with shrinkage of apoptotic cells.

19.4.2.3 Imaging Oxygen Consumption as a Marker of Cell Viability

Finally, novel methods have been developed for the assessment of oxygen consumption by pancreatic islets as a marker of metabolic activity and viability. A recent approach involves the measurement of glucose-stimulated cytochrome C reduction and oxygen consumption in a flow culture system.[125,126] That investigation takes advantage of the differential decay of the phosphorescent emission from oxygen-sensitive dyes. In addition, it has been shown that evaluation of islet respiration through polarographic oximetry is representative of viability and insulin secretion.[127] The results of these studies are encouraging. Still, they do not explore the potential for direct quantitation of viability *in vivo*. Their applicability to magnetic resonance imaging is unclear at this point. However, the conception of strategies for the assessment of oxygen consumption and other markers of metabolic activity in islets is an alternative approach that warrants attention, particularly considering the success of multimodal imaging strategies, combining MRI with near-infrared fluorescence.[128,129]

19.5 IMAGING OF ISLET TRANSPLANTATION

19.5.1 The Promise of Islet Transplantation: Current Status and Challenges

Islet transplantation has emerged as one of the most promising new treatments for diabetes. The earliest reports of islet transplantation in animal models appeared in 1972.[130] The first clinical islet allograft was performed in 1974.[131] Over the next 25 years, numerous attempts were made to achieve normoglycemia in type 1 diabetic patients using islet transplantation, with limited success. Although some functionality was achieved, as evidenced by C-peptide, the majority of cases failed to demonstrate insulin independence or long-term engraftment.[131] With the development in 1999 of an improved protocol for islet transplantation by a group from the University of Alberta, it became possible to attain reproducible success in terms of insulin independence through islet transplantation.[132] The success of this new protocol, named the Edmonton protocol, rests mainly on enhanced immunosupression treatment compared to previous protocols and on the improved islet delivery strategy, namely, intraportal infusion of freshly isolated islets, followed by a second or third infusion of additional islets.[131]

Currently, intraportal infusion is the only method pursued in human trials and proven to lead to insulin independence. As of 2003, there were a total of 282 sites throughout the world actively engaged in islet transplantation. The long-term functionality and viability of transplanted islets are unknown. In a 5-year follow-up report from this group it is emphasized that the results, though promising, still point to the need for improving islet engraftment and preserving islet function.[133] One of the most critical points that need to be resolved, however, is the viability and long-term functionality of the transplant. Even in optimal conditions and in the absence of graft rejection, approximately 60% of transplanted islet tissue is lost 3 days after transplantation, due to cell death.[134,135] Despite the 80% success rate of the Edmonton protocol at 1 year posttransplantation, by 2 years, this number may decline to approximately 65%.[136]

It is unclear at this point to what extent this decline in insulin independence is the result of islet loss and what are the mechanisms behind the deterioration of graft function. Some possibilities include auto- or alloimmune destruction, immunosuppressive toxicity, or a progressive disruption of insulin secretion.[137] Furthermore, since posttransplantation islet mass is a critical element of the potential for glucose stabilization and insulin independence in islet graft recipients, some important relevant questions need to be addressed. For example, it remains unresolved why, even when insulin independence is attained, there are signs of marginal islet mass. The time-course of loss of islet mass in the transplant is also unknown. Successful monitoring of the stability and functionality of the graft would also permit us to test the effectiveness of various immunosuppressive regimens, as well as islet delivery strategies, and ultimately assist further optimization of the islet transplantation procedure.

In addition to these questions, almost nothing is known about the structure and function of islets following transplantation. In a normal pancreas, islet microvasculature is complex. It consists of arterioles splitting into capillaries within the islet core, leaving the islet through the islet mantle and draining into the portal system. During transplantation, islet microvasculature is disrupted. Virtually no information is available concerning islet revascularization and function following transplantation. Besides the observed blood glucose normalization, very little is known about other aspects of islet function, i.e., the participation of glucagon, somatostatin, and pancreatic polypeptide in the regulation of metabolism.[138]

19.5.2 Current Strategies for Islet Graft Assessment

Clearly, an effective approach to islet graft assessment following transplantation is urgently needed. However, currently there are no clinically relevant tools to achieve that goal. The functionality of the graft is evaluated by indirect methods. The major criteria of graft function include indicators of metabolic control, namely, fasting and stimulated glucose levels, oral glucose tolerance testing (OGTT), C-peptide levels, HbA1c levels, mean amplitude glycemic excursions (MAGEs), and insulin secretion. In addition, complications associated with introduction of the graft are assessed by both means of evaluating symptoms of rejection, such as glutamic acid decarboxylase (GAD) antibodies and islet cell autoantibodies, and signs of toxicity or impairment of liver function.[139]

All of the strategies listed above rely on indirect assessment of metabolic parameters. The mechanisms behind islet function, however, represent a finely tuned network of molecular interactions. Because islets upregulate their insulin production in response to need, abnormalities in glucose, C-peptide, and insulin release do not become apparent until most islets have already been destroyed.[140–142] Therefore, these parameters only provide information on the late stage of graft rejection. In view of this, it is critical to develop noninvasive imaging techniques that can determine transplant efficacy and assess its functional status and viability after transplantation.

Direct morphological assessment of transplant fate is currently performed through liver biopsy. From an imaging perspective, modalities such as ultrasound, computerized tomography (CT), and MRI have been used to identify abnormalities in liver tissue following transplantation. Several reports have established the occurrence of hepatic steatosis, or fatty liver, following islet transplantation.[143–145] All of these diagnoses are based on the detection of intrahepatic lesions on CT, MR, and ultrasound scans. The association between graft function and liver steatosis, however, is highly controversial. Markmann et al. concluded that despite the presence of steatosis, graft function at the time of imaging was good, based on the metabolic criteria described above. Since the dissemination of these lesions appears to match the distribution of infused islets, it has been postulated that the appearance of steatosis is the result of local insulin secretion by the graft.[145] It has also been speculated that the appearance of steatosis reflects an abnormal local utilization of insulin, which may indicate or result in graft dysfunction.[143,144]

19.5.3 Potential Strategies for Imaging Transplanted Islets

The strategies employed in these studies do not image islet grafts directly and only provide a functional footprint of graft performance as an indirect tool to follow graft survival over time.

Furthermore, the association between hepatic steatosis and insulin secretion is not absolute. All three techniques are nonspecific and can be affected by various processes, such as excess glycogen accumulation, edema, inflammation, etc.[146] It is still a subject of debate whether steatosis reflects abnormal graft function or is simply a benign consequence of local insulin secretion. It is crucial, therefore, to develop direct methods for imaging of the graft to assess its stability over time, as well as to devise novel, noninvasive, reliable strategies for evaluation of graft viability and function.

The earliest reports of the noninvasive imaging of transplanted islets utilized the bioluminescence optical imaging modality.[147–150] These experiments established proof of principle that isolated rodent or human islets can be genetically engineered to express luciferase without affecting their function and be imaged following transplantation in immunocompromised mice using a cooled charge-coupled device. Since the resulting signal was proportional to the transplanted islet dose, it was suggested that this strategy can be used to noninvasively evaluate islet mass. In addition, the feasibility of long-term monitoring of islets transplanted under the kidney capsule was tested using recombinant lentivirus vectors expressing luciferase under the cytomegalovirus (CMV) promoter. In these studies, luciferase signal emanating from the graft remained stable for at least 140 days, indicating graft stability in this model of transplantation.[149] More recently, Fowler et al.[150] extended the application of bioluminescence imaging to tracking the fate of transplanted islets at the hepatic site, which is the only clinically applied location for islet transplantation.

An interesting approach to imaging early posttransplantation events utilized fluorescence optical imaging. In this investigation, transgenic mice were engineered to express proinsulin II tagged with a live-cell fluorescent reporter protein, Timer.[80] Timer protein is unique because it changes color from green to red in the first 24 h after synthesis. Therefore, insulin synthesis can be easily monitored through the changes in fluorescence over time. Islets from these mice were transplanted into the recipient animals, and the body window device was put in place to monitor insulin synthesis as well as the migration and replication of injected T-cells to these islets. Using this technique, the imaging of both insulin-producing cells and T-cells may be carried out repeatedly for a week or more without repeated surgery, while preserving the life of the studied animal. The combination of Timer monitoring of insulin production with a body window technique holds promise for future investigations that involve the tracking of T-cell graft infiltration, and islet behavior and insulin secretion after transplantation.

Another approach utilized a gene expression reporter labeled with a radiotracer to monitor islet grafts using a different modality — positron emission tomography (PET). Islets were infected with an Adeno-Tkm adenovirus engineered to express a mutant herpes simplex virus type 1 thymidine kinase driven by the CMV promoter.[151] However, the high uptake of its substrate, [^{18}F]FHBG, in the liver may hamper its application in islet imaging.[152] Additionally, adenovirus directs only the transient expression of reporter genes. A major issue in applying the listed imaging methods, however, is that they are based on transgenic modification and represent a highly artificial imaging strategy, lacking an equivalent applicable to human studies and thus preventing direct translation into clinical use.[153]

A clinically relevant approach for the imaging of labeled transplanted islets was utilized to visualize posttransplant early events by positron emission tomography (PET).[154] In this study, rodent islets labeled with 2-[^{18}F]fluoro-2-deoxy-D-glucose (FDG) were implanted in the livers of syngeneic rats and monitored for up to 6 h. A significant problem associated with this method, however, is the short half-life of ^{18}F (110 min), which considerably limits its applicability in a clinical or research setting.

Having these examples in mind, it appears that one of the most logical modalities to explore for imaging of pancreatic islets is MR imaging. MRI does not utilize ionizing radiation, has tomographic capabilities, can deliver the highest-resolution images *in vivo*, and has unlimited depth penetration. Native MR imaging has a low overall sensitivity and is not likely to be capable of distinguishing pancreatic islets from surrounding tissue. However, this drawback can be overcome by the application of contrast agents. As mentioned previously, superparamagnetic iron oxide nanoparticles represent a negative contrast agent whose strong magnetic moment makes them an excellent imaging tool. Nanoparticle presence in tissue is evident primarily by a darkening effect on T2- or T2*-weighted MR images. Remarkably, the T2* effect of superparamagnetic iron oxides results in a hypointensity footprint many times larger than the labeled entity, in essence constituting a powerful signal amplification tool.

As a first step toward unraveling the reasons behind transplanted islet loss and improving transplantation efficiency and graft survival, we recently developed a method for the noninvasive detection and tracking of pancreatic islet grafts by dual-modality fluorescence/magnetic resonance imaging.[129] In that study, we labeled isolated human pancreatic islets with a superparamagnetic iron oxide magnetic nanoparticle (MN) modified with the near-infrared fluorescent Cy5.5 dye (MN-NIRF) and transplanted them under the kidney capsule in immunocompromised mice.

In our initial experiments, we estimated the *in vivo* detection threshold of this method and confirmed the direct correlation between T2 relaxativity in the graft and the number of transplanted islets, corroborating the utility of our strategy for the noninvasive measurement of transplanted islet mass (Figure 19.7C).

In addition, we established proof of principle that our method could be used to track labeled transplanted islets for a substantial period of time, with the goal of monitoring graft longevity. For this purpose, we implanted human pancreatic islets labeled with MN-NIRF under the left kidney capsule of nude mice and unlabeled islets in the right kidney. Imaging was performed up to 188 days after transplantation, demonstrating a significant difference in T2 relaxation times between labeled and unlabeled grafts for the duration of the experiment, as well as preservation of graft-associated T2 relaxation times, reflecting graft stability and persistence of the label (Figure 19.7A and B). Furthermore, we confirmed the functional integrity of the graft in NOD.scid mice (Figure 19.7D) treated with the beta-cell cytotoxic agent streptozotocin. In these mice, labeled grafts were able to restore normoglycemia as efficiently as the unlabeled counterpart.[129]

In order to bring our findings to testing in clinical trials, we employed a preclinical model of islet transplantation at the hepatic site by intraportal infusion, since currently this is the clinically approved protocol. In addition, for islet labeling we utilized the FDA-approved commercially available contrast agent Feridex, which is used clinically for liver imaging. Human pancreatic islets or islet clusters labeled with Feridex could be easily visualized on T2* images as distinct foci of signal loss in the liver parenchyma (Figure 19.8). Furthermore, we were able to track the fate of intrahepatically transplanted human islets during the early posttransplantation period in NOD.scid (immunodeficient) and Balb/c (immunocompetent) mice. We demonstrated a gradual decrease in islet number in both models, with significantly more pronounced loss in immunocompetent mice due to immune rejection.[155] Generally, islet loss after transplantation is associated with immune rejection (in allogeneic transplants), mechanical injury, ischemia, non-alloantigen-specific inflammatory events in the liver after transplantation, and recurrent autoimmunity.[156]

We believe that these new results hold the unique advantage of being directly translatable to a clinical scenario. Potential benefits that could be derived from these studies include the collection of detailed spatial and temporal information regarding location, quantity, and viability of transplanted islets; resolution of many of the dilemmas associated with graft functionality, including the association between transplanted islet mass and insulin independence; elucidation of the mechanisms behind graft failure, as well as monitoring therapeutic intervention during graft rejection; and design of optimized immunosuppressive regimens and transplantation protocols. We are confident that MRI is the modality best suited to address these questions in a clinically relevant context.

19.6 FUTURE OUTLOOK

The capacity for direct cellular imaging of the beta-cell is the first step toward unraveling the etiology and pathophysiology of diabetes and monitoring therapy, particularly islet transplantation.

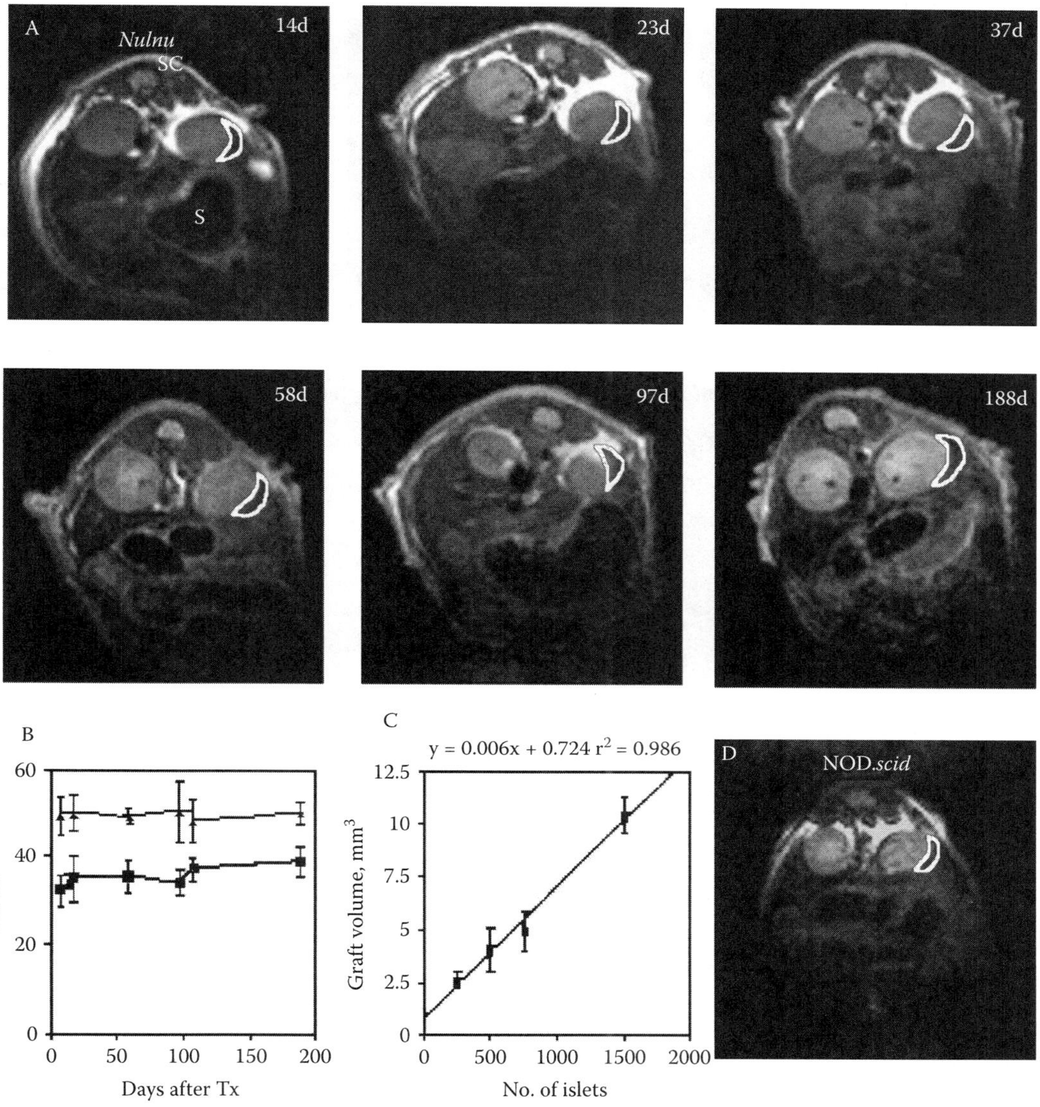

FIGURE 19.7 *In vivo* MRI of islet transplantation under the kidney capsule. (A) Transverse T2-weighted magnetic resonance images of transplanted labeled and nonlabeled human islets 14, 23, 37, 58, 97, and 188 days after transplantation under the kidney capsule in nude (*nu/nu*) mice. The dark area in the left kidney represents a labeled graft. No darkening was reported for the right kidney with unlabeled graft. S, stomach; SC, spinal cord. (B) Analysis of T2 relaxation times of labeled (squares) and unlabeled (triangles) grafts. There was no significant change in T2 values over the period of 188 days in either graft ($p = 0.14$). There was a significant difference in T2 values between labeled and unlabeled grafts ($p = 0.0003$). (C) Detection threshold *in vivo*. There was a direct linear correlation ($r^2 = 0.986$) between the number of islets in the graft and graft volume. (D) A representative magnetic resonance image of a streptozotocin-induced diabetic NOD.scid mouse 2 weeks after transplantation. The delineated areas are associated with a decrease in signal intensity in the left kidney graft on all images. Mice were prone positioned in the magnet with feet first. (Reproduced from Evgenov et al., *Nature Medicine*, 12, 144, 2006. With kind permission from *Nature Medicine*.)

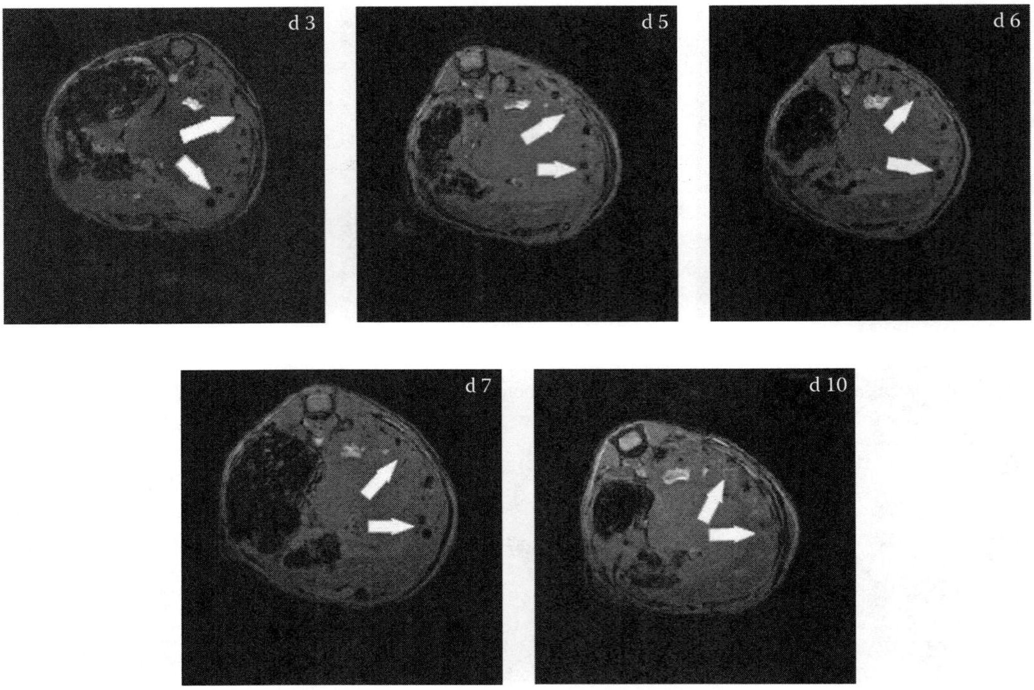

FIGURE 19.8 *In vivo* MRI of intrahepatic transplantation of labeled human islets in NOD.scid mice. A representative time course imaging slice showing islets scattered throughout the liver. Pancreatic islets (0.5 mm slices) appear as hypointense spots on T2*-weighted images (arrows). Islet number declined during the initial posttransplantation period in all animals studied (n = 5).

With the goal of curing diabetes in mind, ongoing and future imaging efforts would most likely involve the identification of biomarkers for the disease, the utilization of these biomarkers in clinically applicable scenarios for early diagnosis, and the image-guided advancement of therapies and improved care. Currently, numerous innovative approaches are being developed, including MRI, PET, optical imaging, gene delivery, bioengineering, and cell targeting. A significant effort has been mounted to discover unique surface markers for the adult and developing beta-cells and provide unique ligands. The location of the pancreas, the fact that beta-cells comprise a small proportion of its volume, and their scattered distribution in the tissue make the task of imaging the diabetic beta-cell and its morphology and function a particular challenge.

The current goals of diabetes cellular imaging can be summarized as follows:

1. Noninvasive molecular imaging of beta-cell morphology, function, and reserve, and their alteration in diabetes
2. Noninvasive measures of beta-cell neogenesis, regeneration, turnover, and apoptosis in the healthy and the diabetic pancreas
3. Noninvasive imaging of pancreatic islet inflammation or autoimmune attack
4. Imaging of transplanted islet engraftment, function, vascularization, and innervation

Although this list is not exhaustive, we believe that it outlines some general directions for the future. The combined progress in these areas certainly has the potential to transform this serious and deadly disease, for which there is currently no established treatment, into a condition amenable to effective curative intervention through early diagnosis.

REFERENCES

1. National Diabetes Fact Sheet, American Diabetes Association, http://www.diabetes.org/diabetes-statistics/national-diabetes-fact-sheet.jsp.
2. Butler, A.E. et al., Beta-cell deficit and increased beta-cell apoptosis in humans with type 2 diabetes, *Diabetes*, 52, 102, 2003.
3. Mathis, D. et al., Beta-cell death during progression to diabetes, *Nature*, 414, 792, 2001.
4. Weir, G.C. and Bonner-Weir, S., Five stages of evolving beta-cell dysfunction during progression to diabetes, *Diabetes*, 53 (Suppl 3), S16, 2004.
5. Bonner-Weir, S. and Orci, L., New perspectives on the microvasculature of the islets of Langerhans in the rat, *Diabetes*, 31, 883, 1982.
6. Laughlin, M., Why image the pancreatic beta cell? *Curr Med Chem Immun Endoc Metab Agents*, 4, 251, 2004.
7. Weir, G. et al., Islet mass and function in diabetes and transplantation, *Diabetes*, 39, 401, 1990.
8. McCullich, D. et al., Correlation of *in vivo* beta-cell function test with beta-cell mass and pancreatic insulin content in streptozocin-administered baboons, *Diabetes*, 40, 673, 1991.
9. Bonner-Weir, S., Anatomy of the islet of Langerhans, in *The Endocrine Pancreas*, Samols, E., Ed., Raven Press, Ltd., New York, 1991, p. 15.
10. Bonner-Weir, S., Regulation of pancreatic beta-cell mass *in vivo*, *Recent Prog Horm Res*, 49, 91, 1994.
11. Moore, A. and Medarova, Z., Approaches for imaging the diabetic pancreas: first results, *Curr Med Chem Immun Endoc Metab Agents*, 4, 315, 2004.
12. Clark, P.B. et al., Neurofunctional imaging of the pancreas utilizing the cholinergic PET radioligand [18F]4-fluorobenzyltrozamicol, *Eur J Nucl Med Mol Imaging*, 31, 258, 2004.
13. Fujibayashi, Y. et al., A new approach toward a pancreas-seeking zinc radiopharmaceutical. I. Accumulation of 65Zn-amino acid and aminopolycarboxylic acid complexes in pancreatic tissue slices, *Eur J Nucl Med*, 11, 484, 1986.
14. Shiue, C.Y. et al., Potential approaches for beta cell imaging with PET and SPECT, *Curr Med Chem Immun Endoc Metab Agents*, 4, 271, 2004.
15. Malaisse, W.J. et al., Pancreatic fate of 6-deoxy-6-[125I]iodo-D-glucose: *in vivo* experiments, *Endocrine*, 13, 95, 2000.
16. Malaisse, W.J. et al., Pancreatic fate of 6-deoxy-6-[125I]iodo-D-glucose: *in vitro* experiments, *Endocrine*, 13, 411, 2000.
17. Malaisse, W.J. et al., Fate of 2-deoxy-2-[18F]fluoro-D-glucose in hyperglycemic rats, *Int J Mol Med*, 6, 549, 2000.
18. Malaisse, W.J. et al., Fate of 2-deoxy-2-[18F]fluoro-D-glucose in control and diabetic rats, *Int J Mol Med*, 5, 525, 2000.
19. Malaisse, W.J. et al., Pancreatic fate of D-[3H] mannoheptulose, *Cell Biochem Funct*, 19, 171, 2001.
20. Ladriere, L. et al., Assessment of islet beta-cell mass in isolated rat pancreases perfused with D-[(3)H]mannoheptulose, *Am J Physiol Endocrinol Metab*, 281, E298, 2001.
21. Malaisse, W.J. and Ladriere, L., Assessment of B-cell mass in isolated islets exposed to D-[3H]mannoheptulose, *Int J Mol Med*, 7, 405, 2001.
22. Malaisse, W.J., On the track to the beta-cell, *Diabetologia*, 44, 393, 2001.
23. Bryan, J. and Aguilar-Bryan, L., Sulfonylurea receptors: ABC transporters that regulate ATP-sensitive K(+) channels, *Biochim Biophys Acta*, 1461, 285, 1999.
24. Proks, P. et al., Sulfonylurea stimulation of insulin secretion, *Diabetes*, 51 (Suppl 3), S368, 2002.
25. Uhde, I. et al., Identification of the potassium channel opener site on sulfonylurea receptors, *J Biol Chem*, 274, 28079, 1999.
26. Ladriere, L. et al., Uptake of tritiated glibenclamide by endocrine and exocrine pancreas, *Endocrine*, 13, 133, 2000.
27. Schmitz, A. et al., Synthesis and evaluation of fluorine-18 labeled glyburide analogs as beta-cell imaging agents, *Nucl Med Biol*, 31, 483, 2004.
28. Schneider, S. et al., *In vitro* and *in vivo* evaluation of novel glibenclamide derivatives as imaging agents for the non-invasive assessment of the pancreatic islet cell mass in animals and humans, *Exp Clin Endocrinol Diabetes*, 113, 388, 2005.

29. Wangler, B. et al., Synthesis and evaluation of (S)-2-(2-[18F]fluoroethoxy)-4-([3-methyl-1-(2-piperidin-1-yl-phenyl)-butyl-carbamoyl]-methyl)-benzoic acid ([18F]repaglinide): a promising radioligand for quantification of pancreatic beta-cell mass with positron emission tomography (PET), *Nucl Med Biol*, 31, 639, 2004.
30. Wangler, B. et al., Synthesis and *in vitro* evaluation of (S)-2-([11C]methoxy)-4-[3-methyl-1-(2-piperidine-1-yl-phenyl)-butyl-carbamoyl]-benzoic acid ([11C]methoxy-repaglinide): a potential beta-cell imaging agent, *Bioorg Med Chem Lett*, 14, 5205, 2004.
31. Alejandro, R. et al., A ganglioside antigen on the rat pancreatic B cell surface identified by monoclonal antibody R2D6, *J Clin Invest*, 74, 25, 1984.
32. Halban, P. et al., Altered differentiated cell surface properties of transformed (RINm5F) compared with native adult rat pancreatic B cells, *Endocrinology*, 123, 113, 1988.
33. Kjaer, T. et al., Interleukins increase surface ganglioside expression of pancreatic islet cells *in vitro*, *Acta Pathol Microbiol Immunol Scand*, 100, 509, 1992.
34. Asfari, M. et al., Establishment of 2-mercaptoethanol-dependent differentiated insulin-secreting cell lines, *Endocrinology*, 130, 167, 1992.
35. Halle, J. et al., Protection of islets of Langerhans from antibodies by microencapsulation with alginate-poly-L-lysine membranes, *Transplantation*, 55, 350, 1993.
36. Ladriere, L. et al., Pancreatic fate of a 125I-labelled mouse monoclonal antibody directed against pancreatic B-cell surface ganglioside(s) in control and diabetic rats, *Cell Biochem Funct*, 19, 107, 2001.
37. Brogren, C. et al., Production and characterization of a monoclonal islet cell surface autoantibody from the BB rat, *Diabetologia*, 29, 330, 1986.
38. Buschard, K. et al., Antigen expression of the pancreatic beta-cells is dependent on their functional state, as shown by a specific, BB rat monoclonal antibody IC2, *Acta Pathol Microbiol Immunol Scand*, 96, 342, 1988.
39. Aaen, K. et al., Dependence of antigen expression on functional state of beta-cells, *Diabetes*, 39, 697, 1990.
40. Moore, A. et al., Non-invasive *in vivo* measurement of beta-cell mass in mouse model of diabetes, *Diabetes*, 50, 2231, 2001.
41. Woods, M. et al., Toward the design of MR agents for imaging beta-cell function, *Curr Med Chem Immun Endoc Metab Agents*, 4, 349, 2004.
42. Sambanis, A. et al., Development and non-invasive monitoring of a pancreatic tissue substitute, in *Imaging the Pancreatic Beta Cell*, Bethesda, MD, National Institutes of Health, 2003.
43. Lu, D. et al., C-13 NMR isotopomer analysis reveals a connection between pyruvate cycling and glucose-stimulated insulin secretion (GSIS), *Proc Natl Acad Sci USA*, 99, 2708, 2002.
44. Zheng, Q. et al., Magnetic relaxation agents for cell labeling and imaging applications, in *Imaging the Pancreatic Beta Cell*, Bethesda, MD, National Institutes of Health, 2003.
45. Gimi, B. et al., NMR spiral surface microcoils: applications, *Magn Res Eng*, 18B, 1, 2003.
46. Hanaoka, K. et al., Design and synthesis of a novel magnetic resonance imaging contrast agent for selective sensing of zinc ion, *Chem Biol*, 9, 1027, 2002.
47. Zhang, S. et al., A paramagnetic CEST agent for imaging glucose by MRI, *J Am Chem Soc*, 125, 15288, 2003.
48. Golman, K. et al., Molecular imaging with endogenous substances, *Proc Natl Acad Sci USA*, 100, 10435, 2003.
49. Golman, K. et al., Molecular imaging using hyperpolarized 13C, *Br J Radiol*, 76, S118, 2003.
50. Gepts, W., Pathologic anatomy of the pancreas in juvenile diabetes mellitus, *Diabetes*, 14, 619, 1965.
51. Bottazzo, G.F. et al., *In situ* characterization of autoimmune phenomena and expression of HLA molecules in the pancreas in diabetic insulitis, *N Engl J Med*, 313, 353, 1985.
52. Imagawa, A. et al., Pancreatic biopsy as a procedure for detecting *in situ* autoimmune phenomena in type 1 diabetes: close correlation between serological markers and histological evidence of cellular autoimmunity, *Diabetes*, 50, 1269, 2001.
53. Bulte, J.W. and Kraitchman, D.L., Monitoring cell therapy using iron oxide MR contrast agents, *Curr Pharm Biotechnol*, 5, 567, 2004.
54. Schulze, E. et al., Cellular uptake and trafficking of a prototypical magnetic iron oxide label *in vitro*, *Invest Radiol*, 30, 604, 1995.

55. Moore, A. et al., Uptake of dextran-coated monocrystalline iron oxides in tumor cells and macrophages, *J Magn Reson Imaging*, 7, 1140, 1997.
56. Weissleder, R. et al., Magnetically labeled cells can be detected by MR imaging, *J Magn Reson Imaging*, 7, 258, 1997.
57. Schoepf, U. et al., Intracellular magnetic labeling of lymphocytes for *in vivo* trafficking studies, *Biotechniques*, 24, 642, 1998.
58. Shen, T. et al., Magnetically labeled secretin retains receptor affinity to pancreas acinar cells, *Bioconj Chem*, 7, 311, 1996.
59. Moore, A. et al., Measuring transferrin receptor gene expression by NMR imaging, *Biochim Biophys Acta*, 1402, 239, 1998.
60. Josephson, L. et al., High-efficiency intracellular magnetic labeling with novel superparamagnetic-Tat peptide conjugates, *Bioconjugate Chem*, 10, 186, 1999.
61. Lewin, M. et al., Tat peptide-derivatized magnetic nanoparticles allow *in vivo* tracking and recovery of progenitor cells, *Nat Biotechnol*, 18, 410, 2000.
62. Bulte, J.W. et al., Magnetodendrimers allow endosomal magnetic labeling and *in vivo* tracking of stem cells, *Nat Biotechnol*, 19, 1141, 2001.
63. Magnitsky, S. et al., *In vivo* and *ex vivo* MRI detection of localized and disseminated neural stem cell grafts in the mouse brain, *Neuroimage*, 26, 744, 2005.
64. Arai, T. et al., Dual *in vivo* magnetic resonance evaluation of magnetically labeled mouse embryonic stem cells and cardiac function at 1.5 t, *Magn Reson Med*, 55, 203, 2006.
65. de Vries, I.J. et al., Magnetic resonance tracking of dendritic cells in melanoma patients for monitoring of cellular therapy, *Nat Biotechnol*, 23, 1407, 2005.
66. Anderson, S.A. et al., Noninvasive MR imaging of magnetically labeled stem cells to directly identify neovasculature in a glioma model, *Blood*, 105, 420, 2005.
67. Hill, J.M. et al., Serial cardiac magnetic resonance imaging of injected mesenchymal stem cells, *Circulation*, 108, 1009, 2003.
68. Kraitchman, D.L. et al., *In vivo* magnetic resonance imaging of mesenchymal stem cells in myocardial infarction, *Circulation*, 107, 2290, 2003.
69. Bulte, J.W. et al., Feridex-labeled mesenchymal stem cells: cellular differentiation and MR assessment in a canine myocardial infarction model, *Acad Radiol*, 12 (Suppl 1), S2, 2005.
70. Heyn, C. et al., *In vivo* magnetic resonance imaging of single cells in mouse brain with optical validation, *Magn Reson Med*, 55, 23, 2006.
71. Shapiro, E.M. et al., *In vivo* detection of single cells by MRI, *Magn Reson Med*, 55, 242, 2006.
72. Heyn, C. et al., Detection threshold of single SPIO-labeled cells with FIESTA, *Magn Reson Med*, 53, 312, 2005.
73. Anderson, S.A. et al., Magnetic resonance imaging of labeled T-cells in a mouse model of multiple sclerosis, *Ann Neurol*, 55, 654, 2004.
74. Sundstrom, J.B. et al., Magnetic resonance imaging of activated proliferating rhesus macaque T cells labeled with superparamagnetic monocrystalline iron oxide nanoparticles, *J Acquir Immune Defic Syndr*, 35, 9, 2004.
75. Kircher, M.F. et al., *In vivo* high resolution three-dimensional imaging of antigen-specific cytotoxic T-lymphocyte trafficking to tumors, *Cancer Res*, 63, 6838, 2003.
76. Kaldany, A. et al., Trapping of peripheral blood lymphocytes in the pancreas of patients with acute-onset insulin-dependent diabetes mellitus, *Diabetes*, 31, 463, 1982.
77. Signore, A. et al., 123I-Interleukin-2: biochemical characterization and *in vivo* use for imaging autoimmune diseases, *Nucl Med Commun*, 24, 305, 2003.
78. Signore, A. et al., New approach for *in vivo* detection of insulitis in type I diabetes: activated lymphocyte targeting with 123I-labelled interleukin 2, *Eur J Endocrinol*, 131, 431, 1994.
79. Signore, A. et al., A radiopharmaceutical for imaging areas of lymphocytic infiltration: 123I-interleukin-2. Labelling procedure and animal studies, *Nucl Med Commun*, 13, 713, 1992.
80. Bertera, S. et al., Body window-enabled *in vivo* multicolor imaging of transplanted mouse islets expressing an insulin-Timer fusion protein, *Biotechniques*, 35, 718, 2003.
81. Moore, A. et al., MR imaging of insulitis in autoimmune diabetes, *Magn Reson Med*, 47, 751, 2002.
82. Billotey, C. et al., T-cell homing to the pancreas in autoimmune mouse models of diabetes: *in vivo* MR imaging, *Radiology*, 236, 579, 2005.

83. Verdaguer, J. et al., Acceleration of spontaneous diabetes in TCRβ-transgenic nonobese diabetic mice by beta cell-cytotoxic CD8+ T-cells expressing identical endogenous TCRα chains, *J Immunol*, 157, 4726, 1996.
84. Verdaguer, J. et al., Spontaneous autoimmune diabetes in monoclonal T cell nonobese diabetic mice, *J Exp Med*, 186, 1663, 1997.
85. DiLorenzo, T. et al., MHC class I-restricted T-cells are required for all but end stages of diabetes development and utilize a prevalent T cell receptor α chain gene rearrangement, *Proc Natl Acad Sci USA*, 95, 12538, 1998.
86. Trudeau, J. et al., Autoreactive T cells in peripheral blood predict development of type 1 diabetes, *J Clin Invest*, 111, 217, 2003.
87. Amrani, A. et al., Progression of autoimmune diabetes driven by avidity maturation of a T-cell population, *Nature*, 406, 739, 2000.
88. Moore, A. et al., Tracking the recruitment of diabetogenic CD8+ T cells to the pancreas in real time, *Diabetes*, 54, 1459, 2004.
89. Lieberman, S. et al., Identification of the b-cell antigen targeted by a prevalent population of pathogenic CD8+ T cells in autoimmune diabetes, *Proc Natl Acad Sci USA*, 100, 8384, 2003.
90. Wraith, D.C. et al., Antigen recognition in autoimmune encephalomyelitis and the potential for peptide-mediated immunotherapy, *Cell*, 59, 247, 1989.
91. Metzler, B. and Wraith, D.C., Inhibition of experimental autoimmune encephalomyelitis by inhalation but not oral administration of the encephalitogenic peptide: influence of MHC binding affinity, *Int Immunol*, 5, 1159, 1993.
92. Liu, G.Y. and Wraith, D.C., Affinity for class II MHC determines the extent to which soluble peptides tolerize autoreactive T cells in naive and primed adult mice: implications for autoimmunity, *Int Immunol*, 7, 1255, 1995.
93. Anderton, S.M. and Wraith, D.C., Hierarchy in the ability of T cell epitopes to induce peripheral tolerance to antigens from myelin, *Eur J Immunol*, 28, 1251, 1998.
94. Karin, N. et al., Reversal of experimental autoimmune encephalomyelitis by a soluble peptide variant of a myelin basic protein epitope: T cell receptor antagonism and reduction of interferon gamma and tumor necrosis factor alpha production, *J Exp Med*, 180, 2227, 1994.
95. O'Brien, B.A. et al., Apoptosis is the mode of beta-cell death responsible for the development of IDDM in the nonobese diabetic (NOD) mouse, *Diabetes*, 46, 750, 1997.
96. Kurrer, M.O. et al., Beta cell apoptosis in T cell-mediated autoimmune diabetes, *Proc Natl Acad Sci USA*, 94, 213, 1997.
97. O'Brien, B.A. et al., Beta-cell apoptosis is responsible for the development of IDDM in the multiple low-dose streptozotocin model, *J Pathol*, 178, 176, 1996.
98. Bar-On, H. et al., Irreversibility of nutritionally induced NIDDM in *Psammomys obesus* is related to beta-cell apoptosis, *Pancreas*, 18, 259, 1999.
99. Donath, M.Y. et al., Hyperglycemia-induced beta-cell apoptosis in pancreatic islets of *Psammomys obesus* during development of diabetes, *Diabetes*, 48, 738, 1999.
100. Federici, M. et al., High glucose causes apoptosis in cultured human pancreatic islets of Langerhans: a potential role for regulation of specific Bcl family genes toward an apoptotic cell death program, *Diabetes*, 50, 1290, 2001.
101. Shimabukuro, M. et al., Role of nitric oxide in obesity-induced beta cell disease, *J Clin Invest*, 100, 290, 1997.
102. Shimabukuro, M. et al., Fatty acid-induced beta cell apoptosis: a link between obesity and diabetes, *Proc Natl Acad Sci USA*, 95, 2498, 1998.
103. Maedler, K. et al., Distinct effects of saturated and monounsaturated fatty acids on beta-cell turnover and function, *Diabetes*, 50, 69, 2001.
104. Piro, S. et al., Chronic exposure to free fatty acids or high glucose induces apoptosis in rat pancreatic islets: possible role of oxidative stress, *Metabolism*, 51, 1340, 2002.
105. Medarova, Z. et al., Imaging beta-cell death with a near-infrared probe, *Diabetes*, 54, 1780, 2005.
106. Fadok, V.A. et al., Exposure of phosphatidylserine on the surface of apoptotic lymphocytes triggers specific recognition and removal by macrophages, *J Immunol*, 148, 2207, 1992.
107. Verhoven, B. et al., Mechanisms of phosphatidylserine exposure, a phagocyte recognition signal, on apoptotic T lymphocytes, *J Exp Med*, 182, 1597, 1995.

108. Vermes, I. et al., A novel assay for apoptosis. Flow cytometric detection of phosphatidylserine expression on early apoptotic cells using fluorescein labelled annexin V, *J Immunol Methods*, 184, 39, 1995.
109. van de Wiele, C. et al., Quantitative tumor apoptosis imaging using technetium-99m-HYNIC annexin V single photon emission computed tomography, *J Clin Oncol*, 21, 3483, 2003.
110. Boersma, H.H. et al., Comparison between human pharmacokinetics and imaging properties of two conjugation methods for 99mTc-annexin A5, *Br J Radiol*, 76, 553, 2003.
111. Hofstra, L. et al., Visualisation of cell death *in vivo* in patients with acute myocardial infarction, *Lancet*, 356, 209, 2000.
112. Kown, M.H. et al., *In vivo* imaging of acute cardiac rejection in human patients using (99m)technetium labeled annexin V, *Am J Transplant*, 1, 270, 2001.
113. Petrovsky, A. et al., Near-infrared fluorescent imaging of tumor apoptosis, *Cancer Res*, 63, 1936, 2003.
114. Schellenberger, E.A. et al., Optical imaging of apoptosis as a biomarker of tumor response to chemotherapy, *Neoplasia*, 5, 187, 2003.
115. Schellenberger, E.A. et al., Annexin V-CLIO: a nanoparticle for detecting apoptosis by MRI, *Mol Imaging*, 1, 102, 2002.
116. Sosnovik, D.E. et al., Magnetic resonance imaging of cardiomyocyte apoptosis with a novel magneto-optical nanoparticle, *Magn Reson Med*, 54, 718, 2005.
117. Zhao, M. et al., Non-invasive detection of apoptosis using magnetic resonance imaging and a targeted contrast agent, *Nat Med*, 7, 1241, 2001.
118. Wedam, S.B. et al., Antiangiogenic and antitumor effects of bevacizumab in inflammatory and locally advanced breast cancer patients, *J Clin Oncol*, 24, 769, 2006.
119. Eichhorn, M.E. et al., Paclitaxel encapsulated in cationic lipid complexes (MBT-0206) impairs functional tumor vascular properties as detected by dynamic contrast enhanced magnetic resonance imaging, *Cancer Biol Ther*, 5, 89, 2006.
120. de Lussanet, Q.G. et al., Dynamic contrast-enhanced magnetic resonance imaging of radiation therapy-induced microcirculation changes in rectal cancer, *Int J Radiat Oncol Biol Phys*, 63, 1309, 2005.
121. Brauer, M., *In vivo* monitoring of apoptosis, *Prog Neuropsychopharmacol Biol Psychiatry*, 27, 323, 2003.
122. Zhao, M. et al., Early detection of treatment response by diffusion-weighted 1H-NMR spectroscopy in a murine tumour *in vivo*, *Br J Cancer*, 73, 61, 1996.
123. Poptani, H. et al., Monitoring thymidine kinase and ganciclovir-induced changes in rat malignant glioma *in vivo* by nuclear magnetic resonance imaging, *Cancer Gene Ther*, 5, 101, 1998.
124. Kauppinen, R.A., Monitoring cytotoxic tumour treatment response by diffusion magnetic resonance imaging and proton spectroscopy, *NMR Biomed*, 15, 6, 2002.
125. Sweet, I.R. et al., Continuous measurement of oxygen consumption by pancreatic islets, *Diabetes Technol Ther*, 4, 661, 2002.
126. Sweet, I.R. et al., Glucose stimulation of cytochrome C reduction and oxygen consumption as assessment of human islet quality, *Transplantation*, 80, 1003, 2005.
127. Zacharovova, K. et al., *In vitro* assessment of pancreatic islet vitality by oxymetry, *Transplant Proc*, 37, 3454, 2005.
128. Moore, A. et al., *In vivo* targeting of underglycosylated MUC-1 tumor antigen using a multimodal imaging probe, *Cancer Res*, 64, 1821, 2004.
129. Evgenov, N.V. et al., *In vivo* imaging of islet transplantation, *Nat Med*, 12, 144, 2006.
130. Hauptman, P.J. and O'Connor, K.J., Procurement and allocation of solid organs for transplantation, *N Engl J Med*, 336, 422, 1997.
131. Stock, P.G. and Bluestone, J.A., Beta-cell replacement for type I diabetes, *Annu Rev Med*, 55, 133, 2004.
132. Bretzel, R.G. et al., Improved survival of intraportal pancreatic islet cell allografts in patients with type-1 diabetes mellitus by refined peritransplant management, *J Mol Med*, 77, 140, 1999.
133. Ryan, E. et al., Five-year follow-up after clinical islet transplantation, *Diabetes*, 54, 2060, 2005.
134. Biarnes, M. et al., Beta-cell death and mass in syngeneically transplanted islets exposed to short- and long-term hyperglycemia, *Diabetes*, 51, 66, 2002.
135. Davalli, A.M. et al., Vulnerability of islets in the immediate posttransplantation period: dynamic changes in structure and function, *Diabetes*, 45, 1161, 1996.

136. Ryan, E. et al., Successful islet transplantation: continued insulin reserve provides long-term glycemic control, *Diabetes*, 51, 2148, 2002.
137. Robertson, R.P., Pancreas and islet transplants for patients with diabetes: taking positions and making decisions, *Endoc Pract*, 5, 24, 1999.
138. Inverardi, L. et al., Islet transplantation: immunological perspectives, *Curr Opin Immunol*, 15, 507, 2003.
139. Ryan, E. et al., Clinical outcomes and insulin secretion after islet transplantation with the Edmonton protocol, *Diabetes*, 50, 710, 2001.
140. Pileggi, A. et al., Factors influencing islet of Langerhans graft function and monitoring, *Clin Chim Acta*, 310, 3, 2001.
141. Castano, L. and Eisenbarth, G., Type-I diabetes: a chronic autoimmune disease of human, mouse, and rat, *Annu Rev Immunol*, 8, 647–679, 1990.
142. Atkinson, M. and Maclaren, N., The pathogenesis of insulin-dependent diabetes mellitus, *N Engl J Med*, 331, 1428–1436, 1994.
143. Bhargava, R. et al., Prevalence of hepatic steatosis after islet transplantation and its relation to graft function, *Diabetes*, 53, 1311, 2004.
144. Eckhard, M. et al., Disseminated periportal fatty degeneration after allogeneic intraportal islet transplantation in a patient with type 1 diabetes mellitus: a case report, *Transplant Proc*, 36, 1111, 2004.
145. Markmann, J.F. et al., Magnetic resonance-defined periportal steatosis following intraportal islet transplantation: a functional footprint of islet graft survival? *Diabetes*, 52, 1591, 2003.
146. Garg, A. and Misra, A., Hepatic steatosis, insulin resistance, and adipose tissue disorders, *J Clin Endocrinol Metab*, 87, 3019, 2002.
147. Kaufman, D. et al., *In vivo*, real-time, non-invasive bioluminescent imaging of transplanted islets in a functional murine model, in *Imaging the Pancreatic Beta Cell*, Bethesda, MD, National Institutes of Health, 2003.
148. Powers, A. et al., Using bioluminescence to non-invasively image and assess transplanted islet mass, in *Imaging the Pancreatic Beta Cell*, Bethesda, MD, National Institutes of Health, 2003.
149. Lu, Y. et al., Bioluminescent monitoring of islet graft survival after transplantation, *Mol Ther*, 9, 428, 2004.
150. Fowler, M. et al., Assessment of pancreatic islet mass after islet transplantation using *in vivo* bioluminescence imaging, *Transplantation*, 79, 768, 2005.
151. Lu, Y. et al., Repetitive microPET imaging of implanted human islets in mice, in *Imaging the Pancreatic Beta Cell*, Bethesda, MD, National Institutes of Health, 2003.
152. Yaghoubi, S. et al., Human pharmacokinetic and dosimetry studies of [(18)F]FHBG: a reporter probe for imaging herpes simplex virus type-1 thymidine kinase reporter gene expression, *J Nucl Med*, 42, 1225, 2001.
153. Massoud, T.F. and Gambhir, S.S., Molecular imaging in living subjects: seeing fundamental biological processes in a new light, *Genes Dev*, 17, 545, 2003.
154. Toso, C. et al., Positron-emission tomography imaging of early events after transplantation of islets of Langerhans, *Transplantation*, 79, 353, 2005.
155. Evgenov, N.V. et al., *In vivo* imaging of immune rejection in transplanted pancreatic islets, *Diabetes*, 55, 2419, 2006.
156. Ricordi, C. and Strom, T., Clinical islet transplantation: advances and immunological challenges, *Nat Rev Immunol*, 4, 259, 2004.

20 Functional Cellular Imaging with Manganese

Vincent Van Meir and Annemie Van der Linden

CONTENTS

20.1 Introduction 369
20.2 The Challenge of Functional Cellular Imaging 370
  20.2.1 Associations between Neuronal Activity and Dynamics of the Plastic Brain 370
  20.2.2 Measurable Correlates of Neuron Function within the Living Brain 370
    20.2.2.1 Localization of Neuronal Activity: Spatial and Temporal Resolution 370
    20.2.2.2 Imaging Neurons *In Vitro*, *In Vivo*, or in a Real-World Environment 372
  20.2.3 Manganese as Contrast Agent 373
  20.2.4 Studying Cardiac Function with Manganese 374
  20.2.5 Manganese Chelates for Human Applications 374
20.3 Mechanisms of Cellular Incorporation of Manganese 374
  20.3.1 Manganese Administration Routes 375
  20.3.2 Molecular and Cellular Correlates of Manganese Uptake and Distribution 376
    20.3.2.1 Chemical Configurations of Manganese 376
    20.3.2.2 Manganese Transport across Membranes and Barriers 376
    20.3.2.3 Intraneuronal Redistribution and Transsynaptic Transport of Manganese 377
    20.3.2.4 Manganese Might Influence Excitable Tissue Activity 378
20.4 Studying Functional Neural Circuitries 378
  20.4.1 Activation-Induced Manganese-Dependent Contrast (AIM) MRI 378
  20.4.2 Activity-Dependent Tract Tracing with MEMRI 380
    20.4.2.1 Olfactory and Visual Pathways 380
    20.4.2.2 Other Brain Circuits 380
    20.4.2.3 Remodelling of Neuronal Circuitries 383
  20.4.3 Activity-Dependent Tract Tracing after Systemic Injections of Manganese, the Least Invasive Approach 383
20.5 Conclusions 384
Acknowledgments 385
References 386

20.1 INTRODUCTION

Manganese has recently emerged as an easily accessible and promising positive contrast agent for magnetic resonance imaging (MRI) that covers a broad spectrum of functional and anatomical applications in a wide variety of small-animal models. This application of manganese is known as manganese-enhanced MRI (MEMRI).[1] The capacity of the Mn^{2+} ion to substitute for Ca^{2+} in excitable tissues such as muscle and nervous tissue, combined with the ability to obtain highly localized contrast changes by using high-resolution scanning techniques, provides a new route to map cellular activity or malfunctions with high accuracy.

The main focus of this chapter will be MRI using Mn^{2+}-induced contrast in the brain and, to a lesser extent, the heart muscle. In the first of three sections we illustrate different levels of cellular function and how they can be highlighted with the application of *in vivo* MEMRI. The advantages and disadvantages of MEMRI are compared to other cellular imaging techniques, and we discuss some practical issues that have to be taken into account when using MEMRI. The second section describes possible administration routes of Mn^{2+}. Distribution and accumulation of excessive Mn^{2+} are discussed at the body, tissue, and cellular levels. Properties of Mn^{2+} distribution, which are specific to brain and neurons, are discussed in detail. The third section reviews applications of MEMRI to visualize anatomy, function, and plasticity of neuronal networks.

20.2 THE CHALLENGE OF FUNCTIONAL CELLULAR IMAGING

20.2.1 Associations between Neuronal Activity and Dynamics of the Plastic Brain

Figure 20.1 represents different levels of localization and timing of neurophysiological events in relation to environmental input and behavior. A simple sensory stimulation or behavioral act leads to activation of the involved neurons. Repetition of a stimulus can change the sensitivity of a neuron. During a learning process, several associations are ultimately captured within a neuronal circuit by changes in neuronal morphology. Cellular activity is usually associated with *short-term events*, such as the generation of action potentials and the intracellular release of calcium. These short-term events can be modulated on an *intermediate term* by experience-dependent mechanisms, regulated by intracellular molecular cascades. In the *long term*, changes in cell function involve morphological adaptations, which are the result of a previous upregulation of protein production and previous extensive neuronal activity. In adult animals, changes in local brain morphology were for a long time believed to occur only at the synaptic and dendrite levels. Recent findings in neurological models, such as the songbirds' song control system, revealed the presence of far more dramatic plasticity mechanisms that take place in brain tissue and deal with neuronal recruitment and axonal sprouting (see Section 20.4.2.2). Similar neuroplastic mechanism will be consulted when, e.g., rewiring the brain after neurological insults. It is therefore obvious that for optimal assessment of cellular activity and function, in normal, diseased, and recovering brains, one needs to unravel the more complete picture of ongoing cellular activity at the background of a changing brain. This requires repeated visualizations of neuronal activity and the mutual interactions between brain regions bridging the gap between localized (*in situ*) measurements of individual neurons and whole brain imaging techniques that allow comparison between distant, albeit related, neuronal populations.

20.2.2 Measurable Correlates of Neuron Function within the Living Brain

20.2.2.1 Localization of Neuronal Activity: Spatial and Temporal Resolution

In modern science, each step, from the initial activation to the ultimate structural modification, can be analyzed using a variety of techniques. However, few techniques combine the potential of long-term *in vivo* observations with sensitivity to cellular activity and network functions that involve multiple brain structures.

Localized cellular activations are mostly studied *in vivo* with electrophysiological techniques (e.g., patch-clamp, intracellular recordings), *in vitro* using fluorescence microscopy, or *postmortem* by means of immunohistochemistry (e.g., immediate early gene expression: *zenk*, *cfos*). Long-term, repeated observations are difficult or impossible because of the invasiveness of the methods.

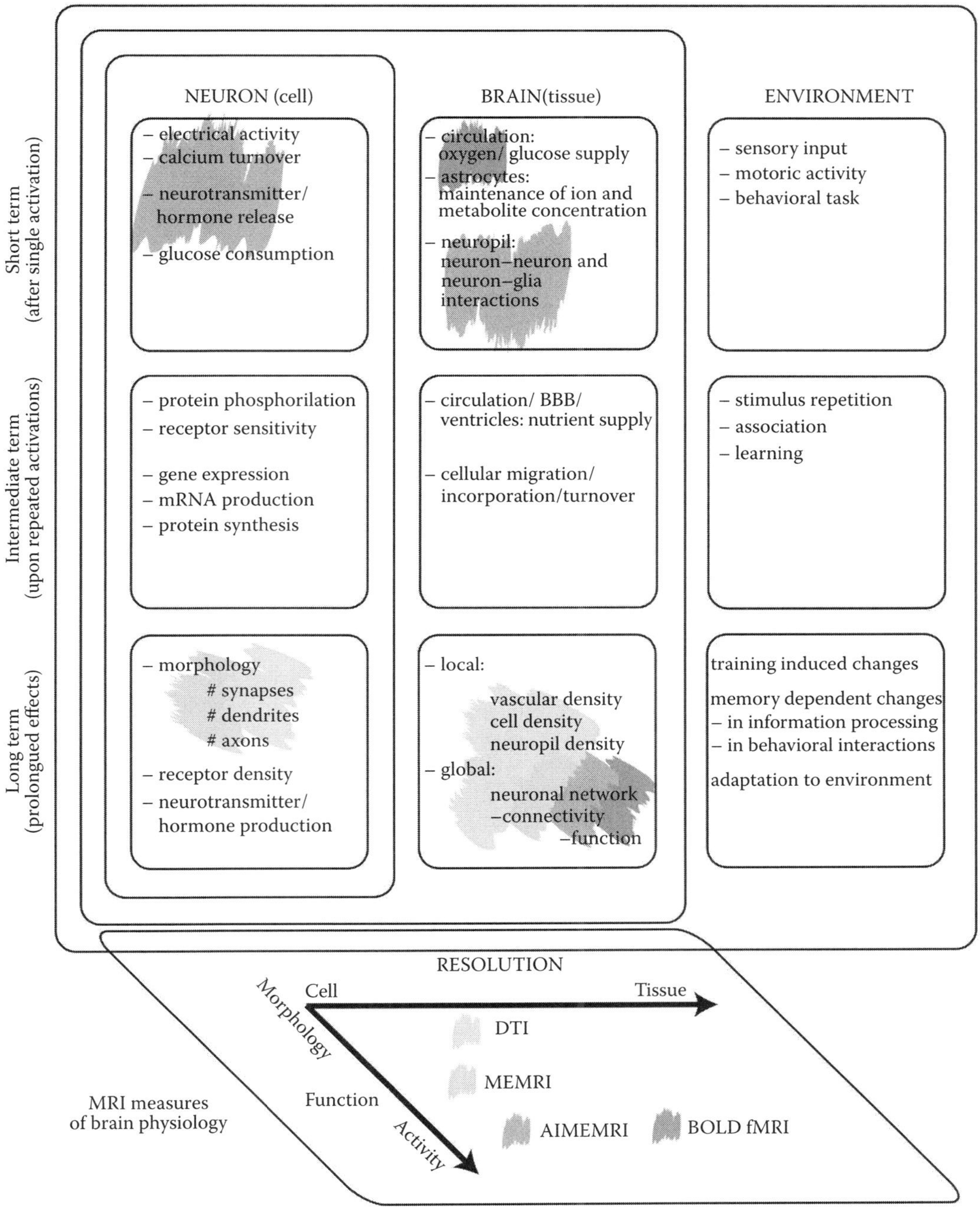

FIGURE 20.1 (please see color insert following page 210) Schematic representation of different levels of localization and timing of brain function. Top: Neuronal activity and molecular and physiological events are represented in relation to environmental input and behavior. Bottom: MRI techniques that provide information on different levels of brain function, indicated on the top panel in their respective colors. MRI, magnetic resonance imaging; MEMRI, manganese-enhanced MRI; AIMEMRI, activity-induced manganese-enhanced MRI; BOLD fMRI, blood oxygenation level-dependent functional MRI; DTI, diffusion tensor imaging.

Alternatively, bioelectricity can be recorded at the surface of the body, and thus can reveal cellular activity in deeper tissues. Examples are electrocardiography (ECG), electroencephalography (EEG), and magnetoencephalography (MEG). The time resolution is comparable to that of electrophysiological recordings, and the techniques are completely noninvasive, but the localization of the signal is very poor. *In vivo* tomographical techniques such as functional MRI (fMRI) and positron emission tomography (PET) are suitable to provide functional measures of large neuron populations and allow localization of the activations of multiple regions on three-dimensional brain reconstructions. Unfortunately, they suffer from either limited resolution (PET) or contrast changes (fMRI). Moreover, fMRI yields an indirect measure of cellular activity by measuring changes in blood oxygenation levels, and accurate localization of the activated cell groups can be obscured by the presence of draining veins. Nevertheless, the *in vivo* technique for soft tissue imaging with the highest spatial resolution today is MRI. Contrary to PET, the intrinsic contrast mechanisms of proton (^1H) MRI are very powerful for measuring tissue properties. The combination of magnetic resonance microscopy (MRM) and magnetic resonance spectroscopy (MRS) provides access to several observable quantities that cannot be determined with microscopical techniques alone (e.g., metabolite concentrations, chemical shifts, spin couplings, T_1 and T_2 relaxation times, and diffusion constants). These quantities have been related to a variety of cellular events as tumor formation, programmed cell death (apoptosis), necrosis, and increased proliferation. MRS has a limited imaging resolution up to 5 to 10 μl (~$2 \times 2 \times 2$ mm^3) for proton spectroscopy at field strengths of 7 to 9.4 T,[2,3] whereas the theoretical and practical limit of *in vivo* MRM at normal physiological body temperature approaches 10-μm resolution.[4]

A popular way of approaching the challenge to obtain cellular information in the context of living tissue has been to combine multiple modalities — MRI, PET, single-photon emission computed tomography (SPECT), ultrasound (US), computed tomography (CT), and bioluminescence imaging (BLI) — to better apply the strengths while overcoming individual modality weaknesses. On the other hand, MRI contrast agents, preferably affecting T_1 relaxation, can also be located with very high precision. Some ions with paramagnetic properties are accumulated within cells in parallel to processes of cellular activity or metabolism, and thus can be used to localize or quantify cell function. Of those ions, Mn^{2+} has demonstrated possession of properties that can be particularly helpful for imaging activity of excitable tissues in general and neurons and neuronal networks in particular. First, the Mn^{2+} ion, which has paramagnetic properties, can substitute for Ca^{2+} ions.[5–11] During cellular activations of excitable tissues, Mn^{2+} can enter through voltage-gated Ca channels. An acute increase of extracellular Mn^{2+} concentrations together with the activation of specific brain regions has been shown to result in an increased local signal intensity in the activated region on T_1-weighted MR images.[12,13] After uptake in the neuronal body, Mn^{2+} is transported to the synapse via fast anterograde axonal transport, a transport mechanism that requires energy consumption and can be activity dependent.[14] Finally, Mn^{2+} can travel across the synapse, and thus can enter a subsequent step of the neuronal network.[15–18] These properties can be extremely useful to image *in vivo* anatomy,[19,20] physiology,[21–23] and activity[24,25] of neuronal networks.

20.2.2.2 Imaging Neurons *In Vitro, In Vivo,* or in a Real-World Environment

Certain features of neuronal function such as topographical organization of the cortex can only be investigated in intact animals.[26,27] Until recently, *in vivo* molecular imaging was limited to anesthetized or restrained animals and was not able to take motion into account, which is critically needed for behavioral studies in normal physiological conditions and dynamic signal transduction visualization. *In vivo* observations of brain function in animal models during the execution of a specific task and the change in activity over different training sessions are severely limited by this restrained environment. This issue has been addressed by building miniature imaging (two-photon/fluorescence microscopy) devices that can be implanted in the region of interest (ROI).[28] Although these techniques are suitable to image local miniature networks at different cortical depths, simultaneously measuring interactions between distant brain regions remains beyond reach. Imaging brain function with tomographical techniques does not readily address this issue either, since it requires restrained highly controlled conditions. In animals, this requires either the use of anesthesia or extensive training protocols.[29–33]

An alternative solution would be to design a contrast agent that enters the cell during the execution of a behavioral task and remains in stable concentrations when this task is finished. Immediately following a training session, high-resolution tomographical techniques can be used to evaluate cell function *in vivo*. Here again, MEMRI would provide an elegant solution since Mn^{2+} can enter cells via activity-dependent mechanisms and is stored within the cell for a prolonged period. Thus, a stimulus or exercise can be performed outside the magnet where the manganese will accumulate at the active sites, and then the animal can be placed in the magnet for imaging. Since the images are not time dependent over a short time, one can spend most of the imaging time averaging to obtain ultrahigh spatial resolution.

20.2.3 Manganese as Contrast Agent

Thus far, manganese has been mainly studied as a nutrient and as a potential toxicant. Manganese is a required trace element that binds to and regulates many enzymes throughout the body and is essential for normal development and body function across the life span of all mammals.[34] Interruption of Mn^{2+} homeostasis has been associated with a variety of disease states in humans, such as liver and heart failure and neurological symptoms resembling Parkinson's disease.[35–37] Despite these phenomena, there has been a growing crescendo of research on biological applications where Mn^{2+} can be used as a contrast agent for MRI.

The Mn^{2+} ion with its five unpaired electrons on its outer shell has paramagnetic properties. This causes the H protons in the vicinity to transfer their energy to the Mn^{2+} ions, which leads to a fast T_1 relaxation, and thus a positive contrast on T_1-weighted MR images in regions with Mn^{2+} accumulation. MR relaxation rates are proportional to the effective concentration of Mn^{2+} in tissue.[38] The required amount and concentration of Mn^{2+} to produce robust and detectable contrast depends on the MR sequence used and magnetic field strength.

Aoki et al.[39] demonstrated that lymphocytes and B cells can be labeled with $MnCl_2$ *in vitro* to a level that allows their detection by T_1-weighted MRI at 11.7 T already 1 h after incubation with 0.05 to 1.0 m*M* $MnCl_2$. Additional investigations revealed that lymphocytes did not undergo apoptosis or necrosis immediately after and 24 h following a 1-h incubation with up to 1.0 m*M* $MnCl_2$, and that NK cells and cytotoxic T cells maintained their *in vitro* killing capacity after being incubated with up to 0.5 m*M* $MnCl_2$.

Volumes of 5 to 10 nl with concentrations in the range of 2 to 10 m*M* were efficient to visualize transsynaptic Mn^{2+} transport after injection in the striatum or amygdala in mice (11.7 T)[17] and after injection in components of the song control system in the canary brain (7 T).[40]

The two common sequences used to image Mn^{2+} distribution are T_1-weighted gradient echo (GE) and spin echo (SE) sequences, which can provide a three-dimensional image of the whole brain in a reasonable amount of time (30 to 120 min). Using inversion recovery (IR) sequences, higher contrast sensitivity is obtained, allowing smaller amounts of Mn^{2+} to be used. T_1 values or T_1 maps can be calculated that allow objective quantification of Mn^{2+}-induced contrast changes.[39,41–43]

Since T_1 relaxation rates decrease with magnetic field, it is important to take this into account to determine the correct imaging parameters. With increasing field strength, repetition time needs to be increased in GE or SE sequences, inversion time and repetition time need to be increased in IR-SE sequences, and optimal flip angles need to be decreased in GE sequences with very short repetition times (~50 msec). The advantage of high magnetic fields is that they yield higher signal-to-noise ratio (SNR), but require longer TR. Since Mn^{2+} is a positive contrast agent, and thus increases the SNR in the regions where it accumulates, imaging at low fields can offer the advantage of faster image acquisitions.[44]

For more in-depth information about MEMRI contrast mechanisms, we refer to Pautler and Fraser[45] for a review on practical considerations of MEMRI, and on both surgical and imaging levels, we refer to Silva et al.[46] and Pautler.[47]

20.2.4 Studying Cardiac Function with Manganese

Manganese ions (Mn^{2+}) can enter viable myocardial cells via voltage-gated calcium channels. Intracellular calcium is a central regulator of cardiac contractility, but despite the importance of Ca^{2+} regulation in the heart, there are currently no established radiological imaging techniques for visualizing Ca^{2+} channel activity. *In vivo* studies show that Ca^{2+} channel blockers inhibit Mn^{2+}-induced MRI signal enhancement and that positive ionotropes, which increase Ca^{2+} influx, further increase Mn^{2+}-induced MRI signal enhancement.[48] Therefore, MEMRI would represent a novel means by which Mn^{2+} uptake can be related to voltage-gated Ca^{2+} channel activity in homogeneous, well-perfused heart with no impaired tissue integrity. Additionally, MEMRI was implemented in a myocardial infarction model[49–53] to simultaneously measure the myocardial infarct size along with left ventricular function at enhanced spatial resolution. The left ventricular functional parameter, myocardial ME signal intensity (SI), was recorded during systole and diastole to assess potential alterations in calcium homeostasis following permanent left anterior descending (LAD) coronary artery ligation. Infarction size can thus be correlated with functional data to gain insight into the relationship between infarction size and decreased cardiac function.

20.2.5 Manganese Chelates for Human Applications

The use of manganese has been limited due to concerns about toxicity, specifically acute cardiovascular depression,[54] as it acts as a transient competitive antagonist to calcium activity, reducing contractility and systemic blood pressure.[54,55] To date it has been approved for clinical imaging only in a chelated form, manganese dipyridoxyl diphosphate (Mn-DPDP).[56] While chelation markedly improves the acute safety profile of manganese,[57] it does so at the expense of many of the properties that make it desirable for cardiac or neuroimaging. This is in particular due to the extracellular distribution of the chelate and the slow release of manganese from the ligand. However, Bremerich et al.[49] and Wendland et al.[58] tested its role as a myocardial viability marker. Another contrast agent, EVP 1001-1 (Eagle Vision Pharmaceutical Corp., Exton, PA),[59] was formulated with the aim of producing an agent with the desired characteristics of free manganese, but without the associated risk of cardiotoxicity. It contains manganese in a readily available form, rather than the chelated form, and addresses the safety issue pharmacologically by the addition of calcium. Storey et al.[60] showed that its pharmacokinetics are similar to those of free manganese, with a short vascular half-life (about 1.5 min) and a long retention time in viable myocardium (greater than 1 h). Storey et al.[60,61] provided evidence of a flow-dependent difference in the uptake of the agent between viable ischemic and normal myocardium, suggesting its application in steady-state imaging of myocardial ischemia, and they provided the T_1-shortening characteristics of this compound in an infarct model. In the future, these studies will allow for assessment of cardiovascular drugs that have the potential to reduce infarction size, improve ventricular function, or improve Ca^{2+} handling.

20.3 MECHANISMS OF CELLULAR INCORPORATION OF MANGANESE

The use of manganese as an MRI contrast agent to highlight activities in excitable tissues largely depends on its properties to enter cells via voltage-gated or even ligand-gated Ca channels. However, Mn^{2+} is a required trace metal with a specific distribution and function in living organisms. As a consequence, tissue enhancement after Mn^{2+} administration can depend on a variety of mechanisms that are not necessarily activity induced. To gain insight into the specificity of activation-induced MEMRI upon different routes of manganese administration, we will discuss the pathways of manganese uptake in tissue and cells. As in the previous section, we will mainly focus on manganese access to the brain.

20.3.1 Manganese Administration Routes

The most natural way of manganese uptake is *oral consumption*. Manganese is absorbed slowly and poorly throughout the length of the small intestine. Only 3 to 4% of the ingested manganese salts are absorbed in rodents. Manganese ions that are absorbed into the portal circulation are almost completely removed by the liver and excreted into the bile.[62] This highly efficient manganese excretory mechanism prevents excess tissue exposure and as such is less interesting as a tool to induce manganese contrast on MR images. The liver may be important as a depot for manganese, with hepatic manganese later delivered to the body. The highest manganese concentrations can be found in the bone, liver, kidney, and pancreas (20 to 50 nmol/g). Concentrations of manganese in the brain, heart, lung, and muscle are typically <20 nmol/g. Natural concentrations in blood and serum are 200 and 20 nmol/l, respectively.[34,63,64]

Following *inhalation*, a considerable amount of manganese is retained in the lungs, providing a depot that slowly releases manganese to the bloodstream. This leads to a prolonged systemic exposure.[41,65] Half-lives for Mn^{2+} leaving the chest ranged from 34 to 187 days. This route in humans is associated with the main cause of Mn^{2+} toxicity due to Mn^{2+}-induced neurological degeneration, resulting in syndromes similar to idiopathic Parkinson's disease.[37] Head levels peaked 40 days after dosing and remained high for 1 year after dosing. In animal experiments, using high magnetic field strengths, however, reasonably low concentrations might be sufficient to induce a long-term contrast enhancement in specific brain regions, without resulting in pathological symptoms.

For the purpose of inducing contrast in the animal brain, thus far mainly *systemic injection* techniques have been used (intraperitoneal (i.p.), subcutaneous (s.c.), or intravenous (i.v.)). This led to contrast enhancements in regions that under natural conditions also contained high Mn^{2+} concentrations.[41,43,66–69] In Sprague-Dawley rats Mn^{2+} distributed under normal conditions among four tested brain regions in the following order: substantia nigra > striatum > hippocampus > frontal cortex in a concentration range of 0.3 to 0.7 μg/g of wet tissue weight.[70] In the healthy human brain, the highest concentrations (dry weight) were found in the pineal gland (~4 μg/g), followed by the olfactory bulb (~3 μg/g) and the caudate nucleus (~2 μg/g).[71–73]

Peripheral administration techniques can be useful to obtain a global manganese delivery to the brain noninvasively. Although a direct access route via the blood exists for most body tissues (e.g., liver, pancreas, and heart), this route is largely cut off to the brain by the existence of a blood–brain barrier (BBB). After a short systemic exposure, Mn^{2+} is cleared relatively fast from the blood, in the range from minutes[74] up to 1.83 h.[75] The subsequently increased influx into the brain is largely dependent on a fast uptake mechanism that exists at the level of the choroid plexus and the ventricular ependyma, rather than direct uptake through the BBB.[76,77]

After *intraventricular injection* Mn^{2+} diffuses into the parenchyma, a process that can last for 24 to 96 h in rats.[42,78] The observed uptake rate and distribution in the brain were similar to those after systemic injection.[43,66,68,79–81] Specific stimulation during this period can alter the normal Mn^{2+} distribution pattern of Mn^{2+}, and thus indicate regions of increased activity.[82]

Alternatively, one can reversibly *break the blood–brain barrier* by applying an osmotic shock during the intravenous infusion of Mn^{2+}.[12] A subsequent stimulation leads to a fast Mn^{2+} accumulation in the activated brain region.[13] Mn^{2+} can also be delivered directly into a brain region of interest via its administration into peripheral sensory organs such as the olfactory epithelium[14,83–86] and the eye,[14,15] or more invasive *focal brain injections*.[16,17,21,87,88] This approach has been mostly used to perform anatomical or activation-induced tract tracing and is based on the fact that Mn^{2+} is taken up into neurons and subsequently transported via anterograde transport toward the synapse.

A variety of cellular and molecular mechanisms are involved in Mn^{2+} uptake into neurons. On many levels these mechanisms have not been completely resolved. Mn^{2+} is often captured or transported by molecules associated with iron (Fe^{3+}) or calcium (Ca^{2+}) metabolism. This diversity of plausible transport systems obscures the molecular picture of important processes such as manganese accumulation in neurons, their subsequent intracellular distribution, and eventual exchange at the synapse. Therefore, the molecular background of these processes is often studied in the function of curing or preventing manganese toxicity. In the next section we will evaluate the multitude of manganese transport mechanisms and discuss their contribution to activity-dependent signal enhancements during MEMRI.

20.3.2 Molecular and Cellular Correlates of Manganese Uptake and Distribution

20.3.2.1 Chemical Configurations of Manganese

In biological fluids and tissues, most metals are present largely as complexes with amino acids, peptides, proteins, phospholipids, and other tissue constituents, rather than as free metal cations. Binding of reactive heavy metals to metallothioneins, ferritin, transferrin, lactoferrin, melanotransferrin, hemosiderin, ceruloplasmin, citrate, ascorbate, glutathione (GSH), cysteine, or other amino acids is a major protective mechanism. Likewise, the biological reactivity of essential metals is regulated by interaction with specific ligands, and in particular with prosthetic groups on proteins.[89] To understand the molecular mechanisms of manganese uptake into cells or neurons, one should consider some of the chemical species under which manganese appears.

Thermodynamic modeling of Mn^{2+} in serum suggests it exists in several forms, for example, as an albumin-bound species (84%), as a hydrated ion (6.4%), and in other low molecular weight ligands (1.8%).[90] The calculations based on this model are consistent with the observation of low molecular weight species slightly larger than the Mn^{2+} ion, in plasma.[91] Approximately 8% of Mn^{2+} in rat blood plasma was found to be in the free fraction, also consistent with the model calculations.[92] According to another model, a large portion (approximately 60%) of divalent manganese in the plasma is reported to be non-protein-bound, probably the free ion.[93] In serum, divalent manganese can be oxidized to trivalent manganese by ceruloplasmin or molecular oxygen.[95] Thermodynamic modeling of Mn^{3+} in serum suggests that it is almost 100% bound to Tf.[94,90]

20.3.2.2 Manganese Transport across Membranes and Barriers

Several transport systems have been reported for *divalent manganese* uptake: calcium channels,[5,8] Na/Ca exchanger,[96] active calcium uniporter,[97] and Na/Mg antiporter.[98] Divalent manganese can also be transported into the cell by the divalent metal transporter DMT1 (also known as DCT1 and NRAMP2).[99,100] *Trivalent manganese*, usually bound to Tf, can be captured by the Tf receptor and subsequently incorporated by endocytosis.

Several studies suggested that Mn^{2+} ion brain *influx* at the *BBB* is carrier mediated,[76,77,101–103] while the identity of the Mn^{2+} carriers is still unknown. Mn^{2+} *efflux* across the BBB, on the other hand, does not appear to occur through a carrier but rather by diffusion.[104] In the brain, capillary endothelia have the highest density of Tf receptors,[105–108] but these seem to play a negligible role in manganese delivery to the brain.[109–111] It was shown recently that the contribution of DMT1 to Mn^{2+} transport across the BBB might also be limited.[92] Alternative pathways for Mn^{2+} transport across the BBB via store-operated Ca channels, as well as another mechanism at the blood–brain barrier, likely play a role in carrier-mediated Mn^{2+} influx into the brain.[112]

Glial cells, particularly *astrocytes*, represent a sink for brain manganese[113] and could contribute significantly to signal enhancements after manganese administration. The kinetics of Mn^{2+} uptake have been studied in cultured astrocytes from rat[114] and mixed glia from chick.[115] Unlike neurons, astrocytes have the ability to concentrate Mn^{2+} at levels 50-fold higher than the culture media.[114,115] The precise transporters for Mn^{2+} into astrocytes are unknown. In both reports, saturable Mn^{2+} transport was found to be both competitively and noncompetitively inhibited by Ca^{2+}. Extracellular Mn^{2+} stimulates the rapid component of manganese efflux from rat astrocytes.[114] An astrocyte-specific manganoprotein critical for ammonia metabolism, glutamine synthetase (GS), accounts for about 80% of the total brain manganese.[113] Areas of high astrocyte density include the hypothalamus and hippocampus, with low astrocyte density in the cerebral cortex, neostriatum, midbrain, medulla oblongata, and cerebellum.[116] The relatively high influx rates for Mn^{2+} (hippocampus) and manganese–citrate (thalamus/hypothalamus) support a correlation, but the high influx despite low astrocyte density in cortical regions for Mn^{2+}, manganese–citrate, and manganese–Tf and in the cerebellum for manganese–citrate decreases the strength of the putative correlation.[117]

Within the brain parenchyma, the Tf receptor is present in *neurons* at only 10 to 20% of the level found in capillary endothelia.[105,107] DMT1 is present in high densities in the pyramidal and granule cells of the hippocampus, cerebellar granule cells, the preoptic nucleus, and pyramidal cells of the piriform cortex.[99] There is the possibility that this transporter is involved in the uptake of manganese ion or low molecular weight ligand-bound manganese by neurons. In excitable cells such as neurons, excessive Mn^{2+} can be incorporated by L-type voltage-gated calcium channels. This was verified by utilizing the drug diltiazem or verapamil, which prevents the uptake of Mn^{2+} into cells by blocking Ca^{2+} channels. This has been verified in the brain as well as the heart.[6,8,9,24] Additional support comes from the accumulation of Mn^{2+} in specific brain areas that contain neuronal populations with high spontaneous activity. For instance, after a focal injection in the hippocampal region of mice, specific Mn^{2+} enhancement in the dentate–CA3 region, as opposed to the CA1–subiculum subfields, seems to reflect the local functional activity in relation to intrahippocampal processing.[118] Extracellular recordings[119] detected spontaneous action potentials from CA3 but not CA1 pyramidal cell populations, while single-channel recordings[120] indicated that low-voltage activated Ca^{2+} channels are particularly abundant on pyramidal neurons in CA3 but not CA1. Important sensory areas in rodents also provide evidence for local activity-induced uptake of Mn^{2+}. The Mn^{2+}-induced MRI signal enhancements in the olfactory and auditory systems of mice after systemic administration are in agreement with 2-deoxyglucose autoradiograms of rat brain, indicating pronounced activity in structures of the olfactory system[121,122] as well as in the inferior colliculus.[123–125]

20.3.2.3 Intraneuronal Redistribution and Transsynaptic Transport of Manganese

The largest subcellular concentration of manganese is found in the mitochondria, although in excess it also accumulates in lysosomes.[126] Manganese is a component of the mitochondrial form of superoxide dismutase,[127] known as MnSOD or SOD2. After experimental exposure *in vivo*, Mn^{2+} has been found to be taken up into mitochondria of the brain.[128] Higher levels of Mn^{2+} were observed in striatal mitochondria than in mitochondria isolated from the cortex. Although it is generally agreed that most minute-to-minute Ca^{2+} buffering in the cell is carried out at submicromolar basal Ca^{2+} concentrations by the endoplasmic reticulum,[129] local spikes of cytosolic Ca^{2+} may reach micromolar levels,[130] capable of triggering mitochondrial influx of both Ca^{2+} and Mn^{2+}.[131] It follows that excitable tissue, experiencing frequent Ca^{2+} spikes, is likely to accumulate mitochondrial Mn^{2+}.

On the other hand, after focal injections into the brain it has been demonstrated that fast axonal transport of Mn^{2+} occurs.[14,21,87,88] Mammalian axons exhibit two major anterograde transport processes with differential speeds. *Slow axonal transport* refers to velocities of 0.01 to 0.33 mm/h, while *fast axonal transport* yields velocities in the range of 2 to 16 mm/h.[132] Since mitochondria are transported via a slow axonal transport mechanism,[133] it has been concluded that the observed fast axonal transport of Mn^{2+} occurs within vesicles along microtubules via an energy-requiring mechanism. This has been demonstrated with the use of the microtubule-disrupting drug colchicine. Upon disruption of the microtubules, transport of the Mn^{2+} was halted.[24,87,88] Interestingly, although the speed of fast axonal transport is constant, the amount of vesicles that can be transported per unit of time can change according to the neuronal activity.[133]

It is likely that Mn^{2+} is sequestered in the endoplasmic reticulum after neuronal uptake, where it is packaged in vesicles for transport. This has been verified utilizing subcellular fractionation obtained through sucrose gradient centrifugation and photometric sensitive Mn^{2+} assays on subcellular fractions of olfactory bulbs exposed to Mn^{2+}.[24] After transport along the microtubules to the synaptic cleft, the Mn^{2+} was then released at the synaptic cleft and taken up by the next neuron in the circuit.[76,77,88,134] Quite interestingly, it has been shown that neurons that have been preloaded with Mn^{2+} co-release the ion with glutamate upon stimulation, indicating the possibility that Mn^{2+} is transported within synaptic vesicles.[135] After release in the synaptic cleft, Mn^{2+} could then be incorporated in the postsynaptic cells by ligand-gated Ca^{2+} channels such as N-methyl-aspartate (NMDA) receptors. Fluorescence quenching techniques showed that Mn^{2+} can enter neurons through NMDA receptors.[9,136] A pharmacologic disruption of normal synaptic transmission by blocking the NMDA-type glutamate receptors with the antagonist APV (DL-2-amino-5-phosphonovaleric acid) caused a depression of the otherwise pronounced enhancement in the dentate–CA3 region of the dorsal hippocampus by APV on a three-dimensional MRI 18 h after injection of Mn^{2+}.[67]

20.3.2.4 Manganese Might Influence Excitable Tissue Activity

Manganese has been found to stimulate adenylate cyclase activity in the brain and other tissues of the body. This is of importance because cyclic-AMP plays a regulatory role in the action of several brain neurotransmitters, by acting as a second messenger within cells in transmitting the messenger hormone.[137] Several researchers have demonstrated that manganese influences synaptic neurotransmission at high doses (~m*M*). Narita et al.[8] suggest that Mn^{2+} can enter nerve terminals through the voltage-dependent Ca^{2+} channel during action potentials in the frog motor nerve, enhancing the release of neurotransmitters. Drapeau and Nachshen[5] also demonstrate that Mn^{2+} permeates presynaptic voltage-dependent Ca^{2+} channels and induces dopamine release from depolarized nerve terminals. In the absence of extracellular Ca^{2+}, Mn^{2+} induces a long-lasting potentiation of acetylcholine release from cardiac parasympathetic nerve terminals following tetanic nerve stimulation.[138] These findings indicate that Mn^{2+} can either substitute for Ca^{2+} in the exocytotic process or induce the release of Ca^{2+} from intracellular stores, possibly the endoplasmic reticulum. Manganese also causes the activation of glutamate-gated cation channels, e.g., N-methyl-aspartate (NMDA) receptor, which contributes to neuronal manganese-induced neurodegeneration.[139] Calcium channels in the heart muscle can be blocked by Mn^{2+} and thereby affect cardiac function. However, a high concentration of Mn^{2+} (1 m*M*) is required to block the calcium channels.[140] Manganese also blocks voltage-dependent calcium channels and nerve-evoked neurotransmitter release *in vitro*, although the effects of manganese on the voltage-dependent calcium channels and the release of neurotransmitters are controversial.[141]

20.4 STUDYING FUNCTIONAL NEURAL CIRCUITRIES

Three specific applications of MEMRI in the animal brain have been demonstrated, and their differences are based on the injection route of Mn^{2+} and the information that is provided. First, due to the fact that manganese ion can enter excitable cells via the voltage-gated calcium channel, protocols have been devised that enable accumulation of Mn^{2+} in active areas of the brain. This technique has been referred to as activation-induced MEMRI. The second use of MEMRI is to image anterograde connections after direct injection of $MnCl_2$ into a specific brain region, the so-called MEMRI tract tracing, which can also be activity dependent. The third use of MEMRI has been as an anatomical contrast agent providing a superior image contrast in the brain after systemic administration of Mn^{2+}.

20.4.1 Activation-Induced Manganese-Dependent Contrast (AIM) MRI

This was first demonstrated by Lin and Koretsky.[12] In AIMMRI, $MnCl_2$ is infused intra-arterially after the blood–brain barrier (BBB) is opened with a hyperosmolar agent. Upon functional stimulation

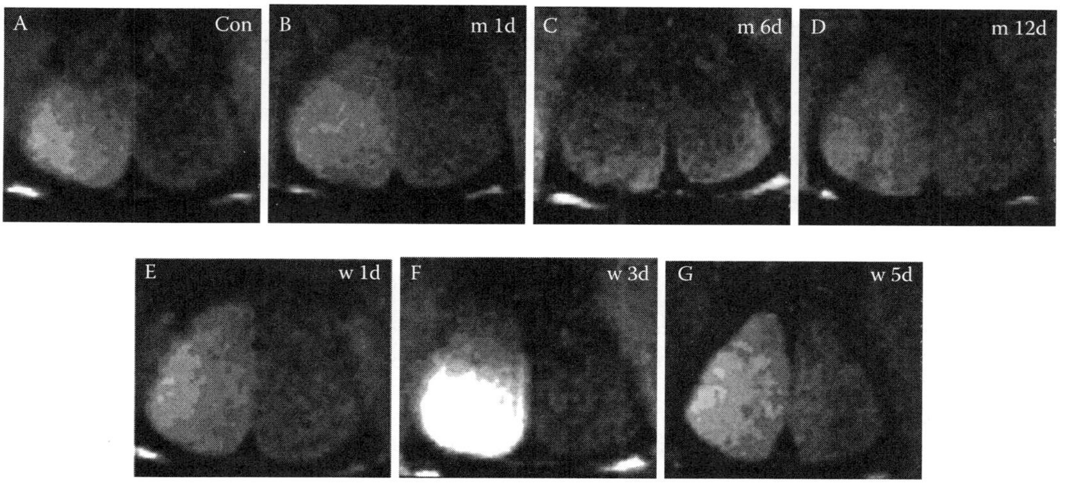

FIGURE 20.2 The differences of the MEMRI signal intensity in the orbitofrontal cortex between the control group, the morphine administration group, and the withdrawal group in rats. (A) The control group. (B–D) The MEMRI signal intensity in the orbitofrontal cortex during morphine administration with morphine administration on (B) the 1st day, (C) the 6th day, and (D) the 12th day. (E–G) The MEMRI signal integrity in the orbitofrontal cortex during the withdrawal period, with withdrawal on (E) the 1st day, (F) the 3rd day, and (G) the 5th day. (From Sun, N. et al., *Neuroscience*, 138, 77, 2006. Copyright 2006, with permission from Elsevier.)

of the brain, Mn^{2+} accumulates in the active regions by entering active cells through voltage-gated Ca^{2+} channels, causing local signal increases in T_1-weighted images. The contrast of AIMMRI depends strongly on the depth of anesthesia, and the low levels used in somatosensory stimulation studies can lead to significant nonspecific accumulation of manganese ion throughout the brain.

In a recent work, Morita et al.[142] using AIMMRI demonstrated that intracarotid arterial injection of hypertonic NaCl elicited a rapid and striking increase in signal intensity in the hypothalamic nuclei involved in central osmotic regulation. These authors also observed consistency with cFos expression, indicating that AIMMRI is a useful technique for investigating the autonomic centers in the hypothalamus.

Aoki et al.[13] published a *dynamic AIM* (DAIM) paradigm, which used sequential MR scans during $MnCl_2$ infusion prior to and following functional stimulation of the brain. Stimulation-specific functional maps were produced using time-course analysis. The new method was tested during glutamate administration and electric stimulation of the rat forepaw. It was shown that DAIM maps are better confined to the specific region of brain activated by somatosensory stimulation, compared to AIMMRI.

Another approach to obtain information on the activity of neurons without the need to disrupt the BBB pharmacologically, but by using Mn^{2+} injections straight into the regions of interest, was introduced by Sun et al.[143] These authors focused on the amount of Mn^{2+} ions that entered excitable cells in the injection area as a measure of the activity of that brain region. They investigated the activity of the orbitofrontal cortex (OFC) as part of the limbic system involved in the reinforcing effects of drug abuse, during opiate administration and withdrawal periods. They used both gamma-band EEG and MEMRI to evaluate changes in neuronal activity. $MnCl_2$ was stereotactically injected into the right OFC 40 min after morphine injection on the 1st, 6th, and 12th days of morphine administration and on the 1st, 3rd, 5th, and 7th days after the withdrawal of morphine (see Figure 20.2). About 5 to 6 h after Mn^{2+} injection rats were anesthetized and perfused with 10% formalin containing 1% potassium ferrocyanide to wash out extracellular Mn^{2+}, and the MRI signal intensity ratios of the ipsilateral ROI and the contralateral OFC were calculated as *a measure for intracellular* Mn^{2+}*, and hence a sign of neuronal activity.* Based on the results of MEMRI and gamma-band EEG, it appears that there is an increase in OFC activity during morphine withdrawal (Figure 20.2F), which was significantly correlated with the intensity of the craving and the power of OFC gamma EEG.

20.4.2 Activity-Dependent Tract Tracing with MEMRI

20.4.2.1 Olfactory and Visual Pathways

In order for the Mn^{2+} to reach the target, the BBB should be circumvented by using either nostril exposure or intravitreal injections of $MnCl_2$ to trace neuronal connections *in vivo*, but this approach is limited to the olfactory[14,24] and visual pathways.[15,144] Watanabe et al.[15] injected $MnCl_2$ into the left vitreous body of the rat and monitored Mn^{2+} enhancement starting at the left retina and proceeding to the left optic nerve, right optic tract, right and left suprachiasmatic nucleus, right lateral geniculate nucleus, right olivary pretectal nucleus, and up to the right superior colliculus 24 h after injection. This *in vivo* mapping of the rat visual pathway is in general agreement with the well-documented projections identified by conventional invasive methods.

One step further is to perform activity-dependent tract tracing as demonstrated in the olfactory system by Pautler and Koretsky.[24] The olfactory system in rodents contains two components, the main olfactory system processing common odors, and the accessory olfactory system involved in perception of pheromones and mediating specific behavioral and endocrine processes. These authors performed a study in which they traced Mn^{2+} from sites of activation in the olfactory epithelium to the olfactory bulb, thereby localizing regions within the olfactory bulb that respond to a particular odor. To that end, a common odor stimulus (amyl acetate) and the odor of mice urine, which is a complicated mixture of common odorants and pheromones, were presented to the nostrils of the mice in combination with aerosoled $MnCl_2$. Subsequently, functional mapping of the olfactory system was done using MEMRI with an anatomical resolution of 100 μm. The high-pheromone-containing solution caused enhancement in the anatomically correct location of the accessory olfactory bulb. Amyl acetate also caused T_1-weighted MRI enhancement in specific regions of the olfactory bulb. The areas showing activation agree well with previous studies, and it is anticipated that manganese-enhanced MRI (MEMRI) could be used to rapidly map a variety of odors.

20.4.2.2 Other Brain Circuits

Brain regions other than the olfactory bulb and the visual system can be reached by injecting $MnCl_2$ immediately in the brain. Small focal injections of manganese ions within the central nervous system combined with *in vivo* high-resolution MRI allow delineation of neuronal tracts originating from the site of injection and using, in some cases, the transsynaptic transport capacity of manganese. This was accomplished in the murine striatum and amygdala,[17] the somatosensory cortex,[19,20] and the spinal cord of the rat,[145,146] as well as the basal ganglia of the monkeys, in combination with simultaneous injection of the common tract tracing agent, wheat germ agglutinin conjugated to horseradish peroxidase (WGA-HRP),[16] and in the brain of songbirds.[40]

The next step is to link this to activity-driven tract tracing as outlined earlier in the case of the olfactory system. This was accomplished in songbirds to investigate neuroplastic changes in the songbird brain[21–23] or the impact of hearing songs on specific neuronal populations.[25] In these studies the dynamics of axonal manganese transport were monitored as manganese-induced signal intensity (SI) enhancement in the projected areas and translated into a Hill plot (function describing a sigmoid curve). This so-called *dynamic manganese-enhanced MRI* (DMEMRI) can then be used as a quantitative tool to monitor the activity of the projecting neurons in the injection area. This has resulted in the discovery of gender-linked[21] or testosterone-induced[22] changes in activity and functional connectivity in two circuits of the song control system of songbirds. These studies as well as the study of Tindemans et al.[25] used the DMEMRI method to segregate the contribution of different neuronal populations from the activated brain region. Although MRI has no spatial resolution to discern activity of each cell separately, DMEMRI seemed able to resolve this, and this will be illustrated by tract tracing studies in the song control circuit. In the songbird brain, a spatially organized neuronal circuitry exists, responsible for the bird's ability to learn and produce songs (Figure 20.3A). HVC (previously called the high vocal center) is an important sensory motor region that serves as a relay within the vocal network connecting the brain areas involved in hearing, song production, and vocal learning. It contains three distinct types of neurons (Figure 20.3B) from which two project to the robust nucleus of the arcopallium (RA) or to Area X of the medial striatum, respectively. The third type can be identified as interneurons. The two types of projection neurons are part of two brain circuits that play a critical role in the production of song (RA-projecting neurons) or its acquisition in juveniles and stability in adults (X-projecting neurons). The song control circuit in songbirds is characterized by a high density of axons connecting HVC with its target nuclei RA and Area X, and by the relatively large volume and clearly defined boundaries of these targets. These characteristics find no match in any mammalian brain circuit. In particular, the existence of dense connections between HVC, on the one hand, and RA and Area X, on the other hand, results in axonal transport highways from the injection site (HVC) to the projection areas (RA and Area X). The rationale behind quantifying the dynamics of the Mn^{2+} uptake in RA and Area X is that, since RA and Area X are monosynaptically connected to HVC, changes in the dynamics of Mn^{2+} accumulation in these nuclei are directly related to the electrical activity of the HVC neurons projecting to these targets or to the density of these projections. To follow the dynamics of Mn^{2+} uptake, Mn^{2+} had to be injected through a permanent nonmagnetic plastic cannula (i.d., 0.39 µm; o.d., 0.69 µm; Plastics One, Inc., Roanoke, VA) implanted in HVC that allowed repeated injections of $MnCl_2$ (typically 200 nl of a 10 m*M* solution) at exactly the same location. The plastic cannulla was connected to a very long and narrow tubing that allowed stereotaxic injection of very small volumes while the bird is in the magnet, so that images could be acquired before, during, and after Mn^{2+} injection. In this way, DMEMRI can be used to investigate fast changes in activity of specific types of neurons that have been traditionally studied by single-cell electrophysiology.

Tindemans et al.[25] used DMEMRI to obtain a global view of the activity changes in two types of HVC neurons in isoflurane-anesthetized male canaries that are affected by auditory stimuli (song from other canaries) and to assess simultaneously their homology. The two HVC cell populations projecting to RA and Area X, respectively, responded differentially to the same set of auditory stimuli. The difference observed in Mn^{2+} accumulation in these two nuclei reflects a difference in the processing of Mn^{2+} ions in the corresponding circuits associated with the activity of a specific neuron type (Figure 20.3G).

The information obtained from the DMEMRI studies in songbirds ultimately led to monitoring manganese-enhanced SI in the projected areas at a crucial time point after injection as a measure of the activity of the projecting neurons at the injection area. This was illustrated in a seasonal study by Van Meir et al.,[23] who combined MEMRI and singing behavior in starling.

Three main conclusions could be drawn from this study:

1. The manganese transport in both circuits was severely affected by the season, with the highest transport rate observed during the season of highest song activity.
2. The individual differences in seasonal changes in manganese transport rate were not correlated to individual differences in singing behavior (song rate, repertoire size, etc.).
3. The study revealed, however, a higher activity in both seasons in the projected area (RA) of the song motor system in birds with the highest singing rate.

At this point, we cannot conclude whether the seasonal changes or individual differences are mainly due to cellular activity or physiological properties of the network.

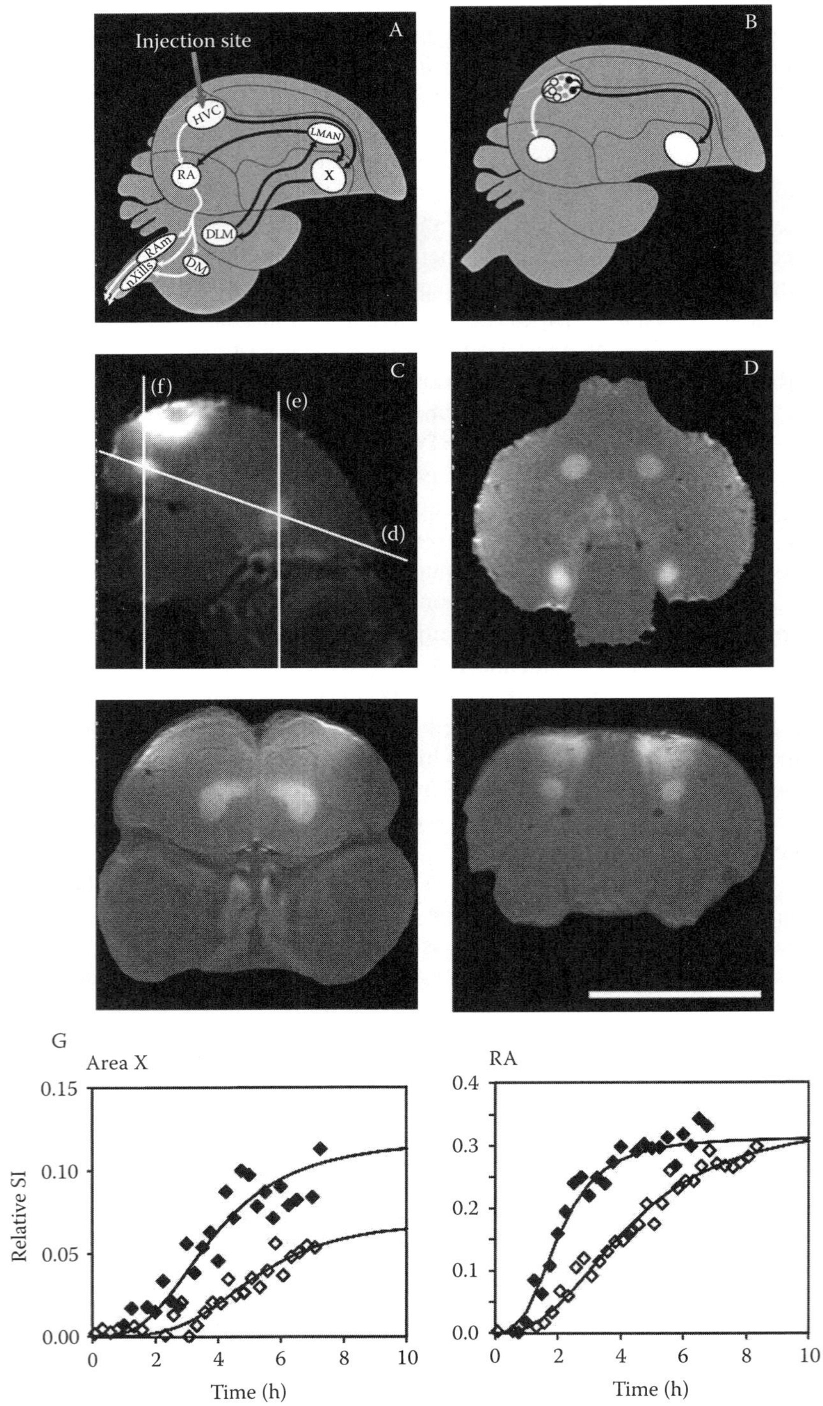

FIGURE 20.3 (A) Schematic overview of the adult songbird brain showing the song control nuclei (SCN) and their connections in the telencephalon. The black arrows represent the anterior forebrain pathway that starts in HVC and projects to Area X DLM lMAN. The white arrows indicate the motor pathway that projects from HVC directly to RA and further down to nXIIts and the syrinx. (B) Schematic overview with the key SCN — HVC, Ra, and Area X — illustrating that the two pathways originate from distinct cell populations within HVC. (C) Sagittal *in vivo* manganese-enhanced MRI of a male starling brain obtained 6 h after $MnCl_2$

FIGURE 20.3 (continued)
injection into HVC. The injection area is indicated by the grey arrow in (A) and by the signal-enhanced area on the corresponding sagittal MRI slice displayed in this panel and the coronal MRI slice in (F). Panel (C) also illustrates the different planes of imaging for panels (D) to (F). On all the images (C–F), the areas enhanced by a brighter signal correspond to the labeling by Mn^{2+}. They correspond exactly to RA and Area X as identified by histological techniques. Image resolution in the coronal plane is 97 μm (pixel size). Magnification bar = 1 cm. (Adapted from Van der Linden, A. et al., *Neuroscience*, 112, 467, 2002. Copyright 2002, with permission from Elsevier.) (G) T_1-weighted images were collected every 15 min starting before the injection and for up to 6 to 7 h after the injection under isoflurane anesthesia with perfectly controlled body temperature conditions. Changes in the mean signal intensity were determined within the two song control nuclei (RA and Area X) as illustrated in (A) and in adjacent control areas in all images collected sequentially. Changes in relative signal intensity were then plotted as a function of time and fitted by nonlinear regression to a sigmoid curve to describe the kinetics of Mn^{2+} accumulation in the areas of interest and reflecting the activity of the respective HVC neuron type. Thick lines and full squares represent data collected when the canary was allowed to listen to canary songs while in the magnet, and thin lines and open squares indicate the control situation, without song stimulation, obtained in the same bird. For more information, see Tindemans et al.[25] (Adapted from Tindemans, I. et al., *Eur. J. Neurosci.*, 18, 3352, 2003. Reprinted with permission of Blackwell Publishing.)

20.4.2.3 Remodelling of Neuronal Circuitries

MEMRI is a very useful method to examine network plasticity and regeneration in songbirds (for review, see Van der Linden et al.[147]). In the earlier described study on canaries (Tindemans et al.[25]), consecutive DMEMRI measurements in the same bird covered a time window in which no morphological changes occurred and the observed modifications in DMEMRI could be entirely assigned to changes in cellular activity. In a different study on starlings from Van Meir et al.,[22] DMEMRI was used to study functional and morphological changes induced by testosterone. In this study, the different DMEMRI parameters allowed the authors to discriminate increased axonal sprouting of HVC-to-RA projecting neurons after 6 weeks of testosterone treatment, while this did not occur in the HVC-to-Area X projecting neurons.

Apart from the information on plasticity within individual networks, Van Meir et al.[23] provided evidence for a putative interaction in the plasticity of the motor pathway through RA and the basal ganglia pathway through Area X. The birds showing the largest seasonal change in manganese enhancement in RA accumulated significantly lower amounts of manganese within Area X in the season of highest song production, whereas this discrepancy was not present in the RA measurements at this time point.

Axonal plasticity is also recognized to be part of several different pathologic processes, such as neural circuitry remodeling as a consequence of hyperactivity during seizures. Sprouting of granule cell axons or mossy fibers is one of the most consistent neuropathologic findings in the hippocampus of animals or humans with temporal lobe epilepsy,[148] providing one of the most extensively characterized examples of *activity-induced axonal plasticity in the brain.* Nairismagi et al.[149] used MEMRI to characterize this activity-dependent plasticity in the mossy fiber pathway after intraperitoneal kainic acid injection. Enhancement of the MEMRI signal in the dentate gyrus and the CA3 subregion of the hippocampus was evident 3 to 5 days after injection of $MnCl_2$ into the entorhinal cortex in both control and kainic acid-injected rats. An increase in the number of Mn^{2+}-enhanced pixels in the dentate gyrus and CA3 subfield of affected rats correlated ($p < 0.05$) with histologically verified mossy fiber sprouting. These data demonstrate that MEMRI can be used to detect specific changes at the cellular level during activity-dependent plasticity *in vivo.*

20.4.3 Activity-Dependent Tract Tracing after Systemic Injections of Manganese, the Least Invasive Approach

Except for the olfactory studies, the above-mentioned MEMRI approaches all required invasive manganese injection in the eye or the brain of the animal. The search for new administration routes has resulted in systemic injections of manganese. It has been demonstrated in mice and rats that an intraperitoneal (i.p.), intravenous (i.v.), or subcutaneous (s.c.) injection of $MnCl_2$ leads to unique MRI contrast revealing the *neuroarchitecture of the brain.*[43,44,66,68,69] Systemic injected manganese will ultimately reach the brain tissue through uptake into the choroid plexus, which then spreads to the cerebral spinal fluid (CSF) spaces in ventricles and periventricular tissues within 2 h of administration.[68]

An interesting application was developed by Wadghiri et al.,[150] who explored an easily implemented approach for contrast-enhanced imaging, using systemically administered manganese to reveal fine anatomical detail in T_1-weighted MR images of *neonatal mouse brains.* In particular, they demonstrated the utility of MEMRI for analyzing early postnatal patterning of the mouse cerebellum. Through comparisons with matched histological sections, it was shown that MEMRI enhancement correlates qualitatively with granule cell density in the developing cerebellum, suggesting that the cerebellar enhancement is due to uptake of Mn^{2+} in the granule neurons. Finally, variable cerebellar defects in mice with a conditional mutation in the Gbx2 gene were analyzed with MEMRI to demonstrate the utility of this method for mutant mouse phenotyping. Taken together and given the importance of genetically modified mice in studies of mammalian brain development and human congenital brain diseases, MEMRI provides an efficient and powerful *in vivo* method for analyzing neonatal brain development in normal and genetically engineered mice.

Besides providing a superior image contrast in the brain, activated areas will be enhanced, and a new approach to study brain activity during motion-accompanied behavior became available. Brain activation in awake small animals can be monitored by performing MRI after the presumed activity has occurred, preceded by a systemic injection of manganese. This minimally invasive approach became feasible due to the low clearance rate from the cell. MEMRI becomes then quite homologue to histological discrimination of immediate early gene expression (e.g., cfos), as it highlights areas with prior activity but probably harbors the same drawbacks in terms of specificity. This method has been proven capable of providing a sensitive and effective method for mapping the mouse auditory brain stem.[82] MEMRI was used to map regions of accumulated sound-evoked activity in awake, normally behaving mice, resulting in high-resolution (100-micron) brain mapping of the tonotopic organization of the mouse inferior colliculus. Systemic (intraperitoneal) administration of $MnCl_2$ allowed longitudinal imaging starting even from early postnatal stages of mouse auditory brain development. The spatial accuracy was tested by generating high-frequency tonotopic maps of mouse inferior colliculus that were in excellent agreement with previously reported electrophysiological measurements (Figure 20.4).

20.5 CONCLUSIONS

At the beginning of this chapter we introduced Figure 20.1, which provided an overview of different levels of localization and timing of neuronal activity and molecular and physiological events in the brain in relation to environmental input and behavior. This overview allowed us to outline how important it is for the assessment of cellular (neuron) activity and functioning to consider and investigate at the same time the broader context of the dynamics in brain physiology and morphology. This requires repeated whole brain imaging of neuronal activity. At this point and in conclusion, we add *two extra layers* to Figure 20.1. The *first layer* displays the potentials of MEMRI for studying cellular function at the different levels of localization and timing of neuronal activity. The *second layer* displays the contribution of other MRI techniques and clarifies to the reader how the application of MEMRI in combination with other MRI tools allows one to unravel brain activity in the neuroplastic brain.

MEMRI can contribute in both short-term and long-term activations. For the latter application, however, the need exists to distinguish these different morphological changes from changes in cellular activity. In this regard, the songbird model has a clear advantage, as it is well documented and can serve to elucidate the different contributors in this highly plastic brain affecting the dynamics of manganese transport. However, MRI also has a clear advantage, as different MRI techniques

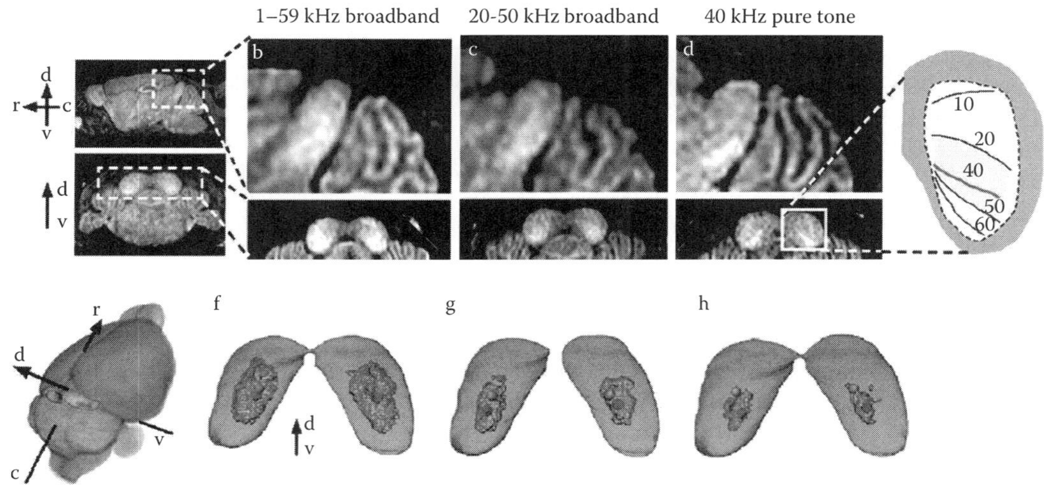

FIGURE 20.4 (please see color insert) MEMRI was used to map the tonotopic organization of the mouse inferior colliculus (IC). (a) Sagittal (upper) and coronal (lower) images of 21-day-old mice with the IC showing obvious differences in mice exposed to defined stimuli. (b) After broadband (1- to 59-kHz) stimulation, enhancement covered most of the rostral-caudal (r-c) and ventral-dorsal (v-d) extent of the central nucleus of the IC. (c) After high-frequency broadband (20- to 50-kHz) stimulation, enhancement was more restricted to the ventral-caudal region. (d) After 40-kHz pure-tone stimulation, enhancement was restricted to an isofrequency band in excellent agreement with electrophysiological maps. (Romand et al., 1990) (inset). (e) Averaged, co-registered images (n = 8) were used to extract whole brain (grey) and IC (green) and to generate three-dimensional maps of MEMRI IC enhancement (red) after stimulation with 1 to 59 kHz (f), 20 to 50 kHz (g), and 40 kHz (h). (From Yu, X. et al., *Nat. Neurosci.*, 8, 961, 2005. Copyright 2006, with permission from Nature Publishing Group.)

used within the same experiment can provide complementary information and help to unravel the more complete picture of ongoing cellular activity at the background of a changing brain. *In vivo* diffusion tensor imaging (DTI) and the quantitative estimation of fractional anisotropy (FA) can become excellent tools for detection of changes in the amount of axonal projections and dendrite branching,[151] while blood oxygenation level-dependent (BOLD) functional MRI is a control for neuronal activity or for incorporation of other brain nuclei in the communication network.[152]

From the content of this chapter, one can conclude that the majority of reported MEMRI applications focus on activity, connectivity, and mapping of somatosensory neuronal circuits and that MEMRI harbors great potential for the study of *neuronal development*, *activity*, and *plasticity* in different small-animal models. Surprisingly, only a few reports exist on MEMRI in neuropathological animal models.[149,150,153] One of the biggest benefits of MEMRI, which finds no comparison in any other method, is its potential as an *in vivo* noninvasive tool to link behavior, performed in a nonrestricted environment, with its neuronal substrate. Given the currently acknowledged ease and importance of behavioral phenotyping in neurodegenerative mice models, to see whether they mimic the human pathology, we are quite close to inserting MEMRI into protocols for phenotyping the neural substrate of the observed modified behavior. Its repeated applicability in the same specimen allows unraveling of how the brain changes behavior, as well as how behavior changes the brain.

ACKNOWLEDGMENTS

Supported by BOF-NOI, GOA funding from the University of Antwerp, and in part by EC-FP6-NoE DiMI, LSHB-CT-2005-512146 and NoE EMIL, and LSHC-CT-2004-503569 to A.V.d.L., and BOF-KP funding from the University of Antwerp to V.V.M., who is a postdoctoral researcher with the National Science Foundation (FWO) Flanders.

REFERENCES

1. Koretsky, A.P. and Silva, A.C., Manganese-enhanced magnetic resonance imaging (MEMRI), *NMR Biomed.*, 17, 527, 2004.
2. Juchem, C., Logothetis, N.K., and Pfeuffer, J., High-resolution (1)H chemical shift imaging in the monkey visual cortex, *Magn. Reson. Med.*, 54, 1541, 2005.
3. Tkac, I., Henry, P.G., Andersen, P., Keene, C.D., Low, W.C., and Gruetter, R., Highly resolved *in vivo* 1H NMR spectroscopy of the mouse brain at 9.4 T, *Magn. Reson. Med.*, 52, 478, 2004.
4. Tyszka, J.M., Fraser, S.E., and Jacobs, R.E., Magnetic resonance microscopy: recent advances and applications, *Curr. Opin. Biotechnol.*, 16, 93, 2005.
5. Drapeau, P. and Nachshen, D.A., Manganese fluxes and manganese-dependent neurotransmitter release in presynaptic nerve endings isolated from rat brain, *J. Physiol.*, 348, 493, 1984.
6. Du, C., MacGowan, G.A., Farkas, D.L., and Koretsky, A.P., Calibration of the calcium dissociation constant of Rhod(2) in the perfused mouse heart using manganese quenching, *Cell Calcium*, 29, 217, 2001.
7. Merritt, J.E., Jacob, R., and Hallam, T.J., Use of manganese to discriminate between calcium influx and mobilization from internal stores in stimulated human neutrophils, *J. Biol. Chem.*, 264, 1522, 1989.
8. Narita, K., Kawasaki, F., and Kita, H., Mn and Mg influxes through Ca channels of motor nerve terminals are prevented by verapamil in frogs, *Brain Res.*, 510, 289, 1990.
9. Simpson, P.B., Challiss, R.A., and Nahorski, S.R., Divalent cation entry in cultured rat cerebellar granule cells measured using Mn2+ quench of fura 2 fluorescence, *Eur. J. Neurosci.*, 7, 831, 1995.
10. Tisch-Idelson, D., Sharabani, M., Kloog, Y., and Aviram, I., Stimulation of neutrophils by prenylcysteine analogs: Ca(2+) release and influx, *Biochim. Biophys. Acta*, 1451, 187, 1999.
11. Wiemann, M., Busselberg, D., Schirrmacher, K., and Bingmann, D., A calcium release activated calcium influx in primary cultures of rat osteoblast-like cells, *Calcif. Tissue Int.*, 63, 154, 1998.
12. Lin, Y.J. and Koretsky, A.P., Manganese ion enhances T1-weighted MRI during brain activation: an approach to direct imaging of brain function, *Magn. Reson. Med.*, 38, 378, 1997.
13. Aoki, I., Tanaka, C., Takegami, T., Ebisu, T., Umeda, M., Fukunaga, M., Fukuda, K., Silva, A.C., Koretsky, A.P., and Naruse, S., Dynamic activity-induced manganese-dependent contrast magnetic resonance imaging (DAIM MRI), *Magn. Reson. Med.*, 48, 927, 2002.
14. Pautler, R.G., Silva, A.C., and Koretsky, A.P., *In vivo* neuronal tract tracing using manganese-enhanced magnetic resonance imaging, *Magn. Reson. Med.*, 40, 740, 1998.
15. Watanabe, T., Michaelis, T., and Frahm, J., Mapping of retinal projections in the living rat using high-resolution 3D gradient-echo MRI with Mn2+-induced contrast, *Magn. Reson. Med.*, 46, 424, 2001.
16. Saleem, K.S., Pauls, J.M., Augath, M., Trinath, T., Prause, B.A., Hashikawa, T., and Logothetis, N.K., Magnetic resonance imaging of neuronal connections in the macaque monkey, *Neuron*, 34, 685, 2002.
17. Pautler, R.G., Mongeau, R., and Jacobs, R.E., *In vivo* trans-synaptic tract tracing from the murine striatum and amygdala utilizing manganese enhanced MRI (MEMRI), *Magn. Reson. Med.*, 50, 33, 2003.
18. Pautler, R.G., *In vivo*, trans-synaptic tract-tracing utilizing manganese-enhanced magnetic resonance imaging (MEMRI), *NMR Biomed.*, 17, 595, 2004.
19. Allegrini, P.R. and Wiessner, C., Three-dimensional MRI of cerebral projections in rat brain *in vivo* after intracortical injection of MnCl2, *NMR Biomed.*, 16, 252, 2003.
20. Leergaard, T.B., Bjaalie, J.G., Devor, A., Wald, L.L., and Dale, A.M., *In vivo* tracing of major rat brain pathways using manganese-enhanced magnetic resonance imaging and three-dimensional digital atlasing, *Neuroimage*, 20, 1591, 2003.
21. Van der Linden, A., Verhoye, M., Van Meir, V., Tindemans, I., Eens, M., Absil, P., and Balthazart, J., *In vivo* manganese-enhanced magnetic resonance imaging reveals connections and functional properties of the songbird vocal control system, *Neuroscience*, 112, 467, 2002.
22. Van Meir, V., Verhoye, M., Absil, P., Eens, M., Balthazart, J., and Van der Linden, A., Differential effects of testosterone on neuronal populations and their connections in a sensorimotor brain nucleus controlling song production in songbirds: a manganese enhanced-magnetic resonance imaging study, *Neuroimage*, 21, 914, 2004.
23. Van Meir, V., Pavlova, D., Verhoye, M., Pinxten, R., Balthazart, J., Eens, M., and Van der Linden, A., *In vivo* MR imaging of the seasonal volumetric and functional plasticity of song control nuclei in relation to song output in a female songbird, *Neuroimage*, 31, 981–992, 2006.

24. Pautler, R.G. and Koretsky, A.P., Tracing odor-induced activation in the olfactory bulbs of mice using manganese-enhanced magnetic resonance imaging, *Neuroimage*, 16, 441, 2002.
25. Tindemans, I., Verhoye, M., Balthazart, J., and Van der Linden, A., *In vivo* dynamic ME-MRI reveals differential functional responses of RA- and area X-projecting neurons in the HVC of canaries exposed to conspecific song, *Eur. J. Neurosci.*, 18, 3352, 2003.
26. Stosiek, C., Garaschuk, O., Holthoff, K., and Konnerth, A., *In vivo* two-photon calcium imaging of neuronal networks, *Proc. Natl. Acad. Sci. U.S.A.*, 100, 7319, 2003.
27. Ohki, K., Chung, S., Ch'ng, Y.H., Kara, P., and Reid, R.C., Functional imaging with cellular resolution reveals precise micro-architecture in visual cortex, *Nature*, 433, 597, 2005.
28. Helmchen, F., Miniaturization of fluorescence microscopes using fibre optics, *Exp. Physiol.*, 87, 737, 2002.
29. Tabuchi, E., Yokawa, T., Mallick, H., Inubushi, T., Kondoh, T., Ono, T., and Torii, K., Spatio-temporal dynamics of brain activated regions during drinking behavior in rats, *Brain Res.*, 951, 270, 2002.
30. Sachdev, R.N., Champney, G.C., Lee, H., Price, R.R., Pickens, D.R., III, Morgan, V.L., Stefansic, J.D., Melzer, P., and Ebner, F.F., Experimental model for functional magnetic resonance imaging of somatic sensory cortex in the unanesthetized rat, *Neuroimage*, 19, 742, 2003.
31. Li, L., Weiss, C., Disterhoft, J.F., and Wyrwicz, A.M., Functional magnetic resonance imaging in the awake rabbit: a system for stimulus presentation and response detection during eyeblink conditioning, *J. Neurosci. Methods*, 130, 45, 2003.
32. Miller, M.J., Chen, N.K., Li, L., Tom, B., Weiss, C., Disterhoft, J.F., and Wyrwicz, A.M., fMRI of the conscious rabbit during unilateral classical eyeblink conditioning reveals bilateral cerebellar activation, *J. Neurosci.*, 23, 11753, 2003.
33. Febo, M., Numan, M., and Ferris, C.F., Functional magnetic resonance imaging shows oxytocin activates brain regions associated with mother-pup bonding during suckling, *J. Neurosci.*, 25, 11637, 2005.
34. Keen, C.L., Ensunsa, J.L., and Clegg, M.S., Manganese metabolism in animals and humans including the toxicity of manganese. In: *Metal Ions in Biological Systems (Vol. 37), Manganese and its Role in Biological Processes*, ed. H. Sigel and A. Sigel, 89–121, New York: Marcel Dekker, 2000.
35. Crossgrove, J. and Zheng, W., Manganese toxicity upon overexposure, *NMR Biomed.*, 17, 544, 2004.
36. Jiang, Y. and Zheng, W., Cardiovascular toxicities upon manganese exposure, *Cardiovasc. Toxicol.*, 5, 345, 2005.
37. Levy, B.S. and Nassetta, W.J., Neurologic effects of manganese in humans: a review, *Int. J. Occup. Environ. Health*, 9, 153, 2003.
38. Gallez, B., Demeure, R., Baudelet, C., Abdelouahab, N., Beghein, N., Jordan, B., Geurts, M., and Roels, H.A., Noninvasive quantification of manganese deposits in the rat brain by local measurement of NMR proton T1 relaxation times, *Neurotoxicology*, 22, 387, 2001.
39. Aoki, I., Takahashi, Y., Chuang, K.H., Silva, A.C., Igarashi, T., Tanaka, C., Childs, R.W., and Koretsky, A.P., Cell labeling for magnetic resonance imaging with the T1 agent manganese chloride, *NMR Biomed.*, 19, 50, 2006.
40. Tindemans, I., Boumans, T., Verhoye, M., and Van der Linden, A., IR-SE and IR-MEMRI allow *in vivo* visualization of oscine neuroarchitecture including the main forebrain regions of the song control system, *NMR Biomed.*, 19, 18, 2006.
41. Newland, M.C., Ceckler, T.L., Kordower, J.H., and Weiss, B., Visualizing manganese in the primate basal ganglia with magnetic resonance imaging, *Exp. Neurol.*, 106, 251, 1989.
42. Liu, C.H., D'Arceuil, H.E., and de Crespigny, A.J., Direct CSF injection of MnCl(2) for dynamic manganese-enhanced MRI, *Magn. Reson. Med.*, 51, 978, 2004.
43. Lee, J.H., Silva, A.C., Merkle, H., and Koretsky, A.P., Manganese-enhanced magnetic resonance imaging of mouse brain after systemic administration of MnCl2: dose-dependent and temporal evolution of T1 contrast, *Magn. Reson. Med.*, 53, 640, 2005.
44. Natt, O., Watanabe, T., Boretius, S., Radulovic, J., Frahm, J., and Michaelis, T., High-resolution 3D MRI of mouse brain reveals small cerebral structures *in vivo*, *J. Neurosci. Methods*, 120, 203, 2002.
45. Pautler, R.G. and Fraser, S.E., The year(s) of the contrast agent: micro-MRI in the new millennium, *Curr. Opin. Immunol.*, 15, 385, 2003.
46. Silva, A.C., Lee, J.H., Aoki, I., and Koretsky, A.P., Manganese-enhanced magnetic resonance imaging (MEMRI): methodological and practical considerations, *NMR Biomed.*, 17, 532, 2004.

47. Pautler, R.G., Biological applications of manganese-enhanced magnetic resonance imaging, *Methods Mol. Med.*, 124, 365, 2006.
48. Hu, T.C., Pautler, R.G., MacGowan, G.A., and Koretsky, A.P., Manganese-enhanced MRI of mouse heart during changes in inotropy, *Magn. Reson. Med.*, 46, 884, 2001.
49. Bremerich, J., Saeed, M., Arheden, H., Higgins, C.B., and Wendland, M.F., Normal and infarcted myocardium: differentiation with cellular uptake of manganese at MR imaging in a rat model, *Radiology*, 216, 524, 2000.
50. Saeed, M., Higgins, C.B., Geschwind, J.F., and Wendland, M.F., T1-relaxation kinetics of extracellular, intracellular and intravascular MR contrast agents in normal and acutely reperfused infarcted myocardium using echo-planar MR imaging, *Eur. Radiol.*, 10, 310, 2000.
51. Wendland, M.F., Saeed, M., Lund, G., and Higgins, C.B., Contrast-enhanced MRI for quantification of myocardial viability, *J. Magn. Reson. Imaging*, 10, 694, 1999.
52. Kim, R.J., Fieno, D.S., Parrish, T.B., Harris, K., Chen, E.L., Simonetti, O., Bundy, J., Finn, J.P., Klocke, F.J., and Judd, R.M., Relationship of MRI delayed contrast enhancement to irreversible injury, infarct age, and contractile function, *Circulation*, 100, 1992, 1999.
53. Rehwald, W.G., Fieno, D.S., Chen, E.L., Kim, R.J., and Judd, R.M., Myocardial magnetic resonance imaging contrast agent concentrations after reversible and irreversible ischemic injury, *Circulation*, 105, 224, 2002.
54. Wolf, G.L. and Baum, L., Cardiovascular toxicity and tissue proton T1 response to manganese injection in the dog and rabbit, *Am. J. Roentgenol.*, 141, 193, 1983.
55. Yanaga, T. and Holland, W.C., Effect of manganese on transmembrane potential and contractility of atrial muscle, *Am. J. Physiol.*, 217, 1280, 1969.
56. Federle, M.P., Chezmar, J.L., Rubin, D.L., Weinreb, J.C., Freeny, P.C., Semelka, R.C., Brown, J.J., Borello, J.A., Lee, J.K., Mattrey, R., Dachman, A.H., Saini, S., Harmon, B., Fenstermacher, M., Pelsang, R.E., Harms, S.E., Mitchell, D.G., Halford, H.H., Anderson, M.W., Johnson, C.D., Francis, I.R., Bova, J.G., Kenney, P.J., Klippenstein, D.L., Foster, G.S., and Turner, D.A., Safety and efficacy of mangafodipir trisodium (MnDPDP) injection for hepatic MRI in adults: results of the U.S. multicenter phase III clinical trials (safety), *J. Magn. Reson. Imaging*, 12, 186, 2000.
57. Elizondo, G., Fretz, C.J., Stark, D.D., Rocklage, S.M., Quay, S.C., Worah, D., Tsang, Y.M., Chen, M.C., and Ferrucci, J.T., Preclinical evaluation of MnDPDP: new paramagnetic hepatobiliary contrast agent for MR imaging, *Radiology*, 178, 73, 1991.
58. Wendland, M.F., Saeed, M., Bremerich, J., Arheden, H., and Higgins, C.B., Thallium-like test for myocardial viability with MnDPDP-enhanced MRI, *Acad. Radiol.*, 9 (Suppl 1), S82, 2002.
59. Harnish, P., Seoane, P., and Vessey, A., Manganese compositions and methods for MRI, U.S. Patent 5980863, 1999.
60. Storey, P., Danias, P.G., Post, M., Li, W., Seoane, P.R., Harnish, P.P., Edelman, R.R., and Prasad, P.V., Preliminary evaluation of EVP 1001-1: a new cardiac-specific magnetic resonance contrast agent with kinetics suitable for steady-state imaging of the ischemic heart, *Invest. Radiol.*, 38, 642, 2003.
61. Storey, P., Chen, Q., Li, W., Seoane, P.R., Harnish, P.P., Fogelson, L., Harris, K.R., and Prasad, P.V., Magnetic resonance imaging of myocardial infarction using a manganese-based contrast agent (EVP 1001-1): preliminary results in a dog model, *J. Magn. Reson. Imaging*, 23, 228, 2006.
62. Ballatori, N., Molecular mechanisms of hepatic metal transport. In: *Molecular Biology and Toxicology of Metals*, ed. R.K. Zalups and D.J. Koropatnick, 346–381, London: Taylor & Francis, 2000.
63. Hurley, L.S. and Keen, C.L., Manganese. In *Trace Elements in Human Health and Animal Nutrition (Vol. 1)*, ed. E. Underwood and W. Mertz, 185–223, New York: Academic Press, 1987.
64. Keen, C.L., Ensunsa, J.L., Lonnerdal, B., and Zidenberg-Cherr, S., Manganese: Physiology, dietary sources and requirements. In: *Encyclopedia of Human Nutrition (Vol. 1)(2nd Edition)*, ed. M. Sadder, B. Caballero, and S. Strain, 217–224, Oxford: Elsevier, 2005.
65. Newland, M.C., Cox, C., Hamada, R., Oberdorster, G., and Weiss, B., The clearance of manganese chloride in the primate, *Fundam. Appl. Toxicol.*, 9, 314, 1987.
66. Watanabe, T., Natt, O., Boretius, S., Frahm, J., and Michaelis, T., *In vivo* 3D MRI staining of mouse brain after subcutaneous application of MnCl2, *Magn. Reson. Med.*, 48, 852, 2002.
67. Watanabe, T., Frahm, J., and Michaelis, T., Functional mapping of neural pathways in rodent brain *in vivo* using manganese-enhanced three-dimensional magnetic resonance imaging, *NMR Biomed.*, 17, 554, 2004.

68. Aoki, I., Wu, Y.J., Silva, A.C., Lynch, R.M., and Koretsky, A.P., *In vivo* detection of neuroarchitecture in the rodent brain using manganese-enhanced MRI, *Neuroimage*, 22, 1046, 2004.
69. Kuo, Y.T., Herlihy, A.H., So, P.W., Bhakoo, K.K., and Bell, J.D., *In vivo* measurements of T1 relaxation times in mouse brain associated with different modes of systemic administration of manganese chloride, *J. Magn. Reson. Imaging*, 21, 334, 2005.
70. Zheng, W., Ren, S., and Graziano, J.H., Manganese inhibits mitochondrial aconitase: a mechanism of manganese neurotoxicity, *Brain Res.*, 799, 334, 1998.
71. Larsen, N.A., Pakkenberg, H., Damsgaard, E., and Heydorn, K., Topographical distribution of arsenic, manganese, and selenium in the normal human brain, *J. Neurol. Sci.*, 42, 407, 1979.
72. Bonilla, E., Salazar, E., Villasmil, J.J., and Villalobos, R., The regional distribution of manganese in the normal human brain, *Neurochem. Res.*, 7, 221, 1982.
73. Duflou, H., Maenhaut, W., and De Reuck, J., Regional distribution of potassium, calcium, and six trace elements in normal human brain, *Neurochem. Res.*, 14, 1099, 1989.
74. Cotzias, G.C., Horiuchi, K., Fuenzalida, S., and Mena, I., Chronic manganese poisoning. Clearance of tissue manganese concentrations with persistence of the neurological picture, *Neurology*, 18, 376, 1968.
75. Zheng, W., Kim, H., and Zhao, Q., Comparative toxicokinetics of manganese chloride and methylcyclopentadienyl manganese tricarbonyl (MMT) in Sprague-Dawley rats, *Toxicol. Sci.*, 54, 295, 2000.
76. Murphy, V.A., Wadhwani, K.C., Smith, Q.R., and Rapoport, S.I., Saturable transport of manganese(II) across the rat blood-brain barrier, *J. Neurochem.*, 57, 948, 1991.
77. Rabin, O., Hegedus, L., Bourre, J.M., and Smith, Q.R., Rapid brain uptake of manganese(II) across the blood-brain barrier, *J. Neurochem.*, 61, 509, 1993.
78. Takeda, A., Sawashita, J., and Okada, S., Localization in rat brain of the trace metals, zinc and manganese, after intracerebroventricular injection, *Brain Res.*, 658, 252, 1994.
79. Dastur, D.K., Manghani, D.K., Raghavendran, K.V., and Jeejeebhoy, K.N., Distribution and fate of Mn54 in the rat, with special reference to the C.N.S., *Q. J. Exp. Physiol. Cogn. Med. Sci.*, 54, 322, 1969.
80. Manghani, D.K., Dastur, D.K., Jeejeebhoy, K.N., and Raghavendran, K.V., Effect of stable manganese on the fate of radiomanganese in the rat with special reference to the CNS, *Indian J. Med. Res.*, 58, 209, 1970.
81. Takeda, A., Akiyama, T., Sawashita, J., and Okada, S., Brain uptake of trace metals, zinc and manganese, in rats, *Brain Res.*, 640, 341, 1994.
82. Yu, X., Wadghiri, Y.Z., Sanes, D.H., and Turnbull, D.H., *In vivo* auditory brain mapping in mice with Mn-enhanced MRI, *Nat. Neurosci.*, 8, 961, 2005.
83. Henriksson, J., Tallkvist, J., and Tjalve, H., Transport of manganese via the olfactory pathway in rats: dosage dependency of the uptake and subcellular distribution of the metal in the olfactory epithelium and the brain, *Toxicol. Appl. Pharmacol.*, 156, 119, 1999.
84. Tjalve, H., Mejare, C., and Borg-Neczak, K., Uptake and transport of manganese in primary and secondary olfactory neurones in pike, *Pharmacol. Toxicol.*, 77, 23, 1995.
85. Tjalve, H., Henriksson, J., Tallkvist, J., Larsson, B.S., and Lindquist, N.G., Uptake of manganese and cadmium from the nasal mucosa into the central nervous system via olfactory pathways in rats, *Pharmacol. Toxicol.*, 79, 347, 1996.
86. Tjalve, H. and Henriksson, J., Uptake of metals in the brain via olfactory pathways, *Neurotoxicology*, 20, 181, 1999.
87. Sloot, W.N. and Gramsbergen, J.B., Axonal transport of manganese and its relevance to selective neurotoxicity in the rat basal ganglia, *Brain Res.*, 657, 124, 1994.
88. Takeda, A., Kodama, Y., Ishiwatari, S., and Okada, S., Manganese transport in the neural circuit of rat CNS, *Brain Res. Bull.*, 45, 149, 1998.
89. Ballatori, N., Transport of toxic metals by molecular mimicry, *Environ. Health Perspect.*, 110 (Suppl 5), 689, 2002.
90. Harris, W.R. and Chen, Y., Electron paramagnetic resonance and difference ultraviolet studies of Mn2+ binding to serum transferrin, *J. Inorg. Biochem*, 54, 1, 1994.
91. Critchfield, J.W. and Keen, C.L., Manganese + 2 exhibits dynamic binding to multiple ligands in human plasma, *Metabolism*, 41, 1087, 1992.
92. Crossgrove, J.S. and Yokel, R.A., Manganese distribution across the blood-brain barrier. III. The divalent metal transporter-1 is not the major mechanism mediating brain manganese uptake, *Neurotoxicology*, 25, 451, 2004.

93. May, P.M., Linder, P.W., and Williams, D.R., Computer simulation of metal-ion equilibria in biofluids: models for low-molecular weight complex distribution of calcium(II), magnesium(II), manganese(II), iron(III), copper(II), zinc(II) and lead(II) in human plasma, *J. Chem. Soc. Dalton Trans.*, 6, 588–595, 1977.
94. Aisen, P., Aasa, R., and Redfield, A.G., The chromium, manganese, and cobalt complexes of transferrin, *J. Biol. Chem.*, 244, 4628, 1969.
95. Gibbons, R.A., Dixon, S.N., Hallis, K., Russell, A.M., Sansom, B.F., and Symonds, H.W., Manganese metabolism in cows and goats, *Biochim. Biophys. Acta*, 444, 1, 1976.
96. Frame, M.D. and Milanick, M.A., Mn and Cd transport by the Na-Ca exchanger of ferret red blood cells, *Am. J. Physiol.*, 261, C467, 1991.
97. Gavin, C.E., Gunter, K.K., and Gunter, T.E., Manganese and calcium efflux kinetics in brain mitochondria. Relevance to manganese toxicity, *Biochem. J.*, 266, 329, 1990.
98. Gunther, T., Vormann, J., and Cragoe, E.J., Jr., Species-specific Mn2+/Mg2+ antiport from Mg2(+)-loaded erythrocytes, *FEBS Lett.*, 261, 47, 1990.
99. Gunshin, H., Mackenzie, B., Berger, U.V., Gunshin, Y., Romero, M.F., Boron, W.F., Nussberger, S., Gollan, J.L., and Hediger, M.A., Cloning and characterization of a mammalian proton-coupled metal-ion transporter, *Nature*, 388, 482, 1997.
100. Conrad, M.E., Umbreit, J.N., Moore, E.G., Hainsworth, L.N., Porubcin, M., Simovich, M.J., Nakada, M.T., Dolan, K., and Garrick, M.D., Separate pathways for cellular uptake of ferric and ferrous iron, *Am. J. Physiol. Gastrointest. Liver Physiol.*, 279, G767, 2000.
101. Aschner, M. and Aschner, J.L., Manganese transport across the blood-brain barrier: relationship to iron homeostasis, *Brain Res. Bull.*, 24, 857, 1990.
102. Aschner, M. and Gannon, M., Manganese (Mn) transport across the rat blood-brain barrier: saturable and transferrin-dependent transport mechanisms, *Brain Res. Bull.*, 33, 345, 1994.
103. Crossgrove, J.S., Allen, D.D., Bukaveckas, B.L., Rhineheimer, S.S., and Yokel, R.A., Manganese distribution across the blood-brain barrier. I. Evidence for carrier-mediated influx of manganese citrate as well as manganese and manganese transferrin, *Neurotoxicology*, 24, 3, 2003.
104. Yokel, R.A., Crossgrove, J.S., and Bukaveckas, B.L., Manganese distribution across the blood-brain barrier. II. Manganese efflux from the brain does not appear to be carrier mediated, *Neurotoxicology*, 24, 15, 2003.
105. Connor, J.R., Iron regulation in the brain at the cell and molecular level, *Adv. Exp. Med. Biol.*, 356, 229, 1994.
106. Jefferies, W.A., Brandon, M.R., Hunt, S.V., Williams, A.F., Gatter, K.C., and Mason, D.Y., Transferrin receptor on endothelium of brain capillaries, *Nature*, 312, 162, 1984.
107. Moos, T., Immunohistochemical localization of intraneuronal transferrin receptor immunoreactivity in the adult mouse central nervous system, *J. Comp. Neurol.*, 375, 675, 1996.
108. Pardridge, W.M., Eisenberg, J., and Yang, J., Human blood-brain barrier transferrin receptor, *Metabolism*, 36, 892, 1987.
109. Dickinson, T.K., Devenyi, A.G., and Connor, J.R., Distribution of injected iron 59 and manganese 54 in hypotransferrinemic mice, *J. Lab. Clin. Med.*, 128, 270, 1996.
110. Malecki, E.A., Devenyi, A.G., Beard, J.L., and Connor, J.R., Transferrin response in normal and iron-deficient mice heterozygotic for hypotransferrinemia; effects on iron and manganese accumulation, *Biometals*, 11, 265, 1998.
111. Malecki, E.A., Cook, B.M., Devenyi, A.G., Beard, J.L., and Connor, J.R., Transferrin is required for normal distribution of 59Fe and 54Mn in mouse brain, *J. Neurol. Sci.*, 170, 112, 1999.
112. Crossgrove, J.S. and Yokel, R.A., Manganese distribution across the blood-brain barrier. IV. Evidence for brain influx through store-operated calcium channels, *Neurotoxicology*, 26, 297, 2005.
113. Wedler, F.C. and Denman, R.B., Glutamine synthetase: the major Mn(II) enzyme in mammalian brain, *Curr. Top. Cell Regul.*, 24, 153, 1984.
114. Aschner, M., Gannon, M., and Kimelberg, H.K., Manganese uptake and efflux in cultured rat astrocytes, *J. Neurochem.*, 58, 730, 1992.
115. Wedler, F.C., Ley, B.W., and Grippo, A.A., Manganese(II) dynamics and distribution in glial cells cultured from chick cerebral cortex, *Neurochem. Res.*, 14, 1129, 1989.
116. Savchenko, V.L., McKanna, J.A., Nikonenko, I.R., and Skibo, G.G., Microglia and astrocytes in the adult rat brain: comparative immunocytochemical analysis demonstrates the efficacy of lipocortin 1 immunoreactivity, *Neuroscience*, 96, 195, 2000.

117. Yokel, R.A. and Crossgrove, J.S., Manganese toxicokinetics at the blood-brain barrier, *Res. Rep. Health Eff. Inst.*, 7, 119, 7–58, 2004.
118. Watanabe, T., Radulovic, J., Spiess, J., Natt, O., Boretius, S., Frahm, J., and Michaelis, T., *In vivo* 3D MRI staining of the mouse hippocampal system using intracerebral injection of $MnCl_2$, *Neuroimage*, 22, 860, 2004.
119. Cohen, I. and Miles, R., Contributions of intrinsic and synaptic activities to the generation of neuronal discharges in *in vitro* hippocampus, *J. Physiol.*, 524 (Pt 2), 485, 2000.
120. Fisher, R.E., Gray, R., and Johnston, D., Properties and distribution of single voltage-gated calcium channels in adult hippocampal neurons, *J. Neurophysiol.*, 64, 91, 1990.
121. Schwartz, W.J. and Sharp, F.R., Autoradiographic maps of regional brain glucose consumption in resting, awake rats using (14C) 2-deoxyglucose, *J. Comp. Neurol.*, 177, 335, 1978.
122. Astic, L. and Saucier, D., Metabolic mapping of functional activity in the olfactory projections of the rat: ontogenetic study, *Brain Res.*, 254, 141, 1981.
123. Sokoloff, L., Reivich, M., Kennedy, C., Des Rosiers, M.H., Patlak, C.S., Pettigrew, K.D., Sakurada, O., and Shinohara, M., The [14C]deoxyglucose method for the measurement of local cerebral glucose utilization: theory, procedure, and normal values in the conscious and anesthetized albino rat, *J. Neurochem.*, 28, 897, 1977.
124. Nelson, S.R., Howard, R.B., Cross, R.S., and Samson, F., Ketamine-induced changes in regional glucose utilization in the rat brain, *Anesthesiology*, 52, 330, 1980.
125. Duncan, G.E., Miyamoto, S., Leipzig, J.N., and Lieberman, J.A., Comparison of brain metabolic activity patterns induced by ketamine, MK-801 and amphetamine in rats: support for NMDA receptor involvement in responses to subanesthetic dose of ketamine, *Brain Res.*, 843, 171, 1999.
126. Suzuki, H., Wada, O., Inoue, K., Tosaka, H., and Ono, T., Role of brain lysosomes in the development of manganese toxicity in mice, *Toxicol. Appl. Pharmacol.*, 71, 422, 1983.
127. Sugaya, K., Chouinard, M.L., and McKinney, M., Induction of manganese superoxide dismutase in BV-2 microglial cells, *Neuroreport*, 8, 3547, 1997.
128. Liccione, J.J. and Maines, M.D., Selective vulnerability of glutathione metabolism and cellular defense mechanisms in rat striatum to manganese, *J. Pharmacol. Exp. Ther.*, 247, 156, 1988.
129. Becker, G.L., Fiskum, G., and Lehninger, A.L., Regulation of free Ca2+ by liver mitochondria and endoplasmic reticulum, *J. Biol. Chem.*, 255, 9009, 1980.
130. Cobbold, P.H. and Rink, T.J., Fluorescence and bioluminescence measurement of cytoplasmic free calcium, *Biochem. J.*, 248, 313, 1987.
131. Unitt, J.F., McCormack, J.G., Reid, D., MacLachlan, L.K., and England, P.J., Direct evidence for a role of intramitochondrial Ca2+ in the regulation of oxidative phosphorylation in the stimulated rat heart. Studies using 31P n.m.r. and ruthenium red, *Biochem. J.*, 262, 293, 1989.
132. Elluru, R.G., Bloom, G.S., and Brady, S.T., Fast axonal transport of kinesin in the rat visual system: functionality of kinesin heavy chain isoforms, *Mol. Biol. Cell*, 6, 21, 1995.
133. Grafstein, B. and Forman, D.S., Intracellular transport in neurons, *Physiol. Rev.*, 60, 1167, 1980.
134. Gallez, B., Baudelet, C., and Geurts, M., Regional distribution of manganese found in the brain after injection of a single dose of manganese-based contrast agents, *Magn. Reson. Imaging*, 16, 1211, 1998.
135. Takeda, A., Ishiwatari, S., and Okada, S., *In vivo* stimulation-induced release of manganese in rat amygdala, *Brain Res.*, 811, 147, 1998.
136. Kannurpatti, S.S., Joshi, P.G., and Joshi, N.B., Calcium sequestering ability of mitochondria modulates influx of calcium through glutamate receptor channel, *Neurochem. Res.*, 25, 1527, 2000.
137. Ashton, B., Manganese and man, *J. Orthomol. Psychiatry*, 9, 237, 1980.
138. Kita, H., Narita, K., and van der, K.W., Tetanic stimulation increases the frequency of miniature end-plate potentials at the frog neuromuscular junction in Mn2+-, CO2+-, and Ni2+-saline solutions, *Brain Res.*, 205, 111, 1981.
139. Brouillet, E.P., Shinobu, L., McGarvey, U., Hochberg, F., and Beal, M.F., Manganese injection into the rat striatum produces excitotoxic lesions by impairing energy metabolism, *Exp. Neurol.*, 120, 89, 1993.
140. Ramos, K.S., Chacon, E., and Daniel Acosta, J., Toxic responses of the heart and vascular systems. In *Casarett and Doull's Toxicology: The Basic Science of Poisons (5th edition)*, ed. C.D. Klaasen, 487–527, New York: McGraw-Hill, 1996.
141. Takeda, A., Function and toxicity of trace metals in the central nervous system, *Clin. Calcium*, 14, 45, 2004.

142. Morita, H., Ogino, T., Seo, Y., Fujiki, N., Tanaka, K., Takamata, A., Nakamura, S., and Murakami, M., Detection of hypothalamic activation by manganese ion contrasted T(1)-weighted magnetic resonance imaging in rats, *Neurosci. Lett.*, 326, 101, 2002.
143. Sun, N., Li, Y., Tian, S., Lei, Y., Zheng, J., Yang, J., Sui, N., Xu, L., Pei, G., Wilson, F.A., Ma, Y., Lei, H., and Hu, X., Dynamic changes in orbitofrontal neuronal activity in rats during opiate administration and withdrawal, *Neuroscience*, 138, 77, 2006.
144. Thuen, M., Singstad, T.E., Pedersen, T.B., Haraldseth, O., Berry, M., Sandvig, A., and Brekken, C., Manganese-enhanced MRI of the optic visual pathway and optic nerve injury in adult rats, *J. Magn. Reson. Imaging*, 22, 492, 2005.
145. Bilgen, M., Dancause, N., Al-Hafez, B., He, Y.Y., and Malone, T.M., Manganese-enhanced MRI of rat spinal cord injury, *Magn. Reson. Imaging*, 23, 829, 2005.
146. Bilgen, M., Peng, W., Al-Hafez, B., Dancause, N., He, Y.Y., and Cheney, P.D., Electrical stimulation of cortex improves corticospinal tract tracing in rat spinal cord using manganese-enhanced MRI, *J. Neurosci. Methods*, 156, 17–22, 2006.
147. Van der Linden, A., Van Meir, V., Tindemans, I., Verhoye, M., and Balthazart, J., Applications of manganese-enhanced magnetic resonance imaging (MEMRI) to image brain plasticity in song birds, *NMR Biomed.*, 17, 602, 2004.
148. Mathern, G.W., Pretorius, J.K., and Babb, T.L., Influence of the type of initial precipitating injury and at what age it occurs on course and outcome in patients with temporal lobe seizures, *J. Neurosurg.*, 82, 220, 1995.
149. Nairismagi, J., Pitkanen, A., Narkilahti, S., Huttunen, J., Kauppinen, R.A., and Grohn, O.H., Manganese-enhanced magnetic resonance imaging of mossy fiber plasticity *in vivo*, *Neuroimage*, 30, 130, 2006.
150. Wadghiri, Y.Z., Blind, J.A., Duan, X., Moreno, C., Yu, X., Joyner, A.L., and Turnbull, D.H., Manganese-enhanced magnetic resonance imaging (MEMRI) of mouse brain development, *NMR Biomed.*, 17, 613, 2004.
151. De Groof, G., Verhoye, M., Van Meir, V., Tindemans, I., Leemans, A., and Van der Linden, A., *In vivo* diffusion tensor imaging (DTI) of brain subdivisions and vocal pathways in songbirds, *Neuroimage*, 29, 754, 2006.
152. Van Meir, V., Boumans, T., De Groof, G., Van Audekerke, J., Smolders, A., Scheunders, P., Sijbers, J., Verhoye, M., Balthazart, J., and Van der Linden, A., Spatiotemporal properties of the BOLD response in the songbirds' auditory circuit during a variety of listening tasks, *Neuroimage*, 25, 1242, 2005.
153. Aoki, I., Naruse, S., and Tanaka, C., Manganese-enhanced magnetic resonance imaging (MEMRI) of brain activity and applications to early detection of brain ischemia, *NMR Biomed.*, 17, 569, 2004.

Part IV

Future Perspectives for Molecular and Cellular Imaging

21 Translating Promising Experimental Approaches to Clinical Trials

Adrian D. Nunn

CONTENTS

21.1 Introduction 395
21.2 General 396
21.3 Nonclinical Safety Assessment 397
 21.3.1 General considerations 397
 21.3.2 Imaging agents 397
21.4 Recent First-in-Man Concepts 398
 21.4.1 Microdosing 399
 21.4.2 Exploratory IND 399
21.5 Discussion 401
References 402

21.1 INTRODUCTION

In most cases (oncology being an exception), traditional phase I clinical trials are performed in healthy volunteers in an attempt to remove variables due to underlying disease and deteriorations in physiology and because of the intensity of the monitoring required. A new chemical entity (NCE) cannot (should not) be introduced into man without an assessment of the risk–benefit ratio, which requires knowledge of its safety profile in animals. As the first clinical studies are traditionally safety studies, the initial administrations are performed using doses that are low relative to the no observable adverse event level (NOAEL) found in animal testing. This requires practical knowledge of toxic effects and an understanding of the dose–response relationship. It should be obvious that in order to accurately translate the animal results into expected behavior in man, the quality of the material used in each case should be known and be similar. It should also be obvious that if the proposed testing is of an existing compound that involves a (new) route of administration or dosage level, or use in a patient population, or other factor that significantly increases the risks (or decreases the acceptability of the risks) associated with the compound, then additional relevant preclinical safety testing may be necessary.

The general requirements for filing an Investigational New Drug (IND) application in the U.S. are laid down in 21 CFR 312,[1] and specific rules for radiopharmaceuticals (only) can be found in 21 CFR 315.[2] There are also a number of guidances addressing all imaging agents. This chapter will focus mainly on the safety aspects of these requirements as elaborated in the various guidances issued by the authorities, with the understanding that the chemical manufacturing and control (CMC) elements are met such that the transfer of those safety aspects from animals to man is not confounded by variations in response due to variations in test material. This chapter will explore the issues relevant to taking a medical imaging agent NCE into clinical trials. Most emphasis will be directed toward the situation in the U.S., as the Food and Drug Administration (FDA) has been more active with respect to imaging agents, but where other regulatory authorities have published relevant documents these will be compared. The general concepts outlined by the position of the FDA may be used as a starting point in the absence of opinions expressed by other authorities.

This chapter is not directed at those professionals in established pharmaceutical companies who have access to experienced toxicology/pathology medical, and regulatory groups, but is instead written for those without these resources. The intent is to provide sufficient information to allow better communication between the various groups, including that with the regulatory authorities.

21.2 GENERAL

The amount and type of information required prior to human testing is determined by the regulatory authorities responsible for the area in which the studies will take place. In the past there was a degree of variability in the requirements from country to country, but in the last decade the authorities have made efforts to develop a common set of requirements for the registration (approval) of new drugs. This process is termed the International Conference on Harmonisation of Technical Requirements for Registration of Pharmaceuticals for Human Use (ICH).[3] Nevertheless, there still exist some regional differences in the requirements. The ICH website contains a list of useful, readable publications covering the whole range of topics pertinent to first administration to man, and extending to ultimate approval and routine use. In brief, the requirements for first use control the quality and reproducibility of the product intended to be used in the clinic for the first time (and to achieve subsequent approval) and define a package of safety data developed in a nonclinical setting that allows an assessment of the risks of administering the NCE to the first group of humans. There is one set of requirements for all drugs, but they are sufficiently flexible to accommodate the wide range in toxicities, desired pharmacologic activity, dosing schedule, duration of use, etc., that may be encountered. Biological products have additional requirements, deriving mainly from their methods of production, but also to reflect their potential for immunogenicity.

Medical imaging agents are generally governed by the same regulations as other drug and biological products. However, because medical imaging agents are by definition used solely to diagnose and monitor diseases or conditions, as opposed to treat them, development programs for medical imaging agents can be tailored to reflect these particular uses, and the regulatory authorities have issued some guidance. Clearly, as we enter further into the field of molecular imaging, the NCEs may have potent pharmacologic activity, such that even at (low) imaging doses they may be little separated from therapeutic drugs with regard to their activity, and the preclinical testing should reflect this.

There is significant pressure from all sides to find ways to speed up the discovery, development, and approval of all new drugs. The FDA has identified imaging methods as a key technology in this regard,[4] but of course the introduction of new imaging agents suffers from the same issues as other drugs. The regulatory authorities are attempting to address these issues. As traditional phase I trials are by definition safety studies performed initially, at least, at low doses, no efficacy data are expected, and this is indeed the case for a therapeutic drug. However, if any imaging is performed during an imaging agent trial, efficacy may be observable. This is a double-edged sword, as it tends to obscure the true need of the traditional phase I trial, which is to collect the safety data. Without the necessary safety data, the regulatory authorities do not allow progression to more patients, higher doses, etc.

The requirements and opinions of the regulatory authorities are discussed below in chronological order, as this best illustrates the movement that has occurred with regard to assisting the passage of NCEs into man.

21.3 NONCLINICAL SAFETY ASSESSMENT

The overall need for the safety assessment of an NCE is to acquire sufficient nonclinical biological data to allow the development of a safety profile for the NCE. This is then used to (1) identify an initial safe dose and subsequent dose escalation schemes in humans, (2) identify potential target organs or tissues for toxicity and for the study of whether such toxicity is reversible, and (3) identify safety parameters for clinical monitoring. The nonclinical safety studies, although limited at the beginning of clinical development, should be adequate to characterize potential toxic effects under the conditions of the supported clinical trial.

21.3.1 General considerations

The ICH document M3[5] lays out the following framework for studies, and the information that must be obtained, prior to human exposure of any drug.

Single-dose (acute) toxicity: A repeated-dose toxicity study in two species (one nonrodent) for a minimum duration of 2 weeks would support phase I (Human Pharmacology) and phase II (Therapeutic Exploratory) studies up to 2 weeks in duration.[5]

Repeated-dose toxicity in two mammalian species (one nonrodent): The duration should be equal to or exceed the duration of the human clinical trials up to the maximum recommended duration of the repeated-dose toxicity studies.[5] In the U.S., may be replaced with a more extensive single-dose study.

Toxicokinetics/pharmacokinetics (absorption, distribution, metabolism, excretion, or ADME) and the generation of pharmacokinetic data to assess systemic exposure: Toxicokinetics is the pharmacokinetics under the conditions of the toxicity studies, which may exhibit different behavior than the expected clinical dosing conditions due to nonlinear kinetics, etc.[6]

Safety pharmacology: The assessment of effects on vital functions, such as cardiovascular, central nervous, and respiratory systems.[7]

Local tolerance studies: The assessment of tolerance to the drug using routes relevant to the drug intended for clinical use, for instance, effects on the immediate vasculature after iv administration (when the highest concentrations may be anticipated) and perivenous or intramuscular effects that may occur after extravasation during administration.

Genotoxicity studies: *In vitro* tests for the evaluation of mutations and chromosomal damage. These are bacterial based, using a standard set of strains and a human lymphocyte chromosomal aberration test. They may be supplemented with a mouse bone marrow micronucleus test.

These requirements and guidelines may be modified based on the type of NCE and on the intended clinical pathology.

21.3.2 Imaging agents

Imaging agents have their own guidance document issued by the FDA.[8] In the referenced document, imaging agents are divided into contrast agents and radiopharmaceuticals. Contrast agents comprise iodinated compounds used in radiography and computed tomography (CT), paramagnetic metallic ions (such as ions of gadolinium, iron, and manganese) linked to a variety of molecules and microparticles (such as superparamagnetic iron oxide) used in magnetic resonance imaging (MRI), and microbubbles and related microparticles used in diagnostic ultrasonography. Radiopharmaceuticals include the final radioactive drugs and the nonradioactive kits from which many are produced. At the time this document was issued, optical imaging agents were not discussed in detail.

The stated position of the FDA is that (CT or MRI) contrast agents may pose safety issues because of the inherently large amounts used for a single administration. It recommends that studies be conducted to address the effects of large mass dose and volume, and osmolality. For MRI agents, the potential for loss of the metal from the chelate or deleterious effects resulting from deposition of the metal in tissues should be examined.[7] Classical high (inorganic) chemical stability determined *in vitro* may not be sufficient to predict behavior in biological systems.

Thus, nonradiolabeled imaging agents generally should be treated like therapeutic agents for the purpose of conducting clinical safety assessments. Inherent in the general guidance, but pointed out in the imaging guidance document is the fact that many medical imaging agents are administered infrequently or as single doses. Thus, adverse events (AEs) that are related to long-term use or to accumulation are less likely to occur than with drugs that are administered repeatedly to the same patient. The nonclinical development programs for such single-use products usually can omit long-term (i.e., 3 months' duration or longer), repeated-dose safety studies. However, where it is possible that the medical imaging agent will be administered to a single patient repeatedly (e.g., to monitor disease progression), repeated-dose studies of 14 to 28 days duration are recommended.

With regard to radiopharmaceuticals, the FDA anticipates the potential for the pharmacologic activity now facing biologically active molecular imaging agents of all sorts and recommends that special safety considerations for diagnostic radiopharmaceuticals include assessment of the mass, toxic potency, and receptor interactions for any unlabeled moiety (in addition to the radiolabeled mass) and assessment of potential pharmacologic or physiologic effects due to molecules that bind with receptors or enzymes.

Also in this document, the FDA has raised the idea that there may be two general categories of imaging agents that could be separated based on their potential for toxicity and that may qualify for different levels of safety assessment *during the early phases of clinical trials.* These two categories were conceived to help drug sponsors identify and differentiate those characteristics that are of greatest interest to the FDA in assessing the potential safety of a medical imaging agent.[9] Generally, a less extensive clinical safety evaluation is appropriate for Group 1 agents. Biological agents are assumed to be Group 2 agents unless they can be demonstrated to lack immunogenicity, and those agents emitting primarily beta or alpha particles are also Group 2. Other agents must meet certain safety requirements, which are a combination of nonclinical and clinical characteristics, to qualify for Group 1. These characteristics are as follows:

- The NOAEL in expanded acute, single-dose toxicity studies and safety pharmacology studies should be at least 100 times greater, and in short-term, repeated-dose toxicity studies, at least 25 times greater, in suitable animal species than the maximal mass dose to be used in human studies.
- The clinical criteria are much less well defined and include whether safety issues were identified during initial human use of the medical imaging agent in appropriately designed studies, and if any AEs occurred that were not predicted from effects observed in animals.

The FDA states that it is willing to consider classifying a medical imaging agent as Group 1 even if its NOAELs are slightly less than the multiples specified above. To do this, the FDA proposes to take into consideration, among other things, how close the NOAELs are to the multiples specified above, the amount of safety information known about chemically similar and pharmacologically related medical imaging agents, the nature of observed animal toxicities, and whether adverse events have occurred during initial human experience, including the nature of such adverse events.

21.4 RECENT FIRST-IN-MAN CONCEPTS

The first attempt to categorize imaging agents into two groups based on their safety profile, with resulting reduced requirements and perceived faster development,[9] occurred at the same time that the need to speed up the development of all drugs was widely recognized.[4] This led to a new set of guidelines being issued that covered not just imaging agents, but all drugs. These guidelines were an acknowledgment that the first human studies resulted in, or led to, a high failure rate, and that ways needed to be found to increase the number of NCEs going into man or to perform better-designed studies. The idea was to allow limited testing in man to determine proof of concept of compounds that had a sufficient set of preclinical safety data to establish the risks of the initial testing, but not necessarily of the higher doses usually seen in the traditional phase I design.

21.4.1 Microdosing

The European Medicines Agency (EMEA) issued a position paper in an attempt to ease the entry of compounds into man and introduced the microdosing concept.[10] This position paper defines modified requirements for the nonclinical safety studies needed to support human clinical trials of a single microdose of a pharmacologically active compound (pre-phase I studies). A microdose is defined as less than 1/100 of the dose calculated to yield a pharmacologic effect based on pharmacodynamic data obtained *in vitro* and *in vivo*, with a maximum dose of ≤100 μg per individual. The nonclinical requirements are very much like those for a Group I drug as described in the FDA's guidance document[9]; however, microdosing does not require prior clinical experience with the compound.

The position paper provides examples of microdosing clinical trails that might comprise the early characterization of an NCE by positron emission tomography (PET) imaging or using some other very sensitive analytical technique. The intent was not to limit the NCEs to imaging agents, but imaging could be a means of characterizing the behavior of a nonimaging (therapeutic) agent. The clinical trials could be conducted with a number of closely related pharmaceutical candidates, to choose the preferred candidate or formulation for further development, provided the total amount of test compounds does not exceed 100 μg per individual. The idea of comparing multiple closely related compounds in a phase I study is not new; mention of it appeared in an FDA publication on single-dose acute toxicity testing in 1996;[11] however, the practicalities were not publicly addressed until the microdosing guidance, and this may have reduced the use of such an opportunity.

21.4.2 Exploratory IND

The FDA recently issued a much more extensive guidance document,[12] elaborating on the concept of a microdose, in the form of an Exploratory IND (E-IND). The guidance describes preclinical and clinical approaches and the CMC information that should be considered when planning E-IND studies in humans.

The E-IND study is intended to describe a clinical trial that is conducted early in phase 1, involves very limited human exposure, and has no therapeutic or diagnostic intent (e.g., screening studies, microdose studies). These studies are conducted *prior to* the traditional dose escalation, safety, and tolerance studies that ordinarily occur in a traditional phase 1. The goal of the microdosing and E-IND trials is fundamentally different from that of the traditional phase I, which is one of safety. Risks must be managed, but the low doses that these new trial designs require or insist on markedly reduce the risks. Anticipated pharmacologic effects may be acceptable during the clinical phase of E-IND studies, but unlike the traditional IND, the studies are not designed to establish maximally tolerated doses (MTDs). Instead, they assess feasibility for further development of the drug or biological product. As for the EMEA's microdosing position, the E-IND provides the option of studying closely related drugs or therapeutic biological products, under a single IND application. Examples of information that an E-IND trial may provide are to:

1. Determine whether a mechanism of action defined in experimental systems can also be observed in humans (e.g., a binding property or inhibition of an enzyme)
2. Provide information on pharmacokinetics (PK)

3. Select the most promising lead product from a group of candidates designed to interact with a particular therapeutic target in humans, based on PK or pharmacodynamic (PD) properties
4. Explore a compound's or group of compounds' biodistribution characteristics using various imaging technologies

In agreement with the EMEA, a microdose as defined by the FDA is less than 1/100 of the dose calculated to yield a pharmacologic effect in man based on animal data and a total dose of ≤100 μg per individual. The maximum dose for proteins is ≤30 nmol due to the high molecular weights. The total dose per individual of all test compounds should not exceed 100 μg, so if multiple compounds are tested or if receptor blocking studies are performed, the total dose should not exceed this.

The guidance describes three types of clinical study that are sufficiently different that they have different entry requirements. The first is clinical studies of pharmacokinetics or imaging. Single-dose toxicity studies may be sufficient, and the text states that the studies should be designed to establish a dose inducing a minimal toxic effect or, alternatively, establishing a margin of safety. A pharmacologic effect (*vide supra*) may or may not be a toxic effect, so the wording here is a little vague, but it would appear that an expected or desired transitory pharmacologic effect may be differentiated from an unexpected, nontransitory pharmacologic effect or a frank toxic effect. To establish a margin of safety, it should be demonstrated that a large multiple (e.g., 100×) of the proposed human dose does not induce adverse effects in the experimental animals. This can present an issue if one is dealing with a nontoxic compound, as large amounts of material may need to be administered to the animals, yet the volume that may be administered is limited.[13] The FDA has recognized this in the sense that it acknowledges that it may be necessary to adjust formulations, dose schedules, etc. Scaling from animals to humans based on body surface area can (should?) be used to select the dose for use in the clinical trial. There is a useful calculator on the FDA website that performs these scaling calculations.[14]

The second type of clinical study involves pharmacologically relevant doses that require more extensive nonclinical testing. At first glance, this appears to be designed for testing therapeutic drugs and not applicable to imaging studies. However, if one considers the potential for pharmacological activity in doses of highly active metabolic imaging agents, or the FDA's previous position on contrast agents, one would be well advised to consult with the FDA early on to avoid any surprises. (Indeed, a desire or need for early consultation is a general theme throughout the document.) Unlike the EMEA's microdosing position paper, which, as written, limits the trial only by dose, the E-IND limits the number of subjects and the dosing period. The duration of dosing in an E-IND study of this type is expected to be limited (e.g., 7 days).

The third example involves studies to determine the mechanism of action of a drug. The animal studies should incorporate endpoints that are mechanistically based on the pharmacology of the new chemical entity and thought to be important to clinical effectiveness. For example, if the degree of saturation of a receptor or the inhibition of an enzyme were considered related to effectiveness, this parameter would be characterized and determined in the animal study and then used as an endpoint in a subsequent clinical trial. Such a study clearly traverses the boundary between a molecular imaging agent and a therapeutic agent.

Table 21.1 attempts to list in a simplified form the various requirements that need to be met before the regulatory authorities will allow the first human administration of an NCE or administration of a previously tested compound administered in a different manner. For the E-IND, only the requirements for a classical imaging study are listed. This is not a simple exercise, because of the degrees of freedom proffered in the various guidances and position papers that themselves reflect changes due to their chronology. It is important to understand that the microdosing or E-IND options are a means to expedite entry of NCEs into early-stage clinical trails. Encouraging results in these early trials do not allow continued progression without meeting the requirements for a regular IND,

TABLE 21.1
Chronological Summary of Imaging Agent Requirements Listed by Regulatory Authorities for First in Man or New Use Studies

Study Type	ICH M3 1997	FDA Imaging Guidance 2004	EMEA Microdosing (Pre-Phase I) 2004	E-IND Imaging (Pre-Phase I) 2006
Safety pharmacology	Yes	Yes	Maybe	No
Toxicokinetic pharmacokinetic	Yes	Yes		Yes
Single-dose toxicity or dose escalation	Two species	Expanded acute single dose	Expanded acute single dose, 2 weeks, one species	Expanded acute single dose, 2 weeks, one species
Short-term (2–4 weeks), multiple-dose toxicity	Yes	No	No	No
Genotoxicity	*In vitro*	*In vitro*	*In vitro*	No
Reproductive and developmental toxicity[a]	Yes	Yes or waived	No	
Dose limits		Group 1 agents *either* <1/100 NOAEL expanded acute single-dose and safety pharmacology *or* <1/25 NOAEL repeated dose *and* clinical data	1/100 pharmacologically active dose; total 100 μg	1/100 pharmacologically active dose; total 100 μg or 30 n*M* protein

[a] There are different requirements depending on whether the first subjects are males or females and, if the latter, of childbearing potential. There are also differences between jurisdictions.

etc., as laid out in ICH M3; indeed, the stated intention of the FDA is that the E-IND should be withdrawn on completion of the intended studies (and an IND opened if warranted).

It is evident that a key change has been made in the switch from the use of the terms *NOAEL* and *toxic effect* to the use of *pharmacologically active dose*. (Although as noted above, these terms are still interchanged.) The problem, though, has not changed — to collect data that allow an accurate risk assessment prior to first-in-man studies. This is a moving target, as it depends on the compound, the goal, and the patient population. For instance, for anticancer compounds focusing on cytotoxic/cytostatic drugs that are presumed to have a direct effect on tumor cells, the risk–benefit ratio is such that the MTD, as determined using single-dose toxicity studies, not the NOAEL, which is generally much lower, is used to establish the appropriate starting dose in phase I studies in cancer patients. One tenth of the MTD is frequently an acceptable starting dose, much higher than would be used with noncancer drugs in healthy subjects.[15]

21.5 DISCUSSION

Imaging agents play a central role in the FDA's view of the future of developing new drugs as outlined in the Critical Path Initiative[4] and further developed in the Report[16] and List.[17] This interest in imaging agents comes at a time when commercial pressures are making the development of imaging agents unattractive.[18] The E-IND, which represents the very latest thinking of the regulatory authorities on the subject of getting NCEs into man, is still a young document. Its development involved a good deal of public debate, some of which has been recorded and is well worth reading.[19]

How will imaging agents fare under these new ideas? The sensitivity of the various imaging techniques would seem to preclude CT agents and gadolinium-based MRI agents from first-in-man trials under the auspices of microdosing, or the E-IND imaging model if imaging is used for detection. More sensitive analytical techniques may allow doses of <100 μg and still provide PK data, but in the age of targeted drugs, this may not have sufficient value. Clearly, the microdosing and E-IND opinions were written with the high-sensitivity imaging techniques of nuclear positron emission tomography or single photon computed tomography (PET/SPECT) and possible ultrasound and optical imaging in mind. Or perhaps it is better to say that these techniques are the ones that are able to meet the low-risk criteria that allow first-in-man studies with limited data.

The superparamagnetic iron particles may meet the criteria in some cases, as they have high sensitivity.[20] The use of cells labeled with such iron particles is a case in point. This has been done in the past using a radiopharmaceutical (indium oxine) approved for radiolabeling cells *ex vivo*. After some years of nonclinical development, iron oxide-labeled dendritic cells were recently administered to melanoma patients and imaged by MRI.[21] This paper raises a number of illustrative points. First, the authors were well aware of the potential safety and regulatory issues of labeling cells using unapproved iron particle preparations, antibodies and transfection techniques, etc., and elected instead to use the normal phagocytic capacity of the immature dendritic cells with an approved iron particle. The particles in question are approved in Europe and the U.S., but not for this indication. The authors administered 7.5×10^6 cells containing 10 to 30 pg of iron per cell, or a total of 1 to 2 μg, which lies well within the mass amount limits of microdosing and the E-IND. The study was performed in stage III melanoma patients being treated by vaccination, so a higher starting dose (10% of the MTD) of iron may well be acceptable if needed. However, although the safety profile of the iron particles after iv administration is established, the toxicity to these particular cells is not, so what is the meaning of a (whole-body) MTD in this case? The authors performed functional assays *in vitro* and observed the behavior *in vivo*, the results of which suggest that the cells have not been grossly altered by their high iron loading. Considering the pivotal role that these cells play in initiating the immune response, it is not unreasonable to ask if this is sufficient. A complicating factor in this study is that the cells were also loaded with tumor-derived antigenic peptides. Similar issues will have to be addressed regarding labeled stem cells where their ability to differentiate into a multitude of cell and tissue types offers at least a theoretical capability to repair, regenerate, or grow almost any type of tissue or organ. How does one establish the long-term toxic effects, if any, of labeling such cells? These are issues that are not touched on in the published regulatory advice. Certainly end-stage cancer patients are an appropriate place to start, but the transition to patients with a greater life expectancy or a worse risk–benefit ratio will require discussion with the regulatory authorities and probably recourse to the traditional IND.

The exchange of information and ideas among the FDA (and other regulatory authorities), drug developers, and users has become much more open and effective in recent years. This is exemplified by the three Critical Path documents that the FDA has issued, in which the major role of imaging in drug development is laid out. There are more formal venues for discussions on imaging, such as the Interagency Council on Biomedical Imaging in Oncology (ICBIO),[22] which makes available FDA and National Cancer Institute staff, among others, to address imaging issues. Finally, there are opportunities to meet formally with the FDA as part of the normal drug approval process. All should be used to clarify the issues that need to be addressed to take an NCE imaging agent into man for the first time.

REFERENCES

1. Investigational New Drug Application. 21 CFR 312. Available at http://ecfr.gpoaccess.gov/cgi/t/text/text-idx?c=ecfr&tpl=/ecfrbrowse/Title21/21cfr312_main_02.tpl. Accessed March 28, 2006.
2. Regulations for *In Vivo* Radiopharmaceuticals Used for Diagnosis and Monitoring. 21 CFR 315. Available at http://ecfr.gpoaccess.gov/cgi/t/text/text-idx?c=ecfr&tpl=/ecfrbrowse/Title21/21cfr315_main_02.tpl. Accessed March 28, 2006.

3. ICH guidance documents. Available at http://www.ich.org/cache/compo/276-254-1.html. Accessed March 28, 2006.
4. Innovation and Stagnation. Challenge and Opportunity on the Critical Path to New Medical Products. FDA. March 2004. pp. 11, 24. Available at http://www.fda.gov/oc/initiatives/criticalpath/. Accessed July 11, 2005.
5. M3 Nonclinical Safety Studies for the Conduct of Human Clinical Trials for Pharmaceuticals. ICH. July 1, 1997; amended November 9, 2000. Available at http://www.ich.org/cache/compo/276-254-1.html. Accessed March 28, 2006.
6. S3A Note for Guidance on Toxicokinetics: The Assessment of Systemic Exposure in Toxicity Studies. ICH. October 27, 1994.
7. S7A Safety Pharmacology Studies for Human Pharmaceuticals. ICH. November 8, 2000.
8. Guidance for Industry Developing Medical Imaging Drug and Biological Products Part 1: Conducting Safety Assessments. FDA. June 2004. Available at http://www.fda.gov/cber/guidelines.htm. Accessed March 28, 2006.
9. Guidance for Industry Developing Medical Imaging Drug and Biological Products Part 1: Conducting Safety Assessments. Part IV, Group 1 and 2, Medical Imaging Agents. FDA, June 2004. Available at http://www.fda.gov/cber/guidelines.htm. Accessed March 28, 2006.
10. Position paper on nonclinical safety studies to support clinical trials with a single microdose. EMEA. June 2004. Available at www.emea.eu.int/pdfs/human/swp/259902en.pdf. Accessed March 28, 2006.
11. Single Dose Acute Toxicity Testing for Pharmaceuticals. FDA. August 1996. Available at http://www.fda.gov/cder/guidance/pt1.pdf. Accessed March 30, 2006.
12. Guidance for Industry, Investigators, and Reviewers. Exploratory IND Studies. FDA. January 2006. Available at http://www.fda.gov/cder/guidance/7086fnl.htm. Accessed March 30, 2006.
13. Hull, R.M. Guideline limit volumes for dosing animals in the preclinical stage of safety evaluation. *Human Exp Toxicol* 14, 305–307, 1995.
14. http://www.fda.gov/cder/cancer/animalframe.htm.
15. Note for guidance on the preclinical evaluation of anticancer medicinal products. EMEA. July 1998. Available at http://www.emea.eu.int/pdfs/human/swp/099796en.pdf. Accessed April 3, 2006.
16. Critical Path Opportunities Report. FDA. March 2006. Available at http://www.fda.gov/oc/initiatives/criticalpath/. Accessed April 3, 2006.
17. Critical Path Opportunities List. FDA. March 2006. Available at http://www.fda.gov/oc/initiatives/criticalpath/. Accessed April 3, 2006.
18. Nunn, A.D. The cost of developing imaging agents for routine clinical use. *Invest Radiol* 41, 206–212, 2006.
19. Exploratory Clinical Studies for Improved Compound Selection. The Toxicology Forum, summer meeting 2004. pdf files of the presentations. Available at http://www.toxforum.org/html/day_2.html. Accessed April 3, 2006.
20. Nunn, A.D., Linder, K.E., and Tweedle, M.F. Can receptors be imaged with MRI agents? *Q J Nucl Med* 41, 155–162, 1997.
21. de Vries, I.J., Lesterhuis, W.J., Barentsz, J.O., Verdijk, P., van Krieken, J.H., Boerman, O.C., Oyen, W.J., Bonenkamp, J.J., Boezeman, J.B., Adema, G.J., Bulte, J.W., Scheenen, T.W., Punt, C.J., Heerschap, A., and Figdor, C.G. Magnetic resonance tracking of dendritic cells in melanoma patients for monitoring of cellular therapy. *Nat Biotechnol* 23, 1407–1413, 2005.
22. Interagency Council on Biomedical Imaging in Oncology. Available at http://www.cancer.gov/dctd/icbio. Accessed April 3, 2006.

22 An Outlook on Molecular and Cellular MR Imaging

Michel M.J. Modo and Jeff W.M. Bulte

CONTENTS

Conclusion 408
References 408

Just as molecular biology revolutionized our understanding of biological processes, molecular and cellular imaging is set to change how we study life.[1] No longer will *in vivo* imaging be limited to visualizing downstream effects of molecular or cellular changes, but it will specifically detect axiomatic elements of ongoing biological mechanisms. As illustrated in this book, scientists and clinicians are no longer addressing questions for which the mere visualization of particular organs is sufficient, but are aiming to provide a more holistic understanding from molecular to functional processes.[2] The technological advances required for cellular and molecular imaging illustrate that these advances in science are largely dependent on the emergence of new technologies.

From the first visualization of cells facilitated by a microscope to the first *in vivo* imaging with x-rays, biological investigations are intimately linked to technological innovations. As with microscope-based histology, the visualization of specific targets by molecular or cellular imaging requires selective dyes and markers, including antibodies, to selectively highlight different types of cells or molecules of interest. High affinity and selectivity of the marker for the target are the chief requirements to ensure imaging specificity. In contrast to light-based histology, detection of these markers by *in vivo* imaging with positron emission tomography (PET) and magnetic resonance imaging (MRI) necessitates contrast agents that contain either detectable radioactive elements or metal ions.

Contrast agents are therefore currently the single-most rate-limiting step for molecular and cellular imaging. PET radioligands have the advantage that radioisotopes can be attached to available compounds that are in routine clinical practice and can therefore be fairly easily and efficiently engineered on-site.[3] As the radioisotopes do not significantly increase the molecular weight of these agents, they generally cross the blood–brain barrier (BBB) and can be detected in the femtomolar range despite the poor spatial resolution of PET.[4] Due to the need of radioisotopes for detection, frequent repeated imaging with PET is more restricted than with MRI. Although the metal ions used for generating exogenous contrast on MR images do not raise the same concerns regarding repeated use, they are larger than PET agents and generally prohibit infiltration across the intact BBB. Great strides have been taken to create functional MR agents that would allow crossing of the BBB. To achieve an active transport across the BBB, specialized peptides can be attached onto the MR contrast agents to ensure CNS penetration.[5–7] However, this process is slower than the CNS availability of PET ligands, which generally diffuse to the brain shortly after injection. Furthermore, attaching additional functional arms, such as antibodies, to the MR contrast agent achieves a selective binding of the agent to the target of interest. Podulso et al.,[8] for instance, described a multifunctional contrast agent that crosses the BBB due to putrescine molecules linked to the gadolinium chelate. Additionally, this agent also presented Aβ peptide fragments that specifically bound the contrast agent to Aβ containing amyloid plaques (a hallmark of Alzheimer's disease). However, this strategy requires that sufficient contrast agent infiltrates the brain and attaches to its targets to provide adequate contrast. Moreover, a delay period during which unbound agent can be cleared needs to be observed. Nevertheless, the versatility of this approach is very promising. Antibodies are available against most targets of interest, and conjugating these to given contrast agents could provide a generic strategy that would be very flexible for various targets and could be prepared on-site in different laboratories. In comparison to PET ligands, the conjugation of the peptide fragments to the MR agent can use fairly simple approaches that are already in daily use in histology laboratories. As MR contrast agents do not have a short half-life like PET agents, it is also conceivable that companies could provide specific kits for molecular MR imaging using this approach. This strategy would provide a technique that could easily be disseminated in a large number of laboratories akin to the availability of immunohistochemistry.

An alternative strategy to visualize, for instance, particular cells or gene expression for preclinical imaging is to regulate specialized genes, such as the ferritin gene, that will change the iron content of the cell (see Chapter 11). As iron is paramagnetic, this change in local iron content can inform about the expression of a particular gene. Especially the study of regional and temporal gene expression during development is likely to benefit from this type of *in vivo* imaging. However, caution needs to be exerted as to not overload cells with iron, as this might have deleterious effects. A conditional gene expression of ferritin might therefore be more desirable, as it would only increase the cellular iron load for brief periods of time, during which *in vivo* imaging could occur. Still, this approach will mainly find fruition in preclinical studies, although applications in cell transplantation/infusion studies could see a clinical translation of this approach.

Most commonly, exogenous MRI contrast relies on perturbing the relaxivity of hydrogen atoms. However, MRI can also detect other nuclei, and it is conceivable that an increased specificity can be gained from developing agents that will rely on other nuclei that are not commonly found in biological tissues (e.g., ^{3}He, ^{19}F, ^{31}P).[9] In principle this approach seems very attractive to visualize molecules that are not very abundant in tissues, as they might not accumulate sufficient contrast agent to alter the hydrogen signal. The main problem lies with the sensitivity of MRI to detect these nuclei against background, and it might be more likely that this approach will find an application in visualizing regions where these molecules are present in abundance. If large quantities of molecules are present in the area to be imaged, contrast agents affecting the hydrogen atoms' relaxation might prohibit proper anatomical MR imaging of this region. Visualizing the molecules by providing contrast agent using, for instance, fluorine could allow anatomical MRI in addition to molecular MRI.[10] Some of these contrast agents also avoid the large metal particles necessary to alter the relaxation of hydrogen. Due to their smaller molecular size, these contrast agents, akin to PET agents, could more easily penetrate the BBB and open up new possibilities for molecular MRI.[11] A substantial increase in imaging flexibility will result from further developing multinuclei imaging.[12] It is conceivable that multiple molecular targets could be visualized simultaneously using this approach, providing a truly unique and enlightening approach to molecular and cellular imaging.

Multinuclei MRI highlights the advantage of adopting imaging to investigate as many independent observations as possible to increase the amount of information available from a single subject at a particular time. Apart from multinuclei MRI, it is also possible to combine MRI with other complementing imaging modalities. As MRI is providing highly resolved anatomical and functional images, PET can offer complementary information regarding particular metabolic or transmitter systems. Implementing both imaging modalities into one machine, such as the PET and NMR Dual Acquisition (PANDA)[13] system, will allow an efficient and informative assessment in many rapidly evolving disease states.[14] For instance, for stroke or tumors, a combined PET and MRI system utilizing multimodal contrast agents[15] could greatly increase the diagnostic assessment of patients. MRI could provide information regarding the functional implications of anatomical changes in grey and white matter, whereas PET could provide details as to the molecular/receptor/metabolic status in the affected areas. However, significant challenges in hardware, contrast agent, and software design remain to be addressed.[16] The needs and applications of dual acquisition systems can already be found in combined computed tomography (CT)/PET or CT/SPECT (single photon emission computer tomography) systems.[17]

Traditional light-based imaging techniques are also increasingly being used for *in vivo* imaging. Foremost, fluorescent imaging modalities such as bioluminescence or optical imaging are used extensively in preclinical investigations. The most simple application of these techniques is to corroborate the presence of contrast agent within cells *ex vivo* at a higher resolution[18] to ensure that the MRI signal is indeed due to the contrast agent rather than imaging artifacts, such as air bubbles or small bleeds. Techniques such as optical imaging can also be readily adapted to visualize targets *in vivo*. Multimodal contrast agents[19,20] with functional fluorescent moieties have already been designed to report on particular molecules of interest. Many fluorescence-based cell assays have been developed to study particular molecular phenomena *in vitro*, and these fluorescent assays could be adapted to link to MRI contrast agents to provide multimodal/functional contrast agents that can bridge the gap between *in vitro* and *in vivo* assessments. For instance, recently Mulder et al.[21,22] were able to visualize the activated endothelium in tumors by MRI and verify this activation by fluorescent histology using the same bimodal agent. Further developments of these approaches promise to provide some interesting insights in relating changes in particular molecules to structure and function. Although this approach is similar to combined PET/MRI systems, it has the added advantage that the fluorescent moieties would easily allow high-resolution *in vitro* assessments that cannot be gained from PET studies.

It is this scaling from the microscopic to macroscopic world that will be essential to develop molecular and cellular imaging into a reliable translatable technology. Advantages, disadvantages, and complementarity will determine which combination of modalities will provide added value. Although the integration of macroscopic images from two different modalities is fairly straightforward, and many software programs are available for this purpose, the integration of particular molecular information with macroscopic images or information from more than two modalities is currently not as readily available. The use of various colors to represent the overlap of a target from independent modalities (e.g., molecular aspects of a tumor by PET and its location and size by MR imaging) can solve some of these issues and help researchers to integrate the disparate information provided by these images. Alternatively, MR images can, for instance, also be annotated to provide an atlas of various structures complemented with molecular information,[23] but fall short of providing quantitative data that would allow group comparisons. Further development is hence needed in the image analysis and integration of multimodal and multiparametric data. Especially the use of contrast agents to detect cells or molecules with MR imaging can greatly benefit from these developments. PET imaging already heavily relies on mathematical models and analyses to provide quantitative assessments. At present, the detection of an MR contrast agent-modulated signal is predominantly identified based on a visual inspection of the image compared to what type of signal would normally be expected in this area. Pre- and postimages are already an improvement on this, as it is possible to calculate a difference image that highlights the area that underwent a signal change due to the contrast agent. However, in molecular and cellular MR imaging it is often not possible to do these pre- and postimages within a single imaging session, and realignment of the images is needed. A difference between pre- and postimages might therefore not solely be due to changes in signal intensity due to the contrast agent, but structural changes could also provide image differences. In structures such as the brain, these changes are often not too dramatic, but other organs, such as the bladder, can dramatically change between two images. Realignment and co-registration of images therefore provide additional challenges to the interpretation of pre- and postimages. The degree or spatial extent of signal change needed to allow a reliable detection of a contrast agent-induced signal change also remains a question of debate. It is possible that image analysis methods would provide more robust detections of smaller signal changes than is currently feasible based on visual inspection. However, at present little research is devoted to this aspect of molecular and cellular imaging.

Image analysis could therefore also be an important aspect in reducing the amount of contrast agent needed to detect a particular target, as it could optimize a reliable detection based on the smallest signal change needed for detection. A reduction in contrast agent would not only possibly provide a better detection, but might also ensure a safer use of contrast agent, as the patient is exposed to a lower amount of metal particles. Although MR contrast agents have an excellent safety record, their use for molecular and cellular MR imaging in many cases is quite different from current practices. At present, MR contrast agents are only temporarily present in the patient's body before being cleared by the reticuloendothelial system (RES). However, labeling of nonphagocytic cells with a contrast agent possibly present for months could exert detrimental effects on cellular functions. The possible degradation of contrast agents in the brain could lead to inner-sphere metal particles being exposed and causing cell death. Rigorous and detailed preclinical studies are therefore essential to ensure the safety of these agents prior to clinical translation (see Chapter 21).

CONCLUSION

Substantial progress in molecular and cellular MR imaging has been achieved in the past few years. Novel avenues for MR imaging have arisen that allow an even better integration with other means of assessments. The ever-increasing range of contrast agents promises exciting new applications that will truly allow researchers to probe the evolution of biological molecules *in vivo* and determine how these elements participate in health and disease. Being able to easily translate these approaches into clinical applications promises to supply the noninvasive diagnostic tools clinicians need to practice molecular medicine.[24] From molecular to functional imaging, the diversity of information provided by MR imaging is unparalleled by other techniques. It is foreseeable that MR imaging will become increasingly the core integrative technology in biomedicine.

REFERENCES

1. Rudin, M. and Weissleder, R., Molecular imaging in drug discovery and development, *Nat Rev Drug Discov*, 2, 123–131, 2003.
2. Massoud, T.F. and Gambhir, S.S., Molecular imaging in living subjects: seeing fundamental biological processes in a new light, *Genes Dev*, 17, 545–580, 2003.
3. Gibson, R.E., Burns, H.D., Hamill, T.G., Eng, W.S., Francis, B.E., and Ryan, C., Non-invasive radiotracer imaging as a tool for drug development, *Curr Pharm Des*, 6, 973–989, 2000.
4. Phelps, M.E., PET: the merging of biology and imaging into molecular imaging, *J Nucl Med*, 41, 661–681, 2000.
5. Pardridge, W.M., Drug and gene targeting to the brain with molecular Trojan horses, *Nat Rev Drug Discov*, 1, 131–139, 2002.
6. Abbott, N.J., Chugani, D.C., Zaharchuk, G., Rosen, B.R., and Lo, E.H., Delivery of imaging agents into brain, *Adv Drug Deliv Rev*, 37, 253–277, 1999.
7. Spellerberg, B., Prasad, S., Cabellos, C., Burroughs, M., Cahill, P., and Tuomanen, E., Penetration of the blood-brain barrier: enhancement of drug delivery and imaging by bacterial glycopeptides, *J Exp Med*, 182, 1037–1043, 1995.
8. Podulso, J.F., Wengenack, T.M., Curran, G.L., Wisniewski, T., Sigurdsson, E.M., Macura, S.I., Borowski, B.J., and Jack, C.R., Molecular targeting of Alzheimer's amyloid plaques for contrast-enhanced magnetic resonance imaging, *Neurobiol Dis*, 11, 315–329, 2002.
9. Hudson, A.M., Kockenberger, W., and Bowtell, R.W., Dual resonant birdcage coils for 1H detected 13C microscopic imaging at 11.7 T, *MAGMA*, 10, 61–68, 2000.
10. Ahrens, E.T., Flores, R., Xu, H., and Morel, P.A., *In vivo* imaging platform for tracking immunotherapeutic cells, *Nat Biotechnol*, 23, 983–987, 2005.
11. Sato, K., Higuchi, M., Iwata, N., Saido, T.C., and Sasamoto, K., Fluoro-substituted and 13C-labeled styrylbenzene derivatives for detecting brain amyloid plaques, *Eur J Med Chem*, 39, 573–578, 2004.

12. Golman, K., Ardenkjaer-Larsen, J.H., Petersson, J.S., Mansson, S., and Leunbach, I., Molecular imaging with endogenous substances, *Proc Natl Acad Sci USA*, 100, 10435–10439, 2003.
13. Marsden, P.K., Strul, D., Keevil, S.F., Williams, S.C., and Cash, D., Simultaneous PET and NMR, *Br J Radiol*, 75, S53–S59, 2002.
14. Jacobs, R.E. and Cherry, S.R., Complementary emerging techniques: high-resolution PET and MRI, *Curr Opin Neurobiol*, 11, 621–629, 2001.
15. Doubrovin, M., Serganova, I., Mayer-Kuckuk, P., Ponomarev, V., and Blasberg, R.G., Multimodality *in vivo* molecular-genetic imaging, *Bioconjug Chem*, 15, 1376–1388, 2004.
16. Cherry, S.R., Multimodality *in vivo* imaging systems: twice the power or double the trouble? *Annu Rev Biomed Eng*, 8, 35–62, 2006.
17. Messa, C., Di Muzio, N., Picchio, M., Gilardi, M.C., Bettinardi, V., and Fazio, F., PET/CT and radiotherapy, *Q J Nucl Med Mol Imaging*, 50, 4–14, 2006.
18. Modo, M., Cash, D., Mellodew, K., Williams, S.C., Fraser, S.E., Meade, T.J., Price, J., and Hodges, H., Tracking transplanted stem cell migration using bifunctional, contrast agent-enhanced, magnetic resonance imaging, *Neuroimage*, 17, 803–811, 2002.
19. Mulder, W.J., Strijkers, G.J., van Tilborg, G.A., Griffioen, A.W., and Nicolay, K., Lipid-based nanoparticles for contrast-enhanced MRI and molecular imaging, *NMR Biomed*, 19, 142–164, 2006.
20. Mulder, W.J., Koole, R., Brandwijk, R.J., Storm, G., Chin, P.T., Strijkers, G.J., de Mello Donega, C., Nicolay, K., and Griffioen, A.W., Quantum dots with a paramagnetic coating as a bimodal molecular imaging probe, *Nano Lett*, 6, 1–6, 2006.
21. Mulder, W.J., Strijkers, G.J., Griffioen, A.W., van Bloois, L., Molema, G., Storm, G., Koning, G.A., and Nicolay, K., A liposomal system for contrast-enhanced magnetic resonance imaging of molecular targets, *Bioconjug Chem*, 15, 799–806, 2004.
22. Mulder, W.J., Strijkers, G.J., Habets, J.W., Bleeker, E.J., van der Schaft, D.W., Storm, G., Koning, G.A., Griffioen, A.W., and Nicolay, K., MR molecular imaging and fluorescence microscopy for identification of activated tumor endothelium using a bimodal lipidic nanoparticle, *FASEB J*, 19, 2008–2010, 2005.
23. Jacobs, R.E., Papan, C., Ruffins, S., Tyszka, J.M., and Fraser, S.E., MRI: volumetric imaging for vital imaging and atlas construction, *Nat Rev Mol Cell Biol*, Suppl, SS10–SS16, 2003.
24. Misgeld, T. and Kerschensteiner, M., *In vivo* imaging of the diseased nervous system, *Nat Rev Neurosci*, 7, 449–463, 2006.

Index

A

α-antiplasmin (α2AP), 50
α-hydroxacids, 62
$\alpha_v\beta_3$ integrins, 143
AAZTA, 40
Abdominal disease, 249
Abdoscan, 24
Acetonitrile, 109
Acid catalyzed exchange, 94
Activity-induced MEMRI (AIMEMRI), 371, 378–380
Acute intraplaque hemorrhage, 164
Adeno-associated virus (AAV), 129, 201
Adenosine, 231
Adenosine triphosphate (ATP), 184, 188
Adenovirus, 126, 201, 205
Adenylate cyclase, 378
AIDS, 351
Albumin, 21, 39, 43, 63, 148, 376
Allosteric effector, 48
Alpha cells, 344
Alzheimer's disease, 1, 128, 183, 260, 330
Amide groups, 86
Amide protein, 107
  transfer (APT), 86, 94, 142
Amino acids, 62
Aminophospholipid, 356
Amphetamine, 222, 228–229, 232
Amphiphilic, 165
Amyloid plaques, 331, 406
Amyl acetate, 380
Anionic maghemite nanoparticle (AMNP), 299, 352
Anemia, 123
Anesthesia, 233, 379
Angina, 162
Angiogenesis, 51
Angiotensin-1, 274
Anionic carboxylate ligating groups, 102
Anisotropic, 63, 67, 385
Annexin V, 63, 143, 174, 312, 356
Antagonism, 229
Antibiotics, 201
Antibodies, 5, 43, 62, 125, 143, 164, 170, 260, 275, 298, 304–305, 402, 406
Antibody-directed prodrug therapy (ADEPT), 204
Antigen-induced arthritis, 263
Antigen presenting cells (APC), 351
Anti-VCAM, 63
Apoferritin, 41, 126, 209
Apolipoprotein, 170
Apoptosis, 117, 144, 162, 174, 183–192, 309, 312, 354–357
Apparent diffusion coefficient (ADC), 186, 277, 356
APV, 378
Arabinogalactan, 62
Archenteron, 286
Arcopallium, 381
Area X, 381
Arginine, 125, 205–206
Arrhenious equation, 112
Arterial
  intima, 161
  stenosis, 162
Aromatic moiety, 39
Arthritis, 263–264
Atherogenic stimulus, 164
Atherosclerosis, 69, 161–177
Atherosclerotic plaque, 77, 162, 170, 176
ATP-binding cassette transporter-1 (ABACA1), 170
Autoimmune recognition, 344, 354
Autonomic system, 235
Autoradiography, 229, 275, 377
Avidin, 43, 144, 191
Avidity, 356
Ayant field fitting, 65
Axonal
  sprouting, 370, 383
  transport, 377

B

β-arylsulfonamide, 47
β-cyclodextrin, 41, 49
β-galactosidase, 50, 124, 143, 199, 203–204
Bacteria, 203
Ballistic gas charge, 303
Banff classification, 266
BANG particles, 299, 304
Basal ganglia, 221, 235, 380
Base catalysed exchange, 94
Basophils, 259
Bcl-2 family, 184
Behavioral activity, 231
Beta cell, 344
  mass (BCM), 345–351
Bidentate anions, 40
Bifunctional ligands, 46
Bile, 375
Binding characteristics, 5
Biocompatibility, 6, 16, 60, 67, 144, 351
Biodistribution, 17, 38, 48, 51, 67, 70, 74, 124, 169, 192, 332, 347, 370, 400
Bioeffects, 283
Bioluminescence, 3, 124, 191, 199, 280, 296, 359, 372, 407
Biomarker, 1, 362
Biomedical research, 2, 199

Biomineralized, 126
Bioreactors, 123
Biomarkers, 277
Biotechnology, 123
Biotin ligase, 206–207
Biopsy, 1, 351
Biovectors, 62
Bis-methylamide-DTPA, 102
Bis-phenylboronate, 115
Black blood sequence, 162
Blastocoel, 286
Blastomer, 203
Blastopore lip, 288
Bloch equation, 85, 88
Bloch model, 104
Blood–brain barrier (BBB), 4, 38, 123, 261, 277, 375, 378, 405
Blood, 48, 259
  acidity, 223
  coagulation, 50
  flow, 344
  half-life, 72, 172
Blood flow autoregulation, 223
Blood oxygen level dependent (BOLD) contrast, 77, 220, 385
Blood pool agent, 4, 43, 75, 77, 302
Bohr magneton, 19
Boltzmann distribution, 105
Boltzmann equation, 14
Bone marrow, 70, 253, 259, 296, 328
Bone sialo protein (BSP), 204
B22956, 43

C

Calcineurin, 264
Calthrin, 297
Canaries, 381
Cancer, 38, 43, 46, 51, 60, 70, 76, 86, 94, 117, 123, 125, 129, 141–155, 183, 203, 331–332, 351, 356, 395, 406
Carbon (^{13}C), 13, 31, 185, 189, 350
Cardiovascular
  disease, 46, 356
  system, 235
Carcinoma, *see* Cancer
Carboxyamides, 41
Carboxydextran, 60, 62
Carboxylates, 41, 46
Carboxypeptidase, 22, 50
Cardiac
  function, 374
  gating, 332
  graft, 266–267
Cardiotoxicity, 302
Cardiovascular depression, 374
Caspases, 183, 191–192
Catalytic activity, 209
Cathecol functionality, 50
Cathepsin B, 263
Catheters, 333
Caudate/Putamen, 228
Caveolin, 297
cDNA, 127, 200
Cell
  differentiation, 296, 303, 306, 313–314
  function, 296, 306
  migration, 306, 328
  proliferation, 303, 313
  replacement, 296
  surface receptors, 143, 206
  survival, 329
  therapy, 124, 130–131, 296
  tracking, 144, 296, 351
  transplantation, 228, 325–335
  viability, 207, 296, 306, 312, 356–357
Cellular distribution, 28
Cellular uptake, 5, 51, 295–316, 325, 374–375
  efficiency, 260
  mechanism, 69
Cerebellum, 262, 377
Cerebral blood flow (CBF), 77, 276
Cerebral blood volume (CBV), 77, 220, 225, 233, 262
Cerebral cortex, 221, 377
Cerebrovascular disease, 51
Ceruloplasmin, 376
CEST spectrum, 106–110, 112
c-fos, 233
Chaltrates, 41
Circular dichroism (CD), 115
cis-hydroxy group, 115
Chemical exchange saturation transfer (CEST), 29, 85–97, 101, 142, 192, 207, 331, 349
  mechanism, 87–89
Chemical exchange rate, 89–93
Chemical manufacturing and control (CMC), 395, 399
Chemical shift, 13, 192, 207
  degeneracy, 101
Chemotherapy, 205, 296, 356
Cholesterol deposition, 161
Cholesterol ester, 188
Choline, 141, 147, 151, 153
Chondrogenesis, 309, 334
Cleft of Brachet, 288
Clinical trial, 306, 314, 329, 335, 395–402
CLIO-NRP, 355
c-myc RNA, 125
Cocaine, 225, 228
Colchicine, 69
Collagen, 162, 248
Colloidal suspension, 299
Combidex, *see* Ferumoxtran
Computer tomography (CT), 276, 358, 372, 397, 401
Concavalin A, 22
Constructed interference in steady state (CISS), 277
Copolymers, 46
Copper (Cu), 40, 296
Coronary artery disease, 51
Correlation time, 39
Cortisol, 234
C-peptide, 344, 357

Creatine, 125, 205–206, 266
Critical micelle concentration (CMC), 168
Critical path initiative, 401
Cross-linked iron oxide (CLIO), 63, 66, 207, 269, 299, 305, 354
Crystal ordering, *see* Spinels
Curie spin relaxation, 20, 24
Cyclen-based complexes, 107, 114
Cyclosporine, 264
Cytochalasin B, 69
Cytochrome C, 184, 357
Cytokines, 141
Cytomegalovirus (CMV), 200, 359
Cytoplasm, 53, 296
Cytoskeleton, 297
Cytosine deaminase (CD), 144, 204–205
Cytosine diphosphocholine (CDP choline), 187
Cytosone triphosphate (CTP), 153
CYP2B1, 209
C184, *see* VSOP

D

DARPP-32, 227, 231
Deep-electrode recording, 235
Delta cells, 344
Demyelinating disease, 327
Dendrimers, 46, 62, 306
Dendritic cells, 335
Dentate gyrus, 333, 383
Dephasing effect, 75
Desferrioxamine, 74
Development, 211, 283–292
Dexamethasone, 263
Dextran, 46, 60
Dextran allergy, 7
Diabetes, 343–362
Diabetogenecity, 264, 354
Diffusion imaging, 142, 149, 185, 356
Diffusion tensor imaging, 330, 385
Dimereaptosuccinic acid (DMSA), 62, 305
Dipolar coupling, 28, 42, 65
Dipolar interaction, 64
Direct cortical stimulation, 235
Displacement, 229
Distearylamide gadopenetic acid microcapsule, 43
DNA-coated nanoparticles, 303
DNA fragmentation, 184
Dopamine, 221
  autoreceptor (DAR), 224, 229–231
  concentration, 228
  release, 228
  transporter, 129
Dopaminergic cells, 228
Dopaquinone, 207
Dose escalation scheme, 397
Dose–response relationship, 395
Dosing schedule, 396
Dotarem, *see* Gadoterate
DOTA-4AmC, 111
DOTA-squaric-acid-ester, 46
DOTA terramide (DOTAM), 31
Drug
  addiction, 229, 231–233
  discovery, 211
DTPA-bisanhydride, 46
DTPA-monoanhydride, 46
DyDOTAM, 107, 111
Dynamic
  AIM (DAIM), 379
  MEMRI (DMEMRI), 380
Dynamic nuclear polarization (DNP), 350
Dysprosium (Dy), 107, 349
3,7-dimethyl-1-propargylaxanthine (DPMX), 231

E

Ectodermal tissue, 288
Edema, 261, 268, 327, 359
EgadMe, 124, 203
Electroencephalography (EEG), 219, 235, 372
Electron paramagnetic resonance (EPR), 24
Electronic g factor, 19
Electronic relaxation rate, 49
Electronic relaxation time, 37
Electron microscopy (EM), 310
Electron-nuclear double resonance (ENDOR) spectroscopy, 22
Electrophoretic mobility, 67
Electroporation, 28, 53, 301
Electrostatic interaction, 299
Elongation factor-1 (EF-1), 200
Embryo development, 274–292
Endarterectomy, 164, 174
Endocytosis, 128, 296, 304, 306, 351
Endogenous anions, 110
Endorem, *see* Ferumoxide
Endorphin, 234
Endosomes, 53, 299, 310
Endothelial cells, 70, 73, 172, 248, 261, 301, 314, 329
Endotoxins, 248
Energy metabolism, 188–189
Enkephalin, 231
Environment sensing agents, 5
Enzyme
  activity, 110, 227
  -responsive agent, 49, 191
  -catalysed reactions, 85
  -linked immunosorbent assay (ELISA), 314
Eosinophils, 259
Epichlorohydrin, 63, 305
Epilepsy, 1
Episome, 130
Erythrocytes, 259, 277
Escherichia coli, 201
E-selectin, 50, 63
Eticlopride, 230
EuCNPHC, 108
EuDOTAM, 103
EuDOTA-4amBBA, 349

EuDOTA-4amCE, 107
EuDOTAMpy, 117
EuDTMA-2PB, 114
Eukaryotic cells, 129
Europium, 30, 49, 103, 349
Eu-S-THP, 109
EVP 1001-1, 374
Everolimus, 266
Exchange correlation spectroscopy (EXSY), 31, 86
Exchange rate, 93
Exchangeable resonance line shape, 86
Exchangeable hydrogens, 28
Exitatory postsynaptic potentials (EPSP), 221
Experimental autoimmune encephalomyelitis (EAE), 261–263
Exploratory IND (E IND), 399
Extracellular matrix (ECM), 141, 144, 146–148, 248
Extravasation, 191

F

Factor XIII transglutaminase activity, 50
Fenton reaction, 127, 207, 303
Feridex, *see* Ferumoxide
Feromoxytrol, 174
Ferrihydrite, 126
Ferritin, 5, 73, 126, 143, 175, 209–210, 406
Ferromagnetism, 5, 24
Ferrous chloride, 311
Ferroxidase activity, 126
Ferucarbotran A (SHU-555A, Resovist), 21, 60, 66, 68, 71, 299
Ferucarbotran C (SHU-555 C, Supravist), 21, 66, 68, 71
Feruglose (NC100150), 60, 66, 71,
Ferumoxtran (AMI-227, Sinerem, Combidex), 24, 60, 62, 66, 68, 71, 170, 174, 299
Ferumoxide (AMI, 25, Endorem, Feridex), 21, 60, 62, 66, 68, 71, 144, 170, 290, 299
Ferumoxytol (C7228), 66, 68, 71
Fetus, 284
Fibrin, 43, 118
Fibronectin, 248
Fibrous cap, 161
FIESTA, 328, 334
Flow field gradient, 61
Flumazenil, 276
Fluorescence
  activated cell sorter (FACS), 313
  imaging, 123, 190, 370, 407
  molecular tomography (FMT), 207
  proteins, 202
  resonance energy transfer (FRET), 301
Fluorine, 6, 13, 31, 43, 125, 143, 170, 189, 192, 204, 325, 331, 359
Fluorobenzyltrozamicol (FBT), 346
Fluorodeoxyglucose (FDG), 184, 189, 359
Food and Drug Administration (FDA), 124, 274, 302, 314, 360, 396
Formation constant, 17
Freed field fitting, 65
Free radicals, 312
Frequency dependence, 90
Frequency-selective saturation pulse, 106
Fructose, 115
  1,6-biphosphate (FBP), 187
Fucoidan, 69
FuGene, 263, 307
Functional circuits, 3, 233, 380
Functionality, 207, 329
Functionalization, 5, 43, 46, 170
Functional magnetic resonance imaging (fMRI), 2, 60, 77, 148–153, 189, 219, 234, 302, 330, 372
Functional recovery, 327
Fusion proteins, 124
4-fluoro-2-nitrophenyl-β-D-galactopyranoside (PFONPG), 204
5-fluorocytosine (5-FC), 144, 205
5-fluorouracil (5-FU), 144

G

GABA, 221
Gadolinium (Gd), 5, 15, 25, 37, 43, 102, 110, 124, 143, 190, 247, 254, 276, 296, 300, 306, 325, 397, 401
Gadobenate (Gd-BOPTA, Multihance), 16, 21, 39, 251–252
Gadodiamide (Gd-DTPA-BMA, Omniscan), 16, 21, 38, 102
Gadofosveset (MS-325, Vasovist), 16, 21, 43
Gadoflurine M, 168
Gadomer, 21
Gadopentate (Gd-DTPA, Magnevist), 16, 21, 38, 102, 143, 148, 166, 169, 277, 300
Gadoterate (Gd-DOTA, Dotarem), 16, 21, 38, 102, 143, 262, 301
Gadoteridol (Gd-HPO3A, ProHance), 16, 21, 28, 38, 53, 301
Gadoxetic acid (Gd-EOB-DTPA, Primovist, Eovist), 21, 39, 252–253
Gd-
  AAZTA, 40
  AAZTAC17, 44
  chitosan, 43
  DOPTA, 348
  DOTMA, 45
  DOTP, 41
  DTPA-BSA, 169
  fullerenol, 42, 302
  G80P, 347
  HOPO, 40
  heanedione (GdH), 43
  incorporated lipid nanoemulsion, 43
  labeled α-LDL
  loaded carbon nanotubes, 42
  loaded liposome, 43
  PCP2A, 40
  PCTA, 40
Galactopyranose, 124, 203
Ganciclovir, 205
Gastric mucosa, 250
Gastrointestinal (GI) tract, 60
Gastrulation, 284

Gene
amplification, 123
expression, 3, 128, 144, 199, 203, 211, 406
gun, 303
manipulation, 201
reporter, 143, 199–211
therapy, 123, 129, 206
Genomolecular studies, 232
Glomerular filtration, 266
GLP-1, 345
Glucagon, 344, 358
Glucose, 115, 345
consumption, 223
homeostasis, 345
GLUT2, 345
Glutamic acid decarboxylase (GAD), 358
Glutamine synthetase (GS), 377
Glycerophosphocholine (GPC), 151
Glycogen accumulation, 359
Glycolytic
activity, 184
enzymes, 147
Glycoprotein, 248
Glycosaminoglycan, 62
Good manufacturing practice (GMP), 303, 314
G-protein-coupled receptors, 226
Graft rejection, 190, 264–266, 315, 356, 360
Granule membrane molecules, 345
Granulocytes, 274
Green fluorescent protein (GFP), 124, 129, 148, 199, 274, 296, 306
GRID, 287, 301
Guanidine, 117

H

Haber-Weiss chemical reaction, 312
HbA1c, 358
Heat shock proteins, 234
Helium, 13, 31
Hepatobillary stem, 32
Hepatocytes, 18, 28, 39, 76, 126, 247
Hemagglutinating virus of Japan (HVJ), 304
Hematopoiesis, 296
Hemodynamic, 220, 223–233
Hemoglobin, 48
Hemophagocytic lymphhistocytosis-associated encephalopathy, 267
Hemosiderin, 73, 175, 267
Hepatitis, 248
Herpes simplex virus (HSV), 129, 201, 205
Her2/scan receptor, 43
Her2/neu receptor, 144
Heteropolymer, 126
Heterotripodic ligand, 41
HEVJ particle, 299
Hippocampus, 377
Holoferritin, 127
Homologuous recombination, 202
Homopolymer, 126
HOPO, 40
Human serum albumin (HAS), *see* albumin
Huntington's disease, 330
Hydrodynamic size, 67, 298
Hydrogenionic concentration, 47
Hydrophillic coating, 60
Hydroxamate, 62
Hydroxyethyl side chains, 107
Hypercholesterolemia, 206
Hyperglycemia, 344
Hyperpolarized nuclei, 13, 31, 350
Hyperventilation, 223
Hypoxia, 146, 261, 284
inducible factor-1 (HIF-1), 147
response element (HRE), 148, 200
6-hydroxy-dopamine (6-OHDA), 228

I

Iatrogenic complications, 1
Image analysis, 407
Image intensity, 101
Imaging half-life, 32
Imaging modalities, 3
Imino groups, 86
Immediate early gene, 227, 231, 370, 384
Immune
cells, 203
deficiency, 123
response, 248, 356
Immunogenecity, 7, 126, 166, 191, 207, 396
Immunosuppression, 264, 315, 357
Implantation guidance, 330
Indium, 296, 335
Inductively couple plasma mass spectrometry (ICP-MS), 72, 312
Infection, 259–269
Inflammation, 161, 183, 259–269, 273
Inflammatory disease, 60
Inhibitory postsynaptic potentials (IPSP), 221
Inner-sphere effect, 17, 38, 166
Insulin
production, 343
secretion, 343, 348, 357
Insulitis, 351
Integrin, 43
Interagency Council on Biomedical Imaging in Oncology (ICBIO), 402
Interferon-gamma, 70, 162, 314
Interferon stimulatory response element (ISRE), 200
Interleukin, 70, 74, 314
Intermolecular exchange, 86
Internal ribosomal entry site (IRES), 206
Internalization, *see* Cellular Uptake
Intracellular targets, 6
Intravital microscopy, 202
Intravoxel dephasing, 27
Investigative new drug (IND), 315, 395
In utero, 284
Invasiveness, 3

Ion channel, 227
Ion transfer, 61
Irbesartan, 274
Iron, 37, 207, 325, 397
IRON method, 220
Iron oxide particles, 5, 24, 37, 46, 171–175, 249, 253, 296, 332, 351
Irradiation frequency, 90
Islet
  langerhans, 343
  transplant, 352, 357–360
  viability, 344
Isostructural complexes, 109
Isothyiocyanate, 46
Isotopes, 31
Isotope exchange processes, 85

K

Ketoheptose D-mannoheptulose, 346
Kidney, 39, 264
Kinetic stability, 17, 38, 49
Knock in/out, 201
Kuppfer cells, 73, 248, 253, 260

L

Labeling strategies, 295–316
Lactate, 113
Langevin function, 65, 67, 172
Lamellarity, 167
Laminin, 248
Lanthanides, 17, 30, 106–110, 114, 164, 348
  shift reagents, 101
Larmor frequency, 14, 19
Lectin, 304
Lentivirus, 129, 201
Leukocyte invasion, 351
Lewis acid type complexes, 101
Ligand–neuronal coupling, 224–227
Ligand–vascular coupling, 223
Lineage diagram, 290
Line shape analysis, 30
Lipid core, 164
Lipid metabolism, 187–188
LIPOCEST, 118
Lipofectamine, 307
Lipofectin, 307
Lipoprotein, 166, 170–171, 297
Liposome, 48, 118, 143, 164, 166–167, 254, 298, 300, 304
Lipovec, 307
Liver, 43, 60, 70, 74, 76, 115, 125, 187, 201, 247–255, 260, 373, 375
  cirrhosis, 248
  sinisoid, 254
  steatosis, 358
Liver activator protein (LAP), 201
LnCNHPC, 120
LnDOTA, 120
LnDOTA-tetraamide, 120
LnDOTMA, 120
LnDOTP, 120
LnDTPA, 120
LnMBDO3AM, 120
LNDTPA-BMA, 120
Longitudinal relaxation rate, 15, 52
Long terminal repeat (LTR), 200
Lovastatin, 70
Lung graft, 266–267
Luciferase, 124, 129, 199, 202–203, 359
Lutethium, 17
Lymphangiography, 302
Lymphatic convective transport, 147
Lymph node, 60, 70, 74, 76, 185, 304, 351
Lymphocytes, 75, 129, 259, 302, 373
Lysine rich protein (LRP), 207
Lysosome, 264, 377

M

Macrocyclic ligand, 17
Macromolecules, 85
Macrophages, 60, 67, 72, 76, 162, 170, 172, 259–269, 273–280, 304, 327, 335
Macropinosomes, 69, 299
Maghemite (Fe_2O_3), 60
Magic angle spinning (MAG) MRS, 152
Magnetite (Fe_3O_4), 60, 171, 266
Magnetic resonance (MR)
  angiography (MRA), 31, 43, 77, 141, 305
  histology, 291
  microscopy (MRM), 285, 352, 372
  relaxometer, 311
  reporter, 5
  spectroscopy (MRS), 4, 86, 104, 125, 141, 148, 170, 184, 187, 189, 192, 203, 205, 284, 348, 372
Magnetic quantum number, 14
Magnetic filtration, 61
Magnetic susceptibility artefact (MSA), 173
Magnetization, 65
  saturation (MS), 67
  transfer, 13, 29, 86, 104, 205
  transfer ratio, 87
Magnetodendrimers, 263, 299, 306
Magnetoencephalography (MEG), 219, 372
Magnetofection, 303
Magnetogyric ratio, 14, 19
Magnetoliposome, 62
Magnetometry, 67
Magnetosome, 126
Magnetotatic bacteria, 126
Magnevist, *see* Gadopentate
Major histocompatiblity complex (MHC), 353
Malignant phenotype, 146–148
Manganese, 5, 15, 22, 37, 48, 296, 300, 302, 348, 369–385, 397
  administration, 375
  chemical configuration, 376
  -enhanced MRI (MEMRI), 369, 378

-mesoporphyrin (MnMeso), 170
toxicity, 373
transport, 376–378
Mangofodipir (Mn-DPDP, Teslascan), 16, 250, 332, 374
Manoheptulose, 346
Massart technology, 62
Matrix Metalloproteinase, 274
Maximally tolerated dose (MTD), 399
Mean amplitude glycemic excursions (MAGE), 358
Medial prefrontal cortex, 227
Medulla, 262
Medullar oblongata, 377
Melanin, 126
Mesendodermal mantle, 286
Mesoscopic inhomogeneities, 27
Metabolites, 86, 150, 187, 297, 349, 372
Metabolic disease, 123
Metabolic imaging, 142
Metabolism, 185, 247, 250, 344, 358, 406
Metal-hydrogen distance, 19
Metal ion concentration, 110
Metal-ligand stability constant, 17
Metalloproteins, 126
Metalloproteinase, 146
Metallo-star, 41
Metastasis, 248
Methotrexate (MTX), 306
Methylprednisolone, 264
Metoxy analog, 347
Micelles, 164, 167–169
Microbubbles, 397
Microdialysis, 231
Microdosing, 399
Microemulsion, 43, 165, 169–170
synthesis, 61
Microhemorrhage, 277, 326
Microinjection, 200, 285
Microparticles, 397
Micron-sized particles of iron oxide (MPIO), 266, 299, 304, 325, 331, 333
Mitochondrial form of superoxide dismutates (MnSOD or SOD2), 377
Mitogen-activated protein (MAP) kinase, 222
Mobile water peak, 28
Molecular
docking (MD), 41, 44
fingerprint, 1
interactions, 5
medicine, 1, 123, 408
switches, 202
therapy, 200
tumbling, 40–41
Molar magnetic susceptibility, 27
Molday coprecipitation assay, 62
Monocyte chemoattractant protein-1 (MCP-1), 274
Monocrystaline iron oxide nanoparticles (MION), 24, 67, 125, 127, 144, 207, 220, 261
Monomagnet, 63
Mononuclear phagocytic system (MPS), *see* Marcophages
Morphogenetic movement, 284
Mössbauer spectroscopy, 65
Motor pathway, 221
mRNA, 203, 233
Multicistronic, 124, 127
Multiconstrast MRI, 162–164
Multidentate ligand, 17
Multifunctionality, 5
Multihance, *see* Gadobenate
Multimodality, 7, 131, 142, 153, 169, 200, 221–223, 279, 292, 296, 357, 372, 406
Multinuclei, 406
Multiple sclerosis, 261–263, 276, 351
Mucin-1 antigen, 144
Muscle, 125, 201
Myelination, 305
Myelin-basic proteins (MBP), 262
Myelin-oligodendrocte glycoprotein (MOG), 262
Myosin light chain-1, 201Myocardial infarction, 51, 162, 183, 332–333
1-methyl-4-phenyl-1,2,3,6,-tetrahydropyridine (MPTP), 228

N

N-acetylaspartate (NAA), 151
Natural killer cells, 259
Near-infrared (NIR), 202
fluorescence (NIF), 263, 356
Necrosis, 183, 273
Néel relaxation, 63, 67
Neovasculature, 38, 43, 51, 77, 142, 146, 167, 174, 329
Nephrotoxicity, 264
N-ethylmaleimide (NEM), 69
Neuroadaptation, 234
Neurochemical imbalance, 3
Neurodegenerative disease, 330
Neuroectoderm, 287
Neuroinflammation, 261
Neurolation, 284
Neuronal
activity, 223, 370
metabolism, 219
sprouting, 228
status, 233
Neuropeptide Y (NPY), 223
Neurophysiological, 370, 384
NeuroAIDS, 268
Neurotransmitter, 219, 346, 406
coupling, 231
-vascular coupling, 227
Neutrophils, 259
N-methyl-aspartate (NMDA), 378
New chemical entity (NCE), 395, 401
NMR spectra, 90
No-observable adverse event level (NOAEL), 395
Nocardia asteroids, 267
Non-frequency-selective observed pulse, 106
Normoglycemia, 360
Nuclear magnetization, 37
Nuclear magnetic relaxation dispersion (NMRD), 44, 64, 67
Nuclear spin, 14

Nucleic acid, 86
Nucleus accumbens (NAc), 222, 227

O

^{17}O, 45, 102
Olfactory bulb, 380
Oligonucleotides, 63
Omniscan, *see* Gadodiamide
Oncology, *see* Cancer
Opsonization, 43, 260
Optical coherence tomography (OCT), 291
Optical imaging, 141, 184, 296, 301, 305, 356, 397, 402, 407
Optical pumping, 31
Oral glucose tolerance testing (OGTT), 358
Organic anion transporter (OAT), 251
Organic siloxane, 62
Organometallic complexes, 85
Osmolality, 38, 398
Osmotic
- regulation, 379
- shock, 375

Outer-sphere effect, 18, 64, 166
Oxidation, 61
Oxidative stress, 74
Oxine, 296
Oxygen, 376
Oxygenation, 187
Oxygen consumption, 223, 276, 356

P

p53, 205
Pancreas, 250,343–362
Pancreatic polypeptide, 344, 358
PARACEST agent, 105, 331, 349
Paramagnetic micelles (pQD), 169
Paramagnetism, 17–22
Paramagnetic chemical exchange saturation transfer (PARACEST), 86, 100–120, 142
Parasympathetic nerves, 346
Parkinson's disease, 128, 228, 330, 373, 375
PCP2A, 40
Pectiniidae, 202
Pegylated starch, 60
Peptide nucleic acid, 125
Peptides, 43, 62, 94, 164, 166, 304, 405
Peptidomimetics, 62
Perfluoro compounds, 14, 31, 118, 143, 170
Perfluorotripropylamine (FTPA)-doped alginate beads, 146
Perfusion imaging, 150
Perisinusoidal fat-storing cells, 248
Permeability, 148, 186, 356
Peroxidase activity, 50
Peroxysome proliferative-activated receptor-gamma (PPAR), 274
PET and NMR dual acquisition (PANDA), 406
pH, 5, 46, 110, 127, 148, 165, 192
- dependence, 93–94

Phagocytic activity, 51, 260, 402
Phagocytosis, 69, 174, 183, 248, 296
Phagosomes, 69
Pharmacokinetic, 24, 31, 70, 166, 252, 374, 399
Pharmacological magnetic resonance imaging (phMRI), 4, 219–235, 330
Pharmacological activity, 396
Pharmacology, 3
Phenylboronate groups, 114
Phenylethanolamine N-methyltransferase (PNMT), 234
Phosphoglycerate kinase (PGK), 200
Phospholipids, 187
Phosphonates, 41
Phosphorescent emission, 357
Phosphorus (^{31}P), 125, 151, 185, 187, 192, 205, 284
Phosphatidylcholine-specific phospholipase D, 153
Phosphatidylethanolamine, 48
Phosphatidylserine (PS), 144, 184, 189–191, 356
Photinus pyralis, 203
Photon correlation spectroscopy (PCS), 67
Photoswitching, 202
Physicochemical properties, 6, 13–33, 65–67, 298
Physiological environment, 146–148
Pinocytosis, 29, 53, 69, 263, 296
Pioglitazone, 274
PK11195, 268, 275
Planck's constant, 14
Plaque
- formation, 161
- instability, 176

Plasma clots, 118
Plasma membrane, 297
Plasticity, 227, 231, 370, 383
Platelets, 259
Polyamido amine (PAMAM), 47, 62, 93, 191, 306
Polyamminocarboxylate ligands, 38
Polyarginine, 117, 120, 298
Polyethylneglycol (PEG), 43, 46, 62
Polyfect, 307
Poly-inosinic acid, 69
Polylysine, 46
Poly-L-arginine, 307
Poly-L-lysine, 263, 307
Polymerase chain reaction (PCR), 221
Polymers, 87
Polynucleotides, 69
Polyornithine, 46, 52
Polyphosophate, 126
Polysaccharides, 69
Porous sinusoidal lining, 248
Porphyrin, 48
Positron emission tomography (PET), 3, 7, 124, 129, 141, 184, 205, 219, 223, 268, 274, 296, 329, 334, 359, 372, 399, 402, 405
Potassium efflux, 347
Potentiometric titrations, 117
Presaturation pulse, 113, 118
Promoters, 200
Propidium Iodide, 312
Protamine sulfate, 307
Proteases, 131

Proteolipid proteins (PLP), 262
Proteolytic enzymes, 147
Protein tracking, 202
Protonation/Deprotonation equilibrium, 85
Proton antenna, 106–110
Proton exchange rate, 88
Proton transfer ratio, 89
Psychiatric disease, 3
ProHance, *see* Gadoteridol
Proton exchange processes, 86
Proton density, 13
Praseodymium (Pr), 111
PrDOTA-4AmC, 111
Pyridine, 40, 116
Pyruvate kinase, 23

Q

Quality control, 396
QuantiChrom Assay, 311
Quantifying exchange using saturation time (QUEST), 89
Quantifying exchange using saturation power (QUESP), 90
Quantum dot, 166, 169
Quenching effect, 53

R

Radiation therapy, 296
Radiofrequncy pulse, 4, 14
Radioisotopes, 4, 17, 184, 190, 205, 229, 405
Rapamycin, 264
Rapid acquisition with relaxation enhancement (RARE), 284
Ras oncogene, 153
Reactive oxygen species, 302, 309, 313
Receptor
  density, 230
  -induced magnetization enhancement (RIME), 19
  -mediated endocytosis, 53, 69, 297, 351
Recombinant DNA, 123
Red fluorescent protein (RFP), 200
Redox
  potential, 46
  state, 48
Regalinide, 347
Regenerative nodules, 248
Regulatory authorities, 396
Relaxation, 14, 19, 105, 211, 308
Relaxivity, 15, 18, 38, 49, 64, 192, 249, 325
Relaxometric ratio, 53
Renila reniformis, 203
Renal
  cortex, 250
  elimination, 252
system, 32
Renography, 266
Repair, 296
Reporter gene, *see* Gene Reporter
Reproducibility, 65
Residence lifetime, 106
Resovist, *see* Ferucarbotran
Response evaluation criteria in solid tumors (RECIST), 152
Reticuloendothelial system (RES), 4, 31, 74, 78, 172, 250, 253–254, 334, 408
Retrovirus, 129, 201
Reverse genetics, 123
Reward expectancy, 234
Risk–benefit ratio, 395, 402
Rotational diffusion, 19
Rous sarcoma virus (RSV), 200
R2D6, 347

S

Saccharomyces cerevisiae, 202
Safety, 6, 38, 315, 334, 395, 408
Samarium-153, 17
Saturation, 106
  bandwidth, 93
  effect, 53
  efficiency, 88
  fields, 93
  time, 88
  transfer, 30, 87
  offset, 108
  power, 90
  pulse, 87, 90, 207
Scaling calculations, 400
Scavenger receptor, 69
Schizophrenia, 228
Scintigraphy, 269, 335
Seasonal changes, 381
Second messenger, 221
Secretagogues, 345
Sendai virus, 304
Sensitivity, 5, 164
Sepsis, 267
Serine proteases, 146
Serological parameters, 345
Serotonin (5-HT), 223
Serum alkaline phosphatase, 50
Side effect, 250
Signal amplification, 191
Simian immunodeficiency virus (SIV), 268
Sinerem, *see* Ferumoxtran
Singing behaviour, 381
Single photon emission computer tomography (SPECT), 7, 141, 205, 296, 402
Site-specific recombinase (SSR) systems, 202
Size exclusion chromatography, 61
Skin, 201
Small interfering RNA (siRNA), 153
Small particles of iron oxide (SPIO), 24, 60, 68, 75, 127, 143, 171, 190, 220, 249, 253, 260, 334, 359
Smart contrast agents, 5, 141, 280
Sodium iodine symporter (NIS), 129
Sodium (^{23}Na)
  extracellular (ECS), 150
  intracellular (ICS), 150
  MRI, 149, 187
  total concentration (TSC), 150

Solubility, 16, 38
Solution technique, 61
Somatostatin, 344, 358
Songbirds, 370
Sonochemistry, 62
Spaces of Disse, 248
Spatial resolution, 75
Species differences, 72
Specificity, 187, 164, 191, 406
Spectrophotometry, 312
Spermidine, 301
Spinal cord, 262, 305, 328, 380
Spinels, 60
Spin exchange, 31
Spin quantum number, 19
Spleen, 43, 70, 74, 253
Splenocyte, 351
Spontaneous exchange, 94
Stability, 17, 398
Staphylococcus aureus, 267
Staphylogenic abscesses, 268
Stem cells, 129, 141, 203, 306, 328, 351
  endogenous, 333
Steroid, 201
Streptozotocin (STZ), 347, 360
Stress-activated protein kinases, 234
Striatum, 377, 380
Stroke, 48, 94, 149, 162, 186, 192, 261, 273–280, 328, 406
Substantia nigra, 127, 330
Subventricular zone (SVZ), 333
Suicide gene, 204–205
Sulfonated styrene-divinylbenzene, 62
Sulfonylurea, 346
Sulfydril, 46
Superconducting quantum interface device (SQUID), 67
Superfect, 307
Superficial intimal erosion, 162
Superparamagnetism, 23–25, 63–65, 127
Supramolecular adducts, 42–43
Supravist, *see* Ferucarbotran C
Surface modification, 62
Susceptibility, 15, 22, 24, 65, 74, 277
Synaptotagmin, 356

T

Targeting vectors, 43
TbDOTA-4AmC, 109
Technetium, 296
Temperature, 5, 46, 165
Temperature decomposition, 62
Terbium (Tb), 109
Teslascan, *see* Mangofodipir
Tetraazamacrocycle chelate, 124
Tetracycline, 201, 207
Tetraphenylsulfonate porphyrin, 48
Tetra-amide-DOTA complexes, 102
Thalamus, 221, 276
Thermodynamic stability, 17, 38, 40, 49
Thrombin-activatable fibrinolysis inhibitor (TAFI), 50
Thrombosis, 162
Thrombotic disorder, 51
Thrombus, 75, 170
Thymidine Kinase (TK), 200, 204–205
Time lapse, 289
Timer protein, 359
Tissue
  contrast model (TCM), 347
  oxygenation, 48
  penetration, 5
TmDOTMA, 118
TmDOTP, 117
Tolbutamide, 345
Tonotopic maps, 384
Toxicity, 16, 31, 126, 130, 205, 260, 296, 315, 334, 358, 373, 397
Tract tracing, 380
Transcription factor, 227
Transcription regulatory elements, 200
Transcytosis, 71
Transfection, 200, 402
  agent (TA), 263, 298, 303, 306–309
Transferrin, 63, 125, 127, 144, 208–209, 298
Transgenic animals, 123, 201–202
Transglutaminase, 51
Transient hydrophilic pores, 53
Transition metal chemistry, 102
Translational aspects, 3, 32, 147, 211, 264, 314, 334–336, 360, 395–402, 408
Transmembrane carrier peptide, 125
Transmetallation, 17, 40
Transmission electronic microscopy (TEM), 65
Transplant, 264
Transverse relaxation rate, 15, 52, 93
Triaciclglycerols, 188
Troglitazone, 274
Trypan Blue, 312
Tumor, *see* Cancer
Tumor necrosis factor (TNF), 183, 274
Twin arginine translocase (TAT), 63, 191, 263, 298, 351
Tyrosine, 126, 207

U

Ultrasmall particles of iron oxide (USPIO), 24, 60, 68, 75, 172, 253, 260, 262, 276, 298, 302–303
Ultracentrifugation, 61
Ultrasound, 141, 358, 397

V

Vacuolar transporter chaperone, 126
Vasa vasorum, 167
Vasodilator, 223
Vasopressin, 234
Vasorestrictor, 223
Vasovist, *see* Gadofosveset
Vascular
  degradation, 277, 279

permeability, 71
regulation, 227
Vascular endothelial growth factor (VEGF), 147
Vectors, 126, 129, 200
Vegetal cell mass, 286
Vegetal blastomere, 289
Ventricular function, 374
Versatile ultrasmall superparamagnetic iron oxide (VUSPIO), 62
Vesicles, 296
Very small iron oxide particle (VSOP, C184), 60, 62, 66, 68, 70, 299, 305
Virus, 304
Vitamin B6, 250

W

Water diffusion, 13
Water exchange (WEX), 19, 25, 38, 86, 92, 103
Water line width, 93
Water residency time, 22
Water soluble iron oxide (WSIO), 299, 305
Weiss domain, 171
Wheat germ agglutinin conjugated to horseradish peroxidase (WG-HRP), 380
White marker techniques, 75, 174–175, 311
World Health Organization (WHO), 152

X

Xenon, 13, 31, 350
Xenopus laevis, 50, 203, 283
X-ray
angiography, 161
diffraction (XRD), 65

Y

YbDOTA-4AmC, 111
YbMB-DO3AM, 113
Yttrium-90, 17, 30
ytterbium, 107

Z

Zeta potential, 67, 299, 308
Zinc (Zn), 40, 116, 344, 348
Z value, 104
Z spectrum, 95, 106, 114